9th International Engineering Research Conference (Eureca 2017)

MATEC Web of Conferences Volume 152 (2018)

Selango, Malaysia
6 December 2017

Editors:

S. N. Namasivayam
M. H. Fouladi
S. W. E. Phang
B. L. Chua
Y. H. Chow
L. C. Yong
A. S. M. Al-Obaidi

ISBN: 978-1-5108-5874-9

Printed from e-media with permission by:

Curran Associates, Inc.
57 Morehouse Lane
Red Hook, NY 12571

CURRAN ASSOCIATES INC. proceedings.com

Some format issues inherent in the e-media version may also appear in this print version.

Printed by Curran Associates, Inc. (2018)

For additional information, please contact EDP Sciences – Web of Conferences
at the address below.

EDP Sciences – Web of Conferences
17, Avenue du Hoggar
Parc d'Activité de Courtabœuf
BP 112
F-91944 Les Ulis Cedex A
France

Phone: +33 (0) 1 69 18 75 75
Fax: +33 (0) 1 69 28 84 91

contact-edps@webofconferences.org

Additional copies of this publication are available from:

Curran Associates, Inc.
57 Morehouse Lane
Red Hook, NY 12571 USA
Phone: 845-758-0400
Fax: 845-758-2633
Email: curran@proceedings.com
Web: www.proceedings.com

TABLE OF CONTENTS

edp sciences

MATEC Web of Conferences

Statement of Peer review

In submitting conference proceedings to MATEC Web of Conferences, I certify to the Publisher that I adhere to the **Policy on Publishing Integrity** of the journal in order to safeguard good scientific practice in publishing.

1. All articles have been subjected to peer review administered by the proceedings editors.
2. Reviews have been conducted by expert referees, who have been requested to provide unbiased and constructive comments aimed, whenever possible, at improving the work.
3. Proceedings editors have taken all reasonable steps to ensure the quality of the materials they publish and their decision to accept or reject a paper for publication has been based only on the merits of the work and the relevance to the journal.

Title, date and place of the conference

Title : 9th Eureca 2017 International Engineering Research Conference
Date : 6th December 2017
Place : Taylor's University, Lakeside Campus, 47500 Subang Jaya, Selangor, Malaysia.

Proceedings editor(s):

1. Ir. Assoc. Prof. Dr. Satesh Namasivayam
2. Assoc. Prof. Dr. Mohammad Hosseini
3. Ms. Eunice Phang Siew Wei
4. Dr. Chua Bee Lin
5. Dr. Chow Yin Hui
6. Dr. Yong Leng Chuan
7. Dr. Abdulkareem Shafiq Mahdi Al-Obaidi

Date and editor's signature

29th January 2017

Solid-Liquid Extraction Kinetics of Total Phenolic Compounds (TPC) from Red Dates

Chua Bee Lin [1,*], and *Chen* Yen Leng [1]

[1]School of Engineering, Taylor's University, 47500 Subang Jaya, Selangor, Malaysia

Abstract. Red dates are one of the most famous herbal plants in making traditional Chinese medicine. They contain large amount of bioactive compounds. The objectives of this research were to optimise the crude extract yield and total phenolic compounds (TPC) yield from red dates using response surface methodology (RSM) and model the extraction kinetics of TPC yield from red dates. Date fruits were dried in an oven under temperatures 50°C, 60°C, 70°C and 80°C until a constant weight was obtained. The optimum drying temperature was 60°C as it gave the highest crude extract yield and TPC yield. Besides that, single factor experiments were used to determine the optimum range of four extraction parameters which were: liquid-solid ratio (10-30 ml/g); ultrasonic power (70-90%); extraction temperature (50-70°C); and extraction time (40-60min). The optimum range of the four parameters were further optimised using the Box-Behken Design (BBD) of RSM. The extraction conditions that gave the highest crude extract yield and TPC yield were chosen. The optimum value for liquid-solid ratio, ultrasonic power, extraction temperature and extraction time were 30ml/g, 70%, 60°C and 60 min respectively. The two equations generated from RSM were reliable and can be used to predict the crude extract yield and TPC yield. The higher the extraction temperature, liquid-solid ratio, and extraction time and lower ultrasonic power, the higher the crude extract and TPC yield. Finally, the results of TPC yield versus time based on the optimum extraction parameters from RSM optimisation were fitted into three extraction kinetic models (Peleg's model, Page's model and Ponomaryov's model). It was found that the most suitable kinetic model to represent the extraction process of TPC from red dates was Page's model due to its coefficient of determination (R^2) was the closest to unity, 0.9663 while its root mean square error (RMSE) was the closest to zero, 0.001534.

1 Introduction

Medical plants have similar properties as pharmaceutical drugs. They can be used for medical purposes as they contain chemical compounds which help humans with their health

[1] Corresponding author: beelin.chua@taylors.edu.my

and prevent diseases. There are 21,000 plants around the world can be used for medical purposes due to their high nutritional values according to World Health Organization [1].

Red dates are described as 'the king of nuts' [2]. They are one of the most famous and useful herbal plants in making traditional Chinese medicine. Red dates contain a high level of Vitamin C which can help improve insomnia, protect the liver and reduce cholesterol [3]. In addition, red dates also contain a large amount of total phenolic compounds (TPC) that increases their antioxidant activity by removing free radicals which cause cancer and other diseases [4].

For an accurate analysis, the optimisation of extraction process is needed. Response Surface Methodology (RSM) is a useful tool to optimise extraction process. It helps to lower the cost of the process and it gives better understanding of the extraction process [5]. RSM has been applied successfully for *Macadamia tetraphylla* and other herbal plants but no article is available regarding the optimisation of crude extract yield and TPC yield of red dates [6].

From an engineering perspective, it is very helpful to do a mathematical modeling of extraction processes. It is used to model the experiment, relate the consumption of energy, time, solvent and other parameters to the extraction kinetics [7]. The widely used extraction kinetic models of antioxidant compounds included Peleg's model, Page's model and Ponomaryoc's model. To find the best fitted model of the extraction process, the correlation coefficient (R^2) and root mean square error (RMSE) were calculated. The closer the R^2 and RMSE to unity and zero respectively, the higher the fit of predicted kinetic models to the experimental data.

To the best of author's knowledge, the optimisation of crude extract yield, TPC yield from red dates was not widely developed. The reported findings regarding the antioxidant activity of TPC yield were inconsistent. There were also three commonly used models were not widely used to describe the process of extraction of TPC from red dates. They were Peleg's model, Page's model and Ponomaryov's model.

One of the objectives of this research was to optimise the crude extract yield and TPC yield of red dates using response surface methodology (RSM). Box-Behnken Design (BBD) with three central points of RSM was used due to its high efficiency and less number of experiments were required thus lowering the total cost [8]. Another objective was to model the extraction kinetics of crude extract yield from red dates. Among the mentioned three kinetic models (Peleg's model, Page's model and Ponomaryov's model), the best fitted extraction kinetic model was chosen to predict the output of extraction process of TPC from red dates.

2 Methodology

The methodology conducted to achieve the objectives of this research was discussed in this section. Also, the analysis and justification of all tasks were included.

2.1 Ultrasonic Extraction of Total Phenolic Compounds from Red Dates

50 g of date fruits with no seeds were washed with tap water and dried in an oven at 50, 60, 70 and 80°C respectively until constant weight was obtained. The dried fruit was grinded to powder using blender. 11g of red dates powder was extracted for 60 min with 165ml of pure ethanol as extraction solvent in an ultrasonic bath under 60°C. The ultrasonic bath was operated at 70% of input power and 37kW. After 60 min, the mixture was placed in a rotary evaporator under 40 °C until crude extract was observed. The crude extract was weighted to determine the crude extract yield.

2.2 Quantification of Total Phenolic Compound Yield from Red Dates

The quantification of TPC yield from red dates was carried out by using gallic acid as a standard [4]. 0.1g of crude extract was dissolved in 1.5ml of distilled water. 5ml of 10 % of Folin-Ciocalteu phenol reagent, 4ml of 7 % of Na_2CO_3 and 1ml of extract were mixed together. After 1 hour, the mixture was poured into 96-well plate. The first row of the 96 well plate was filled with 0.2 ml of pure ethanol to act as blank solution. The 96-well plate was then placed into microplate spectrophotometer to measure its absorbance at 765 nm. The amount of gallic acid from red dates were determined by referring to the calibration curve of gallic acid plotted. Concentration of TPC was calculated using Eq. (1).

$$\text{Concentration of TPC } \left(\frac{\text{g GAE}}{\text{g extract}}\right) = \frac{C_t V_t}{W_1} \times 10^{-6} \quad (1)$$

where C_t is the concentration of gallic acid in sample extract (μg/ml), V_t is the total volume of the sample extract (ml), W_1 is the dry weight of extract (g).

2.3 Selection of Best Drying Temperature and Single Factor Experiments

The best drying temperature of red dates among four different drying temperatures (50, 60, 70 and 80°C) was determined. The drying temperature that gave the highest crude extract and TPC yield was selected as the optimum drying temperature.

Single factor experiments were used to study the four parameters (liquid-solid ratio, ultrasonic power, extraction temperature and extraction time) individually and determine the optimum range of four parameters for RSM. Crude extract yield and TPC yield of red dates were obtained by following the experimental design which was created by varying one factor and others were kept at their midpoint.

2.4 Optimisation of Extraction Conditions

Box-Behken Design (BBD) of response surface methodology was used to optimise extraction conditions based on the optimum ranges determined from single factor experiments. The experiments were carried out by referring to the Design of Experiment (DOE) of BBD which was created using Design Expert software 7.0. The crude extract yield and TPC yield of TPC were determined by following the DOE created. The experiments were repeated for three times. Their average values were used to compare to determine the optimum combination of each parameter. The extraction conditions that give the highest results are chosen as the optimum extraction conditions. Besides that, the validations of optimum extraction conditions were achieved as the difference between the crude extract and TPC yield based on the optimum extraction conditions and the calculated results from the two equations provided from RSM were less than 10 %.

2.5 Mathematical Modeling of Extraction Kinetic Model for Red Dates

Three widely used extraction kinetic models, Peleg's model, Page's model and Ponomaryov's model were used to describe the process of extraction of TPC from red dates mathematically. The equation of Peleg's model, Page's model and Ponomaryov's model were showed in Eq. (2), (3) and (4) respectively. The results of TPC yield under optimum extraction conditions were fitted to the three extraction kinetic models respectively. The constants' values of each models and the coefficient value of R^2 and RMSE were determined using Matlab Curve Fitting Tool 2017. Based on the values of R^2 and RMSE obtained, the best fitted extraction kinetic model to the experiment was chosen. Furthermore, the results of TPC yield of three samples at different time were substituted into the best fitted extraction kinetic models. The validation of the best kinetic model was achieved as the difference of TPC yield calculated using the best fitted model and experimental value was less than 10%.

$$\text{Peleg's model: } C(t) = C_0 + \frac{t}{K_1 + K_2 t} \quad (2)$$

$$\text{Page's model: } C(t) = e^{-kt^n} \quad (3)$$

$$\text{Ponomaryov's model: } 1 - \frac{C_t}{C_e} = b + k_3 t \quad (4)$$

where $C(t)$ is the concentration of TPC at time t (mg/g_{powder}), C_0 is the initial concentration of TPC (mg/g_{powder}), K_1 is Peleg's rate constant (min g_{powder}/mg), K_2 is Peleg's capacity constant (g_{powder}/mg), k and n are Page's constant, C_e is the equilibrium concentration of TPC in the plants (mg/g_{powder}), b is washing coefficient and k_3 is slow extraction coefficient (min^{-1}).

3 Results and Discussions

Based on the methodology described, discussions of results are separated into four sessions, which are determination of the optimum drying temperature, analysis of single factor experiments, analysis of response surface methodology and fitting the extraction kinetic models of red dates. The data was expressed as the average of three replicates ± standard deviation.

3.1 Determination of the Optimum Drying Temperature

Red dates were dried under four different drying temperatures, which were 50, 60, 70 and 80°C. The effects of drying temperature on crude extract and TPC yield of red dates were listed in Table 1. Based on Table 1, the crude extract and TPC yield of red dates increased with higher drying temperature up to 60°C. The heating affects the bioavailability of total phenolic compounds [9]. The heat energy provided during drying process may break the connections between phenolic compounds and the fibre of red dates which is insoluble. Hence, the releasing of phenolic compounds throughout drying process has showed that the bioavailability of total phenolic compounds is increased by heating. Thus, the crude extract and TPC yield increased together with high drying temperature up to 60°C.

However, at drying temperature of 70 and 80°C, the crude extract and TPC yield of red dates started to decrease. The possible reason may be the thermal degradation of total phenolic compounds started from high drying temperature 70°C [10]. Similar to the results of previous findings, in this research, the optimum drying temperature was 60°C as the highest crude extract and TPC yield were obtained under drying temperature of 60°C [11].

Table 1. Effects of drying temperature on crude extract and TPC yield

Temperature (°C)	Crude extract yield (g)	TPC yield (g GAE/g extract)
50	4.44±0.067	0.022±0.00006
60	4.56±0.089	0.023±0.00009
70	4.41±0.022	0.022±0.00004
80	4.37±0.038	0.021±0.00001

3.2 Analysis of Single Factor Experiments

Single factor experiments were conducted to determine the optimum range of each factor for RSM. The four factors involved were liquid-solid ratio, ultrasonic power, extraction temperature and extraction time. The effects of each factor on crude extract and TPC yield were further discussed in section 3.3.1 to 3.3.4.

3.2.1 Effect of Liquid-Solid Ratio on Crude Extract and TPC Yield

Based on Fig. 1, the concentration of TPC increased together with liquid-solid ratio. The concentration of TPC increased from 0.005±0.0001 to 0.0277±0.0007g GAE/g extract and the crude extract yield increased from 1.032±0.0009 to 1.905±0.053 g when the liquid-solid ratio increased from 10 to 30 ml/g. The two graphs are having similar trend. This was because more solutes can be dissolved into solvent as the volume of solvent was larger and the contact area of solute and solvent increased [5]. At the same time, as the contact area of solute and solvent increased, the rate of mass transfer of total phenolic compounds from cell into solvent increased, thus resulting higher TPC yield. Hence, the optimum range of liquid-solid ratio for RSM was 10 to 30 ml/g as they gave the three highest concentration of TPC.

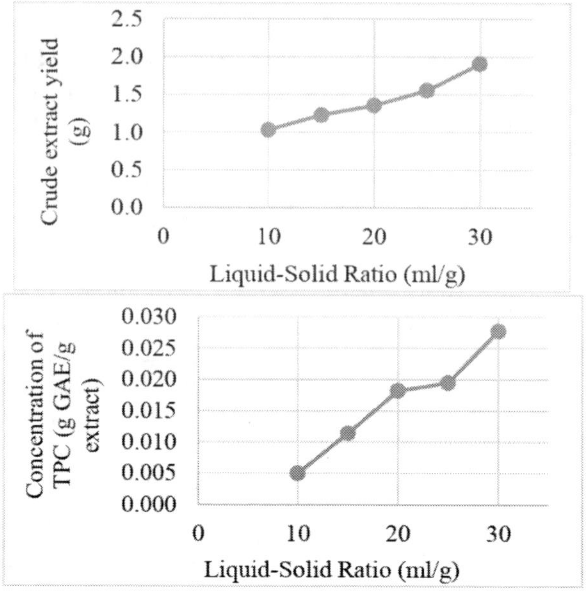

1a. 1b.

Fig. 1. Effects of liquid-solid ratio on crude extract yield (1a) and TPC yield (1b).

3.2.2 Effect of Ultrasonic Power on Crude Extract and TPC Yield

Ultrasonic power is another factor that affects TPC and crude extract yield. Based on Fig. 2, the TPC and crude extract yield were having the similar trend. The concentration of TPC increased from 0.0132±0.0001 to 0.0198±0.00004 g GAE/ g extract and the crude extract yield increased from 1.385±0.023 to 1.889±0.0429 g with higher ultrasonic power up to 80%. This may due to the cavitation shear stress created by ultrasound to break the cell wall of red dates increased, thus, more TPC was able to diffuse out from the cell wall into solvent [12]. As reported from previous research, TPC is having higher stability at higher ultrasonic power [13]. Thus, more TPC was found in the solvent and the crude extract was heavier after removing the solvent.

However, under ultrasonic power of 90%, the concentration of TPC was slightly decreased from 0.0198±0.00004 to 0.0197±0.0005 g GAE/ g extract while the crude extract yield decreased from 1.889±0.0429 to 1.785±0.0247g. The small decrement of the two responses might due to the TPC yield has become saturated as the phenolic compounds have been dissolved into the solvent. Besides that, at 100% of ultrasonic power, a large decrement of concentration can be seen in Fig. 2. The concentration of TPC decreased might due to the degradation of TPC [14]. Hence, the optimum range of ultrasonic power for RSM was 70 to 90%.

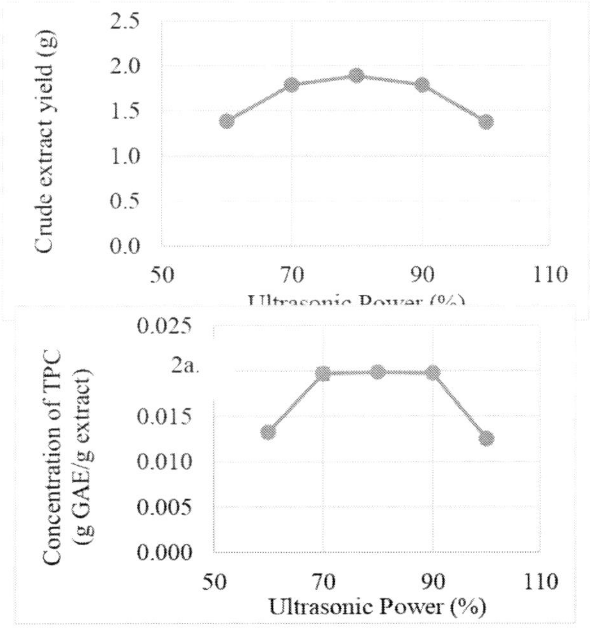

2b.

Fig. 2. Effects of ultrasonic power on crude extract yield (2a) and TPC yield (2b).

3.2.3 Effect of Extraction Temperature on Crude Extract and TPC Yield

Based on Fig. 3, the crude extract yield increased rapidly from 1.465±0.0365 to 1.805±0.0337g and the concentration of TPC increased from 0.0138±0.0003 to 0.0232±0.0003 g GAE/g extract as the extraction temperature was getting higher up to 60°C. The positive effects of extraction temperature on the responses could be due to the molecules gained the energy provided by heating up the extract in the ultrasonic bath. The molecules vibrated violently when they gained the energy and broke the connections between the cell wall and TPC. Thus, TPC was able to diffuse into the solvent more easily. TPC is also readily stable under higher extraction temperature up to 60°C [15]. Apart from that, the solubility of molecules in red dates increased together with extraction temperature [16].

From Fig. 3, the crude extract yield and concentration of TPC dropped from 1.805±0.0337 to 1.775±0.0266g and 0.0232±0.0004 to 0.0198±0.0001 g GAE/g extract respectively when the extraction temperature increased to 70°C. It could be due to the decreasing of cavitation effect by ultrasonic with the increasing of extraction temperature. It caused lesser cavitation bubbles to collapse at the surface of red dates and thus, lesser TPC was extracted. Furthermore, TPC was proved to be thermo-sensitive and they underwent degradation at temperatures beyond 60°C [5]. Thus, the optimum range of extraction temperature for RSM was 50 to 70°C.

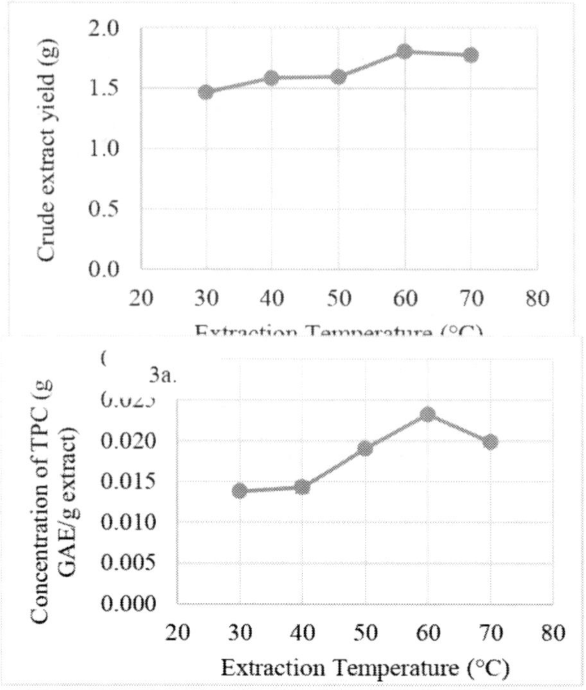

3b.

Fig. 3. Effects of extraction temp. on crude extract yield (3a) and TPC yield (3b).

3.2.4 Effect of Extraction Time on Crude Extract and TPC Yield

Based on Fig. 4, it can be seen that the concentration of TPC increased from 0.0211±0.0004 to 0.0225±0.0002 g GAE/g extract at the beginning but started to decrease from 0.0225±0.0002 to 0.0219±0.0001 g GAE/g extract when extraction time exceeded 50 min. The crude extract yield increased from 1.695±0.0365 to 2.007±0.0744 g for extraction time below 50 min and decreased from 2.007±0.0744 to 1.953±0.0126 g for extraction time more than 50 min.

At extraction time of 40 min, the TPC and crude extract yield and were lower because there wasn't sufficient time for the extraction process to reach equilibrium. The rate of mass transfer of total phenolic compounds from red dates into solvent increased significantly with extraction time until 50 min. The concentration of TPC has reached equilibrium state at extraction time of 50 min as it gave the highest TPC yield. For extraction time longer than 50 min, the TPC yield decreased rapidly. It could be due to the total phenolic compounds were not stable and they started to decompose [17]. Another reason might be more impurities were extracted from red dates by cavitation effect and affected the purity of TPC. Hence, the optimum range of extraction time for RSM was 40 to 60 min.

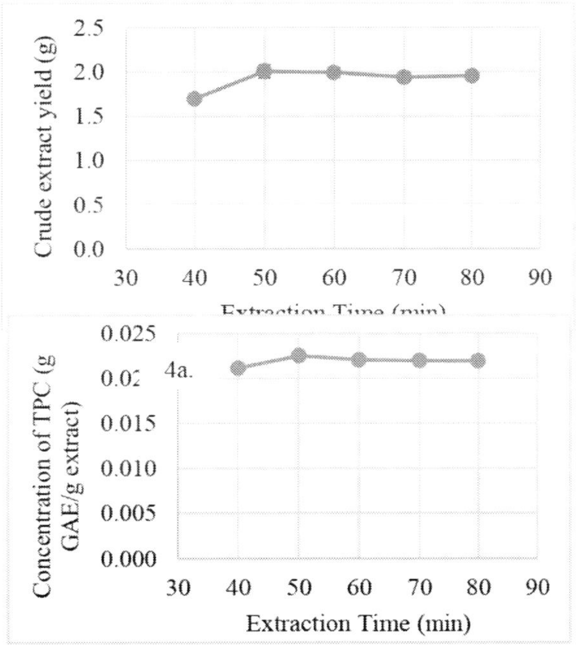

4b.

Fig. 4. Effects of extraction time on crude extract yield (4a) and TPC yield (4b).

3.3 Analysis of Response Surface Methodology

Based on the results obtained from single factor experiments, the range of liquid-solid ratio (A), ultrasonic power (B), extraction temperature (C) and extraction time (D) were 20-30ml/g, 70-90%, 50-70°C and 40-60min respectively. The range of parameters were then used to create design of experiments using Design Expert software 7.0 and the experimental results were listed in Table 2. The results were expressed in the average of three runs ± standard deviation.

Table 2. Experimental crude extract and TPC yield for RSM

Run	A	B	C	D	Crude extract yield (g)	TPC yield (g GAE/g extract)
1	25	70	50	50	2.045±0.032	0.0224±0.0002
2	20	80	70	50	1.755±0.014	0.0195±0.0002
3	20	80	60	60	1.855±0.011	0.0298±0.0006
4	25	70	70	50	2.175±0.074	0.0245±0.0001
5	30	90	60	50	2.025±0.051	0.0302±0.0008
6	20	80	50	50	1.315±0.020	0.0265±0.0002
7	25	80	60	50	1.875±0.024	0.0285±0.0001
8	30	80	70	50	2.065±0.083	0.0187±0.0001
9	25	70	60	40	1.715±0.021	0.0269±0.0006
10	30	80	60	60	2.175±0.022	0.0298±0.0009
11	25	90	60	40	1.625±0.008	0.0273±0.0006
12	25	80	60	50	1.945±0.017	0.0278±0.0003
13	30	80	60	40	1.865±0.026	0.0272±0.0003
14	25	90	50	50	1.545±0.009	0.0258±0.0004
15	25	80	70	60	1.785±0.013	0.0155±0.0003
16	20	80	60	40	1.585±0.016	0.0269±0.0002
17	30	80	50	50	1.745±0.007	0.027±0.0007
18	30	70	60	50	2.065±0.002	0.0299±0.0008
19	25	90	70	50	2.045±0.044	0.0165±0.0004
20	20	70	60	50	1.635±0.059	0.0281±0.0004
21	25	80	60	50	2.075±0.006	0.0277±0.0006
22	25	90	60	60	1.875±0.033	0.032±0.0001
23	25	70	60	60	1.845±0.041	0.0317±0.0001
24	20	90	60	50	1.855±0.005	0.0244±0.0001
25	25	80	60	50	2.075±0.052	0.0286±0.0003
26	25	80	70	40	2.015±0.093	0.0184±0.0006
27	25	80	50	60	1.715±0.021	0.0259±0.0002
28	25	80	50	40	0.845±0.015	0.0194±0.0003

Based on Table 2, the range of crude extract yield was 0.845-2.175g, TPC yield was 0.0165-0.032 g GAE/g extract. However, the experiment with highest TPC yield did not give the highest crude extract yield. Thus, for this research, optimising TPC yield of red dates was the main objective. The optimised conditions determined were 30ml/g of liquid-solid ratio, 70% of ultrasonic power, 60°C of extraction temperature and 60min of extraction time. The validation was achieved as the difference between experimental and predicted TPC yield was 9.45% which was less than 10%.

3.3.1 ANOVA for Crude Extract Yield

ANOVA is a statistical method to analyse the significance of parameters. The parameter is considered significant with P-value less than 0.05. Based on Table 3, liquid-solid ratio (A), extraction temperature (C), extraction time (D) and the interaction of extraction temperature and time (CD) affected the crude extract yield significantly. Ultrasonic power (B) and all the interactions of the four parameters were not significant due to their large P-value. Furthermore, the quadratic model can be used to explain the responses of crude extract yield due to its low p-value 0.0029 ($p < 0.05$). In addition, the p-value of lack of fit ($P = 0.1939$) indicated the lack of fit for this model was not significant. In other words, the quadratic model was highly fitted to the experimental data of crude extract yield.

Table 3. ANOVA for crude extract yield

Source	Sum of squares	df	Mean square	F value	P-value
Model	1.82	14	0.13	5.08	0.0029
A	0.31	1	0.31	12.27	0.0039
B	0.022	1	0.022	0.85	0.3739
C	0.58	1	0.58	22.55	0.0004
D	0.21	1	0.21	8.35	0.0127
AB	0.017	1	0.017	0.66	0.4308
AC	0.0036	1	0.0036	0.14	0.7135
AD	0.0004	1	0.0004	0.016	0.9024
BC	0.034	1	0.034	1.34	0.2680
BD	0.0036	1	0.0036	0.14	0.7135
CD	0.30	1	0.30	11.84	0.0044
A^2	0.017	1	0.017	0.65	0.4356
B^2	0.0007594	1	0.0007594	0.030	0.8658
C^2	0.16	1	0.16	6.29	0.0261
D^2	0.20	1	0.20	7.82	0.0151
Residual	0.33	13	0.026		
Lack of Fit	0.30	10	0.03	3.06	0.1939
Pure Error	0.030	3	0.00982		
Cor Total	2.15	27			

The relationships of linear, interaction and quadratic effects of each parameters were expressed in a quadratic equation, Eq.(5). Eq.(5) can also be used to predict the crude extract yield under different extraction conditions.

$$\begin{aligned}\text{Crude extract yield} = \\ -17.3675 + 0.26733A - 0.0603B + 0.29692C \\ + 0.33183D - 0.0013AB - 0.0006AC + 0.0002AD \\ + 0.000925BC + 0.00003BD - 0.00275CD - 0.0021A^2 \\ + 0.00001125B^2 - 0.0016375C^2 - 0.001825D^2\end{aligned} \quad (5)$$

To simplify the quadratic equation created using Design Expert Software Version 7.0, the non-significant ($P > 0.05$) factors and interactions were removed. The simplified equation, Eq (6) was showed below:

$$\begin{aligned}\text{Crude extract yield} = \\ -19.0265 + 0.0323A + 0.3460C \\ + 0.3526D - 0.00275CD - 0.001555C^2 - 0.001743D^2\end{aligned} \quad (6)$$

The correlation coefficient (R^2) of Eq.(6) was 0.7987 while the R^2 of Eq.(5) was 0.8455. The R^2 of Eq.(5) was higher than Eq.(6). It can be said that the predicted responses value from Eq.(5) were closer to the fitted regression line compared to Eq.(6). Thus, Eq.(5) was suitable to be used to predict crude extract yield instead of Eq.(6).

The parameters with positive coefficient from Eq.(5) gave positive effects to crude extract yield while the parameters with negative coefficient gave negative effects to crude extract yield. Based on Eq.(5), the coefficients of liquid-solid ratio, extraction temperature and time were positive. Thus, the crude extract yield increases together with liquid-solid ratio, extraction temperature and time.

3.3.2 ANOVA for TPC Yield

Based on Table 4, extraction temperature (C), extraction time (D), the interaction of ultrasonic power and extraction temperature (BC), ultrasonic power and extraction time (BD) and quadratic term of extraction temperature (C^2) were significant in affecting TPC yield due to their low p-values ($P < 0.05$). The other parameters and interactions of the four parameters were not significant as their p-values were more than 0.05. In addition, the p-value of quadratic model for TPC yield was 0.0016. The experimental results were highly fitted to the quadratic model due to its low p-value. The lack of fit of the model was also calculated using Design Expert Software Version 7.0 to analyse the error in the model. The value of lack of fit of this quadratic model was 0.5342, which was less than 0.05. It can be said that the quadratic model was very close to the responses of experimental TPC yield.

Table 4. ANOVA for TPC yield

Source	Sum of squares	df	Mean square	F value	p-value
Model	0.0004675	14	0.00003339	5.78	0.0016
A	0.000003715	1	0.000003715	0.64	0.4370
B	0.000003203	1	0.000003203	0.55	0.4697
C	0.00007489	1	0.00007489	12.96	0.0032
D	0.00003571	1	0.00003571	6.18	0.0273
AB	0.000004	1	0.000004	0.69	0.4204
AC	$8.327\text{x}10^{-8}$	1	$8.327\ \text{x}10^{-8}$	0.014	0.9063
AD	0.000001823	1	0.000001823	0.32	0.5839
BC	0.00002704	1	0.00002704	4.68	0.0497
BD	$2.5\ \text{x}10^{-9}$	1	$2.5\ \text{x}10^{-9}$	0.0006891	0.9795
CD	0.00000484	1	0.00000484	0.84	0.3767
A^2	0.000007066	1	0.000007066	1.22	0.2888
B^2	0.000006551	1	0.000006551	1.13	0.3063
C^2	0.0002364	1	0.0002364	40.92	< 0.0001
D^2	$2.226\ \text{x}10^{-7}$	1	$2.226\ \text{x}10^{-7}$	0.039	0.8474
Residual	0.0000751	13	0.00000577		
Lack of Fit	0.00005885	10	0.000005885	1.09	0.5342
Pure Error	0.00001625	3	0.000005417		
Cor Total	0.0005426	27			

A quadratic equation, Eq.(7) was formed to predict the TPC yield under different conditions. The coefficients of extraction temperature and time were positive while the coefficient of liquid-solid ratio ultrasonic power were negative. Hence, the TPC yield increases if lower liquid-solid ratio, lower ultrasonic power, extraction temperature and time are used.

$$\begin{aligned}\text{TPC yield} = \; & -0.19853 - 0.00451A - 0.000663B + 0.00984C \\ & + 0.000688D + 2\times10^{-5}AB - 2.886\times10^{-6}AC + 1.35\times10^{-5}AD \\ & - 2.6\times10^{-5}BC - 2.5\times10^{-7}BD - 1.1\times10^{-5}CD \\ & + 4.34\times10^{-5}A^2 + 1.05\times10^{-5}B^2 - 6.27\times10^{-5}C^2 - 1.93\times10^{-6}D^2\end{aligned} \tag{7}$$

By removing the non-significant factors and interaction, Eq.(7) was simplified to Eq.(8).

$$\text{TPC yield} = 0.028 - 0.002498C + 0.001725D - 0.0026BC - 0.0066C^2 \tag{8}$$

The correlation coefficients (R^2) of Eq.(7) and Eq.(8) were 0.9173 and 0.8042 respectively. The R^2 of Eq.(7) dropped from 0.9173 to 0.8042 after simplifying. By considering more factors and interactions, the predicted values determined using Eq.(7) were closer to the regression line. Thus, Eq. (7) was highly suitable to be used to predict the TPC yield due to its high R^2 value, 0.9173.

3.4 Fitting the Extraction Kinetic Models of Red Dates

In this research, the experimental data was compared with the predicted data using three kinetic models (Peleg's model, Page's model, Ponomaryov's model). Fig. 5-7 show the comparison graphs of experimental data and the three mentioned kinetic models respectively. The results are showed in Table 5.

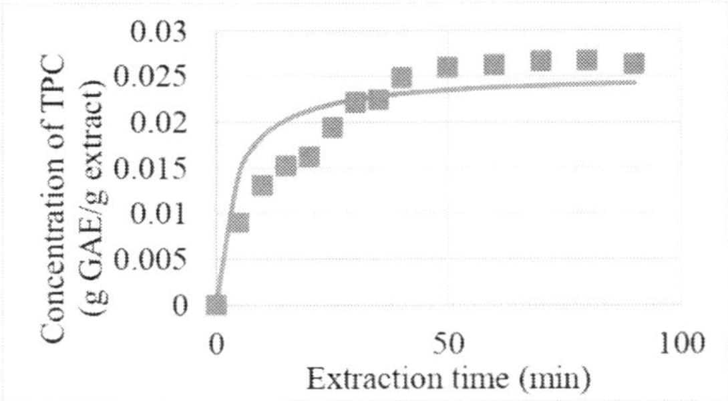

(5a.

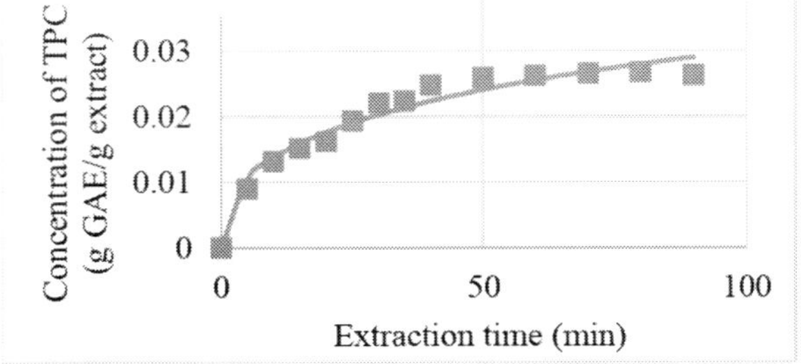

5b.

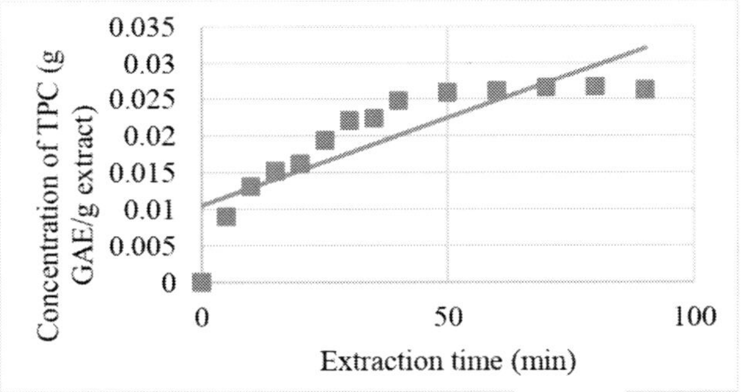

5c.

Fig. 5. Fittings of Peleg's model (5a), Page's (5b) and Ponomaryov's model (5c).

Table 5. Root mean square error (RMSE), coefficient of determination (R^2) and models'constants of kinetic models

Model	Model's constants	R^2	RMSE
Peleg's model	K_1=148.1, K_2=39.65	0.824	0.003506
Page's model	k=5.241, n=-0.08715	0.9663	0.001534
Ponomaryov's model	k_3=0.02623, b=2.157, C_e=-0.009102	0.716	0.004652

The calculated models' constants were showed in Table 5. As shown in Fig. 5, the line of predicted data calculated using Page's model passed through the most experimental data points compared to the rest of kinetic models. In addition, R^2 and RMSE of each model were calculated using Matlab Curve Fitting Tool 2017. The experimental data is highly fitted to the kinetic model if the value of R^2 is close to unity while RMSE is close to zero. Based on Table 5, the model with the highest R^2, 0.9663 and lowest RMSE, 0.001534 was Page's model. Therefore, Page's model was the most suitable to describe the extraction process of TPC from red dates among the mentioned models. The fittings of the mentioned models were in order of: Page's model > Peleg's model > Ponomaryov model.

Validation of extraction kinetic model was carried out to ensure that the most suitable kinetic model (Page's model) to describe the extraction process of TPC from red dates is reliable. The experimental results of TPC yield under optimum extractions at 45 min, 65 min and 85 min were compared with the predicted data calculated using Page's model, Eq. (9). The differences of experimental and predicted data are showed in Table 6.

$$\text{Page's model: } C(t) = e^{-(5.241\times t^{-0.08715})} \quad (9)$$

Table 6. Difference of experimental and predicted TPC yield

Extraction time (min)	TPC yield (g GAE/g extract)		Difference (%)
	Experimental	Predicted	
45	0.0253±0.0002	0.0233	8.20
65	0.0250±0.0004	0.0262	4.83
85	0.0259±0.0002	0.0285	9.97

Based on Table 6, the differences between experimental and predicted TPC yield at extraction time 45, 65 and 85 min were 8.2%, 4.83% and 9.97% respectively. The validation of extraction kinetic model was achieved as the differences between experimental and predicted TPC yield were less than 10%. Therefore, Page's model was most reliable and most suitable to be used to predict the TPC yield of red dates at different extraction times.

4 Conclusion

In this research, the optimised conditions to extract total phenolic compounds from red dates were determined using response surface methodology. The optimum value for liquid-solid ratio, ultrasonic power, extraction temperature and extraction time were 30ml/g, 70%, 60°C and 60 min respectively. Besides that, among three extraction kinetic models, Page's model is the most suitable to be used to describe the extraction kinetics of total phenolic compounds of red dates due to its coefficient of determination (R^2) was the closest to unity, 0.9663 while its root mean square error (RMSE) was the closest to zero, 0.001534.

The author would like to express her gratitude to her supervisor, Dr. Chua Bee Lin for her support, guidance and encouragement. The author also would like to give appreciation to lab assistants from Taylor's University for their advices. The authors would like to acknowledge Taylor's University Lakeside for providing financial support under TRGS/MFS/1/2016/SOE/004.

References

1. Valdivia-Correa, B., Gómez-Gutiérrez, C., Uribe, M. & Méndez-Sánchez, N. Herbal medicine in Mexico: A cause of hepatotoxicity. A critical review. *Int. J. Mol. Sci.* **17,** 48–49 (2016).
2. Norhasidah S, Z.Farahniza, A.G. Maaruf, E.U. Engku Hanisah, B. A. S. Antioxidative properties of edible bird's nest mincroparticulates incorporated into red dates drink. 4–6 (2014).
3. Yao, S. Past , Present , and Future of Jujubes — Chinese Dates in the United States. *HortScience* **48,** 672–680 (2013).
4. Al Harthi, S. S., Mavazhe, A., Al Mahroqi, H. & Khan, S. A. Quantification of phenolic compounds, evaluation of physicochemical properties and antioxidant activity of four

date (Phoenix dactylifera L.) varieties of Oman. *J. Taibah Univ. Med. Sci.* **10,** 346–352 (2015).
5. Yang, Y. C., Wei, M. C., Huang, T. C., Lee, S. Z. & Lin, S. S. Comparison of modified ultrasound-assisted and traditional extraction methods for the extraction of baicalin and baicalein from Radix Scutellariae. *Ind. Crops Prod.* **45,** 182–190 (2013).
6. Dailey, A. & Vuong, Q. Optimum Conditions for Microwave Assisted Extraction for Recovery of Phenolic Compounds and Antioxidant Capacity from Macadamia (Macadamia tetraphylla) Skin Waste Using Water. *Processes* **4,** 2 (2015).
7. Jokic, S. *et al.* Modelling of the process of solid-liquid extraction of total polyphenols from soybeans. *Czech J. Food Sci.* **28,** 206–212 (2010).
8. Anuar, N., Mohd Adnan, A. F., Saat, N., Aziz, N. & Mat Taha, R. Optimization of extraction parameters by using response surface methodology, purification, and identification of anthocyanin pigments in melastoma malabathricum fruit. *Sci. World J.* **2013,** (2013).
9. Kim, H. S. & Chin, K. B. Effects of Drying Temperature on Antioxidant Activities of Tomato Powder and Storage Stability of Pork Patties. *Korean J. Food Sci. Anim. Resour.* **36,** 51–60 (2016).
10. Larrauri, J. A., Rupérez, P. & Saura-Calixto, F. Effect of drying temperature on the stability of polyphenols and antioxidant activity of red grape pomace peels. *J. Agric. Food Chem.* **45,** 1390–1393 (1997).
11. Planinić, M. Influence of Temperature and Drying Time on Extraction Yield of Phenolic Compounds from Grape Pomace Variety 'Portogizac'. *Chem. Biochem. Eng. Q.* **29,** 343–350 (2015).
12. Karki, B. Use of high-power ultrasound during soy protein production and study of its effect on functional properties of soy protein isolate. (2009).
13. S.Rodrigues, F. F. *Advances in Fruit Processing Technologies*. (Taylor & Francis Group, 2012).
14. Ince, A. E., Şahin, S. & Şümnü, S. G. Extraction of phenolic compounds from melissa using microwave and ultrasound. *Turkish J. Agric. For.* **37,** 69–75 (2013).
15. Ruenroengklin, N. *et al.* Effects of various temperatures and pH values on the extraction yield of phenolics from litchi fruit pericarp tissue and the antioxidant activity of the extracted anthocyanins. *Int. J. Mol. Sci.* **9,** 1333–1341 (2008).
16. Qu, W., Pan, Z. & Ma, H. Extraction modeling and activities of antioxidants from pomegranate marc. *J. Food Eng.* **99,** 16–23 (2010).
17. Ibrahim, U. K., Austin, E. A. & Mohd Salleh, R. Effect of Drying Temperature and Time on Antioxidant and Total Phenolic Content in Garcinia mangostana Pericarp. *Adv. Mater. Res.* **1113,** 279–284 (2015).

Drying Kinetics and Optimisation of Pectin Extraction from Banana Peels via Response Surface Methodology

Chua Bee Lin[1,*], and Chong Yek Cze[1]

[1]School of Engineering, Taylor's University, 47500 Subang Jaya, Selangor, Malaysia

Abstract. Banana peels which are the waste in abundance, are used to extract valuable pectin. The gelling ability of the pectin has gained attention in food and pharmaceutical industries. This research aims to select the best drying kinetic model for banana peels and also optimize the pectin extraction process using Box-Behnken response surface design (BBD). Determination of pectin gelling mechanism using degree of esterification (DE) is also focused in this research. In this study, oven drying with temperature 50°C was chosen as the best drying temperature due to highest extraction yield. Furthermore, Page-Two-term model was selected as the best model to describe the drying kinetics of banana peels due to highest R^2 value (0.9991) and lowest RMSE value (0.001). The optimal extraction conditions given by BBD were 75°C extraction temperature, 23 min extraction time and 1:33.3 g/ml solid-liquid ratio. Likewise, the DE for both pectins extracted using unoptimised and optimised conditions were 71.92±1.38% and 76.1±2.07% respectively. Both of the pectins were classified as high-methoxyl pectins. The pectin with higher DE also indicated that the rate of gel formation is higher. The results showed that the pectin yield and gelling time has successfully improved after optimised the pectin extraction process.

1 Introduction

Banana (*Musa*) is one of the popular food crops that can found in the world market. Banana is generally processed into value-added product such as chips and jam. Because of the high demands of banana products, abundant of banana peels are produced every day and it has reported that about 39000 tons of banana peels are generated annually in U.S [1]. Hence, many scientists tend to find out the application of the banana peels to reduce the industrial waste and a valuable compound, pectin is found in the banana peels [2].

Pectin is a polysaccharide that consists of α-(1,4) galacturonic acid [1]. It is widely used as gelling agent for puree and jelly preparation [1]. Pectin is also proved to have wound healing and cholesterol-lowering effects [3]. Pectin is categorized based on the gelling mechanism using degree of esterification (DE) [2]. Pectin can be divided into low-methoxyl (LM) pectin (DE < 50%) and high-methoxyl (HM) pectin (DE > 50%). LM pectin forms

* Corresponding author: beelin.chua@taylors.edu.my

gels under divalent ions whereas HM pectin form gels under high sugar concentration solution [4]. Likewise, the gel formation rate of pectin can be determined by DE. Pectin is classified as slow gelling (DE 58-65%) or fast gelling (DE > 72%) [2]. It is important to analyze the esterification degree of pectin because it is used to determine the suitability of end products [2].

Drying is the primary process prior to pectin extraction process. It is normally performed using natural sunlight or artificial dryer such as oven [5]. Drying is used to preserve the quality and extend the shelf life of the foodstuffs by reducing the moisture content to desired level [6]. Sun drying is the conventional method to preserve foodstuffs as it is renewable and low in cost. Yet, large-scale production is not always preferred to use sun drying [7]. Hence, artificial dryer is the alternative to provide uniform and fast drying process. With regards to this, both methods were studied in this research.

Thin-layer drying kinetic models are the mathematical modelling that used to predict the drying behaviour and time of food products [6]. Many popular drying models such as Page, Logarithmic, Henderson and Pabis models have successfully described the drying kinetics of various plants [5], [8, 9]. Besides, some new models have developed to improve the accuracy [5], [10]. For instances, a new model which combined the Page and Two-term drying models has proved to give a promising results for cocoa beans [5]. These two models are chosen because both of them give favourable fitting in many applications according to past experience [5]. Yet, this new model, Page-two-term model has only been used in cocoa beans [5]. Regarding to this, this model was performed in this research to study the potential of this new model applies on banana peels and also compared with the Page and Two-term models.

Extraction of pectin can be achieved using acid, ultrasound-assisted (UAE) and microwave-assisted extraction (MAE) [11]. However, UAE which used the acoustic energy to break the cell wall, has gained a lot of attention these years [12]. This is because UAE reduces the extraction time and increases the extraction yield with lower energy consumption [12]. With this respect, UAE was selected to extract the pectin from banana peels in this research.

To the best of author's knowledge, no research has been conducted to study both drying kinetics and optimisation of pectin extraction from banana peels using RSM together [13, 14]. Hence, the present study is carried out with the objectives as follows:

1. To select the best drying models to describe the drying process of banana peels.
2. To optimise the pectin yield with different extraction parameters (extraction temperature, extraction time and solid-liquid ratio) using Box-Behnken response surface design (BBD).
3. To determine the gelling mechanism of pectin based on degree of esterification using titrimetric method of Food Chemical Codex 1981.

2 Methodology

This section discusses about the methods that carried out in this research in order to accomplish the objectives.

2.1 Pre-treatment of banana peels

Ripened banana peels were collected from a commercial fruit juice store in Taylor's University, Subang Jaya, Selangor, Malaysia. 100g of banana peels were washed with distilled water to remove dirt on the surface before cutting into slices.

2.2 Drying of banana peels

In this research, the oven drying was compared with the conventional method which is sun drying. The drying method that produced the highest pectin yield was selected for modelling. The initial moisture content of the samples was measured using moisture analyser (Model XM50, Precisa Gravimetrics AG, Switzerland) to estimate the bone-dry weight before drying. All the samples were spread evenly in single layer on an aluminium tray to avoid any overlapping.

For sun drying, the samples were dried under the sun from 9 a.m. to 5 p.m. and tempered from 5 p.m. to 9 a.m. until it reached constant weight [6]. Weighing balance (Model TX423L, Shimadzu, Japan) was used to measure the weight of the samples. The surrounding temperature (32°C), humidity (79.7%) and wind velocity (2.6 m/s) were measured using HVAC datalogger (Model DO2003, Delta OHM, Italy). For oven drying, The peels were dried in an oven (Model UN75, Memmert, Germany) with different temperature 40°C, 50°C, 60°C and 70°C until the weight of samples reached constant.

After drying, the dried banana peels were grinded into powder using a grinder (Model BL-1525 BG OEM, China). The powdered peels were kept in polyethylene bag at room temperature for further analysis.

2.3 Mathematical modeling and statistical analysis

The optimum drying method and temperature that produced the highest pectin yield was used to study the drying kinetic models. 100g of pre-treated banana peels were dried using the optimum drying method/temperature and the weight loss of the banana peels was measured every 5 minutes until constant weight was obtained. The moisture content (MC) and moisture ratio (MR) of banana peels were calculated using Eq. (1) and Eq. (2) [5], [15]

$$MC\text{ (dry basis)} = \frac{w_i\text{-}w_{db}}{w_i} \tag{1}$$

where, w_i is the initial weight of the sample (g) and w_{db} is the bone-dry weight (g).

$$MR = \frac{M\text{-}M_e}{M_o\text{-}M_e} \tag{2}$$

where, M_o is the initial moisture content, M is the moisture content at time and M_e is the moisture content at equilibrium.

The experimental data were fitted into the three drying kinetic models as shown in Table 1. Microsoft Excel SOLVER 2016 was used for non-linear regression analysis [5]. In non-linear regression analysis, statistical indicators, coefficient of determination (R^2) and root mean square error (RMSE) were used to determine the fitting goodness [5], [15]. The formula of R^2 and RMSE are as follow.

$$R^2 = 1\text{-}\frac{\sum_{i=1}^{N}(MR_{Exp.i}\text{-}MR_{Pred.i})^2}{\sum_{i=1}^{N}(MR_{Exp.i}\text{-}MR_{Pred.i})^2} \tag{3}$$

$$\text{RMSE} = \sqrt{\frac{1}{N}\sum_{i=1}^{N}(MR_{Exp.i}\text{-}MR_{Pred.i})^2} \tag{4}$$

where, N is the data number has run, $MR_{exp,i}$ and $MR_{pred,i}$ are the experimental and predicted moisture ratio value respectively.

Table 1. Drying kinetic models

No.	Model	Equation	Ref.
1	Page	$MR = e^{-kt^n}$	[16]
2	Two-term	$MR = ae^{-kt} + ce^{-gt}$	[5]
3	Page-Two-term	$MR = ae^{-kt^n} + ce^{-gt^n}$	[5]

2.4 Ultrasound-assisted extraction (UAE) of pectin

The UAE process was conducted based on publication with minor modification [11, 12]. Four gram of powdered banana peels was added into 0.1 M of citric acid with pH 1.5 and SLR 1:30 g/ml. The mixture was then placed in the ultrasonic water bath (Elmasonic P120H, Elma, Germany) with temperature 80°C and sonicated for 60 min. The power and frequency of sonicator are set to 80 W and 37 kHz respectively. The mixture was left to cool down and then filtered out using a vacuum filtration. The filtrate was precipitated using 95% of ethanol with ethanol-to-filtrate ratio 2:1 v/v. The precipitate and filtrate were then separated using a centrifuge (Scanspeed 1236MG, LaboGene, Denmark) at 6000 RPM for about 30 min. The precipitated pectin was rinsed with 95% ethanol thrice to remove impurities and then dried in the oven with temperature 50 °C for 12 hours. The pectin yield based on the total peels used for extraction was calculated using Eq. (5).

$$\text{Pectin yield (\%)} = \frac{\text{weight of dried pectin (g)}}{\text{initial weight of dried banana peel powder (g)}} \times 100 \qquad (5)$$

2.5 Experimental design using RSM

Box-Behnken Design (BBD) in Design Expert 7.0 Software was used to optimize the pectin extraction conditions from banana peels. Three independent variables (X_1: extraction temperature, X_2: extraction time and X_3: solid-liquid ratio) were employed as these factors were found to be important in affecting the pectin yield based on previous study from publication [17]. For each factor, the level of the factor was chosen by taking the reference from previous work [17]. The three variables with their corresponding coded and actual values is shown in Table 2. By inserting the parameters and levels into BBD, 17 sets of experiment with different extraction conditions were developed as shown in Table 6. The variables in different level combination arranged by the BBD were used for pectin extraction using UAE method as mentioned in Section 2.4.

Table 2. Extraction parameters and levels

Extraction Parameters	Code	Levels		
		-1	0	1
Extraction temperature (°C)	X_1	60	70	80
Extraction time (min)	X_2	20	40	60
Solid-liquid ratio (w/v)	X_3	1:20	1:30	1:40

2.6 Statistical analysis and validation of optimal conditions

Design Expert Statistical 7.0 Software was used to study the experimental data with regression analysis. F-test at P-value (0.05) was used to assess the significance of regression coefficients [7]. The regression model adequacy was evaluated by coefficient of determination, R^2 and adjusted R^2 [7]. Besides that, the significant independent variables in the model were assessed using ANOVA with confidence level 95%. Response surface graphs were then generated using the regression coefficients.

The optimum extraction conditions that obtained from BBD were carried out an experiment. Experimental and predicted values were compared to validate the model and error of less than 5% between both values indicates a good fit [15].

2.7 Determination of degree of esterification (DE)

The esterification degree (DE) of the sample was determined using titration method with a little modification from publication of [4]. 2 ml of ethanol was added to 0.2 g of dried pectin and the mixture was then dissolved in 20ml of distilled water. The mixture was stirred continuously until the pectin was completely dissolved. 3 drops of phenolphthalein were added to the mixture and 0.1M of $NaOH$ was used to titrate the mixture until it turned into pale pink. The titration volume (V_1) was recorded. Next, the mixture was added with 10ml of 0.1M $NaOH$ and stirred continuously for an hour. 10ml of 0.1M HCl was added and the mixture was shaken until the colour of pale pink disappeared. 3 drops of phenolphthalein were added and 0.1M $NaOH$ was used to titrate the mixture until it turned into pale pink. The titration volume (V_2) is recorded. The experiment was repeated for three times. The DE was calculated using Eq. (6).

$$DE(\%) = \frac{V_1}{V_2+V_1} \times 100 \qquad (6)$$

3 Results and discussion

3.1 Selection of drying process and drying temperature

Sun drying (conventional method) and oven drying were compared to select the suitable drying method for banana peels. The drying method was selected based on the highest pectin yield. Since the temperature in the oven can be adjusted, different temperature (40°C, 50°C, 60°C and 70°C) were also tested to select the optimum temperature based on the extraction yield. The pectin yield obtained from sun drying and oven drying was tabulated in the following Table 3. According to the results, oven drying with temperature of 50°C was selected as it gave a highest pectin yield.

Table 3. Pectin yield at different drying process

Method	Temperature (°C)	Pectin Yield (%)
Sun drying	32	1.38±0.07
Oven drying	40	2.84±0.12
	50	2.88±0.07

	60	1.92±0.09
	70	1.78±0.08

Sun drying gave the lowest pectin yield (1.38±0.07%) due to rainy season on September 2017. The drying process became less effective as the average daytime temperature was only about 32°C with low wind velocity of 2.6 m/s and high humidity of 79.7%. These conditions increased the enzymatic degradation since the moisture in the peels could not be removed as soon as possible and this activated the microbiological activity, leading to spoilage of the peels [18]. Thus, the cell structure had destroyed by microbiological activity and this resulted in low pectin yield. The limitation of sun drying is the samples could only dry at day time with sunlight whereas oven drying does not have this limitation. Hence, the samples were able to dehydrate faster than sun drying to prevent microbiological spoilage to produce higher pectin.

For oven drying, results showed that the pectin yield increased from 40°C to 50°C and started to decrease after 50°C. Temperature of 50°C gave the highest pectin yield (2.88±0.07%) while 70°C gave the lowest pectin yield (1.78±0.08%). This could be explained as at low temperature, the heat energy was not strong enough to eliminate the moisture as soon as possible and this led to samples spoilage due to microbiological activity [18]. Yet, high temperature dried and harden the peels surface, blocking the moisture to escape [19]. This is also called as "case hardening" [19]. In other words, high temperature dried the food outside but inside still remained moist. This caused the moisture inside the peels could not be evaporated and the trapped moisture caused bacteria growth and spoilage, resulting in low pectin yield [19].

3.2 Drying kinetics and modeling

The initial moisture content of the banana peels before drying was measured as 92.15%. It is normal to have high initial moisture content because most of the fruits contain more than 90% of water [20]. It is important to measure the initial moisture content because it is used to estimate the bone-dry weight of the sample. The collected moisture content data was converted to moisture ratio and then plotted in graph as shown in Figure 1. Referring to Figure 1, moisture ratio decreased continuously with time and remained constant after 1400 min. The exponential decreasing curve also shows that internal mass transfer had governed by diffusion [15]. The constant zero moisture ratio at the later phase of the process indicated that the moisture of the samples were fully removed. The moisture ratio data was then fitted to the drying models and the following table 4 shows the goodness of fit of each drying model.

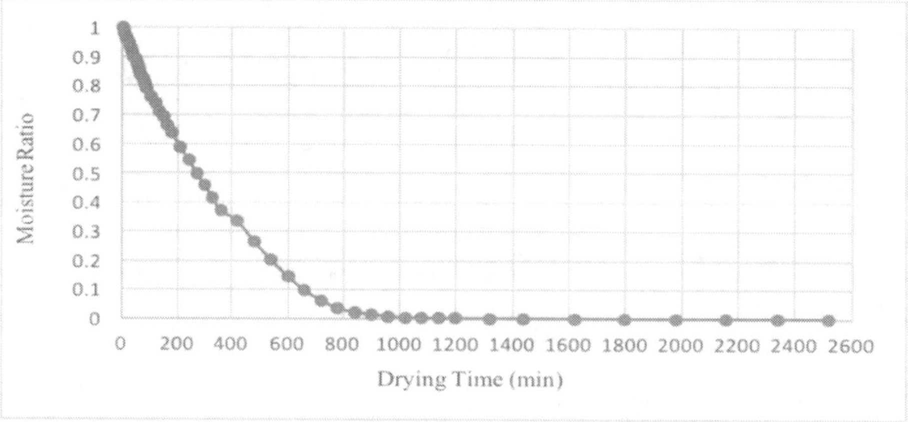

Fig. 1. Graph of moisture ratio against drying time

Table 4. The fitness of different models at 50°C drying temperature

No.	Model name	Coefficients and Constants	R^2	RMSE
1	Page	n = 1.164, k = 0.001098	0.9975	0.0026
2	Two-term	a = 0.0181, k = 1.4819, c = 1.0359, g = 0.00297	0.9947	0.0056
3	Page-Two-term	a = 0.09, k = 0.00405, c = 0.9012, g =0.000207, n = 1.4221	0.9991	0.001

According to Table 4, the fitting goodness of the models according to the statistical indicators (R^2 and RMSE) were in the order of: Page-Two-term > Page > Two-term. Page-two-term was selected as the best-fitted drying model to describe the drying process of banana peels as it gave the highest R^2 value (0.9991) and lowest RMSE value (0.001). Similar result has been obtained in cocoa beans plant in previous study [5]. Since Page-Two-term model is the new model that specially developed to describe the drying kinetics of cocoa beans, it can be concluded that this new model is also applicable in banana peels according to the satisfactory results.

3.3 Experimental design and regression analysis

17 experiments were conducted according to the conditions arranged by BBD as shown in Table 5. The experimental results showed the pectin yield ranging from 2.75% to 6.03%. Experiment 17 (X_1= 70°C, X_2 = 20 min and X_3 = 1:20) gave the maximum pectin yield (6.03%±0.11%) whereas experiment 1 (X_1= 60°C, X_2 = 40 min and X_3 = 1:20) gave the minimum pectin yield with only 2.75%±0.13%. The influence of extraction parameters on pectin yield was discussed in details in Section 3.4.

Table 5. Box-Behnken design with experimental and predicted pectin yield

Std	Run	Temp, X_1 (°C)	Time, X_2 (min)	SLR, X_3 (g/ml)	Pectin Yield, Y (%)	
					Experimental	Predicted
5	1	60	40	1:20	2.75±0.13	2.725

1	2	60	20	1:30	4.50±0.15	4.278
8	3	80	40	1:40	4.50±0.10	4.525
16	4	70	40	1:30	6.00±0.14	5.540
12	5	70	60	1:40	3.75±0.18	3.503
14	6	70	40	1:30	5.20±0.09	5.540
11	7	70	20	1:40	5.75±0.28	5.804
3	8	60	60	1:30	3.50±0.13	3.578
17	9	70	40	1:30	5.50±0.08	5.540
6	10	80	40	1:20	4.85±0.13	4.682
7	11	60	40	1:40	3.00±0.10	3.167
4	12	80	60	1:30	5.00±0.22	5.221
10	13	70	60	1:20	5.00±0.15	4.946
13	14	70	40	1:30	5.60±0.13	5.540
9	15	70	20	1:20	3.83±0.15	4.076
15	16	70	40	1:30	5.40±0.17	5.540
2	17	80	20	1:30	6.03±0.11	5.951

In addition, Table 6 below shows the regression analysis based on the experimental results. ANOVA was used to study the model adequacy and the significant of corresponding variables. Based on rule of thumb, high F-value and p-value lower than 0.05 indicate the model and variable are significant. Referring to the results, quadratic model fitted adequately to the experimental results due to large F-value (19.45) and small p-value (0.004). Likewise, R^2 and adjusted R^2 are used to check for the model adequacy and suitability. The experimental and predicted values showed a good correlation due to high value of R^2 (0.9616) and adjusted R^2 (0.9121) [1]. Besides, adequate precision measures the error related to the predicted response and the desirable value is above 4 [1]. The results showed that the model's adequate precision value was 13.839 which denoted that the model can be used inside the operation region [1].

Likewise, the significant of the corresponding variables on pectin yield (Y) can also be determined by F-value and p-value in Table 6 as shown as below. The higher the F-value and the smaller the p-value, the more significant the variable is. Linear term of extraction temperature (X_1) had the largest effect because of highest F-value and smallest p-value, followed by the quadratic term of extraction temperature (${X_1}^2$) and SLR (${X_3}^2$) and the interaction term of X_2X_3 whilst extraction time (X_2) showed the least significant influence on the pectin yield. However, the results revealed that linear term of SLR (X_3), quadratic effect of extraction time ($X_2^{\,2}$)and interaction effect of X_1X_2 and X_2X_3 did not have significant effects due to p-values more than 0.05. The following Eq. (7) is the polynomial equation that formulated by the surface model to describe the relationship between the extraction parameters and pectin yield.

$$Y\,(\%) = -55.405 + 1.24237X_1 + 0.10112X_2 + 0.85263X_3 - 0.0000375X_1X_2 - 0.0015X_1X_3 - 0.0039625X_2X_3 - 0.00795X_1^2 + 0.00003125X_2^2 - 0.0097X_3^2 \quad (7)$$

Referring to Eq. (7), the coefficient with higher value indicates it has higher impact on the response. Linear variables (X_1, X_2 and X_3) and quadratic variable ($X_2^{\,2}$) had positive effect on the pectin yield due to positive coefficients in the equation. Positive coefficient indicated the variable was directly proportional to the response value. Oppositely, negative coefficients of interactions (X_1X_2, X_1X_3 and X_2X_3) and quadratic variables ($X_1^{\,2}$ and $X_3^{\,2}$) indicated negative effect on the pectin yield. Negative coefficient showed a curvilinear relationship between the variable and response [21]. Curvilinear relationship means the response increases with the variables (negative coefficient) only up to certain point and starts to decrease after that.

Table 6. Variance analysis for quadratic model of response surface

Source	DF	Sum of Square	Mean Square	F-value	p-value
Model	9	16.18	1.80	19.45	0.0004
X_1	1	5.49	5.49	59.47	0.0001
X_2	1	1.02	1.02	11.07	0.0126
X_3	1	0.041	0.041	0.44	0.5285
X_1X_2	1	2.25×10^{-4}	2.25×10^{-4}	2.435×10^{-3}	0.9620
X_1X_3	1	0.090	0.090	0.97	0.3565
X_2X_3	1	2.51	2.51	27.19	0.0012
$X_1^{\,2}$	1	2.66	2.66	28.80	0.0010
$X_2^{\,2}$	1	6.579×10^{-4}	6.579×10^{-4}	7.12110^{-3}	0.9351
$X_3^{\,2}$	1	3.96	3.96	42.88	0.0003
Residual	7	0.65	0.092		
Lack of Fit	3	0.29	0.098	1.12	0.4412
Pure Error	4	0.35	0.088		
Cor Total	16	16.82			

Std. Dev.	0.30				
Mean	4.72				
R^2	0.9616				
Adj-R^2	0.9121				
Adeq Precision	13.839				

3.4 Effect of extraction parameters

The relationship between the extraction variable was studied with the help of three-dimensional (3-D) response surface graphs in Figure 2 and the influence of extraction variables on response was discussed in Section 3.4.1 to 3.4.3.

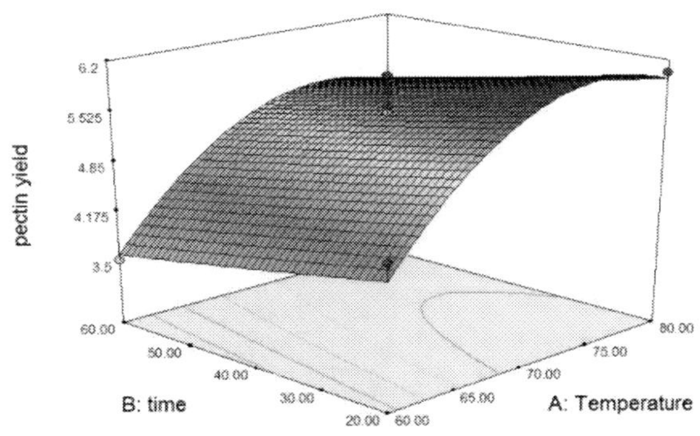

(a)

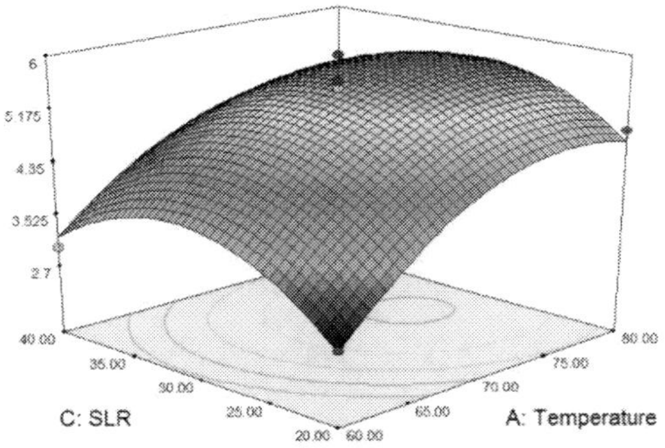

(b)

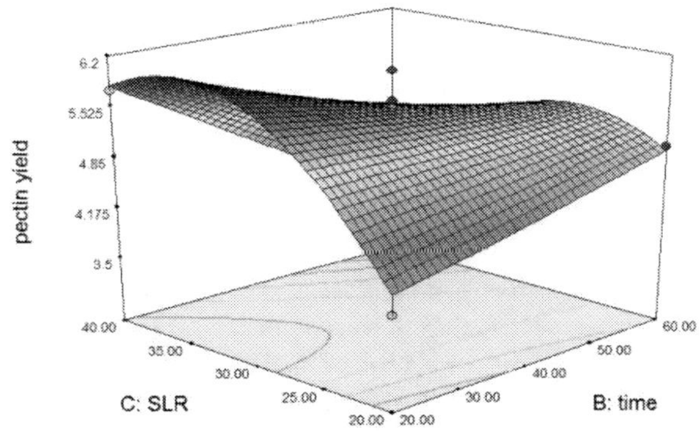

(c)

Fig. 2. Influence of extraction variables on pectin yield. (a) Time vs temperature at 1:30 g/m SLR. (b) SLR vs temperature at 40 minutes. (c) SLR vs time at 70°C temperature.

3.4.1 Effect of extraction temperature on pectin yield

Fig. 2 shows that pectin yield increased significantly with temperature. The pectin yield decreased slowly after reaching the temperature of 76°C. Similar results has been obtained from previous study of [14]. Increasing temperature disrupt the hydrogen bonds and ester linkages, enhancing the solvent penetration to cause higher extraction yield [12]. However, further increasing the temperature decreases the viscosity and surface tension of the extraction solvent [14]. This reduces the cavitation effects and mass transfer intensity, leading to low pectin yield [14]. Moreover, previous study has also reported that high temperature caused thermal degradation of pectin and this resulted in low pectin yield [22].

3.4.2 Effect of solid to liquid ratio (SLR) on pectin yield

Pectin yield was found to increase with SLR until approximately 1:30 g/ml and further increase the solvent volume results in low extraction yield even though SLR was not the significant influence factor according to ANOVA. Similar trend has been found in previous work of [14]. Other than that, 1:30 g/ml of SLR has also been found to produce maximum pectin yield in previous research of [17]. This can be explained as increasing the volume of extraction solvent enhanced the driving force for mass transfer during extraction, leading to higher pectin released into the surrounding solvent [23]. Yet, excess amount of solvent decreased the pectin yield because the solvent was saturated with the solute and this reduced the penetration and mass transfer rate of the pectin into the extraction solvent [24].

3.4.3 Effect of extraction time on pectin yield

Based on the ANOVA results in Table 7 and Fig. 2c, the extraction time and SLR had significant relationship. At low SLR, the pectin yield increased with extraction time whilst the pectin yield decreased with time at high SLR. This can be explained as at low SLR, the driving force is low due to less amount of extraction solvent hence, more reaction time is required for the mass transfer of pectin to the solution [23]. Nonetheless, the pectin yield decreased with longer sonication time at high SLR because the structure of pectin side chain was destructed and decomposed after prolonged exposure of excessive ultrasound waves [22]. Other than that, as mentioned in Section 3.4.2, high SLR saturated the extraction solvent and this reduced the mass transfer between the pectin and the solvent. Therefore, longer extraction time at high SLR resulted in low extraction yield.

3.5 Determination and validation of optimal extraction conditions

The optimum extraction conditions (extraction temperature, extraction time and SLR) generated by the Design Expert 7.0 Software were 74.34°C, 22.73 min and 1:33.3 g/ml respectively with the predicted pectin yield of 6.19%. By considering the operability and limitation of apparatus in real process, the optimal extraction temperature and extraction time were round off to 75°C and 23 min respectively. The suitability of optimized conditions was validated experimentally. The experimental pectin yield obtained was 6.08%±0.2% while the error percentage based on the predicted value was only 1.78 %. The small deviation between the experimental and predicted value might due to the decimal place of the extraction conditions value was round off for operability. Using the unoptimised conditions, only 5% of pectin yield was obtained but the yield had successfully improved by 21.66% after optimisation.

3.6 Esterification degree of pectin before and after optimisation

The degree of esterification (DE) of the pectin before and after optimisation were compared to determine the gelling mechanism. The DE of the pectin before and optimisation were found as 71.92%±1.38% and 76.1%±2.07% respectively. Both of the DE values were acceptable because the tissue pectin is usually in the range of 60 to 90% [2]. However, the DE of pectin after optimisation was slightly higher than the pectin before optimisation because of lower temperature was used in optimised conditions. Previous research has reported that DE values reduced with higher temperature because the methly esterified carboxyl groups that presented in the pectin chains could be destroyed at high temperature [25].

Since both DE values were above 50%, the pectins were classified as high methoxyl (HM) pectins which can form gel with high sugar concentration. In food industry, HM pectin is suitable to process into high sugar jam and canned food [26]. Besides, the pectins were also classified as fast gelling pectins as the DE values were more than 71% [27]. Since the setting rate increased with DE values, the optimized pectin could form gels faster than the pectin before optimisation due to higher DE value [28]. Based on the results, it can be concluded that, the type of gel formation remained unchanged after optimisation yet the gelling speed had improved.

4 Conclusion

In this study, oven drying with 50°C was suggested as the optimum drying temperature for banana peels because it provided the highest pectin yield. Among three drying models (Page, Two-term and Page-Two-term), Page-Two-term Model was selected as the most suitable model to describe the drying kinetic of banana peels with 0.9991 R^2 value and 0.001 RMSE. Three extraction conditions of pectin extraction from banana peels were optimized using Box-Behnken Design and the optimal conditions were 75°C of extraction temperature, 23 min of extraction time and 1:33.3 g/ml of SLR. Using the optimal conditions, the experimental pectin yield was found as 6.08±0.2% which was not much different with the predicted yield of 6.19%. After optimised the extraction process, the isolated pectin was still remained as high-methoxyl pectin (DE>50%) yet the extraction yield and gelling time had successfully improved.

The authors would like to acknowledge Taylor's University Lakeside for providing financial support under TRGS/MFS/1/2016/SOE/004.

References

1. Swamy, G. J. & Muthukumarappan, K. Optimization of continuous and intermittent microwave extraction of pectin from banana peels. *Food Chem.* **220,** 108–114 (2017).
2. Castillo-Israel, K. A. T. *et al.* Extraction and characterization of pectin from Saba banana [Musa 'saba'(Musa acuminata x Musa balbisiana)] peel wastes: A preliminary study. *Int. Food Res. J.* **22,** 202–207 (2015).
3. Freitas de Oliveira, C. *et al.* Extraction of pectin from passion fruit peel assisted by

ultrasound. *LWT - Food Sci. Technol.* **71,** 110–115 (2016).
4. Liew, S. Q., Chin, N. L. & Yusof, Y. A. Extraction and Characterization of Pectin from Passion Fruit Peels. *Agric. Agric. Sci. Procedia* **2,** 231–236 (2014).
5. Hii, C. L., Law, C. L. & Cloke, M. Modeling using a new thin layer drying model and product quality of cocoa. *J. Food Eng.* **90,** 191–198 (2009).
6. Khawas, P., Das, A. J., Dash, K. K. & Deka, S. C. Thin-layer drying characteristics of Kachkal banana peel (Musa ABB) of Assam, India. *Int. Food Res. J.* **21,** 975–982 (2014).
7. Toĝrul, I. T. & Pehlivan, D. Modelling of thin layer drying kinetics of some fruits under open-air sun drying process. *J. Food Eng.* **65,** 413–425 (2004).
8. Onwude, D. I., Hashim, N., Janius, R. B., Nawi, N. M. & Abdan, K. Modeling the Thin-Layer Drying of Fruits and Vegetables: A Review. *Compr. Rev. Food Sci. Food Saf.* **15,** 599–618 (2016).
9. Ertekin, C. & Firat, M. Z. A comprehensive review of thin-layer drying models used in agricultural products. *Crit. Rev. Food Sci. Nutr.* **57,** 701–717 (2017).
10. Fernando, J. A. K. M. & Amarasinghe, A. D. U. S. Drying kinetics and mathematical modeling of hot air drying of coconut coir pith. *Springerplus* **5,** 807 (2016).
11. Bagherian, H., Zokaee Ashtiani, F., Fouladitajar, A. & Mohtashamy, M. Comparisons between conventional, microwave- and ultrasound-assisted methods for extraction of pectin from grapefruit. *Chem. Eng. Process. Process Intensif.* **50,** 1237–1243 (2011).
12. Minjares-Fuentes, R. *et al.* Ultrasound-assisted extraction of pectins from grape pomace using citric acid: A response surface methodology approach. *Carbohydr. Polym.* **106,** 179–189 (2014).
13. Hii, C. L., Law, C. L. & Cloke, M. Modelling of Thin Layer Drying Kinetics of Cocoa Beans During Artificial and Natural Drying. *J. Food Eng.* **3,** 1–10 (2008).
14. Moorthy, I. G., Maran, J. P., Surya, S. M., Naganyashree, S. & Shivamathi, C. S. Response surface optimization of ultrasound assisted extraction of pectin from pomegranate peel. *Int. J. Biol. Macromol.* **72,** 1323–1328 (2015).
15. Kumar, S. k. Drying Kinetics of Banana Peel. *J. Food Process. Technol.* **6,** 10–12 (2015).
16. da Silva, W. P., e Silva, C. M. D. P. S., Gama, F. J. a. & Gomes, J. P. Mathematical models to describe thin-layer drying and to determine drying rate of whole bananas. *J. Saudi Soc. Agric. Sci.* **13,** 67–74 (2014).
17. KulkarniS.G. & Vijayanand, P. Effect of extraction conditions on the quality characteristics of pectin from passion fruit peel (Passiflora edulis f. flavicarpa L.). *LWT - Food Sci. Technol.* **43,** 1026–1031 (2010).
18. Manjarres-Pinzon, K., Cortes-Rodriguez, M. & Rodríguez-Sandoval, E. Effect of drying conditions on the physical properties of impregnated orange peel. *Brazilian J. Chem. Eng.* **30,** 667–676 (2013).
19. Marianski, S. & Marianski, A. *The Art of Making Fermented Sausages*. (2009).
20. *Food Analysis: Theory and Practice - Yeshajahu Pomeranz, Clifton E. Meloan - Google Books*.
21. Lee, W. C., Yusof, S., Hamid, N. S. A. & Baharin, B. S. Optimizing conditions for enzymatic clarification of banana juice using response surface methodology (RSM). *J. Food Eng.* **73,** 55–63 (2006).
22. Xu, Y. *et al.* Effects of ultrasound and/or heating on the extraction of pectin from grapefruit peel. *J. Food Eng.* **126,** 72–81 (2014).
23. Chaharbaghi, E., Khodaiyan, F. & Hosseini, S. S. Optimization of pectin extraction from pistachio green hull as a new source. *Carbohydr. Polym.* **173,** 107–113

(2017).
24. Prakash Maran, J., Sivakumar, V., Thirugnanasambandham, K. & Sridhar, R. Optimization of microwave assisted extraction of pectin from orange peel. *Carbohydr. Polym.* **97,** 703–709 (2013).
25. Huang, X., Li, D. & Wang, L. jun. Characterization of pectin extracted from sugar beet pulp under different drying conditions. *J. Food Eng.* **211,** 1–6 (2017).
26. CP Kelco. *GENU® pectin*. (A HUBER COMPANY, 2010).
27. Sharma, B. R., Naresh, L., Dhuldhoya, N. C., Merchant, S. U. & Merchant, U. C. An Overview on Pectins. *Times food Process. J.* **51,** 44–51 (2006).
28. Guidetti, R. *et al.* Extraction and Characterization of Pectin from Citric Waste. *Chem. Eng. Trans.* **44,** 1–7 (2015).

Characteristics Study of Photovoltaic Thermal System with Emphasis on Energy Efficiency

Chuah Yee Yong[1], *Mohammad Taghi Hajibeigy*[1,*], *Chockalingam Aravind Vaithilingam*[1], *Rashmi Gangasa Walvekar*[1]

[1]Taylor's University Lakeside Campus, Subang Jaya, Selangor, Malaysia

Abstract. Solar energy is typically collected through photovoltaic (PV) to generate electricity or through thermal collectors as heat energy, they are generally utilised separately. This project is done with the purpose of integrating the two systems to improve the energy efficiency. The idea of this photovoltaic-thermal (PVT) setup design is to simultaneously cool the PV panel so it can operate at a lower temperature thus higher electrical efficiency and also store the thermal energy. The experimental data shows that the PVT setup increased the electrical efficiency of the standard PV setup from 1.64% to 2.15%. The integration of the thermal collector also allowed 37.25% of solar energy to be stored as thermal energy. The standard PV setup harnessed only 1.64% of the solar energy, whereas the PVT setup achieved 39.4%. Different flowrates were tested to determine its effects on the PVT setup's electrical and thermal efficiency. The various flowrate does not significantly impact the electrical efficiency since it did not significantly impact the cooling of the panel. The various flowrates resulted in fluctuating thermal efficiencies, the relation between the two is inconclusive in this project.

1 Introduction

Current world energy demand is largely fulfilled by fossil fuel. Fossil fuel has the advantages over renewable energy that it is well documented and has the much more developed technology to harness the energy it provides. However, there are many reasons for us to move on to renewable energies such as solar, wind, geothermal and so on. Conventional coal-fired plants are the top generator all around the world but the process comes with large amounts of greenhouse gases and is unsustainable with the sources depleting. Figure 1 below shows the 2016 statistics of power generation and the methods used.

* Corresponding author: MohammadTaghi.Hajibeigy@Taylors.edu.my

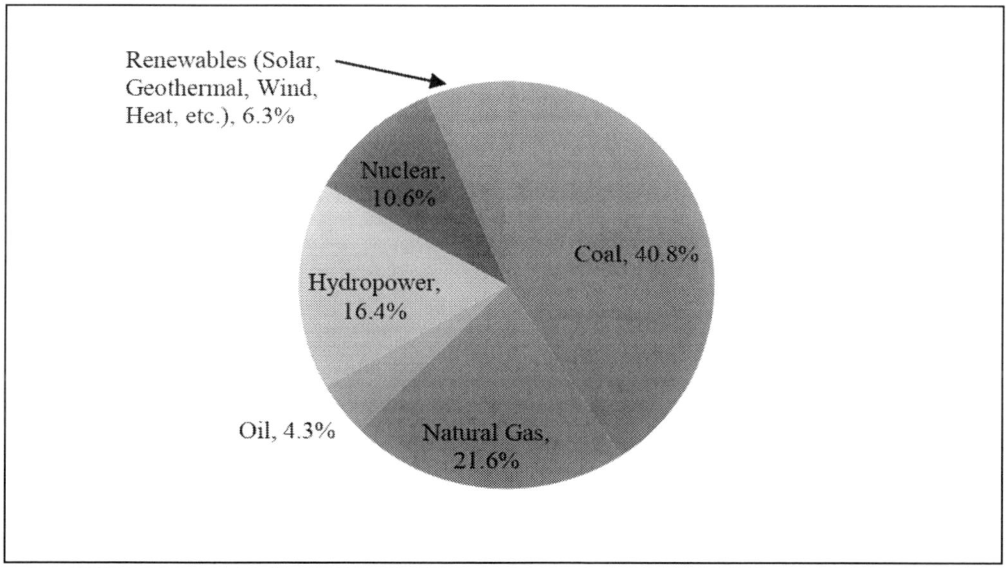

Fig. 1: Sources of Global Electricity Generation [1].

With cutting down the usage of fossil fuel to generate power as a target, this project approaches it through attempting to improve current methods of harnessing solar energy. Present methods of collecting solar energy fall into two categories, photovoltaic (PV) and thermal energy. Photovoltaics uses semiconductor materials which are capable of converting light into electricity, whereas thermal collectors are self-explanatory, where a collector is used to collect heat radiated from the light source.

Only a small amount of the solar energy can be harnessed by the PV panels, which are due to factors such as the current limitations of the technology. Standard monocrystalline panels can operate at roughly 15%-20% efficiency where the remaining solar energy is converted into thermal energy and are wasted. Currently, there are efforts to improve efficiency through methods such as solar tracking where the solar panels are designed to tilt itself towards the direction of the sun, such as passive and active orientation systems [2].

Where the tracking method improves electrical energy through adjusting the angle the sunlight hits the panel, the methods discussed in [3-8] uses the approach of reducing the temperature of the solar panel where high operating temperature causes a reduction in the panels' efficiency. Cooling the solar panel using several fluids were done by [3-6], whereas [7] shows that Phase Change Materials (PCM) can be considered to maintain the low temperature of the solar panel and [8] cools the solar panel by directly spraying water onto the panel surface.

The proposed method to increase solar energy harnessed is to retrofit a thermal system to a standard monocrystalline panel, where the integrated system is expected to operate at a higher efficiency. The design is to attach a heat sink to the back of the solar panel with thermal compound in between to allow heat to transfer from the panel to the heat sink. It is shown that Thermal Interface Materials (TIM) are essential to two contacting surfaces regarding heat transfer, where two flat surfaces can seemingly be tightly contacting while having microscopic air gaps that negatively affects the efficiency of heat transfer [9,10]. The heat sink is submerged into a body of water where it transfers the heat to a water storage tank through a copper tube. The idea of this design is to simultaneously cool the panel, which improves operating efficiency and harness the thermal energy.

2 Methodology

2.1 Setup Design and Fabrication

Solar panels generally are sold in the market without any heat dissipation modifications, where the panels are usually installed individually. That meant the fabrication of heat sinks to suitable specifications are required, SEMS Sdn. Bhd. agreed to arrange a heat sink in the desired dimensions. Due to a limited budget, it is not viable to fabricate a custom heatsink from scratch. Figure 2 shows the solar panel with the heat sink attached to the underside of the monocrystalline panel.

Fig. 2. PVT Panel Setup.

To avoid short circuit of the wirings, the wiring housing is sealed with silicone glue which is waterproof. The space between the panel and the heat sink are filled with thermal compound Kafuter K-5211, that prevents air pockets which reduces the effectiveness of heat transfer between the two layers.

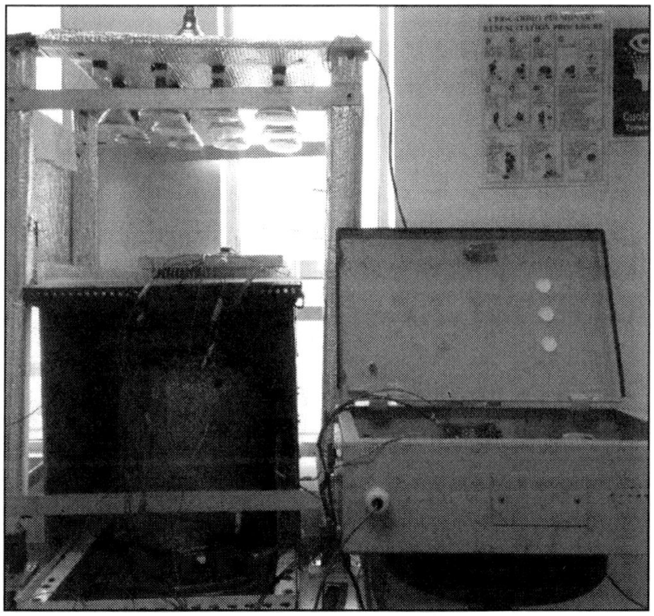

Fig. 3. Overall Experimental Setup.

The experiment will be done indoors to keep factors such as ambient temperature, fluctuating solar energy as consistent as possible. The overall setup can be seen in Figure 3 above, where the lightbulbs are placed directly above the panel that is used to provide consistent light energy. The heat sink is submerged in water within a Perspex container that has water being cycled through a water storage tank. The container on the right of the water storage tank holds the electronics such as the pyranometer and the data-logger.

2.2 Experimental Procedures

The steps below were taken to obtain the electrical and thermal performance:

i. Adjust the power supplied to the water pump at 12V.
ii. Switch on the lights.
iii. Start the data-logger.
iv. Start the EpSolar Monitoring Software.
v. Allow the setup to continue for 10 minutes.
vi. Collect data and switch off the water pump and lights.
vii. Experiment is repeated at different water pump power supply after water storage tank return to ambient temperature.

The panel temperature is observed throughout the experiment to ensure it does not exceed 70°C as it can cause permanent damage to the panel which reduces its operating efficiency.

2.3 Electrical Efficiency

The electrical efficiency of the solar panel was calculated using the equation below:

$$\eta_{elec} = (V_m I_m)/AR \tag{1}$$

Where: η_{elec} = electrical efficiency
V_m = maximum power point voltage, V
I_m = maximum power point current, A
A = panel area, m^2
R = solar radiation, W/m2

2.4 Thermal Efficiency

At steady state, the PVT thermal performance can be calculated using the equation below:

$$\eta_T = \text{(thermal energy collected)/(total amount of solar energy)} \tag{2}$$

$$\eta_T = (mC_p\,(T_o\text{-}T_i))/ARt \tag{3}$$

Where: η_T = thermal efficiency
m = water mass, kg
C_p = specific heat capacity, kJ/kg.K
T_o = outlet temperature, K
T_i = inlet temperature, K
t = time, s

3 Results

3.1 Electrical Efficiency

3.1.1 Difference in Standard PV and PVT Setup

The difference between a standard monocrystalline PV setup and the novel PVT setup fabricated for this project is observed. The factors for both setups such as ambient temperature, distance of panel to light bulbs and more are kept as consistent as possible. The power generated is recorded using the EpSolar Monitoring Software. Figure 4 below shows the irradiance and the power generated by the PV setup throughout the experiment. At the peak of the power generated by the solar panel, the PV setup achieved a 3.77% electrical efficiency. This setup recorded an average efficiency of 1.64%.

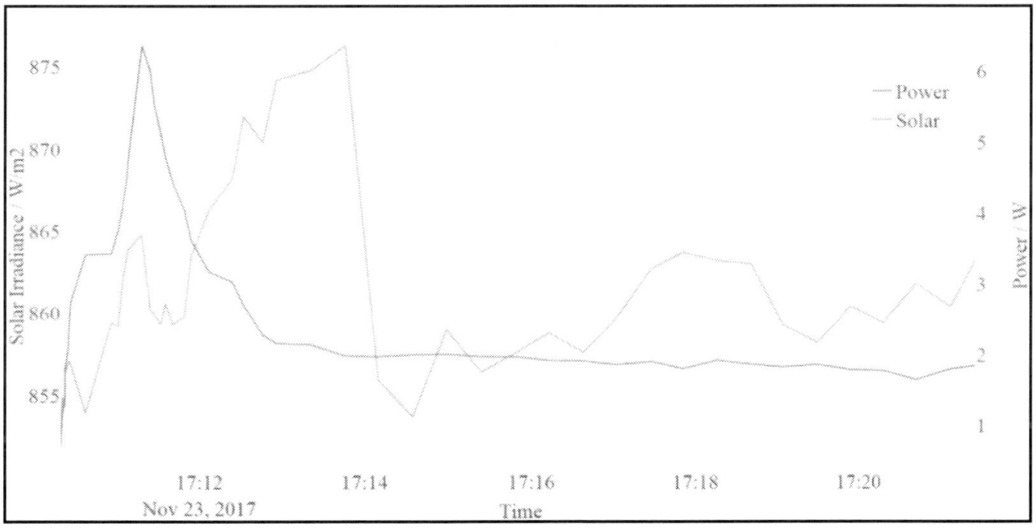

Fig. 4. Real Time Power Graph of PV Panel.

Figure 5 below shows the results of the PVT setup, with the water pump running at 8.4 L/min. It is measured at 4.62% maximum electrical efficiency and 2.15% average electrical efficiency.

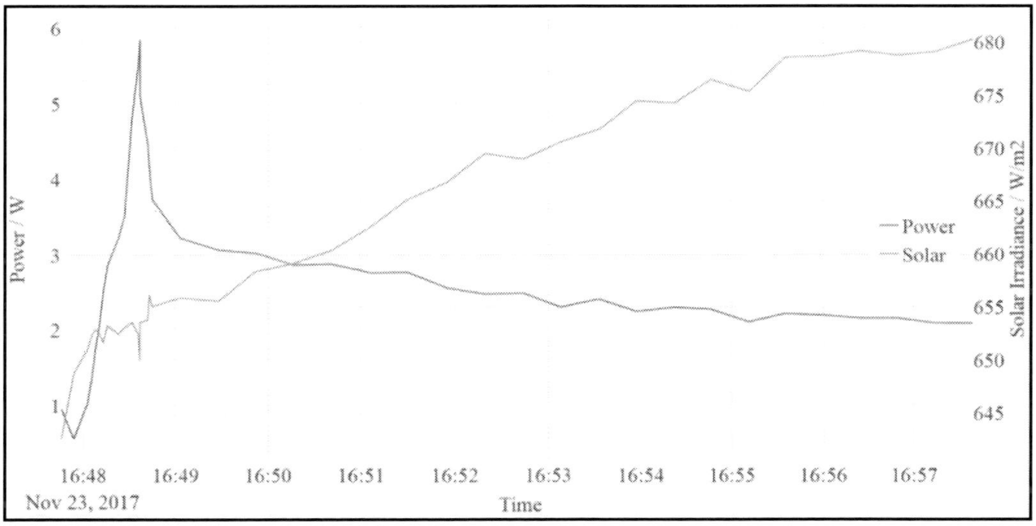

Fig. 5. Real Time Power Graph of PVT Setup.

3.1.2 Difference in Flowrate of PVT Setup

By using an adjustable AC/DC converter, the flowrate flowing into the Perspex container is set to 3.9 L/min, 5.3 L/min, 6.4 L/min and 8.4 L/min. The flowrates are respectively due to the voltage fed to the water pump being 6V, 7.5V, 9V and 12V.

Table 1. Water Pump Flow Rate at the Corresponding Voltages.

Voltage Supplied (V)	Flowrate (L/min)
6.0	3.9
7.5	5.3
9.0	6.4
12.0	8.4

The electrical power generated between the four experiments are recorded and graphed together to be compared in Figure 6 below.

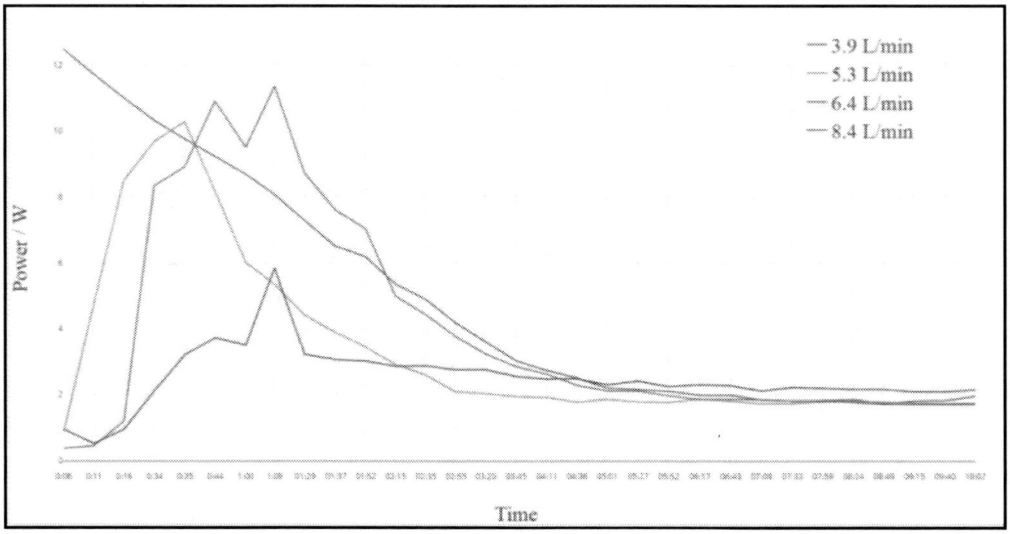

Fig. 6. Real Time Power Graph at Different Flow Rates.

3.2 Thermal Efficiency

The same experimental procedures are used to determine the thermal performance of the PVT setup. The temperature of the water in the storage tank is recorded and the difference between the beginning and the end of the experiment is tabulated as below.

Table 2. The Difference of Water Temperature.

Water Flowrate	Temperature Difference (°C)
3.9 L/min	1.9
5.3 L/min	1.1
6.4 L/min	2.3
8.4 L/min	1.4

By using the formulae mentioned in subsection 2.4, the thermal efficiency of the setup can be calculated. The heat capacity of water is assumed to be at 4184 J/kgK. Table 3 below shows the resulting thermal efficiency of the PVT setup at various flowrates.

Table 3. Different Flowrates and the Corresponding Thermal Performance.

Experiment	Water Flowrate	Average Irradiance	Thermal Efficiency
1	3.9 L/min	704.10	48.25%
2	5.3 L/min	705.57	27.88%
3	6.4 L/min	715.77	57.46%
4	8.4 L/min	672.01	37.25%

3.3 Combined Efficiency

The combined efficiency of the standard PV setup and the fabricated PVT setup are compared. Combined efficiency refers to the sum of both thermal efficiency and the electrical efficiency. It is used to compare whether the integration of thermal collector to the standard PV panel will yield an improvement in harnessing solar energy.

The combined efficiency of the PVT system resulted in 39.4% efficiency while the standard PV setup can only harness electrical energy, thus only operates at 1.64% efficiency in harnessing solar energy.

4 Conclusion

The integration of a thermal system which simultaneously cools the panel and collects thermal energy allowed the solar panel to operate at a higher efficiency. The standard PV setup is rated at 1.64% electrical efficiency while the PVT setup is rated at 2.15%. The decreasing power generated over time can be attributed to the increasing temperature of the panel, illustrated in Figure 7 below where the PVT setup is shown dissipating heat better than the standard PV setup.

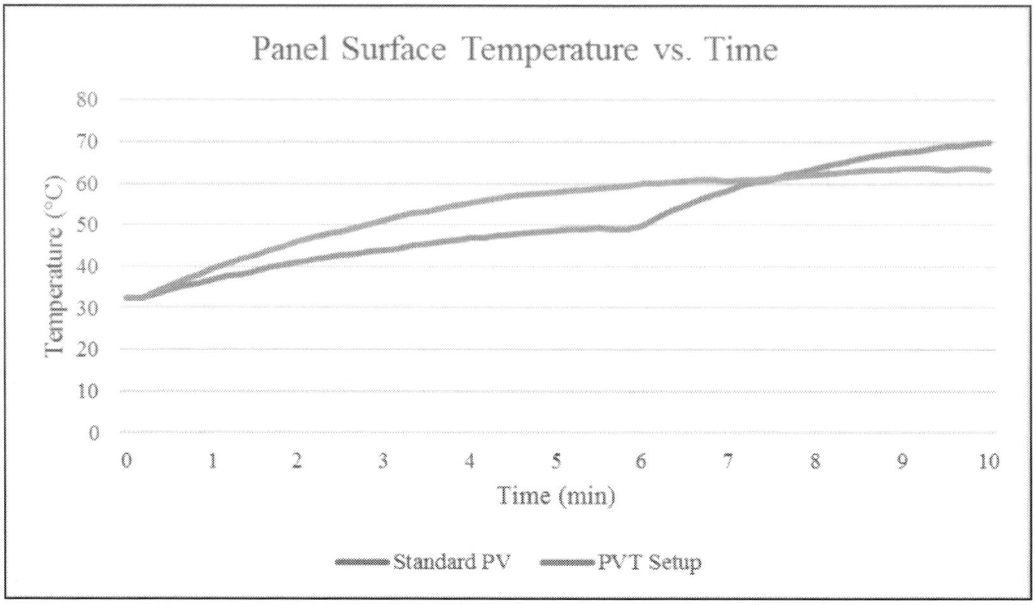

Fig. 7. The temperature of PV Panel Surface over Time.

The PVT setup boasts a much improved combined efficiency in harnessing solar energy, from 1.64% to 39.4% of solar energy collected which deems that integration of thermal system to an existing PV setup successful in terms of improving energy efficiency.

The effects on how the flowrate affect the electrical efficiency and thermal efficiency are also studied in this project. It seems that the flow rate does not significantly impact the electrical efficiency, whereas the thermal efficiency of the PVT setup fluctuates and the relationship between the two could not be defined.

However, it should be noted that the solar panels provided could be unreliable as typical monocrystalline panels operate at roughly 15% efficiency, while the panels used only operated around 2%.

References

1. International Energy Agency, "Key world energy statistics," *Statistics (Ber).*, p. 80, (2016)
2. I. L. Alboteanu, F. Ravigan, and S. Degeratu, "Methods for Increasing Energy Efficiency of Photovoltaic Systems," *Int. J. Power Renew. Energy Syst.*, **vol. 1**, pp. 51–61 (2014)
3. H. Bahaidarah, A. Subhan, P. Gandhidasan, and S. Rehman, "Performance evaluation of a PV (photovoltaic) module by back surface water cooling for hot climatic

conditions," *Energy*, **vol. 59**, pp. 445–453 (2013)

4. R. Mazón-Hernández, J. R.García-Cascales, F. Vera-García, A. S. Káiser, and B. Zamora, "Improving the Electrical Parameters of a Photovoltaic Panel by Means of an Induced or Forced Air Stream" (2013)
5. J. K. Tonui and Y. Tripanagnostopoulos, "Air-cooled PV/T solar collectors with low cost performance improvements," *Sol. Energy*, **vol. 81, no. 4**, pp. 498–511 (2007)
6. C. G. Popovici, S. V. Hudişteanu, T. D. Mateescu, and N.-C. Chereche\$, "Efficiency Improvement of Photovoltaic Panels by Using Air Cooled Heat Sinks," *Energy Procedia*, **vol. 85**, no. November 2015, pp. 425–432 (2016)
7. A. Hasan, S. J. McCormack, M. J. Huang, and B. Norton, "Evaluation of phase change materials for thermal regulation enhancement of building integrated photovoltaics," *Sol. Energy*, **vol. 84**, no. 9, pp. 1601–1612 (2010)
8. K. A. Moharram, M. S. Abd-Elhady, H. A. Kandil, and H. El-Sherif, "Enhancing the performance of photovoltaic panels by water cooling," *Ain Shams Eng. J.*, **vol. 4**, no. 4, pp. 869–877 (2013)
9. R. Prasher and C.-P. Chiu, "Thermal Interface Materials," in *Materials for Advanced Packaging*, D. Lu and C. P. Wong, Eds. Cham: Springer International Publishing, pp. 511–535 (2017)
10. D. L. Saums, "Advances in Thermal Interface Materials for Power LED Applications," in *Thermal Management for LED Applications*, C. J. M. Lasance and A. Poppe, Eds. New York, NY: Springer New York, pp. 299–346 (2014)

Synthesis of A Sustainable Sago-Based Value Chain via Fuzzy Optimisation Approach

Jeffrey Hong Seng Chong[1] , *Yoke Kin* Wan[1,1], and *Viknesh* Andiappan[2]

[1]School of Engineering, Taylor's University, Subang Jaya, Selangor, Malaysia.
[2]School of Engineering and Physical Sciences, Heriot-Watt University Malaysia, 62200, Putrajaya, Wilayah Persekutuan Putrajaya, Malaysia

Abstract. Sago starch is one of the staple foods for human, especially in Asia's Region. It can be produced via sago starch extraction process (SSEP). During the SSEP, several types of sago wastes are generated such as sago fiber (SF), sago bark (SB) and sago wastewater (SW). With the increase in production of existing factories and sago mills, the sago industrial practice in waste disposal management is gaining more attention, thus implementation of effective waste management is vital. One of the promising ways to have effective waste management is to create value out of the sago wastes. In a recent study, sago-based refinery, which is a facility to convert sago wastes into value-added products (e.g., bio-ethanol and energy) was found feasible. However, the conversion of other value added products from sago wastes while considering the environmental impact has not been considered in sago value chain. Therefore, an optimum sago value chain, which involved conversion activities of sago wastes into value-added products, is aimed to be synthesised in this work. The optimum sago value chain will be evaluated based on profit and carbon emissions using fuzzy-based optimisation approach via a commercial optimisation software, Lingo 16.0. To illustrate the the developed approach, an industrial case study has been solved in this work.

1 Introduction

Sago palm (*Metroxylan Sagu*) is one of the underutilised crops that can potentially produce staple food for human, especially in Asia's Region. The sago palm produces sago starch which contains high carbohydrates that can be obtained from the trunks of sago palm. Such palm is an excellent crop for sustainable agriculture due to its ability to thrive in most soil conditions and it requires only one time planting that has the ability of self-sustain by producing its young tree, also known as suckers or tillers [1]. It can also be served as an income for poverty alleviation as it is commonly planted in rural areas [2]. Karim et al. [3] states that, under optimal farming practices, up to 25 metric tonnes of sago starch per hectare annually can be achieved, which is roughly 3 - 4 times the starch yield of staple crops such as rice (6 tonnes/ha/yr) or wheat (5 tonnes/ha/yr) [2], [4]. Sago starch can be produced via sago starch extraction process (SSEP) as shown in Figure 1. As shown, there

[1] Corresponding author: YokeKin.Wan@taylors.edu.my

are several types of sago wastes that are generated during the process, namely sago fiber (SF), sago bark (SB) and sago wastewater (SW)[4].

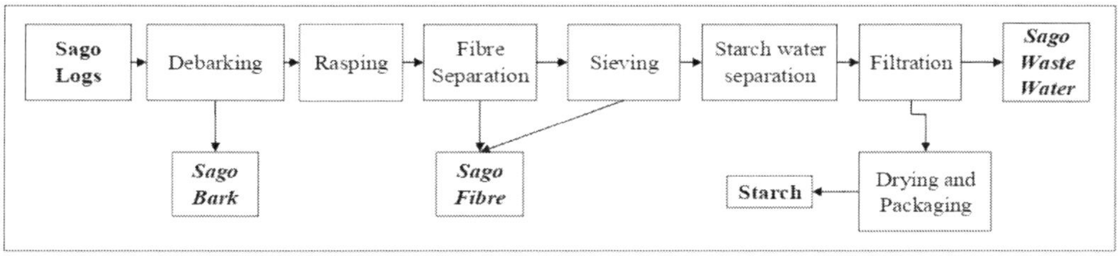

Fig. 1. Block Flow Diagram of SSEP

In current industrial practices, SB is used as timber fuel for firing or will be disposed around the mills to be used as flooring material [4]. SF is used as animal feed, otherwise, it will be released into nearby streams along with SW. According to Adeni et al. [5], SW, combined with SF, represents high organic materials, chemical oxygen demand (COD) and biochemical oxygen demand (BOD), which does not comply to the discharged regulations set in Environmental Quality Act, 1974. The installation of pollution control devices can be costly, approximately 20-50% of the total investment for a large scale production mill, which is not preferred by producers in this relatively low value commodity industry [5]. SF and SB, which are lignocellulosic waste (LCW), contains residue starch which can be converted into value added products such as glucose, which can be further refined to higher value product such as bio-fuels.

To address the aforementioned issues, a viable alternative is to create value out of the sago wastes to improve the current performance of sago value chain. As mentioned, SF and SB could be utilised for the production of value added products (i.e., ethanol, organic acids, enzymes and others)[6]. However, the idea of converting sago wastes into valuable products is still in its infancy. In most recently, Wan et al. [7] studied the feasibility of a sago-based refinery, which is a facility to convert sago wastes into value-added products (e.g., bio-ethanol and energy). The study results showed that it is feasible to convert sago wastes (SB and SF) into bioethanol and energy in sago-based refinery. In addition, Wan and her co-authors [8] also conducted a value chain analysis for sago industry to determine the optimum value chain pathway for sago industry. Value chain is a set of activities that are done to design, produce, market, deliver and support the product created by a party [9]. First proposed by Michael Porter, value chain divides break down a chain of activities into individual parts to scrutinise the dynamics of activities in order to reveal its strength and weaknesses, thus identifying value creation opportunities. Application of value chain in the sago industry has been pioneered by a study done by Wan et al. [8], whom developed a value chain to improve the sustainability of the industry to identify value creation opportunities. However, the conversion of sago wastes into value-added products had not been considered in the sago value chain. Therefore, in order to further improve the economic and environmental performance of the sago value chain, this study aims to expand the previous work done by Wan et al [8], by introducing alternatives towards utilisation of sago wastes to create value added products while considering the environmental impact done to the sago value chain. Figure 2 showed the sago value chain done by Wan et al. [8] and the sago value chain has been expanded in this work by considering value added product conversion technologies (i.e., technology 1, intermediates products, technology 2, and value-added products).

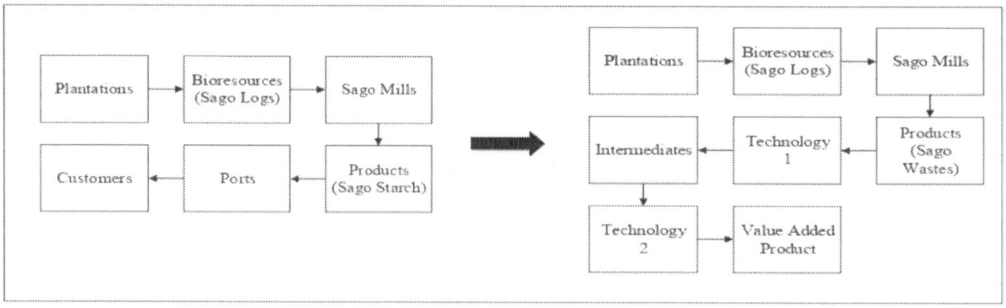

Fig. 2. Proposed Value Chain of Sago Industry in this Study

Douglas et al. [10] adapted fuzzy-based optimisation approach based on 'max-min' aggregation to synthesis of a sustainable integrated bio-refinery. The author's study did not identify the optimal value chain in the sago industry. However, the author has proven that fuzzy-based optimisation approach can be used to obtain an optimal pathway from a superstructure, while considering the economic and environmental impact. Therefore, in order to further improve the economic and environmental performance of the sago value chain, this study aims to expand the previous work done by Wan et al. [8], by introducing alternatives towards utilisation of sago wastes to create value added products to maximise profit, while minimising environmental impact done of the sago value chain. To determine the optimum pathway in this, a fuzzy-based optimisation approach will be developed in this work to find an optimal solution to the multi-objective model in this study. To illustrate the application of proposed model, a real industrial case study has been solved.

2 Methodology

A generic wastewater treatment superstructure is first developed to show all possible pathway to treat wastewater as shown in Figure 1. Based on this generic superstructure, MFCA-based approach has been developed by formulating mathematical equations for each treatment in WWTP as shown in following sections. These equations included mass balance equation, contaminant equation, and cost computation equations. To demonstrate the developed approach, an industry case study, organic wastewater treatment plant is solved.

2.1 Generic superstructure of wastewater treatment

Fig. 3 illustrates the methodology using the fuzzy-based optimisation approach. First step is to determine the superstructure based on the layers of activities from the proposed value chain in this study. Then, the objectives of the study have to be set and relevant data needs to be collected based on the objective set. In this study, the objective of maximising profit and minimising CFP of the sago value chain is set for synthesis of an optimal value chain for sago waste conversion. Profit and CFP objectives are contradictory. Therefore, a fuzzy-based optimisation approach has to be developed. Douglas et al. [10] Following the fuzzy-based optimization approach, the mass balance has to be formulated, as well as the formulation of equations necessary for all objective functions, which in this case, is the profit and CFP equations. The developed formulations are then solved using a commercial optimization software Lingo V.16.

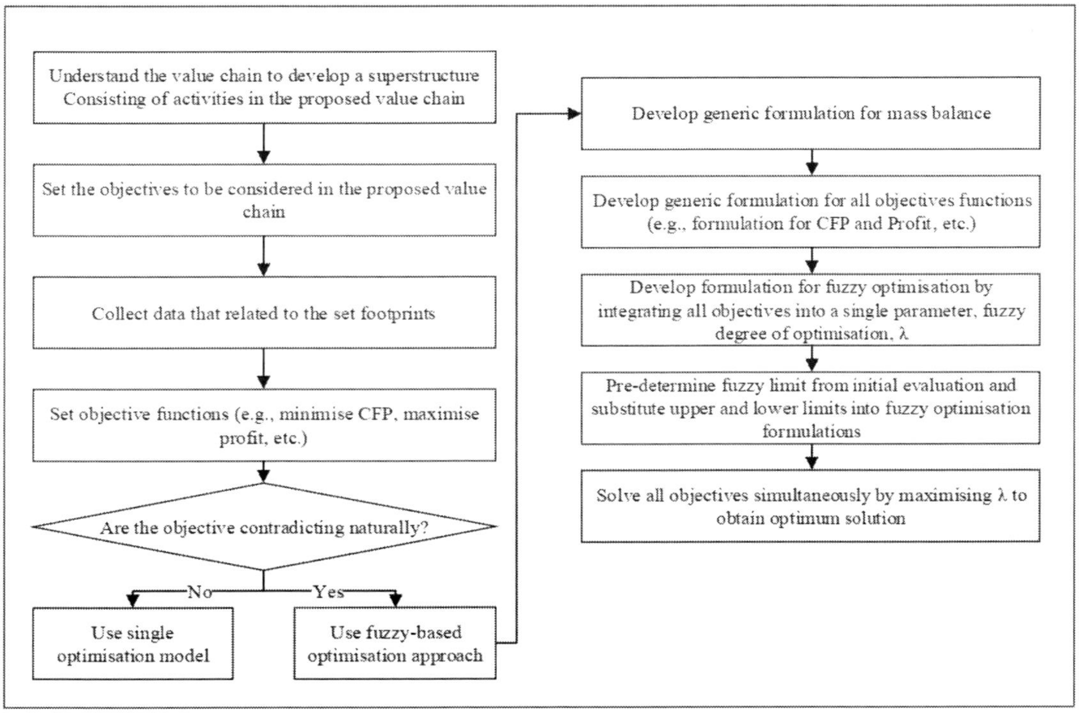

Fig. 3. Methodology Flow Chart for Fuzzy-Based Optimisation Approach

To illustrate the alternatives stages of the value chain, superstructure is first developed. As shown in Figure 4, the superstructure illustrates the seven layers of activities that is considered in the value chain proposed in this study. With reference to the approach used by Wan et al. [8], the activity starts from the harvesting palms from a set of plantation $g \in G$ with annual available palm F_g^{av}, processed and converted to a set of bio-resource $m \in M$, which is limited to only sago logs in this study. The sago logs are then sent to a set of sago mill $f \in F$ with starch production capability $F_f^{starch.lim}$. The sago mill f turns the bio-resource m into a set product $p \in P$ based on the amount of starch produced from the sago mill, which in the case of this study, represents the sago wastes. In the study of Wan et al.[8], the product considered in the author's study is just sago starch in the product p layer. However, in this study, sago starch would not be focused, unlike as shown in the works of Wan et al. [8]. Sago wastes (e.g., SB, SF, SW) will be considered in this study instead.

The sago waste is then sent to a set of conversion technologies 1 $c \in C$ to produce a set of Intermediate product $d \in D$. Intermediate product d is then sent for further processing in a set of technology 2 $e \in E$ to produce a set of end product $z \in Z$. To model this superstructure, mathematical equations have been formulated in following sections.

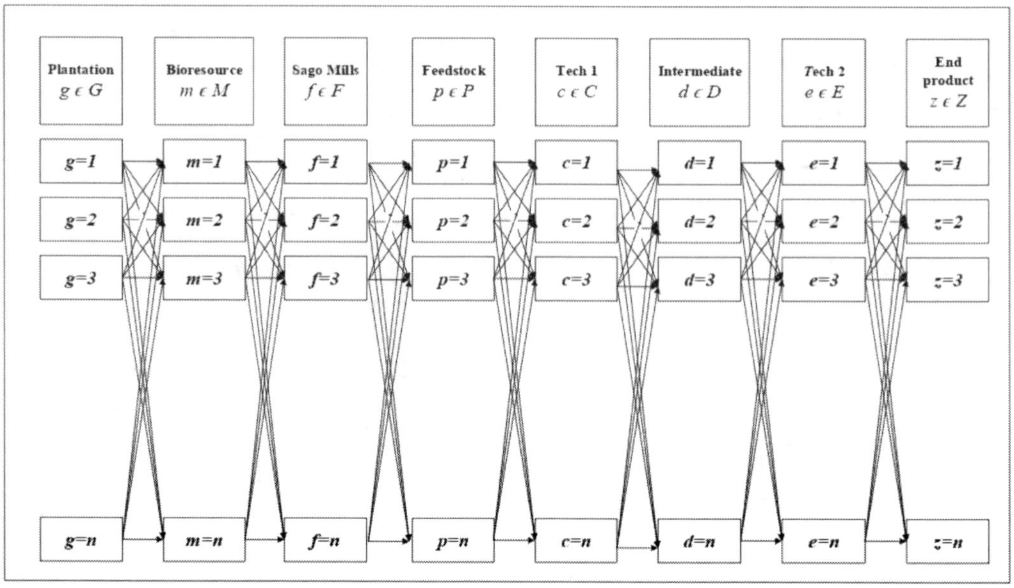

Fig. 4. Generic Superstructure

2.1 Mass Balance Equations

The mass balance for the first four layers of activities illustrated by the value chain in this study is adopted from the work done by Wan et al [8]. The total number of palms harvested F_g^{out} (palm/y) from plantation g has to be less than available palms to ensure a sustainable harvesting process, as:

$$F_g^{av} > F_g^{out} \ \forall g \tag{1}$$

Harvested sago palms from plantation *g* are then converted into bio-resource *m* before its sent to sago mill *f*. Currently, the only bio-resource *m* sent to sago mills are sago logs. Therefore, given the conversion of palms to sago logs $X_{g,\log}$, the total amount of sago logs sent to sago mill *f*, $F_{g,\log}^{in}$ (log/y) can be determined via:

$$F_{g,\log,f}^{in} = F_{g,f}^{out} X_{g,\log} \ \forall g \tag{2}$$

The total sago logs transported to sago mill *f* is the determined via:

$$F_f^{in} = \sum_{g=1}^{G} F_{g,\log,f}^{in} \ \forall f \tag{3}$$

It is noted that sago logs can be transported via roads or river. However, only road transportation is considered in this study, similar to the study done by Wan et al.[8]. Given a lorry capacity $C_{weightlim}^{lorry}$ (t/trip) and weight per log $W_{\log}$ (t/log), the number of trips needed to transport the sago logs from plantation *g* to sago mill *f* $n_{g,f}^{trip}$ (trip) is determined via

$$n_{g,f}^{trip} \geq \frac{F_{g,\log,f}^{in} W_{\log}}{C_{weight\lim}^{lorry}} \quad \forall g \forall f \tag{4}$$

In sago mill *f*, the sago logs are converted into sago starch, producing sago wastes as by-products, labelled as product *p* in this study. To calculate the amount of starch produced from sago mill *f* $Starch^f$ (t/y), Equation (5) is formulated.

$$Starch^f = \sum_{g=1}^{G} F_{g,\log,f}^{in} X_{g,\log,starch} \quad \forall f \tag{5}$$

Given that each sago mill *f* has a production capacity F_{Starch}^{av} (t/y) which cannot be exceeded, hence, the equation below is included in the model

$$Starch^f < F_{Starch}^{av} \quad \forall f \tag{6}$$

To satisfy sago demand, $StarchDemand$ (t/y) which is set at 32,000 t/y in this study, the Equation (7) is added into the model.

$$StarchDemand = \sum_{f=1}^{F} starch^f \tag{7}$$

The amount of product *p* produced from sago mill *f* from SSEP $F_{f,p}^{out}$ is calculated via:

$$F_{f,p}^{out} = Starch^f X_{starch,p} \quad \forall f \forall p \tag{8}$$

Hence, the amount of product *p* produced F_p^{in} (t/y) from all sago mills is calculated as:

$$F_p^{in} = \sum_{f=1}^{F} F_{f,p}^{out} \quad \forall p \tag{9}$$

Product *p* is then sent to technology 1 *c* to be converted into intermediate *d*. The amount of product *p* sent to technology 1 *c* F_c^{in} (t/y) is as

$$F_c^{in} = \sum_{p=1}^{P} F_{p,c}^{in} \quad \forall c \tag{10}$$

In technology 1, the amount of intermediate *d* produced from product *p* is dependent on the conversion $\mathrm{X}_{\mathrm{p,b}}^{c}$, as:

$$F_{c,d}^{out} = F_c^{in} \, \mathrm{X}_{\mathrm{p,d}}^{c} \quad \forall c \forall d \tag{11}$$

Given the conversion $\mathrm{X}_{\mathrm{p,b}}^{c}$, the amount of intermediate *d* produced F_d^{in} (t/y) can be calculated via:

$$F_d^{in} = \sum_{c=1}^{C} F_{c,d}^{out} \quad \forall d \tag{12}$$

Next, intermediate *d* is the distributed to technology 2 *e* via:

$$F_d^{out} = \sum_{e=1}^{E} F_{d,e}^{out} \ \forall d \quad (13)$$

Between these 2 layers of activities, conversion of Intermediate d into suitable unit for Technology 2 e $X_{d,e}^{unit}$ might be needed (e.g., conversion of syngas to power unit GJ in for CHP in technology 2 e). Therefore, the flow into technology 2 e is calculated via:

$$F_{d,e}^{in} = F_{d,e}^{out} \ X_{d,e}^{unit} \qquad \forall d \forall e \quad (14)$$

Hence, the amount received in technology 2 e F_e^{in} (unit/y)can be calculated via:

$$F_e^{in} = \sum_{d=1}^{D} F_{d,e}^{in} \ \forall e \quad (15)$$

Similar to technology 1 c, the amount produced from technology 2 e is dependent on the conversion $X_{d,z}^{e}$

$$F_{e,z}^{out} = F_{e,z}^{out} \ X_{d,z}^{e} \qquad \forall e \forall z \quad (16)$$

Given the conversion $X_{d,z}^{e}$, the amount of end product z produced F_z^{in} (unit/y) can be calculated via:

$$F_z^{in} = \sum_{e=1}^{E} F_{e,z}^{out} \ \forall d \quad (171)$$

2.2 Economic Performance Evaluation Equations

The economics performance of each activities is evaluated to determine the economic performance of the proposed value chain as a whole. In this study, the economic performance is defined as the difference between revenue obtained from production of end product z and the cost associated with the production of end product z. Similar to mass balance formulation at section 2.4, the economic performance of the first four layers of activities in the value chain is formulated with reference to the approach taken by Wan et al in [8]. At plantation g, the cost considered is the harvesting cost, as well as the transportation of bio resource m (sago logs) to sago mill f. the total cost of harvesting $Tot\cos t_g^{Harvest}$ (USD/y), as well as total cost of transportation $Tot\cos t^{Transportation}$ (USD/y) is calculated via:

$$Tot\cos t_g^{Harvest} = \sum_{g=1}^{G} U\cos t_g^{Harvest} F_g^{out} \quad (18)$$

$$Tot\cos t^{Transportation} = \sum_{f=1}^{F} \sum_{g=1}^{G} U\cos t^{km} d_{g,f}^{trip} n_{g,f}^{trip} \quad (19)$$

where $U\cos t_g^{Harvest}$ is the cost of harvesting at plantation g, $U\cos t^{km}$ (USD/palm) is cost of transportation per km USD/km) and $d_{g,f}^{trip}$ (km) is the distance between plantation g and sago mill f. Hence the total cost incurred by all plantations $Tot\cos t^{Plantation}$ (USD/y) is as:

$$Tot\cos t^{Plantation} = Tot\cos t_g^{Harvest} + Tot\cos t_g^{Transportation} \quad (20)$$

In sago mill *f*, the cost considered is the cost of process $Tot\cos t_f^{Process}$ (e.g., electricity, water consumption) as well as raw material cost $Tot\cos t_f^{Material}$ for the production of sago starch. In this study, sago wastes are assumed to be produced from SSEP without cost. The cost of transportation to refinery site for conversion via technology 1 *c* is also not considered in this study. Therefore, the total cost of the sago mill is calculated by summing the cost of process and raw materials as showed in Equations (21) – (23):

$$Tot\cos t_f^{Material} = \sum_{f=1}^{F} U\cos t_f^{Material} Starch^f \quad (21)$$

$$Tot\cos t_f^{Process} = \sum_{f=1}^{F} U\cos t_f^{Process} Starch^f \quad (22)$$

$$Tot\cos t^{mill} = Tot\cos t^{material} + Tot\cos t^{process} \quad (23)$$

where $U\cos t_f^{Material}$ (USD/t) and $U\cos t_f^{Process}$ (USD/t) is the process and raw material unit cost to produce one tonne of sago starch in sago mill *f*. The total cost of production $Tot\cos t^{Prod}$ (USD/y) is the calculated via:

$$Tot\cos t^{Prod} = Tot\cos t^{plantation} + Tot\cos t^{mill} \quad (24)$$

Next, the cost of the product *p* $Tot\cos t_p$ (USD/y) is considered in this study via:

$$Tot\cos t_p = \sum_{p=1}^{P} \mathrm{UCost_p} F_p^{in} \quad (25)$$

where $\mathrm{UCost_p}$ is unit cost of product *p* (USD/t). In order to produce intermediate *d* from Technology 1 *c*, there will be cost associated, which is calculated via:

$$Tot\cos t_b = \sum_{b=1}^{B} \mathrm{UCost}_b F_b^{out} \quad (26)$$

where UCost_b (USD/t) is the unit cost of converting product *p* to intermediate *d*. Similarly, the cost to produce end product *z* via Technology 2 *e* is calculated via:

$$Tot\cos t_e = \sum_{e=1}^{E}\sum_{z=1}^{Z} \mathrm{UCost}_e F_e^{in} \quad (27)$$

where UCost_e (USD/t) is the unit cost of converting Intermediate *d* into End Product Z via technology 2 *e*. Lastly, the total cost of the value chain $Tot\mathrm{Cos}t^{Overall}$ (USD/y) is calculated via:

$$Tot\,\mathrm{Cos}t^{Overall} = Tot\cos t_{production} + Tot\cos t_b + Tot\cos t_c + Tot\cos t_d + Tot\cos t_e \quad (28)$$

Next, the revenue from the proposed value chain $Tot\,Rev_z$ (USD/y) can be calculated via:

$$Tot\,Rev_z = \sum_{z=1}^{Z} Price_z F_z^{in} \tag{29}$$

where $Price_z$ is the unit price for the End Product z produced. Therefore, in order to evaluate the economic performance of the proposed value chain, the overall profit $Tot\,Profit^{Overall}$ (USD/y) is calculated via:

$$Tot\,Profit^{Overall} = Tot\,Rev_z - Tot\,Cost^{Overall} \tag{30}$$

2.3 Carbon Footprint (CFP) Computation Equations

The formulation of CFP is done with reference to CFP formulation done by Wan et al. [8]. The author considered the land use change (LUC) emission from the commercialisation of land for plantation. Given that the yearly emission per metre square from LUC $UCFP_g^{LUC}$ (kg/m^2.y) is 2.33 as reported by Fargione et al. [11]. The total land use change carbon footprint from plantation g $TotCFP^{LUC}$ (kg CO^2/y) can be calculated via:

$$TotCFP_{plantation}^{LUC} = \sum_{g=1}^{G} A_g UCFP_g^{LUC} \tag{31}$$

$$TotCFP_{plantation}^{Fuel} = \sum_{f=1}^{F} UCFP_{km}^{fuel} n_{g,f}^{trip} d_{g,f}^{trip} \tag{32}$$

$$TotCFP_{plantation}^{total} = TotCFP_{plantation}^{LUC} + TotCFP_{plantation}^{Fuel} \tag{33}$$

Next, the total amount of power and fuel consumed in sago mill *f* is determined in order to calculate the CFP emission from sago mill *f*. The CFP from power consumption and fuel consumption is calculated via:

$$TotCFP_{mill} = \sum_{f=1}^{F} UCFP_{KWh}^{Power} E_{f,starch} Starch^{f} \tag{34}$$

Thus the total CFP emission from production process can be calculated via:

$$TotCFP^{Production} = TotCFP_{mill}^{Power} + TotCFP_{plantation}^{total} \tag{35}$$

Considering that product p in this study represents the sago waste produced from SSEP, the CFP emission for production of product p is assumed to be zero. However, the CFP emission from conversion of product p to intermediate d via technology c is considered in this study as:

$$TotCFP_b^{Total} = \sum_{b=1}^{B} UCFP_b F_b^{out} \tag{36}$$

Next, the CFP emission from conversion of intermediate d to end product z via technology 2 e is the sum of the base CFP emission from processing intermediate d and also the specific power consumption of producing end product z via technology 2 e.

$$TotCFP_e^{Base} = \sum_{e=1}^{E} UCFP_e^{base} F_e^{in} \tag{37}$$

$$TotCFP_e^{Specific} = \sum_{z=1}^{Z} \sum_{e=1}^{E} UCFP_{e,z}^{specific} F_{e,z}^{out} \tag{38}$$

$$TotCFP_e^{Total} = TotCFP_e^{Base} + TotCFP_e^{Specific} \tag{39}$$

The total overall CFP $TotCFP^{Overall}$ (kg CO^2/y) is calculated via:

$$\begin{aligned} TotCFP^{Overall} &= TotCFP_{plantation}^{total} + TotCFP_{mill} \\ &+ TotCFP_b^{Total} + TotCFP_e^{Total} \end{aligned} \tag{40}$$

2.4 Fuzzy-Based Optimization Approach Equations

Cost and Environmental objectives are often contradictory, a fuzzy-based optimisation approach is adopted in this study to select the optimum pathway for conversion of sago waste to value –added products. Using the 'min-max' aggregation concept in fuzzy optimisation, the optimum pathway can be obtained by maximising the least satisfied constraints. Fuzzy optimisation combines multiple objectives into a single variable, λ, having a range of 0 to 1. In this study, the objectives are integrated into λ as shown in the equations below:

$$\frac{CFP^{UL} - TotCFP}{CFP^{UL} - CFP^{LL}} \geq \lambda \tag{41}$$

$$\frac{Tot\,Profit - Profit^{LL}}{Profit^{UL} - Profit^{LL}} \geq \lambda \tag{42}$$

where $TotCFP$ and $Tot\,Profit$, are the carbon footprint and profit generated by utilizing the optimum pathway for conversion of sago wastes. Meanwhile, CFP^{LL} and CFP^{UL} are the pre-determined lower and upper limit of the carbon footprint generated, same goes for $Profit^{UL}$ and $Profit^{LL}$ for profit generated. These upper and lower limit are determined by extracting data from evaluation results of the model at different objectives.

It is noted that each objective has a predetermined fuzzy range, represented by a linear function within the lower and upper limit, as shown in Fig 2.

Fig. **2** (a) illustrates the maximization case, where λ approaches 1 when the objective, $Tot\,Profit$ approaches the upper limit, $Profit^{UL}$ of the fuzzy approach. On the contrary, Fig. 2 (b) demonstrates the minimization case, where λ approaches 1 when the objective, $TotCFP$ approaches the lower limit, CFP^{UL} .

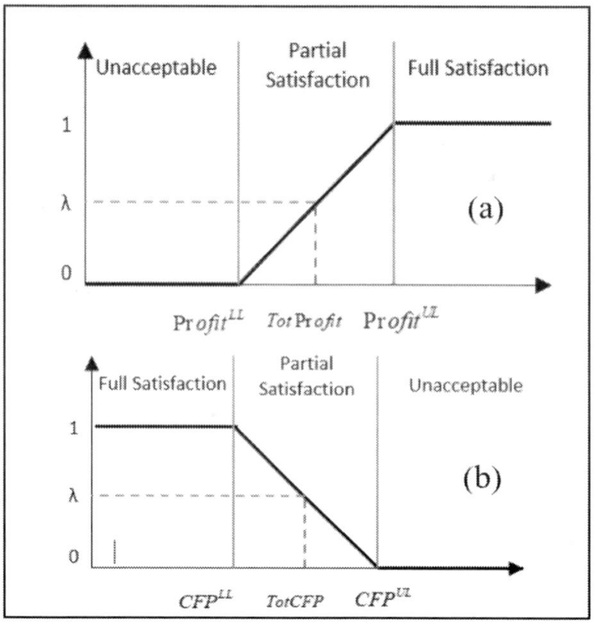

Fig. 2. The Concept of Fuzzy

In order to generate an optimum pathway for the conversion of sago waste while maximizing profit and minimizing carbon footprint, Eq. (40) and (41) are included in the optimization model, and solved simultaneously by maximising the degree of satisfaction, λ. This mixed integer linear programming (MILP) model is solved via Lingo v17.0 in a Lenovo Ideapad 700 with Intel Core i7-6700HQ (2.6 GHz) and 8GB ram running on 64-bit operating system.

3 Case Study

Fig 6 shows the case study superstructure that with all the possible pathways of activities from the proposed value chain in this study. In this case study, plantations from Mukah, Dalat, Saratok and Betong are considered, with sago logs as the only bio-resource produced from the considered sago plantations. The sago mills considered in this study are the ones located around the plantations.

The sago mills products considered in this study are the sago fibers, sago wastes, and sago wastewater, which are produced simultaneously in the sago mill during SSEP. In this study, eight different conversion technology are considered in technology 1 that produces primarily four different intermediates, which will be sent to technology 2 considered in this study. The end product considered in this study consists of high value product from agriculture biomass that has been intensively studied recently. Data from superstructure, such as amount of palms available annually, capacities of sago mill and distance between sago mill and plantation can be found is taken from data reported by Wan et al. [8]. For this study, the conversion rate of palm to sago log $X_{g,log}$ are taken as 10 logs/palm, the weight of log W_{log} is taken as 0.05t/log. Different plantation g produces palms with different starch content due to different soil conditions, which can also be found in [8]. To determine the economic and environmental performance of the value chain proposed in this study, unit costs and CFP emission of producing intermediate d and end product z via technology 1 *c* and technology 2 *e* are estimated from literature and summarized in Table 1 [12]–[18].

Table 1. Conversion, Cost and CFP data

	Feed Type	***Product Type***	***Conversion (product/feed)***	***Cost (USD/Tonne Product)***	***CFP (kg CO_2/tonne Product)***
Alkali Based	Sago Fiber	Glucose	0.4828	650	467.48
	Sago Bark	Glucose	0.16401		
Acid Based	Sago Fiber	Glucose	0.5440	420	503.44
	Sago Bark	Glucose	0.1848		
Steam Explosion	Sago Fiber	Glucose	0.6120	430	521.42
	Sago Bark	Glucose	0.2079		
Liquid Hot Water	Sago Fiber	Glucose	0.6120	790	467.48
	Sago Bark	Glucose	0.2079		
Irradiation Method	Sago Fiber	Glucose	0.2244	370.58	317.15
	Sago Bark	Glucose	0.0762		
Gasification	Sago Fiber	Syngas	0.2271	45,200	2,200
	Sago Bark	Syngas	0.2220		
Pyrolysis	Sago Fiber	Bio-Oil	0.3200	89,630	1,727
	Sago Bark	Bio-Oil	0.3200		
Anaerobic Digestion	Sago Waste Water	Bio Gas	1.7250	4,400	0
Fermentation	Glucose	Butanol	0.0857	1,370	12,652.53
	Glucose	Ethanol	0.2910	874.52	5,977.79
	Glucose	Hydrogen	0.0111	5400	13,055.82
CHP	GJ	Power(GJ)	0.2222	1.55	0
	GJ	Heat(GJ)	0.8000	16.95	0

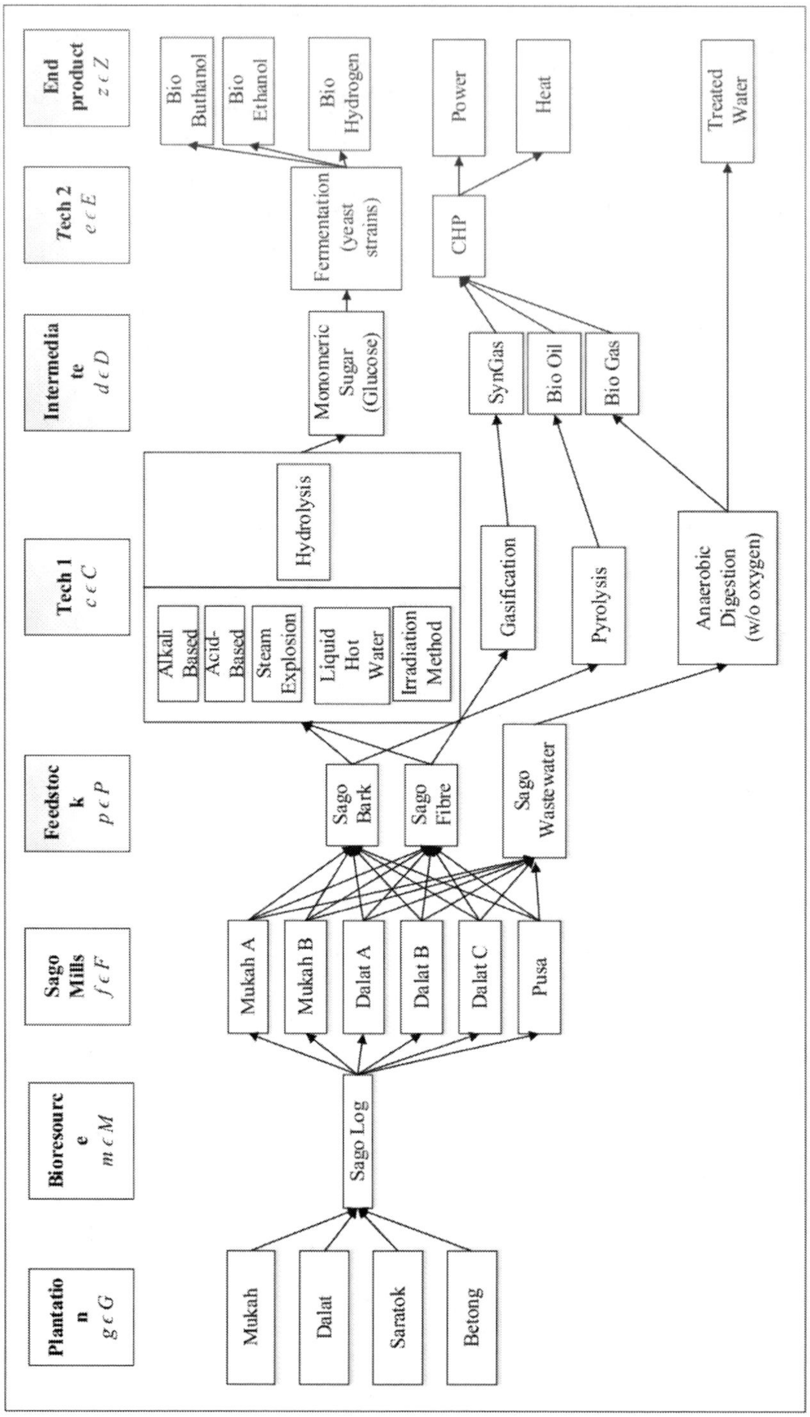

Fig. 6. Case Study Superstructure

3.1 Results and Discussion

To solve the case study, the upper and lower fuzzy limit is pre-determined by solving the objectives individually (e.g., maximize profit, minimize CFP emission) without mutual interaction between objectives. The single objective optimization results are shown in Table 2. Based on the results on, the upper and lower fuzzy limits is coded into the model and solved via Lingo 16.0 in a Lenovo Ideapad 700 with Intel Core i7-6700HQ (2.6 GHz) and 8GB ram running on 64-bit operating system. The CPU time to obtain the maximum global optimal solution for degree of satisfaction λ is approximately within 5s. In addition, Table 3-6 illustrate the flow rate between layers at different objectives.

Table 2. Total Profit and CFP in multiple objectives

Objective Functions	*Total Profit (Million USD/y)*	*Total CFP (tonne CO2/y)*
Max Total Profit	342.90	779,803.00
Min Total CFP	-4.30	768,777.90

Table 3. Mass Flowrates of logs from plantations to sago mills with objective function minimize $TotCFP^{Overall}$, maximize $Tot\,Profit^{Overall}$ and maximize λ.

Flow Rate from sago plantation to sago mill (log/y)						
	minimize $TotCFP^{Overall}$					
	Mukah A	Mukah B	Dalat A	Dalat B	Dalat C	Pusa
Mukah	0	375,000	0	0	200	0
Dalat	0	0	302,400	317,400	204,600	0
Saratok	0	0	0	0	0	220,000
Betong	0	0	0	0	0	0
	maximize $Tot\,Profit^{Overall}$					
	Mukah A	Mukah B	Dalat A	Dalat B	Dalat C	Pusa
Mukah	0	374,200	0	0	0	0
Dalat	0	600	302,400	343,600	178,800	0
Saratok	0	0	0	0	0	220,000
Betong	0	0	0	0	0	0
	maximize λ					
	Mukah A	Mukah B	Dalat A	Dalat B	Dalat C	Pusa
Mukah	0	0	0	0	0	0
Dalat	200	343,600	302,400	343,600	343,600	0
Saratok	0	0	0	0	0	0
Betong	0	0	0	0	0	0

Table 4. Mass Flow Rate of Sago Wastes from sago mills with objective function minimize $TotCFP^{Overall}$, maximize $Tot\,Profit^{Overall}$ and maximize λ.

Production of Sago wastes from Sago mills.									
Objectives	minimize $TotCFP^{Overall}$			maximize $Tot\,Profit^{Overall}$			maximize λ		
	SF	SB	SW	SF	SB	SW	SF	SB	SW
Mukah A	0	0	0	0	0	0	4.8	19.2	96
Mukah B	8,250	33,000	165,000	8,246	32,987	164,936	9,246	32,985	164,928
Dalat A	7,257	29,030	145,152	7,257	29,030	145,152	7,257	29030	145,152
Dalat B	7,617	30,470	152,352	8,246	32,985	164,928	8,246	3,295	164,928
Dalat C	4,914	19,659	98,296	4,291	17,164	85,824	8,246	3,295	164,928
Pusa	3,960	15840	79,200	3,960	1,584	79,200	0	0	0

Table 5. Mass Flow Rate of intermediate *d* into technology 1 with objective function minimize $TotCFP^{Overall}$, maximize $Tot\,Profit^{Overall}$ and maximize λ.

	minimize $TotCFP^{Overall}$			**maximize** $Tot\,Profit^{Overall}$			**maximize** λ		
	SF	SB	SW	SF	SB	SW	SF	SB	SW
Irradiation Method	0	128,000	0	0	0	0	0	128,006	0
Gasification	0	0	0	0	0	0	3,068	0	0
Pyrolysis	32,000	0	0	32,002	128,008	0	28,933	0	0
Anerobic Digestion	0	0	640,000	0	0	640,040	0	0	640,032

Table 6. Amount of Intermediate *d* and End Product *z* produced

	minimize $TotCFP^{Overall}$	**maximize** $Tot\,Profit^{Overall}$	**maximize** λ
Butanol	0	0	0
Ethanol	2,839	0	2,839.
Hydrogen	0	0	0
Power	8,480,067	8,716,897	8,523,204
Heat	0	0	0

An optimum sustainable sago value chain with degree of satisfaction, λ of 0.8596, a total profit of 294.2 million USD/y and CFP emission of 770.32 ton CO_2/y is obtained as shown in Table . Furthermore, Table also shows the flow rate between activities in the optimum sago value chain at optimised solution.

Based on the optimised result, only palms from Dalat is harvested, which is first processed into sago logs before sending into sago mill Dalat A, Dalat B, Dalat C and Pusa SSEP process. The SSEP process then produces sago waste SF, SB and SW. SF is then sent into gasification and pyrolysis to produce syngas and bio-oil respectively, while SB is sent for microwave irradiation pre-treatment, followed by hydrolysis to produce monomeric sugars such as glucose. SW is sent to for anaerobic digestion to produce biogas. The intermediates are then sent for further processing. Glucose undergoes fermentation to produce ethanol, while syngas, bio-oil and bio gas are sent to CHP to generate power. The amount of final products produced in this optimised sago value chain is 2,839.611 t/y of ethanol and 8,523,204 GJ/y of power.

Table 7. Mass Flow Rate of Optimized Sago Value chain

Pathways	*Amount*	*Pathways*	*Amount*
Dalat - Sago Log	133,340	SF- Gasification	3,068
Sago Log - Dalat A	343,600	SF - Pyrolysis	28,933
Sago Log - Dalat B	302,400	SB- Irradiation Method	128,005
Sago Log - Dalat C	343,600	SW- Anaerobic Digestion	640,032
Sago Log - Pusa	343,600	Irradiation Method - Glucose	9,758
Dalat A - SF	8,246	Gasification - Syngas	696

Table 8. Mass Flow Rate of Optimized Sago Value chain (Continued)

Pathways	*Amount*	*Pathways*	*Amount*
Dalat A - SB	32,985	Pyrolysis - Bio Oil	9,258
Dalat A - SW	164,928	Anaerobic Digestion - Bio Gas	1,104,055
Dalat B - SF	7,257	Glucose - Fermentation	9,758
Dalat B - SB	29,030	Syngas - CHP	696
Dalat B - SW	145,152	Bio Oil - CHP	9,258
Dalat C - SF	8,246	Bio Gas - CHP	1,104,055
Dalat C - SB	32,985	Fermentation - Bio Ethanol	2,839
Dalat C - SW	164,928	CHP - Power	8,523,204
Pusa - SF	8,246		
Pusa - SB	32,985	Total CFP	770,324,800
Pusa - SW	164,928	Total Profit	294,206,200

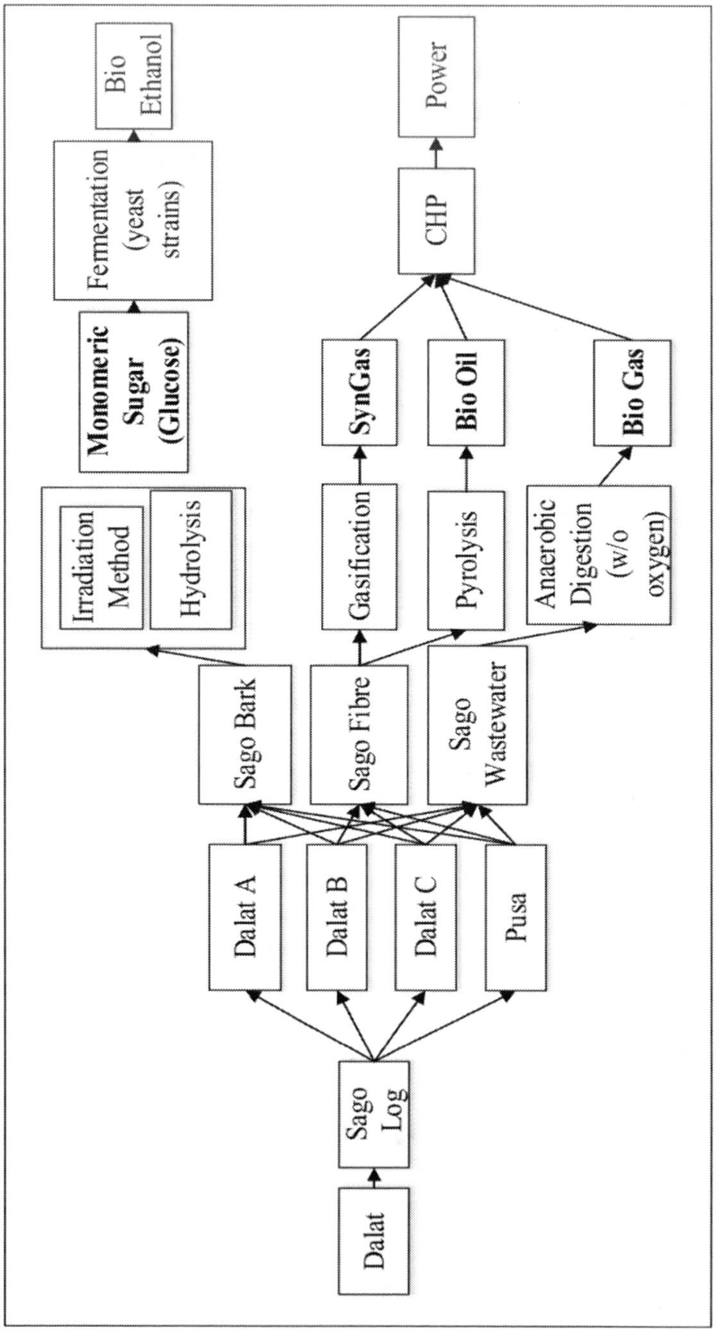

Fig. 7. Optimized Sago Value Chain

3 Conclusions

A fuzzy-based optimisation model approach that considers economic and environmental performance of the proposed sago value chain simultaneously is developed in this study. This model can be used as an analysis tool that aids decision making in selecting the most optimal pathway for conversion of sago waste to value added product while considering the environmental and economic impacts.

This model developed in this study is requires intensive data and is customized for sago industry in this study; however, with some modification, it can be applied in different agriculture industry to evaluate conversion pathways of agriculture biomass into value-added products. Note that data such as yield, cost and CFP of crops might not be readily available and is subjected to the magnitude of attention and research on that specific agriculture industry. Sensitivity to uncertainties or changes in data is not considered in this study via robust optimization, which can be considered in the future.

The limitation of this study is that it only considers the sago wastes produced from the SSEP process in the sago mill, which can also be considered in future researches.

References

1. M. Flach, Sago palm, Diversity, **76**, (1997).
2. FAO, Expert consultation on the establishment of a sago network for Asia and the Pacific, 1–83, (2013).
3. A. A. Karim, A. P. L. Tie, D. M. A. Manan, and I. S. M. Zaidul, Starch from the sago (Metroxylon sagu) palm tree - Properties, prospects, and challenges as a new industrial source for food and other uses, Comprehensive Reviews in Food Science and Food Safety, **7**. 215–228, (2008).
4. D. S. Awg-Adeni, S. Abd-Aziz, K. Bujang, and M. A. Hassan, Bioconversion of sago residue into value added products, African J. Biotechnol., **9**, 2016–2021, (2010).
5. D. S. Awg-Adeni, K. B. Bujang, S. Abd-Aziz, and M. A. Hassan, Recovery of Glucose from Residual Starch of Sago Hampas for Bioethanol Production, 2013, (2013).
6. M. Machines, Chapter 1, 1–11, (1995).
7. Y. K. Wan, J. Sadhukhan, and D. K. S. Ng, Techno-economic evaluations for feasibility of sago-based biorefinery, Part 2: Integrated bioethanol production and energy systems, Chem. Eng. Res. Des., **107**, 102–116, (2016).
8. Y. K. Wan, R. T. L. Ng, D. K. S. Ng, K. B. Aviso, and R. R. Tan, Fuzzy multi-footprint optimisation (FMFO) for synthesis of a sustainable value chain: Malaysian sago industry, J. Clean. Prod., **128**, 62–76, (2016).
9. T. Koc and E. Bozdag, Measuring the degree of novelty of innovation based on Porter's value chain approach, Eur. J. Oper. Res., **257**, 559–567, (2017).
10. D. H. S. Tay, D. K. S. Ng, N. E. Sammons, and M. R. Eden, Fuzzy optimization approach for the synthesis of a sustainable integrated biorefinery, Ind. Eng. Chem. Res., **50**, 1652–1665, (2011).
11. J. Fargione, J. Hill, D. Tilman, S. Polasky, and P. Hawthorne, Land clearing and the biofuel carbon debt., Science, **319**, 1235–8, (2008).
12. N. Nahar, D. Ripplinger, and S. W. Pryor, Process yield and economic trade-offs for enzymatic hydrolysis of alkaline pretreated corn stover, Biomass and Bioenergy, **99**, 97–105, (2017).

13. T. Eggeman and R. T. Elander, Process and economic analysis of pretreatment technologies, Bioresour. Technol., **96**, 2019–2025, (2005).
14. T. S. Kannan, A. S. Ahmed, and F. N. Ani, Energy Efficient Microwave Irradiation of Sago Bark Waste (SBW) for Bioethanol Production, Adv. Mater. Res., 701, 249–253, (2013).
15. T. Pielhop, J. Amgarten, P. R. von Rohr, and M. H. Studer, Steam explosion pretreatment of softwood: the effect of the explosive decompression on enzymatic digestibility, Biotechnol. Biofuels, **9**, 152, (2016).
16. P. Noparat, P. Prasertsan, S. O-Thong, and X. Pan, Dilute Acid Pretreatment of Oil Palm Trunk Biomass at High Temperature for Enzymatic Hydrolysis, in Energy Procedia, (2015), **79**, 924–929.
17. D. Kumar and G. S. Murthy, Impact of pretreatment and downstream processing technologies on economics and energy in cellulosic ethanol production, Biotechnol. Biofuels, **4**, 27, (2011).
18. P. Elaiyaraju and N. Partha, Biogas Production from Sago (Tapioca) Wastewater Using Anaerobic Batch Reactor, Energy Environ., **23**, 631–645, (2012).
19. M. Ben Rabha, M.F. Boujmil, M. Saadoun, B. Bessaïs, Eur. Phys. J. Appl. Phys. (to be published)
20. F. De Lillo, F. Cecconi, G. Lacorata, A. Vulpiani, EPL, **84** (2008)
21. L. T. De Luca, *Propulsion physics* (EDP Sciences, Les Ulis, 2009)

Parametric Study for Devulcanization of Waste Tire Rubber Utilizing Deep Eutectic Solvent (DES)

Rashmi Walvekar[1,*] Kishen Kunju[1], Ricky Saputra[1,†], Khalid Siddiqui[2] and Suganti Ramarad[3]

[1]School of Engineering, Taylor's University, 47500 Subang Jaya, Selangor, Malaysia

[2] Graphene & Advanced 2D Materials Research Group (GAMRG), Research center for nano-materials and energy technology (RCNMET), School of science and technology, Sunway University, Subang Jaya, Selangor, Malaysia

[3]School of Engineering and Physical Sciences, Herriotwatt University Malaysia, 62200 Putrajaya, Malaysia

Abstract. Waste rubber is a polymeric material containing 50% of rubber and is generally referred to as waste tyre rubber. The main purpose of this research is to study ultrasonic devulcanisation of waste rubber utilising deep eutectic solvent (DES) of $ZnCl_2$:Urea by improving process parameters such as sonication time, reaction temperature and rubber: DES mass ratio by effectively cleaving cross-link sulphur bonds. DES was created and prepared by mixing $ZnCl_2$ with urea at 2:7 and 1:4 molar ratios respectively. Physicochemical properties of the prepared DES was measured using DSC, KFT and TGA analysis to find the freezing point, moisture content and degradation temperature, whereby their freezing point below 60°C, moisture content lower than 3.0 wt.% and 200°C degradation temperature average. Rubber to DES mass ratio was varied at 1:20, 1:30 and 1:40 and sonicated for 15 minutes inside ultrasonic water-bath. Samples were placed onto hot plate whereby heating temperature was varied at room temperature, 130°C, 150°C, and 180°C for 15 minutes. Samples were filtered, washed with distilled water and dried in oven for 24 hours. Once dried, samples were taken for analysis using TGA, EDX, FESEM, FTIR and Gel content. Under TGA analysis, most samples have an average degradation temperature of 200°C, hence justifying a successful devulcanisation. EDX analysis shows two occurrences during devulcanisation process which is bond reformation and cleavage. Furthermore, it is determined that heating temperature of 130°C is an important parameter as it is the optimum temperature for $ZnCl_2$:Urea. Under FTIR analysis, it shows that disulphide bond, S-S is the only bond that is being broken while the rest still remains the same. Gel content analysis showed that samples have a lower soluble fraction after devulcanisation process. Finally, FESEM proves that at 130°C and 15

[*] Corresponding author: rashmi.walvekar@gmail.com

minutes is the optimum temperature and time which is illustrated by the smooth surface at that particular point.

1 Introduction

Rubber is a form of elastic and flexible substance which is able to elastically deform through a certain applied force. It mainly composed of polymers exhibiting a long-chain carbon bond, rotating around its axis and twisting the chains into different combinations. Main source of rubber originates from rubber tree known as ***Hevea Brasiliensis*** [1] which grows sufficiently well under cultivation.

Oils, fillers and antioxidants are compounded together with raw rubber to increase the tensile strength of the rubber's life cycle. In order to utilize raw rubber, it needs to be vulcanised in order to increase the rubber's strength and life as well as reducing manufacturing cost. Only after that, vulcanisation rubber process is carried out. It is an irreversible process in which rubber molecules is cross-linked to the sulphur compound, resulting in tires being elastic and strong resistant to bio-degradation and high temperature [2].

In the manufacturing sector, rubber is considered to be the largest component in composing a tire to where it is broadly applied. Throughout the years, demand for rubber has been ever increasing globally from household applications to industrial fields such as gloves, shoes, insulators and especially on tyres. And with the increasing demand of tyres, the number of vehicles on the road has been increasing drastically. This leads to the around 800 million tires being disposed annually and it is said to increase by 2 % every year [3]. 1.3 billion tires are produced globally which is equivalent to nearly 17 million tonnes per year of used tires as reported by Sienkiewicz *et al.* [4] which will ultimately lead the wastage of rubber tyres.

Waste rubber is a polymeric material containing 50 % of rubber in tyres and is generally referred to as waste tyre rubber [5]. There is a large amount of vulcanized residue that comes from the production of rubber materials and it considered to be a major problem. Less efforts were taken into recycling them even after their life cycle has ended, resulting in these tyres being wasted to a point where it becomes a pollution source. This is due to the fact that the rubber tyre has a complex microstructure of a three dimensional cross-linking bonds which makes the material not so easily decomposed and insoluble [5].

Therefore to reduce wastage of rubber tyres, different methods of devulcanisation has been introduced. Devulcanisation method is one of the ways to reduce the impact risks on environment by transforming tyres back into virgin rubber or original rubber that can be used for re-vulcanisation purposes. Rubber with a complex structure of three-dimensional network arrangement has carbon-sulphur (-C-S-) bond and sulphur-sulphur (-S-S-) bond attached to it. This -C-S- bond is longer compared to -C-C- bond as sulphur is much larger than carbon. During the devulcanisation process, -C-S- and -S-S- bonds are broken down resulting in a three dimensional structure to obtain plasticity whereby -C-S- bond is broken down under the presence of high temperature, where it gains enough energy to in order to break the -C-S- bond. It is then re-vulcanised again by heating to 120-150°C to form better products.

Some of the currently available methods used for devulcanisation are microwave heating [7], continuous ultrasonic [8], microbial action [9], and many other methods. These methods used are definitely capable for devulcanisation of rubber material, however up to a certain limit. This is due to the fact that few of the solvents used before this are not suitable as it is costly and hazardous. In addition to that, end product from using from those methods mentioned has a poor mechanical strength which is not ideal for re-vulcanisation process.

Recent study carried out by V. V. Rajan *et al.* [11] shows experimental analysis where ionic liquid (IL) was used as a solvent for devulcanisation of rubber through microwave irradiation method and it showed positive results. Devulcanisation of rubber tyre was carried out with the combination of microwave radiation and IL which achieved to 50% removal of sulphur. However, further research found that ionic liquids was not very convenient to be used due to its limitations whereby it has high toxicity level, poor degradability, combustible material and not to mention highly cost. Therefore, these limitations can be avoided by finding a more suitable candidate to replace ILs.

One of the emerging alternative solvent which shows similar properties to ILs is known to be Deep Eutectic Solvent (DES). It's been proven that ILs is able to devulcanise waste tyre rubber and with that justifies DES is an excellent alternative to IL which can devulcanise the waste tyre rubber whilst avoiding all the limitations and problems encountered. It is a mixture of two or more components having a lower melting point than each individual component. By mixing ammonium halide salt which is a hydrogen bond acceptor (HBA) and hydrogen bond donor (HBD) molecule, DES is obtained, resulting in significant decrease in freezing point. Since it has similar properties to ILs such as easy regeneration and non-flammable, it shares similar advantages as well in terms of low vapour pressure, biodegradability, chemical stability, cheap material and easily obtained [13]. Thus, DES is considered as a green solvent in being an important alternative to volatile organic solvents.

Similar research area [12] has been done within the petrochemical industry whereby for desulphurisation process, DES was used to remove sulphur from petrochemical products such as oil. Based on the research done by Wang, Liu, Li *et al.* [10] , it was found that the removal efficiency of sulphur is as high as 99.48% which further proves that DES is a suitable solvent in removing sulphur and it can be used for devulcanisation process.

Many different combinations of HBA and HBD have been performed since the first experimental analysis of DES was reported. One of the common combinations is with choline chloride, ChCl and zinc chloride, $ZnCl_2$ salt with other types of HBDs such as urea. ChCl salt is commonly picked because it is low in cost and toxicity, biodegradable, and its biocompatibility [10]. On the other hand, zinc chloride, $ZnCl_2$ is chosen due to its biocompatibility and biodegradability of the salt [10].

Hence, currently new approach for devulcanisation process using the DES alongside with ultrasonic treatment under controlled condition of temperature and sonication of water bath was introduced. With the use of DES as a cheaper and greener solvent, this can help reduce energy consumption during the devulcanisation process as well improving its recovery process. Thus, for the interest of the study, the chosen DES used alongside ultrasonic devulcanisation of rubber samples is $ZnCl_2$:Urea at various molar ratios of 1:4 and 2:7.

This study aims to synthesize and characterize $ZnCl_2$:Urea as a deep eutectic solvents (DESs) of a various molar ratio and to optimise the sonication time, heating temperature and rubber:DES mass ratio on devulcanisation of the waste rubber.

2 Methodology

Based on the research project that will be carried out, quantitative study will be used throughout the entire project to which the justification would be made based on the removal amount of sulphur from waste tyre rubber sample during devulcanisation process.

2.1 DES preparation samples

The selection and synthesis process for two types of DESs solvent chosen was carried out. The selection of DES is made based on boiling point of DES whereby it must be higher than the temperature of devulcanisation which is about 180°C. Based on thesis done by Ricky S. [39], it was determined that zinc chloride: urea and choline chloride: urea is the best selection for DES synthesis to be carried out.

In order to prepare DES, zinc chloride (salt) as the hydrogen bond acceptor and urea as the hydrogen bond donor was ordered and purchased under Sigma Aldrich Sdn. Bhd. Preparation of DES was made based on 1:4 and 2:7 molar ratio of $ZnCl_2$:Urea. By following the steps to prepare DES based on a research from [14], mass of $ZnCl_2$ and urea were weighed accordingly and both compounds were added into a 100 ml Schott bottle alongside a magnetic stirrer. The bottle was sealed with a parafilm to ensure minimum moisture content. This is to prevent the crystallization of DES from occurring. It was then placed onto a hot plate placed inside fume hood running at 300 rpm and at 130-170°C, in order to synthesize DESs solvent. The solvent is left approximately for 2-4 hours. It is considered successful if the solution obtained is clear and homogeneous. If it is not, then the process has to be repeated [13]. The reason for the DES samples to be conducted in a fume hood and sealing the cap with parafilm is to ensure no moisture goes into solvent. Table 1 depicts the molar ratio used for $ZnCl_2$:Urea solvent.

Table 1. Preparation of DES molar ratio

Ammonium/Halide Salt	HBD	Molar Ratio	Freezing Point (K)
$ZnCl_2$	Urea	1:4	278.15
$ZnCl_2$	Urea	2:7	275.15

2.2 Physicochemical properties of DES

In order to determine whether the DES prepared is suited to be used as a devulcanising agent, Karl-Fischer Titration, Differential Scanning Calorimetric (DSC), Thermogravimetric Analysis (TGA) will be performed upon the prepared DES samples.

2.2.1 Karl-fischer titration

The moisture content of the prepared DES sample is done through KFT analysis whereby the moisture must not exceed 3 wt % as declared by C. Acids *et al.* [16]. To find out the accuracy and precision of the analysis instrument, 1.00 mg/g of Hydranal water standard was utilised whereby it was purchased and ordered from Sigma-Aldrich Malaysia.

2.2.2 Differential scanning calorimetric analysis

Freezing point of the synthesized DES is determined using the DSC available at Taylor's University School of Bioscience, where it will be carried out in nitrogen surrounding. Before conducting the experiment, the equipment is calibrated correctly to the standards values so that the accuracy of the readings obtained will be high. A portion of DES sample was extracted out and weighed between 5-10 mg in a small pan. The small pan was then placed onto Perkin Elmer – DSC 8000 instrument analysis. The sample was first heated from 30°C to 60°C at a rate of 10 K/min, then cooled from 60°C to -70°C at a rate of 5 K/min and finally heated from -70°C to 0°C at rate of 5 K/min. The suitable freezing point of the synthesized DES should lower than the constituent components and also, the preferred melting point of DES should be below 60°C [12].

2.2.3 Thermogravimetric analysis

The degradation temperature is determined through TGA analysis. The experiment is conducted at temperature of 180°C. A portion of synthesized DES sample was extracted out and weighed between 5-10 mg into a small pan. The small pan was then placed onto Perkin Elmer – TGA 8000 instrument analysis allowing the analysis to be carried out from ambient temperature to 500°C, with 10°C increment every minute under constant nitrogen purge gas of 40 ml/min environment [15]. The amount of liquid is measured as a temperature function.

2.3 Rubber: DES sample preparation

Once physiochemical properties of DES sample were measured and gave good results, devulcanisation of rubber samples were carried out next. Under this process, 4 variables were fixed which consist of sonication time, sonication frequency, DES molar ratios and heating time. Apart from that, 2 variables were varied which are heating temperature and rubber: DES mass ratio.

DES samples as shown in Table 1 were selected for the fact that they are amino-compound-based DESs which are suitable for desulphurisation process [28] and thus, it is considered to be the same condition for devulcanisation process. Rubber:DES mass ratio was varied at 1:20, 1:30 and 1:40 whereby the total mass of the rubber and DES should be equivalent to 50 grams. These ratios are specifically picked as to avoid sluggish mixture being formed which could render the devulcanisation process during ultrasonic treatment [3]. Rubber and DES were weighed accordingly to mass ratio and inserted inside the Schott bottle where it will be mixed thoroughly.

2.4 Ultrasonic devulcanization & heating treatment

Once it is mixed thoroughly, the rubber sample mixtures were placed inside the ultrasonic water bath to undergo ultrasonic treatment at sonication time of 15 minutes at 37 kHz [17]. During treatment, rubber: DES samples experienced and received the ultrasonic frequency wave in all possible directions. With the rubber and DES being exposed to ultrasonic wave, the cross-link bonds are overstressed which allows the sulphur-sulphur bonds as well cross-link bonds to become detached. Thus, penetration of DES is easier which leads to cleavage of cross-link bonds [3]. Through sonication, it is predicted that small fragments of the cross-link bond are broken. Therefore to further improve recovery, rubber: DES samples are subjected to heating treatment at 4 different temperatures for each sample after ultrasonic treatment.

Under this treatment, samples mixtures are transferred onto conical flask. It is then preceded by heating on hot plate in a fume hood environment at room temperature, 130°C, 150°C and 180°C for 15 minutes. Heating temperatures was selected based on the flash point of rubber which is nearly 205°C [18-20]. Thus, heating temperature is recommended to be below the flash point temperature for safety aspects. A controlled sample was used to which it does not undergo ultrasonic water bath treatment and heating.

As soon as the heating process is completed, the rubber mixture sample was filtered out using a tea bag instead of a filter paper. This is due to the DES high viscosity which makes it very difficult for the rubber to be filtered out. The rubber mixture which was filtered was washed thoroughly with distilled water [18] to ensure there is no contamination of DES in the rubber mixture samples. Finally, the samples were placed inside the oven for approximately 24 hours at 70°C for it to dry.

Therefore, the total number of samples required for two different molar ratios, three different rubber:DES mass ratios as well four different heating temperatures is 30. The following Table 2 and 3 depict the formulation and experimentation of the samples of various molar ratios.

Table 2. The formulation and experimentation of the sample 2:7 molar ratio

<table>
<tr><th>DES molar ratio</th><th>Rubber:DES</th><th>Sonication time / min</th><th>Heating Temperature / °C</th><th>Experimental run</th></tr>
<tr><td rowspan="15">2:7</td><td rowspan="5">1:20
(2.38 gr : 47.2 gr)</td><td rowspan="4">15</td><td>30</td><td>1A</td></tr>
<tr><td>130</td><td>2A</td></tr>
<tr><td>150</td><td>3A</td></tr>
<tr><td>180</td><td>4A</td></tr>
<tr><td colspan="2">Controlled sample</td><td>5A</td></tr>
<tr><td rowspan="5">1:30
(1.61 gr : 48.39 gr)</td><td rowspan="4">15</td><td>30</td><td>6A</td></tr>
<tr><td>130</td><td>7A</td></tr>
<tr><td>150</td><td>8A</td></tr>
<tr><td>180</td><td>9A</td></tr>
<tr><td colspan="2">Controlled sample</td><td>10A</td></tr>
<tr><td rowspan="5">1:40
(1.22 gr : 48.78 gr)</td><td rowspan="4">15</td><td>30</td><td>11A</td></tr>
<tr><td>130</td><td>12A</td></tr>
<tr><td>150</td><td>13A</td></tr>
<tr><td>180</td><td>14A</td></tr>
<tr><td colspan="2">Controlled sample</td><td>15A</td></tr>
</table>

Table 3. The formulation and experimentation of the sample 1:4 molar ratio

DES molar ratio	Rubber:DES	Sonication time / min	Heating Temperature / °C	Experimental run
1:4	1:20 (2.38 gr : 47.2 gr)	15	30	1B

			130	2B
			150	3B
			180	4B
		Controlled sample		5B
	1:30 (1.61 gr : 48.39 gr)	15	30	6B
			130	7B
			150	8B
			180	9B
		Controlled sample		10B
	1:40 (1.22 gr : 48.78 gr)	15	30	11B
			130	12B
			150	13B
			180	14B
		Controlled sample		15B

2.5 Sample analysis

Analysis on the rubber sample mixture can be carried out once the rubber is completely dry after being left in the oven for 24 hours. Fourier Transform Infrared Spectroscopy (FTIR) was performed on the rubber sample mixture since it revolves around the presence of the carbon-sulphur bonds.

2.5.1 Fourier transform infrared spectroscopy

The FTIR method used for this particular research study is the FTIR-ATR Spectrum 100. To make further confirmation, FTIR equipment allows the user to determine the specific bond that is broken down after devulcanisation process. This is done by analysing the spectrum generated from FTIR based on before and after data results of the rubber sample mixtures. In order to understand spectrum generated from FTIR, it is best to refer to journal by Garcia and co-workers *et al.* [13] so that the spectrum can be determined as to which compound it corresponds to specifically. Through this analysis, it is expected that there will be decrease in the intensity of the spectrum based on sulphur cross-link bonds during devulcanisation process. It is also expected that C-S and S-S bonds will decrease after the completion of devulcanisation process.

2.5.2 Thermogravimetric analysis

Thermogravimetric analysis (TGA) analysis was carried out in with Shimadzu system form Japan. As reported by F. Endres *et al.* [21], it is heated within a certain temperature range, ranging from 20°C-600°C, with an increment of 10°C per minute. The entire analysis was carried out in a nitrogen controlled environment. Through this analysis, degradation temperature of the rubber samples can be determined.

Experiment was carried out at temperature of 180°C, in which small amount of DES is inserted into TGA tool. Analysis is carried out from ambient temperature to 500°C, having an increment of 10°C per minute under a controlled nitrogen environment. Based on the TGA curve obtained from [35], a decreasing slope produced indicates there is a sudden drop in mass which indicates the degradation temperature of the rubber sample.

3 Results and discussion

The data obtained was analysed carefully and comparison was made based on previous research carried out on this topic. Results of the analysis will consist of DES characterization, FTIR and TGA results on the rubber sample mixtures.

3.1 DES analysis of physicochemical properties

In order to prepare DES, zinc chloride (salt) as the hydrogen bond acceptor and urea as the hydrogen bond donor was ordered and purchased Sigma Aldrich Sdn. Bhd. Preparation of DES was made based on 1:4 and 2:7 molar ratio of $ZnCl_2$: Urea. A homogeneous solution is obtained which has a very high viscosity at ambient temperature. Based on these, its physiochemical properties of the prepared DES samples such as moisture content, freezing point and degradation temperature is measured and determined.

The moisture content present inside the prepared DES samples must not exceed more than 3 wt % as it will affect and change its properties which could lead to crystallisation during devulcanisation process [22]. Thus 2 batches of different molar ratio of DES consisting of 1:4 and 2:7 were used to determine its respective moisture content. Based on Table 4, it can be seen that the moisture content for both the DES molar ratio does not exceed than the maximum allowable moisture content, thus they are suited to be used as a devulcanising agent.

Table 4. Moisture content of prepared DES sample

DES molar ratio	Sample abbreviation	Moisture content (%)
1:4	DES 1	3.0000
2:7	DES 2	0.0717

Freezing point is another important physiochemical property to be measured off from the prepared DES samples in order to show the liquidity range of the solvent and is ensured to be less than 60°C so that it can be applicable in industrial fields. By looking at the freezing point of DES obtained, its freezing point is below than its constituent component depicted in Table 5. This confirms a successful prepared synthesis of DES. Another confirmation of a successful synthesis of DES is the experimental freezing point obtained from the DSC analysis is compared with the literature value from [13, 21], whereby it show a small percentage error between the readings obtained.

Table 5. Moisture content of prepared DES sample

Solvent	Freezing point (K)		Percentage error (%)
	Literature	Experimental	
DES 1	278.15	288.15	3.47
DES 2	275.15	278.81	1.31

Degradation temperature of the prepared DES sample is an important requirement to be known so that DES prepared earlier does not degrade at 180°C which is operating temperature of devulcanisation. The measurement is carried out accordingly as mentioned in sub-section 2.2.3. The following two figures below depicts the weight loss of $ZnCl_2$:Urea for two different molar ratio vs. temperature.

Based on the TGA curve reported by Seghar *et al.* [15], a decreasing slope produced indicates there is a sudden drop in mass. This point at which it has a sudden drop is the degradation temperature of the DES. Referring to Figure 1a and 1b, it can be seen that the degradation temperature was found to be at 187.35°C and 324.36°C respectively for 2:7 and 1:4 molar ratio. The values obtained clearly states that devulcanisation temperature must not be allowed to reach to the degradation temperature to avoid the DES degradation. Therefore, the prepared DES samples are suited to be used for this particular devulcanisation research study. In addition to that, the operating temperature of the devulcanisation used is below the DES degradation temperature, making it a suitable operating temperature for devulcanisation process.

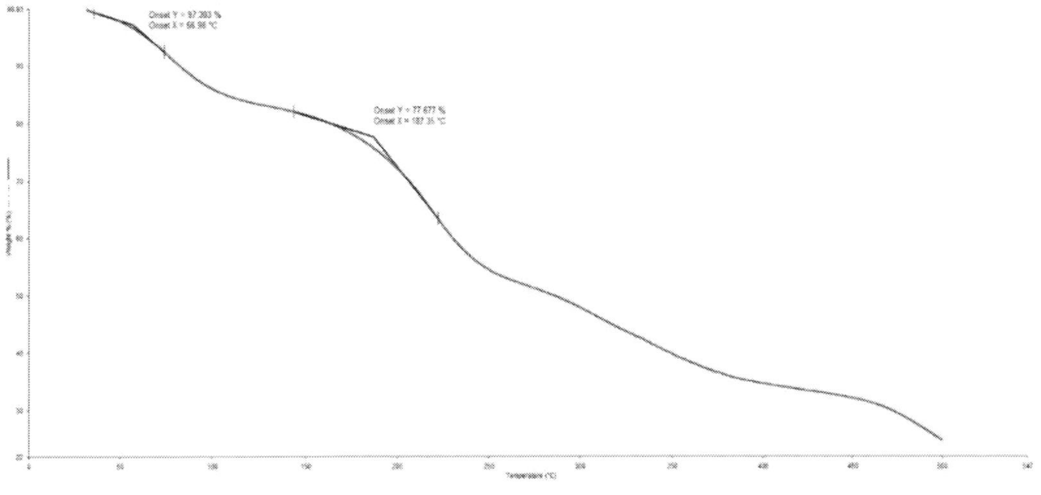

(a)

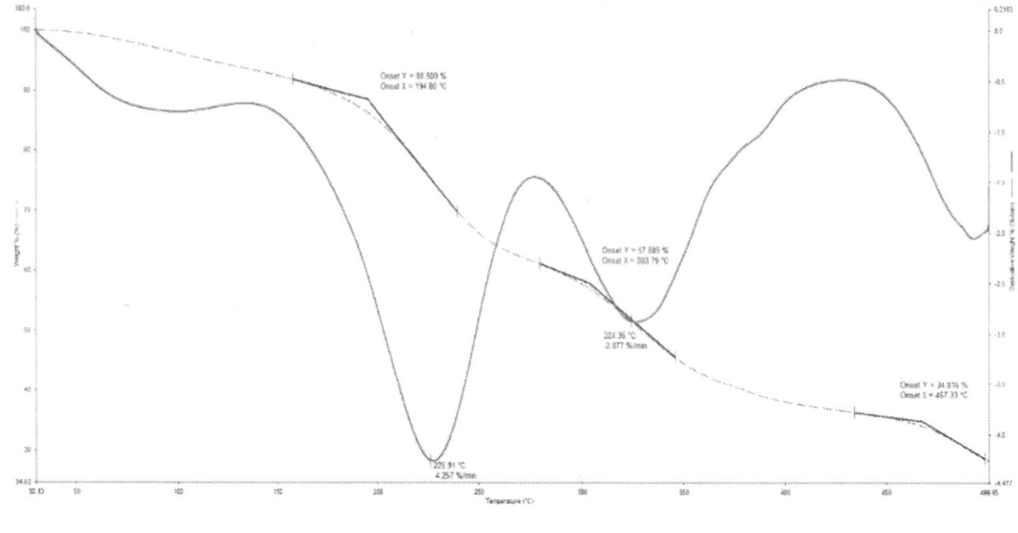

(b)

Fig. 1. TGA analysis for DES molar ratio of (a) 2:7 and (b) 1:4

3.2 Analysis of devulcanized rubber

Analysis is performed on the devulcanised rubber mixture samples on its structural bonds and temperature degradation using FTIR and TGA respectively. Then the analysis is followed by composition, morphological structure and gel content analysis.

3.2.1 Fourier transform infrared spectroscopy analysis

FTIR graph for the first three samples treated with DES 1 is shown in Figure 2a, b and c.

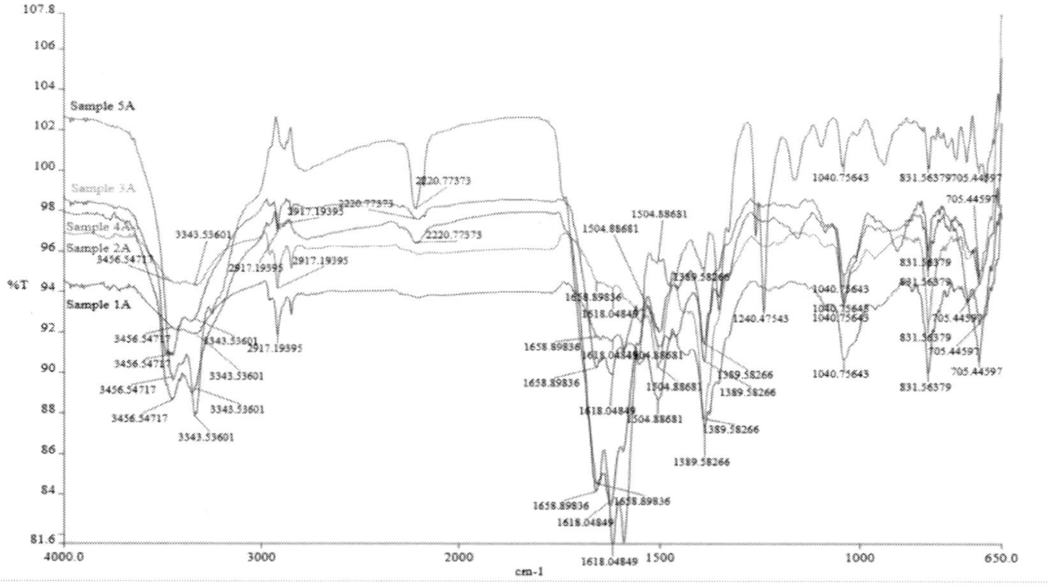

(a)

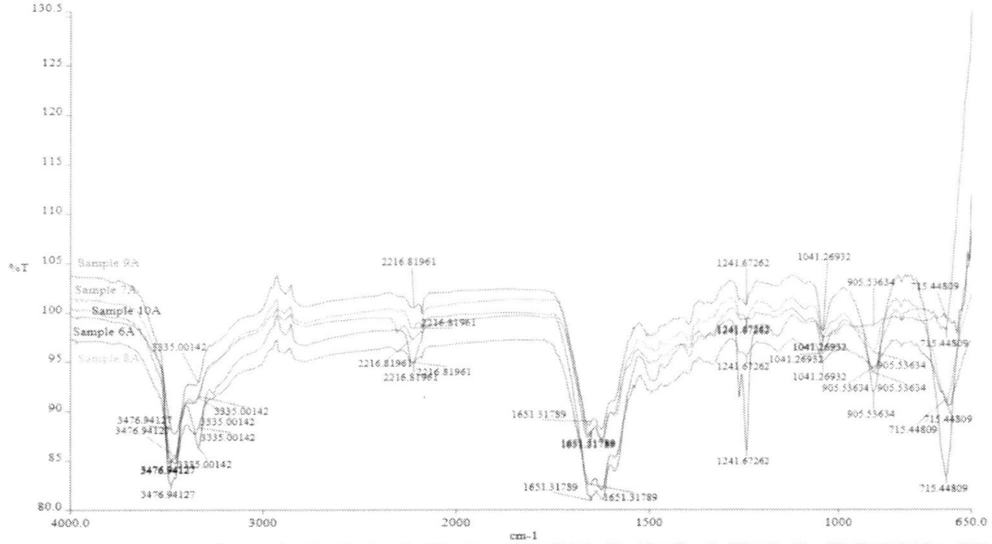

(b)

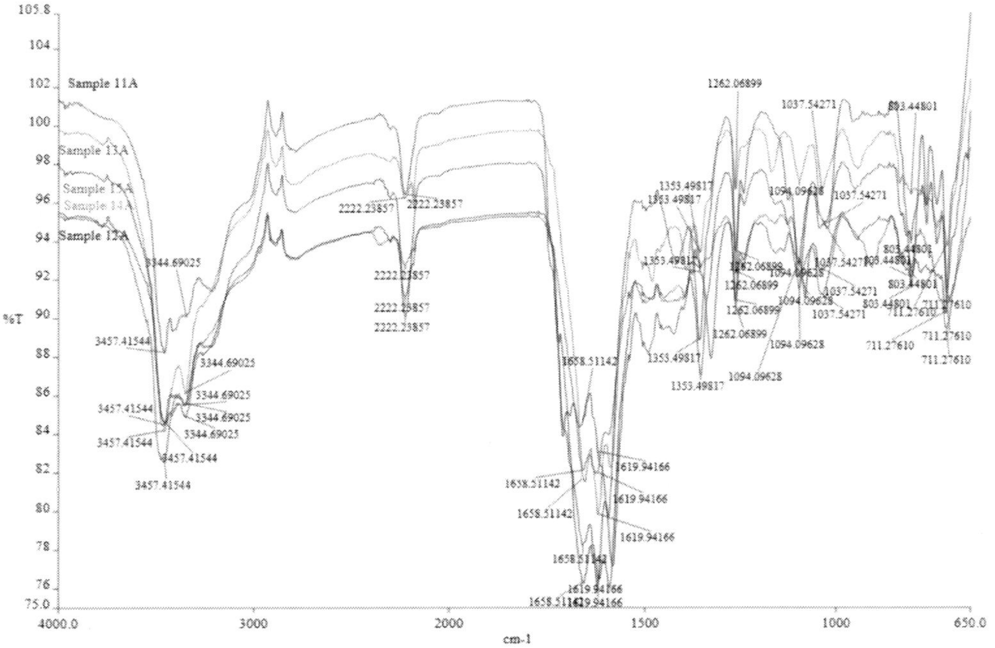

(c)

Fig. 2. FTIR analysis for rubber:DES mass ratio at (a) 1:20, (b) 1:30 and (c) 1:40

Referring to Figure 2a, it can be seen that above 3550 cm^{-1}, there is an appearance of a broad peak especially for Sample 1A and 2A, which indicates that the presence of O-H group. This also indicates water presence which might still remain in the sample. Presence of water is due to the fact the rubber sample mixture was washed previously with distilled water so droplets of water might still be present. Thus, it can be said that there is a low amount of rubber degradation happening.

In Figure 2a and 2c, wavelength for -CH_3 functional group makes an appearance in the region of 1389.58266 cm^{-1} and 1353.49817 cm^{-1} respectively for all of the samples. However, for rubber: DES mass ratio at 1:20 in Figure 2b, there was no presence of -CH_3 functional group, in other words there is no structural change. Aside from that, functional group of -CH- appears at the region of 2917.19395 cm^{-1} for mass ratio at 1:20 as shown in Figure 2a. Despite that, there was no peak appearance of the functional group –CH- for mass ratio at 1:30 and 1:40.

–S-S- functional group or bond was analysed at wavelength between 700-550 cm^{-1} and it can be identified that sample 1A, 2A, 3A, 4A, 6A, 7A, 8A, 9A, 11A, 12A, 13A, and 14A have a peak appearance indicating bond reformation is occurring. On the other hand, sample 5A, 10A and 15A does not have any peak appearance since it is a controlled sample, therefore there is no reformation occurring. Another bond that was analysed is the –C-S- bond for which it falls in the wavelength region of 700-520 cm^{-1}. Since there is still peak appearance in Figure 2a, b and c for all the samples, hence it can be said that bonds have not been broken down. Thus, it can be concluded that ultrasonic treatment was unable to break the bond which has higher bond energy than –S-S- bond.

Another analysis was made based on bonds through FTIR analysis as shown in Figure 3a, b and c for devulcanisation process of $ZnCl_2$:Urea for 2:7 molar ratio at different mass ratio.

Based on FTIR curve plotted from Figure 3a, 3b and 3c, it can be analysed that for mass ratio of 1:30, there is disappearance of –S-S- peak whereas for mass ratio of 1:20 and 1:40, there is an appearance of –S-S- peak. Carbon-sulphur bond for this particular treatment is also not broken down since it has a low supply of energy, resulting in zero cleavage of sulphur cross-link bond. From Figure 3b, it also can be seen that there is a disappearance of -S-S- peak for sample 8B. This occurs because of the uneven dispersion particles of rubber or rubber mixture sample composition during devulcanisation process.

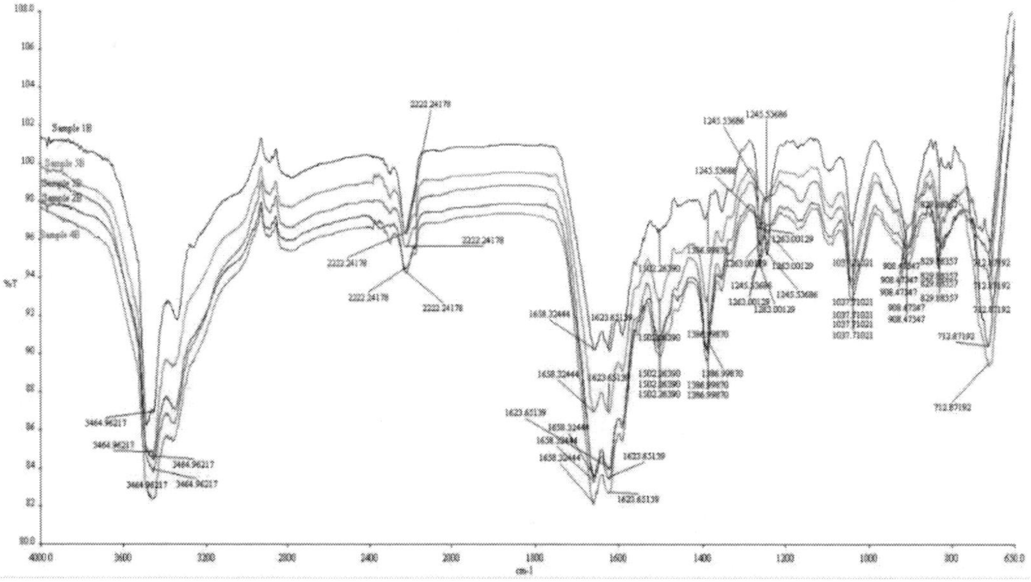

(a)

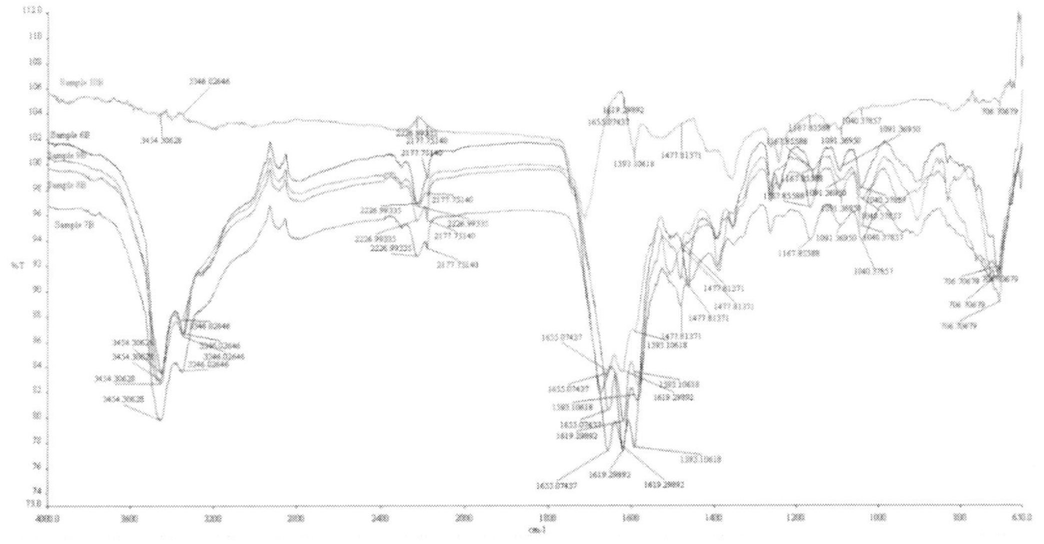

(b)

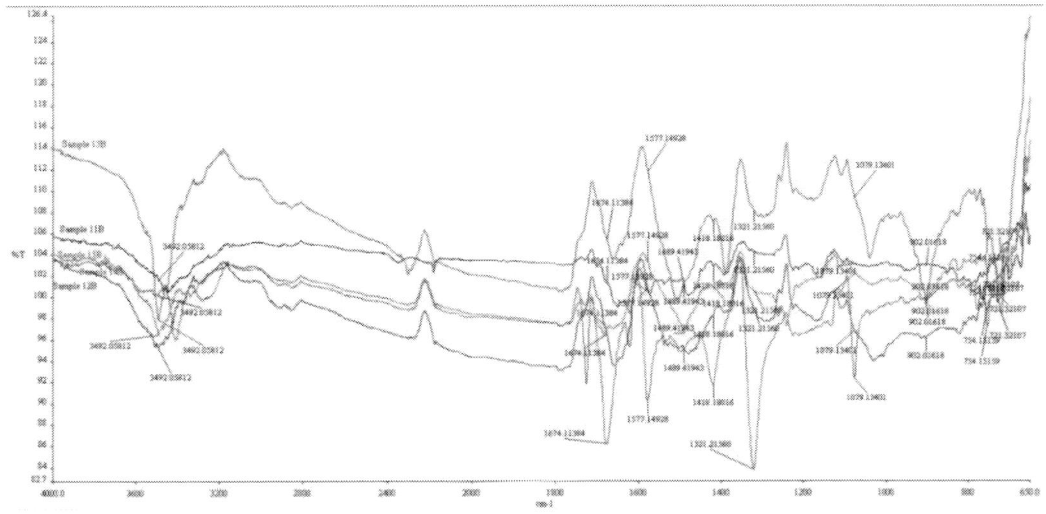

(c)

Fig. 3. FTIR analysis for rubber:DES mass ratio (a) 1:20, (b) 1:30 and (c) 1:40

From Figure 2 and 3, it can be said that wavelength range which are 1300 and 3500 cm-1, there is no appearance of a peak other than broad peak which is apparent at above 3000 cm-1 wavelength range. Therefore, this shows there is still presence of water inside the rubber mixture sample. Thus, it can deduced that there is low amount of degradation rubber which contains the -O-H group.

All in all, various conclusion can be deducted that through FTIR analysis. It was determined through FTIR analysis that –S-S- or sulphur-sulphur bond is the only bond which is leaved during devulcanisation process which proves a successful devulcanisation process at an optimum heating temperature of 130°C. Other bonds are not broken or cleaved as there was insufficient energy to cleave them. Hence, the main carbon chain is still maintained even after devulcanisation process was carried out which further proves that devuclanised rubber has excellent physical properties.

3.2.2 Thermogravimetric analysis

Based on the TGA curve obtained from S. Seghar *et al.* [15], a decreasing slope produced indicates there is a sudden drop in mass. This point at which it has a sudden drop is the degradation temperature of the rubber samples mixture. Referring to Figure 4, it can be seen that the degradation temperature for sample 1A, 2A, 3A, 4A, 5A was found to be at 288.0°C, 298.29°C, 325.23°C, 332.95°C and 112.59°C respectively. Degradation temperature for sample 4A is highest among the rest of the samples. However degradation for sample 1A is comparably much smaller than the rest.

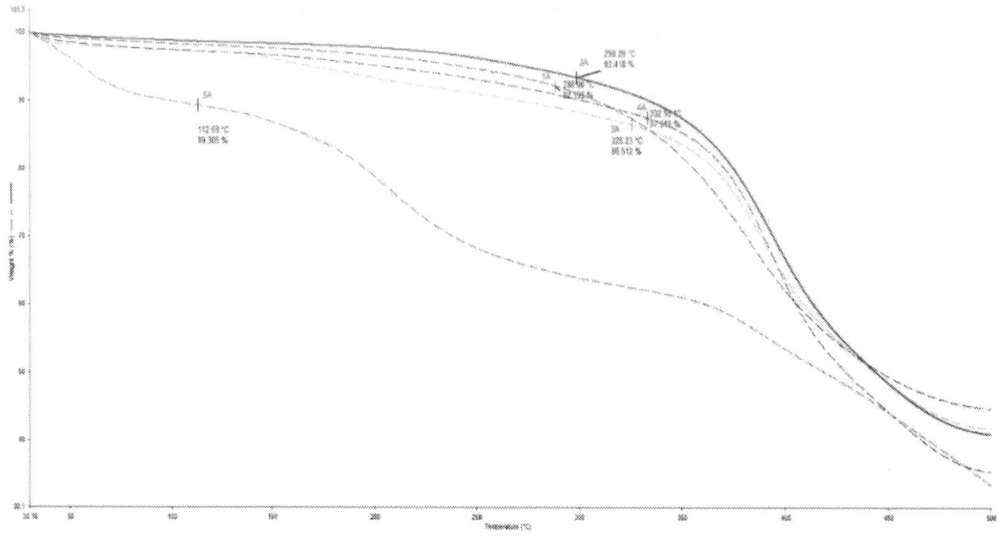

Fig. 4. TGA analysis for samples 1A to 5A at 1:20 mass ratio

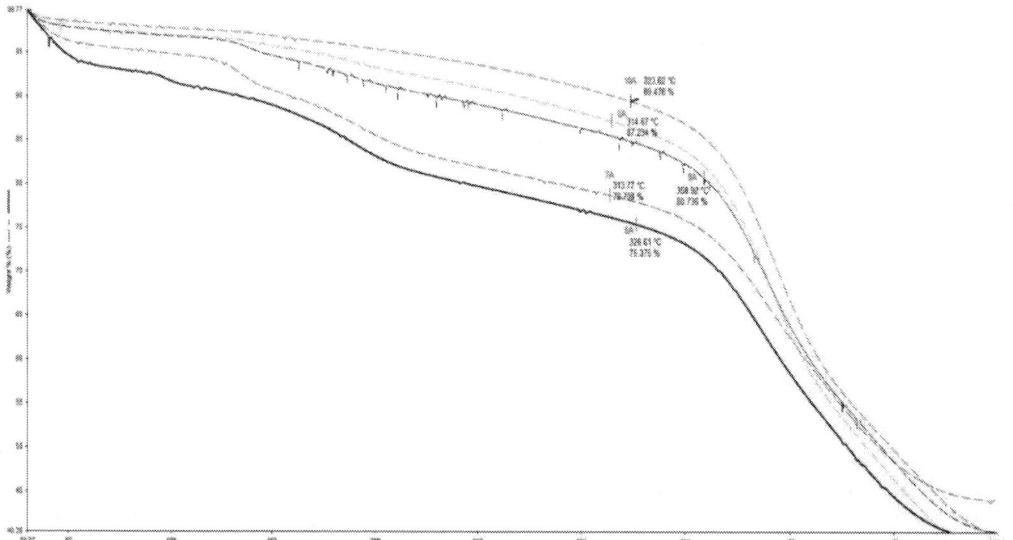

Fig. 5. TGA analysis for samples 6A to 10A at 1:30 mass ratio

Next, with respect to Figure 5, it can be seen that the degradation temperature for sample 6A, 7A, 8A, 9A, 10A was found to be at 326.61°C, 313.77°C, 314.67°C, 358.92°C and 323.62°C respectively. Degradation temperature for sample 9A is highest among the rest of the samples. However degradation for sample 7A is comparably much smaller than the rest. Similarly observation from Figure 6 gives the fact that the degradation temperature for sample 11A, 12A, 13A, 14A, 15A are 313.45°C, 302.51°C, 314.95°C, 312.88°C and 350.30°C respectively whereby the highest degradation temperature is possessed by sample 15A while the lowest is on sample 12A.

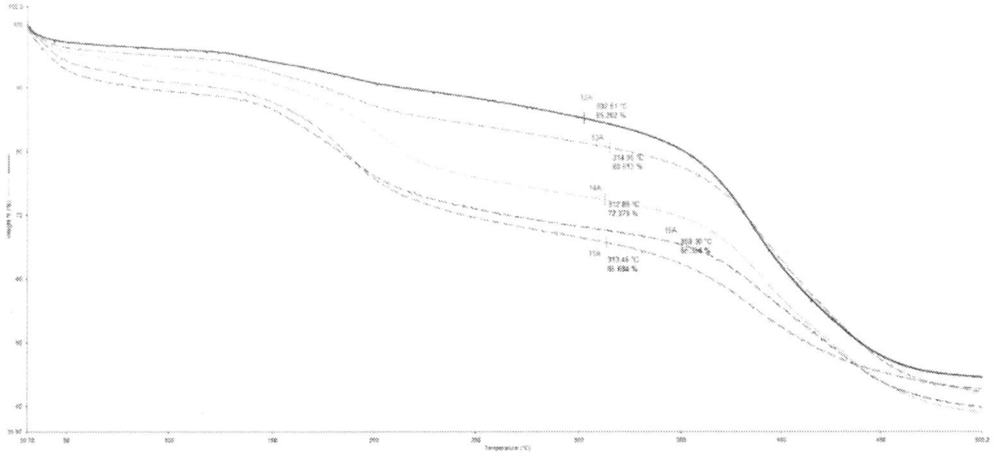

Fig. 6. TGA analysis for samples 11A to 15A at 1:40 mass ratio

With respect to Figure 4-6 for treatment of rubber with $ZnCl_2$:Urea at 2:7, the trend that can be observed is that the controlled sample possesses the highest degradation temperature as –S-S- and/or –C-S- bonds still intact into the rubber. However, as more –S-S- and/or –C-S- extracted, degradation temperature also reduces as expected. Thereby, the most optimum devulcanization heating temperature at fixed sonication time and frequency as well as heating time of 15 minutes for treatment with $ZnCl_2$:Urea (2:7) is 130°C.

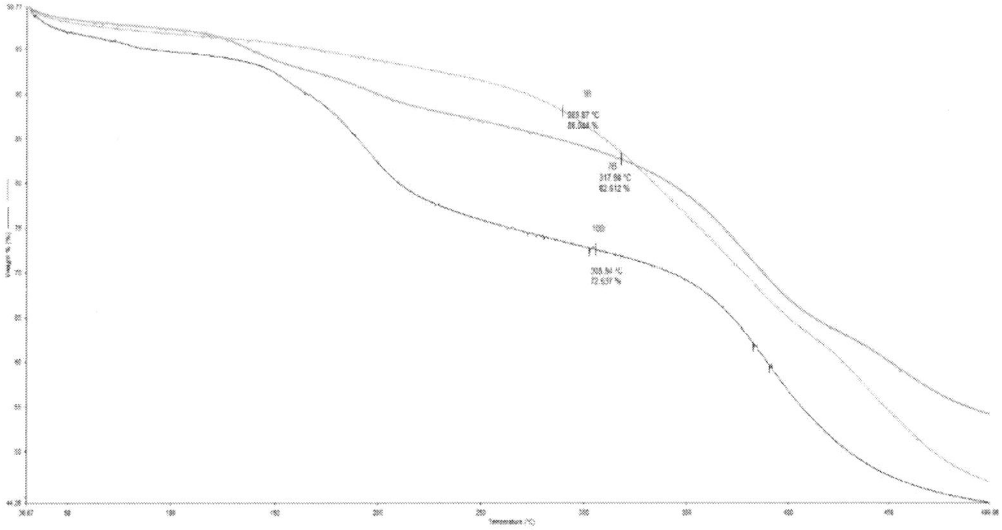

Fig. 7. TGA analysis for samples 3B, 7B and 10B

Continue on to analysis of treatment with ZnCl:Urea at 1:4, Figure 7 that shows the degradation temperature of sample 3B, 7B and 10B which are at 289.97°C, 317.98°C, and 305.94°C respectively whereby it was found that sample 7B has the highest degradation temperature while sample 3B is the lowest with respect to controlled sample 10B.

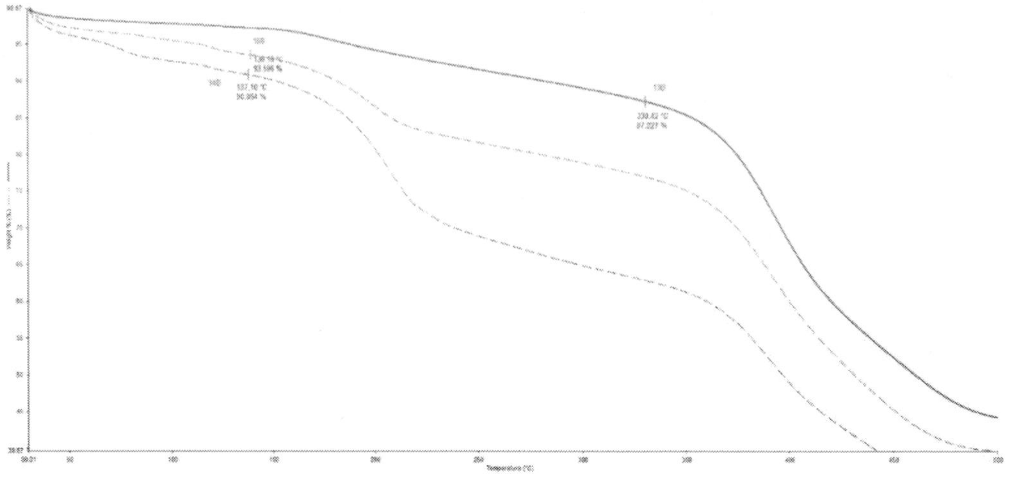

Fig. 8. TGA analysis for samples 13B, 14B and 15B

Moving on to Figure 8, it can be seen that the degradation temperature for sample 13B, 14B, and 15B was found to be at 330.42°C, 137.10°C, and 138.16°C respectively. In a similar manner, degradation temperature for sample 13B is the highest among the rest of the samples, while 14B is the lowest in comparison with controlled sample 15B.

Conclusively for treatment with $ZnCl_2$:Urea at 1:4 molar ratio, it is found that the degree of devulcanization increases as the heating temperature increases which indicated by lower degradation temperature with the minimum temperature setting of 150°C.

3.2.3 EDX analysis

Results acquired for composition of rubber sample from before and after devulcanisation process through EDX analysis is depicted in Figure 9 for $ZnCl_2$:Urea for 2:7 molar ratio, whereby the graph is plotted in terms of sulphur removal percentage against heating temperature.

Referring to Figure 9, it can be observed that, for mass ratio of 1:30 and 1:40, the sample is having an increase in devulcanisation process efficiency as the temperature increases from 150°C to 180°C. Thus, this shows that cross-link network is being broken and soluble fraction is increasing. In fact, at temperature 130°C, it has the highest percentage removal of sulphur which indicates that devulcanisation efficiency is particularly high at that temperature. Reduction in devulcanisation efficiency can be seen for rubber sample under mass ratio of 1:20 as the temperature increases from 150°C to 180°C. This is due to the fact that bond reformation is occurring during heating treatment as the sulphur cross-link bonds are being reformed again.

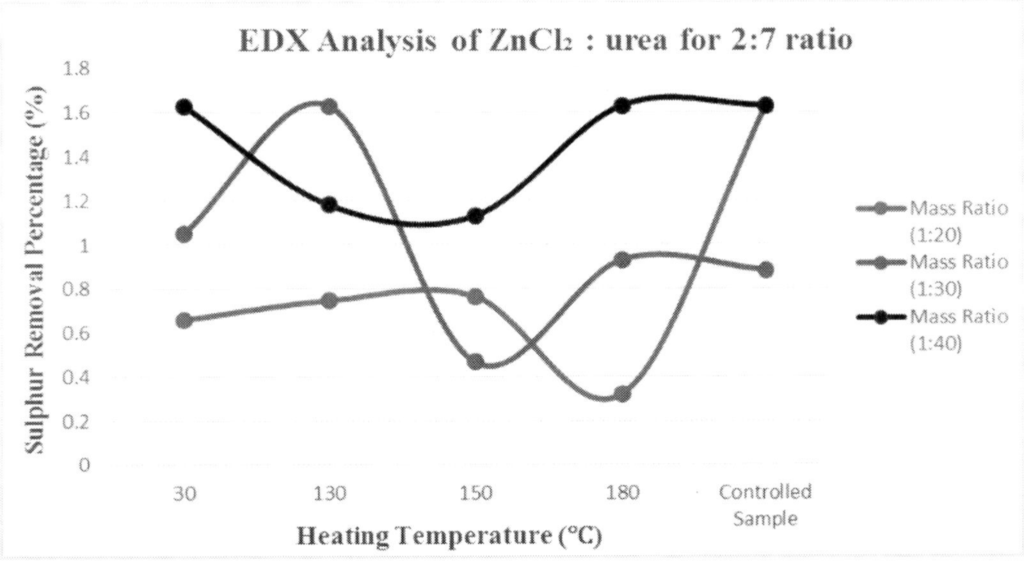

Fig. 9. EDX analysis for $ZnCl_2$:Urea for 2:7 molar ratio

Based on the results acquired, it can be deduced that during devulcanisation, bond reformation and bond cleaving of the sulphur-sulphur bond or carbon-sulphur bond can simultaneously occur as proven by [7]. Therefore, heating temperature is key process parameter in order to have a successful devulcanisation process. Hence, for this case, the optimum heating temperature is 130°C for rubber samples at 1:20 and 1:30 mass ratio at maximum removal of sulphur content

With bond reformation happening, it can explained that as bonds are cleaved or broken down, sulphur still remains inside the rubber sample together with free electrons which are detached from the -C-S- bond. Thus, this leads to carbon active sites formation which allows the sulphur to reform the sulphur cross-link bonds again by utilising those free electrons released. At a temperature of 130°C, rubber samples will have the highest removal of sulphur content if the heating treatment is stopped at moment. Despite that, temperature above 130°C will result in decrease in devulcanisation efficiency and bond reformation, allowing removed sulphur to reform back again with carbon backbone by utilising those free electrons. Therefore, it is important to stop at the optimum temperature in order to avoid bond reforming from occurring and to also ensure the maximum removal of sulphur content from the rubber sample mixture.

Based on Figure 9, in terms of mass ratio, it can be shown that using higher amount of DES will result in higher devulcanisation efficiency. This means that degree of cleavage on the cross-link bonds will be greater. This is because DES capacity is higher allowing the bond to break more easily. Nonetheless, it is an inconclusive hypothesis due to the uneven trend displayed for the rubber sample at 1:30 mass ratio.

The cause of such uneven trend displayed by the rubber sample might be due to non-homogenous of rubber sample mixture. This is because the sample mixture might not been completely well mixed, thus leading to such trend as shown in Figure 9. Another cause to it might be due to the rubber not completely mixed with DES as this DES has a very high viscosity, making it hard to mix both components. This will eventually leads in forming a two layered mixture due to density difference whereby devulcanisation process under ultrasonic treatment will be affected entirely.

3.2.4 FESEM Analysis

For this analysis, study on the morphology of rubber before and after devulcanisation was carried out using FESEM analysis. The findings of the results were discussed in detailed in the following section below.

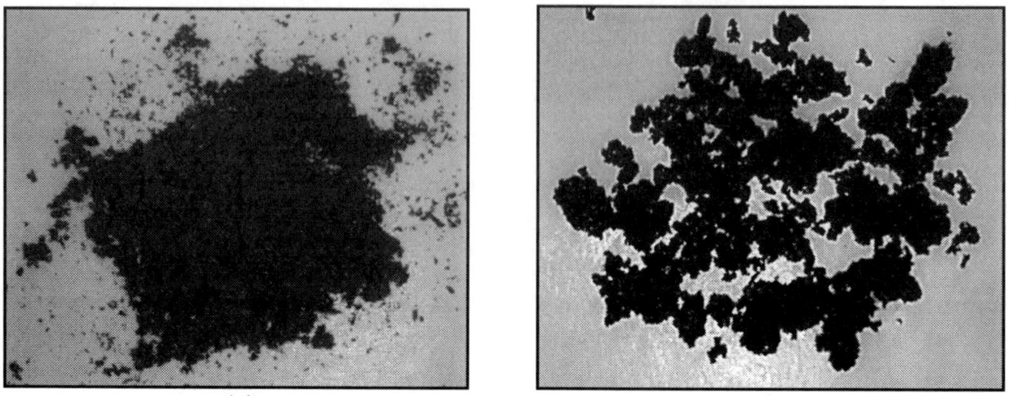

(a) (b)

Fig. 10. (a) Vulcanised rubber sample and (b) Devulcanised rubber sample

Properties of devulcanised rubber as shown in Figure 15 consist of soft, sticky in nature and feels rougher compared to unvulcanised rubber. Displaying these kind properties indicates that devulcanisation process was a success. FESEM analyses are carried out for each rubber sample through the use of different solvent agent. To make a confirmation on whether rubber has been vulcanised efficiently, it depends entirely on the gaps presence and smooth surface of the rubber morphology sample. In Figure 16, it shows the FESEM analysis on devulcanised rubber for all of the samples based on $ZnCl_2$:Urea for 2:7 molar ratio.

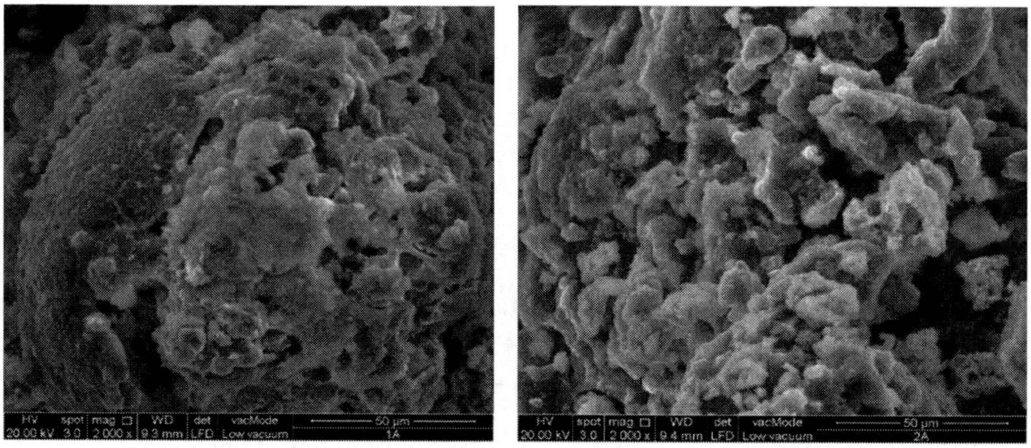

(a) (b)

HV spot mag WD det vacMode 50 µm
20.00 kV 3.0 2 000 x 9.1 mm LFD Low vacuum 3A

(c)

HV spot mag WD det vacMode 200 µm
20.00 kV 3.0 500 x 9.2 mm LFD Low vacuum 4A

(d)

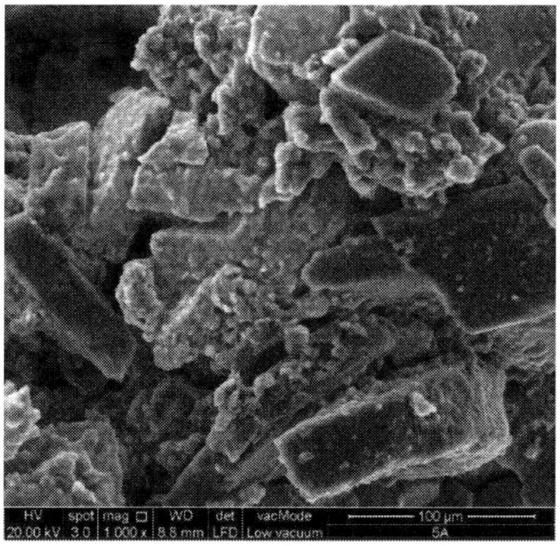
HV spot mag WD det vacMode 100 µm
20.00 kV 3.0 1 000 x 8.8 mm LFD Low vacuum 5A

(e)

HV spot mag WD det vacMode 200 µm
20.00 kV 3.0 500 x 9.2 mm LFD Low vacuum 6A

(f)

HV spot mag WD det vacMode 100 µm
20.00 kV 3.0 1 000 x 9.4 mm LFD Low vacuum 7A

(g)

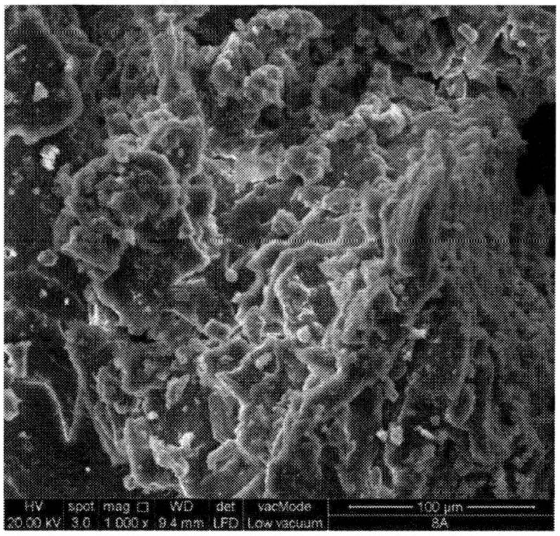
HV spot mag WD det vacMode 100 µm
20.00 kV 3.0 1 000 x 9.4 mm LFD Low vacuum 8A

(h)

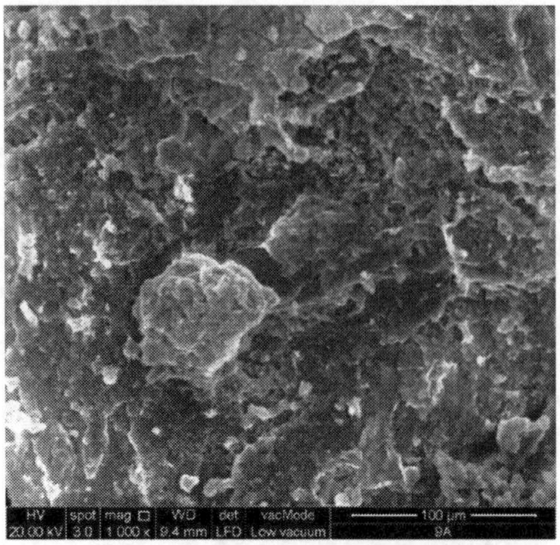
HV spot mag WD det vacMode 100 µm
20.00 kV 3.0 1 000 x 9.4 mm LFD Low vacuum 9A

(i)

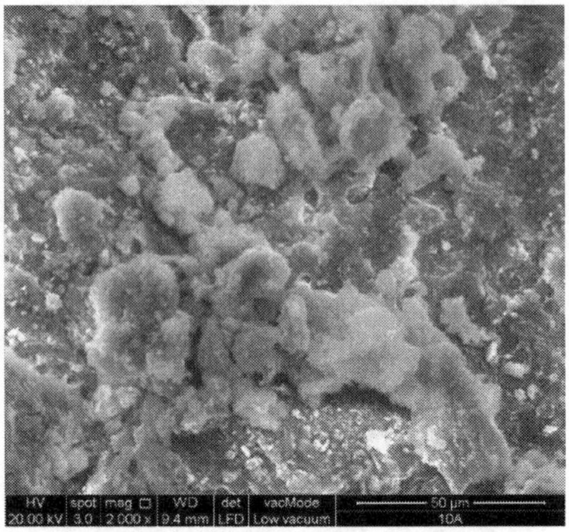
HV spot mag WD det vacMode 50 µm
20.00 kV 3.0 2 000 x 9.4 mm LFD Low vacuum 10A

(j)

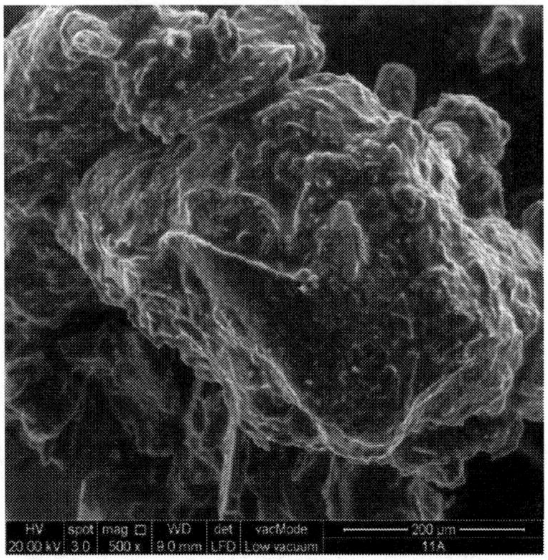
HV spot mag WD det vacMode 200 µm
20.00 kV 3.0 500 x 9.0 mm LFD Low vacuum 11A

(k)

HV spot mag WD det vacMode 100 µm
20.00 kV 3.0 1 000 x 9.4 mm LFD Low vacuum 12A

(l)

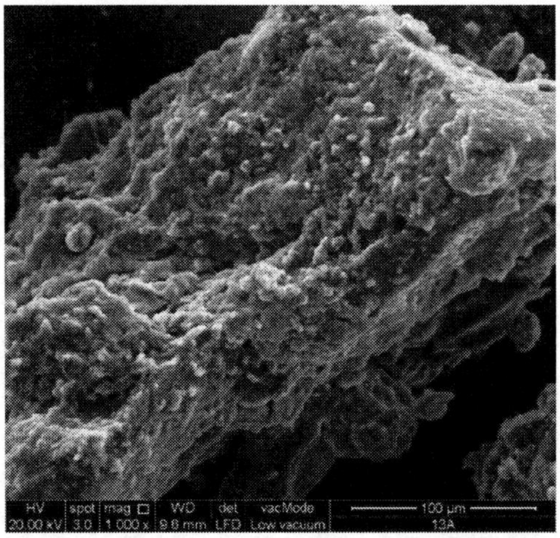
HV spot mag WD det vacMode 100 µm
20.00 kV 3.0 1 000 x 9.6 mm LFD Low vacuum 13A

(m)

HV spot mag WD det vacMode 200 µm
20.00 kV 3.0 500 x 9.3 mm LFD Low vacuum 14A

(n)

(o)

Fig. 11. FESEM analysis on rubber sample (a) 1A, (b) 2A, (c) 3A, (d) 4A, (e) 5A, (f) 6A, (g) 7A, (h) 8A, (i) 9A, (j) 10A, (k) 11A, (l) 12A, (m) 13A, (n) 14A and (o) 15A for $ZnCl_2$:Urea at 2:7

Referring and comparing Figure 11a to 11o, the morphology of the samples were evaluated. Along with sample 1A, 2A, 3A, 4A and 5A under mass ratio of 1:20, it can be noticed that Sample 2A as the smoothest surface out of all the other samples, followed by 1A, 3A, 4A and finally 5A. From a microscopic view, it can be observed that sample 1A is still rough in surface and stiff compared to sample 2A. From the micrograph, it is clearly seen that sample 1A is still very rough and rigid in surface in comparison to sample 2A. Despite that sample 3A has a similar surface as to sample 2A, hence the result tallies with EDX analysis result where the amount of sulphur removed is nearly similar. Such findings is said to be in equilibrium reformation for which if the heating temperature is more than optimum temperature, bond reformation will occur. A perfect example is sample 3A. Furthermore, small particles can been on top of the surface of the rubber sample. This could be due to contamination or left over of DES solution.

With sample 6A, 7A, 8A, 9A and 10A under mass ratio of 1:30, it can be shown clearly that sample 7A has the smoothest surface out of all the other samples, followed by 6A, 8A, 9A and finally 10A. Sample 7A has almost similar surface area to sample 5A. Hence, it is presumed that both these samples will have similar characteristics as shown in EDX analysis whereby the percentage of sulphur removal is almost the same.

For sample 11A, 12A, 13A, 14A and 15A, under the mass ratio of 1:40, it can be observed clearly that sample 12A has the smoothest surface out of all the samples since its surface is less rigid compared to others. Similar deduction is made for which the optimum heating temperature for devulcanisation is 130°C.

All of the respective figures shown above for morphology structure, neither of the figures showed any failure, damage on quality on the rubber samples. This indirectly comes to conclusion that devulcanised rubber properties are maintained without destroying the backbone chain of the rubber and the results is in agreement with FTIR result.

3.2.5 Soluble content analysis

Gel content samples were carried out and determined in accordance to ASTM D2765 [42].The degree of devulcanisation process of the sample rubber was determined by obtaining the soluble and gel fraction through Soxhlet extraction apparatus using boiling toluene. Then, the samples were collected and immediately dried inside an oven at a temperature of 70°C once a steady mass was obtained. Hence, the lower the content of the gel, the higher the efficiency of the devulcanisation process [21]. Soluble fraction is determined from the following equation;

$$\textit{Soluble Content} = [(Wo - W_i)/Wo] \times 100\% \quad (1)$$

Whereby, Wo is dry weight of devuclanised rubber before extraction process and W_i is dry weight of devulcanised rubber after extraction process. From there, the gel fraction is determined from the expressed equation. The following Table 6 and 7 show the gel content analysis tabulation and the analysis was thoroughly discussed.

Table 6. Gel content analysis on rubber treated with 2:7 molar ratio of $ZnCl_2$:Urea

Sample	Wo	W_i	Soluble fraction (%)	Sample	Wo	W_i	Soluble fraction (%)
1A	0.7276	0.7437	102.2127543	9A	0.8875	0.8448	95.18873239
2A	0.6779	0.6788	100.1327629	10A	0.7573	0.741	97.84761653
3A	0.8463	0.8252	97.50679428	11A	0.8667	0.7962	91.86569747
4A	0.8647	0.8505	98.35781196	12A	0.8865	0.8518	96.0857304
5A	0.9188	0.8171	88.93121463	13A	0.8565	0.7715	90.07589025
6A	0.8801	0.8182	92.96670833	14A	0.8659	0.783	90.42614621
7A	0.8069	0.7198	89.20560169	15A	0.9059	0.8201	90.52875593
8A	0.8681	0.8487	97.76523442				

With respect to Table 6, the trend that can be seen is that from sample 5A to 1A, soluble fraction increases whereby sample 1A has the highest soluble fraction. Next, from sample 10A to 6A, it can be seen that the highest soluble fraction is achieved by sample 8A. And lastly, from sample 15A to 11A, similar observation is seen where sample 12A has the highest soluble fraction. High percentage of soluble fraction declares the fact that devulcanisation is efficient and it was experienced by sample 1A, 8A and 12A. These results are tally with FTIR, TGA and EDX results earlier.

In a similar manner, observation from Table 7 shows that from sample 5B to 1B, soluble fractions are approximately the same but sample 3B has the highest value. Next, in between sample 10B to 6B, sample 7B and 9B are on par in having the highest value. Lastly, from sample 15B to 11B, it is clear that sample 13B has the highest soluble fraction. Thereby, treatment by $ZnCl_2$:Urea at 1:4 shows that the highest degree of devulcanisation was experienced by sample 3B, 7B and 9B as well as 13B which again tally with previous TGA, FTIR and EDX results.

Table 7. Gel content analysis on rubber treated with 1:4 molar ratio of $ZnCl_2$:Urea

Sample	Wo	W_i	Soluble fraction (%)	Sample	Wo	W_i	Soluble fraction (%)

1B	0.8582	0.834	97.18014449	9B	0.8994	0.8686	96.57549477
2B	0.8802	0.8577	97.44376278	10B	0.9056	0.8176	90.28268551
3B	0.8645	0.8562	99.03990746	11B	0.8865	0.7762	87.55781162
4B	0.7617	0.7445	97.74189313	12B	0.8376	0.8069	96.334766
5B	0.8975	0.8714	97.09192201	13B	0.9141	0.9039	98.88414834
6B	0.8654	0.7756	89.62329559	14B	0.8686	0.7829	90.13354824
7B	0.8817	0.8499	96.39333106	15B	0.8641	0.7941	91.89908575
8B	0.8821	0.8251	93.5381476				

4 Conclusion

DES as an alternative to IL was utilized to devulcanise rubber tyre due to its property of being cheap, non-toxic and biodegradable. Amino-compound-based deep eutectic solvents were chosen as they have been proven to be efficient in attracting sulphur in desulphurisation process. Particularly in this study, $ZnCl_2$:Urea DES was chosen and its molar ratio was varied to 1:4 and 2:7. Characterisation found that their moisture contents are below the maximum allowable limit with freezing point of < 60°C and degradation temperature that is higher than 200°C.

Rubber tyre were then treated with $ZnCl_2$:Urea DES at 1:20, 1:30 and 1:40 mass ratios (rubber:DES) at variable temperature of 30,130, 150 and 180°C. Finding shows that the optimum heating temperature is at 130°C for devulcanisation process for treatment with $ZnCl_2$:Urea (2:7). And when the temperature were increased to higher than 130°C, it was observed that decrease in devulcanisation efficiency and bond reformation took place, which allowing removed sulphur re-bonded back to carbon backbone by utilising free electrons. Therefore, it is important to stop at the optimum temperature in order to avoid bond reforming from occurring and to also ensure the maximum removal of sulphur content from the rubber sample mixture. FESEM analysis showed that most samples at 130°C gives the smoothest surface of rubber sample thus indicating the devulcanisation process was a success. In addition, FTIR analysis showed that –S-S- bond is the only bond which was cleaved during the process which proves a successful devulcanisation. On the other hand, for samples treated with $ZnCl_2$:Urea (1:4), high degree devulcanization is achieved at temperature >130°C. Thereby, indeed DES has great capability in devulcanising tyre.

References

1. O. E. Ikwuagwu, I. C. Ononogbu, and O. U. Njoku, "Production of biodiesel using rubber seed oil," vol. 12, pp. 57–62, 2000.
2. S. Rooj, G. C. Basak, P. K. Maji, and A. K. Bhowmick, "New Route for Devulcanization of Natural Rubber and the Properties of Devulcanized Rubber," *J. Polym. Environ.*, vol. 19, no. 2, pp. 382–390, 2011.

3. P. J. H. van Beukering and M. A. Janssen, "Trade and recycling of used tyres in Western and Eastern Europe," *Resour. Conserv. Recycl.*, vol. 33, no. 4, pp. 235–265, 2001.
4. M. Sienkiewicz, J. Kucinska-Lipka, H. Janik, and A. Balas, "Progress in used tyres management in the European Union: A review," *Waste Manag.*, vol. 32, no. 10, pp. 1742–1751, 2012.
5. S. Ramarad, M. Khalid, C. T. Ratnam, A. L. Chuah, and W. Rashmi, "Waste tire rubber in polymer blends: A review on the evolution, properties and future," *Prog. Mater. Sci.*, vol. 72, pp. 100–140, 2015.
6. A. Joseph, B. George, M. K N, and R. Alex, "The Current Status of Sulphur Vulcanization and Devulcanization Chemistry," *Rubber Sci.*, vol. 29, no. 1, pp. 62–100, 2016.
7. K. Bredberg, B. Erik Andersson, E. Landfors, and O. Holst, "Microbial detoxification of waste rubber material by wood-rotting fungi," *Bioresour. Technol.*, vol. 83, no. 3, pp. 221–224, 2002.
8. a. Zanchet, L. N. Carli, M. Giovanela, J. S. Crespo, C. H. Scuracchio, and R. C. R. Nunes, "Characterization of Microwave-Devulcanized Composites of Ground SBR Scraps," *J. Elastomers Plast.*, vol. 41, no. 6, pp. 497–507, 2009.
9. X. Zhang, C. Lu, and M. Liang, "Properties of natural rubber vulcanizates containing mechanochemically devulcanized ground tire rubber," *J. Polym. Res.*, vol. 16, no. 4, pp. 411–419, 2009.
10. H. Wang *et al.*, "Multifunctional TiO2 nanowires-modified nanoparticles bilayer film for 3D dye-sensitized solar cells," *Optoelectron. Adv. Mater. Rapid Commun.*, vol. 4, no. 8, pp. 1166–1169, 2010.
11. V. V. Rajan, W. K. Dierkes, R. Joseph, and J. W. M. Noordermeer, "Science and technology of rubber reclamation with special attention to NR-based waste latex products," *Prog. Polym. Sci.*, vol. 31, no. 9, pp. 811–834, 2006.
12. V. A. Online, "Deep eutectic solvents based on N -methylacetamide and a lithium salt as suitable electrolytes for lithium-ion batteries," no. 3, pp. 20054–20063, 2013.
13. G. Garcia, S. Aparicio, R. Ullah, and M. Atilhan, "Deep Eutectic Solvents: Physicochemical Properties and Gas Separation Applications," *Energy & Fuels*, vol. 29, pp. 2616–2644, 2015.
14. Q. Zhang, K. De Oliveira Vigier, S. Royer, and F. Jérôme, "Deep eutectic solvents: syntheses, properties and applications," *Chem. Soc. Rev.*, vol. 41, no. 21, p. 7108, 2012.
15. S. Seghar *et al.*, "Devulcanization of styrene butadiene rubber by microwave energy: Effect of the presence of ionic liquid," *Express Polym. Lett.*, vol. 9, no. 12, pp. 1076–1086, 2015.
16. C. Acids, C. Florindo, and A. M. Fernandes, "Insights into the Synthesis and Properties of Deep Eutectic Solvents Based on Cholinium Chloride," ACS Sustainable Chemistry Engineering, vol. 2, pp. 2416-2425, 2014.
17. A. I. Isayev, S. P. Yushanov, S.-H. Kim, and V. Y. Levin, "Ultrasonic Devulcanization of Waste Rubbers: Experimentation and modeling," Rheol. Acta, vol. 35, no. 6, pp. 616–630, 1996.
18. S. Seghar, N. A. Hocine, V. Mittal, S. Azem, B. Schmaltz, and N. Poirot, "Devulcanization of styrene butadiene rubber by microwave energy : Effect of the presence of ionic liquid," vol. 9, no. 12, pp. 1076–1086, 2015.
19. R. U. N. D. Anwendungen, "Devulcanization of Whole Passenger Car Tire Material," pp. 20–25, 2013.

20. P. S. Garcia, F. D. B. De Sousa, J. A. De Lima, S. A. Cruz, and C. H. Scuracchio, "Devulcanization of ground tire rubber : Physical and chemical changes after different microwave exposure times," vol. 9, no. 11, pp. 1015–1026, 2015
21. F. Endres, D. McFariane and A. Abbot, Electrodeposition from Ionic Liquid, Germany: Wiley-VCH, 2008.
22. K. Shahbaz, F. S. Mjalli, M. A. Hashim, and I. M. AlNashef, "Prediction of deep eutectic solvents densities at different temperatures," *Thermochimica Acta,* vol. 515, pp. 67-72, 2011.
23. K. Aoudia, S. Azem, N. A. Hocine, M. Gratton, V. Pettarin, and S. Seghar, "Recycling of waste tire rubber: Microwave devulcanization and incorporation in a thermoset resin," *Waste Manag.*, vol. 60, pp. 471–481, 2017.
24. A. I. Isayev and B. Sujan, "Nonisothermal vulcanization of devulcanized GRT with reversion type behavior," *J. Elastomers Plast.*, vol. 38, no. 4, pp. 291–318, 2006.
25. E. T. Thostenson and T.-W. Chou, "Microwave processing: fundamentals and applications," *Compos. Part A Appl. Sci. Manuf.*, vol. 30, no. 9, pp. 1055–1071, 1999.
26. C. Guo, L. Zhou, and J. Lv, "Effects of expandable graphite and modified ammonium polyphosphate on the flame-retardant and mechanical properties of wood flour-polypropylene composites," *Polym. Polym. Compos.*, vol. 21, no. 7, pp. 449–456, 2013.
27. L. Landini, S. G. de Araújo, A. B. Lugão, and H. Wiebeck, "Preliminary analysis to BIIR recovery using the microwave process," *Eur. Polym. J.*, vol. 43, no. 6, pp. 2725–2731, 2007.
28. V.YU.Levin, S.H.Kim, and A.I.Isayev, "Ultrasound Devulcanization of Sulfur Vulcanized SBR Crosslink Density and Molecular Mobility," *Rubber Chemistry and Technology*, vol. 69, no. 1. pp. 104–114, 1996.
29. A.Pelofsky, "Rubber reclamation using ultrasonic energy," US3725314 A, 1971.
30. M. Okuda and Y. Hatano, "Devulcanization of cured rubber," US5891926 A, 1998.
31. N. Senapati and D. Mangaraj, "Ultrasonic vulcanization," US4548771 A, 1984.
32. G. Garcia, S. Aparicio, R. Ullah, and M. Atilhan, "Deep Eutectic Solvents: Physicochemical Properties and Gas Separation Applications," *Energy & Fuels*, vol. 29, pp. 2616–2644, 2015.
33. A. R. Ferreira, M. G. Freire, J. C. Ribeiro, F. M. Lopes, J. G. Crespo, and J. A. P. Coutinho, "Ionic liquids for thiols desulfurization: Experimental liquid-liquid equilibrium and COSMO-RS description," *Fuel*, vol. 128, pp. 314–329, 2014.
34. V. A. Online, "Deep eutectic solvents based on N -methylacetamide and a lithium salt as suitable electrolytes for lithium-ion batteries," no. 3, pp. 20054–20063, 2013.
35. S. Seghar *et al.*, "Devulcanization of styrene butadiene rubber by microwave energy: Effect of the presence of ionic liquid," *Express Polym. Lett.*, vol. 9, no. 12, pp. 1076–1086, 2015.
36. Q. Zhang, K. De Oliveira Vigier, S. Royer, and F. Jérôme, "Deep eutectic solvents: syntheses, properties and applications," *Chem. Soc. Rev.*, vol. 41, no. 21, p. 7108, 2012.
37. M. H. Chakrabarti *et al.*, "Prospects of applying ionic liquids and deep eutectic solvents for renewable energy storage by means of redox flow batteries," *Renew. Sustain. Energy Rev.*, vol. 30, pp. 254–270, 2014.
38. L. Bahadori, M. H. Chakrabarti, F. S. Mjalli, I. M. Alnashef, N. S. A. Manan, and M. A. Hashim, "Physicochemical properties of ammonium-based deep eutectic solvents and their electrochemical evaluation using organometallic reference redox systems," *Electrochim. Acta*, vol. 113, pp. 205–211, 2013.

39. R. Saputra, "Devulcanization of Waste Rubber Tyre Utilizing Deep Eutectic Solvents and Ultrasonic Energy," Taylor's University Lakeside Campus, 2016.
40. S. Ghose and A. I. Isayev, "Recycling of unfilled polyurethane rubber using high-power ultrasound," *J. Appl. Polym. Sci.*, vol. 88, no. 4, pp. 980–989, 2003.
41. T. Liang, "Continuous Devulcanization of Ground Tire Rubber of Different Particle Sizes Using an Ultrasonic Twin-Screw Extruder," 2013.
42. S. Ramarad, C. T. Ratnam, M. Khalid, A. L. Chuah, and S. Hanson, "Improved crystallinity and dynamic mechanical properties of reclaimed waste tire rubber/EVA blends under the influence of electron beam irradiation," *Radiat. Phys. Chem.*, vol. 130, pp. 362–370, 2017.

Design of Biomass Combined Heat and Power (CHP) Systems based on Economic Risk using Minimax Regret Criterion

Wen Choong Ling[1], *Viknesh* Andiappan[2], and *Yoke Kin* Wan[1,1]

[1]School of Engineering, Taylor's University, Subang Jaya, Selangor, Malaysia.
[2]School of Engineering and Physical Sciences, Heriot-Watt University Malaysia, 62200, Putrajaya, Wilayah Persekutuan Putrajaya, Malaysia

Abstract. It is a great challenge to identify optimum technologies for CHP systems that utilise biomass and convert it into heat and power. In this respect, industry decision makers are lacking in confidence to invest in biomass CHP due to economic risk from varying energy demand. This research work presents a linear programming systematic framework to design biomass CHP system based on potential loss of profit due to varying energy demand. Minimax Regret Criterion (MRC) approach was used to assess maximum regret between selections of the given biomass CHP design based on energy demand. Based on this, the model determined an optimal biomass CHP design with minimum regret in economic opportunity. As Feed-in Tariff (FiT) rates affects the revenue of the CHP plant, sensitivity analysis was then performed on FiT rates on the selection of biomass CHP design. Besides, design analysis on the trend of the optimum design selected by model was conducted. To demonstrate the proposed framework in this research, a case study was solved using the proposed approach. The case study focused on designing a biomass CHP system for a palm oil mill (POM) due to large energy potential of oil palm biomass in Malaysia.

1 Introduction

Global energy demands are projected to increase 48% from 2012 by 2040 [1, 2]. To meet this projected increase in energy demand, biomass energy is expected to play an imperative role over the coming years. In fact, global biomass energy demands from 2000 to 2015 have seen a significant increase as compared to stable hydropower demands [3]. Such increase indicates that biomass is gaining attention in the renewable energy market. Biomass can be categorised into lignocellulose biomass, municipal waste and animal manure. These biomass resources can be obtained from agricultural residues, biomass plantations and organic wastes from residential areas [4]. Approximately 70% of these

[1] Corresponding author: YokeKin.Wan@taylors.edu.my

biomass are lignocellulosic biomass from plantations such as cereal, oil and sugar crops [5]. Intergovernmental Panel on Climate
Change Special Report on Renewable energy (IPCC-SRREN) reported that solely on biomass from plantations are estimated to have potential energy up to 700 x 10^{12} MJ/year whereas other sources of biomass such as food and municipal waste had only 300 x 10^{12} MJ/year of potential energy [4]. Due to the high availability of lignocellulose biomass, recent studies on biomass utilisation focus on converting biomass into bioenergy and biofuels.

In Malaysia, one major source of lignocellulose biomass is oil palm biomass from the palm oil industry which provides approximately 82.61wt% of total biomass in Malaysia [6]. The palm oil industry generates biomass such as empty fruit bunches (EFBs), palm mesocarp fibre (PMF), palm kernel shell (PKS) and palm oil mill effluent (POME). These oil palm biomass are wastes produced during the extraction of crude palm oil (CPO) and crude palm kernel oil (CPKO) from fresh fruit bunches (FFB). Based on Malaysia Palm Oil Board [7], amount of FFB processed in Malaysia palm oil mill in 2016 is approximately 60 million tonnes. Solid oil palm biomass such as EFB, PMF and PKS are generated at about 28.42 million tonnes while POME is produced at about 40.04 million tonnes during palm oil extraction annually in 2017 [6–8]. These biomass have high calorific value that can be converted into heat and power using biomass combined heat and power (CHP) systems. This provides an opportunity for the industry to harness high amount of bioenergy from sources that were initially disposed as waste by-products.

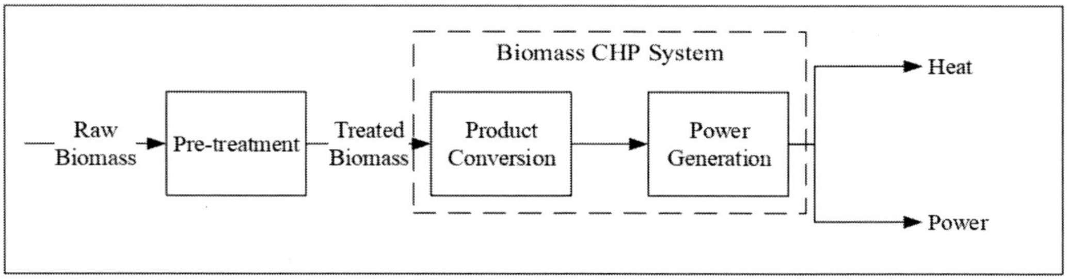

Fig. 1.1. Simple Illustration of Biomass CHP system

Figure 1.1 above illustrates a biomass CHP is a system to generate power and thermal energy from biomass sources for industrial and/or residential needs [9]. A biomass CHP system commonly consists of pre-treatment unit, product conversion unit and power generation unit. Pre-treatment units such as shredder and dryer are required to reduce size and moisture content in biomass to allow efficient conversion into products through conversion units such as (but not limited to) boilers, gasifiers, digesters [6, 10, 11]. Biomass undergoes combustion in boilers to produce steam where steam produce can be used to produce energy using a steam turbine. Gasifiers convert biomass into syngas and digesters convert biomass into biogas. Syngas consists of carbon monoxide and hydrogen that could be combusted in gas engines, gas turbines and micro-turbines. Similarly, biogas consist of mostly methane gas can be combust in similar power generation units to produce power [10, 11].

The Malaysian government had announced a program known as Small Renewable Energy Power (SREP) program back in 2001 to encourage industrial players to utilise agricultural wastes to generate power. With support from United Nation Development Programme and Global Environment Facility and SREP, the first demonstration of 13 MW CHP plant using EFB was built in Bahau, Negeri Sembilan [8]. Currently in Malaysia, most palm oil mills have CHP plants installed to meet internal energy demands. However, the power generated from these CHP plants are insufficient to supply to national power grid. This is due to some challenges met by these power plants that was outlined by several work from Umar et. al.[11–13] and it is summarised as following.

- Varying energy demand and biomass supply – Subjected to energy demand, palm oil mill will have sufficient biomass supply to meet energy demand for operation of the palm oil mill. However, feeding any power generated to national grid will have difficulties to ensure stable supply of biomass to fed into CHP systems due to seasonal nature of palm oil production to cope with varying energy demand.

- Non-favourable government program and schemes – SREP and Feed-in-Tariff (FiT) had proposed by the government in hope of palm oil mill owners to utilise their oil palm biomass to generate electricity and feed into gridlines. However, respondents from these work expressed that current FiT rates are not attractive enough to them. Hence, most of the power generated from biomass CHP plant were used for self-consumption in palm oil mill instead of feeding it into power grid.

- Location of palm oil mill from gridlines – Approximately 62.3% of palm oil mills in Malaysia are situated at least 10 km from any electricity network system. Connecting these CHP systems in palm oil mill to the gridlines requires intensive amount of capital cost. A heuristic value regarding cost required for connection of grid is given as approximately RM 1 to 1.5 million per km of gridline [14]. Hence, palm oil mill owners are reluctant to invest such capital cost when there are risk of having unstable power generation with varying biomass supply.

- Perspective of palm oil mill owner – Respondents considers biomass energy systems as a highly intensive capital invest with long payback period. Since the palm oil industry focuses on edible oil production as their main traditional business, most mill owners prefer not to risk participating in the renewable energy industry. Furthermore, mill owners have expressed that challenges in financial capability to participate in such capital-intensive projects.

Due to uncertainty of biomass supply in meeting varying energy demands, palm oil mill owner possess a lack confidence to invest in a capital-intensive biomass CHP system. Others factors such as distance of palm oil mill from gridlines and unattractive government subsidy or tariff programs contribute further to this lack of confidence. Therefore, there is a need of a framework to design CHP system that cope with varying energy demands and subsequently provides an optimal design with low economical risk in order to gain confidence from palm oil mill owners.

In a similar area of research, several research works have been proposed to analyse economic risk of renewable energy technologies (RET) using an approach known as Monte Carlo Simulation (MCS) approach. MCS considers uncertainty in variables behaviour through probability distribution to quantify the possibility of a certain condition will occur [15]. Arnold et al. [16] assessed economical risk for renewable energy infrastructures using

MCS to compare different boundary conditions such as climate, capital and operating cost. However, it is stated by the author that this MCS model requires a matching tool to select suitable RET for each case study of interest. In addition, Edinaldo et al. [15] presented a similar mathematical model to help in justifying whether the selected RET is worth investing through MCS approach. However, MCS approach used in both Arnold and Edinaldo's work only analyses a certain RET potential economic risk without providing clear suggestion on selection of optimal RET. Therefore, there is a need for a framework that is able to analyse potential economic risk and subsequently synthesise an optimal RET system.

Recent studies by Yokoyama [17, 18] used Minimax Regret Criterion (MRC) approach to evaluate economic risk and subsequently providing optimal design of energy supply system. However, Yokoyama's study only focuses on conventional energy supply system. Besides, challenges faced in utilising biomass CHP such as varying biomass supply and energy demand were not addressed in recent work. Therefore, in this work, a mathematical optimisation model was developed to synthesise an optimal design for biomass CHP systems based on potential loss of profitable opportunity using MRC approach at varying energy demand. The objective of this developed framework are as follows:

1. To determine the optimal design for the biomass CHP system based on maximum gross profit.
2. To assess potential economy risk for a given biomass CHP design based on varying energy demand using MRC approach.

Such model will enable designers to systematically design a biomass CHP with minimal economic risk using MRC approach.

2 Methodology

As described previously, a mathematical model framework was developed to design an optimal biomass CHP system to meet certain energy demand with minimal economic risk through Minimax Regret Criterion (MRC) approach. A case study based on palm oil mill biomass was conducted using the framework generated. To carry out the case study, a palm oil mill case study superstructure was developed to show possible biomass CHP designs to convert palm oil mill biomass into heat and power. With the superstructure developed, mathematical model equations were formulated for each biomass CHP technology unit. These equations correlates biomass mass flow rates, efficiency of CHP systems, heat and power produced and capital expenditure (CAPEX) of equipment. The mathematical model will ultimately provide biomass CHP design based on maximum gross profit at different energy demands. Following this, MRC approach was utilised to select the optimum biomass CHP design which has the lowest maximum regret value. The term "maximum regret value" is defined as the as maximum amount of profitable opportunity loss by making a wrong decision in design [19]. A sensitivity analysis of Feed-in Tariff (FiT) rates on selection of optimum biomass CHP design was conducted to analyse the impact of the change of FiT rates on the selection for optimum design of biomass CHP. Lastly, design analysis on optimum biomass CHP design was conducted to study the trend of selection for biomass CHP design by the framework base on case study.

2.1 Generic Superstructure of Biomass CHP Design

Resources i represents biomass supply with flow rates F_i can be used to convert into product $p \in P$ through technology $j \in J$. The product p can be further converted to energy $e \in E$ via power generation technology $j' \in J'$. Figure 2.1 below shows the generic superstructure of Biomass CHP design. With the generic superstructure, equations formulated in the following sections are based on Figure 2.1.

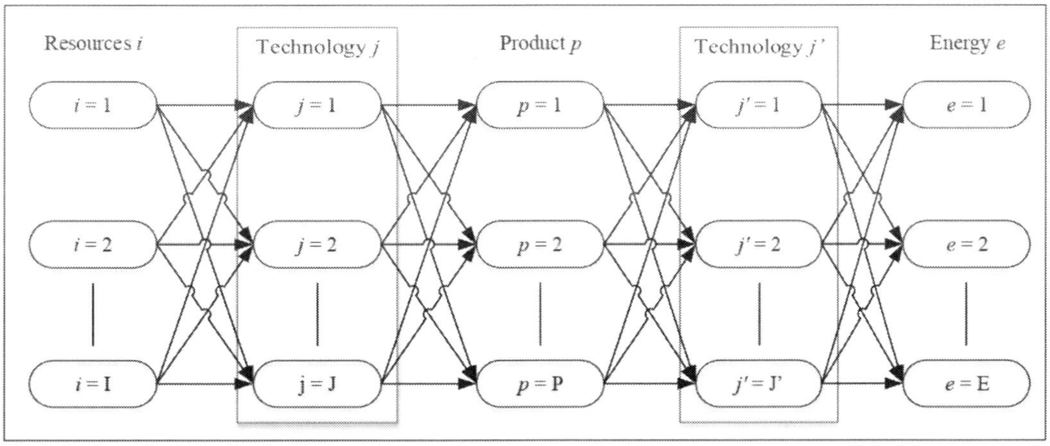

Fig 2.1. Generic Case Study Superstructure

2.2 Mathematical Model Equations

Referring to Figure 2.1, following mathematical modelling equations were formulated based on data collected for the respective process unit to convert biomass to heat and power. Mass and energy balance equation was developed first for each biomass CHP system to convert biomass to heat and power. Following this, the equation for capital expenditure (CAPEX) of each equipment was developed based on the equipment capacity parameters such as flow rates and energy produced. With CAPEX of equipment and energy production equations, the optimum biomass CHP design that provide maximum gross profit was generated using LINGO programming software. The data of gross profit and optimum biomass CHP design at different power demand will be utilised in Minimax Regret Criterion (MRC) approach to solve for optimum biomass design that provide maximum gross profit with minimal economical risk.

2.2.1 Mass and Energy Conversion Mathematical Equations

Eq. (1) shows the component balance for biomass i where f_i and x_i are the component mass flow rates and composition respectively based on Figure 2.1 where F_i represents the total flow rate of biomass i.

$$f_i = F_i x_i \qquad \forall i \qquad (1)$$

Biomass i is allocated to potential technology j as shown in in Eq. (2) where F_{ij} is the flow rate entering technology j.

$$F_i = \sum_{j=1}^{J} F_{ij} \qquad \forall i \forall j \qquad (2)$$

Technology j converts entering biomass F_{ij} into product p. Eq. (3) shows the conversion of F_{ij} into mass flow rate of product F_{jp} with a conversion factor of k_{jp} for each technology.

$$F_{jp} = k_{jp} F_{ij} \qquad \forall i \forall j \forall p \qquad (3)$$

Total rate of production for product p for all technologies j is shown as Eq. (4).

$$F_p = \sum_{j=1}^{J} F_{jp} \qquad \forall j \forall p \qquad (4)$$

The product p is then distributed to power generation technology j' to convert into energy e. The distribution of product p is as shown in Eq. (5).

$$F_p = \sum_{j'=1}^{J'} F_{pj'} \qquad \forall j' \forall p \qquad (5)$$

Eq. (6) shows distributed product p converts into power $E_{j'e}$ through power generation technology j'. $n_{j'p}$ is the conversion efficiency of each power generation technology j'.

$$E_{j'e} = n_{j'p} F_{pj'} \qquad \forall e \forall j' \forall p \qquad (6)$$

The total power generated by technologies j' is calculated as Eq. (7) below.

$$E_T = \sum_{e=1}^{E} E_{j'e} \qquad \forall e \forall j' \qquad (7)$$

2.2.2 Capital Expenditure (CAPEX) of Equipment Mathematical Equations

The capital expenditure (CAPEX) equations of each technologies shown in Figure 2.1 were formulated based on data collected. CAPEX of each technologies C_j and $C_{j'}$ are determined via Eq. (8) and (9). k_{cj} and $k_{cj'}$ are the cost factor of each technologies correlated to its technology size factor M_j and $M_{j'}$ respectively. Example of size factor M_j are flow rates through the equipment (F_{ij} or F_{jp}) and power capacity of the equipment $E_{j'e}$.

$$C_j = k_{Cj} M_j \qquad \forall j \qquad (8)$$
$$C_{j'} = k_{Cj'} M_{j'} \qquad \forall j' \qquad (9)$$

The total capital expenditure (CAPEX) of the technologies is as shown in Eq. (10).

$$CAPEX = \sum_{j=1}^{J} C_j + \sum_{j'=1}^{J'} C_{j'} \qquad \forall j \forall j' \qquad (10)$$

2.2.3 Solving Case Study by Minimax Regret Criterion (MRC) Approach

With the total *CAPEX* of equipment and energy produced from biomass CHP design calculated using Eq. (10), gross profit based on energy produced through CHP system were calculated with power tariff data at the location of the palm oil mill. Eq. (11) below shows the gross profit P_{gross} calculation of the CHP system. AOT is the annual operation time of the plant and (A/P) is the capital recovery compound interest factor for uniform payment of the biomass CHP plant.

$$P_{gross}=E_T \times Tariff \times AOT - CAPEX \times (A/P) \quad (11)$$

By maximising gross profit from Eq. (11) at different power demand, the data of gross profit are subsequently used in Minimax Regret Criterion (MRC) approach to decide on the optimum design of biomass CHP with minimal regret. The term "regret" refers to a quantifiable value on loss of profitable opportunity due to a decision made in design. For example, if the design selected have much higher power capacity than the power demand, the biomass CHP plant will be running on low capacity which will generate lower gross profit than its potential gross profit. MRC approach is commonly used in business and finance model to make decision that minimises the possible maximum regret [13, 18]. Generic tables used for MRC approach are shown in Table 2.1 and 2.2 below.

Table 2.1. Generic gross profit table used for MRC approach

Factor 1	**Factor 2**		
	j_1	j_2	j_3
i_1	P_{11}	P_{12}	P_{13}
i_2	P_{21}	P_{22}	P_{23}
i_3	P_{31}	P_{32}	P_{33}

Where i_n is the levels of factor 1, j_n is the levels of factor 2 and P_{ij} as stated in the table are the gross profit of the system.

Table 2.2. Generic Regret Table used for MRC approach

Factor 1	**Factor 2**			**Maximum Regret Value**
	j_1	j_2	j_3	
i_1	R_{11}	R_{12}	R_{13}	$R_{max,1}$
i_2	R_{21}	R_{22}	R_{23}	$R_{max,2}$
i_3	R_{31}	R_{32}	R_{33}	$R_{max,3}$

Where R_{ij} is the regret value of each level combination of the factors and $R_{max,i}$ is the maximum regret value of each row.

The MRC approach calculations based on Table 2.1 and 2.2 with i = 1, 2, 3 …, I and j = 1, 2, 3…, J are as following:

For gross profit table,

1. $P_{i,j}$ is gross profit calculated with Eq. (11) at energy demand and capacity of the CHP designed are equal to each other. (diagonal row of the table from top left to bottom right)
2. $P_{i,i+j}$ are equal to $P_{i,j}$ since the profit is limited by the maximum capacity of the CHP designed where energy demand is more than designed CHP capacity. (columns on top of diagonal row of the table)
3. $P_{i+j,j}$ is calculated with Eq.(11) at energy demand is lesser than capacity of CHP designed. (Columns below of diagonal row of the table)

For regret value table,

1. $R_{i,j}$ (diagonal row from top left to bottom right) equal to zero since energy demand occurred is equal to capacity of CHP designed.
2. $R_{i,i+j}$ (columns on top of diagonal row of the table) are calculated by Eq. (12).

$$R_{i,i+j}=R_{i+j,i+j} - R_{i,i} \tag{12}$$

3. $R_{i+j,j}$ (columns below of diagonal row of the table) are calculated by Eq. (13)

$$R_{i+j,j}=R_{i+j,j} - R_{j,j} \tag{13}$$

4. Determine the maximum regret value for each row.
5. Select the design criteria with the smallest maximum regret value.

2.3 Palm Oil Mill Case Study Superstructure

Figure 2.2 below shows the palm oil mill case study superstructure developed for the mathematical model.

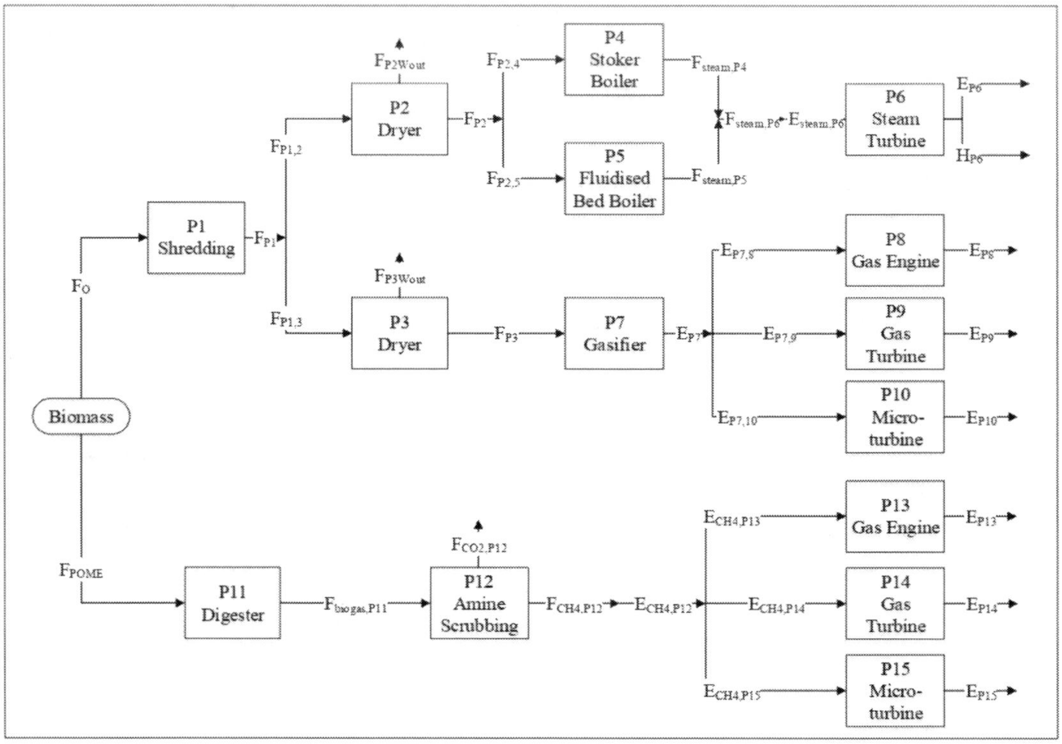

Fig. 2.2. Palm Oil Mill Case Study Superstructure

The amount of biomass supply for this case study is taken as 10000 kg/hr of mix solid biomass and palm oil mill effluent (POME) of 10 m^3/day. Varying the power demand as the manipulating variable, power demand range from 250 kW to 13 MW were input into the model to identity the optimum biomass CHP technology pathway.

2.4 Sensitivity Analysis of Feed-in Tariff (FiT)

Feed-in tariff (FiT) system is a distributed premium license to individual or company to legally sell renewable energy at the FiT rate to companies that distribute electricity [20]. FiT rates for power generation using biogas and solid biomass are different rates. Basic FiT rates varies with installed capacity and bonus FiT rates applies when certain criteria of the installed CHP technology is met. To ease the calculation in sensitivity analysis, the FiT rates used to calculate the gross profit of CHP plant was taken as average of RM 0.2985 per kWh for biomass power generation and RM 0.2886 per kWh for biogas power generation [20]. For the sensitivity analysis conducted, the FiT rates are changed at ±10, ±30 and ±50% from its original tariff. The purpose of doing this is to analyse the impact of the change of FiT rates on the selection for optimum design of biomass CHP technologies.

2.5 Analysis of Optimum Biomass CHP Design

Optimum biomass CHP design that are recommended by model through maximising gross profit at varying power demand for the case study was analysed to study the trend of optimum selection for biomass CHP design. Besides, the analysis identifies which biomass CHP design is compulsory across a certain range of power demand. These "must-have" design is compulsory to have in order to achieve the required power demand. The analysis also identifies "optional" equipment that can be invested in if process designers are ambitious to meet higher power demands and equipment that are never selected throughout the power demand range appear as a "must-avoids". This analysis helps plant designers to narrow down their selection of biomass CHP design and subsequently decide on the optimum selection of biomass CHP design.

3 Results and Discussion

As described previously, the methodology for this work was carried out to determine an optimum biomass CHP design for palm oil mill using Minimax Regret Criterion (MRC). In this section, the results obtained from the case study are analysed and discussed. Following this, a sensitivity analysis on Feed-in Tariff (FiT) rates was performed to analyse the impact of FiT rates on the selection of biomass CHP design. Lastly, a design analysis was done on the optimum design of biomass CHP to study the trend of recommended design by the mathematical model. In the design analysis, "must-have", "optional" and "must-avoid" designs were identified to ease plant designer decision on selecting the optimum selection of biomass CHP design.

3.1 Optimum Biomass CHP Design using MRC Approach

To select the optimum design of CHP plant throughout the wide range of power demand, Minimax Regret Criterion (MRC) approach was utilised as described in methodology. Taking the power demands are 3000, 5000, 8000 and 10000 kW, Table 3.1 below shows the gross profit table of optimum biomass CHP designs at different power demand.

Table 3.1. Gross Profit Table of Case Study in unit RM/year

Designed Capacity (kW)	Power Demand (kW)			
	3000	**5000**	**8000**	**10000**
3000	4,470,740.00	4,470,740.00	4,470,740.00	4,470,740.00
5000	1,481,316.30	6,154,717.00	6,154,717.00	6,154,717.00
8000	(2,213,380.58)	2,460,019.42	9,964,220.00	9,964,220.00
10000	(3,897,126.95)	776,273.05	8,280,473.05	12,938,480.00

As shown in Table 3.1, values in parentheses represent loss of profit. The mathematical model provided the biomass CHP design at its respective designed capacity. The equipment that were selected while the gross profit of biomass CHP design is maximised will have CAPEX value shown in the mathematical model while solving gross profit for each designed capacity. Hence, the biomass CHP design for each designed capacity identified are listed below in Table 3.2.

Table 3.2. Biomass CHP Design at each Designed Capacity

Biomass CHP Design	**Plant Power Capacity (kW)**
Biogas with GE	3000
Syngas + Biogas GE	5000
Boiler with Steam Turbine	8000
Combined Boiler with Steam Turbine and Biogas with GE	10000

The regret table was developed to identify the maximum regret value of each biomass CHP technology. The biomass CHP technology with the smallest value between the maximum regret values was selected as the optimum biomass CHP technology. The regret table was developed as below.

Table 3.3. Regret Table of Case Study in unit RM/year

Designed Capacity (kW)	**Power Demand (kW)**				**Max Regret Value**
	3000	**5000**	**8000**	**10000**	
3000	-	1,683,977	5,493,480	8,467,740	8,467,740
5000	2,989,423	-	3,809,503	6,783,763	6,783,763
8000	6,684,120	3,694,697	-	2,974,260	6,684,120
10000	8,367,866	5,378,443	1,683,746	-	8,367,866

Selecting the smallest maximum regret value in Table 3.3, the optimum design capacity of biomass CHP plant is 8000 kW. Based on Table 3.2, the optimum biomass CHP technology at 8000 kW is steam generation using boiler with steam turbine to generate power. Therefore, boiler with steam turbine power generation is the recommended optimum biomass CHP technology that provide maximum gross profit with minimal regret/loss of opportunity.

3.2 Sensitivity Analysis of Feed-in Tariff (FiT) Rate on Selection of Optimum Biomass CHP Design

The optimum design recommended by the developed model through MRC approach at different FiT rates are shown in Table 3.4 below using similar case study assumptions in Section 2.3. Lowest maximum regret value, potential minimum and maximum gross profit values between different biomass CHP designs are included to show their changes with FiT rates.

Table 3.4. Optimum Design of Biomass CHP Plant at Different FiT Rates

FiT Rates (%)	Optimum Design	Minimax Regret Value (RM/yr)	Potential Minimum Gross Profit (RM/yr)	Potential Maximum Gross Profit (RM/yr)
-50	Biogas GE	96,457.00	1,007,540.00	1,103,997.00
-30	Syngas + Biogas GE	3,137,180.00	2,392,820.00	5,811,957.00
-10	Syngas + Biogas GE	5,567,543.00	3,778,100.00	10,562,280.00
0	Boiler Steam Turbine	6,684,120.58	4,470,740.00	12,938,480.00
10	Boiler Steam Turbine	6,684,120.58	5,163,380.00	15,314,680.00
30	Boiler Steam Turbine	6,684,120.58	6,548,660.00	20,067,090.00
50	Boiler Steam Turbine	6,684,120.58	7,933,940.00	24,819,490.00

As the FiT rates decreases, the optimum design of biomass CHP plant changed into combined syngas and biogas with gas engine (GE) as power generation technology. At 50% reduction of FiT rates, the optimum design of biomass CHP plant is small capacity biogas generation. The lowest maximum regret value reduces significantly as FiT rates decreases. Although the magnitude of regret value reduces, the potential minimum and maximum gross profit reduces significantly as well. Lower profit margin as FiT rate reduces causes smaller scale biomass CHP design such as biogas and syngas generation with GE more economical compared to boiler with steam turbine. As FiT rate increases, the selection of optimum design for biomass CHP plant remain the same as boiler with steam turbine. Increasing FiT rates will not change the selection on boiler with steam but will significantly increase potential gross profits. This is due to higher revenue on feeding in power to grid thus giving a higher potential of profit as FiT rate increases.

In summary, the impact of FiT rate on selection of biomass CHP design based on Table 3.4 is summarised as following:

1. Decrease in FiT rates will economically favour smaller scale biomass CHP design such as biogas and syngas generation with GE due to lower profit margin.
2. Decrease in FiT rates reduces lowest maximum regret significantly; however, potential profit from biomass CHP plant decreases significantly as well.
3. Higher revenue is generated on feeding in power to grid with higher plant power capacity and FiT rate which eventually provide higher gross profit. Hence, increase in FiT rates will favour larger scale biomass CHP design such as boilers with steam turbine which is able to provide higher power capacity.

3.3 Analysis of Optimum Biomass CHP Design

Table 3.5 below shows the optimum biomass CHP technology that provide maximum gross profit at different power demand with their respective usage of biomass using the developed model. The amount of biomass available was taken as the same as case study as described in Section 2.2.

Table 3.5. Optimum Biomass CHP Design at Different power Demand of Case Study

Power Demand (kW)	POME Consumption (m^3/day)	Usage of Solid Feed (kg/hr)	Optimum Biomass CHP Design
250	0.58	0.00	Biogas with GE
500	1.16	0.00	
750	1.75	0.00	
1000	2.33	0.00	
3000	6.98	0.00	
4000	9.31	0.00	
5000	10.00	1691.00	(Syngas + Biogas) with GE
6000	0.00	7472.37	Boiler with Steam Turbine
7000	0.00	8479.30	
8000	0.00	9486.24	
9000	1.14	10000.00	Combined Boiler with Steam Turbine and Biogas with GE
10000	3.47	10000.00	
13000	9.55	10000.00	

As shown in Table 3.5, biogas with GE engine is used for lower power demand from 250 to 4000 kW where only POME is utilised. This shows that biogas generation through anaerobic digestion of POME is preferred at lower power demand rather than utilising solid biomass such as empty fruit bunches (EFB), palm kernel shell (PKS) and palm mesocarp fibre (PMF). This occurs because utilisation of solid biomass requires pre-treatment equipment such as dryers and shredder, which requires more capital cost. Since the objective function in model focuses on maximising profit and minimising capital costs, the model favours a cheaper route, which is anaerobic digestion equipment. Meanwhile, when the power demand reaches 5000 kW, available POME supply of 10 m^3/day is completely consumed, leaving the remaining amount of power required to be met by syngas generation. Syngas generation was selected to top up remaining power required rather than utilising boiler technology at low usage of solid feed for power generation. This occurs as boilers are more economically viable at larger capacities and high usage of solid feed while syngas is more viable at lower capacities. However, as the power demand increases to range of 6000 to 8000 kW, only solid biomass is utilised through boiler with steam turbine technology without utilising any POME for biogas generation. This shows that boiler steam generation is a favoured selection of CHP technology at higher power demand comparing to syngas and biogas generation. This is because solid biomass that used in boiler can generate higher amount of energy to satisfy higher range of power demand at 6000 to 8000kW. Alternatively, biogas and syngas can be generated to meet this demand. However, due to the high power demand, high amount of biogas and syngas would be required. This will cause the equipment to increase in size much more than of boilers as both biogas and syngas are in gaseous state. Hence, biogas and syngas generation is not economically viable at higher capacity. As the solid biomass are completely utilised as shown in power demand above 9000 kW, POME is utilised to make up to the power demand required which there are not much selection for the model to utilise both biomass. The data stops at power demand of 13000 kW due to insufficient amount of biomass to produce the required power demand. In addition, only GE is selected as power generation technology for biogas and syngas shows that it is a much better selection than micro-turbine and gas turbine. This is due to GE is a widely used and developed power generation technology compare to micro-turbine. Gas turbine is commonly used in natural gas CHP plants but it is only economically viable at very high power demand due to its large power capacity. Therefore, GE is a more economical option than micro-turbine and gas turbine for biomass CHP plant as shown in Table 3.5.

Results from Table 3.5 also indicate other important analysis. It is found that there are common equipment selected for a range of power demands. These equipment are considered as the "must-haves", as without them, the design would be incapable of meeting the required power demands. On top of this, there are certain equipment in the CHP design that appear as "optional". "Optional" equipment are equipment that can be invested in if process designers are ambitious to meet higher power demands. On the other hand, equipment that are never selected throughout the power demand range appear as a "must-avoid". To allow designers to analyse these options, Figure 3.1 summarises the "must-have", "optional" and "must-avoid" equipment in the case study.

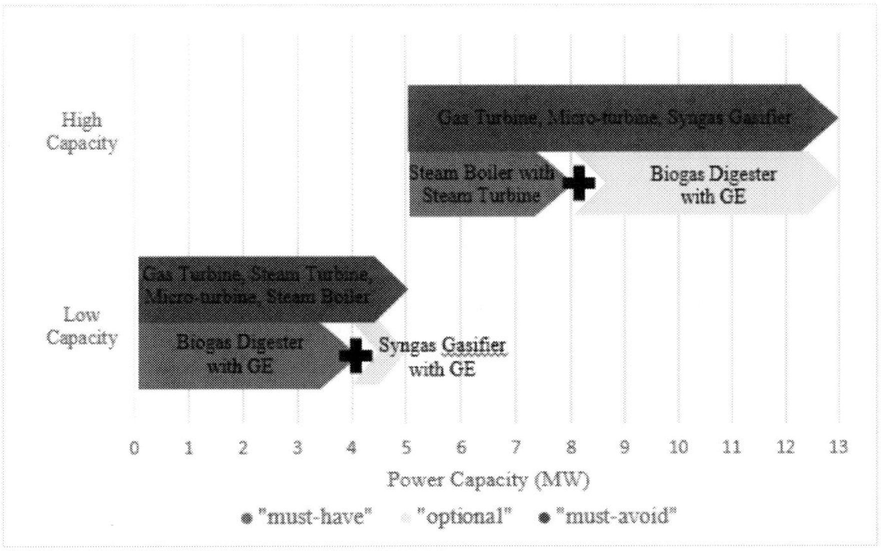

Fig 3.1. "Must-have", "optional" and "must-avoid" Design

For low capacity biomass CHP plant ranges up to 5 MW, the "must-have" design for the CHP plant is biogas generation with GE as shown in Figure 3.1. If the designer would like to invest for a higher power demand at this range, the designer could combine "optional" syngas generation on top of biogas generation to meet higher power demand of 5 MW. For higher capacity biomass CHP plant that ranges from 5 to 13 MW, the "must-have" design for the CHP plant is boiler with steam turbine. Similarly, a higher power demand of 8 to 13 MW could be achieved by combining "optional" biogas with GE on top of boiler with steam turbine. The "must-avoid" design of biomass CHP plant are gas turbine and micro-turbine as these design are not economically viable compare to GE as described.

In a nutshell, Table 3.5 and Figure 3.1 shows a trend of biomass CHP technology selection as below:

1. Biogas generation utilising POME is selected for low power generation.
2. Combined syngas and biogas generation is selected at full usage of POME with low usage of solid biomass at mid power generation.
3. Steam generation using boiler is utilised for high power generation.

4. Internal combustion engine (GE) is used for both syngas and biogas combustion rather than micro-turbine and gas turbine.
5. The "must-have" design for low capacity biomass CHP design ranges up to 5 MW is biogas generation with GE. "Optional" syngas generation design can be invested and combined to achieve a higher power demand from 4 to 5 MW.
6. The "must-have" design for high capacity biomass CHP design ranges from 5 to 13 MW is boiler steam generation with steam turbine. "Optional" biogas generation design can be invested and combined to achieve a higher power demand from 8 to 13 MW.

4 Conclusion

This research work had presented a linear programming systematic framework to design biomass CHP system based on potential loss of profitable opportunity due to varying energy demand. Then, Minimax Regret Criterion (MRC) approach was used to assess potential economic risk of the given biomass CHP design based on energy demand. Based on MRC approach for the case study, the recommended optimum biomass CHP design is steam generation through boiler with steam turbine generating power. This design is selected as the optimum design due to its lowest maximum regret. This means that boiler with steam turbine biomass CHP design will provide minimum potential loss of profitable opportunity across the power demand range of 3000 to 10000 kW. Sensitivity analysis of FiT rates on the selection of biomass CHP design was conducted. Boilers with steam turbine is generally favoured as the FiT rates increases based on the sensitivity analysis of FiT rates at ±10, ±30 and ±50% from its original tariff. This is due to higher revenue is generated on feeding in power to grid with higher plant power capacity and FiT rate which eventually provide higher gross profit. On the other hand, decrease in FiT rates will economically favour smaller scale biomass CHP design such as biogas and syngas generation with GE due to lower profit margin. Besides, design analysis on the trend of the optimum design selected by model was conducted. Biogas with GE utilising POME is favoured for lower power capacity ranges from 250 to 5000 kW. Boilers with steam turbine is favoured for high power capacity ranges from 5000 to 13000 kW instead. Certain equipment will be essential for certain power demand range as described as a "must-have" design. These equipment had been outlined to ease the decision in final selection of process designer. In overall, a systematic framework to design biomass CHP system had been established to recommend an optimum biomass CHP design which minimise the potential loss of profitable opportunity while maximising gross profit. The case study had shown that the framework was able to provide a preliminary design. The model also highlights optional design that allow process designer to invest in to achieve higher power capacity. It is worth mentioning that the mathematical model can be easily revised and reformulated to take in account of other aspects such as variation in biomass supply, variation in tariff rates, and changes in government schemes which will be considered as prospects of future work.

The author would like to acknowledge Prof. Denny K. S. Ng for generous advice and recommendation that had provide insightful analysis towards this work.

References

1. EIA, "World energy demand and economic outlook EIA's handling of non-U.S. policies in the International Energy Outlook", *U.S. Energy Inf. Adm.*, vol. **2016**, no. May 2015, pp. 7–17 (2016)
2. T. W. John Twidell, *Renewable Energy Resources*, 3rd ed. New York : Routledge (2015)
3. EIA, "World Renewable Energy Consumption," *U.S. Energy Inf. Adm,* (2016)
4. O. Edenhofer *et al., IPCC, 2011: Summary for Policymakers. In: IPCC Special Report on Renewable Energy Sources and Climate Change Mitigation*. (2011)
5. F. Rosillo-Calle, "A review of biomass energy - shortcomings and concerns," *J. Chem. Technol. Biotechnol.*, no. March, pp. 1933–1945, (2016)
6. S. Mekhilef, R. Saidur, A. Safari, and W. E. S. B. Mustaffa, "Biomass energy in Malaysia: Current state and prospects," *Renew. Sustain. Energy Rev.*, vol. **15**, no. 7, pp. 3360–3370, (2011)
7. Malaysia Palm Oil Board, "FFB Processed by Mill 2017." [Online]. Available: http://bepi.mpob.gov.my/index.php/my/statistics/sectoral-status/179-sectoral-status-2017/806-ffb-processed-by-mill-2017.html. [Accessed: 18-Jun-2017].
8. Y. L. Chiew and S. Shimada, "Current state and environmental impact assessment for utilizing oil palm empty fruit bunches for fuel, fiber and fertilizer - A case study of Malaysia," *Biomass and Bioenergy*, vol. **51**, pp. 109–124, (2013)
9. K. Darrow, R. Tidball, J. Wang, and A. Hampson, "Catolog of CHP Technologies," no. March, p. 131, (2015)
10. U. S. E. P. Agency, "Biomass Combined Heat and Power Catalog of Technologies," *Biomass*, no. September, pp. 2000–2003, (2007)
11. M. S. Umar, P. Jennings, and T. Urmee, "Strengthening the palm oil biomass Renewable Energy industry in Malaysia," *Renew. Energy*, vol. **60**, pp. 107–115, (2013)
12. M. S. Umar, P. Jennings, and T. Urmee, "Generating renewable energy from oil palm biomass in Malaysia: The Feed-in Tariff policy framework," *Biomass and Bioenergy*, vol. **62**, pp. 37–46, (2014)
13. M. S. Umar, P. Jennings, and T. Urmee, "Sustainable electricity generation from oil palm biomass wastes in Malaysia: An industry survey," *Energy*, vol. **67**, pp. 496–505, (2014)
14. ICESN, "Biogas Asia Pacific Forum 2017", no. 15–19th May, pp. 17–18, (2017)
15. E. J. da S. Pereira, J. T. Pinho, M. A. B. Galhardo, and W. N. Macêdo, "Methodology of risk analysis by Monte Carlo Method applied to power generation with renewable energy," *Renew. Energy*, vol. **69**, pp. 347–355, (2014).
16. U. Arnold and Ö. Yildiz, "Economic risk analysis of decentralized renewable energy infrastructures – A Monte Carlo Simulation approach," *Renew. Energy*, vol. **77**, pp. 227–239, (2015)
17. R. Yokoyama, K. Fujiwara, M. Ohkura, and T. Wakui, "A revised method for robust optimal design of energy supply systems based on minimax regret criterion," *Energy Convers. Manag.*, vol. **84**, pp. 196–208, (2014)
18. R. Yokoyama, R. Nakamura, and T. Wakui, "Performance Comparison of Energy

Supply Systems Under Uncertain Energy Demands Based on a Mixed-Integer Linear Model," *Energy*, (2016)

19. A. S. Christian and W. Winston, *Data Analysis and Decision Making*, 4th Edition. Mason, USA: South Western Cengage Learning, (2010)
20. "SEDA PORTAL." [Online]. Available: http://seda.gov.my/. [Accessed: 23-Oct-2017].

Devulcanization of Waste Tire Rubber Using Amine Based Solvents and Ultrasonic Energy

Rashmi Walvekar[1*], *Zulkefly* Mohammad. Afiq[1], *Suganti* Ramarad[2], *and Siddiqui* Khalid[3]

[1] School of Engineering, Taylor's University, 47500 Subang Jaya, Selangor, Malaysia
[2] School of engineering and physical sciences, Heriot Watt University Malaysia, 62200 Putrajaya, Malaysia
[3] Graphene & Advanced 2D Materials Research Group (GAMRG), Research Centre for Nano-materials and Energy Technology (RCNMET), School of Science and Technology, Sunway University, No. 5, Jalan Universiti, Bandar Sunway, 47500 Subang Jaya, Selangor, Malaysia

Abstract. This research project focuses on an alternative pathway of devulcanizing waste tire rubber by using amine based chemicals. Waste tire rubbers are known to be as toxic, non-degradable material due to their vulcanized crosslink carbon structure, and disposing of such waste could impose hazardous impacts on the environment. The current rubber recycling methods that are practiced today are rather uneconomical, non-environmentally friendly, and also producing recycled rubber with low quality due to the alteration in the main polymeric chains of waste rubber. This project aims to answer the question of whether the usage of amine can produce high quality rubber, where the properties of recycled rubber is almost the same as new/virgin rubber. With known potential of amine, it is a challenge for the chemical to selectively cleave the sulfur bonds without affecting the main carbon backbone chain in the rubber structure and diminishing much of the rubber properties. To study this research, amine-treated rubber must undergo devulcanisation process by applying heat and sonication energy. Then, the properties of the amine-treated rubber were determined through a set of characterization tests and analysis which are: gel content test to determine the weight of rubber before and after devulcanization, the thermogravimetric analysis (TGA) to determine the thermal degradation and stability of rubber, and Fourier Transform Infrared Spectroscopy (FTIR) to determine any structural change of the rubber. In this research so far, the first two preliminary analysis tests have been performed. The gel content test has shown that tertiary amine samples possessed a lower gel content (%) of (77 – 63 %), compared to primary amine samples (falls within the range of 80%), as well as the TGA test in which tertiary amine samples degrade faster than primary amine samples (suggesting a higher degree of rubber structure breakdown). For each type of amine, the concertation of amine did not play a major role in affecting the degree of devulcanization (as the concentration increased, the degree of devulcanization decreased for some samples). FTIR analysis showed that only sulphur-sulphur bonds were cleaved during the devulcanization process, leaving the carbon-sulphur bonds unaffected.

[1] Corresponding author: rashmi.walvekar@gmail.com

1 Introduction

Rubber is an elastomer with a potential to return to its original state when deformed. External conditions (such as external force), the structure and the molar mass of the rubber are factors which can influence the deformation rate of rubber. Usually, rubbers possess regular and long macromolecule chains which can restore their initial position once the external deforming forces are removed. There are natural rubbers and synthetic rubbers used in industrial applications. Natural rubbers are usually obtained from a plant called Hevea Brasiliensis, while synthetic rubbers are mainly derived by constructional polyreactions of chain or grade nature [1]. For rubber to be used in commercial purposes, raw rubber is required to undergo treatment processes in order to increase the rubber's strength and life. Treatment processes of rubber may involve the addition and mixing of the rubber with industrial oils, additives, and carbon black filler [2].

Vulcanization is a curing process which transforms the rubber into a strong, elastic, and rubbery hard state. This process involves the crosslinking of rubber macromolecules which results in the foundation of three dimensional network of rubber matrix as shown in Figure 1 [1]. This network pattern makes the rubber more durable and stronger which contributes to improved wear and durability of the rubber [2].

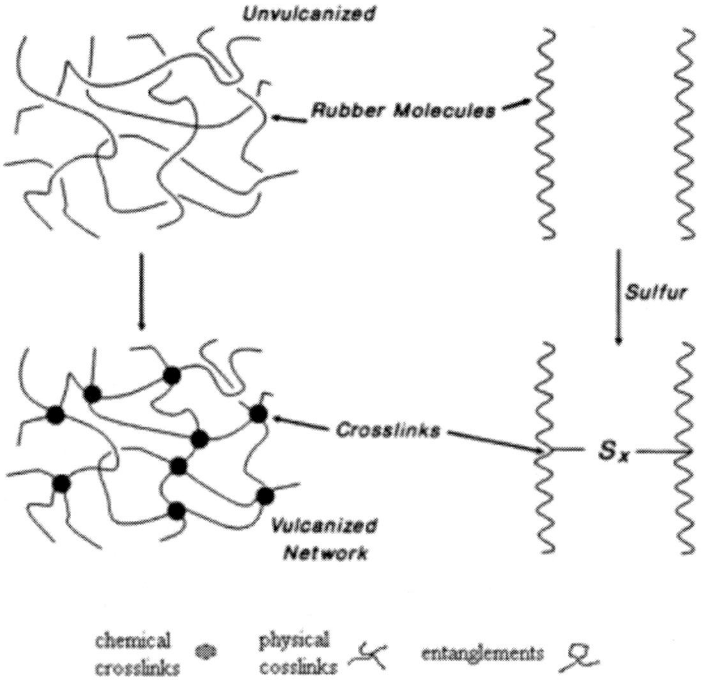

Fig. 1. Three dimensional rubber matrix [1].

Different types of chemical-based crosslinking agents, such as peroxides, sulfur, metal oxides, and resins, are involved in the process of rubber macromolecules crosslinking formation. These agents are able to react with suitable functional rubber groups and to create cross-links between them. Typical composition of car and lorry tires are shown in Table 1 below. Out of all these agents, sulfur remains the most successful and economical cross linking agent even today. This is due to the fact that sulfur is low in toxicity and cost, compatible with other additives, and able to predictably provide the desired vulcanisation properties. The typical vulcanisation system consists mainly of sulfur, metal oxide (usually zinc oxide), a fatty acid (to make the metal oxide soluble), and one or more organic accelerators, where the chemical reaction between sulfur and the rubber hydrocarbon occurs mainly at the C = C (double bonds) [3]. Vulcanization is generally considered irreversible. In other words, after it has been altered, the once long, complex, and twisted rubber molecule cannot return to its original form [3].

Table 1: Composition of a tire [4].

Ingredient	Passenger Car Tire	Lorry Tire
Rubber/Elastromers	41 – 48%	41 – 45%
Carbon black	22 – 28%	20 – 28%
Metal	13 – 16%	20 – 27%
Textile	4 – 6%	0 – 10%
Zinc Oxide	1%	2%
Sulphur	1%	1%
Additives	10 – 12%	7 – 10%

Tires produced by vulcanization are non-degradable material, which means that they could potentially stick around indefinitely. It is estimated that over 250 million tires are discarded annually. The amount of discarded waste tire rubber around the world is increasing at an alarming rate and only a small part of these wastes is recycled. The remaining are either buried, landfilled or stockpiled [4]. The disposal area of the tires contributes to the reduction of biodiversity since tires release toxic components to the environment. Tires are also very susceptible to be burned when subjected to high temperature. Once they are ignited, they will release toxic fumes and produce oil that will eventually contaminate soil and water [5]. Waste tires could also pose a threat to the humans by becoming a home ground for rodents and mosquitos that are responsible for carrying many types of disease [6]. Hence, the recycling of rubber wastes is vital in order to maintain a safe and healthy environment.

Although the rubber industry has a long time experience dealing with rubber, the perfect solution for disposal and recycling method for rubber products is yet to be found. Devulcanization is a technique where the crosslink structure, specifically the S-S and/or C-S bonds are cleaved. This can be done by various methods such as mechano-chemical, chemical, biological, and using ultrasonic energy, which are available and practiced today [7]. Common current scission processes of S-S and S-C bonds are demonstrated in Figure 2.

However, the optimum efficiency of devulcanization is yet to be discovered as the current existing processes produce uncontrolled polymer rubber chain degradation, where the breakage of main macromolecule (C-C bonds) could occur. In actuality, the breakage of C-C bonds is called a reclamation process. So in light of this, the current physical or chemical processes of converting used rubber into reusable form should be called reclamation rather than devulcanisation. Currently, when used rubber is recycled through any one of the existing processes, the main polymer chain (C-C bonds) is broken, leading to poor mechanical properties due to structural changes, where the rubber cannot be reused or re-vulcanized [8]. A proper devulcanisation process should only involve the cleavage of the S-S and/or C-S bonds, which basically means the reverse process of vulcanization, where the devulcanizates resemble the original material in quality and structure (high quality rubber).

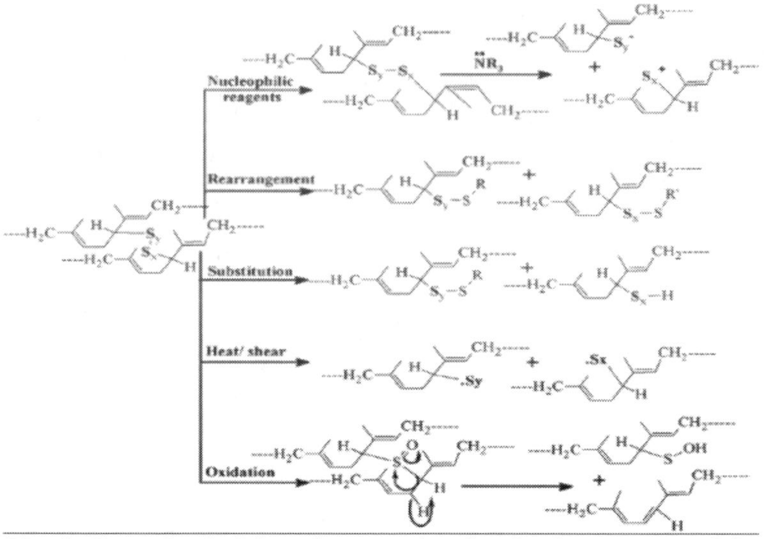

Fig. 2. General principles of current devulcanization processes [9].

Today, finding and developing a more selective breakdown process in recycling used rubber is a global challenge. Some of the latest works where the introduction of amine usage as an alternative to devulcanise rubber have been initiated. Several types of amine function as degradation stabilizers for coatings and plastics. Most degradation processes such as devulcanization at high temperature, would have a free radical character, where amine could facilitate in the process reaction by serving as radical scavengers [10]. Works performed by Verbruggen et al. [10], show that amines (eg. Primary, secondary, and tertiary) could be an effective devulcanization agent. This may be due to the ability of amines to reduce the crosslink density mainly by selective reaction with sulfur bridges known as nucleophilic reaction. Figure below demonstrates the reaction of amine with rubber crosslinks.

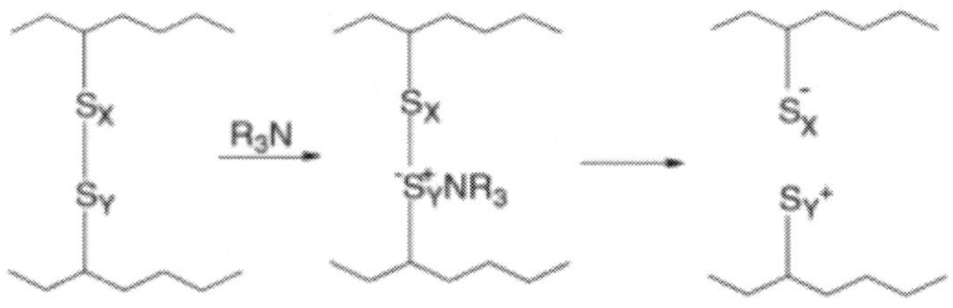

Fig. 3. Amine reaction with sulphur crosslinks [11].

As such, in this research, the devulcanisation of waste tire rubber in the presence of amine to produce high quality recycled rubber is studied. Different conditions will be performed to aid the devulcanisation process, which includes varying temperature and also by sonication radiation as these two are the most effective commercial techniques. Thus, this study aims to compare the effect of different types of amines on the degree of rubber devulcanization and to assess the effect of amine concentration on the degree of rubber devulcanization.

2 Methodology

Quantitative research approach was applied where analysis was done based on experimental data that was collected along the way. This involved the process of analyzing the degree of devulcanization where the amount of sulfur removed was studied based on the experimental and numerical results.

2.1 Amine Preparation

Two types of amines, which are primary [(3-AminoPropylTrimethoxysilane), $C_6H_{17}NO_3Si$] and tertiary [(n-Diethyl-3-AminoPropyl) Trimethoxysilane), $C_{10}H_25NO_3Si$] amines, were bought from Nano Life Quest Sdn. Bhd. and used as the de-vulcanizing agent for ground waste tire rubber. To compare the effect of concentrations, the primary amine was prepared at three different concentrations while the tertiary amine was prepared at two different concentrations based on molarity, where at each concentration of amine, the sample was mixed with 50 ml of water as base. Every amine sample was then added to 20 g of ground waste tire rubber. In this study, the different in concentrations were referred to as different in R:A ratio. Based on preliminary study, the ratios of the amount of rubber to amine are as follows:

Table 2: Rubber (R) to Amine (A) ratio for Primary Amine.

R:A ratio	Amount of Rubber	Amine Volume
1:0.5	20 g	10 ml
1:0.25	20 g	5 ml
1:0.125	20 g	2.5 ml

Table 3: Rubber (R) to Amine (A) ratio for Tertiary Amine.

R:A ratio	Amount of Rubber	Amine Volume
1:0.25	20 g	5 ml
1:0.125	20 g	2.5 ml

On top of that, two types of control samples were also prepared as to compare the results after the devulcanization process. One solely contains rubber and the other contains a mixture of rubber and 50 ml of water, as to investigate if water would play a part in the devulcanization process.

2.2 Devulcanization Process and Rubber Sample Preparation

After the preparations of amine and rubber samples were done, the next step was to perform the devulcanization process. In this stage, every sample prepared in Schott bottle was allowed to experience and undergo the ultrasonication process. In this study, three variables were fixed; they were the sonication time and sonication frequency. In addition to that, another two variables were varied; they were the sonication temperature, and the rubber to amine ration, as shown in Table 6 and 7 for primary and tertiary amine, respectively.

One reason of why amine was chosen was because of the fact that amine is known to be an agent, suitable for desulphurization processes [12], and assumed to possess the same ability for the process of devulcanization. According to, an acceptable heating treatment for rubber, for it to not have any effects on the main polymer chain (carbon-carbon bonds), is within the range of 130 to 180°C. The heating treatment in this devulcanization process was paired together with the sonication. Since amines were involved during the process of devulcanization, the temperature chosen must be at or lower than 80°C as the boiling point of the primary amine used, is at 80°C. On top of that, since the ultrasonic water bath was used, temperatures lower than 100°C were decided on.

The temperatures chosen to be the heating environment during the sonication were 30, 50 and 80°C, while the sonication frequency of 37 Hz was also chosen as the Ultrasonic water bath equipment was only limited to this frequency. In addition to that, 37 Hz was also selected as the sonication frequency was because of the fact that it is the optimum condition with minimum degradation of rubber properties, according to Suganti R. et al. [13]. Due to safety purposes, the experiment was conducted at a pressure of 1 atm.

Varying temperatures and applying a frequency of 37 Hz were expected to agitate the rubber crosslink structure and weaken the Sulphur bonds, in order to allow the amine to penetrate through the crosslink structure. Since there were three different set temperatures, each rubber to amine ratio would then have three replications so that one rubber to amine ratio sample would experience different temperature than the other samples in the same ratio. The number of samples are summarized in Tables 4, 5 and 6.

Table 4: Rubber control samples.

Sonication Temperature (°C)	Amount of Rubber (grams)	Amount of Water (mL)	Sample Number
30	20	-	1
50			2
80			3
30		50	4
50			5
80			6

Table 5: Primary Amine Samples.

R:A ratio	Sonication Temperature (°C)	Amount of Rubber (grams)	Amine Volume (mL)	Sample Number
1:0.5	30	20	10	7
	50			8
	80			9
1:0.25	30		5	10
	50			11
	80			12
1:0.125	30		2.5	13
	50			14
	80			15

Table 6: Tertiary Amine Samples.

R:A ratio	Sonication Temperature (°C)	Amount of Rubber (grams)	Amine Volume (mL)	Sample Number
1:0.25	30	20	5	16
	50			17
	80			18
1:0.125	30		2.5	19
	50			20
	80			21

As shown in the tables above, there are a total of 21 samples including the control samples. Each devulcanization process (application of 37Hz sonication frequency in the respective temperature) took place for 15 minutes. Afterwards, the rubber from every sample was then filtered out by using the gravity filtration method [14]. Then, every sample was washed thoroughly by using distilled water at least three times to ensure no presence of amine in the rubber samples. Lastly, the washed rubber was then dried in the oven at 70°C for a period of 24 hours, to ensure minimum to none moisture content present in rubber samples, before characterization can take place.

2.3 Rubber Characterization

Characterization of the rubber test samples was the next stage of this research. There were three different techniques in dealing with the determination of the amount of Sulphur removed in the rubber samples, they are: gel content test, thermogravimetric analysis (TGA), and Fourier Transform Infrared Spectroscopy (FTIR).

2.3.1 Gel Content Test

The gel content test samples were determined according to ASTM D2765. 0.2±0.05g of every rubber sample was measured and placed in a stainless steel wire mesh pouch of 120 mesh size. Three pouches were ready from each sample in order to attain an average result. In order to dissolve the soluble content, the samples placed in wire mesh were extracted in boiling toluene by using a Soxhlet apparatus for 24 hours. Then, test samples were collected and dried in an oven for 24 hourse at 70°C until a constant mass is achieved. The gel content was then calculated by using the following Eq. (1) [13].

$$\text{Gel content (\%)} = \frac{w_1}{w_0}\ x\ 100 \qquad (1)$$

Where w_0 is the weight of dried sample before extraction and w_1 is the weight of dried sample after extraction. The soluble content was calculated by using the following Eq. (2):

$$\text{Soluble \%} = 100 - \text{Gel content (\%)} \qquad (2)$$

2.3.2 Thermogravimetric Analysis (TGA)

TGA was responsible for the determination of the test samples' thermal degradation and stability. This was done by utilizing a computerised thermo gravimetric analyser (Perkin Elmer – TGA 8000). To assess the thermal stability, dynamic TGA experiments was carried out. Rubber sample, weighted around 5 to 10 mg was placed on a sample pan before placing it in the instrument. The experiment was conducted by heating the samples from room temperature up to a temperature of 600°C to collect the weight loss vs temperature thermogram. Every analysis was done with test samples in Nitrogen atmosphere (with flow rate of 20 ml/min) with a rate of heating of 10°C/min. Before analysis can be done, a least-squares averaging technique was used to smoothen out the curve of normalised weight loss vs temperature. T_{max} was denoted as the temperature at the maximum rate of weight loss. T_{max} will be identified from the peak of the derivative (dW/dT) curve, where W is weight loss and T is temperature. With all these temperatures, they then was referred to and used to determine the thermal degradation and stability of the test samples [15,16].

2.3.3 Fourier Transform Infrared Spectroscopy

This method allows the evaluation of any structural change on the rubber based on which bonds are broken after the devulcanization process. Data analysis can be done based on the graphs generated by FTIR (Perkin Elmer FTIR Spectrum 100). ATR method was used for this FTIR analysis. As aforementioned, waste tire rubber with and without the help of amine as the devulanizing agent during devulcanization were examined in order to detect the degree of devulcanization.

This test analysis involved the testing of control samples and rubber samples with amine, in order to observe and detect if amine would actually impose any additional impacts in cleaving the sulphur bonds, reducing the crosslink structure of rubber during the devulcanization process. In addition to that, studies on which other bonds or compounds were removed during the devulcanization and comparisons between different samples can be made, which can be done by identifying the spectra produced.

3 Results and Discussion

As per discussed previously, the first preliminary stage of screening the rubber samples involved two tests which are: gel content test and thermogravimetric analysis (TGA).

3.1 Gel Content Test Analysis

Figure 4 shows the average gel content value for every sample at three different sonication temperatures. The results obtained portray an irregular pattern as it does not have a proper trend. At 30°C sonication temperature, all of seven samples fall within the range of 75 to 88% of gel content.

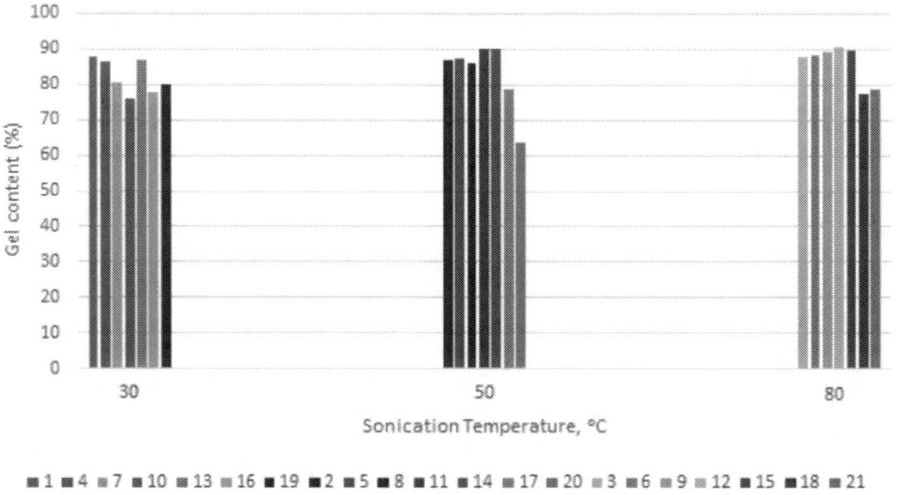

Fig.4. Gel content of rubber samples at different sonication temperature.

It is common for rubber to show gel content between 50% and 90% due to the presence of crosslinked structure of molecules. This can be seen from the control samples 1 – 6. The sonication temperature has a little effect on the gel content of rubber as all control samples project gel content % that is almost similar in value. However, at 30°C, as 10 ml of primary amine was added into sample 7, the gel content decreased in value from 87 to 80%, which shows that amine had an effect in aiding the breakdown of the rubber structure. As the amine volume was decreased to 5 ml in sample 10, the gel content however decreased even more up to 75%. However, when the amine concentration was further decreased to 2.5 ml in sample 13, the gel content shows a value of that near the control sample.

A feasible explanation to this results is that, as the amount of amine concentration decreases, it could become more reactive relative to the amount of rubber in one sample. When the concentration was further decreased, the sulphur bonds could have been reformed leading to the higher gel content. The results also show that as the temperature of sonication was increased, it did not have much effect of the degree of crosslinked structure cleavage. This can be seen for instance, from samples 7, 8 and 9. The gel content tend to increase rather than decrease with an increase in temperature. One explanation to this occurrence is that, the primary amine's boiling temperature used in this experiment, is at 80°C. Increasing the sonication temperature higher near to its boiling point, may have agitated the amine too much, leading it to evaporate, and leaving no chemicals left to aid in the process of cleaving sulphur bonds through nucleophilic reaction. This may have been a factor influencing the efficiency of the devulcanization, as the temperature was increased. This can be seen from the increasing pattern of gel content in every rubber to amine ratio samples, as the temperature was increased. In addition to that, according to [4], the acceptable temperature where the crosslinked structure would be reduced effectively is within the range of 130 to 180°C. This shows that the involved heating treatment in the ultrasonicator was not sufficient to break down the rubber structure.

Similarly, tertiary amine results also portray an irregular pattern. However, the gel content analysis shows that the gel content of rubber which was treated by tertiary amine was lesser than those treated by primary amine. The gel content was managed to be reduced and has an average range of 77 to 63%. The most effective treatment was observed from sample 20, in which the rubber was treated by 2.5 ml of tertiary amine and heated at 50°C during the sonication process.

Based on Figure 4, Samples 16, 17 and 18 display an almost the same amount of gel content, even all these three samples were under different temperatures using the heating and sonication treatments. This results may have been due to the uneven distribution of amine in all samples or unequal distribution of rubber particle during the process of devulcanization. On top of that, it may have also have been because of the discrepancies in the mixture of rubber sample composition. On the other hand, the results show that the gel content decreased in value to 63% as the temperature was increased to 50°C, and increased back to around 78% as the temperature increased to 80°C. This result shows that the efficiency of tertiary amine in helping breaking down the crosslink structure of rubber increased to a certain temperature, in this case was 50°C, and would decrease when the temperature was further increased to 80°C. The increase of gel content from sample 20 to 21, must have been due to the reformation of sulphur bonds with the main backbone chain, as the amine failed to capture the free radicals during the devulcanization process.

The results from this gel content experiment show that tertiary amine was more effective and reactive towards breaking down the rubber structure than the primary amine. This is contrary to the initial expectation as primary amine is usually more reactive than tertiary amine.

3.2 Thermal Analysis

High temperature TGA (30 - 600°C) plots of samples 1, 4, 7, 10, 13 and 19 are shown in Figure 5. Based on Figure 12, all TGA plots show a one-step degradation. Different degradation temperatures calculated from Figure 5 can be seen in Table 7.

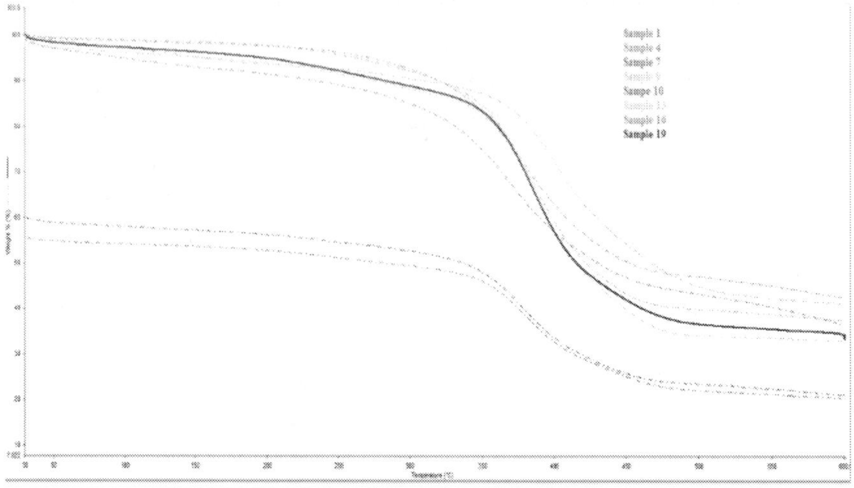

Fig.5. TGA plots of Samples 1, 4, 7, 9, 10, 13, 16, and 19.

Table 7: TGA parameters

Sample Number	T_1 (°C)	T_{40} (°C)	T_{80} (°C)
1	333.96	408.18	496.09
4	349.58	398.62	494.26
7	325.59	389.73	439.50
8	363.25	387.87	440.13
9	371.12	401.72	468.13
13	357.71	393.85	448.97
19	349.58	394.96	460.54

Based on Table 7, the onset degradation (T_1) of the control sample 4 started at 349.58°C. Forty percent degradation (T_{40}) and 80% degradation of the same control sample started at 398.62 and 494.26°C respectively. For the sample which was treated with primary amine, sample number 7, the onset degradation (T_1) occurred at lower temperature (325.59°C). The T_{40} and T_{80} of sample 7 occurred at 389.73 and 439.50°C respectively. For the sample which was treated with a lower primary amine concentration, sample number 13, the onset degradation (T_1) occurred at higher temperature (357.71°C). The T_{40} and T_{80} of sample 13 occurred at 393.85 and 448.97°C respectively. For the sample which was treated with tertiary amine, sample number 19, the onset degradation (T_1) occurred at a slightly lower temperature (349.58°C). The T_{40} and T_{80} of sample 7 occurred at 394.96 and 460.54°C respectively.

The delayed degradation of control sample 4 compared to sample 7 suggested that the treated sample 7 with the devulcanizing agent had a cross-linking structure breakdown due to the presence of amine. The control sample 4 was more stable towards thermal degradation because of the extra-cross-linking due to the absence of a devulcanizing agent during the devulcanization process. Cross-linking increased the rigidity of the system, which in turn increased the thermal stability. When the primary amine had a lower concentration as in sample 13, the rubber took longer to degrade, in comparison to sample 7. As discussed in the previous analysis section, this may due to the reformation of Sulphur bonds, which lead to a stronger crosslinked structure.

Based on Table 7, Sample 8, which was treated with primary amine at a sonication temperature of 50°C had an onset degradation (T_1) which occurred at temperature (363.25°C). The T_{40} and T_{80} of sample 8 occurred at 387.87 and 440.13°C respectively. For the sample which was also treated with primary amine, but at a higher sonication temperature of 80°C, sample number 9, the onset degradation (T_1) occurred at higher temperature (371.12°C). The T_{40} and T_{80} of sample 9 occurred at 401.72 and 468.13°C respectively. This is an indication of a more rigid structure of sample 9 compared to 8, and sample 8 compared to sample 7. As previously discussed, the gel content of sample 9 is the highest due to the amine evaporated from the rubber samples. Hence, it took sample 9 the longest for it to be degraded.

On top of that, sample 19 was treated with the same amine volume as sample 13, and it showed a lower degradation temperature than sample 13. This data also tallied with the gel content analysis, portraying that the tertiary amine had helped more in the reduction of rubber crosslinked structure than the primary amine.

3.3 Fourier Transform Infrared Spectroscopy (FTIR) Analysis

Based on Figure 6 above, it can be observed that the peaks for sulphur-sulphur bonds (in the range of 700-550 cm-1 and carbon-sulphur bonds (in the range of 720-570 cm-1) still exist based on the peaks generated from the plot. This means that during the devulcanization process, the sonication energy at a frequency of 37 Hz and at three different temperatures did not possess enough energy to cleave the sulphur bonds. It also means that water did not perform any additional effects on the rubber structure during the devulcanization process. Though there is some variation in graph patterns for devulcanization at different temperature environment, lines A and B marked on the graph still indicate that the sulphur bonds still exist in the rubber structure.

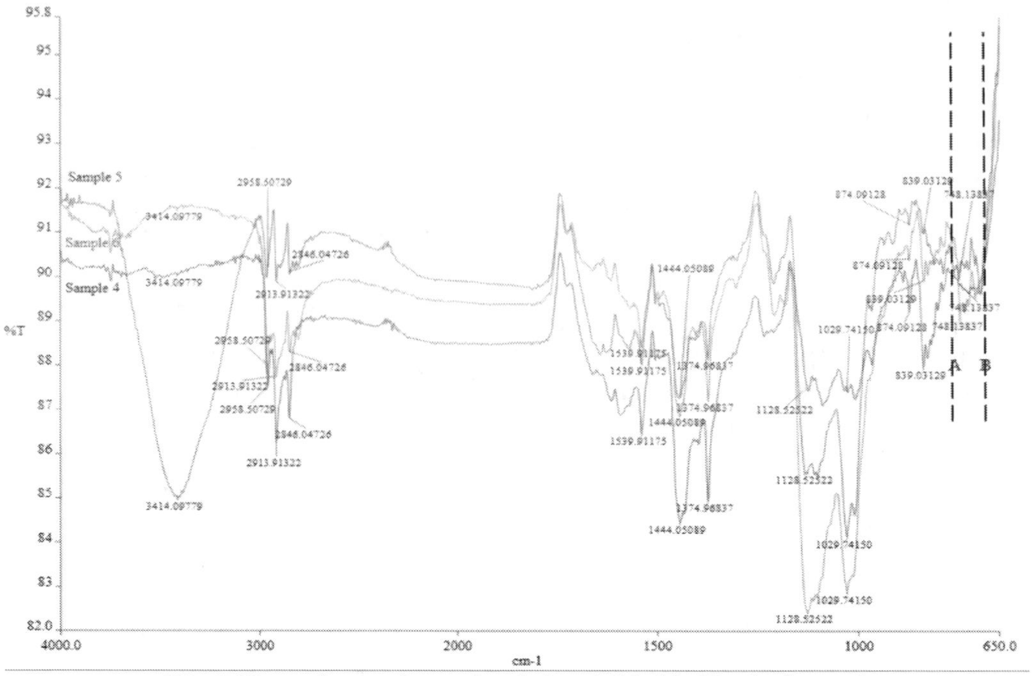

Fig.6. Plots of control samples 4, 5 and 6 (with 50 ml of water)

Based on Figure 7(a),(b),and, (d) lines A and B represent the regions for carbon-sulphur bonds and sulphur-sulphur bonds respectively. As seen from the plots, it can be seen that sulphur bonds are still present even with after going through devulcanization with the presence of primary amine. From Figure 7(a), where the amine concentration is the largest, at rubber to amine ratio of 1:0.5, there seems to be an increasing pattern in sulphur-suphur bonds. That is, the peaks seem to be appearing as the sonication temperature was increased, from Samples 7 to 9. This is an indication that the at higher the sonication temperature, the less effective the amine is to cleave the sulphur bonds. This may due to the fact that as the temperature increases to 80°C the amine must have evaporated slowly out of the rubber mixture, as the boiling temperature of the primary amine used is 80°C. Similarly, for the other two rubber to amine ratio, as the data presented in Figures 7(b) and (d), the sulphur-sulphur bonds tend to be more in terms of amount, as the sonication temperature increased. This pattern can be seen from Samples 10 to 12 in Figure 7(b), for rubber to amine ratio of 1:0.25, and Samples 13 to 15 in Figure 7(d), for amine ratio of 1.0.125. The explanation to this is may also be to the fact that as the temperature increases, it gets near to the primary amine's boiling temperature. Hence, the amount of amine to allow more cleavage of sulphur bonds would be less, and devulcanization becomes less effective. On top of that, there is no indication of carbon-sulphur bonds cleavage throughout the curves for the samples. This is an indication of the energy exerted on the rubber samples were not strong enough to break the carbon-sulphur bonds.

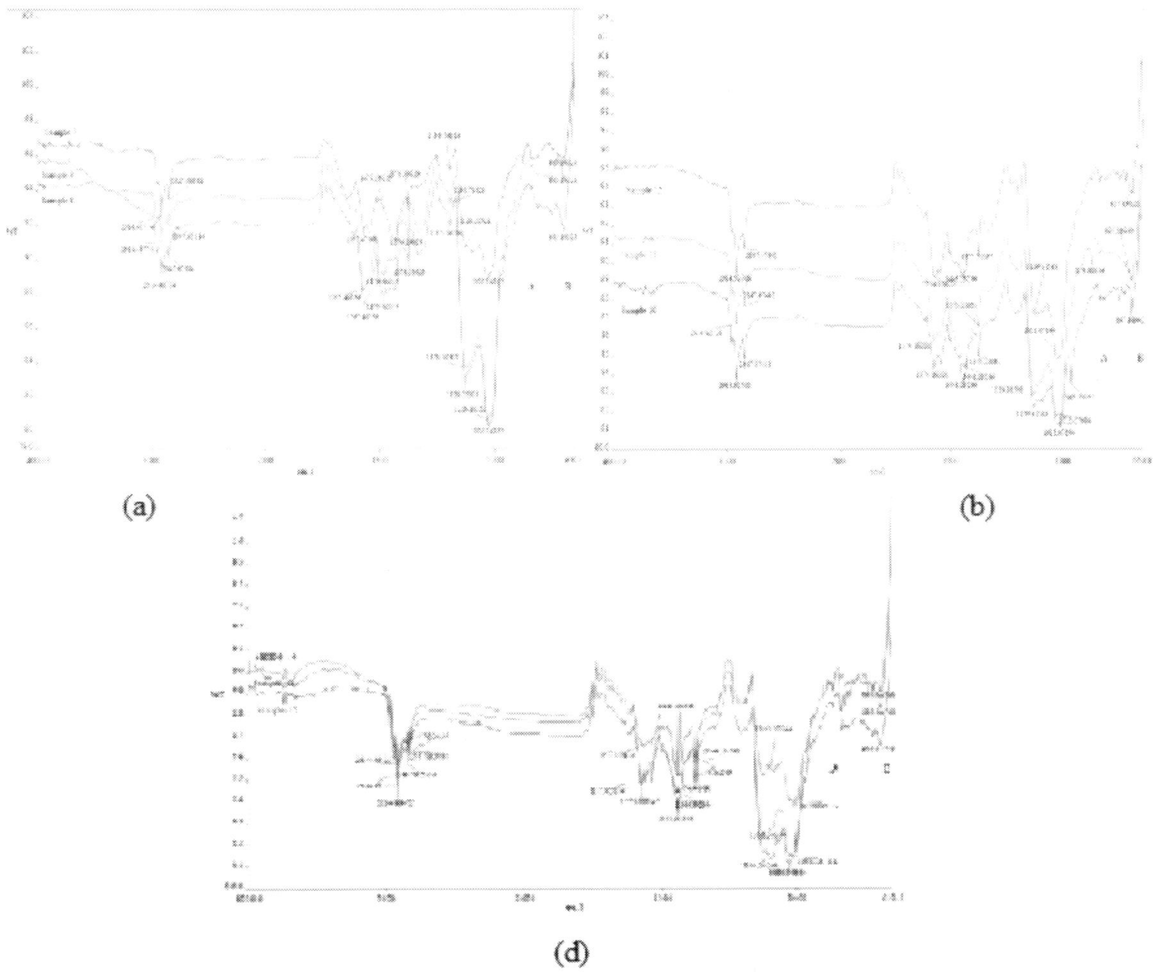

Fig.7. FTIR plots for primary amine, R:A of 1:0.5 (a), 1:0.25 (b), & 1:0.125 (d)

These results show that primary amine helps in breaking sulphur bonds to a certain degree, when compared to the results of primary amine to the control samples. The energy was sufficient enough to break sulphur-sulphur bonds, but not carbon-sulphur bonds as carbon-sulphur bonds would require higher energy to be broken. However, the findings also show that, the effectiveness of primary amine decreases as the temperature increases. This may due to the fact that, the primary amine may have escaped as the temperature gets higher, near to its boiling temperature, which leads to the lower efficiency of devulcanization

Based on Figure 8(a) and (b) the curves shown are the results generated for tertiary amine at rubber to amine ratio of 1:0.25 and 1:0.125 respectively. Samples 16, 17 and 18 curves marked at B, show an even distribution of plots, which may indicate that the amount of sulphur-sulphur bonds existing in these three samples may be almost the same. This explains as why these rubber samples possess almost the same amount of gel/soluble fraction, as previously recorded in the gel content test. Based on Sample 19 plot from the Figure 8(b), there seems to be some sulphur-sulphur bonds marked in the region B. But the peaks seem to be disappearing in Sample 20, which indicates the disappearing of sulphur-sulphur bonds in Sample 20. However, the peaks reappear in Sample 21, which indicates the reformation of sulphur bonds as the sonication temperature increases. The explanation to this phenomenon is that, as the temperature increases, the tertiary amine participated in the process of cleaving the sulphur-sulphur bonds even more as it gets more excited. However, after some time, the sulphur bonds tend to reappear which shows that to a certain degree, when the temperature increases, the sulphur bonds tend to reform and adhere back to the rubber structure. These results may enhance the findings from the gel content test analysis where from Sample 19 to 20, the gel fraction tends to decrease. While it increases from Sample 20 to 21. Similarly, the carbon-sulphur bonds during this treatment were unaffected, which can be seen from the peaks in the graphs. Indicating that the energy applied onto the rubber was not sufficient enough to break the chain.

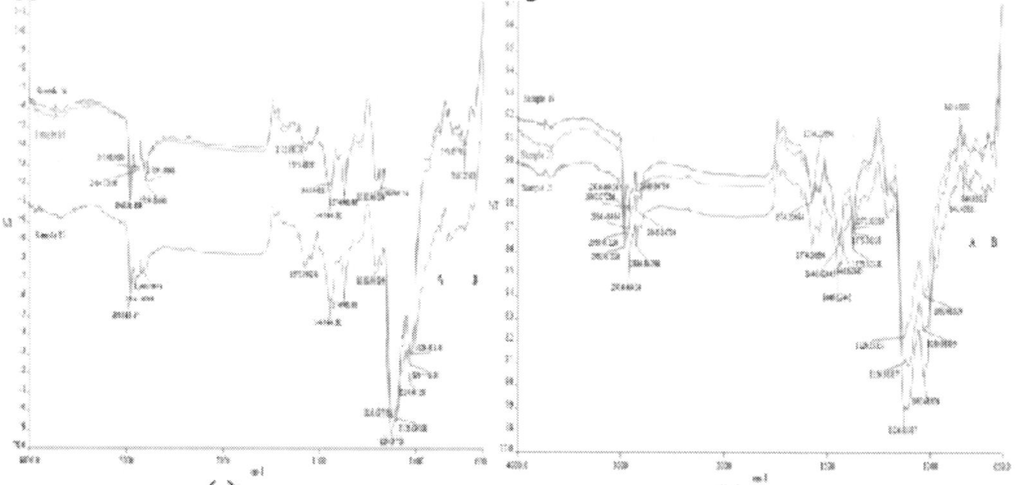

(a) (b)

Fig.8. FTIR plots for tertiary amine, R:A of 1:0.25 (a) and 1:0.125 (b)

Based on all the FTIR figures shown previously, the control samples are the only samples which generated some peaks beyond the range of 3000 and 1700 cm-1, which may indicate the presence of water remained in the samples. These peaks are not observed in all the other FTIR curves for any amine + rubber samples. This is an indication that there is a small degradation of bonds containing groups –OH and –C=O [17].

In conclusion, both types of amines could undergo nucleophilic reaction, due to the existence of free radicals during the process of devulcanization. Both types of amines were capable of only cleaving sulphur-sulphur bonds, where tertiary amine showed a more promising result than primary amine. The efficiency of primary amine was compromised, because of the high sonication temperature which was near to its boiling temperature. In addition to that, no other bonds were broken as there was not enough energy to break them. The main carbon backbone chain was not affected during the dvulcanization process, which is an indication of if, the devulcanized rubber were to be vulcanized, it could result in rubber with good physical property.

4 Conclusion

Based on the preliminary tests, the usage of amine as the devulcanizing agent, coupled with heating and sonication treatments has shown that there is a structural change in the rubber structure. However, the degree of structural breakdown is not substantial. One of the factors may be due to the fact that the devulcanizing environment, such as temperature, is not sufficient to encourage the scission of the Sulphur bonds in the rubber crosslinked structure. Amine is able to undergo the nucleophilic reaction, but only to a certain extent. As oppose to the previous expected outcome, these two analysis tests have proven that tertiary amine is more efficient in breaking down rubber structure than the primary amine. This is proven by the gel content test which has shown a lower gel content (%) of (77 – 63 %), compared to primary amine samples (falls within the range of 80%), as well as the TGA test in which tertiary amine samples degrade faster than primary amine samples (suggesting a higher degree of rubber structure breakdown).

The effects of amine concentration were difficult to be studied as the results generated did not have a regular pattern, as the sonication temperature also played a role during the process of devulcanization within the 15 minutes timeframe. However, the results have portrayed that the higher the concentration of amine is, the higher the breakdown of structure takes place. However, for some types of amine such as tertiary, under a certain condition, the lower the concentration is, the higher the degree of devulcanization. This can be seen from sample 20, with rubber to amine ratio of 1:0.125, had a more structural breakdown than sample 17, whose rubber to amine ratio was 1:0.25. Hence, amine may have the capacity to cause crosslinked structure breakdown of rubber, but it still may not be efficient enough to be used on industrial levels

References

1. J. A. Brydson "Rubber Chemistry," pp. 194–205, 1978.

[2] T. Liang et al., "Continuous Devulcanization of Ground Tire Rubber of Different Particle Sizes Using an Ultrasonic Twin-Screw Extruder," pp 200-230, 2013.

[3] G. S. Whitby and H. E. Simmons, "Vulcanization Accelerators—I.," *Ind. Eng. Chem.*, vol. 17, no. 9, pp. 931–935, 1925.

[4] S. Ramarad, M. Khalid, C. T. Ratnam, A. L. Chuah, and W. Rashmi, "Waste tire rubber in polymer blends: A review on the evolution, properties and future," *Prog. Mater. Sci.*, vol. 72, no. March, pp. 100–140, 2015.

[5] F. P. Torgal, A. Shasavandi, and S. Jalali, "Tyre Rubber Wastes Based Concrete: A review," *WASTES Solut. Treat. Oppor.*, p. 3, 2011.

[6] Rinkesh, "Tire Recycling," *Conserv. Energy Futur.*, 2015.

[7] M. Hess, H. Geisler, and R. H. Schuster, "Devulcanization as an opportunity to recycle rubber," *Chem. List.*, vol. 103, no. 13, 2009.

[8] S. Seghar *et al.*, "Devulcanization of styrene butadiene rubber by microwave energy: Effect of the presence of ionic liquid," *Express Polym. Lett.*, vol. 9, no. 12, pp. 1076–1086, 2015.

[9] J. Anu Mary, G. Benny, K. N. Madhusoodanan, and A. Rosamma,

"The current status of sulphur vulcanization and devulcanization chemistry: devulcanization," *Rubber Sci.*, vol. 29, no. 1, pp. 62–100, 2016.

[10] P. Sutanto et al., "Development of a Continuous Process for EPDM Devulcanization in an Extruder," p. 124, 2006.

[11] S. Saiwari et al., "Post-consumer tires back into new tires: de-vulcanization and re-utilization of passenger car tires". 2013.

[12] A. Bocoum, S. Hosseinnezhad, and E. Fini, "Investigating effect of amine based additives on asphalt rubber rheological properties," *Asph. Pavements*, no. September, pp. 921–931, 2014.

[13] S. Ramarad, C. T. Ratnam, M. Khalid, A. L. Chuah, and S. Hanson, "Improved crystallinity and dynamic mechanical properties of reclaimed waste tire rubber/EVA blends under the influence of electron beam irradiation," *Radiat. Phys. Chem.*, vol. 130, pp. 362–370, 2017.

[14] Univeristy of calgary, "Filtration Methods, Organic Laboratory Techniques," pp. 1–5, 2012.

[15] W. J. Sichina, "Characterization of Polymers Using TGA," pp. 1–5, 2000.

[16] ASTM International, "Standard Test Method for Compositional Analysis by Thermogravimetry," vol. i, no. C, pp. 9–13, 2012.

[17] S. Rodsuk, S. Ritsuar, and K. Wattanakul, "Characterization the effect of disulfide compound on the devulcanization of thermoplastic vulcanizate," *AIP Conf. Proc.*, vol. 1482, pp. 240–245, 2012.

Removal of Dye in Wastewater by Adsorption-Coagulation Combined System with *Hibiscus sabdariffa* as the Coagulant

Ho Nicholas Jian Hoong[1] , and *Nurhazwani* Ismail[1,*]

[1] School of Engineering, Faculty of Built Environment, Engineering, Technology & Design, Taylor's University, Subang Jaya, Selangor DE, Malaysia

Abstract. The conventional process to treat dye wastewater is the physicochemical treatment such as coagulation, flocculation and adsorption process. A new approach has been demonstrated to treat Congo red dye wastewater, which is the adsorption-coagulation hybrid process. Natural coagulant extracted from *Hibiscus sabdariffa* seeds is used as the coagulant while activated carbon is used as the adsorbent in this case study. The objective of this experiment is to study the significant factors that will affect the efficiency of dye removal. Then, the optimum conditions for the hybrid process is determined using Respond Surface Methodology (RSM). The variables are pH, initial dye concentration, coagulant dosage and adsorbent dosage while the response of experiment is the dye removal percentage. A three-level and four-variable Box-Behnken design (BBD) is used for the RSM. A total of 27 sets of experimental results is required to determine the optimum conditions. Jar test is used to conduct the experiment with the addition of coagulant and adsorbent simultaneously. Based on the regression model analysis and ANOVA, the highly significant factors that contribute to the dye removal efficiency through adsorption-coagulation hybrid process are pH of solution and initial dye concentration. The RSM results shows that the optimised process parameters for adsorption-coagulation hybrid process with *Hibiscus sabdariffa* seeds as the coagulant and activated carbon as the adsorbent are pH 2, initial dye concentration of 385 ppm, coagulant dosage of 209 mg/L and adsorbent dosage of 150 mg/L. The dye removal reaches up to 96.67% under optimum parameters.

1 Introduction

Synthetic dyes are commonly used in many industries, such as textile, paper, plastic, ink cartridge and paint manufacture factories. During the washing and finishing stage in the manufacture process, they consume and discharge a huge amount of dye contaminated water. Based on research studies, it is estimated that 280,000 tons of textile dyes are

* Corresponding author: nurhazwani.ismail@taylors.edu.my

discharged annually in the textile industry as effluent worldwide, ranking the first place in the usage of dyes among various industries [1]. Direct discharge of dye wastewater into the environment can be hazardous since synthetic dyes contain carcinogenic substances and heavy metals such as chromium (Cr), lead (Pb) and copper (Cu) [2].

Most dyes are synthetic compounds with complex, highly-branched aromatic molecular structures [3]. Thus, dye wastewater is very resistant to biodegradation treatment process. The most conventional way of treating it is the physicochemical treatment. This method involves the coagulation and flocculation process which use up various coagulants such as alum, lime and ferric [4]. Chemical coagulants are capable of dissociating polyelectrolytes when in solution. The polyelectrolytes will neutralise the negatively charged dye molecules to form particle-polymer-particle complexes [4]. These contaminants will be precipitated in the form of chemical sludge when the treated water is further added with flocculants. However, the generated sludge contains a considerable amount of chemical residue such as iron and aluminium salts. These components are hazardous to human health and possess environmental impact. According to [5], aluminium-based coagulants attribute the development of Alzheimer's disease in human beings. Thus, there is an emerging trend in wastewater treatment industry using natural coagulants such as Moringa oleifera, Strychnos potatorum (Nirmali), Cactus opuntia etc [6]. Natural coagulants possess a few crucial advantages compared to chemical coagulants, such as environmentally friendly (sustainable and biodegradable), toxic-free and safer for application.

The usage of *Hibiscus sabdariffa* seeds as natural coagulants was first reported to clarify high turbidity water (above 300 NTU) in the Indian Journal of Environmental Health, 1974. The chemical polyelectrolyte is substituted by natural polyelectrolyte in the form of polysaccharide and proteins [7]. According to [8] and [6], these structures usually exhibit higher molecular weight compared to synthetic polymer. This greatly increase the unoccupied binding sites for the coagulation and partial adsorption process. *Hibiscus sabdariffa* seeds also contain coagulation proteins, mainly composed of cationic peptides, such as glutamic acid, aspartic acid and leucine [9]. Based on previous experiment case studies, the removal of dye using *H. sabdariffa* as coagulant typically ranges from 81.2-93.3% [6,10].

Adsorption process is also one of the most effective and low-cost treatment process to remove dyes in wastewater. The most conventional adsorbent used in industries is activated carbon. The efficiency of adsorption using activated carbon is relatively high due to its microporous structure, large surface area and high adsorption capacity [11]. In fact, it shows a stronger affinity for dyes removal compared to zeolite [11]. In this view, activated carbon can be potentially used as the adsorbent in this case study.

Till date, the combination compromising of more than one wastewater treatment process as a single system has shown to be more effective when treating wastewater [12]. Based on previous studies, adsorption-coagulation hybrid process has been used to treat real wastewater of complex dyes manufacturing plant and wastewater containing high concentration of humic and tannic acids [12,13]. Results show that hybrid process is proved to have 90% efficient in colloidal particles removal than coagulation as an individual process [13]. In this method, the coagulant and adsorbent are added simultaneously at one single unit for reaction. The hybrid process enhances and improves the performance of dye removal because the dye molecules are eliminated from the water, and not because of the partial decomposition of dye molecules, which can potentially lead to the formation of more harmful and toxic aromatic compounds [4]. Meanwhile, the activated carbon will adsorb the colour and odour in the solution and the forming sludge.

Thus, with respect to the study that has been done extensively, the hybrid process is used to treat Congo red dye solution with *Hibiscus sabdariffa* seeds as coagulant and activated carbon as adsorbent. Congo red is targeted in this study based on a few reasons. The molecular structure is also shown in Figure 1.

(i) Congo red is a typical benzidine-based azo dye. It is widely found in the effluent of industries. After releasing it to the environment, Congo red is metabolised to benzidine, a well-known carcinogen that can cause allergic diseases in human. Thus, an effective and feasible method is needed to treat the Congo red in the effluent [3].

(ii) The potentiality of adsorption-coagulation hybrid process of Congo red removal has not yet been studied. With this in view, in this present study, an attempt has been made to identify the coagulation potential of coagulation agent extracted from *Hibiscus sabdariffa* seeds with activated carbon, for the removal of Congo red dye from aqueous solution.

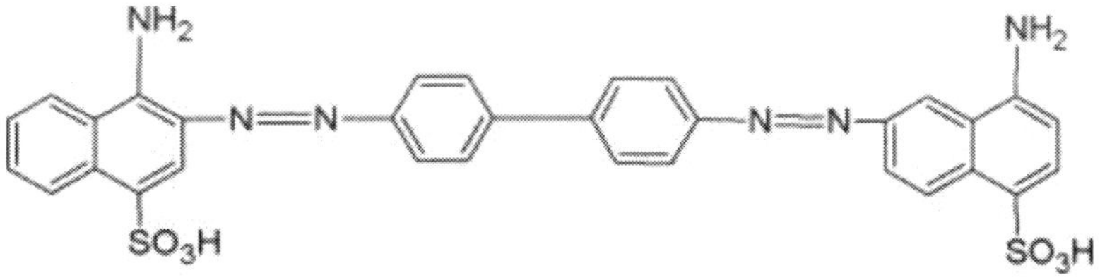

Fig. 1. Molecular structure of Congo red dye [14]

The objectives of conducting this research are to study the significant factors that will affect the performance of adsorption-coagulation hybrid process with *Hibiscus sabdariffa* seeds as coagulant and activated carbon as adsorbent on Congo red dye removal. Besides that, the optimum process parameters (pH, initial dye concentration, coagulant dosage and adsorbent dosage will be determined using Respond Surface Methodology (RSM).

2 Methodology

In this research studies, quantitative method will be implemented extensively to collect objective measurements, then numerical, mathematical and statistical analysis of data will be conducted to verify and validate the results. Lastly, the results will be used for RSM to determine the optimised working parameters for the adsorption-coagulation hybrid process.

2.1 Preparation of Synthetic Dye Wastewater

The stock solution is prepared by dissolving 1 gram of Congo red dye powder in 1 L of distilled water. Theoretically, the concentration of stock solution will be 1001ppm [14]. The stock solution is then diluted using distilled water to prepare the required concentration. The pH of dye solution is adjusted by adding 1.0 M NaOH solution and 1.0 M HCl solution. Noted that the colour of Congo red changes from red to blue when inorganic acid is added into the solution. This is due to the resonance between charged canonical structures [15]. The pH value is monitored using a pH meter. Table 1 simply summarise the physicochemical properties of Congo red dye.

Table 1. Physicochemical properties of Congo red

Parameter	Value (unit)
Molecular Weight	696.66 g/mol

Molecular Formula	$C_3H_{22}N_6Na_2O_6S_2$
Absorption Maxima	498nm

2.2 Preparation of Natural Coagulants

2.2.1 Pre-treatment of Hibiscus sabdariffa Seeds

Hibiscus sabdariffa flowers are collected and the seeds are removed manually from their dried capsules. The seeds are quickly washed with distilled water to remove any fibrous traces and tower dried. The clean seeds are then dried in an oven at 60°C for 2 hours to preserve the protein in the seed. Drying temperature higher than 60°C will cause denature of the proteins [16].

Fig. 2. Dried *H. sabdariffa* seeds

2.2.2 Extraction of Coagulating Agent

In this present work, sodium chloride solution is selected to be the extraction agent. since it can be obtained easily. During the extraction, the interaction between the peptides and the molecules of sodium chloride produces H-bonds with the adjacent molecules. The peptides will dissolve in the solution when the protein surface turns fully hydrophilic [17].

The concentration of sodium chloride solution was investigated. When the concentration of NaCl solution increases above 0.5 M, the binding sites of solvents are saturated and all protein fractions are extracted [18]. Research studies also show that there will be no significant changes of the extracted proteins when the concentration of NaCl solution increases from 0.6 M to 1.0 M, [18]. Thus, 0.5 M of NaCl solution is used as the solvent.

The seed is measured 0.5 gram and mixed with 100 mL of 0.5 M NaCl solution. The mixture is then blended in a household food blender for 2 minutes [10]. The mixture is sieved through a muslin cloth to remove any traces of seed residues. The filtrate is the extracted coagulating agent and it is milky in colour. The coagulating agent is used immediately after extraction to avoid deterioration.

2.3 Preparation of Calibration Graph

The stock solution is used to prepare various dye concentration solutions, which are 200, 400, 600, 800 and 1000 ppm. The samples are transferred into the cuvettes and placed into the UV Spectrophotometer for the measurement of absorbance value. After taking the measurements, a calibration graph of absorbance value against dye concentration (ppm) is plotted. The calibration graph is used to determine the dye removal percentage of each treated samples. The dye removal percentage can be determined using Eq (1) as shown below [10].

$$Dye\ Removal\ (\%) = \frac{Abs_o - Abs}{Abs_o} \times 100\% \quad (1)$$

Where Abs_o and Abs are the absorbance value before and after treatment process respectively.

2.4 Design of Experiment

2.4.1 Jar Test

Jar test apparatus is used to conduct the experiment in a few batches. The apparatus can accommodate a maximum of 6 beakers in a batch. Each batch will involve rapid mixing, slow mixing and sedimentation process. The dye solutions are stirred for 4 minutes at 100 rpm to allow floc formation. Then, the speed of mixer is reduced to 40 rpm and agitating process is continued for 25 minutes for flocculation process. At the final stage, the precipitate is allowed to settle at the bottom of jar for 30 minutes [10].

2.4.2 Factors

There are four factors being manipulated in this experiment, which are pH of dye solution, initial dye concentration, coagulant dosage and adsorbent dosage. The range and level of each factor is summarised in Table 2.

Table 2. Range and level of each factor

Variable	Description	Code	Range and Levels		
			Low (-1)	Central (0)	High (1)
1	pH	A	2	7	12
2	Initial Dye Concentration (ppm)	B	200	400	600
3	Coagulant Dosage (mg/L)	C	120	260	400
4	Adsorbent Dosage (mg/L)	D	50	100	150

2.4.3 Response of Experiment

The dye removal percentage is the response of the experiment. After the dye solution is being treated, the supernatant liquid is obtained without disturbing the precipitated flocs and transferred into a cuvette. It is then placed into UV spectrophotometer for the measurement of absorbance value. The dye removal percentage is then calculated using E.q (1).

2.4.4 Optimisation of Process Parameters using RSM

RSM is a mathematical statistics technique method to build models, evaluate the complex relationships between independent factors and experiment response and to determine the optimum working conditions [19]. In this experiment, a three-level and four-variable Box-Behnken design (BBD) is demonstrated. In order to further analyses the optimum combination of working parameters, a 27-runs of experiment is conducted. The experimental results for the 27 sets of experiment are the input for aggression model analysis, ANOVA and RSM for optimisation purpose. The results and justifications are discussed at the part below.

3 Results and Discussion

3.1 Model Fitting

There are four variables tested in this experiment, which are pH of dye solution, initial dye concentration, coagulant dosage and adsorbent dosage. The relative contribution of each variables to the response (dye removal) is predicted from the regression coefficient in the fitted model [19]. The regression coefficient for dye removal is obtained using a statistical software, Minitab 18.0 by least square method. Table 3 simply summarises the obtained result from the software.

Table 3. Predicted regression coefficient of the polynomial model for dye removal

Independent Variable	Regression Coefficient	t-Value	p-Value	Significant Level
Constant	1.07	0.47	0.650	
Linear				
A	-46.040	-39.986	0.000	***
B	4.526	3.931	0.002	***
C	-1.866	-1.620	0.131	
D	2.422	2.103	0.057	
Square				
A*A	51.274	29.687	0.000	***
B*B	-4.678	-2.709	0.019	**
C*C	2.202	1.275	0.226	
D*D	1.055	0.611	0.553	
2-Way Interaction				
A*B	5.375	2.695	0.019	**
A*C	-0.926	-0.464	0.651	
A*D	0.539	0.270	0.792	
B*C	-0.069	-0.034	0.973	
B*D	0.258	0.130	0.899	
C*D	-0.593	-0.297	0.771	

** $p < 0.05$; *** $p < 0.01$

Table 4. Model Summary

R^2	R^2 (adjusted)	R^2 (predicted)
99.57%	99.06%	97.56%

Based on the analysis, the pH of CR solution (A), initial dye concentration (B) and second-order effect of pH (A*A) are the highly significant variables on dye removal. Theses independent variables have p-values smaller below 0.01 ($p<0.01$). Besides that, the second-order effect of initial dye concentration (B*B) and the interaction of pH and dye concentration (A*B) are significant on dye removal. Both of the variables have p-values below 0.05 ($p<0.05$). A quadratic polynomial equation that describes the behaviour of dye removal to optimise the process can be determined from the regression coefficients of all the significant factors. The regression equation in term of coded variables is shown below.

$$Y_2 = -46.040A + 4.526B + 51.274A_2 - 4.678B_2 + 5.375AB \tag{4}$$

Coefficient of determination (R_2) is a measurement on how close the collected data are to the fitted regression line. It simply describes how well the model fits the data. Meanwhile, adjusted coefficient of determination (adjusted R_2) is an adjustment for R_2 to include the number of variables in a data set [19].

The analysis shows a R_2 value of 99.57%. This indicates that 99.57% of the total variation is explained by the model and the remaining 0.43% of the variation is unexplained. The adjusted R_2 is determined to be 99.06% and is in good agreement with the R_2. If a new set of data is fitted to the model for dye removal analysis, it shows 99.06% similar value with the R_2. Thus, the second-order regression model obtained for dye removal is satisfactory and E.q. (4) is fitted to represent the model.

3.2 Analysis of Variance (ANOVA)

Next, the fitted models obtained from section 3.1 will be further discussed by diagnostic checking. Table 5 shows the extracted results obtained from Minitab 18.0.

Table 5. ANOVA of the factors as linear, square and interaction on the effect dye removal

Source	Degree of Freedom	Sum of Square	Mean Square	F-Value	P-Value
Model	**14**	**43857.3**	**3132.7**	**196.910**	**0.000**
Linear	4	25794.5	6448.6	405.341	0.000
Square	4	17941.0	4485.2	281.929	0.000
2-Way Interaction	6	121.8	20.3	1.276	0.338
Error	**12**	**190.9**	**15.9**		
Lack-of-Fit	10	183.7	18.4	5.085	0.175
Pure Error	2	7.2	3.6		
Total	**26**	**44048.2**			

The results of ANOVA test indicate that the regression model for the data on dye removal is highly significant as the p-value is small ($p < 0.01$). The p-value of LOF for this model is large ($p>0.05$), this shows that the LOF is insignificant. According to [35], the sum of square due to pure error (SSPE) measures the inherent variability of repeated measurements which cannot be explained by any model. However, a small value of pure error is still preferred. In this model, the pure error of is very small, which is 7.2%.

In short, the ANOVA test shows that the regression model is fitted to describe the actual behaviours of dye removal based on the relationships of independent variables and thus Eq. (4) is able to represent the model. Th interaction between variables for dye removal is essential and is discussed below.

3.3 Contour Plots and Optimisation of Parameters

The model fitting analysis and ANOVA test as shown in section 3.1 and 3.2 have clearly illustrated the relationships between the independent variables and response. Thus, RSM method can be used to evaluate the optimisation of process parameters. The regression model analysis has shown that the significant interaction between variables are A: pH and B: Initial dye concentration. The relationships can be further illustrated in the three-dimensional respond surface plot and contour plot, from Figure 3 to 5.

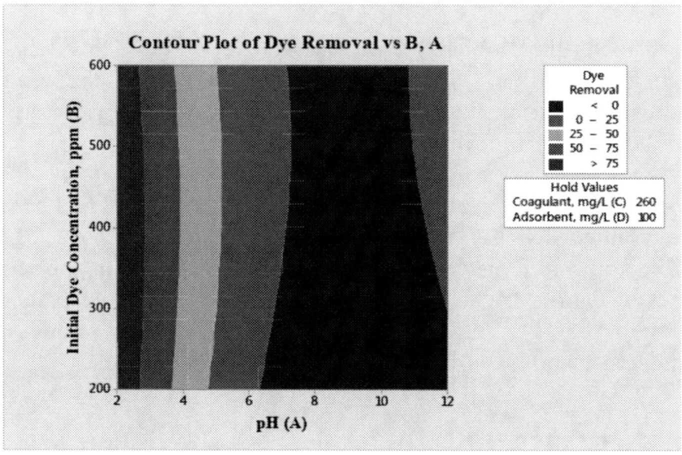

Fig. 3. Contour and respond surface plot of dye removal against pH and initial dye concentration

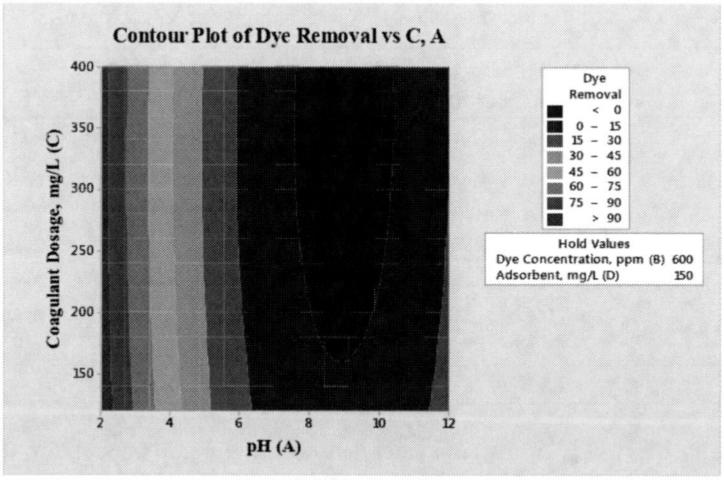

Fig. 4. Contour and respond surface plot of dye removal against pH and coagulant dosage

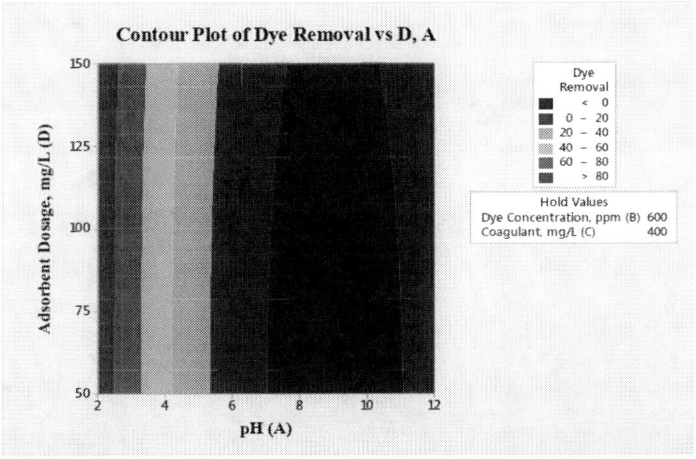

Fig. 5. Contour and respond surface plot of dye removal against pH and adsorbent dosage

The optimisation of the process variables was analysed using the response optimiser function in Minitab 18.0. The optimum conditions for the adsorption-coagulation hybrid process on dye removal are pH 2, initial dye concentration of 385 ppm, coagulant dosage of 209 mg/L and adsorbent dosage of 150 mg/L. With these optimum process parameters, the predicted dye removal percentage fits up to 96.67%. Meanwhile, to further validate the predicted optimised conditions, an experiment using the stated values of variables was conducted and the experimental result shows a 98.71% of dye removal. Thus, the obtained optimised combination of process parameters is accepted.

3.3.1 Influence of pH on Dye Removal

Since natural coagulants are used in this case study, the pH is manipulated from 2 to 12 during the adsorption-coagulation hybrid process [10]. From the contour plot shown in Figure 3, the highest range of dye removal is established when the pH is between 2 and 3. This shows that acidic condition is favourable for *Hibiscus sabdariffa* seeds to perform its coagulation process. Conversely, the dye removal is reduced significantly when the pH increases from 6 to 12. Based on [9], the coagulant protein extracted from *Hibiscus sabdariffa* seeds consist of glutamic acid (27%) and aspartic acid (11%), both are the cationic amino group. When the coagulation condition is acidic, it further enhances the positively-charged environment by increasing the density of positive charges (H+) around the coagulants hydrolysates. Charge neutralisation and complex reaction between the coagulant proteins and dissociated organic compounds of Congo red dominate the coagulation mechanism [20]. At the isoelectric point, the surface colloid particle is electroneutral and the combined effect of bridging and charge neutralisation results in higher dye removal efficiency [20]. This may explain for the coagulants to achieve optimum dye removal effect at pH 2 to 3.

According to [16], *Hibiscus sabdariffa* seeds also contain anionic peptides Under neutral and alkaline conditions, the dye removal percentage is rather low. This is because not all colloid particles will coprecipitate with the coagulant hydrolysates. In fact, most of the colloid particles and organic compounds of dyes merely absorbed on the surface of coagulants under neutral conditions when they collide with each other [20]. The absorption occurs due to weak chemical bonding but not attraction of charges. However, there is also a slight increase in dye removal when the dye solution reaches pH 12. The possible reason is that alkaline condition (OH-) is favourable for the dissociation of small amount of coagulant protein, arginine. This coagulating agent coagulates the dye colloid thus increases the dye removal [9].

Fig. 6. Sample of 400 ppm dye solutions – pH, coagulant dosage, adsorbent dosage (Left: 2, 120, 100; Centre: 2, 260, 50; Right: 7, 260, 150)

3.3.2 Influence of Initial Dye Concentration on Dye Removal

From the aggression model analysis, the initial dye concentration is also tested to be one of the highly significant factors that will attribute the dye removal percentage. In this case study, the dye concentration is manipulated from 200 ppm to 600 ppm.

From the contour plot shown in Figure 3, the dye removal increases when the dye concentration increases. As the dye concentration increases, there are more free dye colloid particles readily existed to form particle-protein-particle complexes with the coagulant and thus increases the dye removal efficiency [21]. However, according to [10], the dye removal will drop or remain unchanged when the dye concentration reaches a saturation point. At this point, the coagulating agent is said to be saturated, there is no free amino group with active site available to coagulate the colloid particles. Therefore, the dye removal percentage will drop or remain unchanged. In this case study, the dye concentration is said to be optimised at 385 ppm, which means beyond this indicated concentration, the dye removal does not increase significantly. However, Figure 2 shows that the dye removal continues to increase even the dye concentration exceeds 385 ppm. This is because dye removal is resulted from the adsorption of Congo red dyes onto the activated carbon surfaces. Thus, the dye removal increases up to 600 ppm dye concentration.

3.3.3 Influence of Coagulation Dosage on Dye Removal

In this case study, the coagulant dosage extracted from *Hibiscus sabdariffa* seeds is tested from 120 mg/L to 400 mg/L [10]. As shown in Figure 4, the dye removal increases gradually from 120 mg/L to 400 mg/L at pH 2 to 3. This is because as the cationic amino group increases, the peptides form more and stronger hydrogen bonding with the aromatic hydroxyl groups on dye particles. When abundant of coagulative components are available, it also neutralises the charges found on the surface of particles so that the particles can attract with one another more efficiently [5]. Thus, this explains for the increment of dye removal as the coagulant dosage increases.

However, it is also reported that the coagulation activity of *Hibiscus sabdariffa* seeds is decreased when the dosage reaches a certain point. This declination is caused by charge reversal which results the re-dispersion and re-stabilisation of dye colloidal particles due to overdosing of coagulant proteins [22]. However, this phenomenon will only occur when the coagulation dosage reaches up to 709 mg/L [23].

3.3.4 Influence of Adsorption Dosage on Dye Removal

The adsorbent dosage is manipulated from 50 mg/L to 150 mg/L [12]. As shown in Figure 5, It can be seen that the increase in the adsorbent dosage increases the dye removal percentage of Congo red. This is because higher adsorbent dosage provides a greater surface area with more adsorption sites and thus increase the adsorption of dye.

The adsorption of dye particles at pH 2 to 3 also shows higher removal percentage. When the density of positive charges in the solution increases, it avoids the adsorbents becoming more aggregated due to electrostatic interactions and steric repulsion force [12]. Thus, the stabilised adsorbent provides more adsorption sites. The excessive positive charges will also affect the surface charge of the adsorbent. As a result, the highly branched aromatic compounds of Congo red will gradually adsorb on the highly porous surface of the activated carbon and balances the dispersion status [24].

In short, pH and initial dye concentrations play an important role to maximise the performance of adsorption-coagulation hybrid process. The solution must be in between 2 to 3 so that the hybrid process can perform effectively (as shown in Figure 2 to 4).

The optimisation of the hybrid process and the interaction between process variables had been discussed above. The analysis shows that the dye removal can reach up to 96.67% when the pH of CR solution is 2, initial dye concentration is 385 ppm, coagulant dosage is 209 mg/L and adsorbent dosage is 150 mg/L.

3.4 Summary

Shah et al. had demonstrated the adsorption-coagulation process to remove pollutants in the textile dye wastewater from a manufacturing plant in Pakistan. In his research work, commercial coagulants such as ferric chloride ($FeCl_3$), ferrous sulphate ($FeSO_4$) and alum [$Al_2(SO_4)_3$] are used for comparison purpose and bottom-based coal ash (BBCA) is used as an adsorbent in the hybrid process. According to the findings from Shah et al., the hybrid process was optimised when 1.68 g/L of $FeCl_3$ and 4 g/L of BBCA was used to treat the dye effluent. The colour removal achieves up to 82%, turbidity removal 84% and total suspended solid (TSS) 94%.

The hybrid process performed in this current research study shows a 96.67% of dye removal. As compare with the colour removal reported from Shah et al., it shows that the properties of wastewater are important for removal efficiency. Shah et al. described the dye wastewater as effluent with high suspended organic compounds, because it consists of dyes, pigments, textile chemicals and binder emulsion. In contrast with the synthetic CR wastewater, the dye molecules are the only component presented in the solution, thus it can achieve high dye removal percentage.

In this research work, the optimum coagulant dosage is 209 mg/L, which is much lower than 1.86 g/L. This is because of the composition of wastewater and the dosage of natural coagulant is normally lower than chemical coagulant [25]. Besides that, the optimum adsorbent dosage in this research work is 150 mg/L, which is lower than 4 g/L. According to [11], the dosage of synthetic adsorbent such as activated carbon is normally lower than the dosage of natural adsorbent in water treatment process. This is because the manufacturing process of synthetic adsorbent which requires strict temperature and pressure control produces highly porous activated carbon. This greatly increase the adsorption surface area of activated carbon. Note that the dye molecules adsorbed on the surface of activated carbon is relatively small compared with the organic pollutants in real wastewater. Thus, the dosage of activated carbon is lower than the dosage BBCA.

Shah et al. had also reported that higher dosing of coagulant and adsorbent more than 1.68 g/L and 4g did not improve the removal efficiency in terms of the reduction of colour and concentration of pollutants. This is due to the occurrence of steady state condition with higher dosing of coagulant and adsorbent. This is true based on the analysis of the current experimental result. When the coagulant and adsorbent dosage reach 209 mg/L and 150 mg/L respectively in the CR solution with dye concentration of 385 ppm, the removal efficiency will drop below 96.67%.

Shah et al. also suggested that among the three coagulants used, the combination of alum-BBCA has the lowest removal efficiency, with 77% of colour removal, 79% of turbidity removal and 93% of TSS removal. This may be due to the pH of effluent that influence the coagulation activity of alum as lime is added into the water sample for pH maintenance. Based on [4], the optimum pH for alum is 5.3. Since lime is an alkaline, it will affect the performance of coagulation in the hybrid process and thus reduce the removal efficiency. In this research work, the analysis shows that the coagulation of *H. sabdariffa* seeds is efficient at pH 2. Thus, it can remove the dye efficiently.

From this summary, it can be concluded that the adsorption-coagulation hybrid process with *H. sabdariffa* seeds as coagulant and activated carbon as adsorbent can potentially achieve the removal efficiency obtained from RSM if it is scaled up or used to treat real dye wastewater. The process condition must be optimised so that the interaction of adsorbent and coagulant with the pollutants can be improved and thus increase removal efficiency.

4 Conclusion

In conclusion, this case study shows that *Hibiscus sabdariffa* seeds are workable natural coagulants. It also serves as an alternative to remove Cong red dye wastewater in real practice. Based on the aggression model analysis and ANOVA, the highly significant factors that contribute to the dye removal efficiency through adsorption-coagulation hybrid process are pH of solution and initial dye concentration. The RSM results shows that the optimised process parameters for adsorption-coagulation hybrid process with *Hibiscus sabdariffa* seeds as the coagulant and activated carbon as the adsorbent are pH 2, initial dye concentration of 385 ppm, coagulant dosage of 229 mg/L and adsorbent dosage of 50 mg/L. The dye removal reaches up to 98.68% under optimum parameters.

Acknowledgment

The author would like to offer a token of appreciation to Ms. Nurhazwani Ismail, Ms. Nur Farihah A. Malek and Ms. Lithnes Kalaivani for their kind support throughout the research studies until it is successfully completed.

References

[1] B. Frindt, J. Mattusch, T. Reemtsma, A. G. Griesbeck, A. Rehorek, Multidimensional monitoring of anaerobic/aerobic azo dye based wastewater treatments by hyphenated UPLC-ICP-MS/ESI-Q-TOF-MS techniques, Environ. Sci. Pollut. Res., **24**, 12, 10929–10938 (2017)
[2] A. Pirkarami, M. E. Olya, "Removal of dye from industrial wastewater with an emphasis on improving economic efficiency and degradation mechanism," J. Saudi Chem. Soc., **21**, 179–186 (2017)
[3] C. Jiang, B. Fu, H. Cai, T. Cai, Efficient adsorptive removal of Congo red from aqueous solution by synthesized zeolitic imidazolate framework-8, Chem. Speciat. Bioavailab., **28**, 1, 199–208 (2016)
[4] A. K. Verma, R. R. Dash, P. Bhunia, A review on chemical coagulation/flocculation technologies for removal of colour from textile wastewaters, J. Environ. Manage., **93**, 1, 154–168 (2012)
[5] T. . V. K. Vijayaraghavan, G. Sivakumar, Application of Plant Based Coagulants for Waste Water Treatment, Int. J. Adv. Eng. Res. Stud., **1**, 1, 88–92 (2011)
[6] C. P. Pise, S. A. Halkude, A New Technique for Purification of Water Using Natural Coagulant, Int. J. Eng. Technol., **6**, 6, 2564–2572 (2015)
[7] C. Y. Yin, Emerging usage of plant-based coagulants for water and wastewater treatment, Process Biochem., **45**, 9, 1437–1444 (2010)
[8] U. A.Y, I. T.S., The Role of Green Coagulants in Wastewater Treatment : A Review, Wastewater Remediat. Using Plant Tech.(2016)
[9] A. N. Jones, Investigating the potential of Hibiscus seed species as alternative water treatment material to the traditional chemicals (2006)
[10] M. Y. Yong, N. Ismail, Optimisation of *Hibiscus sabdariffa* as a natural coagulant to treat Congo red in wastewater, J. Eng. Sci. Technol., **11**, 153–165 (2016)
[11] A. Bhatnagar, A. K. Minocha, Conventional and non-conventional adsorbents for removal of pollutants from water - A review, Indian J. Chem. Technol., **13**, 3, 203–217 (2006)
[12] C. Jung, N. Phal, J. Oh, K. H. Chu, M. Jang, Y. Yoon, Removal of humic and tannic acids by adsorption-coagulation combined systems with activated biochar, J. Hazard. Mater., **300**, 808–814 (2015)
[13] A. Shah, A.K. Memon, A. Metlo, G. Peerzada, Simultaneous Coagulation-Adsorption Process Optimization for the Treatment of Real Wastewater of Complex Dyes Manufacturing Plant, **45**, 3, 506–513 (2013)
[14] G. Vijayaraghavan, S. Shanthakumar, E. E. Division, B. Sciences, Removal of Congo Red Dye Using a Natural Bio- Polymeric Coagulant, Int. Conf. Recent Adv. Chem. Environ. &Energy Eng., **17**, 1–5 (2014)
[15] B. Meroufel, O. Benali, M. Benyahia, Y. Benmoussa, M. A. Zenasni, Adsorptive removal of anionic dye from aqueous solutions by Algerian kaolin: Characteristics, isotherm, kinetic and thermodynamic studies, *J. Mater. Environ. Sci.*, **4**, 3, 482–491 (2013)
[16] S. M. N. Halimatul, I. Amin, N. Mohd-Esa, A. G. Nawalyah, M. Siti Muskinah, Protein quality of roselle (*Hibiscus sabdariffa L.*) seeds, Int. Food Res. J., **14**, 2, 131–140 (2007)
[17] K. C. Duong-Ly, S. B. Gabelli, Salting out of proteins using ammonium sulfate precipitation, Methods Enzymol., **541**, 85–94 (2014)
[18] S. Eldein, B. Ahmed, Some Physicochemical and Nutritional Studies on Karkade (*Hibiscus sabdariffa*) Seed Proteins, Rev. Lit. Arts Am (1998)
[19] T. Amir, N. Ismail, F. Alkarkhi et al., Optimization of Coagulation Process for Landfill Leachate Pre-Treatment Using Response Surface Methodology (RSM), J. Sustain. Dev., **2**, 2, 159 (2009)

[20] B. C. Cao, B. Y. Gao, C. H. Xu, Y. Fu, X. Liu, Effects of pH on coagulation behavior and floc properties in Yellow River water treatment using ferric based coagulants, Chinese Sci. Bull., **55**, 14, 1382–1387 (2010)

[21] S. Shamsnejati, N. Chaibakhsh, A. R. Pendashteh, S. Hayeripour, "Mucilaginous seed of *Ocimum* basilicum as a natural coagulant for textile wastewater treatment," Ind. Crops Prod., **69**, 40–47 (2015)

[22] N. Fathinatul, R. Nithyanandam, Wastewater Treatment by using Natural Coagulant, *2nd Eureca*, 2–3 (2014)

[23] C. Y. Yong, Removal of Congo red by Using *Hibiscus sabdariffa* for Coagulation and Flocculation Process (School of Engineering, Taylor's University, Malaysia, 2016)

[24] K. Zare et al., "A comparative study on the basis of adsorption capacity between CNTs and activated carbon as adsorbents for removal of noxious synthetic dyes: a review," J. Nanostructure Chem., **5**, 2, 227–236 (2015)

[25] R. Sanghi, B. Bhattacharya, A. Dixit, and V. Singh, *I. dasysperma* seed gum: An effective natural coagulant for the decolorization of textile dye solutions, J. Environ. Manage., **81**, 1, 36–41 (2006)

Potential of *Hibiscus Sabdariffa* and *Jatropha Curcas* as Natural Coagulants in the Treatment of Pharmaceutical Wastewater

Sheena Sibartie[1], and *Nurhazwani* Ismail[1*]

[1] School of Engineering, Faculty of Built Environment, Engineering, Technology & Design, Taylor's University, Subang Jaya, Selangor DE, Malaysia

Abstract. Pharmaceutical wastewater is one of the most difficult wastewater to treat due to the presence of pharmaceutical compounds resulting in high concentration of organic matter, high turbidity and Chemical Oxygen Demand (COD). Chemical- based coagulation is a common method used to treat wastewater. However, the issue that has been raised with the use of chemical coagulants is their presence in water after treatment that can cause risks to the human health such as Alzheimer and cancer. Natural coagulants can be used as a safe alternative to these chemicals instead. Therefore, the objective of this experiment was to study the effect of *H. Sabdariffa* and *J. Curcas* as natural coagulants, separately and as a combination, on the treatment of pharmaceutical wastewater. Jar test experiment were carried out where beakers of 0.5L wastewater were mixed with the coagulants. The pH of the wastewater was varied from 2 to 12 while the coagulant dosage was varied from 40 to 200 mg/L. It was found that *H. Sabdariffa* works best at pH 4 and at a coagulant dosage of 190 mg/L with a highest turbidity removal of 35.8% and a decrease of COD by 29%. *J. Curcas* was found to perform best at pH 3 and with a coagulant dosage of 200 mg/L with a highest turbidity removal of 51% and a decrease of COD by 32%. When *J. Curcas* and *H. Sabdariffa* were used in combination, the optimum composition was found to be 80% *J. Curcas* and 20% *H. Sabdariffa* by weight with a maximum turbidity removal of 46.8% and a decrease in COD by 46%. In comparison between the two natural coagulants, *J. Curcas* is found to be a better and more suited coagulative agent for the treatment of pharmaceutical wastewater. The same experiment was carried with alum at pH 6 and coagulant dosage of 750 mg/L and a turbidity removal of 48% and a decrease in COD by 38% were recorded. In comparison with alum, *J. Curcas* was a better coagulant in treating the pharmaceutical wastewater. This shows that natural coagulants can be used to replace chemical coagulants in the treatment of pharmaceutical wastewater.

1 Introduction

In the last decade, traces of pharmaceuticals in the water cycle including drinking water have become a subject of discussion among the public, governments and the drinking-water regulators. Their presence in water, even at low concentrations, has raised concerns regarding its risks to human health from consuming these pharmaceuticals in drinking

* Corresponding author: Nurhazwani.Ismail@taylors.edu.my

water [1], [2]. Therefore, it is essential to remove these pharmaceuticals from the wastewater before consumption. As a result, effective treatment methods must be applied to efficiently treat the contaminated water. Current pharmaceutical wastewater plants use methods such as activated sludge, membrane bioreactor and coagulation [3], [4].

Coagulation is a well-known technique used to reduce the pollutant contents in the water body that are present as turbidity, colour and organic matters [5], [6] . It is used as a treatment aiming to reduce the amount of pharmaceuticals such as paracetamol in the wastewater before it leaves as effluents [7]. Coagulants work by neutralising the charges caused by the colloidal particles. The suspended particles accumulate, forming flocs that then settle at the bottom and separate from the water suspension [6]. The most common coagulants used are chemicals such as the aluminium and ferric salts which are highly effective and cheap. However, recent studies have shown that traces of the aluminium based coagulants in the treated water can cause Alzheimer and other diseases such as intestinal constipation, abdominal colic and convulsions [6], [8]. Regarding the use of synthetic polymers, the presence of its residues in the water can be toxic and have strong carcinogenic properties [8]. Besides that, they are also ineffective in low- temperature water, produce large sludge volumes and significantly affect the pH of the treated water [9]. Therefore, researchers are focusing more on using natural coagulants which can be obtained from plants, animals or microorganisms as replacement to the chemical ones. They are non-toxic, biodegradable and can help to achieve the same results as the chemical coagulants [10].

Natural coagulants have been used for domestic households in the earlier days for traditional water treatment. However, due to the cheap cost and effectiveness of chemical coagulants, the use of natural coagulants was overlooked upon [10]. Furthermore, the use of natural coagulants have not been sufficiently investigated [10]. Therefore, this paper is focused on the study of *Hibiscus Sabdariffa* and *Jatropha Curcas* as natural coagulants. These plants are known as natural coagulants due to the presence of proteins in the seeds that act as an active coagulative agent [10]. These two plants are easily and readily accessible in various countries. The *H. Sabdariffa* seeds consist of 29.9% of proteins while *J. Curcas* seeds was found to contain 27.2% of proteins [11], [12]. Since the proteins inside the seeds act as a coagulative agent, both of these plants have the potential to treat wastewater.

H. Sabdariffa, most commonly known as Roselle, is used in herbal drinks, as food and as a flavouring agent in the food industry. The flower, known as calyx, is mostly used for these purposes. The seeds are most often discarded since till now, there is no use for it [11]. However, these seeds can be used to treat wastewater due to its coagulative properties. *J. Curcas* is a shrub or small tree belonging to the *Euphorbiaceous* family. Its main use is to extract oil from its fruit which is then used in the cosmetic industry, in soap making and as a substitute to kerosene or diesel. The seed and the press cake which is the waste after the oil extraction are normally discarded. However, these contain the protein which is the active coagulating agent that is efficient in the wastewater treatment [13], [14]. These wastes and the seeds can be used for coagulation, reducing the amount of waste and promoting sustainable development [8].

This paper investigates the efficiency of *H. Sabdariffa* and *J. Curcas* as natural coagulants, individually and as a combination, in the treatment of pharmaceutical wastewater. The effect of pH and coagulant dosage on the coagulation process are also studied in this research. The performance of the natural coagulants is compared with the effectiveness of the chemical coagulant alum.

2 Materials and Methods

2.1 Materials

The materials used for this experiment include: good quality dry seeds of *H. Sabdariffa* and Jatropha Curcas, pharmaceuticals (namely paracetamol, ibuprofen, naproxen and caffeine) to simulate the wastewater, 1M hydrochloric acid and 1M sodium hydroxide (to adjust the pH of the wastewater), 0.5M sodium chloride solution (to extract the proteins from the seeds).

2.2 Preparation of synthetic wastewater

Synthetic pharmaceutical wastewater was prepared by dissolving pharmaceuticals in distilled water to simulate the pharmaceutical wastewater properties in terms of the concentration of pharmaceuticals present in the wastewater. The pharmaceuticals were crushed into a fine powder using a mortar and pestle. 62.5 mg of each pharmaceutical (paracetamol, ibuprofen, naproxen and caffeine) was added to 1L of distilled water to prepare the stock solution of the synthetic wastewater. The solution was mixed for 1 hour at 50 rpm to achieve uniform dilution of the pharmaceuticals. The synthetic wastewater was then allowed to settle for 24 hours for complete hydration before it was filtered and ready for the experiment.

pH of the wastewater was adjusted using 1M hydrochloric acid and 1M sodium hydroxide before the jar tests.

2.3 Preparation of the *H. Sabdariffa* and *J. Curcas* seed powder

Good quality seeds were selected and washed with tap water to remove any dust or impurities on the shells. They were then dried in an oven at 60 ± 2°C for 3 hours to remove all the moisture content from the seeds. The dried seeds were crushed into a fine powder using a mortar and pestle.

2.4 Preparation of the H. Sabdariffa and J. Curcas coagulant

To prepare the coagulants, sodium chloride was used as the solvent to extract the proteins from the seed powder. As studied by [13], sodium chloride was found to be the best solvent compared to distilled water and sodium hydroxide for the extraction of the coagulative agent from the seeds.

0.5g of the seed powder was dissolved in 100 ml of 0.5M sodium chloride. The mixture was blended for 2 minutes using a food processor. The solution was filtered through a muslin cloth and the filtrate was used in the following jar tests.

2.5 Jar Tests Experiment

The coagulation experiments were carried out by varying the pH, coagulant dosage and wastewater concentration in terms of the concentration of pharmaceuticals in the water. The pH was varied from 2 to 12 while the coagulant dosages were varied from 40 to 200 mg/L. The initial wastewater concentration was investigated at 62.5 mg/L of each pharmaceutical. The temperature of the wastewater was kept constant at room temperature.

The jar test was used to conduct the coagulation experiments. Six 0.5L beakers were each filled with 500 mL of wastewater. The optimum pH was first tested by varying the pH while keeping the coagulant dosage constant at 200mg/L. Upon addition of the coagulant, the mixture was stirred rapidly at 100 rpm for 4 minutes followed by slow mixing at 40 rpm for 25 minutes. The rapid mixing is to promote the collision between the pharmaceuticals and the coagulant while the slow mixing is to allow the growth of the flocs. The mixture was then allowed to settle for one hour for sedimentation, after which 50 ml of the liquid was carefully collected without disturbing the flocs settled at the bottom of the beakers. The liquid was then filtered and the residual turbidity (RT) and COD of the filtrate were measured. From the results, the pH with the best turbidity and COD removal was chosen as the optimum pH.

The same procedures were repeated while experimenting with different coagulant dosages while keeping the pH constant at the optimum value. The COD and turbidity are measured and the dosage with the best turbidity and COD removal was taken as the optimum. Next, the initial wastewater concentration was varied while the pH and the dosage was kept to their optimum values. The initial wastewater concentration with the best turbidity and COD removal was selected. The same coagulation test without the addition of coagulants was carried out as a control and the residual turbidity (RTc) and the COD (CODc) were measured.

The jar coagulation tests were carried out for both J. Curcas and H. Sabdariffa individually and as a combination.

2.6 Analytical Method

When the pharmaceutical wastewater is put under study, different parameters are always considered to have a better indication of the characteristics of the wastewater. The parameters measured are pH, COD, Biological Oxygen Demand (BOD), turbidity, total suspended solids (TSS) and total dissolved solids (TDS) [15].

As mentioned earlier in Section 1, coagulation is used to reduce the amount of colloidal matter in the water. Turbidity measures the cloudiness of the water, and therefore measures the amount of colloidal matter present inside the water.

Coagulation is a treatment used prior to the biological treatment which involves activated sludge where microorganisms are used to treat the water [16]. During coagulation, a reduction of COD is favored so that when the treated water undergoes the biological treatment, less sludge is produced from the microbial activity. As a result, the two parameters measured in this study are turbidity and COD.

Turbidity

It was measured using a turbiditimeter (EUTECH Instruments TN-100). The turbidity removal was calculated as shown by Eq.1:

$$\text{Turbidity removal} = \frac{RT_c - RT}{RT_c} \times 100\ \% \tag{1}$$

Where: (RT_c) is the turbidity of the wastewater before undergoing coagulation, (RT) is the turbidity of the wastewater after undergoing coagulation.

Chemical Oxygen Demand (COD)

COD was measured using reaction cells of range 25-1500 mg/L which contained sulphuric acid, potassium dichromate and mercury (II) sulphate as reagents. 3 mL of the sample to be tested was carefully added to the reaction cells inside a fume hood. The cell was then vigorously shaken and heated at 150 °C for 2 hours using a thermoreactor. The cells were then removed and let to cool to room temperature. The COD was then measured using a photometer (Prove 300 UV/VIS Spectrophotometer). The COD removal was calculated as shown by Eq.2:

$$\text{COD removal} = \frac{COD_c - COD_i}{COD_c} \times 100\ \% \tag{2}$$

Where: (COD_c) is the COD of the wastewater before undergoing coagulation, (COD_i)is the COD of the wastewater after undergoing coagulation.

3 Results & Discussion

The synthetic wastewater was prepared to simulate the properties of the real pharmaceutical wastewater. The properties of the synthetic wastewater prepared are shown in Table 1 below:

Table 1: Characteristics of synthetic wastewater.

Characteristics of the synthetic wastewater	
Temperature (°C)	30
pH	4.95
Turbidity (NTU)	9.96
COD (mg/L)	665

The pharmaceutical compounds used to simulate the wastewater are acidic in nature and as a result, the wastewater becomes slightly acidic with an average pH of 4.95. The real wastewater is also acidic due to the acidic nature of the pharmaceutical compounds present [15].

3.1 Coagulation activity using *H. Sabdariffa*

3.1.1 Effect of pH on coagulation activity

Fig. 1 and Fig. 2 shows the effect of pH on the turbidity and COD of the treated water after jar tests. pH is an important factor to consider since the solubility of the matter depends on the pH level of the wastewater. The wastewater consists of mainly negative charges and the pH at which the particles become stable is known as the isoelectric point [6]. The pH is varied from 2 to 12 while the coagulant dosage is kept constant at 200 mg/L. From the results obtained, it can be seen that the lowest turbidity is obtained at pH 4 with a turbidity removal of 44.7% and a decrease in COD by 22.9%. This is in agreement with a previous study done by [6] in which it was found that *H. Sabdariffa* is efficient under acidic conditions, especially at pH 4.

This phenomena can be explained by studying the amino acids present in the *H. Sabdariffa* seeds and their isoelectric point. The coagulative agent present in the seeds consists mainly of glutamic acid and aspartic acid which are categorised as acidic amino acids [17]. While the isoelectric point of glutamic acid and aspartic acid are 3.2 and 2.77 respectively, the seeds also contain significant amount of the other amino acids with isoelectric point ranging from 3 to 6 as shown earlier in table 3. As a result, the coagulative agent works best in acidic conditions, especially at pH 3 and 4.

Also, from Fig. 1, pH 4 gives the lowest turbidity of 5.51 NTU with a turbidity removal of 44.7%. It can also be seen that at some pH values, the turbidity increases above its initial turbidity of 9.96 NTU. This is because the pharmaceutical wastewater can only be treated effectively at its optimum pH and at other pH far from its optimum pH, the coagulation activity fails and is unable to produce any flocs. As a result, it causes an increase in the turbidity of the water.

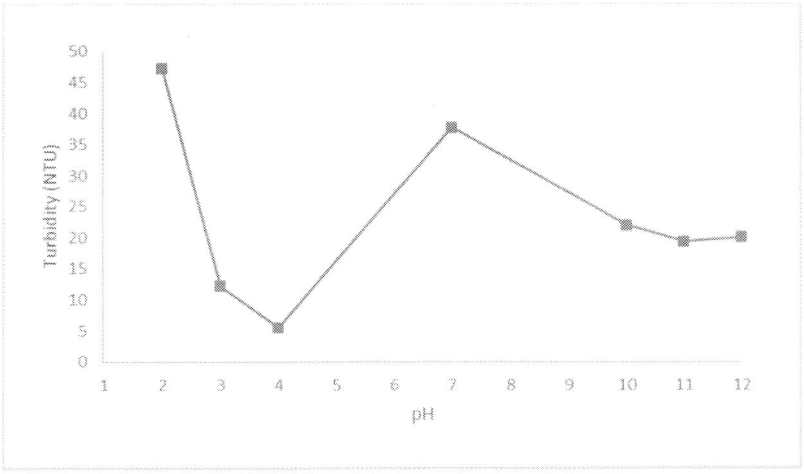

Fig. 1. The effect of pH on the turbidity of the wastewater after coagulation.

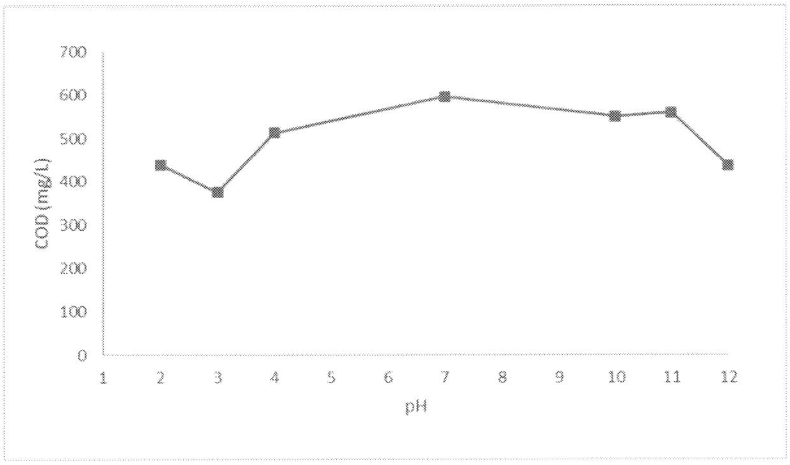

Fig. 2. The trend in COD when varying the pH of the wastewater after coagulation.

Fig. 1 and 2 show the relationship between turbidity and COD. The graph for COD shows a similar trend to the graph of turbidity. This shows that pH also has an effect on the COD of the wastewater. At pH 4, the % COD removal is 22.9%.

3.1.2 Effect of coagulant dosage on coagulation activity

The formation of flocs depends on the coagulant dosage. The charges of the coagulants work by neutralising the charged ions in the synthetic wastewater. The floc strength depends on the attraction and repulsion between the particles. When the dosage is low, the particles do not have enough energy and therefore form weak and small flocs [18]. Overdosing also result in poor performance of coagulation and also directly increases the cost of dosing and sludge formation [19]. Therefore, the coagulant dosage was varied from 40 to 200 mg/L.

From Fig.3, it is found that at 190 mg/L, the coagulant gives the highest turbidity removal of 28.4% with a COD removal of 28.3%. It can be seen that at lower dosages, the turbidity removal is low due to insufficient dosages producing weak flocs that are unable to settle down. This is also in correlation with the findings of [20]. As from 90mg/L of coagulant, the COD remains stable at 476 mg/L. This shows that the dosage does not have much of an effect on the COD of the water.

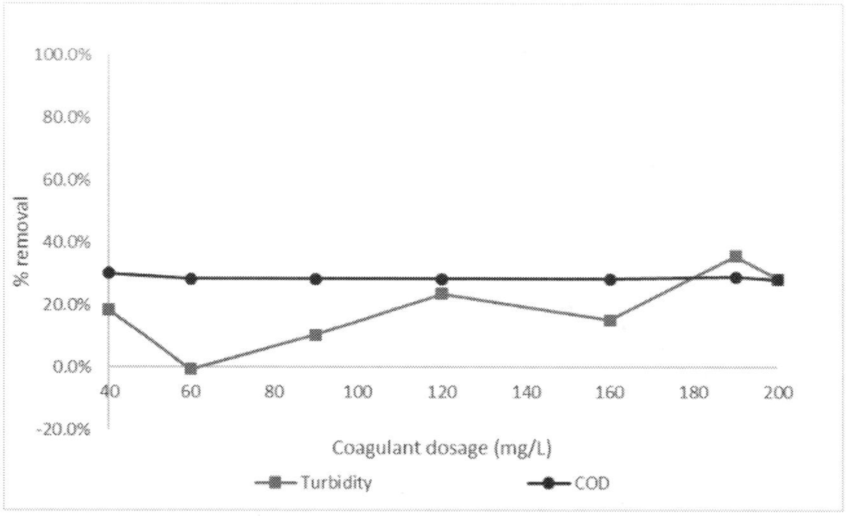

Fig. 3. Percentage removal of turbidity and COD with increasing coagulant dosage.

3.2 Coagulation activity using *J. Curcas*

3.2.1 Effect of pH on coagulation activity

Fig.4 and Fig.5 show a similar trend with the effect of increasing pH. The lowest turbidity is observed at pH 3 with a turbidity removal of 41% and COD removal of 34.9%. The turbidity is reduced from 9.96 NTU to 5.88 NTU. This is in agreement with [13] where it proves that Jatropha works best in acidic conditions and works best at pH 3.

It can also be observed that only at pH 3 and 4, the coagulative agent is able to reduce the turbidity. This is because at pH 3 and 4, the positive charges on the amino acids that make up the protein molecules dominate. The composition of amino acids in the seeds of *J. Curcas* was studied by [21]. Glutamic acid and aspartic acid were abundantly found in the seeds. Since these amino acids are acidic and have an isoelectric point at 3.22 and 2.77 respectively, they favour coagulation in acidic conditions, especially at pH 3 [22]. The *J. Curcas* protein is expected to work well as a cationic coagulant agent. As an amphoteric molecule, the charge of the amino acids is dependent on the pH of the environment. As a result, this is why we can observe flocs at pH 3 and 4 [23]. However, at pH greater than 3, the positive and negative charges of the different amino acids in the protein may have reduced the cationic potential of the coagulant and as a result, no coagulation can be observed at these pH [23].

However at other pH values, *J. Curcas* is unable to neutralise the charges in the wastewater and unable to produce any flocs. This may be due to the presence of oil in the jatropha seed powder which was not extracted prior to coagulation. The oil when mixed with the sodium chloride solvent produces a milky emulsion that causes an increase in turbidity at pH 2 and at alkaline conditions [14].

Similar to the coagulation activity with *H. Sabdariffa,* the COD of the treated water shows a similar trend as the turbidity of the water. When the turbidity decreases, the COD of the water also reduces. Therefore, this shows that pH also has an effect on the COD of the water.

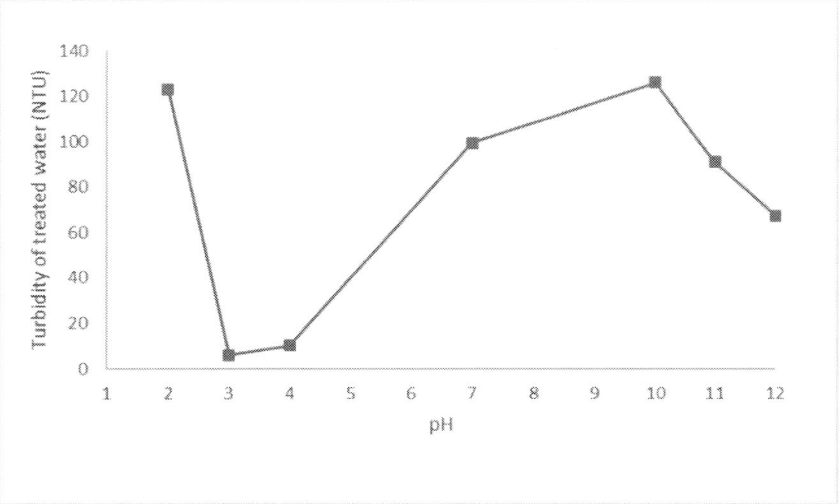

Fig. 4. The effect of pH on the turbidity of the wastewater after coagulation.

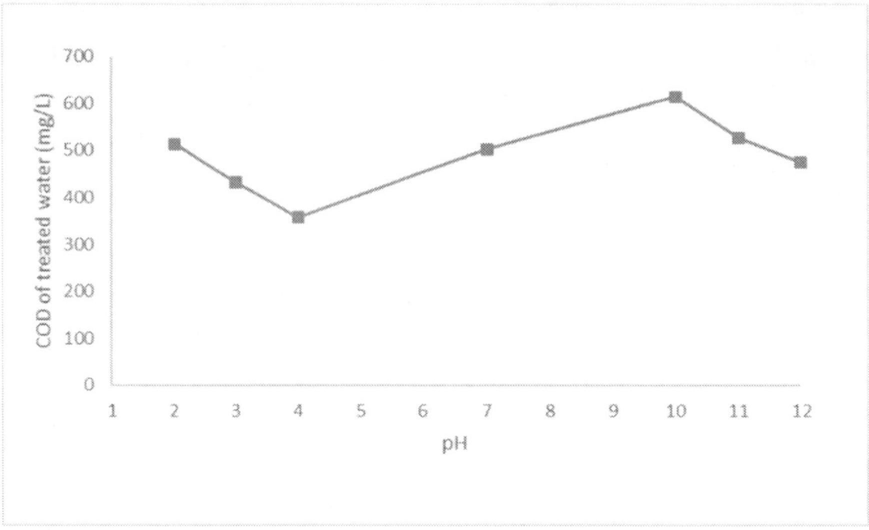

Fig. 5. The effect of pH on the COD of the wastewater after coagulation.

3.2.2 Effect of coagulant dosage on coagulation activity

From Fig. 6, it can be seen that the turbidity increases when the coagulant dosage is below 80 mg/L. This is because the dosage is not enough to produce any flocs and therefore the milky emulsion of the jatropha coagulative agent causes the turbidity of the water to increase. From 90mg/L onwards, the turbidity considerably decreases and reaches its lowest turbidity of 4.81 NTU at 200mg/L. The highest turbidity reaches a 51% removal with a 32% decrease in COD.

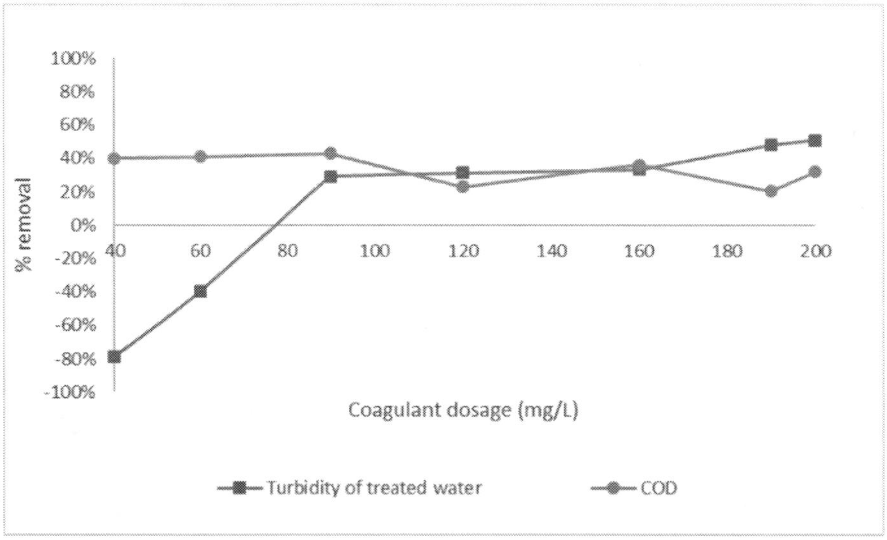

Fig. 6. The effect of coagulant dosage on the turbidity and COD % removal.

3.3 Coagulation activity using a combination of *H. Sabdariffa* and *J. Curcas*

3.3.1 Effect of pH on coagulation activity

The effect of pH is studied on treatment of water by using a combination of both seeds. As mentioned in Section 3.1.1, H.Sabdariffa works best at pH 4 while in Section 3.2.1, it was found that *J. Curcas* works at an optimum pH of 3. For a combination of both, it is expected that the coagulants will work under acidic conditions as well. Therefore, the pH is varied from 2 to 7 while the combination is kept to 200 mg/L with equal amounts of *J. Curcas* and *H. Sabdariffa* (50:50%).

Fig. 7 studies the effect of pH on the turbidity of wastewater after coagulation. It can be observed that the turbidity considerably decreases at pH 3 and 4. It reaches its lowest turbidity of 7.59 NTU at pH 3 with a turbidity removal of 24% and a 32.3 % decrease in COD. Fig. 8 shows that pH also affects the COD of the wastewater. The highest decrease in COD is observed at pH 3 and pH 4.

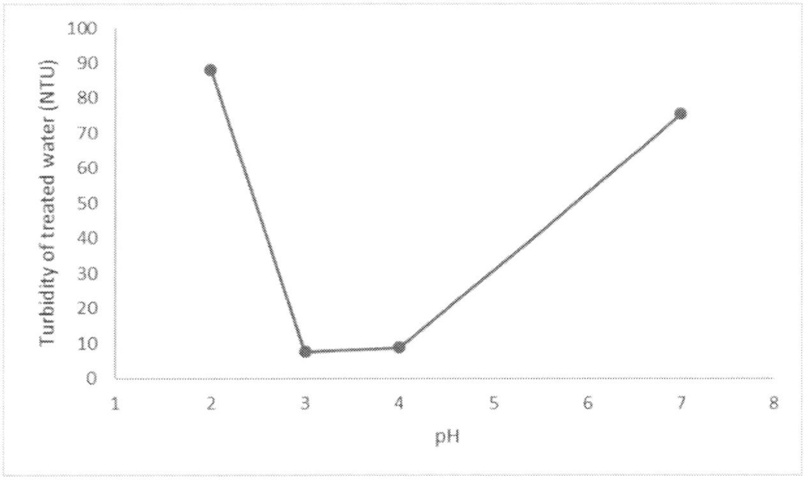

Fig. 7. The effect of pH on the turbidity of treated water after coagulation.

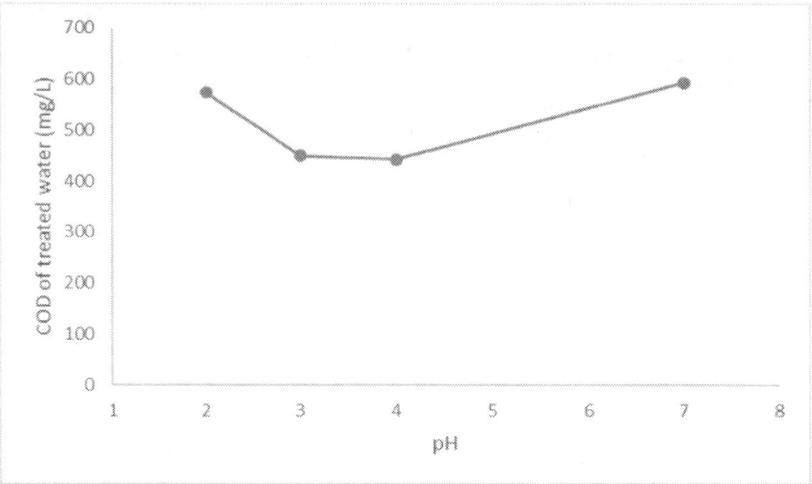

Fig. 8. The effect of pH on the COD of treated water after coagulation.

3.3.2 Effect of coagulant dosage on coagulation activity

Fig. 9 shows the effect of increasing the weight of *J. Curcas* in the mixture of both coagulants on the turbidity and COD of the treated water after coagulation. It can be observed that the % turbidity removal gradually increases from 38.8% to 46.8% when the % weight of *J. Curcas* increases. However, the COD remains almost constant at a COD removal of 46%. This shows that increasing the amount of *J. Curcas* does not affect the COD of the water. The optimum composition is 80% *J. Curcas* and 20% *H. Sabdariffa* by weight with a maximum turbidity removal of 46.8% and a decrease in COD by 46%. This shows that *J. Curcas* has a higher effect than *H. Sabdariffa* in the treatment of the wastewater.

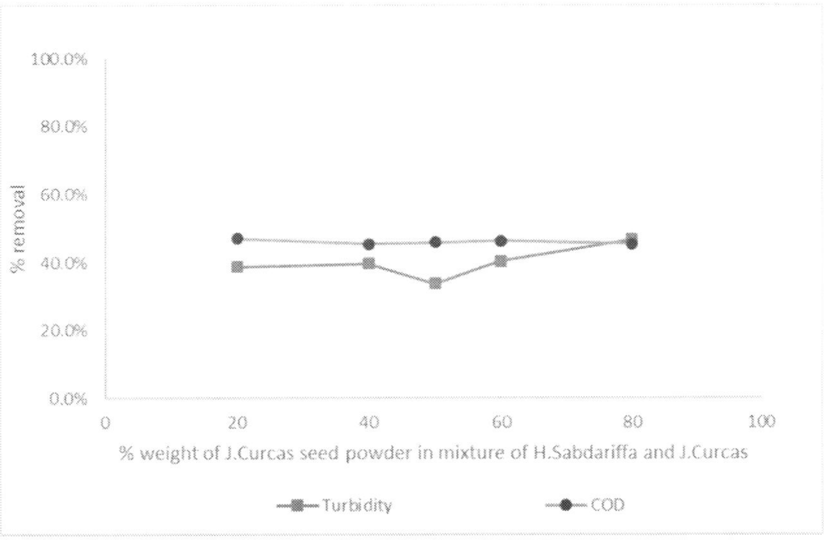

Fig. 9. The effect of coagulant dosage on the turbidity and COD of the wastewater.

3.4 Coagulation with Alum

The results are compared with the conventional chemical coagulant which is alum. The effect of alum in the treatment of pharmaceutical wastewater was studied by [7]. It was found that alum worked best at pH 6 with a dosage of 750mg/L.

The synthetic wastewater was treated with alum with the optimum conditioned mentioned above. The same experiment was carried out, with only alum being used as the coagulant. The chemical coagulant was successful in treating the pharmaceutical wastewater with a turbidity removal of 48% and a COD removal of 53%.

3.5 Comparison of results for all different coagulants experimented

The results obtained for all different scenarios tested in this research are compared. The graph below shows the results obtained for all 4 scenarios.

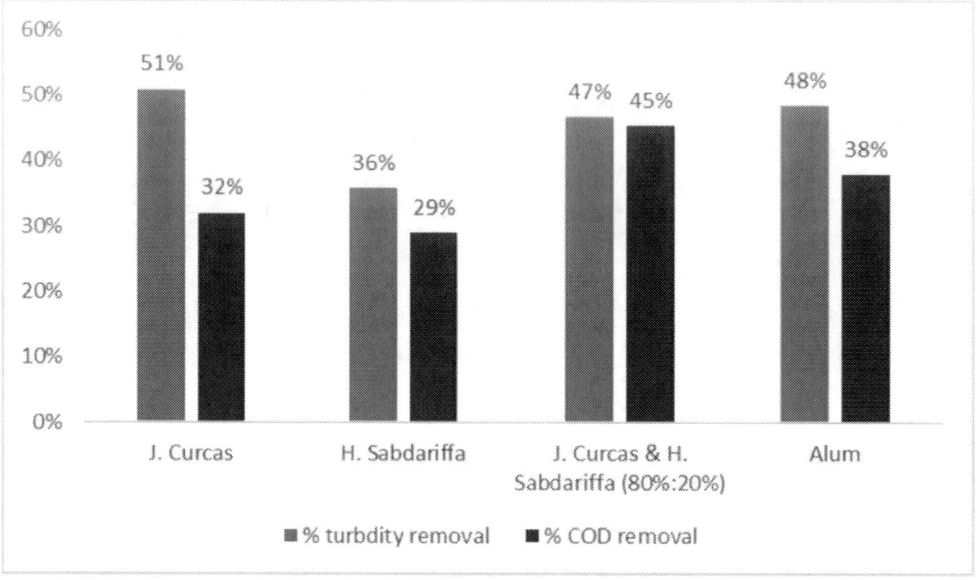

Fig. 10. Result comparison of all the different scenarios tested.

From Fig. 10, it can be seen that in terms of % turbidity removal, *J. Curcas* gives the best results compared to the others. It is followed by Alum with a 48% turbidity removal. *H. Sabdariffa* gives the lowest results among the 3 coagulants. This shows that the natural coagulants can be used to replace chemical coagulants in the treatment of pharmaceutical wastewater since the results obtained are comparable and close to each other. By comparing both natural coagulants, *J. Curcas* is the most suitable one to be used as it is more effective in treating the wastewater.

4 Conclusion

In this study, it was found that both *H. Sabdariffa* and *J. Curcas* were able to effectively treat pharmaceutical wastewater. It was found that *H. Sabdariffa* works best at pH 4 and at a coagulant dosage of 190 mg/L with a highest turbidity removal of 35.8% and a decrease of COD by 29%. *J. Curcas* was found to perform best at pH 3 and with a coagulant dosage of 200 mg/L with a highest turbidity removal of 51% and a decrease of COD by 32%. In comparison, *J. Curcas* is found to be a better and more suited coagulative agent for the treatment of pharmaceutical wastewater. Even a combination of both coagulants only gave a turbidity removal of 46.8%. This study also showed that only pH has a distinctive effect on the COD of the wastewater while both pH and dosage has an effect on the turbidity of the wastewater. In comparison with alum, *J. Curcas* gave a better result in the treatment of the wastewater. This research was successful in showing that natural coagulants can be used to treatment of pharmaceutical wastewater and therefore can replace the use in alum in treatment plants.

For future work, the oil can be extracted from the *J. Curcas* seeds prior to coagulation. Also, coagulation of higher concentrations of pharmaceutical wastewater can be studied. High Performance Liquid Chromatography (HPLC) can be used to detect the pharmaceuticals still present in the wastewater after coagulation and its corresponding concentration.

Acknowledgements

I would like to extend my utmost gratitude towards my supervisor, Ms. Nurhazwani Binti Ismail, for her guidance throughout this project and for her words of encouragement. I would also like to thank the laboratory technicians for providing training for the different apparatus required and for assisting us when having difficulty with any equipment.

I would also like to thank my friends for supporting me, for providing their help whenever needed. My special thanks to Rikesh Ramsurn for always encouraging me in the tough times.

Last but not least, my sincere gratitude to my parents and family for the words of encouragement and for their guidance.

References

[1] "Pharmaceuticals in Drinking-water Public Health and Environment Water, Sanitation, Hygiene and Health," Switzerland, 2011.

[2] A. Shraim, A. Diab, A. Alsuhaimi, E. Niazy, M. Metwally, M. Amad, S. Sioud, and A. Dawoud, "Analysis of some pharmaceuticals in municipal wastewater of Almadinah Almunawarah," *Arab. J. Chem.*, **vol. 10**, pp. S719–S729, (2012).

[3] N. Deziel, "Pharmaceuticals in Wastewater Treatment Plant Effluent Waters," *Univ. Minnesota, Morris*, **vol. 1**, no. 2, pp. 1–20, (2014).

[4] X. Li and G. Li, "A Review : Pharmaceutical Wastewater Treatment Technology and Research in China," *Asia-Pacific Energy Equip. Eng. Res. Conf. (AP3ER 2015)*, pp. 345–348, 2015.

[5] Hendrawati, I. R. Yuliastri, Nurhasni, E. Rohaeti, H. Effendi, and L. K. Darusman, "The use of Moringa Oleifera Seed Powder as Coagulant to Improve the Quality of Wastewater and Ground Water," *IOP Conf. Ser. Earth Environ. Sci.*, **vol. 31**, p. 11, (2016).

[6] N. Fathinatul and R. Nithyanandam, "Wastewater Treatment by using Natural Coagulant," (2014).

[7] M. I. Ashraf, M. Ateeb, M. H. Khan, N. Ahmed, Q. Mahmood, and Zahidullah,

"Integrated treatment of pharmaceutical effluents by chemical coagulation and ozonation," *Sep. Purif. Technol.*, **vol. 158**, pp. 383–386, (2016).

[8] G. Vijayaraghavan, T. Sivakumar, and a Vimal Kumar, "Application of plant based coagulants for wastewater treatment.," *Int. J. Adv. Eng. Res. Stud.*, **vol. 1**, no. 1, pp. 88–92, (2011).

[9] H. D. Beyene, T. D. Hailegebrial, and W. B. Dirersa, "Investigation of coagulation activity of cactus powder in water treatment," *J. Appl. Chem.*, **vol. 1**, no. 5, p. 9, (Nov. 2016).

[10] M. Schiban, M. Antov, and M. Klasnja, "Extraction and Partial Purification of Coagulation of Coagulation Active Components from Common Bean Seed," University of Novi Sad, (2006).

[11] I. Da-Costa-Rocha, B. Bonnlaender, H. Sievers, I. Pischel, and M. Heinrich, "Hibiscus sabdariffa L. - A phytochemical and pharmacological review," *Food Chem.*, **vol. 165**, pp. 424–443, (2014).

[12] J. M. Nzikou, L. Matos, F. Mbemba, C. B. Ndangui, N. P. G. Pambou-Tobi, A. Kimbonguila, T. Silou, M. Linder, and S. Desobry, "Characteristics and Composition of Jatropha curcas Oils, Variety Congo-Brazzaville," *Res. J. Appl. Sci. Eng. Technol.*, **vol. 1**, no. 3, pp. 154–159, (2009).

[13] Z. Z. Abidin, N. S. Mohd Shamsudin, N. Madehi, and S. Sobri, "Optimisation of a method to extract the active coagulant agent from Jatropha curcas seeds for use in turbidity removal," *Ind. Crops Prod.*, **vol. 41**, no. 1, pp. 319–323, (2013).

[14] K. Openshaw, "A review of Jatropha curcas: An oil plant of unfulfilled promise," *Biomass and Bioenergy*, **vol. 19**, no. 1, pp. 1–15, (2000).

[15] S. Kumar Gupta, Y.-T. Hung, and S. Kumar Gupta, *Treatment of Pharmaceutical Wastes*, (October 2005).

[16] Hlthman, "Waste water treatment 2 2.1," *Waste Water*, **vol. 20,** no. 30, pp. 1–18, (2002).

[17] K. I. E. Hainida, I. Amin, H. Normah, and N. Mohd.-Esa, "Nutritional and amino acid contents of differently treated Roselle (Hibiscus sabdariffa L.) seeds," *Food Chem.*, **vol. 111**, no. 4, pp. 906–911, (2008).

[18] Z. Ma, J.-J. Qin, C.-X. Liou, L. Zhang, and S. Valiyaveettil, "Effects of Coagulation, pH and Mixing Conditions on Characteristics of Flocs in Surface Water Treatment," National University of Singapore, (2012).

[19] H. Patel and R. T. Vashi, "Removal of Congo Red dye from its aqueous solution

using natural coagulants," *J. Saudi Chem. Soc.*, **vol. 16**, no. 2, pp. 131–136, (2012).

[20] M. Y. Yong and N. Ismail, "Optimisation of hibiscus sabdariffa as a natural coagulant to treat Congo red in wastewater," *J. Eng. Sci. Technol.*, **vol. 11**, no. Special Issue onthefourtheureca2015, pp. 153–165, (2016).

[21] J. Xiao and H. Zhang, "Comparative evaluation of Jatropha curcas L. seed meals obtained by different methods of defatting on toxic, antinutritional and nutritive factors," *J. Food Sci. Technol.*, **vol. 51**, no. 6, pp. 1126–1132, (2014).

[22] A. E. Ugbogu, E. I. Akubugwo, F. O. Uhegbu, C. G. Chinyere, O. C. Ugbogu, and K. A. Oduse, "Quality assessment profile of Jatropha curcas (L) seed oil from Nigeria," *Int. Food Res. J.*, **vol. 21**, no. 2, pp. 735–741, (2014).

[23] Z. Z. Abidin, N. Ismail, R. Yunus, I. S. Ahamad, and A. Idris, "A preliminary study on Jatropha curcas as coagulant in wastewater treatment," *Environ. Technol.*, **vol. 32**, no. 9, pp. 971–977, (Jul. 2011).

Synthesis and Optimisation of A Sustainable Wastewater Treatment Plant via Material Flow Cost Account (MFCA)-based Approach

Run Ting Siew[1] , *Yoke Kin* Wan[1,*], and *Viknesh* Andiappan[2]

[1]School of Engineering, Taylor's University, Subang Jaya, Selangor, Malaysia.
[2]School of Engineering and Physical Sciences, Heriot-Watt University Malaysia, 62200, Putrajaya, Wilayah Persekutuan Putrajaya, Malaysia

Abstract. In this work, MFCA-based approach has been developed to synthesise an optimum wastewater treatment plant (WWTP). To develop the approach, a series of mathematical equations have been generated based on a generic superstructure that presented all possible pathways of WWTP. In this work, four continuous treatment stages (pre-treatment, chemical treatment, biological treatment and tertiary treatment) with various treatment technologies have been considered. In addition, raw material cost, energy cost, labour cost, and waste management cost as well as the hidden cost and carry-forward cost have also been considered in this approach. In this work, hidden cost is referred to the cost that associated with the process stream, while the carry-forward cost is referred to the cost that is carried forward from one process to another process. Furthermore, pollutants (e.g., TSS, COD, BOD and O&G) have also been considered to ensure the discharged water is complied with discharged regulations. To illustrate the developed approach, an industrial case study, has been solved. As results, an optimum sago wastewater treatment process with minimum waste generation cost is synthesised via a commercial optimisation software, LINGO.

1 Introduction

Due to the dramatic rise of human population and the rapid growth of industries, waste production rate is getting higher. Generally, wastes can be divided into organic waste (e.g., municipal solid waste (MSW), sewage wastewater, food wastes, etc.) and inorganic waste (e.g., electronic waste, plastics, pesticide, cans, metals, chemical wastes, etc.). Organic waste refers to the waste that consist of biodegradable materials meanwhile inorganic wastewater refers to the waste that consist of relative stable substances which cannot easily degradable by organisms [1]. According to Hoornweg et al. [2], three million tonnes of wastes was produced per day in year 2000, and this amount is expected to hit 6 million tonnes by year 2025. Thus, waste management is a crucial step to reduce environmental issues and to increase economic performance of an industry. Waste management involves

* Corresponding author: YokeKin.Wan@taylors.edu.my

various types of treatment processes to ensure the treated wastes complied with discharged regulations before being discharged to the environment.

Among the wastes above-mentioned, organic wastewater is the most common waste generated from industries and households daily. This is because huge amount of water is being used daily for cleaning purpose in industrial processes and household's activity. Numerous research works have been conducted for wastewater treatment technologies. Gobel et al. [3] evaluated elimination of pollutants in secondary (e.g., activated sludge, fludized bed reactor (FBR) and membrane bioreactor (MBR)) and tertiary treatment (e.g., sand filter). Performance of upflow anaerobic sludge blanket (UASB) also being evaluated by Manjunath et al. [4]. Besides that, Mamais et al. [5] and Ekama et al. [6] had researched removal of colloidal matter by using flocculation-precipitation method. Chen et al. [7] and Edzwald et al. [8] had reviewed electrochemical technologies and DAF treatment in WWTP respectively. Other than that, Lefebure et al. [9] also had reviewed physico-chemical treatments. Sustainability of treatment technologies such as mechanical, lagoon and land treatment system in WWTP also being evaluated by Muga et al. [10]. Microbial fuel sells (MFC) hybrid process being proposed by Li et al. [11] for sustainability of wastewater treatment process. In optimisation of WWTP technology, Finger et al. [12] had evaluated grit removal performance by varying operating conditions, whereas Al-Mutairi et al. [13] conducted research on coagulant selection and sludge conditioning on slaughterhouse WWTP. These showed that there are various different treatment technologies can be used to treat organic wastewater.

Apart from the conventional technologies, some research works have been done on new technologies. Gogate et al. [14] highlighted on newer technologies of biological oxidation processes (cavitation, photocatalytic oxidation, Fenton's chemistry, ozonation and hydrogen peroxide) and the comparison among different WWTP using Life Cycle Approach (LCA) had been done by Kalbar et al. [15]. Furthermore, different modelling on WWTP also had been done. Wilas et al. [16] based on plant capacities calculation and effluents standards for selection of treatment technology, Woods et al. [17] evaluated economic viable of phosphorus recovery processes, Gernaey et al. [18] focused on modelling of WWTP using white-box model, black-box, stochastic grey-box, and hybrid modelling. Apart from that, Albino et al. [19] used enterprise input-output model. Modelling on UASB performance being conducted by Singh et al. [20]. This showed that modelling is one of the vital tool for treatment technologies selection. In economic aspect, cost modelling (operating cost and maintenance cost) for wastewater treatment process had conducted by Hernendez-Sancho et al. [21]. For optimisation of wastewater treatment plant control, Durrenmatt et al. [22] had built software sensors based on SCADA system of plant. In environmental aspect, Corominas et al. [23] had evaluated Greenhouse Gas Emission (GHG) of entire WWTP by comparing different modelling approaches (1. simple comprehensive model from empirical assumptions and 2. sophisticated model consist of Anaerobic Digestion Model 1 and also biological reactor).

As above-mentioned, many researches had been done on wastewater treatment technologies in aspect of removal efficiency, selection based on performance and optimum operating condition. Meanwhile, development and research on new wastewater treatment technologies also getting more interested on research field. All this imply that wastewater treatment are getting more important. Research on selection of wastewater technologies by modelling also had been done by using input-output model which take consideration on material and energy flows of production process. Apart from that, modelling on economic aspect, environmental aspect and optimisation on process control had also been done. Note that although many researches had done on WWTP technologies and modelling of WWTP processes, however selection of optimum wastewater treatment pathway based on MFCA concept which considered hidden cost of waste stream and product stream haven't be applied.

MFCA is Environmental Management Accounting (EMA) for material efficiency analysis of the process [24]. It focused on material flow, energy used and cost accounting [25]. Material cost, energy cost, system cost (e.g., labour cost, depreciation cost) and waste management cost being considered in cost accounting [26]. Significant of MFCA is hidden profits from waste generated can be determined so that further improvement can be done [26]. This is due to the fact that material cost is roughly similar around the world whereas labour cost, energy cost vary significantly by different countries [25]. Thus, MFCA can be the useful tool for cost reduction, higher material and energy management [25]. By doing this, value of corporate can be enhanced. Other than that, MFCA can be the management tool for green productivity since higher energy efficiency also imply that lesser CO_2 or waste will be generated from process so environmental performance can be maximised [26]. By concluding benefits above, MFCA account for both economic and environment aspects which lead to sustainability performance [26]. In September 2011, MFCA had become international standard (ISO 14051) and its application is expected to be increased in future [25].

MFCA had been successfully applied on different fields. These fields included lens manufacturing process, printing process, ceramics tiles manufacturing, Canon firm manufacturer and many more. In most recently, Wan et al. [27] developed MFCA-based approach for prioritisation of waste recovery. Two new costs terms were introduced by Wan et al. [27], which is hidden cost and carry-forward cost using the concept of MFCA. MFCA is a concept that focuses on distribution of imputing cost to waste streams. The main purpose of MFCA is to minimise environmental impact concurrently improve economic performance with the selection of minimum total hidden cost of waste stream [27]. Since MFCA concept focused on tracking in amount, quality and cost aspects of input and output material flow in processes, there are 4 types of cost considered in MFCA which are material, system, energy and waste management costs. Waste management is also taken into consideration since waste is considered as one of the by-product in MFCA concept. Attribution of energy, material and system cost are based on the distribution percentage of material toward product and waste stream. These costs were used to determine the hidden cost and carry-forward cost. Hidden cost is the summation cost of processing cost and carry-forward cost. Material, energy and system cost attributed to processing cost. Cost associated to recycle stream or intermediate stream was identified as carry-forward cost. In conventional cost accounting concept, hidden cost that allocated to waste which consists significant cost is overlooked which might affect the final analysed result. Thus, in MFCA, each waste stream has its associated hidden cost which represents its cumulative cost invested to generate these waste. Consideration of hidden cost make MFCA concept more suitable for waste stream prioritisation selection than other accounting prioritisation approaches.

Hence, this research aimed to develop a MFCA-based approach to synthesise an optimum wastewater treatment process with minimum waste generation cost using the concept of MFCA. The significant of this study is the developed model can be used as decision making tool for technology selection to synthesize an optimum wastewater treatment plant. Meanwhile, it is to ensure the treated water is comply with the discharged regulations. In addition, the MFCA-based approach can be used for any type of wastewater by modifying the parameters. To illustrate the developed approach, an industrial case study is solve. In this work, organic sago wastewater treatment is selected as case study. The rest of this paper is organised as follows: research methodology is first presented which is followed by mass balance equations, contaminant balance equations, and cost computation equations of the MFCA-based approach. Then, case study on sago wastewater treatment is solved to illustrate the proposed approach. Finally, conclusion and future work are given at the end of paper.

2 Methodology

A generic wastewater treatment superstructure is first developed to show all possible pathway to treat wastewater as shown in Figure 1. Based on this generic superstructure, MFCA-based approach has been developed by formulating mathematical equations for each treatment in WWTP as shown in following sections. These equations included mass balance equation, contaminant equation, and cost computation equations. To demonstrate the developed approach, an industry case study, organic wastewater treatment plant is solved.

2.1 Generic superstructure of wastewater treatment

A generic superstructure of wastewater treatment plant is shown in Figure 1 below. As shown, Feedstock *WW* represents wastewater supply with volumetric flow rate F^{in}_{WW} is treated through different treatment included pre-treatment $p \in P$, chemical treatment $c \in C$, biological treatment $b \in B$ and tertiary treatment $t \in T$ before convert into product *TW* which is treated water. With the aid of generic superstructure, a series of equations are formulated in the following sections.

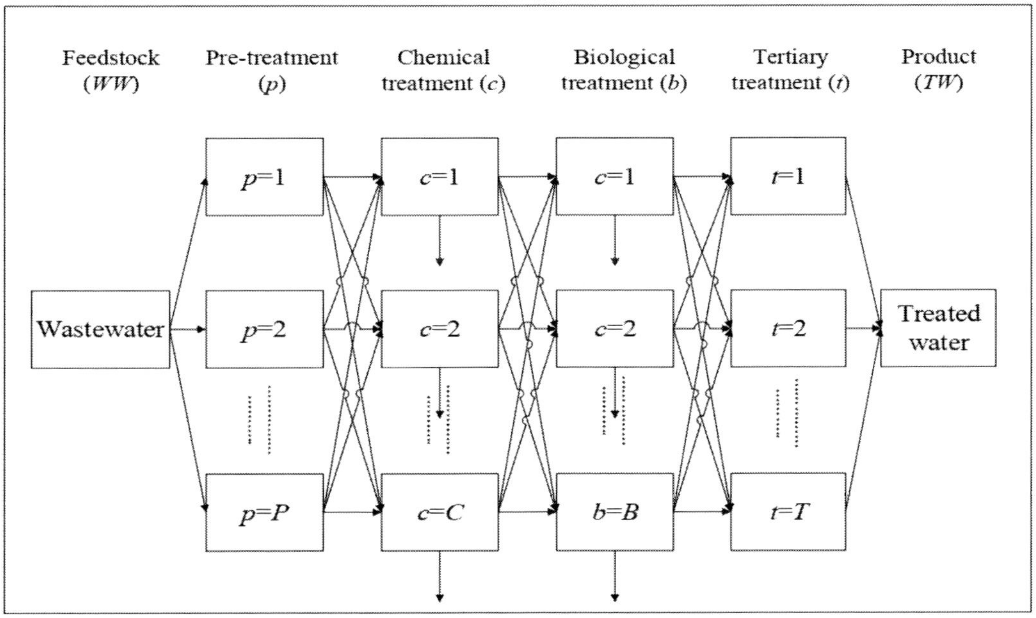

Fig. 1. General superstructure for wastewater treatment.

2.2 Mathematical equations

2.2.1 Mass balance equations

As shown in generic superstructure (Figure 1), the total flow rate of feedstock, *f* F^{in}_{WW} will be distributed to pre-treatment, *p* with flowrate of $F^{out}_{WW,p}$ as shown in Eq. (1).

$$F^{in}_{WW} = \sum_{p=1}^{n} F^{out}_{WW,p} \qquad \forall WW \qquad (1)$$

In other words, the total flowrate that are received by pre-treatment, *p* F^{in}_{p} is shown as Eq. (2),

$$F^{in}_{p} = \sum_{WW=1}^{n} F^{out}_{WW,p} \qquad \forall p \qquad (2)$$

where $F^{out}_{WW,p}$ is the feedstock flowrate distributed to pre-treatment, *p*. As a generic rule of mass balance, the inlet flowrate, F_p^{in} must be equal to outlet flowrate F_p^{out} as shown in Eq. (3).

$$F^{in}_{p} = F^{out}_{p} \qquad \forall p \qquad (3)$$

In pre-treatment *p*, since it do not produce sludge. Thus, total inlet flow rate of pre-treatment, *p* will be distributed to chemical treatment, *c* with flowrate of $F_{p,c}^{out}$ as shown in Eq. (4) below. So, total flowrate that received by chemical treatment *c*, F_c^{in} is shown on Eq. (5). Eq. (6) shown a constraint equation, inlet flowrate, F_c^{in} must be equalled to outlet flowrate, F_c^{out}.

$$F^{in}_{p} = \sum_{c=1}^{n} F^{out}_{p,c} \qquad \forall p \qquad (4)$$

$$F^{in}_{c} = \sum_{p=1}^{n} F^{out}_{p,c} \qquad \forall c \qquad (5)$$

$$F^{in}_{c} = F^{out}_{c} \qquad \forall c \qquad (6)$$

For chemical treatment *c*, since there was additional sludge produced. So other than distribution to biological treatment, *b* with flowrate of $F_{c,b}^{out}$, another sludge produced from chemical treatment, *c* with flowrate of F_c^{sludge} is considered in Eq. (7).

$$F_c^{in} = \sum_{b=1}^{n} F_{c,b}^{out} + F_c^{sludge} \qquad \forall c \qquad (7)$$

Since there is only one stream that flow through biological treatment b, total flow rate received by biological treatment b, F_b^{in} can be defined as Eq. (8). Similarly to chemical treatment c, a constraint equation as shown in Eq. (9) is set,

$$F_b^{in} = \sum_{c=1}^{n} F_{c,b}^{out} \qquad \forall b \qquad (8)$$

$$F_b^{in} = F_b^{out} \qquad \forall b \qquad (9)$$

where F_b^{in} is inlet flowrate of biological treatment b and F_b^{out} is outlet flowrate of biological treatment. Besides, there was sludge produced on biological treatment b. Thus, total outlet flowrate of biological, F_b^{out} is given as Eq. (10),

$$F_b^{out} = \sum_{t=1}^{n} F_{b,t}^{out} + F_b^{sludge} \qquad \forall b \qquad (10)$$

where $F_{b,t}^{out}$ is the flowrate of biological treatment, b that will be distributed to tertiary treatment, t and F_b^{sludge} is the sludge flowrate produced in biological treatment. Similar to previous case, total flowrate which received by tertiary treatment t, F_t^{in} is shown in Eq. (11). Eq. (12) shown generic rule of mass balance which inlet flowrate of F_t^{in} must be equaled to outlet flowrate of F_t^{out}. In addition, Eq. (13) shown the treated water, TW that discharged from tertiary treatment, t with flowrate of F_{TW}.

$$F_t^{in} = \sum_{b=1}^{n} F_{b,t}^{out} \qquad \forall t \qquad (11)$$

$$F_t^{in} = F_t^{out} \qquad \forall t \qquad (12)$$

$$F_{TW} = \sum_{t=1}^{n} F_t^{out} \qquad (13)$$

2.2.2 Contaminant balance equations on COD

First, COD concentration at feed inlet is represented in $C^{COD,in}{}_{WW}$ in ppm unit. COD removal efficiency for pre-treatment p, chemical treatment c, biological treatment b and tertiary treatment t are given as x_p^{COD}, x_c^{COD}, x_b^{COD}, x_t^{COD} respectively. For sludge yield in chemical c and biological treatment b are represented as $Y^{COD,sludge}{}_c$ and $Y^{COD,sludge}{}_b$, both with unit of kg sludge/kg COD removed. Sludge density for chemical and biological treatment are represented by ρ_c^{sludge} and ρ_b^{sludge}.

In order to determine the mass of COD, $m_p^{COD,in}$ at inlet of pre-treatment, p total flow rate received by pre-treatment that determined Eq. (2) is multiplied with COD concentration at feed inlet as shown in Eq. (14).

$$m_p^{COD,in} = F_p^{in} \times C_{WW}^{COD,in} \qquad \forall p \qquad (14)$$

By using removal efficiency x_p^{COD}, outlet COD mass, $m_p^{COD,out}$ and removal COD mass, $m_p^{COD,Rev}$ from pre-treatment, p can be determined using Eq. (15) and (16).

$$m_p^{COD,out} = m_p^{COD,in} \times (1 - x_p^{COD}) \qquad \forall p \qquad (15)$$

$$m_p^{COD,Rev} = m_p^{COD,in} \times x_p^{COD} \qquad \forall p \qquad (16)$$

Then, concentration of COD, $C_p^{COD,out}$ at pre-treatment outlet can be determined via:

$$C_p^{COD,out} = \frac{m_p^{COD,out}}{F_p^{out}} \qquad \forall p \qquad (17)$$

Next, mass of COD $m_c^{COD,in}$ at inlet stream to chemical treatment, c can be calculated by using COD concentration obtained above multiplied with distribution of flowrate from pre-treatment p to chemical treatment c, $F_{p,c}^{out}$ which as shown in Eq. (18).

$$m_c^{COD,in} = C_p^{COD,out} \times F_{p,c}^{out} \qquad \forall c \qquad (18)$$

By using removal efficiency x_c^{COD} for chemical treatment c, outlet COD mass $m_c^{COD,out}$ and removal COD mass $m_c^{COD,Rev}$ can be determined as shown on Eq. (19) and (20).

$$m_c^{COD,out} = m_c^{COD,in} \times (1 - x_c^{COD}) \qquad \forall c \qquad (19)$$

$$m_c^{COD,Rev} = m_c^{COD,in} \times x_c^{COD} \qquad \forall c \qquad (20)$$

Based on amount of COD removed in chemical treatment c, sludge volumetric flow rate F_c^{sludge} can be determine by using sludge yield $Y_c^{COD,sludge}$ and density ρ_c^{sludge} for chemical treatment c as shown in Eq. (21) below.

$$F_c^{sludge} = \frac{Y_c^{COD,sludge}}{\rho_c^{sludge}} \times m_c^{COD,Rev} \qquad \forall c \qquad (21)$$

For biological treatment *b* and tertiary treatment *t*, same procedure can be used to determine COD outlet mass and COD removal from each treatment. Similar to chemical treatment *c*, additional sludge volumetric flow rate need to be account in biological treatment *b*. All equations involved can be represented by Eq. (22) to (30) below.

$$C_c^{COD,out} = \frac{m_c^{COD,out}}{F_c^{out}} \qquad \forall c \qquad (22)$$

$$m_b^{COD,in} = C_c^{COD,out} \times F_{c,b}^{out} \qquad \forall b \qquad (23)$$

$$m_b^{COD,out} = m_b^{COD,in} \times (1 - x_b^{COD}) \qquad \forall b \qquad (24)$$

$$m_b^{COD,\mathrm{Rev}} = m_b^{COD,in} \times x_b^{COD} \qquad \forall b \qquad (25)$$

$$F_b^{sludge} = \frac{Y_b^{COD,sludge}}{\rho_b^{sludge}} \times m_b^{COD,\mathrm{Rev}} \qquad \forall b \qquad (26)$$

$$C_b^{COD,out} = \frac{m_b^{COD,out}}{F_b^{out}} \qquad \forall b \qquad (27)$$

$$m_t^{COD,in} = C_b^{COD,out} \times F_{b,t}^{out} \qquad \forall t \qquad (28)$$

$$m_t^{COD,out} = m_t^{COD,in} \times (1 - \mathrm{x}_t^{COD}) \qquad \forall t \qquad (29)$$

$$m_t^{COD,\mathrm{Rev}} = m_t^{COD,in} \times \mathrm{x}_t^{COD} \qquad \forall t \qquad (30)$$

where $C_c^{COD,out}$ is concentration of COD at chemical treatment outlet, F_c^{out} is outlet flowrate of chemical treatment, $m_b^{COD,in}$ is mass of COD at inlet stream to biological treatment, $F_{c,b}^{out}$ is distribution of flowrate from chemical treatment to biological treatment, $m_b^{COD,out}$ is outlet COD mass, $m_b^{COD,Rev}$ is removal COD mass, x_b^{COD} is COD removal efficiency for biological treatment, F_b^{sludge} is sludge flowrate of biological treatment, $C_b^{COD,out}$ is concentration of COD at biological treatment outlet, F_b^{out} is outlet flowrate of biological treatment, $m_t^{COD,in}$ is COD mass at inlet stream of tertiary treatment, $F_{b,t}^{out}$ is distribution of flowrate from biological to tertiary treatment, $m_t^{COD,out}$ is outlet COD mass, $m_t^{COD,Rev}$ is removal COD mass and x_t^{COD} is COD removal efficiency at tertiary treatment.

In order to comply with COD discharge standard, outlet COD concentration for tertiary treatment *t*, $C_t^{COD,out}$ which is treated water need to be convert from mass that shows on Eq.

(31). Eq. (32) below shown calculated COD concentration must comply with $C^{Std,COD}$ which is standard concentration for COD.

$$C_t^{COD,out} = \frac{m_t^{COD,out}}{F_t^{out}} \qquad \forall t \qquad (31)$$

$$C_t^{COD,out} < C^{Std,COD} \qquad (32)$$

2.2.3 Contaminant balance equations on BOD, TSS, O&G

Formulation of contaminant balance equation for BOD, TSS and O&G is same as the formulation of COD balance equation above but just need to replace respective index value into all COD balance equations above.

2.2.4 Cost computation equations

Operating cost for technology in each treatment take account of raw material cost, energy cost and labour cost. Thus, operating cost of pre-treatment p $Cost_p^{opt}$ can be determine as Eq. (33) below.

$$Cost_p^{mat} + Cost_p^{energy} + Cost_p^{labour} = Cost_p^{opt} \qquad \forall p \qquad (33)$$

where $Cost_p^{mat}$ is raw material cost, $Cost_p^{energy}$ is energy cost and $Cost_p^{labour}$ is labour cost. Based on calculated pre-treatment p operating cost, hidden unit cost for pre-treatment $UCost_p^{Hidden}$ can be determined via:

$$UCost_p^{Hidden} = \frac{Cost_p^{opt}}{F_p^{in}} \qquad \forall p \qquad (34)$$

Carry-forward cost from pre-treatment p to chemical treatment c, $CFC_{p,c}$ can be determined using Eq. (35) below.

$$CFC_{p,c} = UCost_p^{Hidden} \times F_{p,c}^{out} \qquad \forall p \forall c \qquad (35)$$

Similar to operating cost determination of pre-treatment p, operating cost of chemical treatment c, $Cost_c^{opt}$ is determined on Eq. (36) below with an additional carry-forward cost from Eq. (35). By using calculated chemical treatment c operating cost, hidden unit cost of chemical treatment c $UCost_c^{Hidden}$ is calculated using Eq. (37) below.

$$Cost_c^{mat} + Cost_c^{energy} + Cost_c^{labour} + \sum_{p=1}^{n} CFC_{p,c} = Cost_c^{opt} \qquad \forall c \qquad (36)$$

$$UCost_c^{Hidden} = \frac{Cost_c^{opt}}{F_c^{in}} \qquad \forall c \qquad (37)$$

where $Cost_c^{mat}$ is raw material cost , $Cost_c^{energy}$ is energy cost and $Cost_c^{labour}$ is labour cost for chemical treatment c.

In chemical treatment c, instead of just carry forward cost from chemical treatment c to biological treatment b $CFC_{c,b}$, there is another hidden cost for sludge generation HC_c^{sludge} as shown in Eq. (38) and (39).

$$CFC_{c,b} = UCost_c^{Hidden} \times F_{c,b}^{out} \qquad \forall c \forall b \qquad (38)$$

$$HC_c^{sludge} = UCost_c^{Hidden} \times F_c^{sludge} \qquad \forall c \qquad (39)$$

For biological treatment b and tertiary treatment t same procedure can be used to determine operating cost, hidden unit cost and carry-forward from each treatment. Similar to chemical treatment c, additional sludge hidden cost need to be account in biological treatment b. All equations involved can be represented by Eq. (40) to (46) below.

$$Cost_b^{mat} + Cost_b^{energy} + Cost_b^{labour} + \sum_{c=1}^{n} CFC_{c,b} = Cost_b^{opt} \qquad \forall b \qquad (40)$$

$$UCost_b^{Hidden} = \frac{Cost_b^{opt}}{F_b^{in}} \qquad \forall b \qquad (41)$$

$$CFC_{b,t} = UCost_b^{Hidden} \times F_{b,t}^{out} \qquad \forall b \forall t \qquad (42)$$

$$HC_b^{sludge} = UCost_b^{Hidden} \times F_b^{sludge} \qquad \forall b \qquad (43)$$

$$Cost_t^{mat} + Cost_t^{energy} + Cost_t^{labour} + \sum_{b=1}^{n} CFC_{b,t} = Cost_t^{opt} \qquad \forall t \qquad (44)$$

$$UCost_t^{Hidden} = \frac{Cost_t^{opt}}{F_t^{in}} \qquad \forall t \qquad (45)$$

$$CFC_{t,TW} = UCost_t^{Hidden} \times F_t^{out} \qquad \forall t \forall TW \qquad (46)$$

where $Cost_b^{mat}$, $Cost_b^{energy}$, $Cost_b^{labour}$ and $Cost_b^{opt}$ are raw material cost, energy cost, labour cost and operating cost for biological treatment b. $UCost_b^{Hidden}$ is hidden unit cost for biological treatment b. Carry-forward cost from biological treatment b to tertiary treatment t and sludge hidden cost for biological treatment b are $CFC_{b,t}$ and HC_b^{sludge} respectively. Raw material cost, energy cost, labour cost and operating cost involved in tertiary treatment t are represented by $Cost_t^{mat}$, $Cost_t^{energy}$, $Cost_t^{labour}$ and $Cost_t^{opt}$ respectively. $UCost_t^{Hidden}$ is hidden unit cost for tertiary treatment t, $CFC_{t,TW}$ is carry-forward cost from tertiary treatment t to product.

Once cost accounted on all streams around each treatment by using all cost computation equations above, total hidden cost for product and total sludge produced can be determined by Eq. (47) and (48) below.

$$HC^{TW} = \sum_{t=1}^{n} CFC_{t,TW} \tag{47}$$

$$HC^{Total,sludge} = \sum_{c=1}^{n} HC_c^{sludge} + \sum_{b=1}^{n} HC_b^{sludge} \tag{48}$$

Hence total hidden cost HC^{Total} and total operating cost of technology involved in each treatment $Cost^{Total\ opt}$ for entire wastewater treatment can be calculated via:

$$HC^{Total} = HC^{TW} + HC^{Total,sludge} \tag{49}$$

$$Cost^{Total,opt} = \sum_{p=1}^{n} Cost_p^{opt} + \sum_{c=1}^{n} Cost_c^{opt} + \sum_{b=1}^{n} Cost_b^{opt} + \sum_{t=1}^{n} Cost_t^{opt} \tag{50}$$

Finally, optimization objective of this modelling is to minimize HC^{Total} as shown in Eq. (51) below.

$$Min = HC^{Total} \tag{51}$$

In order to demonstrate approach that mentioned above, a case study is solved in the following section.

2.3 Case study

A sago wastewater treatment is select as case study in this research. Thus, Figure 2 below shows superstructure of organic wastewater treatment with specific technologies listed on different treatment. Based on comprehensive literature review done on organic wastewater treatment technology, typical wastewater treatment technologies used in organic wastewater treatment is adopted as treatment technologies in each treatment stage which presented in Figure 2 below.

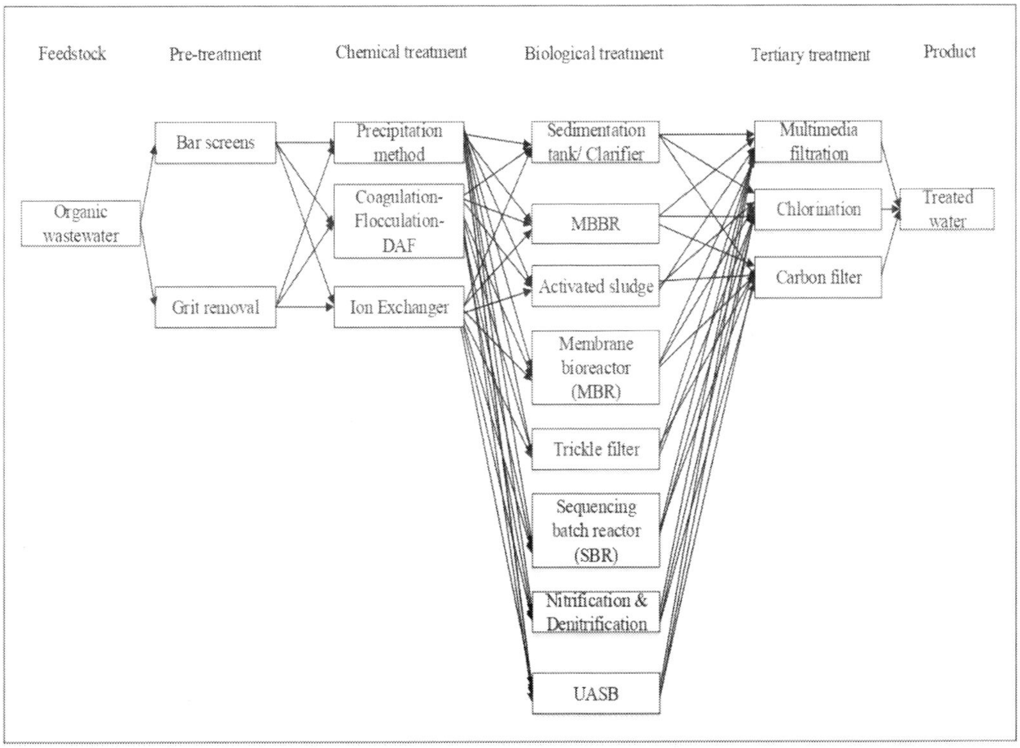

Fig. 2. Superstructure of organic wastewater treatment.

As shown, organic wastewater treatment begins with pre-treatment which consists of bar screen and grit removal. Next is chemical treatment with technologies of precipitation method, coagulation-flocculation-DAF and ion exchanger. Then follow by biological treatment which involved sedimentation tank/clarifier, MBBR, activated sludge and others. Tertiary treatment is the final treatment stage with multimedia filtration, chlorination and carbon filter. Through selection of technology in each treatment stage, treated water that complied with discharged regulation can be generated.

Contaminants characteristic of sago wastewater and discharged regulations are tabulated in Table 1 below. Contaminants characteristics of sago wastewater is based on sago mill industry in Sarawak [28]. Besides that, contaminant removal efficiency of each technology on different treatment stage are summarised in Table 2. Apart from that, in Table 3, operating cost such as raw material cost, energy cost and labour cost in each technology is tabulated.

For energy cost for cost computation listed in each technology on Table 3, it included electricity consumption of pumps, compressors, clarifier and other equipment which involved in that particular treatment.

Table 1. Sago wastewater contaminants characteristic and discharged regulations [28, 29]

Concentration (ppm)				
	TSS	COD	BOD	O&G
Sago wastewater	4800	11650	5750	300
Discharged regulations (Standard A)	50	50	20	10

Table 2. Contaminant removal efficiency of technologies.

Pre-treatment [30, 31]				
	Removal efficiency (%)			
	TSS	COD	BOD	O&G
Bar screen	50	20	35	85
Grit removal	50	20	30	90
Chemical treatment [32, 33, 34, 35]				
	Removal efficiency (%)			
	TSS	COD	BOD	O&G
Precipitation method	90	80	80	50
Coagulation-Flocculation-DAF	85	80	70	95
Ion exchanger	0	38	38	20
Biological treatment [33, 36, 37, 38, 39]				
	Removal efficiency (%)			
	TSS	COD	BOD	O&G
Sedimentation tank	70	20	35	20

Table 2. Contaminant removal efficiency of technologies (Cond.).

Biological treatment [33, 36, 37, 38, 39]				
	Removal efficiency (%)			
	TSS	COD	BOD	O&G
MBBR	0	85	88	0
Activated sludge	61	72	90	0
MBR	99	97	98	0
Trickle filter	95	85	90	0
SBR	98	85	98	79
Nitrification/De nitrification	0	81	89	0
UASB	80	80	85	0
Tertiary treatment [40, 41]				
	Removal efficiency (%)			
	TSS	COD	BOD	O&G
Multimedia filtration	86	69	67	0
Carbon filter	0	30	50	0
Chlorination	0	10	20	0

Table 3. Operating cost of technologies.

Pre-treatment [30, 31]				
	Cost (USD/day)			
	Raw material	Energy	Labour	Opt. cost
Bar screen	0	1668.48	174	1842.48
Grit removal	0	973.52	174	1147.52
Chemical treatment [33, 34, 35, 38, 42]				
	Cost (USD/day)			
	Raw material	Energy	Labour	Opt. cost
Precipitation method	0.18	1264.9	632	1897.18
Coagulation-Flocculation-DAF	240.87	30.6	696	967.47
Ion exchanger	472.22	97.22	175	1094.44
Biological treatment [33, 34, 42, 43]				
	Cost (USD/day)			
	Raw material	Energy	Labour	Opt. cost
Sedimentation tank	0	184.252	175	359.252
MBBR	0	695	875	1570
Activated sludge	0	9147.3	1427	10574.3
MBR	0	8295	1095	9390
Trickle filter	0	5840.565	1095	6935.57
SBR	0	833.33	875	1708.33

Table 3. Operating cost of technologies (Cond.)

Biological treatment [33, 34, 42, 43]				
	Cost (USD/day)			
	Raw material	Energy	Labour	Opt. cost
Nitrification/De nitrification	5039	702	875	6616
UASB	0	4740	875	5615
Tertiary treatment [40, 41, 43]				
	Cost (USD/day)			
	Raw material	Energy	Labour	Opt. cost
Multimedia filtration	0	1043.5	875	1918.5
Carbon filter	67	373	875	1315
Chlorination	208.33	373	875	1456.33

Based on sago wastewater treatment plant with flowrate of 79,000 m^3/day and the contaminants characteristic as shown in Table 1, mass of TSS, COD, BOD and O&G are determined as 379,200, 920,350, 454,250 and 23,700 kg/day, respectively. By input respective contaminant removal efficiency and operating cost of each technology into previous mass and cost computation modelling equations, selection among treatment technology has been done by using LINGO software.

As results, technology that gives minimum hidden cost meanwhile comply with discharge regulations is selected in each treatment stage. Table 4 shows concentration of contaminants on selected technology in each treatment stage. Apart from that, in order to ensure the final treated water that discharged from the selected treatment technologies is complied with discharged regulation (standard A), Table 5 is presented to show the comparison between contaminants concentration of treated water with discharged regulation. It can be clear identified that all contaminants concentration of treated water was lower that discharged regulation so it is safe to be discharged to water bodies.

Next, cost computation is being considered using MFCA concept. Results of unit cost, hidden cost and carry-forward cost for each treatment technology selected by solving Eq. (33) - (46) are summarized in Table 6. Since additional sludge being produced in chemical and biological treatment, so additional hidden cost being calculated via Eq. (39) and (43). Besides that, total hidden cost of entire WWTP also shown in Table 6.

As shown, minimum total hidden cost (THC) of 12,819.98 USD/day can be achieved by selecting grit removal, coagulation-flocculation-DAF, MBR and carbon filter as treatment technology in entire WWTP. THC is calculated by summation of carry-forward cost from tertiary treatment to treated water, hidden cost for sludge generation of chemical and biological treatment. By referring to cost computation equations in section 2.2.4, calculation of unit cost, carry-forward cost and hidden cost are based on operating cost and volumetric flow rate around each treatment technology. Due to conservation of mass, inlet and outlet volumetric flow around each technology must be equal. Hence, operating cost of technology stands out to be one of the main factor in determine value of hidden cost and carry-forward cost. On other word, higher operating cost will lead to higher hidden cost and carry-forward cost.

Table 4. Contaminants concentration on selected technologies.

Treatment	**Technology**	**Stream**	**Contaminant concentration (ppm)**			
			TSS	COD	BOD	O&G
Pre-treatment	Grit removal	Inlet	4800	11650	5750	300
		Outlet	2400	9320	4025	30
Chemical treatment	Coagulation - Flocculation -DAF	Inlet	2400	9320	4025	30
		Outlet	389	2014	1305	1.6
Biological treatment	MBR	Inlet	389	2014	1305	1.6
		Outlet	4.1	64.0	27.6	1.7
Tertiary treatment	Carbon filter	Inlet	4.1	64.0	27.6	1.7
		Outlet	4.1	44.8	13.8	1.7

Table 5. Comparison between contaminants concentration of treated water with discharged regulations.

	Concentration (ppm)			
	TSS	COD	BOD	O&G
Treated water	4.12	44.8	13.8	1.72
Discharged regulations	50	50	20	10

Table 6. Results of unit cost, hidden cost and carry-forward cost of each treatment technology.

Treatment	Technology	Cost (USD/day)		
		Unit cost	Carry-forward cost	Hidden cost
Pre-treatment	Grit removal	0.0145	1,147.52	-
Chemical treatment	Coagulation-Flocculation-DAF	0.02677	1,957.30	157.69
Biological treatment	MBR	0.1552	10,709.91	637.38
Tertiary treatment	Carbon filter	0.1743	-	12,024.91
Total hidden cost				12,819.98

In pre-treatment, grit removal is being selected instead of bar screen. This is due to both technologies have comparable contaminants removal efficiency however grit removal provide lower operating cost than bar screen which can be clearly identified on Table 3.

Among three technologies in chemical treatment, ion exchanger is eliminated due to low contaminants removal efficiency. Although contaminants removal efficiency of precipitation method is similar to coagulation-flocculation-DAF but high operating cost in precipitation method which makes it uneconomical. Thus coagulation-flocculation-DAF is selected as chemical treatment technology.

Even though operating cost for MBR ranked as second highest among technologies in biological treatment, it still selected as technology for biological treatment. The main reason is MBR provide high contaminants removal efficiency especially in COD. Although highest COD removal technology being selected in pre-treatment, chemical treatment and tertiary treatment, up to 87% of COD removal efficiency technology must be selected for biological treatment in order to meet discharged regulations This imply that technologies in biological treatment having COD removal efficiency lower than 87% would not be considered as selection. Based on Table 2, it can be clearly identified that all technologies have COD removal efficiency lesser than 87% except MBR which make MBR necessary to be selected as biological treatment technology. In biological treatment, the main factor for technology selection is technology contaminants removal performance instead of operating cost.

In tertiary treatment, since final treated water is just to discharge into water bodies. So multimedia filtration which provide high purity on treated water is not necessary to be selected. Moreover, operating cost of multimedia filtration is the highest among tertiary treatment technologies. By comparing among chlorination and carbon filter, since chlorination operating cost is higher than carbon filter meanwhile chlorination treatment is specific on disinfection which is not necessary for sago wastewater treatment. Thus, carbon filter is selected as tertiary treatment technology.

3 Conclusion

A novel MFCA-based approach is developed in this work for optimum pathway selection of wastewater treatment plant. Hidden cost associated to treatment waste stream is considered using this approach for WWTP pathway selection. Case study of sago wastewater is solved to illustrate the proposed approach. Trends of wastewater treatment pathway selection also analysed. In case study, there are several factors such as discharged wastewater quality, technology operating cost and sludge yield that will affect selection of wastewater treatment pathway. To determine minimum THC of WWTP, these factors are traded off by using MFCA based-approach. Thus, this approach can be used as selection tools for decision maker in selection of wastewater treatment pathway so that both economic and environmental aspect can be monitored and improved. This approach could be further extended for systhesis and optimisation of wastewater treatment plant by considering sludge handling technologies and cost.

References

1. N. K. S. Lawrence K. Wang, Yung-Tse Hung, *Handbook of Advanced Industrial and Hazardous Wastes Treatment - Google Books*. Taylor & Francis Group (2010).
2. D. Hoornweg, P. Bhada-Tata, and C. Kennedy, "Environment: Waste production must peak this century," *Nature*, **502**, no. 7473, pp. 615–617 (Oct. 2013).
3. A. Göbel, C. S. McArdell, A. Joss, H. Siegrist, and W. Giger, "Fate of sulfonamides, macrolides, and trimethoprim in different wastewater treatment technologies," *Sci. Total Environ.*, **372**, no. 2–3, pp. 361–371 (2007).
4. N. T. Manjunath, I. Mehrotra, and R. P. Mathur, "Treatment of wastewater from slaughterhouse by DAF-UASB system," *Water Res.*, **34**, no. 6, pp. 1930–1936 (2000).
5. D. Mamais, D. Jenkins, and P. Prrr, "A rapid physical-chemical method for the determination of readily biodegradable soluble COD in municipal wastewater," *Water Res.*, **27**, no. 1, pp. 195–197 (1993).
6. Ekama GA; Marais GvR, "Theory, design and operation of nutrient removal activated sludge processes," (1984).
7. G. Chen, "Electrochemical technologies in wastewater treatment," *Sep. Purif. Technol.*, **38,** no. 1, pp. 11–41 (2004).
8. J. K. Edzwald, "Principles and applications of dissolved air flotation," *Water Sci. Technol.*, **31**, no. 3–4, pp. 1–23 (1995).

9. O. Lefebvre and R. Moletta, "Treatment of organic pollution in industrial saline wastewater: A literature review," *Water Res.*, **40**, no. 20, pp. 3671–3682 (2006).
10. H. E. Muga and J. R. Mihelcic, "Sustainability of wastewater treatment technologies," *J. Environ. Manage.*, **88**, no. 3, pp. 437–447 (2008).
11. W.-W. Li, H.-Q. Yu, and Z. He, "Towards sustainable wastewater treatment by using microbial fuel cells-centered technologies," *Energy Environ. Sci.*, **7**, no. 3, pp. 911–924 (2013).
12. R. E. Finger, J. Parrick, and R. E. Finger, "Optimization removal grit a wastewater of at treatment of," **52**, no. 8, pp. 2106–2116 (2015).
13. N. Z. Al-Mutairi, M. F. Hamoda, and I. Al-Ghusain, "Coagulant selection and sludge conditioning in a slaughterhouse wastewater treatment plant," *Bioresour. Technol.*, **95**, no. 2, pp. 115–119 (2004).
14. P. R. Gogate and A. B. Pandit, "A review of imperative technologies for wastewater treatment I: Oxidation technologies at ambient conditions," *Adv. Environ. Res.*, **8**, no. 3–4, pp. 501–551 (2004).
15. P. P. Kalbar, S. Karmakar, and S. R. Asolekar, "Assessment of wastewater treatment technologies: Life cycle approach," *Water Environ. J.*, **27**, no. 2, pp. 261–268 (2013).
16. B. M. Practices, B. Treatment, S. Treatment, and Z. D. Options, "Wastewater treatment technologies," *Facilities*, **4**, pp. 33–43 (1996).
17. N. C. Woods, S. M. Sock, and G. T. Daigger, "Phosphorus recovery technology modeling and feasibility evaluation for municipal wastewater treatment plants," *Environ. Technol.*, **20**, no. March 2015, pp. 663–679 (1999).
18. K. V. Gernaey, M. C. M. Van Loosdrecht, M. Henze, M. Lind, and S. B. Jørgensen, "Activated sludge wastewater treatment plant modelling and simulation: State of the art," *Environ. Model. Softw.*, **19**, no. 9, pp. 763–783 (2004).
19. V. Albino, E. Dietzenbacher, and S. Kühtz, "Analysing materials and energy flows in an industrial district using an enterprise input-output model," *Econ. Syst. Res.*, **15**, no. 4, pp. 457–480 (2003).
20. K. P. Singh, N. Basant, A. Malik, and G. Jain, "Modeling the performance of 'up-flow anaerobic sludge blanket' reactor based wastewater treatment plant using linear and nonlinear approaches-A case study," *Anal. Chim. Acta*, **658**, no. 1, pp. 1–11 (2010).
21. F. Hernandez-Sancho, M. Molinos-Senante, and R. Sala-Garrido, "Cost modelling for wastewater treatment processes," *Desalination*, **268**, no. 1–3, pp. 1–5 (2011).
22. D. J. Dürrenmatt and W. Gujer, "Data-driven modeling approaches to support wastewater treatment plant operation," *Environ. Model. Softw.*, **30**, pp. 47–56 (2012).
23. L. Corominas, X. Flores-Alsina, L. Snip, and P. A. Vanrolleghem, "Comparison of different modeling approaches to better evaluate greenhouse gas emissions from whole wastewater treatment plants," *Biotechnol. Bioeng.*, **109**, no. 11, pp. 2854–2863 (2012).
24. M. NAKAJIMA, "APO e-Learning Course on Green Productivity and Material Flow Cost Accounting (MFCA) Module," (2012).
25. A. W. Hiroshi Tachikawa, "(NPO)Module1_Day1_Overview_MFCA _Mr.Tachikawa_ForDistribution (1).pdf," *Propharm Japan Co., Ltd.* [Online]. Available: https://www.dropbox.com/home/FYP/Concept of MFCA?preview=(NPO)Module1_Day1_Overview_MFCA_Mr.Tachikawa_ForDistribution+(1).pdf. (Accessed: 04-May-2017).
26. Y. Furukawa, "APO e-Learning Course on Green Productivity and Material Flow Cost Accounting," no. C (2012).

27. Y. K. Wan, R. T. L. Ng, D. K. S. Ng, and R. R. Tan, "Material flow cost accounting (MFCA)-based approach for prioritisation of waste recovery," *J. Clean. Prod.*, **107**, pp. 602–614 (2015).
28. Yoke Kin Wan, "Syhthesise A Sustainable Sago Industry," (2008).
29. E. Q. A. EQA, "Sewage Characteristics and Effluent Discharge Requirements, Volume4, Section 3.," *Malaysian Sewerage Ind. Guidel.*, **4** (1974).
30. US EPA, "Combined Sewer Overflow Technology Fact Sheet," *Environ. Prot.* (1999).
31. United States, Environmental Protection, and Agency, "Wastewater Technology Fact Sheet," (1999).
32. P. Molahalli, "Chemical Pre-Precipitation of Municipal Wastewater Treatment," pp. 52, (2011).
33. A. O. Tanyi, "Comparison of Chemical and Biological Phosphorus Removal in Wastewater - a Modelling Approach," pp. 72–78 (2013).
34. I. Hamawand and N. Centre, "Review of wastewater treatment chemicals and organic chemistry alternatives for abattoir effluent," pp. 1–67 (2015).
35. F. A. Brinson *et al.*, "Evaluating Nanofiltration , Reverse Osmosis , And Ion Exchange To Meet Consumptive Use Constraints And Finished Water Quality Goals For Broward County Scope of Analysis," (2002).
36. M. M. Socías-viciana, "Coagulation-flocculation of domestic wastewater by moroccan clays : Removal of organic matter and suspended solids Moroccan Clays : Removal Of Organic Matter," (April 2016, 2014).
37. A. G. Kapagiannidis, I. Zafiriadis, and A. Aivasidis, "Upgrading the efficiency of an external nitrification BNR system – The modified Dephanox process," *Chem. Eng. J.*, **175**, pp. 124–135 (2011).
38. K. Kim, "The characteristic of the sequencing batch reactor (SBR), anaerobic sequencing batch reactor (ASBR) and sequencing batch biofilm reactor (SBBR) Koobum Kim," *Most* (2011).
39. A. Aygun, B. Nas, and A. Berktay, "Influence of High Organic Loading Rates on COD Removal and Sludge Production in Moving Bed Biofilm Reactor," *Environ. Eng. Sci.*, **25**, no. 9, pp. 1311–1316 (2008).
40. T. M. Huggins, A. Haeger, J. C. Biffinger, and Z. J. Ren, "Granular biochar compared with activated carbon for wastewater treatment and resource recovery," *Water Res.*, **94**, pp. 225–232 (2016).
41. M. L. Gulhane and A. U. Charpe, "Domestic Wastewater Treatment using Multimedia Filter Technology," **3**, no. 1, pp. 1119–1121 (2015).
42. F. Jiang, M. Beck, R. Cummings, and K. Rowles, "Estimation of costs of phosphorus removal in wastewater treatment facilities: construction de novo," *Water Policy Work.*, no. June, pp. 1–28 (2004).
43. A. Wendland and Y. Ozoguz, "Operation Costs of Wastewater Treatment Plants," (2013).

Preliminary Study on Effect of Chemical Composition Alteration on Elastic Recovery and Stress Recovery of Nitrile Gloves

Kai Yang Tan[1], *Siew Wei* Phang[1]*, *Chee Kin* Phang[2], *Jing Lang* Choh[2], *Ching Pang* Goh[2], *Chong Ban* Wong[2], *Tiam Ting* Tee[2,3]

[1] School of Engineering, Taylor's University, Subang Jaya, Selangor, Malaysia
[2] Top Glove R&D Centre, GMP Medicare Sdn Bhd, Factory 25, 41050 Klang, Malaysia
[3] Faculty of Lee Kong Chain Engineering & Science, UTAR, Malaysia

Abstract. Nitrile gloves are widely used in the medical and automobile field due to its superiority in hypo-allergic component and chemical resistance over natural latex gloves. However, poor elastic recovery of nitrile glove to compressive force also creates an aesthetic issue for customers with high levels of wrinkling after removing from glove box. This paper demonstrates the preliminary study on the varies chemical composition such as crosslinking agents, sulphur and zinc oxide, the accelerator agent added during curing process, and the rubber filler Titanium Dioxide, on the elastic recovery and stress relaxation in nitrile gloves manufacturing. These chemical were studied at different concentration level comparing the high and low level versus the normal production range. Due to the inconsistency in the analysis technique on the surface imaging, the elastic recovery result was unable to be quantified and was not conclusive at this point. The cross linking agents, sulphur and zinc oxide, and the accelerator agent, played a significant role in the mechanical strength of the gloves. Increment of these chemicals result in higher tensile strength, but a reduction in the elasticity of the materials in which causes a lesser elongation at break percentage for the gloves. Both cross-linkers demonstrate different behaviour where higher sulphur content, provide higher stress relaxation (SR%) yet zinc oxide shows otherwise.

1. Introduction

Nitrile Butadiene Rubber (NBR) gloves, as commonly known as nitrile gloves, generally manufactured from a synthetic polymer that exhibits rubberlike characteristics such as elasticity and tensile strength when vulcanized [1]. With the increased reports of allergic reactions to natural latex products, NBR gloves are now the highly recommended alternative glove material in the health care industry [2,3]. NBR has excellent oil resistance. However, shows no self-reinforcing effect, as there is no crystallinity, but when used in combination with reinforcing fillers, vulcanises with excellent mechanical properties can be obtained from NBR. Vulcanisation occurs by a chemical agent, such as sulphur or zinc oxide [4]. Studies have shown that both vulcanized nitrile and natural latex gloves are

[1] Corresponding author: eunicepsw@gmail.com

comparable or some shown better physical properties in terms of susceptible to material breakdown, chemical resistance and mechanical behaviour [5,6]. This makes nitrile examination glove to mainstay option for medical use.

Physical properties between NBR and NR latex have contrasting difference due to their difference in molecular constituents. Natural latex (NR) rubber generally have superior elasticity, resilience and softness compared to NBR, which in nature has higher stiffness, poor elasticity, poor resilience but better tensile strength [5,6]. The unique characteristics of NBR is linked to the presence of Acrylonitrile (ACN) monomer in its structure, increase in ACN content increases polarity of NBR molecule and glass transition temperature (Tg) properties. Although NBR gloves possess better mechanical properties compared to NR gloves, the stiffer and less elastic nature of NBR gloves do bring about several issues. Poor elastic recovery from compressive stress in NBR gloves results in forming surface wrinkles when subjected to compressive force when folded. Besides, the low stress relaxation property of NBR gloves makes it less comfortable for the user.

Sulphur (S) is the most common vulcanizing agent used in the rubber industry, where it provides sulphur cross-links between long chain polymer molecules which would determine the produced glove properties. Zinc Oxide (ZnO) act as a vulcanizing activator where it would enhance cross-linking between polymer chains and form ionic cross linkage between polymer molecules. Addition of ZnO enhances vulcanizing efficiency and acts as an activator for sulphur crosslinking. Accelerators are also added into NBR latex as vulcanizing catalyst to greatly increase speed of vulcanisation with process taking a much shorter time while reducing the use of sulphur in the system. Composition of these three chemicals involved in rubber curing would hold a crosslink density of produced rubber and subsequently the mechanical properties of produced glove [7]. Titanium Dioxide (TiO_2) acts as a rubber filler which reinforces rubber structure for better mechanical properties. Preliminary study will be conducted to test significance of composition change of each chemical towards stress relaxation and elastic recovery of produced latex film. Three of the four chemicals with most significant effect will be studied for optimization to obtain resultant latex with best stress relaxation and elastic recovery properties.

This project is a collaboration study between Top Glove Sdn Bhd and Taylor's University to address the two stated issues regarding their nitrile glove product line. Customer feedback on the high degree of wrinkling when removed from glove box, which is aesthetically unpleasing as well as causing inconvenience for customers to "open" the gloves up from wrinkled state. The poor stress relaxation of nitrile gloves makes it less comfortable when donned for long periods as nitrile glove would not conform to the user's hand. Therefore, a glove that has higher stress relaxation properties would experience deformation (loss of stress) over time and conform to shape of user's hand giving a better comfort when worn for long durations. This research studies the effect of the vulcanisation chemical agents such as Sulphur, zinc oxide, vulcanising catalyst (accelerator) and titanium dioxide and their composition in the intention to improve the stress relaxation and elastic recovery properties of the product.

2. Methodology

2.1 Materials and sample preparation

2.1.1 Materials

All compounding chemicals including coagulant mixture, NBR mixture, accelerators (Brand name and chemical mixture non-disclosed due to production confidentiality), industrial grade vulcanizing agent such as sulphur, zinc oxide, and titanium dioxide were provided by production line. Compounding of the chemicals was done according to the composition shown at section 2.1.2 at the laboratory for 24 hours prior to glove samples dipping.

2.1.2 Preparation of latex solution and formulations

Based on the four target chemicals discussed, composition of chemicals was adjusted to extreme highs and lows to test significance of each chemical towards targeted response (stress relaxation and elastic recovery). Table 2.1 shows the composition of chemicals tested at three levels. Medium level composition represents normal phr values used in nitrile glove production. With this variation, a total of nine batches of latex samples were prepared as shown in Table 2.2.

Table 2.1 Phr values used for each chemical for significance test

	Normal level (phr)	Low level (phr)	High level (phr)
Accelerator	0.5	0.1	3.0
Sulphur	1.5	0.2	3.0
Zinc Oxide	1.5	0.1	5.0
Titanium Dioxide	1.6	0.2	3.0

Table 2.2 Phr values for all target chemicals for each batch of variation in single factor preliminary study.

No.	Sample name	Accelerator (phr)	Sulphur (phr)	ZnO (phr)	TiO_2 (phr)
1	**Normal**	0.5	1.5	1.5	1.6
2	**High ZnO**	0.5	1.5	**5.0**	1.6
3	**Low ZnO**	0.5	1.5	**0.1**	1.6
4	**High Sulphur**	0.5	**3.0**	1.5	1.6
5	**Low Sulphur**	0.5	**0.2**	1.5	1.6
6	**High accelerator**	**3.0**	1.5	1.5	1.6
7	**Low accelerator**	**0.1**	1.5	1.5	1.6
8	**High TiO_2**	0.5	1.5	1.5	**3.0**
9	**Low TiO_2**	0.5	1.5	1.5	**0.2**

One kg NBR latex solutions were mixed with the chemicals as shown in Table 2.2. The solution was constantly stirred at 600 RPM for 24 hours at room temperature to ensure homogenization of the solution, in order to avoid settlement of high density solid particles. Calculation of weight of each chemical requires individual chemical's parts per hundred (phr) information and Total solid content (TSC%). The information can be subbed into (Eq.

1) shown below to calculate chemical weight (g) of each individual chemical. Latex weight was calculated based on 1000g of final solution.

$$\text{Chemical weight (g)} = \frac{\text{chemical phr} \times \text{Latex TSC (\%)} \times \text{Latex weight (g)}}{\text{Chemical TSC}} \quad (1)$$

2.1.3 Plate-type glove sample preparation

Coagulant solution was obtained directly from the production and was maintained at 55°C on a hotplate. Lab scale plate-type former was used to prepare the samples. The former was heated up to 60°C ± 2°C, followed by a dipping at the coagulant solution for about 2 minutes and dried at the 120°C for another 2 minutes. The formers were air-cooled to the range of 60 – 65°C and were dipped in the latex solution prepared from section 2.1.2 for 15 seconds followed by a 15 minutes drying at 120°C.

2.2 Elastic recovery testing

At this moment, there is no developed industrial standardized method of measurement of surface wrinkles available. A preliminary testing was carried out via surface imaging with a simple MATLAB® (version R2016b) coding on elastic recovery for thin rubber films. A simple test method was carried out via folding the latex samples (uniformly and random crumping as shown in Fig. 1) followed by adding a 2.5 kg-weight compression on the samples for 18 hours period. The samples were then scanned through a table top scanner and the image were analysed through the MATLAB® coding.

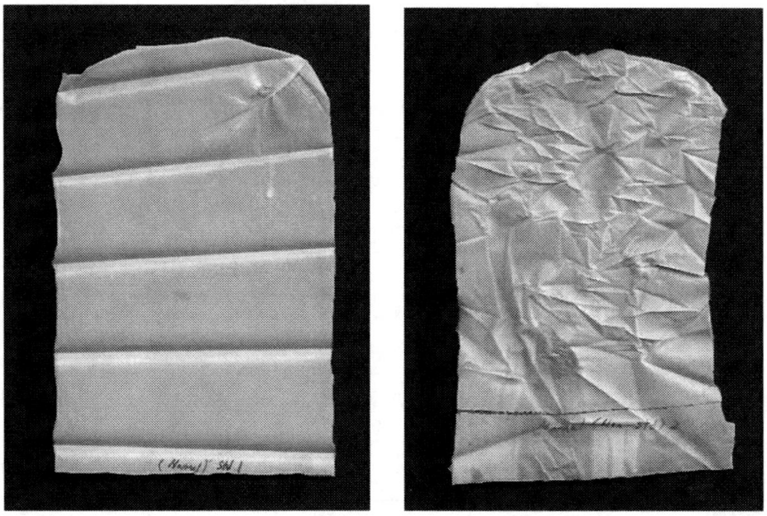

Fig. 1. Standardised folding method (Left) and non-uniform crumpling method (Right)

Surface area selection for wrinkled/folded latex samples were manually adjusted using rectangular box as shown in **Fig. 2.** Surface imaging analysis was done on area measured 13 cm from top; where surface below the 13cm line was excluded from analysis. The samples were scanned the moment it was removed from the weighted-compression, and were scanned again at an interval of 10 minutes for recovery. Several manual adjustments were required to ensure no surface area outside of the sample was included in analysis, to avoid creating additional analytical error. Image processing using MATLAB® results in contrasting of wrinkled surface (white region) and unwrinkled surface (black region) shown in Fig. 2. Wrinkled surface area (white region) over total selected surface was analysed and displayed as wrinkled percentage (%). Calculation of elastic recovery for 10 minute duration is shown in Equation (2)

$$\textit{Elastic Recovery (\%)} = (\% \textit{ of wrinkles}, t=0) - (\% \textit{ of wrinkles}, t=10) \quad (2)$$

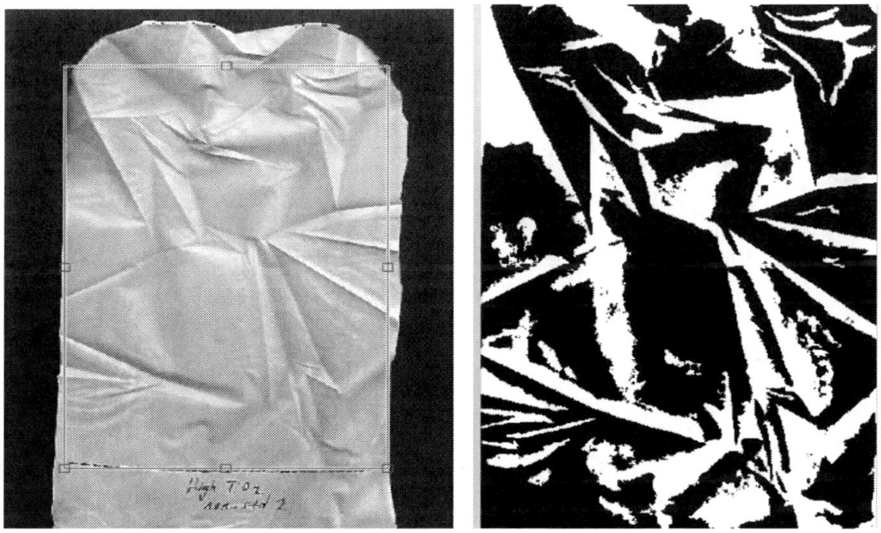

Fig. 2. Selection of analysis area (Blue box) (left) and scanned image (right) for MATLAB®®

2.3 Stress relaxation and mechanical property test

2.3.1 Preparation of tensile test piece

All latex test pieces were cut based on ASTM D412 standard as shown in Fig. 3. Dimensions shown are in units of millimetres (mm).

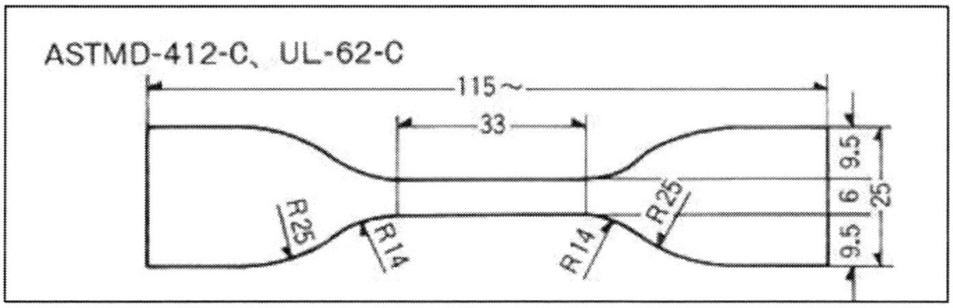

Fig. 3. ASTM D412 Type C die cutter sample

2.3.2 Tensile strength test

Dumbbell shaped test samples were gripped on onto clips of the Universal tensile machine **(Fig.2)**. Standard ASTM D882 tensile test before aging standard was used, with upward tension force of 500mm/min set for all test samples.

Universal tensile machine was used for conducting stress relaxation. Samples were gripped at both ends to clips on the test machine and elongated upwards for a length of 25.414mm. Samples were left in elongated state for 6 minutes. Stress decay of samples throughout 6 minutes with initial stress and final stress values were recorded in test report with calculated Stress recovery (%). Eq. 3 shows the formula of calculation for stress relaxation (%).

$$Stress\ Relaxation\ (\%) = \frac{Fo-Ft}{Fo}\ x\ 100\% \tag{2}$$

where,

Fo = Initial stress value at t=0

Ft = Stres value after certain duration, t

3. Results and Discussion

3.1 Elastic Recovery

Table 3.1 shows the elastic recovery (%) of latex films measured at t = 0 and t = 10 minutes for both standardized and non-standard folding method. Unfortunately, the outcome of this testing were inconclusive due to high experimental errors caused by inconsistency in manual adjustment of the scanned area. Although the calculated recovery % was not able to differentiate elastic recovery in 10 minutes, but the minor differences can be seen through visual judgement based on scanned images as shown in **Fig 4.** It was observed that there are certain amounts of recovery in wrinkling depth between both samples. The scanned image after t = 10-minute of recovery (right) shows a larger surface area compared to sample image at t = 0-minute due to the minor elastic recovery in wrinkle depth, causing latex sample to slightly "expand" in size as it recovers. Although the samples displayed a higher degree of elastic recovery after a long period of time, but that defeat the purpose as end users desired a fast recovery of the product as soon as it is immediately removed from glove box.

Table 3.1 Elastic recovery percent of latex film for different folding method.

	Wrinkle percentage %					
Sample Name	**Standard folding**			**non-standard folding**		
	***t* = 0**	***t* = 10**	**Recovery%**	***t* = 0**	***t* = 10**	**Recovery%**
High Sulphur 1	34.27	26.34	7.9279	34.14	40.38	error
High Sulphur 2	53.98	54.10	error	54.27	53.58	0.6939
High Sulphur 3	30.54	31.51	error	44.30	44.77	error
High Sulphur 4	16.75	22.13	error	54.68	50.44	4.2341
Low Sulphur 1	51.03	57.83	error	45.92	45.14	0.7731
Low Sulphur 2	51.33	52.93	error	46.70	46.89	error
Low Sulphur 3	38.07	32.78	5.2911	47.12	46.16	0.9601
Low Sulphur 4	30.05	25.89	4.1584	39.07	39.13	error

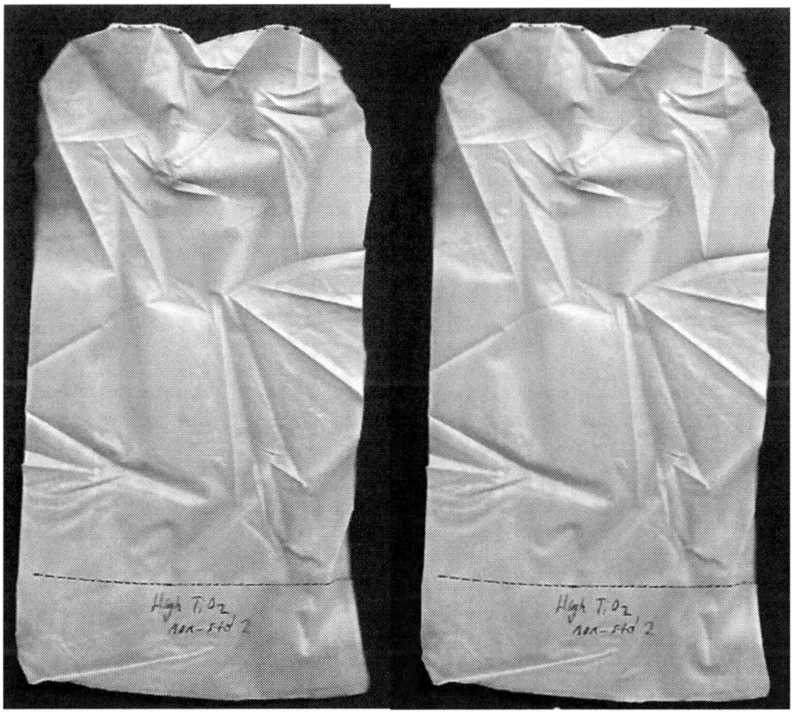

Fig. 4 Scanned images of latex surface at t = 0 minutes (left) and t = 10 minutes (right)

There were some limitations in the MATLAB®® analysis coding, as it was only able to contrast wrinkled (white) and unwrinkled regions (black) as shown in Fig. 2. The calculation was simulated on estimating the total surface area of wrinkled regions (white) over total selected surface area. But based on visual observation shown in Fig. 4, regions of wrinkle lines and regions remains largely the same for samples at 0-minute and 10-minute; yet a change in depth of wrinkles was observed. At this moment, the coding created was not able to distinguish the differences at depths, hence creating insignificant results.

Besides the coding limitation, huge error due to inconsistency on manual adjustment for selected surface area was observed. The repeatability of the scanning surface selection was questionable, that creates most invalid data in Table 3.1. Future improvements in coding to automatically fixed or to select area of scanning and improvement in depth analysis will significantly improve this preliminary scanning test method.

3.2 Significance test for stress relaxation and mechanical strength

As a preliminary study to select the most influencing chemicals for production optimization, the vulcanising agents concentration were tested at an extreme condition as shown in Table 2.1. Table 3.2 provides the result in stress relaxation (%) for each chemical variation. All data shown for stress relaxation and tensile strength were averaged out between data obtained from 4 test samples of the same composition batch. Table 3.3 below shows the tensile strength, elongation at break (%) and elastic modulus at 300% elongation data for the same samples.

Table 3.2 Stress relaxation (%) after 6 minutes of constant strain

	Stress at 0 min, *Fo* (MPa)	Stress after 6 min, *Ft* (MPa)	Stress Relaxation (%)	Difference (%)
High Sulphur	1.861	0.856	42.58	3.40
Low Sulphur	1.696	0.722	45.98	
High TiO_2	1.664	0.734	44.10	0.47
Low TiO_2	1.405	0.626	44.57	
High Accelerator	1.708	0.789	41.38	4.81
Low Accelerator	1.553	0.643	46.19	
High Zinc Oxide	0.921	0.338	36.76	13.03
Low Zinc Oxide	0.500	0.250	49.79	

Table 3.3 Tensile strength, Elongation at break(%) and Elastic modulus result for test samples

	Tensile Strength (MPa)	Difference between High and Low formulation (MPa)	Elongation at break (%)	Difference between High and Low formulation (%)	Modulus at 300% Elongation (MPa)	Difference between High and Low formulation (MPa)
High Sulphur	46.70	23.71	460.8	73.9	12.05	6.49
Low Sulphur	22.99		534.7		5.56	
High TiO_2	21.44	1.45	481.37	26.93	6.64	1.18
Low TiO_2	22.89		508.3		5.46	
High Accelerator	23.62	2.95	473.8	94.37	7.37	2.42
Low Accelerator	20.67		568.17		4.95	
High Zinc Oxide	24.60	8.18	304.37	213.68	22.17	20.08
Low Zinc Oxide	14.19		518.05		2.09	

Titanium dioxide (TiO_2) that acts as reinforcement filler, has not demonstrated a significant effect in stress relaxation (%).The insignificant difference for high and low samples of TiO_2 in stress relaxation (%) can be backed up by mechanical test data shown in Table 3.3, which shows the least difference across the board for tensile strength, elongation at break (%) and elastic modulus at 300% elongation. However, the other three vulcanising agents were observed to significantly affect the mechanical behaviour of the product. Mainly as they were directly involved in rubber curing process, that affect the crosslink density of produced latex [8]. Data shown in the Table 3.3 further proves that a direct influence in crosslink density can affect mechanical strength and overall stiffness of a latex sample.

An additional observation can be made in Table 3.2, stress relaxation (%) high and low concentration of Zinc Oxide (ZnO) shows an exceptionally higher difference compared to other chemicals, with difference of 13.30% in stress relaxation. This may hint that ionic crosslinking in rubber molecules can significantly impact mechanical properties of produced rubber, in which, suggested that ZnO is the most significant factor involved in enhancing stress relaxation in produced latex. By referring to results in Table 3.3, elongation at break (%) High and low Zinc Oxide samples show the highest difference of 213.68% between samples from both batches. The same trend is observed with elastic modulus at 300% elongation. High ZnO samples have given a 10x higher elastic modulus comparing to low ZnO samples. This meant that low ZnO samples were exceptionally soft, with much lesser load (N) needed to stretch samples to 300% elongation. A summary between the relationships of mechanical properties is shown in Table 3.4.

Table 3.4 Relationship between mechanical properties in gloves

	Tensile strength (MPa)	**Elongation at break (%)**	**Elastic Modulus – Rubber Stiffness (MPa)**
Higher Stress Relaxation (%)	Lower	Higher	Lower Stiffness
Lower Stress Relaxation (%)	Higher	Lower	Higher Stiffness

Results from preliminary study of significance in effect of chemicals can conclude two major statements: Titanium dioxide (TiO_2) was the least significant chemical towards stress relaxation (%). It will be excluded for full factorial design and composition optimization. Zinc Oxide (ZnO) is the most significant factor affecting stress relaxation (%) and overall softness of produced latex.

4.0 Conclusion

The preliminary measurement setup for quantifying the glove surface wrinkle has room for improvement, and the current result was inconclusive due to high measurement error on both human and equipment error. The limitation on the analysis coding for detecting surface wrinkle was only able to determine the number of wrinkles / wrinkled regions of selected area but unable to distinguish the depth of surface wrinkles on nitrile latex surface at this moment. Elastic recovery of wrinkled sample had shown recovery in terms of wrinkled depth but little to no recovery from number of wrinkled. Manual selection of analysed surface area also caused huge inconsistency in repeatability of the selected scanned area, resulting in most samples yielding a negative elastic recovery value. Titanium Dioxide (TiO_2) showed least significant changes for SR% and mechanical properties between high-low concentrations and between normal formulation samples. Increment in sulphur and accelerator concentration caused an increment in SR%, possibly due to increased sulphur crosslinking in samples. But increment of Zinc Oxide caused decrease in SR% due to ionic crosslinking between rubber molecules. Tensile strength results show an expected outcome with an increment in mechanical strength with higher concentrations of sulphur, accelerator and zinc oxide due to higher density in crosslinking. But enhanced mechanical strength had a resultant effect of decreased elasticity of produced samples, with elongation at break (%) showing decreased value with higher concentrations. Sulphur, accelerator and Zinc Oxide are chosen for full factorial study for chemical composition optimization for next phase of the study.

References

1. A. Rego and L. Roley, *Am. J. Infect. Control*, **27**, 405–410 (1999).
2. N. Chaiear, S. Sadhra, M. Jones, P. Cullinan, I.S. Foulds, and P.S. Burge, *Occup. Environ. Med.*, **58**, 386–91 (2001).
3. E. Aldlyami, A. Kulkarni, M.R. Reed, S.D. Muller, and P.F. Partington, *J. Arthroplasty*, **25**, 27–30 (2010).
4. K.F. El-Nemr, *Mater. Des.*, **32**, 3361–3369 (2011).
5. M.D. Fisher, V.R. Reddy, F.M. Williams, K.Y. Lin, J.G. Thacker, and R.F. Edlich, *J. Emerg. Med.*, **17**, 1011–8 (1999).
6. J. Sawyer and A. Bennett, *Ann. Occup. Hyg.*, **50**, 289–296 (2005).
7. Z.Q. Cao, D.G. Wang, C.B. Cong, Y.F. Wang, and Q. Zhou, *Tribol. Int.*, **69**, 141–149 (2014).
8. J. Tao, *Soft nitrile medical gloves having improved glove relaxation properties*, US 2002/0193488 A1, (2002).

Process Debottlenecking and Retrofit of Palm Oil Milling Process via Inoperability Input-Output Modelling

May May Tan[1], *Viknesh* Andiappan[2], and *Yoke Kin* Wan[1]*

[1]School of Engineering, Taylor's University, Subang Jaya, Selangor, Malaysia.

[2]School of Engineering and Physical Sciences, Heriot-Watt University Malaysia, No. 1 Jalan Venna P5/2, Precinct 5, 62200 Putrajaya, Wilayah Persekutuan Putrajaya.

Abstract. In recent years, there has been an increase in crude palm oil (CPO) demand, resulting in palm oil mills (POMs) seizing the opportunity to increase CPO production to make more profits. A series of equipment are designed to operate in their optimum capacities in the current existing POMs. Some equipment may be limited by their maximum design capacities when there is a need to increase CPO production, resulting in process bottlenecks. In this research, a framework is developed to provide stepwise procedures on identifying bottlenecks and retrofitting a POM process to cater for the increase in production capacity. This framework adapts an algebraic approach known as Inoperability Input-Output Modelling (IIM). To illustrate the application of the framework, an industrial POM case study was solved using LINGO software in this work, by maximising its production capacity. Benefit-to-Cost Ratio (BCR) analysis was also performed to assess the economic feasibility. As results, the Screw Press was identified as the bottleneck. The retrofitting recommendation was to purchase an additional Screw Press to cater for the new throughput with BCR of 54.57. It was found the POM to be able to achieve the maximum targeted production capacity of 8,139.65 kg/hr of CPO without any bottlenecks.

1 Introduction

Crude palm oil (CPO) is a commodity with large potential and is the most versatile vegetable oil, with the ability to be processed into a range of products with different characteristics [1]. The global demand for palm oil has been increasing at a rapid speed due to the increase in consumption all over the world. The reason for the vast economic growth of palm oil is that it is cheap to produce and has a production rate 10 times that of soy oil, which is its direct competitor [2]. In a POM, CPO is extracted from fresh fruit bunches

1 Corresponding author: YokeKin.Wan@taylors.edu.my

(FFBs), which are harvested from oil palm trees. The process block diagram (PBD) of a typical POM is as depicted in Fig. 1.

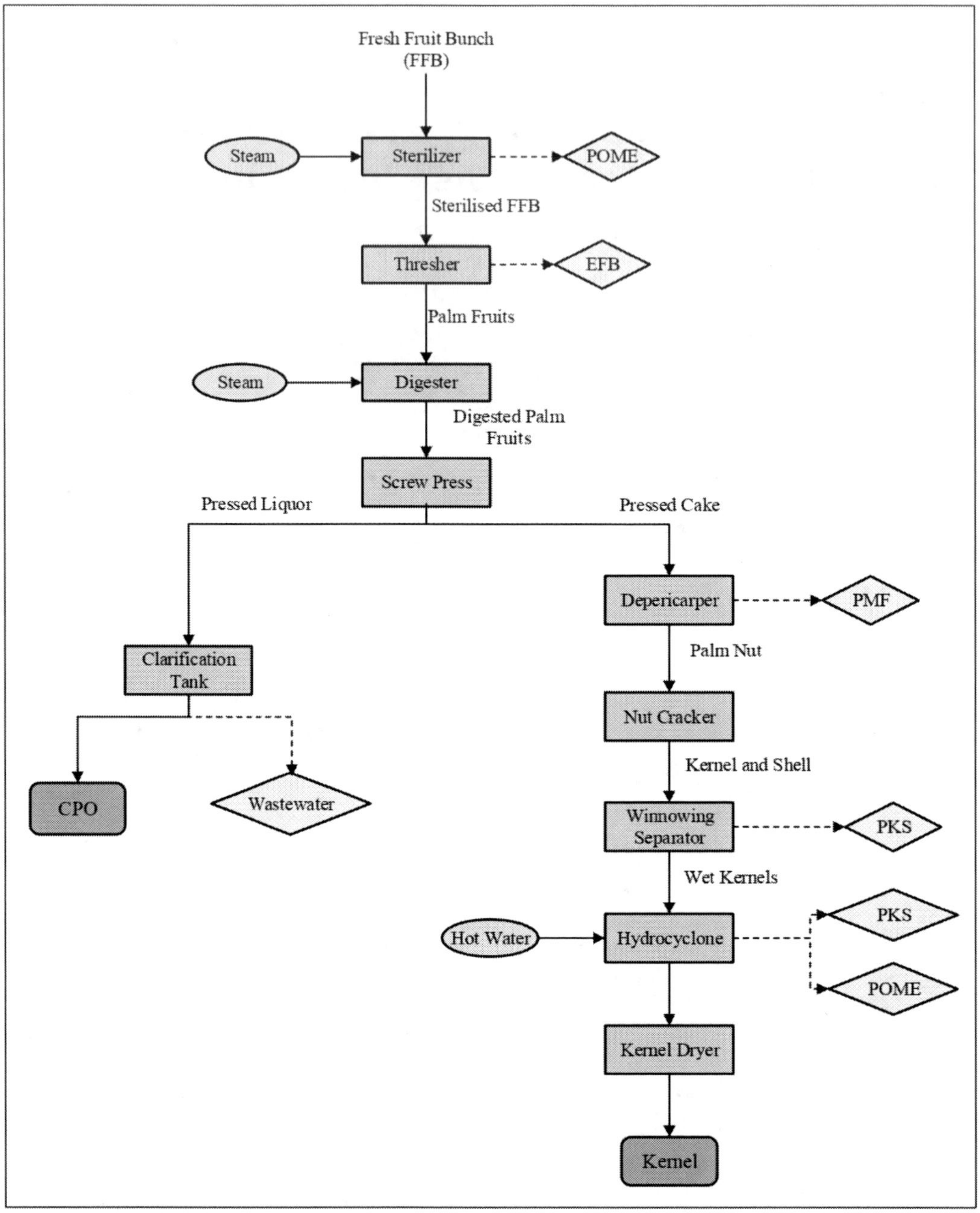

Fig. 1. Block flow diagram (BFD) of a typical POM process [3].

Each equipment in the POM has very specific functions and are designed to operate at their optimum capacities. When there is a need to increase CPO production, some of its equipment may be limited by their fixed design capacities, creating a process bottleneck. A process bottleneck is a situation whereby the available capacity of a process unit is insufficient to meet the increased production rates. The challenge becomes increasingly complex when a POM experiences a drop in equipment efficiency. To solve the mentioned challenges, production capacity of a POM can be increased by retrofitting a plant (e.g. building new production facilities, debottlenecking an existing plant or by designing a plant that meets the production capacity requirement prior to building it) [4]. However, since POM plants are very well-established, it is usually more economical to retrofit an existing POM by debottlenecking it [5].

Retrofitting refers to a procedure to implement modifications to an existing process or system that may improve its energy efficiency, decrease its energy demand, increase its capacity and etc [6]. Several research works have been presented in the area of retrofitting, proving the importance of increasing the productivity of a process to fit increase in demand of a product [7, 8]. In literature, retrofitting and debottlenecking approaches have been proposed in areas including using computer-aided simulation tools, mathematical modelling and programming tools, and systematic frameworks [4, 8–12]. Previous works related to process retrofitting and debottlenecking were focused on the industries such as heat exchanger networks, pharmaceutical, biorefinery and bioenergy [4, 9, 11, 13]. However, there is limited work done on a retrofit and debottlenecking framework of a POM to cater for increased CPO production. At present, work has only been focused on palm oil-based biomass utilisation in a biorefinery and biomass tri-generation system from a palm oil mill (POM), but not the operations of the POM itself [4, 13].

Hence, this research presents a debottlenecking framework, which serves as a stepwise guideline and decision-making tool for engineers to identify bottlenecks and debottleneck the process when there is a desire to increase production capacity. To illustrate the application of the framework, an existing POM was solved in this work, by maximising its production capacity. The identification of bottlenecks was carried out by first developing the mathematical model of the process using an algebraic approach, known as Inoperability Input-output Modelling (IIM) adapted from the work of Kasivisvanathan et al. [4]. IIM assumes that each equipment is a 'blackbox', meaning that the operating conditions of the equipment are not considered. The model was then solved using a commercial optimisation tool, known as LINGO v17.0 [14]. The results of the model were analysed to determine the equipment which have become process bottlenecks. Upon identifying the bottlenecks in the process, the framework entails the procedures on retrofitting a POM process to cater for the increased production capacity. Following this, once the POM was free from bottlenecks, benefit-to-cost ratio (BCR) was performed on the debottlenecking methods to analyse the economic feasibility. Based on the economic performance, the decision to approve the debottlenecking solution is then made.

With this framework, process engineers are able to analyse, and make informed decisions on the debottlenecking a POM process in a systematic manner. The challenge of process bottlenecking can surface in any process. Hence, the purpose of this approach is not limited to a POM production process, it can also be applied to any process that may face potential bottlenecks.

2 Research Methodology

2.1 Debottlenecking Framework

The debottlenecking framework methodology in Figure 2 essentially acts as a step-by-step guide and a decision-making tool for engineers to debottleneck a process. Section 2.2 to 2.3 explains the framework steps in detail.

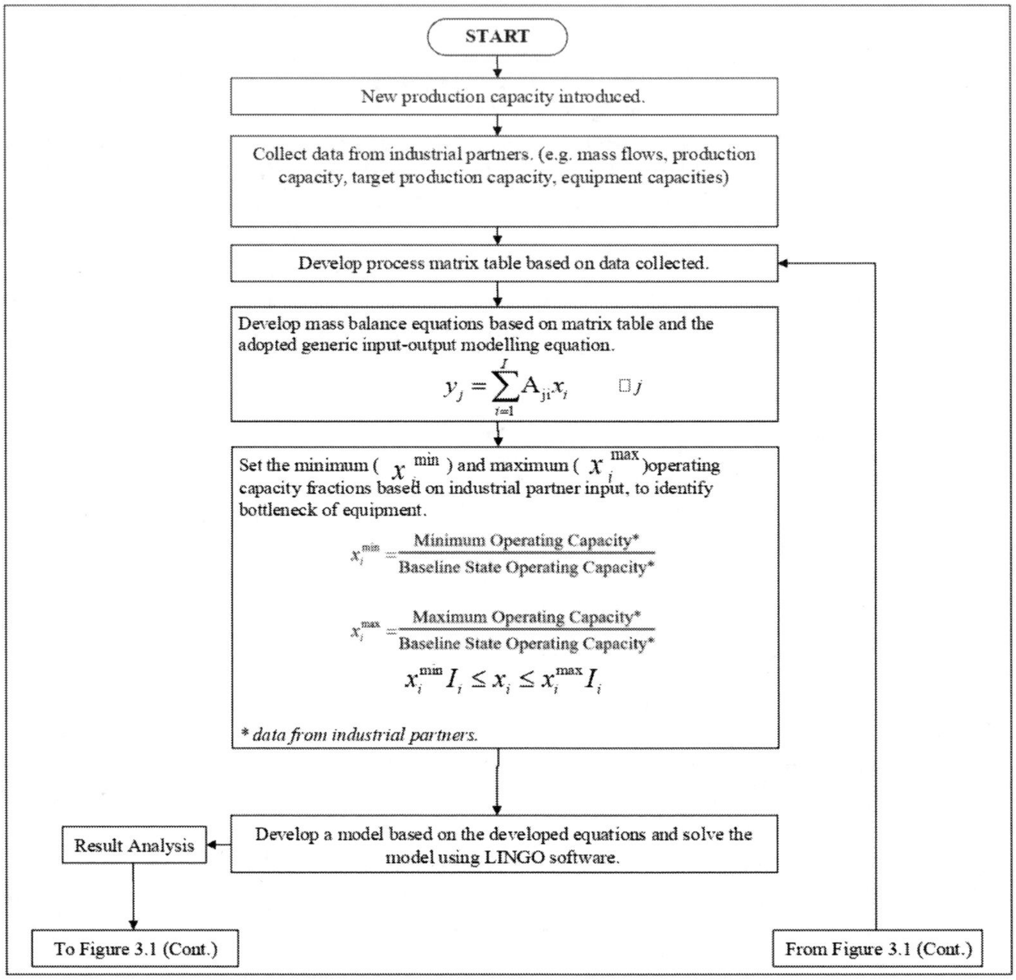

Fig. 2. Debottlenecking framework methodology.

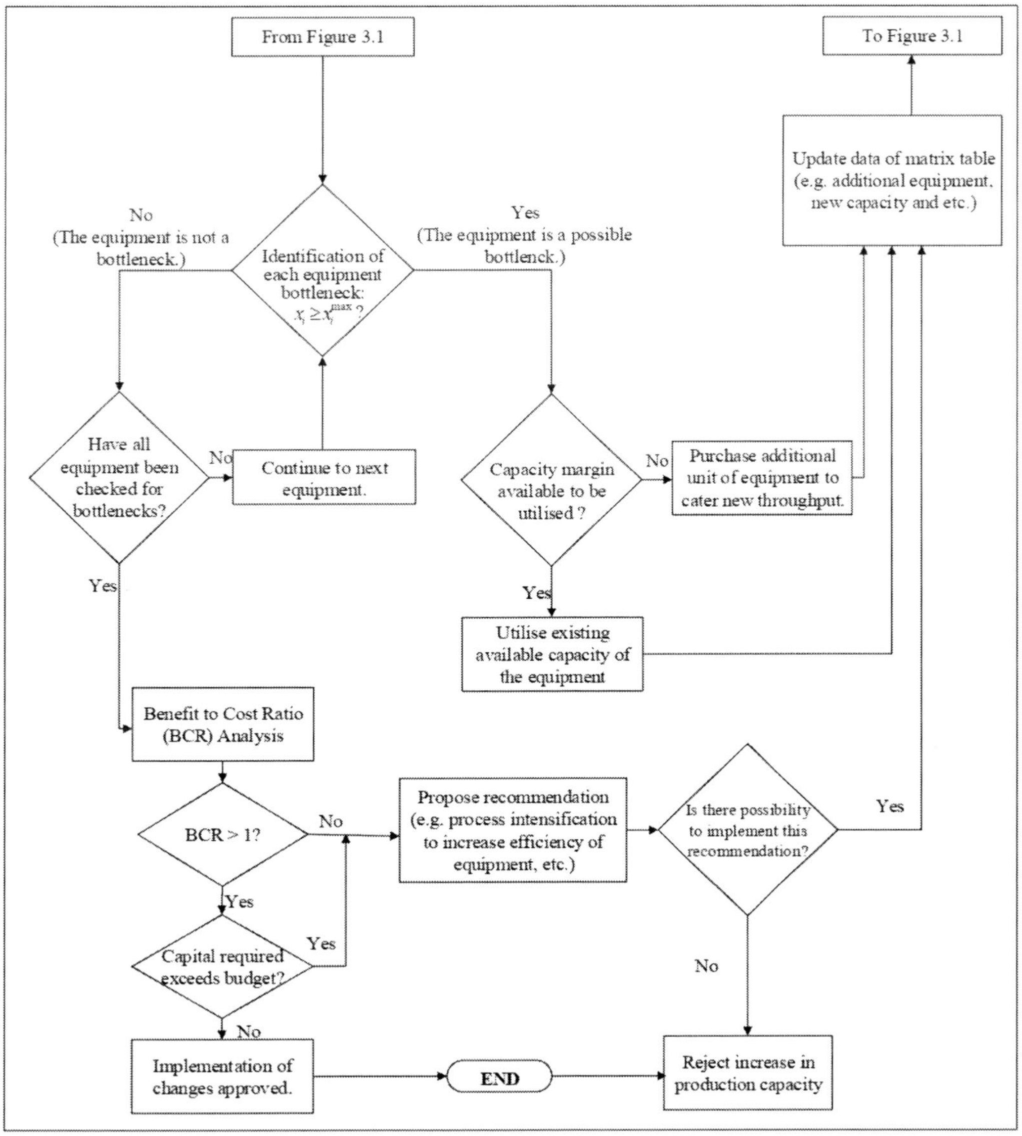

Fig. 3. Debottlenecking framework methodology (Cont.).

As shown in Figure 2, when the need to increase production capacity arises, the first task in the framework is to collect information of the process and equipment from industrial partners for study and analysis as shown in Table 1. For any confidential data that was unavailable, assumptions can be made based on similar, heuristic values or theoretical calculations.

Table 1. Data required for process and specification required for process unit.

Process	Process Unit
Target capacity for increased production	Design capacities
Number of equipment used of each process step	Minimum and maximum operating capacity
Process flow, Mass flow	Capital cost of equipment
Production capacity during normal operation	Utility consumption

Moving down the framework to the next task, a process matrix table needs to be constructed. A general example of a matrix table is shown in Table 2. It shows the mass flows of components, j (j=1,2,3…J), through each process unit, i (i=1,2,3…I), for baseline state operation. In this work, the baseline state operation is referred to the normal existing operating capacity of the plant, before there is an increase in production capacity. Next, mathematical model needs to be formulated based on the matrix table shown in Table 2. The detailed explanation is presented in Section 2.1.

Table 1. General example of a process matrix table.

Equipment, x_i / Component, y_j	x_1	x_2	x_3	…	x_i
y_1	$A_{1,1}$	$A_{1,2}$	$A_{1,3}$	…	$A_{1,j}$
y_2	$A_{2,1}$	$A_{2,2}$	$A_{2,3}$	…	$A_{2,j}$
y_3	$A_{3,1}$	$A_{3,2}$	$A_{3,3}$	…	$A_{3,j}$
…	…	…	…	…	…
y_i	$A_{i,1}$	$A_{i,2}$	$A_{i,3}$	…	$A_{i,j}$

2.2 Formulation of Model

Based on the matrix table in Table 2, the process is then modelled using an algebraic approach known as input-output modelling. A general expression for input-output modelling adapted from the work of Kasivisvanathan et al.[4] is shown in Eq. (1),

$$y_j = \sum_{i=1}^{I} \mathrm{A}_{ji} x_i \qquad \forall j \qquad (1)$$

where y_j is the net flowrate of a given component, which can be an intermediate or a product. It is noted that y_j is denoted by a negative sign for inputs, a positive sign for outputs and is equal to zero for all intermediates. i (i = 1, 2, …, I) represents a process unit index, j (j= 1, 2, ..., J) represents the type of component. A_{ji} showed in the matrix table represents the input and output mass flows to and from a certain process unit i. Meanwhile, x_i is an input-output variable of the equation, known as the operating capacity fraction for a process unit (where baseline state is denoted by 1).

Based on input-output modelling, the net output of the intermediate products in the process are fixed to zero as shown in Eq. (2) below.

$$y_j = 0 \tag{2}$$

This is to ensure that all the intermediate products will be completely converted to produce the final product at the capacity that the plant is designed for [4]. Other than the formulation of the modelling equations, the constrains required to solve the model (i.e. minimum and maximum operating capacity fractions, x_i^{min} and x_i^{max}) are formulated as well, which has been further discussed in Section 2.3.

2.3 Set the Minimum and Maximum Operating Capacity Fractions

The next task in the framework presented in Figure 2 is to set the minimum and maximum operating capacity fractions, x_i^{min} and x_i^{max}, of each process unit. As seen in Eq. (1), the input-output variable; x_i, is the operating capacity fraction of a particular process unit. Each process unit has a minimum and maximum limit of operating capacity fraction whereby it is safe to operate. This is required to limit the model, in order to properly identify bottlenecks. x_i^{min} and x_i^{max} of each process unit has to be determined using Eqs. (3) and (4) based on the information provided by industrial partners.

$$x_i^{min} = \frac{\text{Minimum Operating Capacity*}}{\text{Baseline State Operating Capacity*}} \tag{3}$$

$$x_i^{max} = \frac{\text{Maximum Operating Capacity*}}{\text{Baseline State Operating Capacity*}} \tag{4}$$

*** provided by industrial partners.***

The formulation of the range of operating capacity fraction that is to be modelled in LINGO is as shown in Eq. (5).

$$x_i^{min} I_i \le x_i \le x_i^{max} I_i \qquad \forall\, i \tag{5}$$

where x_i is the operating capacity fraction of process unit i (i = 1,2,3..., I) and I_i is a binary number (I_i = 0 or I_i = 1). For better understanding, during baseline state operation, the operating capacity fraction of an equipment, x_i, is equal to 1. When a new production capacity is introduced, x_i might move above or below 1. The model is able to compute the value of x_i of the process unit. If it is running below the normal capacity, x_i is less than 1, and is not a possible bottleneck; if it is running over the normal capacity, x_i is larger than 1, then it may be at risk of being a bottleneck. If x_i is equal or larger than x_i^{max} , then the equipment is most definitely a bottleneck.

Moving down in the framework (Figure 2), the model could be solved using a commercial optimisation software known as LINGO and data below is extracted from the results:

1. The operating capacity fraction, x_i, of all process units,
2. The production capacity of the plant (i.e. amount of product produced).

Following the solved model, Section 2.3 entails the analysis of the result obtained from the model to identify of bottlenecks in the process.

2.4 Identification of Bottlenecks

After the model is solved, the process is checked for any bottlenecks which may have occurred by analysing the results of the model. Beginning from the 'Result Analysis' box onwards in Figure 2, the analysis will be limited to only one process unit before moving to the next unit. The loop continues until all process units have been checked for bottlenecks. For example, if the equipment is a bottleneck, further analysis is carried out to determine if there is existing capacity margin available to be utilised; this will be discussed in Section 2.5. If the equipment is not a bottleneck, then the next equipment will be checked for bottlenecks. This is done until all equipment have been checked for bottlenecks, then the economic analysis is carried out to assess the feasibility of the debottlenecking method proposed as discussed in Section 2.6.

2.5 Debottlenecking

From Section 2.4, if the particular process unit is a possible bottleneck, the framework will lead to the next diamond box shown in Figure 2, to be checked for additional capacity available. Some process steps may have more than one process unit operating in parallel; meaning that if one unit is a bottleneck, there may be capacity available in another unit, allowing for the utilisation of available existing capacity in the other unit. This can be checked by first calculating the capacity required for the process step and the available capacity for the process step, using Eq. (6) and (7) which is obtained by rearranging Eq. (3) and (4).

$$\text{Capacity Required for Process} = x_i \times \text{Baseline State Capacity} \quad (6)$$

$$\text{Available Capacity} = \text{Capacity of Equipment} \times \text{Number of Equipment} \quad (7)$$

Following this, based on Figure 2, if there is margin available for adjustments, the bottleneck can be removed by utilising the available capacity in the other equipment which is operating in parallel. If there is no margin available, the debottlenecking solution is to purchase additional unit of equipment. Once the debottlenecking method for a particular process unit is decided, the data in the matrix table needs to be updated and the formulations in the model are then updated accordingly. The entire process is then reanalysed for bottlenecks again. For example, if an additional unit of equipment is purchased, the operating capacity fraction, x_i, for that particular equipment is introduced into the matrix table, then the steps described from Section 2.2 onwards are carried out again. This is done until the entire network is free from any other equipment bottlenecks. Once there are no more bottlenecks and all equipment have been checked, the economic feasibility of the debottlenecking solution is assessed in Section 2.6.

2.6 Benefit-to-Cost Ratio (BCR) Analysis

Benefit-to-cost ratio (BCR) analysis of the proposed debottlenecking solutions is performed to assess the feasibility of the modifications. BCR is to determine the ratio of profits gained by implementing the recommendation to the initial capital investment of implementing the recommendation. BCR can be calculated using Eq. (8).

$$BCR = \frac{C_j(\Delta y_j)}{C_{eq}} \tag{8}$$

where j is the valuable product of the process which brings in profits, C_j is the price of component j and Δy_j is the increased flow rates of component j after debottlenecking is performed. C_{eq} is the annualised capital cost and annual operating costs of the equipment purchased to solve the process bottleneck, calculated using Eq. (9).

$$C_{eq} = \sum_{i=1}^{I}\left(C_i x_i^{max} + C_{AOC,i}\right) \tag{9}$$

where i (i = 1,2,3 or…I) refers to the process units that have been added to the process, C_i is the annualised capital cost of the process unit at maximum operating capacity fraction, x_i^{max} and $C_{AOC,i}$ is the annual operating cost of the process unit.

Referring to the framework in Figure 2, the BCR is less than 1, this means that the cost outweighs the benefits. In this case, the process bottlenecks would have to be re-assessed and overcome using different methods (e.g. purchase smaller equipment or process intensification to increase yield of a process) if applicable. On the contrary, when BCR is more than 1, this shows that the benefits of the modification outweigh the costs. In line with BCR, a secondary evaluation is done on the allocated budget. If the investment on the debottlenecking solution falls below the budget, the decision is made to approve the solution. However, if both BCR and/or budget constraints are not satisfied (i.e. BCR < 1 and/or Capital Cost > Allocated Budget), new debottlenecking solutions will have to be proposed and the process will be reanalysed for bottlenecks. For example, a possible recommendation to reduce the initial capital investment is to reduce the maximum operating capacity of the equipment. The plant is then to be re-analysed to determine if these modifications would bring about other bottlenecks. The cycle continues until it reaches the end stage of the framework in Figure 2. In the case where all proposed recommendations have to be rejected due to not achieving the desired results for economic analysis, the entire project will have to be abandoned. This means that the plant is not suited for increase in production capacity or the budget provided is insufficient. To demonstrate the application of the developed framework (Figure 2), the bottleneck identification and debottlenecking of a palm oil mill (POM) case study is presented in Section 3.

3 Palm Oil Mill (POM) Case Study

Figure 3 illustrates a process block diagram (PBD) of an existing and operating palm oil mill (POM) case study that will be used to demonstrate the debottlenecking framework steps described in Section 2. The POM owner is interested in **maximising** the crude palm oil (CPO) production capacity of the existing POM with a budget allocation of US$ 500,000. This increase of CPO production may cause some of the equipment in Figure 3 to reach their maximum capacities and result in process bottlenecks. In this case, the proposed framework serves as a stepwise guideline and decision-making tool to analyse the plant for increased production and to debottleneck accordingly. Based on the framework (Figure 2), the data of the POM were first collected from industrial partners. Some information of the equipment specifications of the POM at baseline state are shown in Table 3. Table 4 is the process matrix table, which shows the capacities of the equipment in the POM at baseline state, also obtained from industrial partners. In Table 4, the negative values are the inputs into an equipment and the positive values are the outputs of an equipment.

Table 2. Current specifications of existing palm oil mill (POM) at baseline state.

Fresh fruit bunches (FFB) Consumed (kg/hr)	30,000
CPO Production Capacity (kg/hr)	7611.42
Steam Consumption (kg/hr)	15435.85
Electricity Consumption (kWh)	419.8

Table 3. Equipment specifications in Palm Oil Mill (POM).

Equipment	Unit	Maximum Capacity	Minimum Capacity	Steam Consumption (kg/kg FFB)	Electricity (kWh/unit)
Steriliser	kg FFB/unit	20000	8,000	0.25	30
Thresher	kg SFB/unit	50,000	20,000	-	28
Steam Injection Digester	kg SFB/unit	12,500	5,000	0.13	18
Screw Press	kg DF/unit	20,000	8,000	-	45
2 -Phase Decanter	kg PL/unit	10,000	4,000	-	40
Depericarper	kg PC/unit	5,000	2,000	-	57.3
Nut Cracker	kg PF/unit	8,000	3,200	-	26.4
4-stage Winnowing Column	kg cracked mixture/unit	15,000	3,000	-	20
Hydrocyclone	kg cracked mixture/unit	5,000	2,000	Water: 1.5 kg/kg cracked nut	30
Kernel Dryer	kg WK/unit	5,000 kg wet kernet/unit	1,000	-	37.1

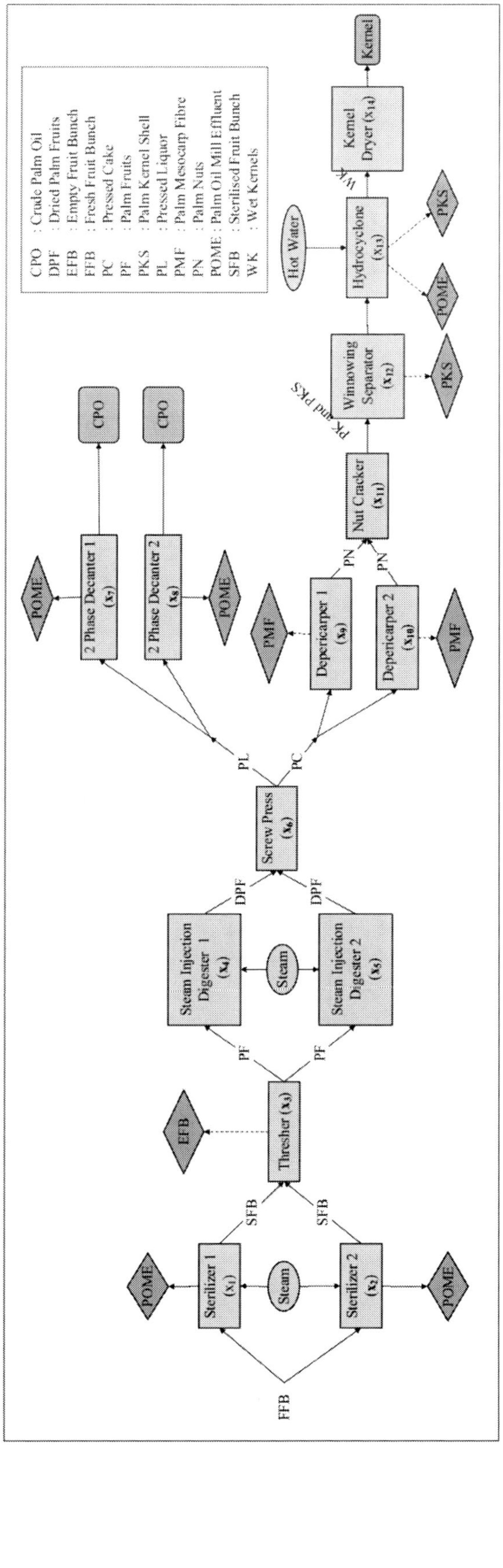

Fig. 4. Process block diagram (PBD) of the palm oil mill (POM) Case Study.

Table 4. Process matrix table at baseline state.

Component, y_j (ton/h)	x_1	x_2	x_3	x_4	x_5	x_6	x_7	x_8	x_9	x_{10}	x_{11}	x_{12}	x_{13}	x_{14}
FFB	-15.0	-15.0												
SFB	12.2	12.2	-24.3											
PF			18.0	-9.0	-9.0									
EFB			6.3											
DPF				9.4	9.4	-18.7								
PL						10.3	-5.1	-5.1						
CPO							3.8	3.8						
PC						8.4			-4.2	-4.2				
PMF									2.0	2.0				
PN									2.4	2.4	-4.9			
CN											4.6	-4.6		
UN											0.2	-0.2		
PKS												1.6	1.6	
PK												3.3	-3.3	
WK													1.5	-1.5
DK														1.4
POME	4.3	4.3					1.3	1.3					4.6	
S	-4.1	-4.1		-1.3	-1.3								-4.6	

Abbreviations

FFB	Fresh Fruit Bunch
SFB	Sterilised Fruit Bunch
PF	Palm Fruit
EFB	Empty Fruit Bunch
DPF	Digested Palm Fruit
PL	Pressed Liquor
CPO	Crude Palm Oil
PC	Pressed Cake
PMF	Palm Mesocarp Fibre
PN	Palm Nut
CN	Cracked Nut
UN	Uncracked Nut
PKS	Palm Kernel Shell
PK	Palm Kernel
WK	Wet Kernel
DK	Dry Kernel
POME	Palm Oil Mill Effluent
S	Steam
E	Electricity

Legend

x_1	Steriliser 1
x_2	Steriliser 2
x_3	Thresher
x_4	Digester 1
x_5	Digester 2
x_6	Screw Press
x_7	2 Phase Decanter 1
x_8	2 Phase Decanter 2
x_9	Depericarper 1
x_{10}	Depericarper 2
x_{11}	Nut Cracker
x_{12}	Winnowing Separator
x_{13}	Hydrocyclone
x_{14}	Kernel Dryer

The next step in the framework is to formulate the input-output model of the POM based on Eq. (1), using the mass flow values in Table 4. Based on Eq. (2), the intermediates of the process are set to zero. Next, the minimum and maximum operating capacity fractions were calculated using based Eq. (3) and (4), and the information in Table 3. The constrains were formulated using Eq. (5). After the completion of these steps, the model is ready to be solved using LINGO v17.0. The results of the model are presented and discussed Section 3.1.

3.1 Results and Discussion

In this case study, the crude palm oil (CPO) production capacity was maximised and the palm oil mill (POM) was analysed for bottlenecks. The POM model is solved for the objective function to maximise the production capacity of CPO shown in Eq. (10).

$$\max CPO \tag{10}$$

Upon modelling and solving the POM case study, the results showed that the existing POM was able to achieve a maximum CPO production capacity of 8139.65 kg/hr and the results are tabulated in the fifth column of Table 6. Based on the framework, the operating capacity fractions of each equipment upon introducing the increased capacity (in the fifth column of Table 4) is to be compared with the existing maximum operating capacity fraction of the equipment (in the fourth column of Table 4).

Table 5. Identification of possible bottlenecks.

Equipment	Input-Output Variable, x_i	Minimum Operating Capacity, X_i^{min}	Maximum Operating Capacity, X_i^{max}	Operating Capacity Fraction to Maximise CPO, x_i	Possible bottleneck ?
Steriliser 1	x_1	0.5333	1.3333	1.3330	**YES**
Steriliser 2	x_2	0.5333	1.3333	0.8058	NO
Thresher	x_3	0.8230	2.0576	1.0694	NO
Digester 1	x_4	0.5561	1.3903	0.7485	NO
Digester 2	x_5	0.5561	1.3903	1.3903	**YES**
Screw Press	x_6	0.4278	1.0694	1.0694	**YES**
Decanter 1	x_7	0.7778	1.9444	0.7778	NO
Decanter 2	x_8	0.7778	1.9444	1.3610	NO
Depericarper 1	x_9	0.4753	1.1883	0.9505	NO
Depericarper 2	x_{10}	0.4753	1.1883	1.1883	**YES**
Nut Cracker	x_{11}	0.6556	1.6390	1.0269	NO
Winnowing Separator	x_{12}	0.6146	3.0731	1.0269	NO
Hydrocyclonne	x_{13}	0.6146	1.5369	1.0269	NO
Kernel Dryer	x_{14}	0.6884	3.4422	1.0269	NO

The comparisons suggested that the possible equipment bottlenecks were the Steriliser 1, Digester 2, Screw Press and Depericarper 2. These are the processes that may be at risk of becoming bottlenecks, should there be a further increase in capacity. This is probably because the Digester and any process relating to palm nut (PN) processing (i.e. Depericarper) had smaller operating capacities compared to other equipment (i.e. Capacity of the Digester is 12,500 kg palm fruit/hr for one unit and capacity of Depericarper is 5000 kg pressed cake/hr for one unit).

Moving down the next diamond box in the framework, the possible bottlenecks identified were then checked if there was available existing margin available to be utilised. This means that the particular process step may have more than one of the same equipment running in parallel, which has excess capacity that can be utilised (i.e. from Figure 3, the Steriliser, Digester, Depericarper and 2-phase Decanter). The calculations were made using Eq. (6) and (7). The results tabulated in Table 7 suggested that the all the possible bottlenecks had available existing capacity to be utilised, except for the Screw Press, which was approaching its maximum available capacity. The conclusion made from this step was that the Steriliser, Digester and Depericarper had available capacity margin to be utilised, because hese processes have two process units which operate in parallel. Hence, when one is at its maximum capacity, the other still had excess capacity to be utilised. However, the Screw Pressing process had only one equipment in the existing POM, leaving no capacity available to be utilised. Therefore, the bottleneck identified and validated for the case study was the Screw Press.

Table 6. Checking for available existing capacity.

Possible Bottlenecks	Capacity Required (kg/hr)	Available Capacity (kg/hr)	Available Existing Capacity Margin to be Utilised?
Steriliser 1	32,081.99	40,000.00	YES
Digester 2	19,229.95	25,000.00	YES
Screw Press	19,999.15	20,000.00	**Approaching Max**
Depericarper 2	8,999.62	10,000.00	YES

Based on the framework (Figure 2), the subsequent step was to debottleneck the Screw Press. In this research, all equipment was assumed to be a 'blackbox', meaning that operating conditions and efficiencies are not taken into consideration and are assumed to be at optimum. However, in reality this may not be the case. Each equipment has its own range of operating conditions at which it has the best performance. For example, a Screw Press may have an optimum operating capacity fraction, x_6, between 0.8 to 0.9, where it may be able to have a higher efficiency. However, since this research is conducted with the assumption of a 'black box', this optimum range is not taken into consideration. Hence, based on the framework (Figure 2), the debottlenecking method selected was by purchasing an additional Screw Press. to cater for the new capacity. Following the decision of the debottlenecking method, the next step is to update the process matrix table, equipment constraints and mathematical model accordingly based on the modifications made. The modifications made (i.e. purchase an additional Screw Press) will allow the POM to have a new maximum CPO capacity. However, since a comparison would like to be made between existing POM and debottlenecked POM, the constraint in Eq. (11) below, has to be defined in the model.

$$CPO = 8139.65 \quad (11)$$

This was done because based on the previous LINGO results mentioned, the maximum CPO capacity of the existing plant was found to be 8139.64 kg/hr. Hence, by introducing Eq. (11) to the model, a comparison can be made to check if the bottlenecks that occurred from the said capacity have been removed. The results are tabulated in Tables 8 to 11. Table 8 shows that Screw Press 1 is at its maximum capacity and is a possible bottleneck. However, the Screw Press 2 that was just added is at zero capacity. Hence, there is existing margin available to be utilised for the Screw Press process, as shown in Table 9. With the additional Screw Press 2, the Screw Pressing process is now able to handle a larger throughput, allowing the plant to achieve the maximised CPO capacity without any bottlenecks.

Table 7. Identification of bottlenecks after debottlenecking.

Equipment	Input-Output Variable, x_i	Minimum Operating Capacity, x_i^{min}	Maximum Operating Capacity, x_i^{max}	Operating Capacity Fraction to Maximise CPO, x_i	Possible bottleneck ?
Screw Press 1	x_6	0.8556	2.1389	2.1388	**YES**
Screw Press 2	x_{6a}	0.8556	2.1389	0.0000	NO

Table 8. Check for capacity margin available for adjustment after debottlenecking.

Bottlenecks	Capacity Required (kg/hr)	Available Capacity (kg/hr)	Available margin for adjustment?
Screw Press	19999.15	40000.00	YES

According to Table 10, upon debottlenecking the existing POM for maximised CPO production capacity, it is now able to produce 8139.65 kg CPO/hr, by consuming 32,082.00 kg FFB/hr, 16,507.08 kg steam/hr and 522.1 kWh of electricity. By comparing the CPO production capacity to the baseline state shown in Table 3 (i.e. 7611.42 kg CPO/hr), the CPO capacity of the debottlenecked plant is has increased by 7%. With the increase in production, the amount of raw material (i.e. fresh fruit bunch (FFB) and steam) consumed has also increased by 7%. A trend is observed in the interactions of the mass flows, whereby an increase in production of CPO leads to the same percentage of increase in consumptions of raw material and utilities required in the process. This is because the process units in the POM are interconnected by a linear programming model. This high interdependency among individual process units results in disruption of the entire system when there are any changes made (i.e. increase production capacity of CPO). The electricity consumption was found to be the same because the additional Screw Press added into the process was not being utilised, as shown in Table 8, where the operating capacity fraction was equal to zero.

Table 9. Production capacity and process requirements after debottlenecking.

New CPO Production Capacity (kg/hr)	8139.65
New FFB Capacity (kg/hr)	32,082.00
Steam Consumption (kg/hr)	16507.08

Electricity Consumption (kWh)	419.8

Based on the framework (Figure 2), after the entire process is free from bottlenecks, the next step is to perform benefit-to-cost ratio analysis (BCR). The BCR was calculated using Eq. (8) and Eq. (9). The economic parameters used, information on the capital cost and annual operating cost of the equipment, along with the calculated values for BCR can be found in Table 9. From Table 9, by purchasing an additional Screw Press, the BCR obtained for the case study was 54.57, which is a lot more than 1. This shows that the additional income gained from increased CPO production far outweighs the cost required to purchase an additional Screw Press. This also shows that one process bottleneck could lead to a big loss in potential revenue that could be earned. Other than that, the BCR could be very high because there was only one equipment bottleneck. After debottlenecking, the POM now has a production capacity 7% higher than the baseline. If for example, the POM owners would like to increase capacity by 50%, a lot more equipment would become bottlenecks and required to be debottlenecked. Different equipment have different costs, some have higher costs than others. These differences would reflect in the BCR results. Since in this case study, there is only one bottleneck involved, the BCR yields a value much larger than 1.

The following step in the framework was to verify if the cost to carry out the project exceeds the allocated budget of the project. In line with BCR, the initial capital investment of the Screw Press (i.e. US$ 100,000) is lesser than the budget allocated of US$500,000. Therefore, the debottlenecking solution can be justified economically and can be approved for construction.

Table 10. Benefit-to-cost ratio analysis for case study.

Total Operating Hours (hrs/yr)	6000
Additional CPO from Increased Capacity, Δy_{CPO} (kg/hr)	528.23
(kg/yr)	3,169,380.00
Selling Price of CPO, C_{CPO} (US$/kg)	0.6565
$C_{CPO}\Delta y_{CPO}$ (US$/yr)	**2,080,697.97**
Annualising Factor (/yr)	0.1315
Capital Cost of Screw Press (US$)	100,000.00
Annualised Capital Cost of Screw Press, C_{6a} (US$/yr)	13,150.00
Maximum Operating Capacity Fraction of Screw Press, x_{6a}^{max}	2.1389
$C_{6a}\, x_{6a}^{max}$ (US$/yr)	28,126.42
Annual Operating Cost of Screw Press (US$/yr)	10,000
Cost of Additional Equipment, C_{eq} (US$/yr)	**38,126.42**
BCR	**54.57(ACCEPTABLE)**

4 Conclusion

This research work presents a framework on identifying bottlenecks and debottlenecking in a palm oil milling (POM) process in the case where production capacity of crude palm oil (CPO) is required to increase. To identify the bottlenecks, an algebraic approach known as input-output modelling is adapted to first formulate the modelling equations of the process. The model was then solved in an optimisation tool, known as LINGO, to identify the bottlenecks. In addition to identifying the bottlenecks, the framework then proposes debottlenecking methods to free the process from bottlenecks. Benefit-to-cost ratio (BCR) analysis was performed on the debottlenecking methods to analyse its economic feasibility. To illustrate the developed framework, an existing POM was presented, given the objective to maximise CPO production capacity. It was found that the equipment at high risk of becoming possible bottlenecks when capacity increases were the Steriliser, Digester, Screw Press and Depericarper. The equipment which reached its maximum design capacity was identified as the Screw Press. The debottlenecking solution selected was to purchase an additional Screw Press to cater for the increase in capacity. After debottlenecking, the plant was able to produce an additional 528.23 kg CPO/hr, which was a 7% increase from the baseline state. The BCR was found to be 54.57. This validates that the benefits of purchasing the additional Screw Press far outweighs the cost.

References

[1] Green Palm, “Why is Palm Oil Important?,” 2015. [Online]. Available: http://greenpalm.org/about-palm-oil/why-is-palm-oil-important. [Accessed: 24-Oct-2017].

[2] J. Mark, “What’s Fueling the Demand for the Palm Oil Destroying the Rainforests of Indonesia?,” 2013. [Online]. Available: http://www.earthisland.org/journal/index.php/elist/eListRead/whats_fueling_the_demand_for_the_palm_oil_destroying_the_rainforests_of_ind/. [Accessed: 11-Jun-2017].

[3] E. Lau, Y. Man, and A. Baharum, “A Qualitative Approach of Identifying Major Cost Influencing Factors in Palm Oil Mills and the Relations towards Production Cost of Crude Palm Oil,” *Am. J. Appl. Sci.*, **8**, no. 5, pp. 441–446, (2011)

[4] H. Kasivisvanathan, R. R. Tan, D. K. S. Ng, M. K. Abdul Aziz, and D. C. Y. Foo, “Heuristic framework for the debottlenecking of a palm oil-based integrated biorefinery,” *Chem. Eng. Res. Des.*, (2014)

[5] Malaysia Palm Oil Council, “The Oil Palm Tree.” [Online]. Available: http://www.mpoc.org.my/The_Oil_Palm_Tree.aspx. [Accessed: 17-Jun-2017].

[6] T. Pinto-Varela, A. Barbosa-Póvoa, and A. Carvalho, “Sustainable batch process retrofit design under uncertainty—An integrated methodology,” *Comput. Chem. Eng.*, (2016)

[7] O. Alshekhli, D. C. Y. Foo, C. L. Hii, and C. L. Law, “Process simulation and debottlenecking for an industrial cocoa manufacturing process,” *Food Bioprod. Process.*, **89**, pp. 528–536, (2011)

[8] R. R. Tan *et al.*, “An algebraic approach to identifying bottlenecks in linear process models of multifunctional energy systems,” *Theor. Found. Chem. Eng.*, **46**, no. 6, pp. 642–650, 2012.

[9] N. Tahouni, R. Khoshchehreh, and M. H. Panjeshahi, “Debottlenecking of condensate stabilization unit in a gas refinery,” *Energy*, **77**, pp. 742–751, (2014)

[10] V. M. Enríquez-Gutiérrez, M. Jobson, L. M. Ochoa-Estopier, and R. Smith, “Retrofit of heat-integrated crude oil distillation columns,” *Chem. Eng. Res. Des.*, **99**, pp. 185–198, (2015)

[11] H. Sugiyama *et al.*, “Multiobjective Retrofitting Methodology for Manufacturing Processes of Active Pharmaceutical Ingredients Considering Environment, Health and Safety Risks,” in *Computer Aided Chemical Engineering*, **38**, pp. 79–84, (2016)

[12] L. T. Biegler and P. Saderne, "A mixed integer flowsheet optimization strategy for process retrofits - the debottlenecking problem,", **13**, no. 8, pp. 947–957, (1989)
[13] V. Andiappan, D. S. Ng, and R. R. Tan, "Design operability and retrofit analysis (DORA) framework for energy systems," *Energy*, (2017)
[14] Lindo Systems Inc., "LINGO," (2017) [Online]. Available: http://www.lindo.com/index.php/products/lingo-and-optimization-modeling. [Accessed: 22-Nov-2017].

Coagulant Dipping Time and Temperature Optimisation for Latex Glove Uneven Coating Investigation

Loong Than[1], *Siew Wei* Phang[1,*], *Kin Nam* Ho[2]

[1] School of Engineering, Taylor's University, Subang Jaya, Selangor, Malaysia
[2] TG Medical Sdn Bhd, Factory 14, Klang, Selangor, Malaysia.

Abstract. Powdered glove is produced by natural latex with the aid of coagulant. The composition of the coagulant contains wetting agent, calcium nitrate as coagulant and anti-tack agent. Generally wetting agent causes foam and results in uneven wall thickness in glove production. Hence, the objective of this research was to select a suitable wetting agent sample that could be used by the production, as well as optimising the parameter such as temperature of coagulant, weight percentage of wetting agent and coagulant dipping time to reduce the rejection of uneven thickness. This paper presented the preliminary testing including foam height test, cloud point test, wetting agent characterisation and the samples mechanical analysis. Two setting agents (sample T and sample E) were chosen due to its availability in the production line. Throughout the wetting agent characterisation, sample T was preferred compared to sample E due to similar bonding of the wetting agent. The surface tension of coagulant and tensile strength of glove samples meet the glove industry standard at the variation of wetting agent amount. The optimise parameter were at 0.0175 wt% of wetting agent and 5 seconds dipping time at 63 °C. The average thickness of the gloves at this parameter was at 0.09 ± 0.004 mm which is at the acceptable level comparing to 0.1mm thickness of ASTM standard.

1. Introduction

In order to remain competitive in the market, organisations look for ways to improve their production and management processes, including but not limited to reduction in production cost, enhance productivity and improve product quality. In other words, organisations must utilise all the available resources efficiently and effectively to cater their customers with high quality products at a low price. Uneven wall thickness has been an issue in the glove manufacturing industry. It has led to several severe rejections in the production line including glove pin-hole, reduction in mechanical properties, and the thin area defect or normally known as "fish-eye" defect as shown in Fig. 1. The uneven thin area defect is mainly due to the present of uneven adherence of latex on the mould former.

[1] Corresponding author: eunicepsw@gmail.com

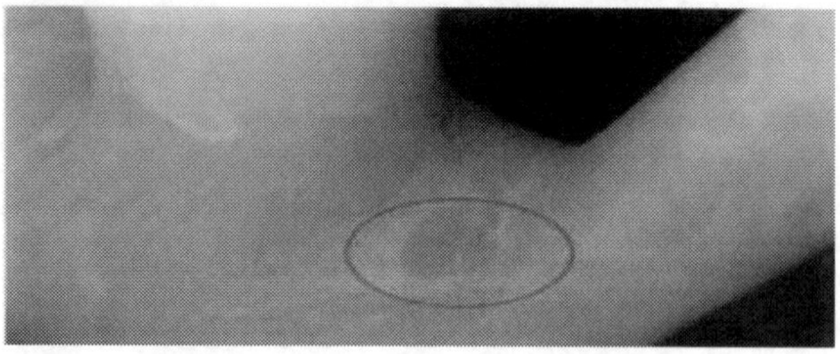

Fig. 1 Fisheye defect located in powdered glove.

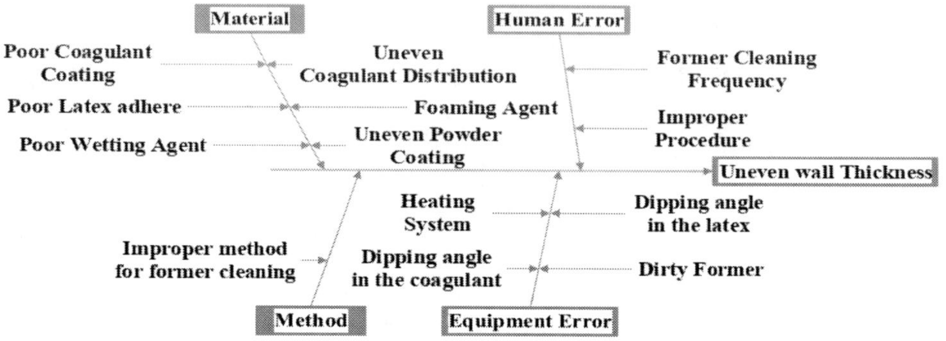

Fig.2 Ishikawa diagram for uneven thickness of glove [1].

Fig. 2 shows the potential causes of the uneven thickness of the gloves, in which includes human error, the materials (mainly chemicals), and / or the improper process method. In the "material" factor, poor coagulant wetting on formers causes the jagged coagulant allocation, that leads to poor latex pick up during dipping in latex process, which resulted in forming a curve shape in the product after inflation and during usage [1]. Besides, insufficient cleaning of the former may also lead to similar defect [2].

Wetting agent is also known as surfactant, where the concentration of the chemical affects the thickness of the glove [3]. The main purpose of the wetting agent is to reduce the surface tension of the coagulant to ensure even coating on the former. However, most wetting agents form layers of foam during the coagulant mixing process due to stirring at high RPM. This foam layer counters the main purpose of the wetting agent on even coating on the former by create areas of air pockets during the coagulant coating process. As a result, create more rejection of uneven thickness in gloves especially the thin area defect as shown in Fig. 1 [4]. This paper presented a preliminary study on the foam creation by different wetting agents, namely sample T and sample E (actual chemical mixture non-disclosed due to company confidentiality), the chemical coagulant dipping time and concentration, prior to process optimisation to reduce the thin area defect.

2. Methodology

2.1 Pre-screen Test – Foam Heights Test

The first screening test was to determine the root cause of the foam creation during the coagulant mixing process. The prepared coagulant was stirred at 600 RPM for at least 24 hours to ensure the homogeneity of the chemicals dispersion as well as avoid anti-tack agent, calcium carbonate's sedimentation. Small amount of wetting agent has known to be more effective in reduction of the coagulant's surface tension [4,5].

Coagulants samples were collected from the production line and different amount of the wetting agent, 0.005 wt%, 0.03 wt% and 0.05 wt%, were added to the samples. One sample without wetting agent was taken as a control sample. The mixtures were then place in the mechanical stability tester, which stirred at 14000 rpm, and the foam heights were recorded.

2.2 Pre-screen Test – Cloud Point Test

Cloud point test was to investigate the temperature when the coagulant turned cloudy and lose its functionality as a coating agent. In nature, additional of anti-tack powder in the coagulant causes the mixture to be cloudy, hence, this test were conducted without additional of the anti-tack powder, to ensure the solutions were remain transparent at the beginning of the testing. Same amount of wetting agent were added to the coagulant mixture collected from the production line, with one without wetting agent as controlled sample. The samples were heated on the hot plate and the temperature of the solutions was recorded till the solution turned cloudy.

2.3 Fourier Transform Infrared Spectroscopy (FTIR)

Two wetting agent samples, sample E and sample T were tested using FTIR to studied its functional group. Since both of this sample are alcohol ethoxylate that consist of general chemical structure of R-$(CH_2H_2O)_n$-H; where R was referring to an alkyl group, which is hydrophobic in nature. While -$(CH_2H_2O)_n$- was referring to the ethylene oxide, a hydrophilic functional group. The higher number of ethylene oxide group present in the sample, it has a better efficiency in reducing the surface tension in the solution, but at the same time, creating more foams.

2.4 Full Factorial Design

Full factorial design was use a design of experiment tool to optimise the temperature of coagulant (X_1), weight percentage of wetting agent (X_2) and mould dipping time in the coagulant (X_3) to minimise the uneven thickness of the glove. A 3^3 full factorial, which lead to a total of 27 formulations were designed.

2.5 Glove Sample Preparation

By refer to section 2.4, glove samples were prepared in lab dipping scale, which consist of four dipping tank (volume for 7kg each). The first tank was filled with diluted acid; second tank was filled with coagulant with the designed wetting agent amount; 3rd tank of pure latex, and 4th tank was filled with corn-starch. Lab scale hand former was used to prepare the samples. The former was dipped in the acid tank to clean up the residue and rince with water. Later, it was heated up to 60°C ± 2°C, followed by a dipping at the coagulant solution for about 2 minutes and dried at the 120°C for another 2 minutes. The formers were air-cooled to the range of 60 – 65°C and were dipped in the latex dispersion obtained from production for 15 seconds followed by a 15 minutes drying at 120°C [6]. Lastly, the formers were dipped in the last tank, coated with corn starch and left for cooling down to room temperature. Latex glove samples were then stripped from the mould.

2.6 Glove Sample Characterisation

2.6.1 Tensile Strength Tester and Sample Cutter

Tensile testing was done following the ASTM D 412-80 test method using dumb bell shaped samples at a cross-head speed of 500 mm/min with a Universal testing machine [7]. Table 1 shows the ASTM standard for the samples.

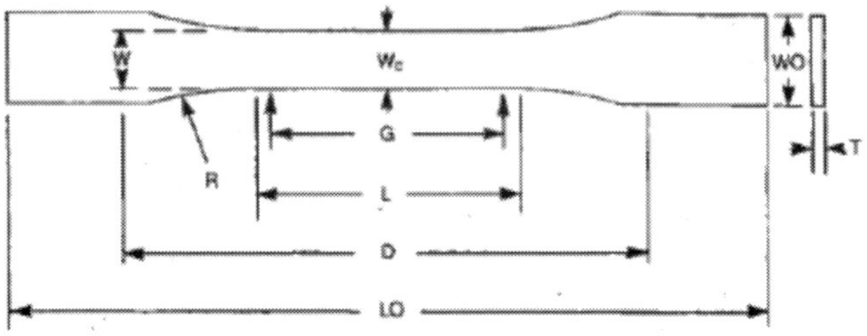

Fig. 3 ASTM plate shape cutting.

Table 1 ASTM standard.

Dimension	Unit, mm
Width of narrow section (W)	6
Length of narrow section (L)	33
Width overall (WO)	19
Length overall (LO)	115
Gage length (G)	25
Distance between grips (D)	65
Radius of fillet (R)	14
Outer Radius (RO)	25

2.6.2 Thickness Gauge Test

The thickness of the samples was measured using Vernier Caliper. Based on industrial glove standard, the average thickness of the glove is at 0.1mm. The measurement was at the area of the palm, areas between fingers such as between thumb with first finger, between first finger with second finger, between second finger with third finger and between third fingers with fourth finger.

2.7 Surface Tension Test

The coagulant of each combination with wetting agent was collected during the glove sample preparation, and was analysed with surface tension meter DST-60.

3. Result and Discussion

3.1 Analysis of First Pre – Screen test

Fig. 4 shows the result of the foaming heights measurement at different wetting agent wt%. It shows that increment in amount of wetting agent in coagulant causes higher the foam height formed. No foam forming in the controlled sample without the wetting agent. The uneven thickness between fingers was found caused by webbing foam on the area during coagulant dipping. When the webbing foam rupture, it leads to no or too little coating of calcium nitrate on the former, which produce thin area of the latex film.

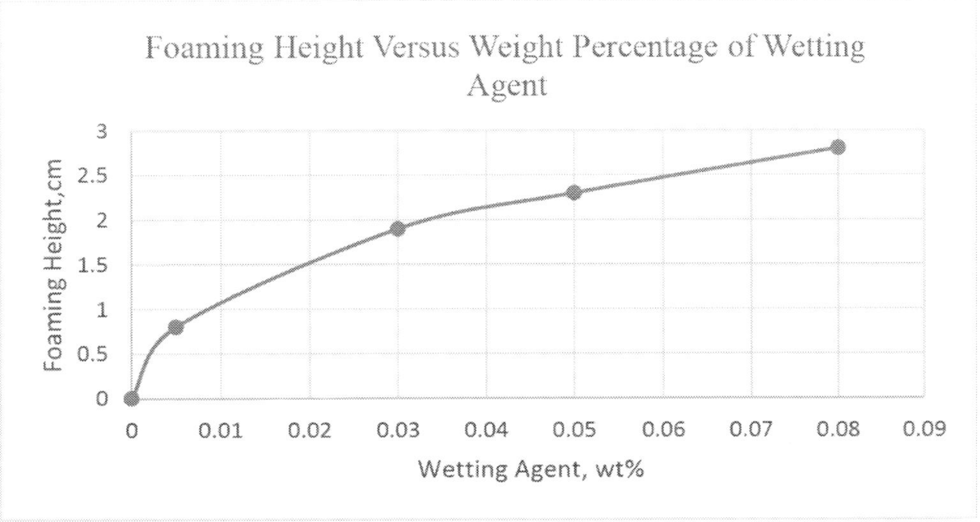

Fig. 4 Foaming Height versus Weight Percentage of Wetting Agent.

3.2 Analysis of Second Pre – Screen test

The cloud point analysis was to determine the effective coagulants' temperature range for the optimisation process, as when the chemical reaches its cloud point, the coagulant loses its functionality as the coating agent for latex solution. Therefore the design temperatures have to be lowered than the tested cloud point temperature. From Table 2, sample E and sample T both proven to have the cloud point above 80°C. This concludes that the designed temperature in between 58°C to 68°C was safe to be tested.

Table 2 Cloud Point Test result

Chemical	Weight Percentage of wetting agent, %	Cloud Point for Sample E, °C	Cloud Point for Sample T, °C
Coagulant	0	No Cloud Point	No Cloud Point
Coagulant	0.005	More than 80 °C	More than 80 °C
Coagulant	0.03	More than 80 °C	More than 80 °C

Coagulant	0.05	More than 80 °C	More than 80 °C
Coagulant	0.08	More than 80 °C	More than 80 °C

3.3 Analysis of Third Pre – Screen test

This test was to determine the bonding of the wetting agent. Since both sample E and T are alcohol ethoxylate, there is possibility of the wetting agent to be having similar bonding. The screening is necessary as 27 runs of full factorial design consume a large amount of time for glove sample preparation.

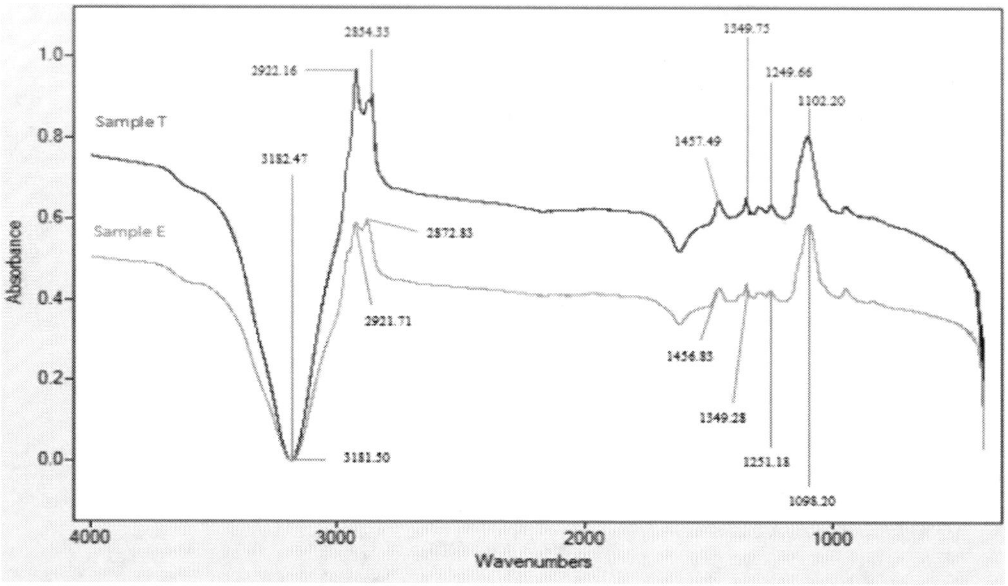

Fig. 5 FTIR Spectrum of both samples

Based on Fig. 5, the first zone section of sample T, there is U shape peak that located about 3182.47 cm^{-1} which in between 2650cm^{-1} to 3200cm^{-1}. This U shape peak is broad and strong that would represent the OH- group. On the zone 2, second peak is about 2872.79cm^{-1} and 2854.33cm^{-1} is refer to alkyl group or CH_3- which is in between 2960cm^{-1} to 2850 cm^{-1}. On the fingerprint zone, peak at 1457.49cm^{-1}, 1349.75cm^{-1} and 1249.66cm^{-1} are methylene or $-CH_2$ [8].

While peak at 1102.20 cm^{-1} is refer to ether group or -C-O-. While based on spectrum of sample E, in the first zone section, the U shape peak that located about 3181.50 cm^{-1} that lies in between 2650 cm^{-1} to 3200 cm^{-1}. This peak is broad and strong that represent the OH- group. While on the zone 2, second peak is about 2921.71 cm^{-1} and 2872.83 cm^{-1} is refer to alkyl group or CH_3- that lies in 2960 cm^{-1} to 2850 cm^{-1}. Lastly on the fingerprint zone, peak at 1456.83 cm^{-1}, 1349.28 cm^{-1} and 1251.18 cm^{-1} are methylene or $-CH_2$. While peak at 1098.20 cm^{-1} is refer to ether group or -C-O-. With both of the sample spectrum is shown in Fig. 5, it shows that both of the samplse of wetting agent is similar in term of bonding. This shows that sample T have more of hydrophilic and hydrophobic bonding compare to Sample E due to absorbance value is higher which mean the finger print zone of sample T have more ethylene oxide group. Ethylene oxide group in alcohol ethoxylate is the role that promotes foam and surface tension reduction properties. As a result, sample T promote more foam and reduce more surface tension compare to sample E. Besides, sample T was more readily available in market compared to sample E. Hence, sample E was removed.

3.4 Analysis of Surface Tension

Based on the coagulant with varies of wetting agent amount, Fig. 6 below shows the decreases trends of the surface tension on coagulant with the increase of wetting agent. This test was to verifying the coagulant that needs to be within the standard of the typical surface tension of the coagulant. The typical surface tension of coagulant is about 0.001 N/m to 0.040N/m. Based on the Fig. 6, the coagulant without the wetting agent is having greater surface tension which is 0.0486N/m. This means that the coagulant is not suitable for glove sample preparation. While with the small amount of the 0.005 wt% wetting agent added, the surface tension is about 0.04 N/m.

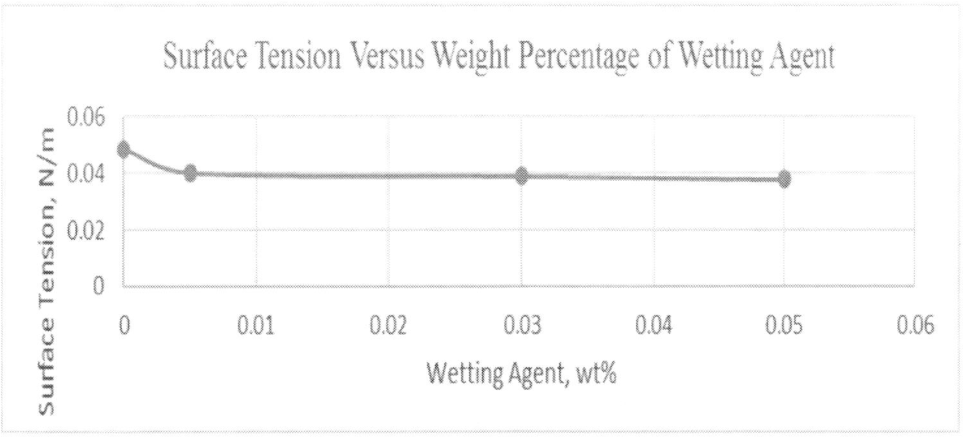

Fig. 6 Surface Tension versus Weight Percentage of Wetting Agent

3.5 Analysis of Glove Sample Characterisation

After the glove sample that have prepared, glove weight, glove thickness and its tensile strength are require to be studied so that optimisation on the parameter can be determined. With the result being collected, it can make use of Matlab® software to plot the surface plot with colour bar as parameter to be studies. With the response surface methodology, optimisation of the production can be done.

3.5.1 Tensile Strength Analysis

Based on Fig. 7, the tensile strength against temperature and dipping time met the requirement of the standard at 18MPa which shown based on the colour bar.

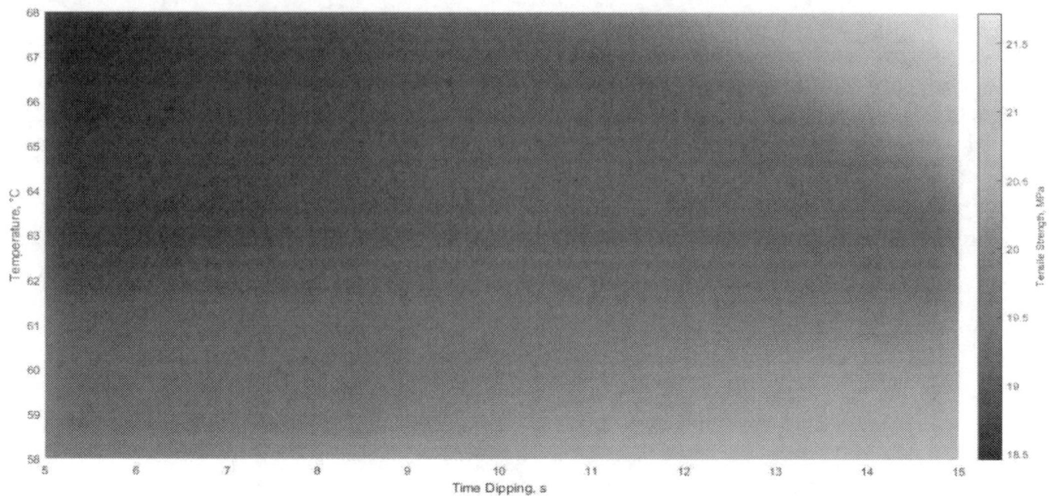

Fig. 7 Contour of temperature and time against tensile strength of glove for 0.005 wt% of wetting agent

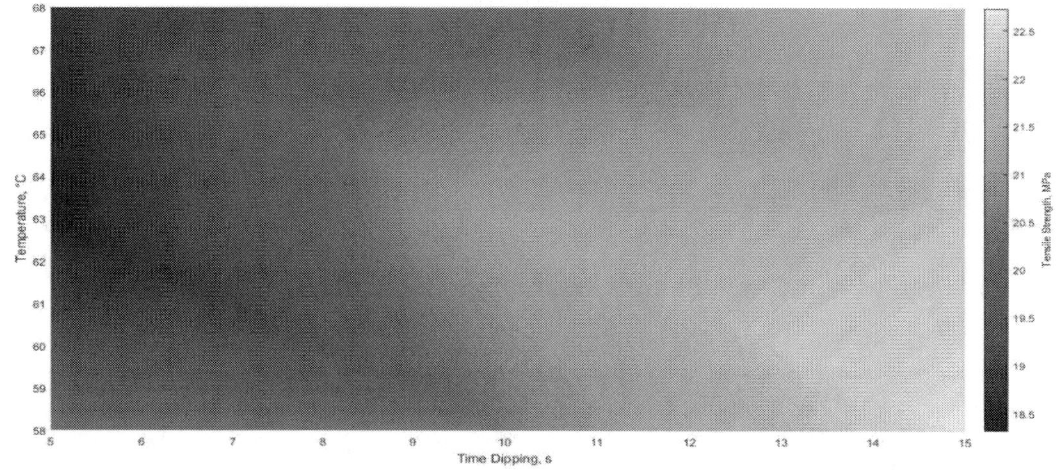

Fig. 8 Contour of temperature and time against tensile strength of glove for 0.03 wt% of wetting agent

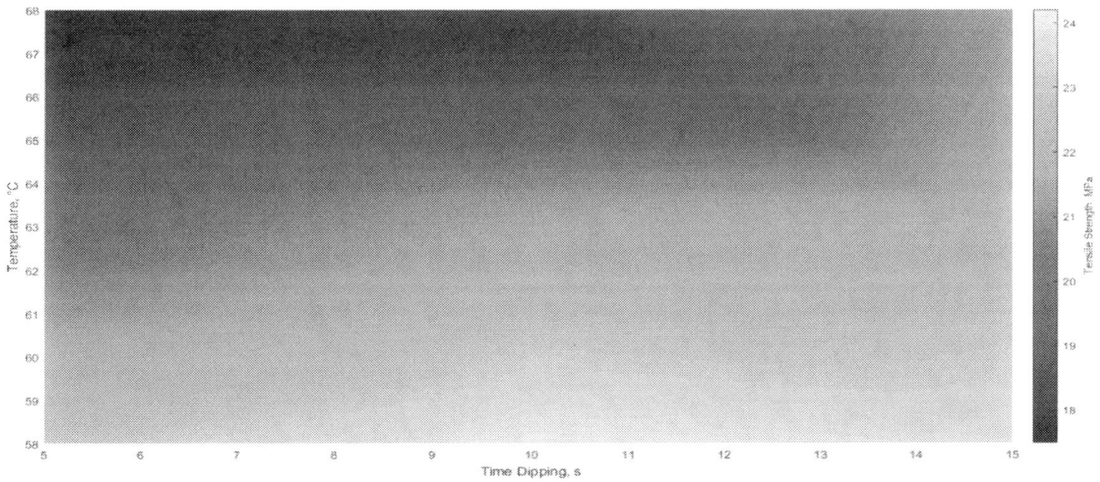

Fig. 9 Contour of time and temperature against tensile strength of glove for 0.05 wt% of wetting agent

In Fig. 7, the glove sample prepared at 0.005 wt% of wetting agent. The glove that produce on at temperature range from 58°C to 68°C as well as the range of time dipping of 5 seconds to 15 seconds meet the standard requirement of the tensile strength. While at temperature of 63°C and 15 seconds, tensile strength should have about 20MPa and above. Based on observation, whenever the temperature of the coagulant is low such as 58°C, the coagulant is hardly evaporating on the surface of the mould. This resulted the coagulant to be flowing on the mould instead of dried where the mould is having smooth surface. If the mould does not have smooth surface which allow forming of thin film without any restrain and would not dripped off. This provides coagulant to adhere on the surface with a thicker film and provide higher tensile strength as well. By referring to Fig. 8, coagulant is hardly evaporating at 58°C and this allows the area on the mould that to have lesser amount of coagulant. Hence, the glove tensile strength should be lowered at lower temperature. This means that there is fluctuation in reading where it is actually require more glove sample so that the glove reading would be consistent.

While based on Fig. 8, the glove sample is tested at 0.03 wt% of wetting agent. Based on the colour of the contour, it shows that the tensile strength against temperature and dipping time is again meeting the requirement of the standard at 18MPa. Since the tensile strength does not varies much compare from 58°C to 68°C, more glove sample are required to be studied so that to allow consistent reading. Based on the Fig. 9, the tensile strength of the glove at 58°C is greater than 68°C. This is basically not supposed to be happened theoretically and the result should be other way around. The reason of this difference of tensile strength is due to the dipping time for latex is not consistent that may due to allow the glove tends to be thicker. This resulted at 58°C to have higher tensile strength.

3.5.2 Thickness Glove Analysis

On the other hand of glove thickness, it is the most crucial part as it determines the uneven wall thickness. The glove manufacturing standard of the powdered glove thickness is at 0.1mm.

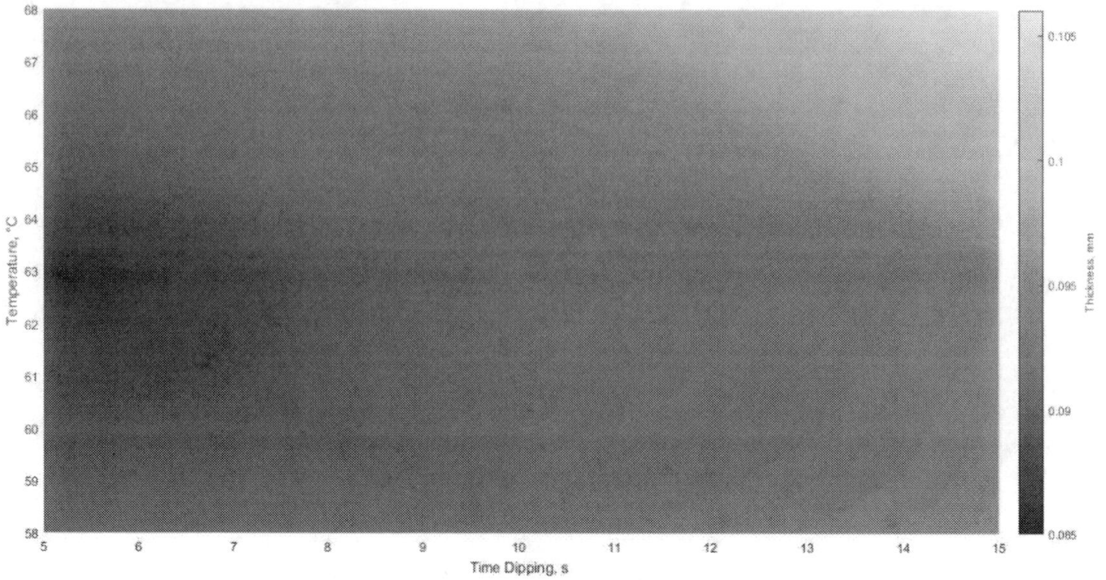

Fig. 10 Contour of time and temperature against glove thickness for 0.005 wt% of wetting agent.

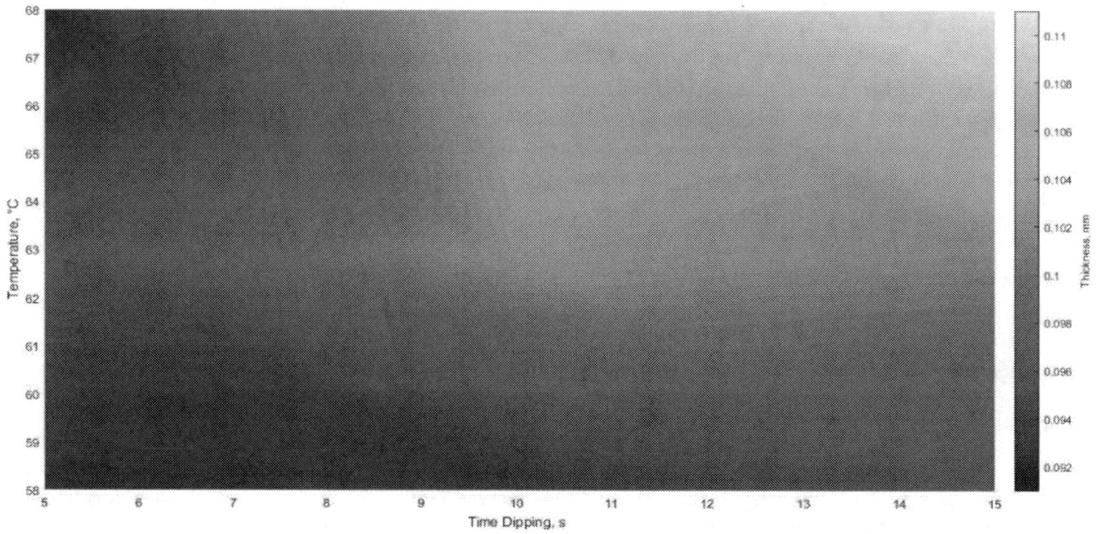

Fig. 11 Contour of time and temperature against glove thickness for 0.03 wt% of wetting agent

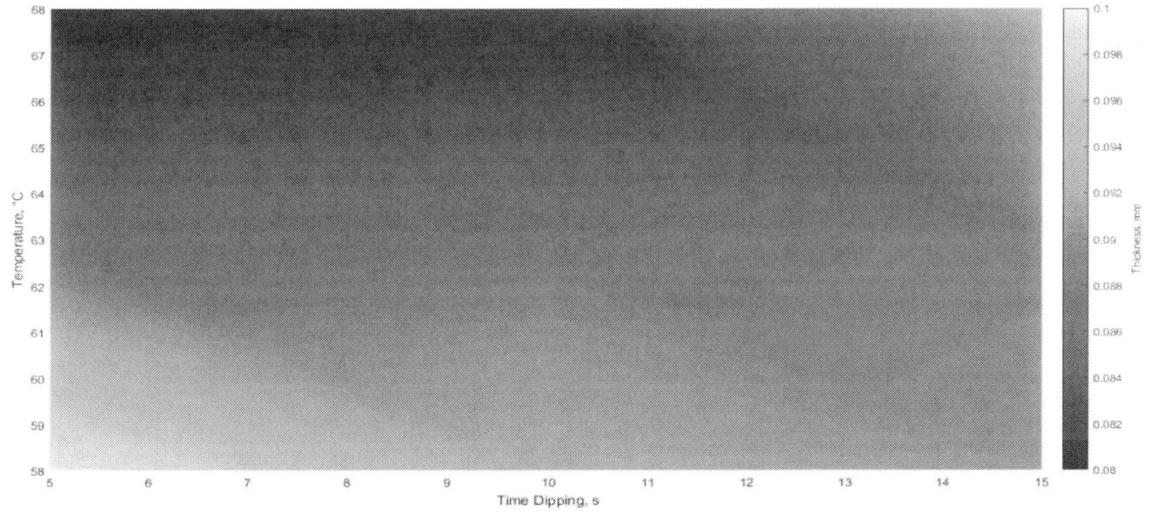

Fig. 12 Contour of time and temperature against glove thickness for 0.05 wt% of wetting agent

Refer to the Fig. 10, the colour bar shows the coagulant temperature at about 63°C with 6 seconds dipping time. This enable to achieve the 0.1mm. Besides, at 68°C with 8 seconds dipping time also able to achieved the 0.1mm. Moreover, this Fig. 11 shows that the trend where thickness increases with the dipping time of the coagulant which is synchronise with the concept that mention previous section. Based on Fig. 15, temperature at 66°C until 68°C with 12 to 15 seconds able to achieve thickness of 0.1mm. This result is not able to be applied in production line. This is because process line operate at high speed where 12 to 15 seconds is not able applicable. By comparing Fig. 11 and Fig. 12, Fig. 12 should have better performance as 0.03 wt% wetting agent basically able to coat on the mould better than 0.005 wt%.

However, more amount of wetting agent would create more foam and bubble on the mould. Based on observation, the coagulant of wetting agent at 0.03 wt% and 0.05 wt% are having large amount of bubble compared to 0.005 wt% wetting agent. This can be shown in Fig. 13. The reason that the coagulant need to be agitate constantly due to avoid the anti-tack agent to be sediment.

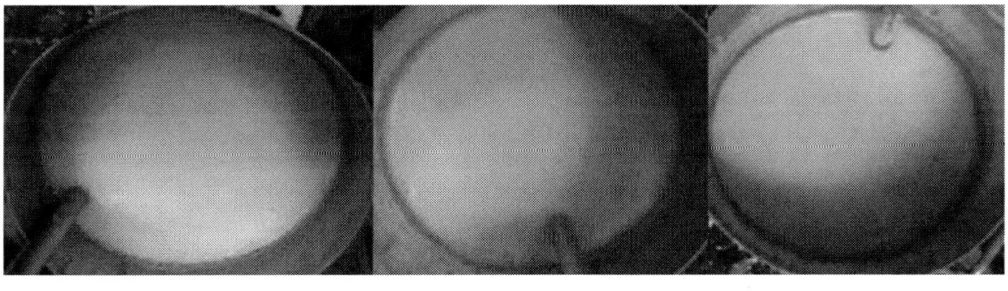

Fig.13 Wetting agent at 0.005 wt% (left), 0.03 wt% (middle) and 0.05 wt% on the (right)

The Fig. 14 show the foam on the mould that affect the on the latex dipping. With this reason, the tensile strength of the glove is also affected by this foam that can compare Fig. 8 and Fig. 11. Also, higher temperature of the coagulant able to break the foam compared to the lower temperature as shown in Fig. 10 and Fig. 11. As a result, lower surface tension liquid would allow bubble foam of liquid would be larger. Hence, increase in liquid drainage and lead bubble to rupture during withdrawing of mould after coagulant dipping. While at lower temperature, it would have higher viscosity that allow the bubble to retain is shape for a longer period. Besides, the bubble foam would having smaller bubble compare to higher temperature and liquid drainage tendency would be lower. As a result, the foam does not rupture. While this foam would only rupture during the drying process of coagulant in oven. This means that the rupture of foam in the certain area in oven will have less deposited of calcium nitrate. Hence, thickness of glove is affected.

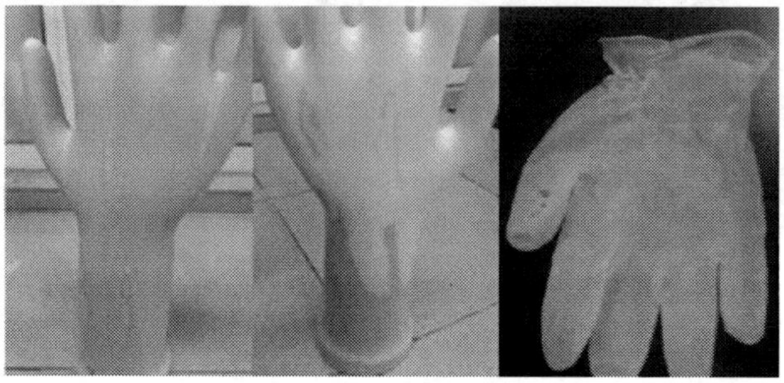

Fig.14 Foam on the mould (Left) and Latex uneven coating (Middle) and Sticky Glove at the cuff area (Right)

Based on Fig. 14, the foam height at 0.05 wt% is greater than 0.005 wt% and 0.03 wt% where it could be refer to Fig. 4. This shows 0.05 wt% have the foam issue that similar to wetting agent at 0.03 wt% that can be based on Fig. 14. But 0.05 wt% wetting agent is having more serious issue where there is presence of webbing that happen between the thumb and first finger. Even though higher temperature expected to reduce the foam creation but there was too much wetting agent in this sample. A higher temperature is required to break more foam and more glove sample is require for consistent reading.

At 0.005 wt% wetting agent, the glove sample having issue of sticky where anti-tack agent is not disperse well enough. This is due to wetting agent is used to ensure a better coating effect on the mould. While wetting agent at 0.03 wt% and 0.05 wt%, the glove also the issue of sticky glove where located at the foam area during coagulant dipping.

3.6 Analysis of Response Surface Methodology Result

Since the result of the tensile strength on Fig. 7, Fig. 8 and Fig. 9 show a positive result. It narrowed the optimisation range of the glove thickness. The best parameter of the glove sample was at 63°C with 6 seconds dipping time and 68°C with 8 seconds dipping time. Both of this parameter gave 0.1mm thickness with 0.005 wt%. But at 0.005 wt%, the glove tends to be sticky. Hence, more wetting agent is require so that anti-tack agent able to disperse evenly. Based on Fig. 15 below, by choosing the parameter 63°C with 6 seconds dipping time would be wise as lower temperature would have lower operating cost for the coagulant. While lesser dipping time that is 6 seconds would have a better profit in glove production compare to 8 seconds. This is due to lesser time require to produce the glove sample. As a result, fixing the temperature at 63 °C and making variation of the wetting agent amount and dipping time would the final optimisation. The optimisated parameter was at 0.0175 wt% of wetting agent and about 5 seconds dipping time at 63 °C.

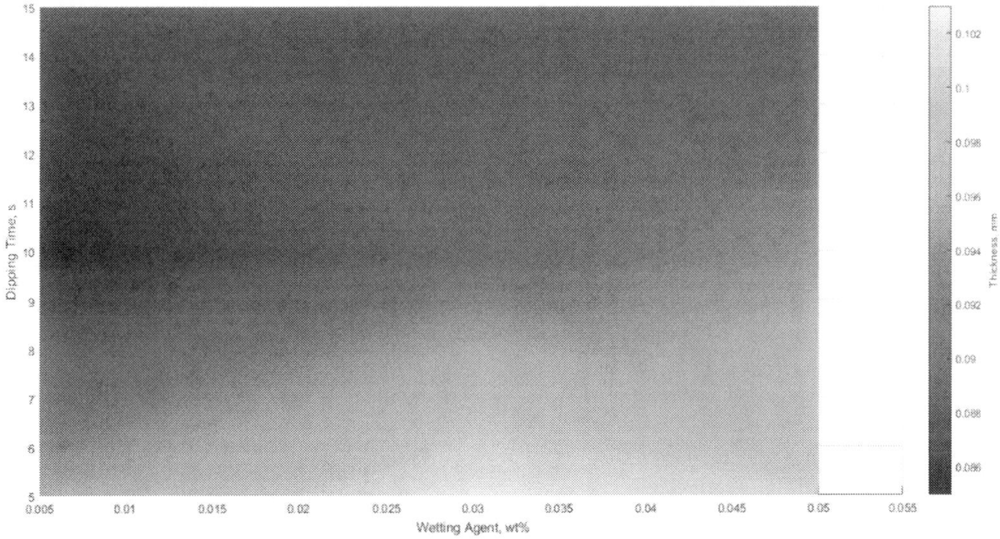

Fig. 15 Predicted time and wetting agent amount against glove thickness for 63°C

3.7 Verification Result

Table 3 below shows the average result of the thickness and tensile strength as well as error percentage of the result. By referring to Table 3, the average thickness gloves that produced from the experiment was in the range of $0.09 \pm 0.004\ mm$. The result did not achieve the targeted value of 0.1mm for ASTM standard. Besides, there is having error about 11.33% lies in glove thickness result compared to targeted thickness. The reason of this error is mainly due to human error.

While the average tensile strength of glove produce at 19.480MPa is higher than ASTM standard of 18MPa. This means that the gloves meet the requirement standard as the parameters suitable. Unfortunately, the optimised-parameters' latex samples was having a sticky surface, due to the inefficiency in dispersing the anti-tack agent throughout the mould. In order to avoid this sticky issue, adding more anti-tack agent into coagulant would be able to negate this issue.

Table 3: Error percentage compared to targeted thickness.

Average Thickness, mm	Targeted Thickness, mm	Error Percentage, %	Average Tensile Strength, MPa
0.09	0.1	11.33	19.480

4. Conclusion

In conclusion, 5 seconds dipping time at 63 °C were chosen compared with 8 seconds dipping time at 63 °C. With lower operating cost on time dipping and temperature, this parameter is expected to able to produce mass amount of glove with low number of uneven wall thickness issue. This would indirectly increasing the profit due to more glove can be sold instead of dispose or sold in lowered price for uneven wall thickness glove. As a result, the optimise parameter from this investigation are set at 0.0175 wt% of wetting agent and about 5 seconds dipping time at 63 °C . The produced latex glove gave an average thickness of 0.09 ±0.004 mm. Both sample E and sample T shows comparable functional group, and with an easier access in the market, sample T was chosen to conducted the optimisation test. However, this data will be further trial run in production line.

References

1. M.C. Ng, H. Ab-Samat, and S. Kamaruddin, *Int. J. Eng. Sci.*, 2319–1813 (2013).
2. P. Cacioli, *Rev. Fr. d'Allergologie d'Immunologie Clin.*, **37**, 1173–1176 (1997).
3. E. Yip and P. Cacioli, *J. Allergy Clin. Immunol.*, **110**, S3–S14 (2002).
4. V.R. Gundabala and A.F. Routh, *J. Colloid Interface Sci.*, **303**, 306–314 (2006).
5. L. Chen and F. Wu, *J. Saudi Chem. Soc.*, **18**, 545–550 (2014).
6. L. Sikong, M. Masae, K. Kooptarnond, W. Taweepreda, and F. Saito, *Appl. Surf. Sci.*, **258**, 4436–4443 (2012).
7. J. George, K.T. Varughese, and S. Thomas, *Polymer (Guildf).*, **41**, 1507–1517 (2000).
8. A. Elena, I. Gozescu, A. Dabici, P. Sfirloaga, and Z. Szabadai, "Organic Compounds FT-IR Spectroscopy", in: *Macro To Nano Spectrosc., InTech*, (2012).

Assessment of Natural Deep Eutectic Solvent Pretreatment on Sugar Production from Lignocellulosic Biomass

Yoon Li Wan[1]*, and *Yuen* Jun Mun[1]

[1] School of Engineering, Taylor's University, Subang Jaya, Selangor, Malaysia

Abstract. Before the conversion of lignocellulosic biomass into fuel such as ethanol, the biomass needs to be pretreated and the yield of ethanol is highly dependent on the pretreatment efficiency. This study investigate the performance of deep eutectic solvent (DES) in pretreating sago waste which is a type of starchy biomass. The suitable type of DES in sago waste pretreatment was selected based on three criteria, which is the structural characteristic, the sugar yield during enzymatic hydrolysis and the amount of sugar loss during pretreatment. In this study, three types of DES namely Choline Chloride-Urea (ChCl-Urea), Choline Chloride-Citric acid (ChCl-CA) and Choline Chloride-Glycerol (ChCl-Glycerol) was investigated. The effect of temperature and duration on DES pretreatment was also investigated. All DES reagents were able to disrupt the structure and increase the porosity of sago waste during pretreatment. ChCl-Urea was selected in this study as it shows apparent structural disruption as examined under Scanning Electron Microscope (SEM). The highest glucose yield of 5.2 mg/mL was derived from enzymatic hydrolysis of ChCl-Urea pretreated sago waste. Moreover, reducing sugar loss during ChCl-Urea pretreatment was low, with only 0.8 mg/mL recorded. The most suitable temperature and duration for ChCl-Urea pretreatment is at 110°C and 3 hr. In a nutshell, the application of DES in pretreatment is feasible and other aspects such as the biodegradability and recyclability of DES is worth investigating to improve the economic feasibility of this pretreatment technique.

1 Introduction

Lignocellulosic biomass has the potential to replace existing petroleum-dependant energy source due to its abundant and renewable nature [1]. Many studies have shown that lignocellulosic biomass can be potentially converted to biofuel such as bioethanol and biodiesel [2]. In common practice, bioethanol is produced by following a pathway of biomass pretreatment, enzymatic hydrolysis, fermentation and distillation [3]. Pretreatment is the first process in lignocellulosic biomass conversion and hence, the efficiency of pretreatment process is the key to effective enzymatic hydrolysis and related downstream processes [3]. Pretreatment is necessary due to remove the lignin content in the biomass which resists solubilization and hamper enzymatic hydrolysis process [4]. In addition to that, factors such as cellulosic compositions, crystallinity structure, rigid nature of cellulose,

* Corresponding author: LiWan.Yoon@taylors.edu.my

formation of inhibitors and particle size of biomass could contribute to the enzymatic digestibility of biomass [4].

Many approaches such as physical, chemical and biological pretreatment methods have been applied. Pretreated substrate would normally have a structure with increased porosity and reduced crystallinity [4]. During pretreatment, modification or removal of lignin content takes place to improve the accessibility of enzymes to the biomass substrate. The conversion of complex carbohydrates into simple sugars and subsequently fermented into bioethanol could be improved when pretreated substrate is applied [3]. However, as good as conventional pretreatments may be beneficial to sugar production from biomass, these methods have their respective drawback. For example, acid pretreatment produce fermentation inhibitors that could reduce the efficiency of fermentation.

The desire for green solvents in pretreatment has led to discovery of deep eutectic solvents (DES) and ionic liquids (IL) [5]. DES is made up of a hydrogen bond donor (HBD) and a hydrogen bond acceptor (HBA). Both DES and IL pretreatment techniques are affected by combination of chemicals used to produce the solvents, the type of biomass feedstock, size of particles and substrate loading [6 - 9]. These advanced pretreatment techniques are gaining attentions as extreme operating conditions such as high pressure and temperature is not required.

Further advancement of DES has been made due to its potential as green solvents to replace IL. Due to low costs and easily accessible raw materials, DES are being studied extensively in various fields such as electrochemistry and pharmaceutical [5]. DES has similar physio-chemical properties to IL with extra benefits such as easier preparation, non-complicated purification processes and it is recyclable [9]. Majority of the DES are Choline Chloride-based (ChCl), which is applied as food additive. With that being said, ChCl is acknowledged as safe compound for human consumption by relevant authorities [10]. Many researches have evaluated degradability of DES and have classified DESs as readily biodegradable compounds [9]. Combination of ChCl into DES has shown significant enhancement of biodegradability. Recently, the potential application of DES in pretreatment of biomass has been proven positive as the enhancement of sugar production from DES-pretreated biomass was observed in a few studies [7 - 9].

To date, no research has been done on the effect of DES pretreatment on starchy biomass. Starchy biomass such as sago waste which has approximately 36% starch, 40% cellulose and 13% hemicellulose serves as a promising feedstock for glucose production. In this study, three types of DES namely Choline Chloride-Urea, Choline Chloride-Citric acid and Choline Chloride-Glycerol was investigated for the pretreatment of sago waste. These DES were selected as they are commonly applied as pretreatment agents in other study [7 – 9]. The best DES was selected based on sugar production from enzymatic hydrolysis while investigating the sugar loss during pretreatment. Structural characteristic was also serves as one of the criteria for selection of DES. The effect of temperature and duration on DES pretreatment on sago waste was also investigated.

2 Methodology

2.1 Preparation of Deep Eutectic Solvent (DES)

Three types of DES, namely choline chloride with urea (ChCl-urea), citric acid (ChCl-CA) and glycerol (ChCl-Glycerol) were prepared at molar ratio of 1:2, 1:2 and 1:1 respectively [7 – 9]. The mixture was heated by using hot plate with stirring rate of 150 rpm at 80°C. The heating and stirring process was halted when a homogenous and clear solution was obtained. The solution was cooled down before it was stored in a tightly-sealed bottle. The DES solution was kept and stored in a dessicator with silica gel.

2.2 Preparation of Sago Waste

Sago waste was collected from Hak Mong Sago Mill, Batu Pahat, Johor, Malaysia. The sample was washed and dried at 60°C for 24 hr. Sago waste was grinded and sieved to particle size of approximately 500 μm prior to be used in pretreatment.

2.3 DES pretreatment

Pretreatment was carried out with 5% solid loading by mixing 1 g of sago waste and 20 g of DES respectively in a Schott bottle. The samples were heated at 110 °C for 3 hours in the oven (UN-75, Memmert). The biomass/DES mixtures were then transferred to centrifuge tubes with the addition of 5 ml deionized water. The contents were centrifuged at 3500 rpm for approximately 30 minutes. The supernatant was collected into sampling bottle and stored at 4°C. Reducing sugar concentration in the supernatant collected from pretreatment was subjected to DNS analysis. Washing and rinsing of biomass was then carried out by centrifugation. The pretreated biomass was then dried at 50°C and kept in desiccator before subjecting it to hydrolysis process and scanning electron microscopy (SEM) analysis.

2.4. Enzymatic hydrolysis

Enzymatic hydrolysis was conducted in 2% solid loading by mixing 0.2 g of pretreated biomass 10 mL of citrate buffer solution (pH 4.8 and 50 mM) added with 0.1 g of cellulase (Cellulysin, *Trichoderma viride*, Merck). The hydrolysis process was carried out at 50 °C for 48 hours at 100 rpm in an incubating shaker (YLHDER, LM-400D). After hydrolysis, the samples were centrifuged at 3500 rpm for 15 min. The supernatant was collected and glucose content was analysed by using High Performance Liquid Chromatography (HPLC).

2.5 Effect of pretreatment temperature and duration on sugar production

The DES with the best performance was selected to carry out this study. The pretreatment duration and solid loading was fixed at 3 hr and 5% when the temperature was varied from 90°C, 110°C and 130 °C. To examine the effect of duration, DES pretreatment was carried out for 1, 3 and 5 hr at 5% solid loading and the most suitable temperature.

2.6 Analytical method

2.6.1 3, 5 Dinitrosalicyclic Acid (DNS) analysis

DNS reagent with 1% (w/v) NaOH (Merck), 1% (w/v) 3,5-dinitrosalicyclic acid (DNS, Merck), 0.2% (w/v) phenol (Merck) and 1% (w/v) of sodium metabisulfite (Merck) was prepared. During the analysis, 3 mL of DNS solution was then added into the test tube which contain 1.5 mL of sample. The mixture was boiled for 5 min, placed in a cold water bath before 1 mL of 40 % Rochelle salt (Potassium sodium tartrate tetrahydrate, Merck) solution was added. 0.2 mL of reaction mixture was then diluted by adding 2.5 mL of ultrapure water and the absorbance of the solution was measured at 540 nm by UV-Vis spectrophotometer (Genesys10, Thermofischer Scientific). The reducing sugars concentration in the sample solution was determined from the standard curve constructed by using glucose solution.

2.6.2 High Performance Liquid Chromatography (HPLC)

Glucose concentration was examined by using HPLC (LC1220, Agilent Technology USA). Deionized water was used as the mobile phase of HPLC with flowrate of 0.6 ml/min. The Agilent H+ column was used while column temperature is fixed at 65 °C. Before conducting HPLC analysis, the sample is filtered using Nylon filter with particle size of 0.2 μm. Glucose concentration was determined from a calibration curve in the range of 1 g/L to 30 g/L glucose.

2.6.3 Scanning Electron Microscopy (SEM) analysis

The morphology of both untreated and DES-pretreated sago waste was observed with scanning electron microscope (Quanta 200 FESEM, FEI, USA). The images were acquired under 10kV acceleration voltage and magnification of 1600X.

2.7 Statistical analysis

All the experiments were performed in duplicate. The significance of the results were identified by using 2-way Analysis of Variance (ANOVA) method at 95% confidence level.

3 Results and discussion

3.1 Selection of suitable DES to pretreat sago waste

3.1.1 Structural analysis - Scanning Electron Microscopy (SEM) Analysis

Fig. 1. shows the SEM images of untreated and pretreated sago waste by using different DES. From the SEM analysis, all the pretreatments were successfully carried out to make the samples more susceptible for sugar extraction via enzymatic hydrolysis. Raw and untreated sago hampas as shown in Fig. 1.(a) was less porous and its structure was less disrupted. The fibers in untreated sago waste were aligned in higher degree of order as compared to its pretreated counterpart. After pretreatment, the rigid and smooth surface was disrupted and the sago waste has demonstrated higher porosity as depicted in Fig. 1.(b), (c) and (d). The disruption is crucial to make biomass more susceptible to enzymatic attack in hydrolysis process, therefore resulting higher yield of sugar. However, the morphology of ChCl-Glycerol pretreated sago waste consists of a large portion of smooth surface which indicates the fibers are still intact, thus making the enzymes less accessible to cellulose.

Fig. 1. SEM images of (a) untreated sago waste, (b) ChCl-Urea pretreated sago waste, (c) ChCl-CA pretreated sago waste, (d) ChCl-Glycerol pretreated sago waste at 1600x magnification.

In general, all pretreatment methods has increase the porosity of biomass and some DES might shows the ability in lignin removal [11]. A few studies have reported similar results whereby the morphology of biomass became more porous, less crystalline and more vulnerable to enzymatic hydrolysis after using ionic liquid pretreatment which has similar characteristic with DES [12]. The result from SEM analysis is aligned with biomass pretreated by using other reagents such as NaOH, HCl and ionic liquids which concluded that alteration of biomass structure occurs in favor to microbial attack and enzymatic hydrolysis [12].

3.1.2 Enzymatic hydrolysis of DES-pretreated sago waste

Sago wastes was subjected to pretreatment with different types of DES at 110 °C, 3 hr and 5% solid loading before enzymatic hydrolysis. The performance of these DES pretreatments was analysed and justified based on yield of reducing sugar obtained from enzymatic hydrolysis and sugar loss during pretreatment besides examining the structural characteristics as outlined in Section 3.1.1.

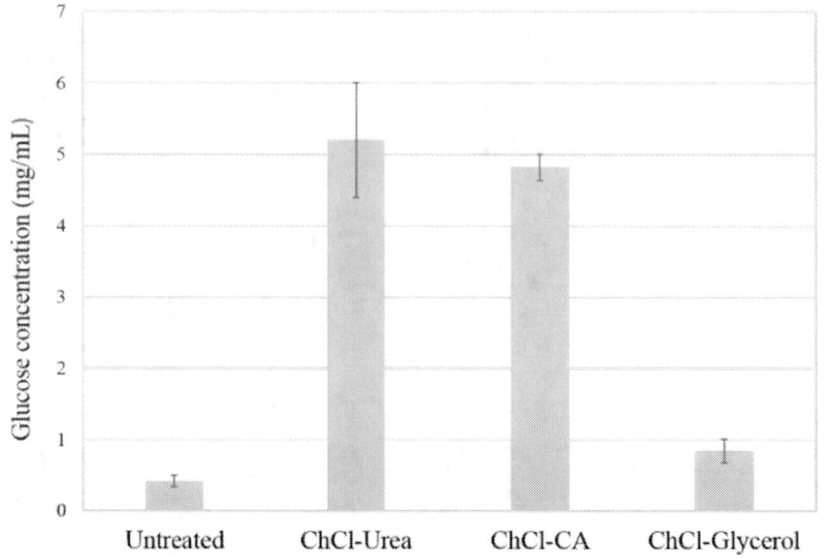

Fig. 2. Glucose yield from enzymatic hydrolysis of sago wastes pretreated by different types of DES

From Fig.2., ChCl-Urea pretreated sago waste gave the highest amount of glucose yield at 5.2 mg/ml (which is 0.26 g/g of pretreated biomass) compared to ChCl-Glycerol and ChCl-CA pretreated biomass. It was found that sago waste pretreated by ChCl-Glycerol yielding a similar glucose yield with the untreated sample which renders this DES the least effective in pretreatment of sago waste. From 2-way ANOVA analysis, the glucose yield obtained from different DES pretreatment was significant (F = 36.56) whereas the batch-to-batch difference (F = 0.22) was insignificant under 95% confidence level.

Based on the nature of each DES, urea is slightly alkaline with pH approximately 9 whereas both glycerol and citric acid are acidic with pH at 6.3 – 6.9 and 3.2 respectively. From the results obtained, it can be concluded that sugar yield increases when wither acidic or alkaline reagent was used in pretreatment. ChCl-Urea might act just like the reagent in conventional alkaline pretreatment to break down lignin and facilitates the access of enzymes to biomass. On the other hand, ChCl-glycerol, which is having neutral pH, yielded the lowest amount of glucose with 0.9 mg/ml (which is 0.04 g/g of pretreated biomass). This finding aligns with a research done which stated that the maximal enzymatic glucose and xylose yields from the biomass, wheat straw fraction were noticed at the most extreme pH values (pH 1 and pH 13) after pretreatment [13]. This trend can be further explained as the surface response models showed significant interactions between pretreatment temperature and pH on glucose and xylose [13]. Acidic pretreatment was favorable at high petreatment temperature compared to alkaline pretreatment, which is not as temperature dependent. From research, alkaline pretreatment generally proven to have better glucose yield even at lower pretreatment (110 °C) temperature as compared to acidic pretreatment [13]. Therefore, the results obtained from Fig.2. were found to be agreeable when slightly lower yield of reducing sugar from was obtained from sago waste pretreated by acid-based DES than alkali-based DES.

3.1.3 Reducing sugar loss during pretreatment

To further confirm the suitability of using DES-pretreated sago waste for glucose production via enzymatic hydrolysis, reducing sugar loss during DES pretreatment was examined and the results are tabulated in Table 1. As such, DES was collected after pretreatment and the reducing sugar content in DES solution was analysed. DES pretreatment which contributes to a higher sugar loss is unfavourable in this case.

Table 1. Reducing sugar loss during pretreatment

Types of DES	Reducing sugar concentration in DES (mg/mL)
ChCl-Urea	0.9 ± 0.14
ChCl-Glycerol	2.0 ± 0.28
ChCl-CA	0.8 ± 0.14

From Table 1, during ChCl-Glycerol pretreatment, the amount of reducing sugar loss was recorded to be twice as much as ChCl-Urea and ChCl-CA pretreatments. Similar to strong acid and some ionic liquids, DES might also be able to depolymerise polysaccharides to compounds with low molecular weight [14], and this causes reducing sugar loss was observed. Based on Table 1, ChCl-Glycerol can be said to have the highest depolymerisation power followed by ChCl-Urea and lastly, ChCl-CA. The mechanism of ChCl-glycerol in depolymerisation of biomass remained unknown and worth investigating. Since reducing sugar loss is unfavourable as it is more beneficial to retain the sugar within the pretreated biomass for subsequent enzymatic hydrolysis, ChCl-urea which produced the highest amount of sugar during enzymatic hydrolysis and shows low reducing sugar loss was chosen in this study. This was further supported by the structural analysis whereby ChCl-urea pretreated sago waste has a more porous and disrupted structure. Based on the DES chosen, a preliminary study on suitable temperature and duration for pretreatment of sago waste was then conducted.

3.2 Effect of temperature on DES pretreatment

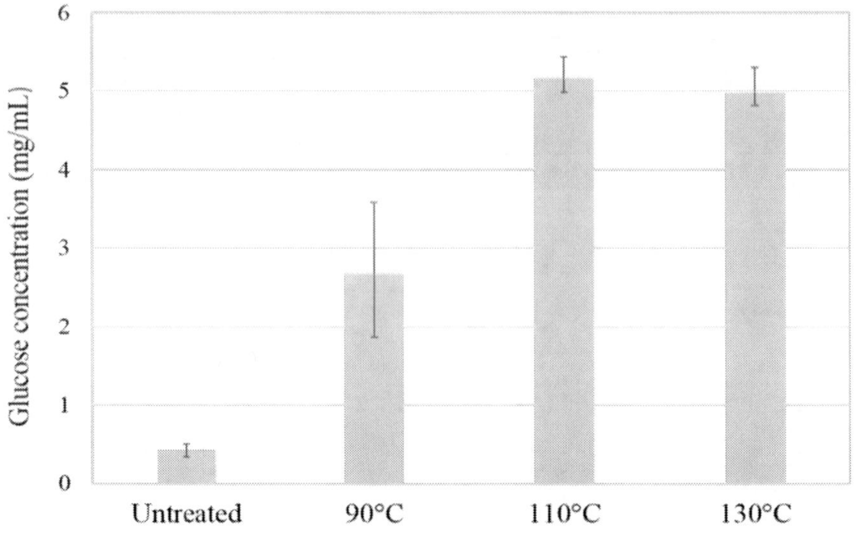

Fig. 3. Glucose concentration at 3 hr under different temperature

Based on Fig. 3., glucose concentration increases from 90°C to 110°C and reduced slightly at 130°C. The result shows the optimum temperature would be at 110 °C with 5.2 mg/ml (which is 0.26 g/g of pretreated biomass) of glucose yield. The trend is aligned to a DES pretreatment study using ChCl-lactic acid to pretreat corncob with varying pretreatment temperature as efficiency of glucose extraction started to decline at certain temperature. The losses of polysaccharides could be due to the balanced effects to improve digestability of hydrolysis [9]. Also, high temperature may contribute to degradation of biomass samples, thus lead to sugar loss. This is further explained by higher pretreatment temperature increases the deconstruction of biomass components and improves amount of sugar produced [10]. Moreoever, at higher temperature, increment in pore size of biomass along with reduction of lignin is expected to happen within the biomass [10].

3.3 Effect of duration on DES pretreatment

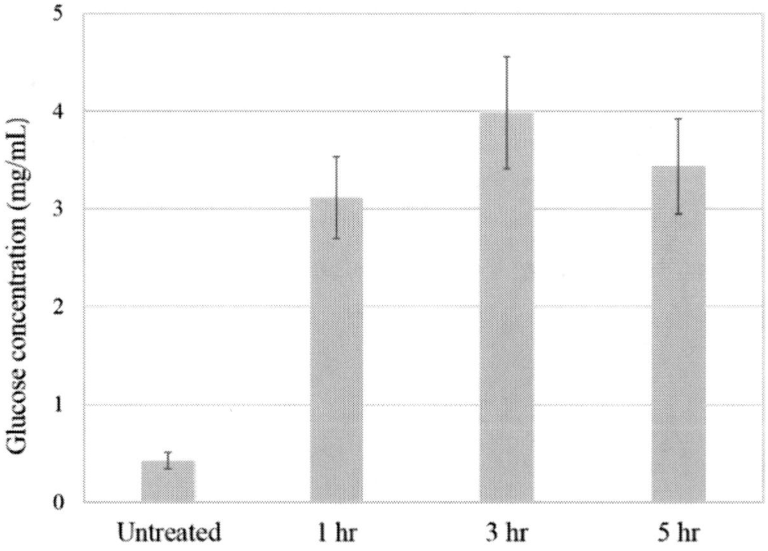

Fig. 4. Glucose concentration at 110 °C under different duration

Based on Fig. 4., it can be observed that glucose concentration of 4.0 mg/mL (which is 0.2 g/g of pretreated biomass) was the highest when pretreatment was applied for 3 hr. This was followed by 5 hr and 1 hr with glucose concentration of 3.4 mg/ml and 3.1 mg/ml respectively. Glucose yield was seen declining at the 5th hr. The sugar loss that might happen at a more profound rate and the degradation of biomass are two possible factors that would cause a decrease in glucose yield when pretreatment duration is extended to 5 hr. This finding agrees well with recent research in which glucose yield increases significantly with pretreatment duration and gives better hydrolysis efficiency until the approximately the 3rd hour before a consistent glucose yield was obtained thereafter [9]. The experiment done by using corncob under different pretreatment duration found that better interaction between biomass and DES-based solvent was expected under prolonged duration [9]. From Fig.4., the most suitable pretreatment duration is 3 hr because the highest amount of sugar yield was obtained.

This study highlights the use of DES as potential green solvents for pretreatment process. At current stage, limited study is available as reference on the application of DES in pretreatment of lignocellulosic biomass. This study could serves as a reference on how DES pretreatment affects the sugar production and structure from starchy biomass. The development in this field would promote and lead to application of eco-friendly pretreatment technology in the future.

4. Conclusion

This preliminary study has demonstrated the potential application of DES in pretreating starchy biomass. DES pretreatment was able to produce biomass with higher porosity and lower crystallinity which is beneficial for the subsequent enzymatic hydrolysis process. It was found that alkaline or acidic-based DES are more suitable in pretreating sago waste as compared to DES with neutral pH. This is because a higher glucose yield were obtained from enzymatic hydrolysis of sago waste pretreated by alkaline and acidic-based DES. Reducing sugar loss during pretreatment was found to be lower when ChCl-urea (alkaline-based DES) and ChCl-CA (acidic-based DES) were ussed. ChCl-Urea was chosen to pretreat sago waste in this study as it demonstrates a high glucose yield from enzymatic hydrolysis with low reducing sugar loss during pretreatment besides the favorable structural changes. The best operating condition for sugar production from ChCl-Urea pretreated sago waste via enzymatic hydrolysis is at 110°C and 3 hr. Further investigation on the compositional analysis of pretreated biomass as well as the optimization of pretreatment condition could be conducted to further elucidate the effectiveness of the pretreatment process.

References

1. F. Isikgor, C. R. Becer, Polym. Chem. **6**, 4497 – 4559 (2015)
2. S. S. da Silva, A. K. Chandel, S. R. Wickramasinghe, J. M. G. Domínguez, J. Biomed. Biotechnol. **2012**, 1-2 (2012)
3. A. Limayem, S. C. Ricke, Prog. Energy Combust. Sci. **38**, 449–467 (2012)
4. A. K. Kumar, S. Sharma, Bioresour. Bioprocess. **4**, 7 (2017)
5. Y. P. Mbous, M. Hayyan, A. Hayyan, W. F. Wong, M. A. Hashim, C. Y. Looi, Biotechnol. Adv. **35**, 105–134 (2017)
6. K. Ellinor B. Heggset, Kristin.S, Biomass Bioenergy **93**, 194–200 (2016)

7. A.K. Kumar, B.S. Parikh, M. Pravakar, Environ. Sci. Pollut. Res. **23**, 9265 – 9275 (2016)
8. A. Procentese, E. Johnson, V. Orr, A. Garruto Campanile, J.A. Wood, A. Marzocchella, L. Rehmann, Bioresour. Technol. **192**, 31 – 36 (2015)
9. C.W. Zhang, S.Q. Xia, P.S. Ma, Bioresour. Technol. **219**, 1–5 (2016)
10. A. N. Gunny, D. Arbain, E. M. Nashef, P. Jamal, Bioresour. Technol. **181**, 297–302, (2015)
11. N. Nor, W. Mustapha, O. Hassan, Procedia Chem. **18**, 147–154 (2016)
12. L. W. Yoon, T. N. Ang, G. C. Ngoh, A. S. M. Chua, Biomass Bioenergy **36**, 160–169 (2012)
13. M. Pedersen, K. S. Johansen, A. S. Meyer, Biotechnol. Biofuels **4**, 11 (2011)
14. D. Di Marino, D. Stöckmann, S. Kriescher, S. Stiefel, M. Wessling, Green Chem. **18**, 6021–6028 (2016)

Suppression of Ground Borne Vibration Induced by High-Speed Lines

Ali Mohamed Rathiu, *Mohammad* Hosseini Fouladi*, *Satesh* Narayana Namasivayam, *Hasina* Mamtaz

School of Engineering, Taylor's University, Subang Jaya, Selangor, Malaysia

Abstract. Vibration of high-speed lines leads to annoyance of public and lowering real estate values near the railway lines. This hinders the development of railway infrastructures in urbanised areas. This paper investigates the vibration response of an isolated rail embankment system and modifies the component to better attenuate ground vibration. Mainly velocity response is used to compare the responses and the applied force is of 20 kN at excitation frequencies of 5.6 Hz and 8.3 Hz. Focus was made on ground-borne vibration and between the frequency range of 0 and 250 Hz. 3D Numerical model was made using SolidWork software and frequency response was produced using Harmonic Analysis module from ANSYS Workbench software. For analytical modelling MATLAB was used along with Simulink to verify the mathematical model. This paper also compares the vibration velocity decibels (VdB) of analytical two-degree of freedom model mathematical model with literature data. Harmonic excitation is used on the track to simulate the moving load of train. The results showed that modified analytical model gives the velocity response of 75 VdB at the maximum peak. Changes brought to the mass and spacing of the sleeper and to the thickness and the corresponding stiffness for the ballast does not result in significant vibration response. Limitations of two-degree analytical model is suspected to be the cause of this inactivity. But resonance vibration can be reduced with the aid of damping coefficient of rail pad. Statistical analysis methods t-test and ANOVA single factor test was used verify the values with 95% confidence.
Keywords: Vibration, Rail, Embankments, Analytical model, Numerical model.

1 Introduction

Rail transportation was a novel mode of transportation created during the enlightened era of industrial revolution in 18th century [1]. Back then, the trains were driven by coal powered steam engines which were loud, dirty and puffed out nauseating thick black smoke. Nevertheless, this mode of transportation was popular among people because, compared to

* Corresponding author: mfoolady@gmail.com

available alternative modes of transportation, trains had many appealing benefits. These include regular intervals of travelling to a predetermined location safely and cheaply. The benefits have increased over the years with technological advancement in engines which has allowed it to travel much faster and efficiently with cleaner and more powerful diesel engines

This popularity has been retained, despite remarkable advances in automobile technology in recent years. Moreover, train usage is raising in urbanised areas across the world for both developed and developing countries. Abe *et al.* [2] investigated rail travel demand in Tokyo from 1950 to 2010 and found that the usage was still increasing, with more travellers using underground trains nowadays. Also, a study made on the market competition of the European rail freight trains, which are large cargo trains, by Laroche *et al.* [3] similarly found that the market was growing despite high initial start-up cost in European Union countries. Furthermore, Chen *et al.* [4] stated that since the beginning of 2000, Chinese government has started to build a massive rail system infrastructure and the effects of this decision has been beneficial to the economy and livelihood.

Therefore, rail travel is still a booming and essential industry for developed countries and even greater for developing countries. But despite the advances in technology, the current railways still have many issues. The most recognisable problems with the railway are vibration and noise. This paper will focus on the ground-borne vibrations which are caused by the movement of the wheel on the track, which propagates vibration from the track to the surrounding areas. These vibrations can be an annoyance to humans who live in nearby buildings. This becomes more problematic in urbanised cities, as Connolly *et al.* [5] states that, the economical repercussion due to high vibrations in city areas are more serious as these vibrations can reduce real estate values and can entirely disrupt the work in sensitive workplaces such as record studios, tv studios and theatres. The paper states that even though these sensitive buildings are fewer compared to residential buildings, the high number of reported complaints from these places shows that there is a direct detriment to the working condition of these places due to railway lines. It also cites that ground vibration is also a growing concern in most European countries due to increasing railway lines and soaring frequency of trips per day.

This research investigates and proposes a modified track embankment to suppress the vibrations. To achieve that the components of the embankment needs to be investigated and the influence of the parameters on the vibration needs to be identified through literature review. Referring to the parameters influence on vibration, the components will be modified to form a mathematical model. Analytical calculations using this model will be conducted using MATLAB and numerical calculations using Finite Element analysis will be made using ANSYS software. The analytical and numerical calculations are compared with each other and also verified using literature data.

1.1 Objectives

Vibration induced by high speed train is a growing urban issue and mitigating methods needs to be put in place to attenuate the vibrations. Generally, the methods to lower the effects of vibration can be classified into two, which is active mitigation, where modification is applied to the railway support and passive mitigation, where nearby buildings are modified with better vibration damping properties. This paper focuses on active mitigation as this is the most sought out method in real life [4].

To modify the railway support or embankment, the properties of the components which makes up the railway embankment is investigated. Next the properties are altered and the result of that quantified using the research methods of analytical and numerical modelling.

1. Investigate the parameters which influence ground vibration induced by high speed lines.
2. To modify the embankment components parameters to achieve greater vibration suppression.
3. To verify the accuracy of the analytical and numerical modelling with each other and literature data.

1.2 Literature review

1.2.1 Railway components

A typical rail embankment has basic four components, track, sleeper, rail pad and ballast, as shown in Figure 1. In addition to these four, external dampers are used in some railways as well. The track is the steel bar on which the train travels and the sleepers are orderly arranged blocks of concrete blocks on which the track is placed. Ballast is the floor which is made from small stones on which the sleeper is placed and half buried in. The rail pad is the placed on the sleeper and acts as the contact to the track. Dampers are external damping structures placed on the track and sleeper to absorb vibrations. The three components to which the parameter changes are made are sleepers, ballast and dampers.

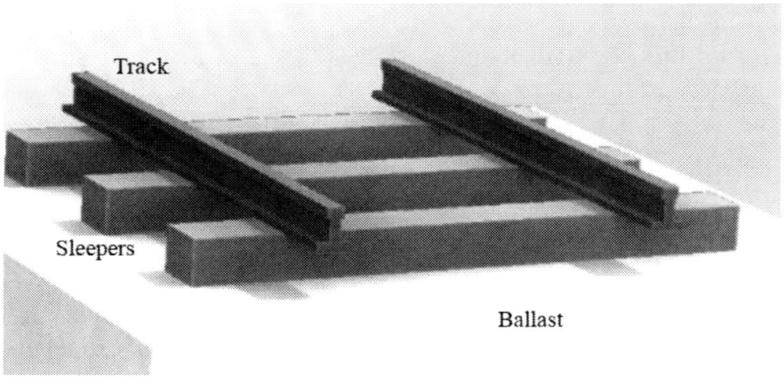

Fig. 1. Rail, sleeper and ballast of a railway embankment.

1.2.2 Types of Vibration

Vibration is a periodically repeating motion. When a train moves on a rail embankment, different types of vibrations occurs. Abe *et al.* [6] states that the main causes for vibration is due to the roughness between the track and the wheel and due to differing stiffness due to periodically placed sleepers. A vibration caused by this sleeper placement is called the pinned-pinned resonance and it occurs when the vibration wavelength is same as the sleeper spacing [7]. Static loading on the track also causes vibration and in Auresh et al [8] it is modelled as a quasi-static load as the load is moved along the track. But it is the wheel/track irregularities which causes most significant vibrations. Nielsen *et al.* [9], investigates the effect on the track as the wheel becomes more flattened due to skidding and grinding. The results showed that the level of ground vibration was much higher for trains with flattened wheels.

1.2.3 Selected parameters

For sleeper, the parameters selected are spacing between sleepers and its mass. Abe *et al.* [6] states that by reducing the spacing the pinned-pinned vibration can be reduced. It also a study by Dwight *et al.* [10] found that mass of the sleeper effects the wave frequencies of the track. The dimension of the sleeper is also investigated because changing the height decreases the bending movement as stated by Kollo *et al.* [11] and this could influence the vibration as well.

Thickness and stiffness is selected for the ballast as these two parameters can be easily altered by increasing the height by adding more stones and by changing the stiffness by altering the size of the stone. Benedetto *et al.* [12] investigated the response of a rubber mixed ballast which would have low stiffness and the results should increase in vibration in low frequency bands of 10 - 30 Hz. Zhang *et al.* [13] found that the increase in thickness, decreases particle velocity. This would decrease the vibration levels as well.

The influence of rail pad on the vibration and flexural modes of the sleeper was investigated by Kaewunruen *et al.* [14] with results showing a relation between the rail pad stiffness and the bending mode with generally high stiffness of rail pad, reducing translation and rotational modes. Results also showed that most damage occurs to the sleepers in the resonant frequencies due to the modes of vibration. It is also cited in Abe *et al.* [6] that the stiffness reduction of the rail pads can influence a lower vibration on the rail system. This component is a focus of this paper due to the ease of adjustability. Changing the rail pad is economically viable compared to changing larger components such as track, sleeper and ballast. Furthermore stiffness and damping are the only variable parameter available for this component since changes to the dimension and mass are limited by the track and sleeper dimensions.

2 Modelling Methodology

2.1 One-dimensional Analytical model and Harmonic excitation

The literature review showed that the type of model used for study reflected the objectives of that study. For example, Zhang *et al.* [15] used two-dimensional analytical model as the effect of one unsupported sleeper on the other sleeper is required to be known. For investigation into wheel and track interactions as in the case of Ferrara *et al.* [16] where a new numerical model to simulate the wheel/track irregularity vibrations is presented and in the case of Nielsen *et al.* [9] where the vibration and the force impacts caused by damaged wheels are investigates, a non-linear Hertzian theory is used would also mean a two-dimensional calculation. As for Dwight *et al.* [10] which compares the Timoshenko continuous beam theory with semi-analytical finite element (SAFE) calculations uses three-dimensional modelling as the lateral in addition to vertical movement of the beam in investigated.

To comply with the objectives of this study, the model uses a one-dimensional model as the effects of direct harmonic load is investigated. Harmonic forces are useful for the investigation as they are continuously applied forces which stimulates the effects of a moving wheel and the consequent stiffness variation in addition to allowing easier understanding of resonances or parametric excitations which provides the maximum values [8]. These extremes are useful to gauge the effectiveness of parameter changes.

Therefore, for this study series of increasing degrees of freedom models up to three degrees of freedom are formed using spring, mass and damping blocks. Only one side of the rail track is used, hence the mass of the ballast and sleeper and excitation force are halved. Further changes to the model are made based on the properties of the components.

2.2 Properties of Embankment Components

The important three components are the mass, damping and stiffness of the components. While mass is an easy property to determine, stiffness and damping are more challenging. Damping of materials is very difficult to discern unless by experimentation [18] and while the stiffness of a component might be simpler in comparison, stiffness changes with the dimension of the component. This is challenging for the ballast where a small volume is required to be isolated to study the effects of vibration. The effort was simplified with the analytical data from Ferrara *et al.* [16] and Zhai *et al.* [19]. Table 1 shows the properties of different components. The data is taken from sources which has similarity to parameters which is studied in this research. This includes high-speed trains at least 100 km/s of speed, concrete sleeper and at-grade railways (not underground subways or LRTs).

To calculate the stiffness the ground directly below the sleeper is taken in to account in a diverging geometry similar to a trapezoid [19]. This geometry takes in account of the load distribution. Stiffness is calculated by the formula provided by Ferrara *et al.* [16].

$$k = \frac{2(l-b)\tan\emptyset}{\ln\left[\frac{l(2h \times tan\emptyset + b)}{b(2h \times tan\emptyset + l)}\right]} \times E \qquad (1)$$

Table 1. Data for the components.

Component	Property	Value	Units
Rail track	Mass	51	kg/m
	Length	850	Mm
	Young Modulus	207 [16]	GPa
Rail pad	Stiffness	20 – 3500 [14]	MN/n
	Damping coefficient	15 [9], 20 [14] 75 [19]	Ns/m
	Size	148 × 180 × 10 [20]	Mm
Sleeper	Mass (half)	150 [21, 19]	Kg
	Spacing	600 [6, 10]	Mm
	Dimension	1250 × 250 × 200 [8]	Mm
	Poisson Ratio	0.15 [8]	
	Density	2500 [8], 2750 [15]	kg/m^3
Ballast	Stiffness	280.4	MN/m
	Damping coefficient	132 [9], 165 [15] 58.8 [19],	kNs/m
	Modulus of Elasticity	0.11 [19]	GPa
	Thickness	300* (200 mm subballast) [21]	Mm
	Angle	35 [19]	Degree

2.3 Properties of Embankment Components

The excitation force is the axle load placed on the track from the wheels of the train. This force is not the weight or the static load of the train itself, but the downward force on the track due to the movement of train. In fact, axle load of a train changes with the speed of the train [8]. Similarly, to the embankment properties, the axial force should also match with the experimental excitation. The type of train from which the measurements were taken by Connolly *et al.* [21] is a high-speed TGV *Reseau* (TGV), train which operates at an average speed of 294.7 km/h or 81.89 m/s. The axle load on the track only increases up to a certain speed and from then onwards the axial load decreases [8]. Therefore, fast trains such as Sweden's X2000 train which operates at 200 km/h [22] had a lower axial load of 18.25 kN than Spain's Talgo train which operates at 160 km/h [23] had a maximum axle load of 75 kN. As the axial load of TGV was not provide, an axle load of 20 kN is selected as this value is closer to the axle load of X2000.
The excitation frequency is selected to be 5.6 Hz and 8.3 Hz for speed of 200 km/h and 300 km/h respectively. This is for an axle distance of 10 m.

2.4 Analytical model using One, Two and Three Degrees of Freedom

The equation of motion for a single degree mass with damping and stiffness is given by Equation 2 and the harmonic excitation by Equation 3. [24]

$$m_T\ddot{x} + c_{eq}\dot{x} + k_{eq}x = 0 \tag{2}$$

$$F\cos(\omega t) \tag{3}$$

Where, m_T is the mass. $\ddot{x}$ is the acceleration. $c_{eq}$is the equivalent damping of the system. $\dot{x}$ is the velocity. k_{eq} is the equivalent stiffness of the system and x is the displacement. For the Eq. 2 the force is represented by F and ω is the driving frequency and t is time.

The amplitude of the steady force response is given by the Equation 4 and the phase, φ is given by the Equation 5.

$$X = \frac{F}{\sqrt{[(k_{eq}-m_T\omega^2)^2+(c_{eq}\omega)^2]}} \tag{4}$$

$$\varphi = \tan^{-1}\left(\frac{2\tau\omega_n\omega}{\omega_n^2-\omega}\right) \tag{5}$$

Where, τ is the damping ratio.

The Equations 4 and 5 are used to find the harmonic response of the system. Equation 6 shows the final form of the equation. The homogenous or transient response is not included as the motion begins from zero initial velocity and displacement, therefore these responses will be zero.

$$x(t) = X\sin(\omega t - \varphi) \tag{6}$$

The analytical model is also verified using MATLAB Simulink model shown in Figure 2. The results from Equation 6 is verified by the Simulink model which is a second order model calculated using ordinary differential equation of Dormand-Prince method.

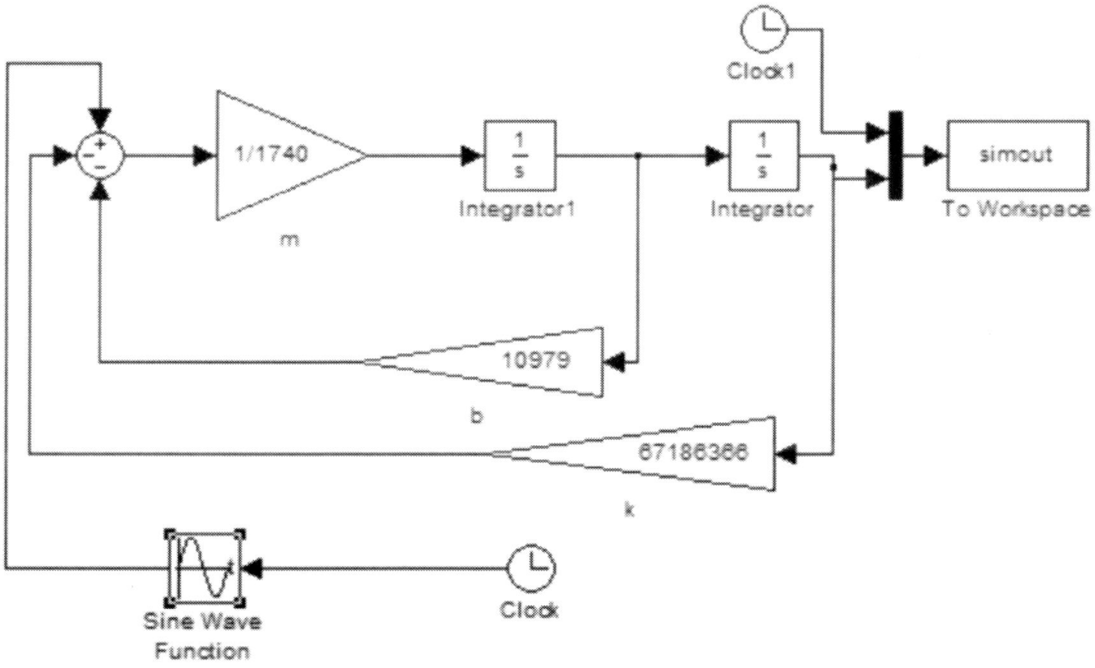

Fig. 2. Simulink model for harmonic excitation for a spring damper system.

Multi-degree of freedom models is more complicated than one degree of freedom as each mass are coupled with the subsequent masses. Therefore, decoupling is required to measure the individual response of the mass. Equations 7 and 8 shows that general two-degree and three-degree of freedom system under harmonic force.

$$\begin{bmatrix} m_1 & 0 \\ 0 & m_2 \end{bmatrix}\begin{Bmatrix} \ddot{x}_1 \\ \ddot{x}_2 \end{Bmatrix} + \begin{bmatrix} c_1 + c_2 & -c_2 \\ -c_2 & c_2 \end{bmatrix}\begin{Bmatrix} \dot{x}_1 \\ \dot{x}_2 \end{Bmatrix} + \begin{bmatrix} k_1 + k_2 & -k_2 \\ -k_2 & k_2 \end{bmatrix}\begin{Bmatrix} x_1 \\ x_2 \end{Bmatrix} = \begin{Bmatrix} 1 \\ 0 \end{Bmatrix} F e^{i\omega t} \quad (7)$$

$$\begin{bmatrix} m_1 & 0 & 0 \\ 0 & m_2 & 0 \\ 0 & 0 & m_3 \end{bmatrix}\begin{Bmatrix} \ddot{x}_1 \\ \ddot{x}_2 \\ \ddot{x}_3 \end{Bmatrix} + \begin{bmatrix} c_1 + c_2 & -c_2 & 0 \\ -c_2 & c_2 + c_3 & -c_3 \\ 0 & -c_3 & c_3 \end{bmatrix}\begin{Bmatrix} \dot{x}_1 \\ \dot{x}_2 \\ \dot{x}_3 \end{Bmatrix} + \begin{bmatrix} k_1 + k_2 & -k_2 & 0 \\ -k_2 & k_2 + k_3 & -k_3 \\ 0 & -k_3 & k_3 \end{bmatrix}\begin{Bmatrix} x_1 \\ x_2 \\ x_3 \end{Bmatrix} = \begin{Bmatrix} 1 \\ 0 \\ 0 \end{Bmatrix} F \cos\omega t \quad (8)$$

Beginning from three-degree of freedom, Equation 8 shows the general system of Figure 3. The masses from top to bottom with the subscript 1,2 and 3 are of track, sleeper and ballast. The stiffness and damping coefficient of subscript 1 are of the rail pad, then for the ballast for subscript 2 and 3. As the force is applied on the track, only the first matrix for force is placed by a harmonic excitation while the rest are zero.

For the two-degree of freedom, in Equation 7 the mass (m_1) combines the mass of track and the sleeper. Also stiffness and the damping are for the rail pad and for ballast. The property of ballast is not duplicated as in the three-degree of freedom equation.

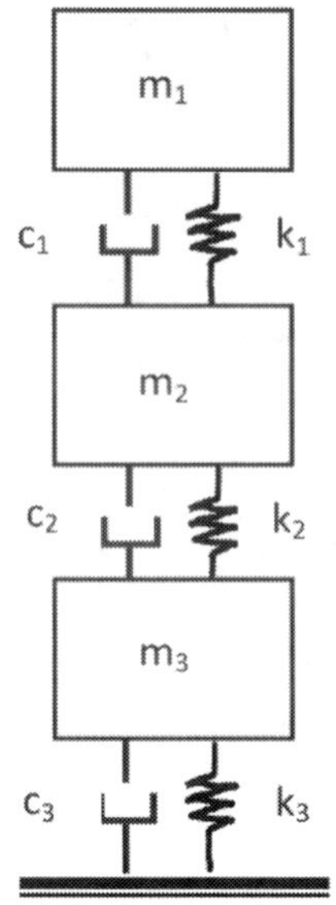

Fig. 3. Three degree of freedom system.

Two separate methods have been used for solving the Equations 7 and 8. For two-degree of freedom system, mechanical impedance $Z(i\omega)$ is used to find the response for the two-degree of freedom displacement, velocity and acceleration as shown in Equations 9 – 13.

$$x_1(t) = X_1 e^{i\omega t} \tag{9}$$

$$x_2(t) = X_2 e^{i\omega t} \tag{10}$$

Equations 9 and 10 are steady state displacement solution for the amplitudes X_1 and X_2 with the forcing frequency of ω. The values for the amplitudes are obtained from the mechanical impedance Equations 12 and 13.

$$Z_{rs}(i\omega) = -\omega^2 m_{rs} + i\omega_{rs} + k_{rs} \tag{11}$$

The subscript rs corresponds to the location of the property in the matrices in Equation 7.

$$X_1 = \frac{Z_{22}(i\omega)F_{10}}{Z_{11}(i\omega)Z_{22}(i\omega) - Z_{12}^2(i\omega)} \tag{12}$$

$$X_2 = \frac{-Z_{12}(i\omega)F_{10}}{Z_{11}(i\omega)Z_{22}(i\omega) - Z_{12}^2(i\omega)} \tag{13}$$

But substituting the values of X_1 and X_2 to Equations 9 and 10, the displacement due to force can be calculated. To find the velocity and the acceleration the Equations 9 and 10 are differentiated once for velocity and twice for acceleration.

An easier route has been taken for three-degree of freedom equation by using MATLAB ordinary differential equation 23 for calculation. Appendix A shows the differential equations used where the three equations from the matric in Equation 8 is set as six first order differential equations.

Another important formula for quantifying the level of vibration is decibels formula Equation 14. Decibels helps to standardise results so that it can be compared with data from other sources [25].

$$L_v = 20\, log_{10} \left(\frac{v}{v_{ref}}\right) \tag{14}$$

L_v is the velocity level represented in decibels and v_{ref} is the reference velocity. This velocity is an accepted value for reference agencies such as International Organisation for Standardisation (ISO) and Federal Transit Administration (FTA). The value used in this paper is $2.54 \times 10^{-8} m/s$ which is the value used by the FTA [25].

2.5 Numerical modelling

Similar to the analytical model, the numerical model uses a single track, sleeper and ballast is shown in Figure 4. But in addition to the analytical modelling a rail pad is modelled and placed as well. Furthermore, the ballast was modelled in a trapezoid geometry for model A and model B uses two sleepers to test the response due to varying the sleeper distance. The components where modelled from SOLIDWORK and was imported to ANSYS Workbench. The bodies were given the material properties and then Harmonic Response module was selected to apply a force of 10000N on the track surface. The boundary condition is set as frictionless support and the frequency range is selected from 0 to 250 Hz with 10 Hz increment. Same properties from Table 1 is used for the material data.

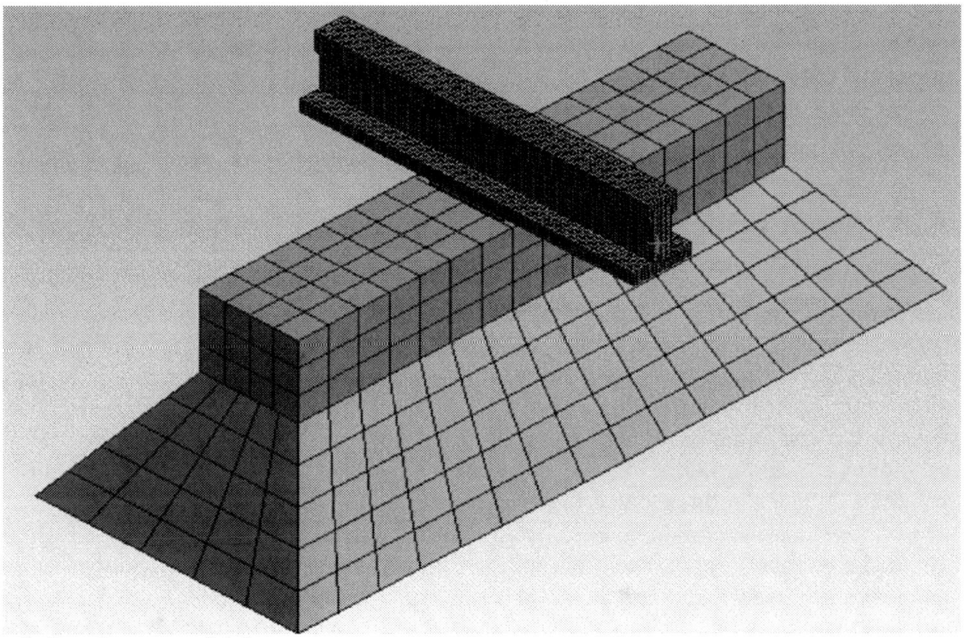

Fig. 4. One sleeper model A.

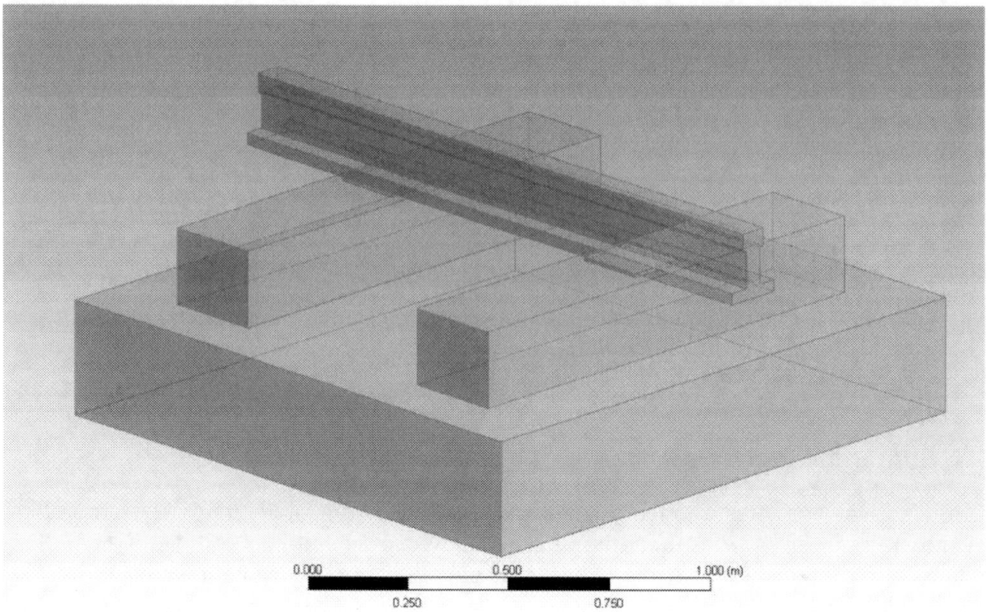

Fig. 5. Numerical model of the track, sleeper and ballast with the force being applied to the track in negative y-direction.

Table 2. Mesh Property of Model A and B

Mesh Property	Model A, one sleeper	Model B, two sleepers
Element size	For track 0.01 m Rest of the body Automatic	For track 0.01 m Rest of the body Automatic
Meshing Method	Hexagonal	Hexagonal
Number of Elements	61714	1339267
Element Quality	0.955	0.962
Skewness	0.0574	0.0486

3 Results and Discussion of Analytical and Numerical Modelling

The analytical model is done using formulas and calculations. Therefore, an error might occur in the calculations which might offset the results. To verify that there is no such error, Simulink model is made to calculate the harmonic motion of a second order differential equation. The results showed that there is observable correlation, and this is statistically proven with the Table 3. Pearson correlation, which indicates strong relation between the two means as the value is close to value one, and the value of the statistical test, 0.9733 is very close to one. Furthermore, the t-critical is much higher than the t-statistical value, hence there is 95% confidence that the mean is similar.

There are many disadvantages of using the one-degree of freedom as the effects of individual components properties cannot be discussed as the total and equivalent values are used. Therefore, the second-degree of freedom model from Equations 9 and 10 are used to provide the results. The results are formed with velocity response for two masses. The first mass is the total mass of track and sleeper and the second mass is that of the ballast. These results are also given in frequency domain as well to compare with the numerical results for harmonic excitation. The three-degree of freedom is not used because second-degree od freedom model is easier to convert to velocity response and the mass of track and sleeper is not large compared to ballast, so they can be combined.

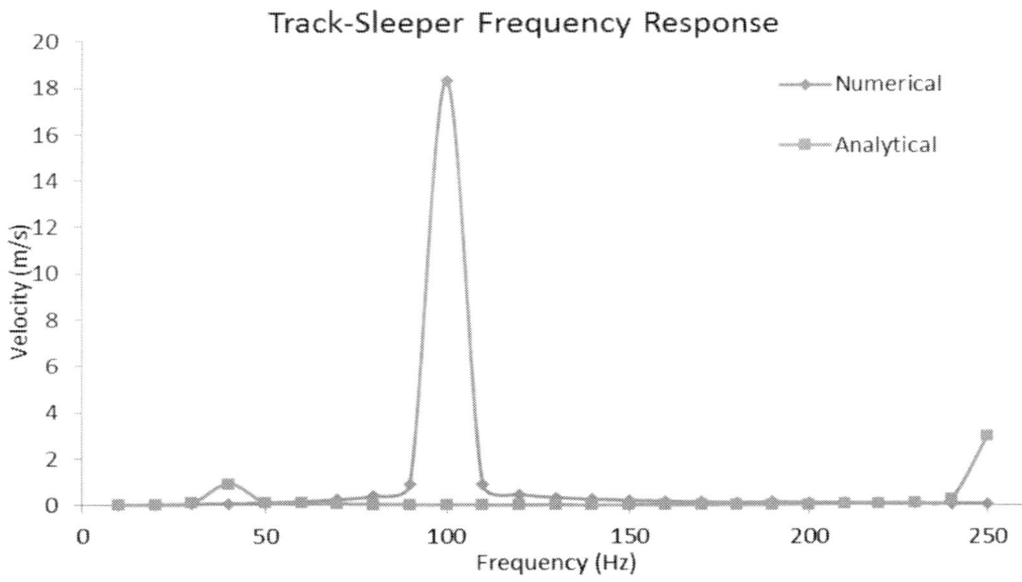

Fig. 6. Velocity frequency of numerical and analytical second-degree freedom model for track-sleeper.

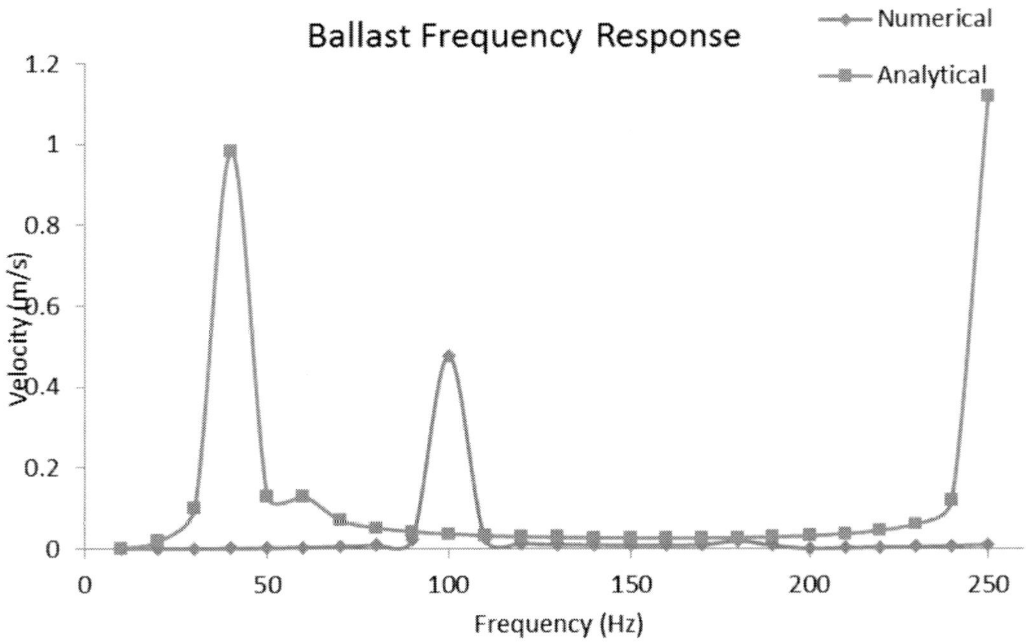

Fig. 7. Velocity frequency of numerical and analytical second-degree freedom model for ballast

While, the visual comparison between the numerical and the analytical model, Figures 6 and 8 might seem apparent because of the peaks, the lower values remain close to each other for both figures. Provides 95% confidence that the means are the same for both of the models. The t-statistical is lower than the t-critical for both the tests but the Pearson Correlation is low. This means that while correlation is not very high, the means are nevertheless same. The peaks for numerical model is at 100 Hz and for analytical at 40 and 250 Hz. The modal simulation using ANSYS Workbench modal analysis module shows that at 100 Hz the natural frequency of the rail pad is reached as shown in Figure 8. This causes the high vibration response. Therefore, the excitation frequency must not reach this 100 Hz while using the rail pad with this particular stiffness form Table 1 of 20 MN/m.

Fig. 8. Modal analysis for numerical Model A using modal analysis

Table 3. Paired Two Sample for Means t-test for Analytical and Simulink results

	Simulink	*Analytical*
Mean	1.48E-05	7.56E-06
Variance	1.44E-07	1.82E-07
Observations	21	21
Pearson Correlation	0.973328	
Hypothesized Mean Difference	0	
Degree of Freedom	20	
t Stat	0.317091	
P(T<=t) two-tail	0.754461	
t Critical two-tail	0.686954	
Mean	0.973328	

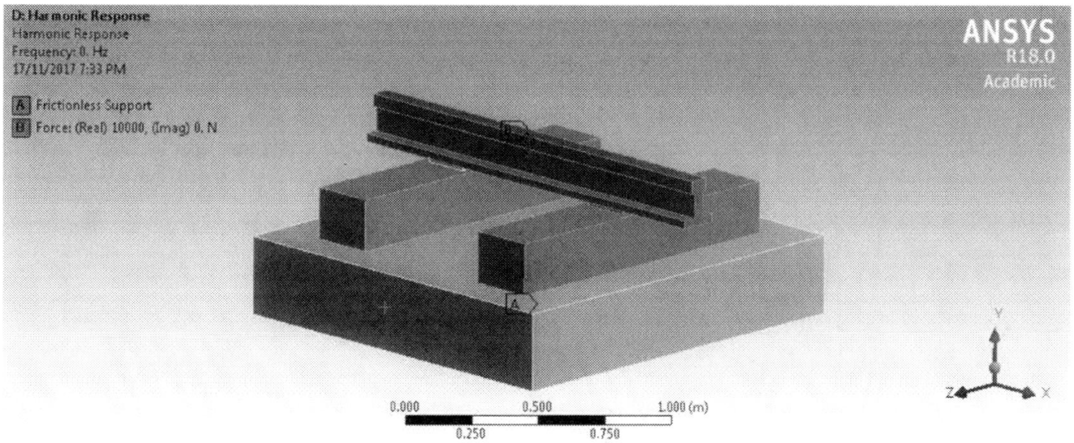

Fig. 9. Model used to study the vibration response to different sleeper spacing

Numerical model to investigate the relation between acceleration response due to sleeper distances of 50, 60 and 70 cm was simulated using the model as shown in Figure 9. The results provided in Appendix B gave insufficient difference to the response. This is validated using ANOVA test in Table 4. where the F-critical 3.12 is greater than F-statistics of 1.88. The reason for using ANOVA is that means of three sets needed to be compared.

Table 4. Show the summary of variance of the sleeper spacing and the F-test

SUMMARY						
Groups	*Count*	*Sum*	*Average*	*Variance*		
50cm	25	14.24983	0.569993	0.805951		
60cm	25	208.6954	8.347815	578.3901		
70cm	25	77.36599	3.09464	48.01165		
ANOVA						
Source of Variation	*SS*	*df*	*MS*	*F*	*P-value*	*F crit*
Between Groups	787.2018	2	393.6009	1.882634	0.159604	3.123907
Within Groups	15052.99	72	209.0692			
Total	15840.19	74				

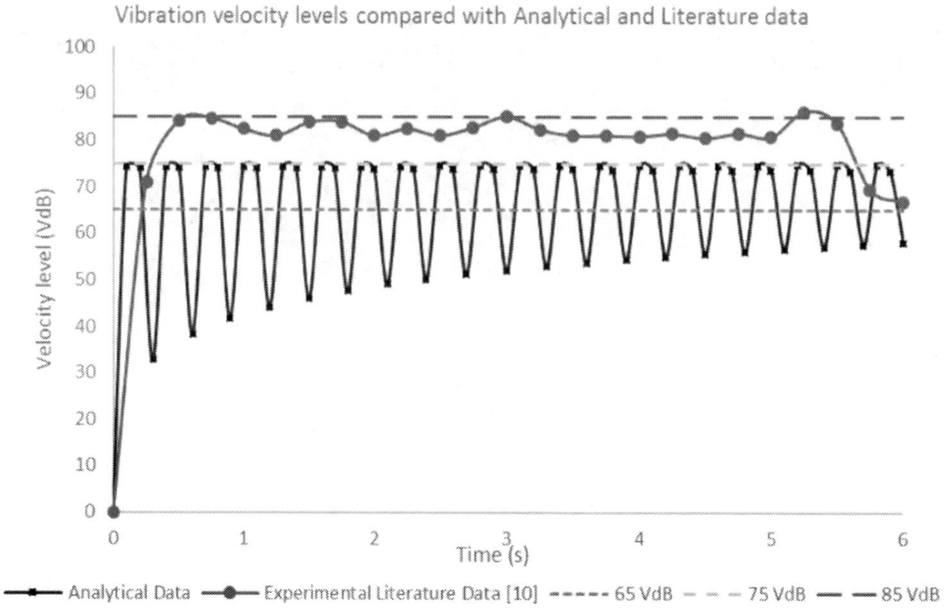

Fig. 10. Actual literature data compared with analytical data with benchmarking of 65 VdB, 75 VdB and 85 VdB

It was found out by analytical calculations that rail pad stiffness produces high response on the vibration of the system. Therefore, using Matlab, the lowest possible value of vibration was produced with the provided pad stiffness from Table 1. The as the stiffness increases the vibration of the system decreased. Therefore, the highest stiffness value of 3500 MN/m value was used. This material corresponds to high density Ethylene vinyl acetate polymer (EVA) [14]. The maximum value from the analytical data was compared for the frequency using alternatively 5.6 and 8.3 Hz. As 8.3 Hz had higher magnitudes it was selected.

The lowest analytical vibration response was then converted to velocity decibels which has the reference value of $2.54 \times 10^{-8} m/s$ [25] to make it easier to compare with the benchmark values. In Figure 10, three dashed lines of values 65, 75 and 85 VdB are the three bands of vibration levels which will be felt by humans with the intensity increasing with the values. This is stated in the "Transit Noise and Vibration Impact Assessment", released by American Federal Transit Administration (FTA) [25], that while 65 VdB is barely precipitable for humans and 75 VdB can be annoying for some, at 85 VdB people will find this level disruptive and is acceptable if only infrequent amount of trips.

Changes to the rail pad stiffness is effective for suppressing vibrations which is shown in Figure 10. This statistically proven with ANOVA single factor test of 95% confidence in Table 5 as the F-statistical values of 8.47 larger than 3.95 F-critical. ANOVA single factor was used due to the difference in count for the results.

Table 5. Show the summary of variance of the analytical and literature data and the F-test

SUMMARY						
Groups	*Count*	*Sum*	*Average*	*Variance*		
Analytical Data	61	3976.862	65.19446	211.7338		
Literature Data	29	2181.412	75.22112	278.9576		
ANOVA						
Source of Variation	*SS*	*Df*	*MS*	*F*	*P-value*	*F crit*
Between Groups	1976.05	1	1976.05	<u>8.476418</u>	0.004558	<u>3.949321</u>
Within Groups	20514.84	88	233.1232			
Total	22490.89	89				

4 Conclusions

In this study, a simplified analytical model and a numerical model is made to understand the vibration response due to a train load which is simulated by harmonic excitation. The numerical modelling showed that a resonance does occur 100 Hz range which belongs to the rail pad. List of components of the rail embankment has been reviewed using journal articles and their vibration influencing parameters in addition to their other properties have been complied and refined to make sure they meet the scope of the project. Then analytical and numerical models where created using ANSYS and Matlab for numerical and analytical respectively. A model very closed to the actual literature data was made and the vibration was supressed by increasing the rail pad stiffness. Comparison between the modified analytical model gives the velocity response of 75 VdB as the maximum peak.

Two-dimensional modelling is required to investigate more complex vibrations and also take in the influence of geometry of the components such as sleeper spacing. Also scope of the project needs to be more focused on particular train or trains from certain region, as different regions uses different types of trains and embankment. Most importantly an acoustic assessment is required to understand the noise generated under the new rail pad stiffness. If the stiffness had worsened the noise, other methods needs to be explored such as using external dampers.

References

1. H. Williams, *Road and rail transportation,* **1st ed.** New York: Facts on File, 50-52(2004).
2. R. Abe and H. Kato, What led to the establishment of a rail-oriented city? Determinants of urban rail supply in Tokyo, Japan, 1950–2010, TRANSPORT POLICY, **58**, 72-79 (2017).
3. F. Laroche, C. Sys, T. Vanelslander, E. Van de Voorde, Imperfect competition in a network industry: The case of the European rail freight market, TRANSPORT POLICY, **58**, 53-61 (2017).
4. Z. Chen, J. Xue, A. Rose, K. Haynes, The impact of high-speed rail investment on economic and environmental change in China: A dynamic CGE analysis, Transp Res Part A Policy Pract, **92**, 232-245 (2016).
5. D. Connolly, G. Marecki, G. Kouroussis, I. Thalassinakis, P. Woodward, The growth of railway ground vibration problems — A review, Sci Total Environ, **568**, 1276-1282 (2016).
6. K. Abe, Y. Chida, P. E. Balde Quinay, K. Koro, "Dynamic instability of a wheel moving on a discretely supported infinite rail," J. Sound Vib., **333**, no. 15, 3413–3427 (2014).
7. N. Persson, Predicting Railway-Induced Ground Vibrations, Post Graduate, Lung University (2016).
8. L. Auersch, The excitation of ground vibration by rail traffic: Theory of vehicle-track-soil interaction and measurements on high-speed lines, J. Sound Vib., **284**, no. 1–2, 103–132 (2005).
9. J. C. O. Nielsen, G. Lombaert, S. François, A hybrid model for prediction of ground-borne vibration due to discrete wheel/rail irregularities, J. Sound Vib., **345**, 103–120 (2015).
10. W. Li, R. A. Dwight, T. Zhang, On the study of vibration of a supported railway rail using the semi-analytical finite element method, J. Sound Vib., **345**, 121–145 (2015).
11. S. A. Köllő, G. Köllő, A. Puskás, Nexus of the Load Bearing Capacity of Rails and the Stiffness of the Optimized Sleepers," Procedia Technol., **22**, 312–318 (2016).
12. A. Benedetto, F. Tosti, L. Bianchini Ciampoli, A. Calvi, M. G. Brancadoro, A. M. Alani, Railway ballast condition assessment using ground-penetrating radar – An experimental, numerical simulation and modelling development, Constr. Build. Mater., **140**, 508–520 (2017).
13. Z. Zhang, X. Zhang, H. Qiu, M. Daddow, Dynamic characteristics of track-ballast-silty clay with irregular vibration levels generated by high-speed train based on DEM, Constr. Build. Mater., **125**, 564–573 (2016).
14. S. Kaewunruen and A. M. Remennikov, Sensitivity analysis of free vibration characteristics of an in situ railway concrete sleeper to variations of rail pad parameters, J. Sound Vib., **298**, no. 1–2, 453–461 (2006).
15. S. Zhang, X. Xiao, Z. Wen, X. Jin, Effect of unsupported sleepers on wheel/rail normal load, Soil Dyn. Earthq. Eng., **28**, no. 8, 662–673 (2008).
16. R. Ferrara, G. Leonardi, F. Jourdan, "SIIV - 5th International Congress - *Sustainability of Road Infrastructures Numerical Modelling of Train Induced Vibrations*, Procedia - Soc. Behav. Sci., **53**, no. 0, 155–165 (2012).
17. D. Inman, *Engineering vibration*, **4th ed**. Boston: Pearson, 16 – 58, (2014).

18. W. Zhai, K. Wang, J. Lin, Modelling and experiment of railway ballast vibrations, JSV, **270**, no. 4-5, 673-683 (2004).
19. J. Stead and M. Kerr, *Spc 232 concrete sleepers*, (2012).
20. D. P. Connolly, G. Kouroussis, P. K. Woodward, P. Alves Costa, O. Verlinden, M. C. Forde, Field testing and analysis of high speed rail vibrations, Soil Dyn. Earthq. Eng., **67**, 102–118 (2014).
21. S. A. Suhairy, Prediction Of Ground Vibration From Railways, Undergraduate Graduate Thesis, Swedish National Testing and Research Institute (2000).
22. I. Crespo-Chacón, J. García-de-la-Oliva, E. Santiago-Recuerda, On the Use of Geophones in the Low-Frequency Regime to Study Rail Vibrations, *Procedia Engineering*, **143**, 782-794 (2016).
23. S. Rao, *Mechanical vibration*, **5th ed**. Reading, Massachussetts: Addison Wesley, 257 – 374 (2011).
24. Federal Transit Administration, *Transit Noise and Vibration Impact Assessment*, (2006).

Numerical Analysis and Geometry Optimisation of Vertical Vane of Room Air-conditioner

Abdulkareem Sh. Mahdi Al-Obaidi[1,*], *Chan Wei* Yin[1] , *Felicia Wong Yen* Myan[1] and *Yan Yao* Chen[1]

[1]School of Engineering, Faculty of Built Environment, Engineering, Technology & Design, Taylor's University, Subang Jaya, Selangor DE, Malaysia

Abstract. Vertical vanes of room air-conditioners are used to control and direct cold air. This paper aims to study vertical vane as one of the parameters that affect the efficiency of dissipating cold air to a given space. The vertical vane geometry is analysed and optimised for lower production cost using CFD. The optimised geometry of the vertical vane should have the same or increased efficiency of dissipating cold air and have lesser mass compared to the existing original design. The existing original design of vertical vane is simplified and analysed by using ANSYS Fluent. Efficiency of wind direction is define as how accurate the direction of airflow coming out from vertical vane. In order to calculate the efficiency of wind direction, 15° and 30° rotation of vertical vane inside room air-conditioner are simulated. The efficiency of wind direction for 15° rotation of vertical vane is 57.81% while efficiency of wind direction for 30° rotation of vertical vane is 47.54%. The results of the efficiency of wind direction are used as base reference for parametric study. The parameters investigated for optimisation of vertical vane are focused at length of long span, tip chord and short span. The design of 15% decreased in vane surface area at tip chord is the best optimised design of vertical vane because the efficiency of wind direction is the highest as 60.32%.

1 Introduction

Vertical vane is part of air-conditioners that can set horizontal air flow direction to left or right. Currently, vertical vane is rarely been studied as part of parameters that affect the efficiency of dissipating cold air to a given space. Most studies and researches have been done on horizontal louver's parameters that will affect the performance of air-conditioners. However, the studies that investigate the horizontal louvers can be used as reference for vertical vanes since the boundary conditions for both are the same in CFD simulation. With the assist of CFD software, geometries changes of vertical vane can be analysed without wasting money to manufacture it [1]. Moreover, CFD software can save time by simulating the case study instead of conducting physical testing with equipment such as wind tunnel. Panasonic is one of the companies that concern about any improvement can be made from design of vertical vane.

* Corresponding author: abdulkareem.mahdi@taylors.edu.my

Nakanishi et al. [2] investigated the air flow passing through louver. The change of louver angle that cause pressure is studied. Three different louver models respectively with louver angle of 45°, 35° and 25° were tested. Laser Doppler Velocimetry (LDV) and smoke wire methods have been conducted to visualise the pattern and velocity of airflow. After that, CFD++ software is used to simulate the result. The result of experiment and CFD were compared where the compressible Navier-Stokes equation and k-ε turbulence model are used in CFD simulation. The CFD could have good agreement with the experimental result from LDV and smoke wire methods. It is concluded that the larger the louver angle, the larger the vortexes. The velocity of airflow is increased due to the narrowed flow path by large vortexes. As a result, the pressure loss is increased with larger louver angle.

A similar study from Noh et al. [3] was conducted which was based on the research of Nakanishi et al. [2]. This research investigates how the shape of airfoil and flat louver affect the pressure drop across weather hood. Moreover, different angles of attack which were 25°, 35° and 45° were studied at both shape and flat louvers. The numerical method was conducted by using SolidWorks Flow Simulation software. Since the inlet velocity of airflow did not reach 0.3 Ma (Mach number), incompressible flow was applied [4]. Incompressible flow was applied and standard k-ε turbulence model were used in CFD simulation. It is concluded that higher louver angle will induce greater pressure drop due to the increase in velocity of airflow. Other than that, airfoil louvers show less pressure drop compared to the flat louvers. In result, airfoil louvers with louver angle of 25° contributed the lowest pressure drop compared to flat louvers and other louver angles.

CFD used as reliable tool to predict the indoor airflow in recent years [5-8]. For instance, Karimipanah et al. [6] conducted several times of researches about the indoor air quality and comfort conditions in classroom. The initiative of investigating indoor air quality is due to the reason of caring the pupil's health and those children who have allergy or asthma. In this research, Karimipanah et al. compared the results calculated by RNG (Re-Normalisation Group) turbulence model and standard k-ε turbulence model respectively with measured result. It is concluded that RNG turbulence model can simulate velocity of airflow with higher agreement with measured result. In contrast, standard k-ε turbulence model shows better capability of predicting the temperature distribution over the classroom compared to RNG turbulence model. This research gives a good idea of choosing suitable turbulence model to get desired simulation result which is approximated to measured result closely.

Sadrizadeh and Holmberg [9] conducted a very detail study about different type of turbulence models applied for indoor airflow prediction. There is nine type of turbulence models are analysed and validated with the experimental result. The experiment is carried out in a box-shaped wind tunnel. A mannequin was built to model a people sitting inside a room (wind tunnel). The locations of measured temperature and velocity were at both; front of and behind the, mannequin. Most of the Reynolds averaged Navier-Strokes approach (RANS) were selected to apply in the CFD simulation in this study. RNG k-ε (Re-normalisation Group) is the best prediction of velocity of airflow out of RANS turbulence models.

This research aims to analyse and optimise the geometry of the vertical vane in order to minimise its mass for lower production cost. Besides that, optimising geometry of vertical vane shall not bring negative impact to the performance and efficiency. Firstly, the method of defining the efficiency for simplified vertical vane is studied and analysed. Next, different geometries and dimensions of vertical vane will be analysed using ANSYS Fluent ver. 18 (CFD software) for a given set of flow conditions. The research process will be follow by optimisation to select the vertical vane shape which should have same efficiency of dissipating cold air compared to the existing original vertical vane.

2 Research methodology

The method of doing numerical analysis of vertical vane is written and explained in this section. Besides that, the calculation of efficiency for vertical vane is defined.

2.1 3D CAD Modelling

The geometry of existing original design of vertical vane is studied and the assumptions of 3D CAD modelling is made for geometry clean-up. The simplified vertical vane is drawn by using SolidWorks ver. 2016. The existing original design and simplified vertical vane is shown in Fig. 1. The simplified vertical vane is redrawn based on a few assumptions made.

Firstly, the tip of vertical vane where it attached to air-conditioner is assumed no air flow into the hole. The reason is the tip part will later be locked with a nut and the volume will be filled after assembly. Thus, the tip of vertical vane is simplified to have filled volume. Besides that, the hook of vertical vane where the connecting bar attached to is simplified to become fillet instead of having irregular shape of hook. Changing the geometry of hook will not affect the air flow that much but it helps to reduce complexity of geometry in CFD simulation afterward. The circumference length of simplified vertical vane follows exactly the dimension of existing original design. For the upper surface of vertical vane, it is simplified to have same thickness as lower surface. The reason is the thickness of vertical vane contributes very small effect to velocity of air flow.

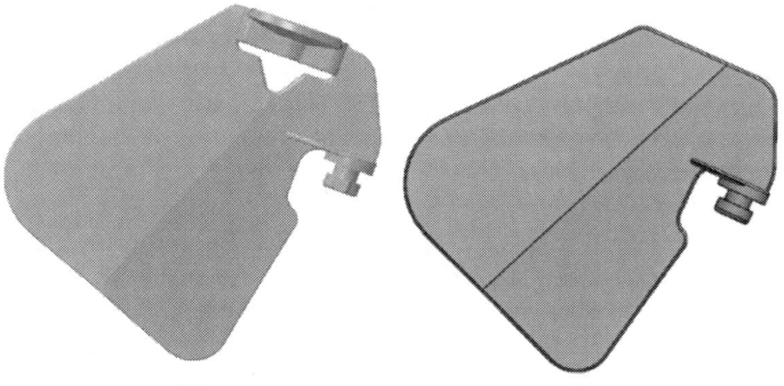

(a) (b)

Fig. 1. Vertical vane (a) existing original design (b) simplified.

2.2 Numerical modelling

2.2.1 Geometry modelling

Geometry model for CFD is based on a condition that air-conditioner is installed in a 6 m×5 m×3.5 m of room. Usually, the fluid domain should build with 5 meter square base with height of 3.5 m for this model. However, back flow of air is not the outcome that wanted to focus in simulation. Thus, the fluid domain drawn for this model is much bigger than size of the room. It can increase the accuracy of simulated result. Shell of room air-conditioner is drawn and vertical vanes are placed inside the shell. Following by that, a half sphere shape of fluid domain is drawn and shown in Fig. 2. The half sphere is drawn with 10 m diameter.

Two different rotations of simplified vertical vane are studied. First case is to simulate the 0° of vertical vane in Fig. 3 (a) and the second case is to simulate the 30° angle of vertical vane in Fig. 3 (b). The case of 0° angle of vertical vane is used as base reference for meshing size. The 30° is the maximum rotation that allows rotating for the product of room air-conditioner.

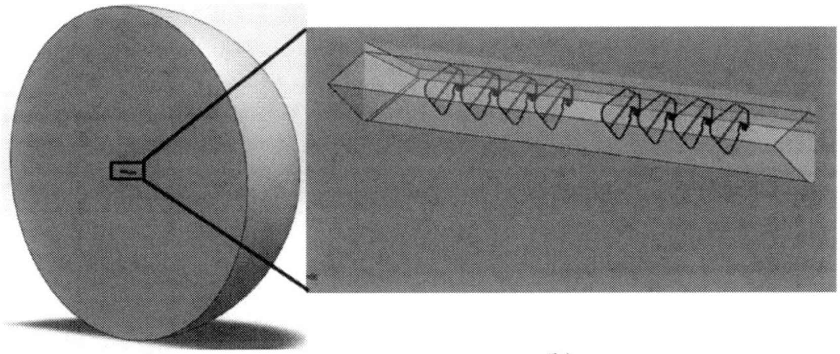

(a) (b)

Fig. 2. (a) The half sphere of fluid domain and (b) location of vertical vane in transparent view.

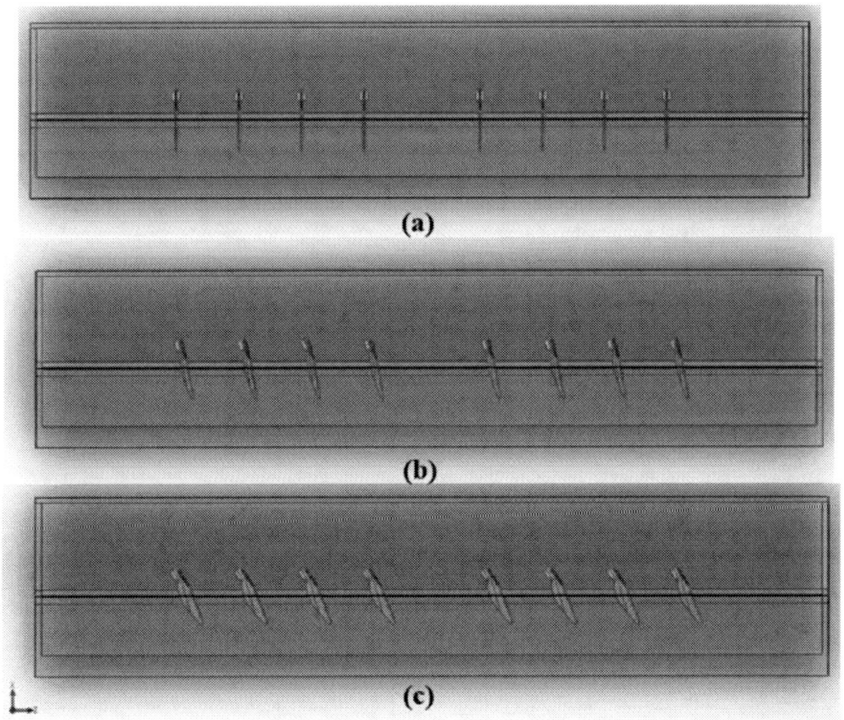

Fig. 3. Rotation of simplified vertical vane (a) 0° (b) 15° (c) 30°.

2.2.2 Meshing generation

The mesh quality of model is very important as it will affect the accuracy of solution generated in post-processing. First of all, unstructured type of mesh is applied for this model as the fluid domain has complicated geometry of vertical vane. By choosing the unstructured mesh, the meshing quality is improved and the computing time is also increased to solve for simulation [7].

Tetrahedral type of meshing element is selected due to the fillets and curvatures of vertical vanes and enclosed domain. After that, the edge sizing is added to shell of room air-conditioner. It is to increase the number of mesh element, so the velocity of air flow is captured with more accurate result. Moreover, inflation is added to the surfaces that bounded to the air flow inlet. The first layer thickness of 0.000643 m with 1.1 growth rate is set for the inflation. It is to ensure that the boundary layer effect is captured. The first layer of thickness is calculated based on the wall Y+ value of 30. The air flow that inside the boundary layer is named as viscous flow as it will form no slip condition where velocity of air flow equal to zero. Lastly, face meshing is added to the shell of room air-conditioner. It helps to organise the element equally over the surfaces.

However, meshing setup for 15° is slightly different from that that. In order to reach the better convergence during post-processing, another inflation is added to the surfaces of vertical vanes. When the vertical vane is rotated to 15° angle, it will cause more turbulent energy. Thus, mesh refinement is required to do for the surface of vertical vane. The meshing of model is shown in Fig. 4.

Although the meshing method applied to 15° angle of simplified vertical vane is not same as 0° angle of simplified vertical vane, it does not affect the selection of using same mesh size because meshing size is independent of grid refinement [7]. Hence, the meshing size will use the same after the grid independence.

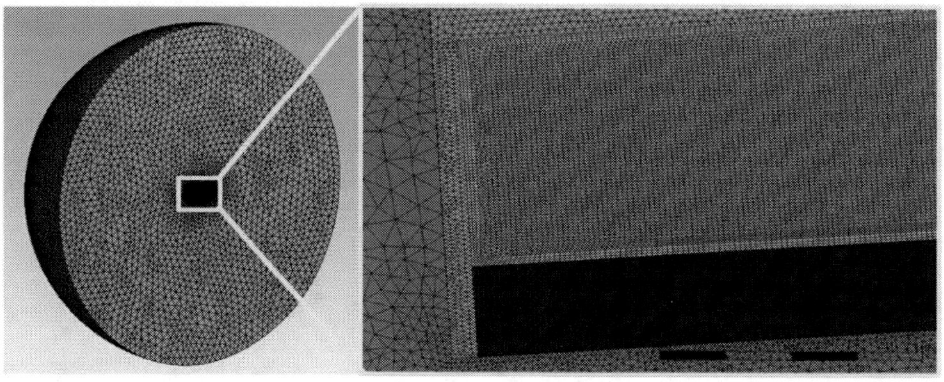

Fig. 4. The meshing of CFD model.

2.2.3 Grid independence analysis

Three level of meshing size is classified into coarse, medium and fine with respectively 0.008 m, 0.004 m and 0.002 m sizing. The purpose of doing grid independence analysis is to determine the maximum mesh size that can produce the same result as the smallest mesh size. The smallest mesh size for this model is 0.002 m as the element quality doesn't increase more after this value. The minimum growth rate for this mesh size is 1.7. Any value below 1.7 growth rate costs high increase in computing time of solving result. The advantage of doing grid independence analysis will save time on solving solution during post-processing.

As mentioned previously, 0° angle of vertical vane is used as base reference. The following result presented in table is the result of base reference. From the results of Table 1, it is not enough data to judge which mesh size to use. As the outlet velocity is the main focus in this study, the velocity of air flow should be measured and compared between these three different mesh sizes.

Table 1. Result of different mesh size

Level	Sizing (m)	Growth rate	Meshing Quality	Node	Elements	Iteration	Time (hours)
Coarse	0.008	1.15	0.80003	733666	3140164	1408	1.7
		1.17	0.79756	717586	3044681	597	0.6
Medium	0.004	1.15	0.80015	733224	3137247	2154	2.6
		1.17	0.79718	717363	3043687	831	1.0
Fine	0.002	1.15	0.79992	733761	3140411	> 4000	>10
		1.17	0.79729	717637	3043217	770	0.9

2.2.4 Numerical setup

RNG k-ε turbulent model is derived from instantaneous Navier-Stokes equations or renormalization group which is using statistical technique. RNG k-ε turbulent is applied to the CFD model because it has been proved to have good prediction for velocity of airflow at indoor environment. [6, 8]. Next, standard wall function is used for this conditions. For the outlet surface, it is defined as pressure outlet.

Lastly, the boundary conditions that have been set in ANSYS Fluent ver. 18 are shown in Table 2. Since the inlet velocity is 5.8 m/s which is less than the 0.3 Ma (Mach number), incompressible flow is applied. On the other hands, the volume of simplified vertical vane is 2.4 X 10-6 m3. This information is important to check whether the new design of vertical vane is minimised or not as compared to the simplified vertical vane.

Table 2. Boundary setting for CFD setup

Inlet velocity (m/s)	Temperature (°C)	Density of air (kg/m^3)	Kinematic viscosity	Viscosity of air (kg/ms)
5.8	20	1.205	1.511×10^{-5}	1.82076

2.2.5 Assumptions

The assumptions of this research's study have been made as following statements:

i. The velocity inlet for the room air-conditioner is uniform and constant at the inlet surface.
ii. The temperature of room which is the enclose domain is constant.
iii. No backflow is occurred at the end of wall surface (outlet).

2.2.6 Calculation

The method of analysing simplified vertical vane will be defined by calculating the distance of how far the airflow can be achieved. First of all, any velocity of airflow that is below 0.25 m/s in velocity contours is considered no wind according to the standard requirement from Panasonic. From this point, the distance where the velocity of airflow is higher than and equal to 0.25 m/s will be taken into calculation.

Furthermore, simulation result of different angle of vertical vane is analysed for efficiency of how accurate the direction of airflow coming out from vertical vane. The distance, where velocity of airflow is equal to and higher than 0.25 m/s, in *z*-axis is taken into calculation by applying trigonometry method. For distance X, it is measured 2 m away from the room air-conditioner. For the measured line of starting point will be located at 1.13 m (Y axis) below the room air-conditioner. The calculation of angle of wind is illustrated in Fig 5.

$$\tan\theta_{wind} = \frac{\text{distance in } z \text{ axis (velocity} \geq 0.25\,\text{m/s)}}{\text{distance in } x \text{ axis (velocity} \geq 0.25\,\text{m/s)}} \quad (1)$$

The efficient of wind direction is defined as equation below.

$$\text{Efficiency of wind direction} = \frac{\theta_{wind}}{\theta_{vertical\ vane}} \times 100\% \quad (2)$$

where θ_{wind} is the angle of wind in simulation and $\theta_{vertical\ vane}$ is the angle of rotation of vertical vane inside room air-conditioner.

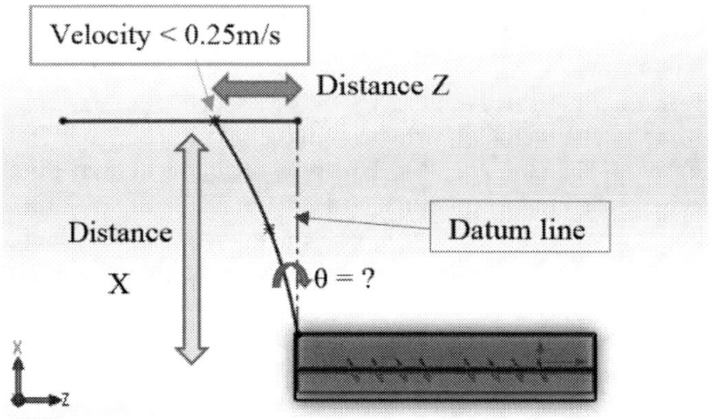

Fig. 5. Calculation of angle of wind.

The equation of efficiency of wind direction is used to define the efficiency of dissipating cold air. It can be applied as long as the location line is within the pathway of airflow in simulation result. In order to make the efficiency of wind direction is comparable with different angles and new designs of vertical vane; the location of measured line is fixed for all case. The location of measured line that determines the velocity of airflow from starting point to ending point is shown in Table 3.

Table 3. Distance from datum line

Location	*x*-axis (m)	*y*-axis (m)	*z*-axis (m)
Starting point	2	-1.13	0
Ending point	2	-1.13	-1.5

3 Result and discussion

3.1 Grid independence analysis

The result of 0° angle of vertical vane simulated from 0.002 m, 0.004 m and 0.008 m mesh size is analysed. The velocity of airflow along the boundary edge and line 1 are measured. The measured location is shown in Fig. 6 and the graph of velocity against the location is plotted in Fig. 7. Observing the graph from Fig. 7, the fine mesh size has higher velocity along the boundary edge as compared to medium and coarse mesh size. However, the velocity result generated by medium mesh size is more reasonable. The reason is the velocity at the middle of boundary edge reached uniform velocity of airflow where this location has big spacing between each vertical vane. Thus, the velocity of airflow is expected to have higher velocity compared to other location. Besides that, the velocity of airflow is almost symmetrical. The medium mesh size can be chosen as constant variable for meshing setting.

In additions, the trend line of medium mesh has the same behaviour as the fine mesh in Fig 8. For the coarse mesh, it only has same pattern of trend line at left side of line 1. Hence, the grid independence analysis is done and the medium mesh which is 0.004 m is used as constant mesh size to all case studies. Lastly, the growth rate of 1.7 is selected as constant variable since its result is almost same as result generated with 1.5 growth rate.

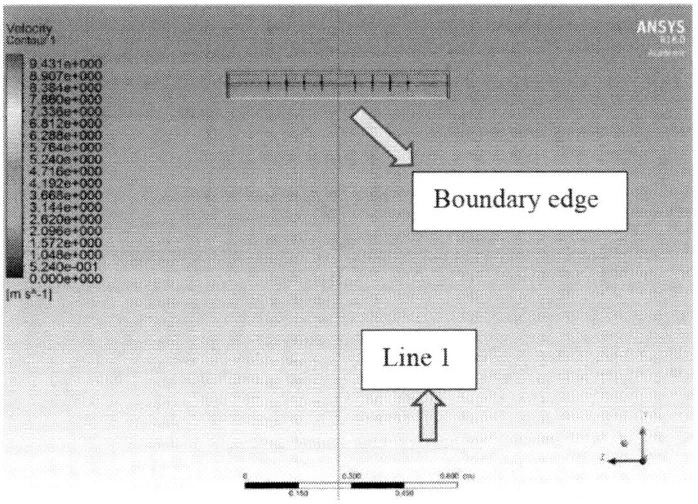

Fig. 6. Location of boundary edge and line 1.

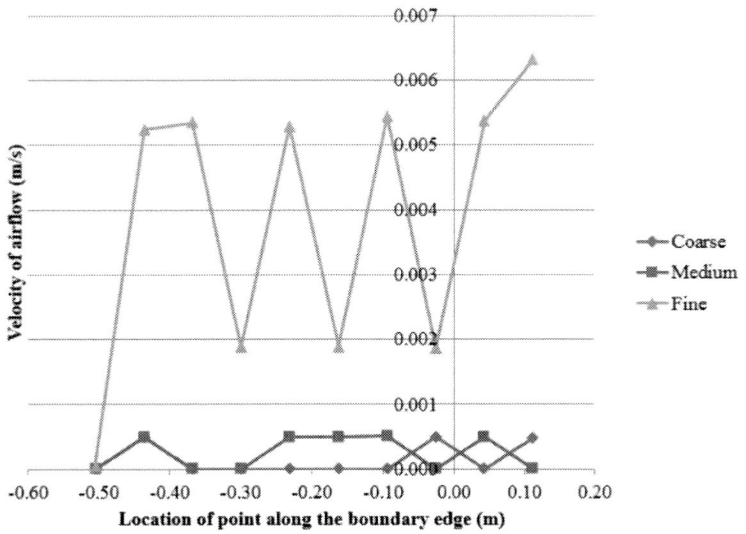

Fig. 7. Graph of velocity against location of point for different mesh size.

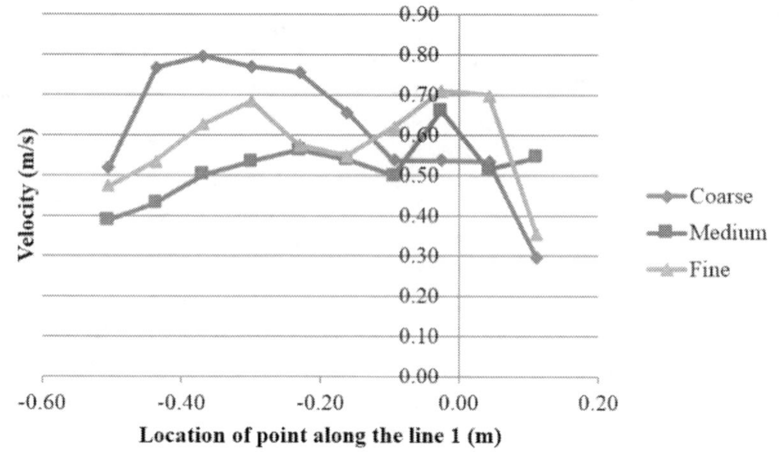

Fig. 8. Graph of velocity against location of point for different mesh size.

3.2 Efficiency of wind direction

The velocity streamline of 15° and 30° of vertical vane is plotted in Fig. 9. Besides that, the result of efficiency of wind direction is calculated and shown in Table 4. Comparing efficiency of wind direction of 15° and 30° of vertical vane, it can be concluded that the efficiency of wind direction is decreased as the angle of vertical vane increased. This result explained the reason why the angle of vertical vane of this room air-conditioner has been set a limit to rotate. The higher the angle of vertical vane to rotate the lesser the efficiency of wind direction and dissipating cold air. These result will used as based reference in order to check whether new designs of vertical vane has made improvement on it.

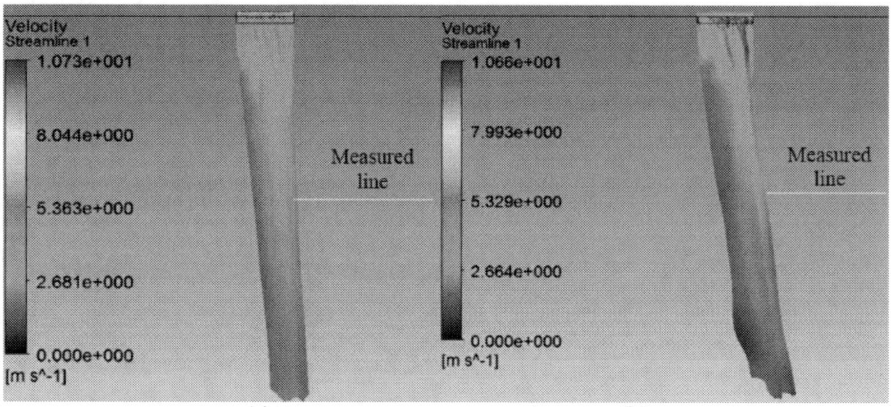

(a) (b)

Fig. 9. Velocity streamline of vertical vane (a) 15° and (b) 30°.

Table 4. Efficiency of wind direction

Angle of rotation of vertical vane (°)	Distance Z (m)	Distance X (m)	Angle of wind (°)	Efficiency of wind direction (%)
15	0.30503	2.0	8.67	57.81
30	0.50839	2.0	14.26	47.54

The velocity contours at mid-plane of vertical vane are shown in Figs. 10 and 11. Based on these figures, it proved that the higher the angle of rotation of vertical vane will lead to shorter distance of the airflow can be achieved at that angle. It also explained 100% efficiency of wind direction can't be obtained since the minimum velocity of airflow, 0.25 m/s is indicated before reaching the measured line.

Fig. 10. Velocity contour at mid-plane of 15° of vertical vane.

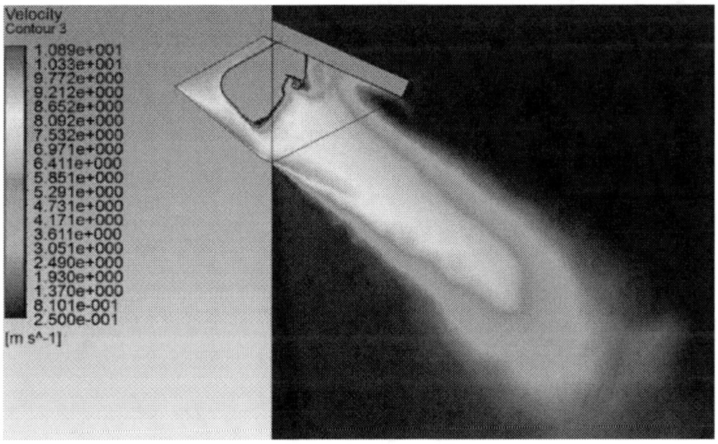

Fig. 11. Velocity contour at mid-plane of 30° of vertical vane.

3.3 Different contour analysed at 30° of vertical vane

The pressure contour at mid-plane of 30° of vertical vane is plotted and shown in Fig. 12. It is used to investigate which parts of vertical vane are main parameters of affecting the efficiency of wind direction. According to the Fig. 12, high increase in gauge pressure is clearly determined at leading edge of vertical vane compared to Fig. 11. Based on Bernoulli's principle, increase in velocity will cause decrease in pressure [4]. In this study, the low velocity of airflow at leading edge induces the high increase in pressure of airflow.

Another velocity contour in Fig. 13 is plotted perpendicular to the inlet and is aligned to leading edge of vertical vane. It is plotted to observe how the velocity of airflow varies when airflow is passing the vertical vane. Referring to the top view of velocity contour, that perpendicular to inlet in Fig. 14, the velocity of airflow passing by the left surface of vertical vane is lower than the velocity of airflow at right surface of vertical vane. In additions, the rear part of vertical vane induce vortex. Vortex induced at rear part of vertical vane caused decreased in velocity of airflow.

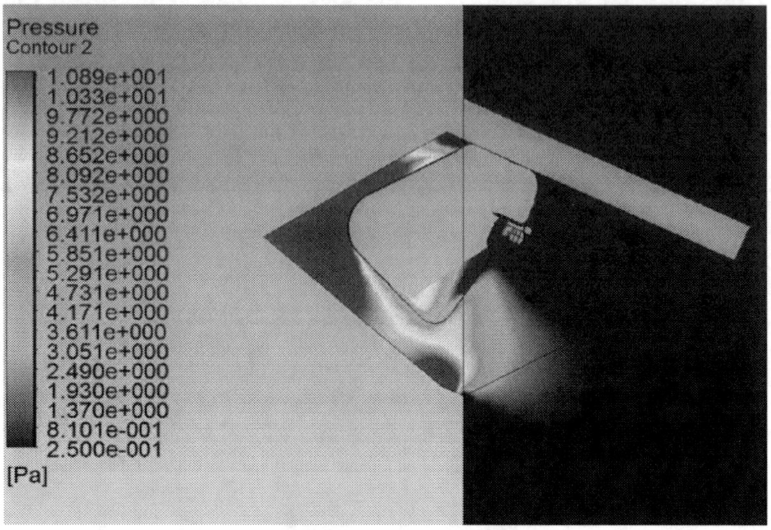

Fig. 12. Pressure contour at mid-plane of 30° of vertical vane.

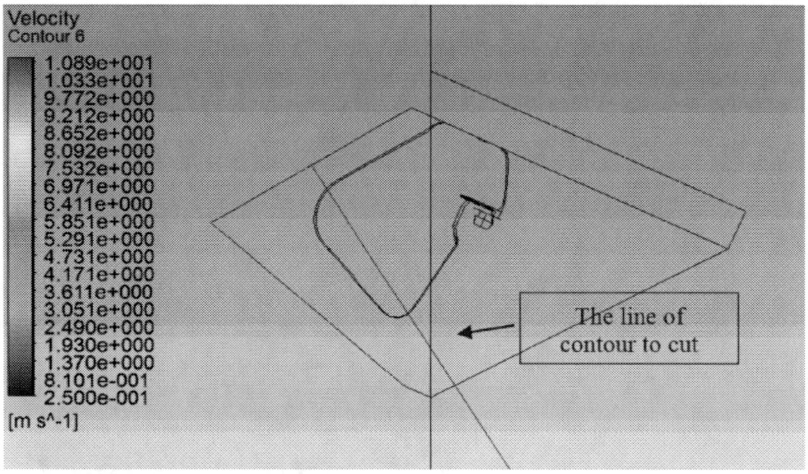

Fig. 13. Position of velocity contour cut from inlet.

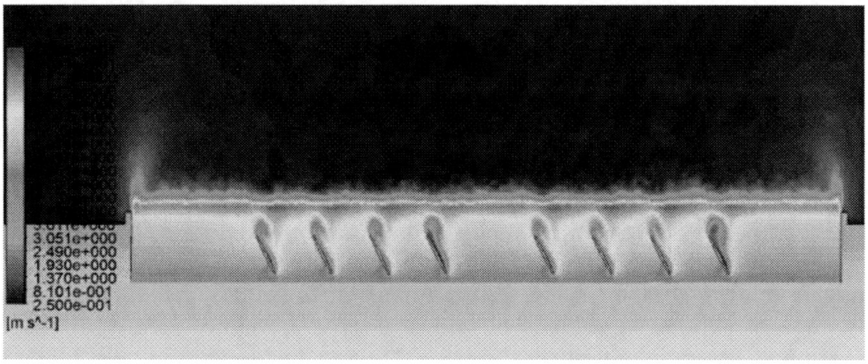

Fig. 14. Top view of the velocity contour cut perpendicular to inlet

Thus, two parameters are selected to study for geometry optimisation of vertical vane. The parameters studies will be the length of span and length of tip chord. The name of length is shown in Fig. 15. The span length and tip chord length of simplified vertical vane are 0.04436 m and 0.04478 m respectively.

The restriction set for parameter studies are as following:

1. The surface of root chord that is attached to room air-conditioner is constant.
2. The total height of vertical vane is maximum length. (due to size of product)
3. The total area of vertical vane is constant.
4. The material of vertical vane is constant.

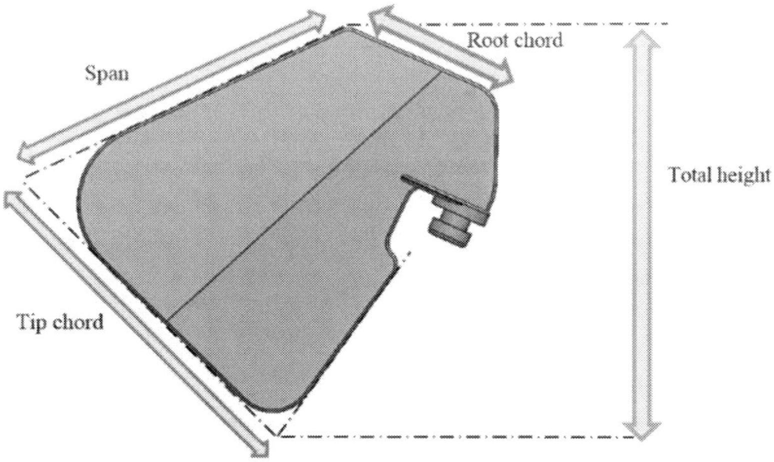

Fig. 15. Name of different length of vertical vane.

3.4 Parametric studies for geometry optimisation of vertical vane

Since long span, tip chord and short span are chosen as main parameter studies, new designs of vertical vane are drawn based on 5%, 10% and 15% decreased in vane surface area. A basic sketching is illustrated in Fig. 16 shown position of each parameter study with different percentage decreased in vane surface area.

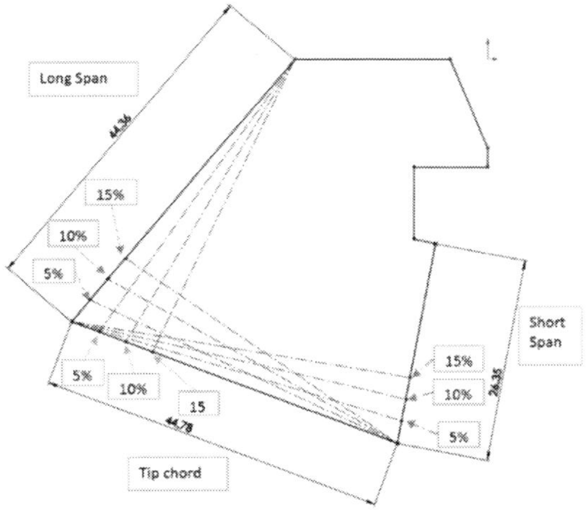

Fig. 16. The position of each parameter study with different percentage decreased in vane surface area

In this paper, the simulation is only run for one side (turning left side) of direction because the design of CFD model is symmetrical with respect to centre of shell of room air-conditioner. Moreover, only 15 degree angle of vertical vane is studied as it is sufficient to identify which parameter is important to analyse. Furthermore, the result of new vertical vane can be compared with base reference (efficiency of 15° and 30° of simplified vertical vanes). In previous discussion, it is found that the higher the angle of vertical vane, the lower the efficiency of wind direction. As long as the result of new vertical vane has same trend as base reference, the reliability of simulation result is justified. The CAD drawing of new vertical vane based on each parameter study at different percentage decreased in vane surface area is shown in Fig 16. The simulation result of new vertical vane is tabulated in Table 5.

According to the Table 5, the changes in efficiency of wind direction for long span are significantly small. Besides that, the trend of decreasing the vane surface area of long span gives result of dropping in efficiency of wind direction. Hence, it can be assumed that long span can't be decreased but it could be increased for improving efficiency of wind direction.

Next, all the result of short span shows negative impact to efficiency of wind direction. So, it is better to keep length of short span as constant as it can't be exceed the length of total height.

Lastly, the result of 5% and 10% decreased in vane surface area at tip chord show negative impact to efficiency of wind direction. However, the 15% decreased in vane surface area at tip chord is actually higher than the efficiency of base reference which is 57.81%. The design of this new vertical vane is selected best configuration of design since its efficiency of wind direction is 60.32% which is increased by 2.51% as compared to efficiency of base reference.

Table 5. Result of each parameter study at different percentage decreased in vane surface area

Parameter studied	Percentage decreased in vane surface area (%)	Distance from fix point (mm)	Distance Z (m)	Distance X (m)	Angle of wind (°)	Efficiency of wind direction (%)
Long Span	5	3.6480	0.27601	2	7.86	52.38
	10	7.2020	0.27146	2	7.73	51.53
	15	10.6955	0.26691	2	7.60	50.68
Tip chord	5	3.7599	0.26236	2	7.47	49.82
	10	7.4350	0.28966	2	8.24	54.94
	15	11.0610	0.31847	2	9.05	60.32
Short Span	5	2.9729	0.20776	2	5.93	39.54
	10	5.8646	0.22293	2	6.36	42.40
	15	8.6985	0.21231	2	6.06	40.40

4 Conclusions

The efficiency of wind direction for different angles for vertical vane is studied. The efficiency of wind direction for 15° of vertical vane is 57.81% while the efficiency of wind direction for 30° is 47.54%. The results of the efficiency of wind direction are used as base reference. The parameters studied for optimisation of vertical vane are focused at length of long span, tip chord and short span. The design of 15% decreased in vane surface area at tip chord is the best optimised design of vertical vane because the efficiency of wind direction is the highest as 60.32%. In addition to that, the mass of vertical vane of this design is lower than base reference and thus it is optimised without bringing any negative effect to the efficiency of dissipating cold.

References

1. B. Xia, D.-W., Sun, Comput. Electron. Agric. **34** (1-3) 5-24 (2004)
2. T. Nakanishi, T. Nakamura, W. Youichirou, H. Katsumasa, K. Takahiro, *Technical Paper*, **53**(160), 1–9 (2007)
3. M. H. M. Noh, H. Rashid, A. H. A. Hamid, and M. F. Iskandar, Procedia Eng, **41**, 1761-1768 (2012).
4. Y. A. Cengel, J. M.Cimbala. Fluid mechanics: *Fundamentals and applications*, (3rd Ed. McGraw-Hill 2014)
5. B. S. Hurak, Undergraduate Honors Thesis, The Ohio State University (2011)
6. T. Karimipanah, H. B. Awbi, M. Sandberg, C. Blomqvist, Build. Environ, **42**(2) 647-655 (2007)
7. N. Mao, M. Song, S. Deng, Appl. Energy **180**, 536-545 (2016)
8. S. Lin, B. T. Tee, C. F. Tan, *7th Int. Conf. on Cooling & Heating Technologies*, Bristol, United Kingdom (2015)
9. S. Sadrizadeh, S. Holmberg, *Proceedings of 9th International Conference on Indoor Air Quality Ventilation & Energy Conservation in Buildings*, 1-7 (2016)

The Fabrication of Non-Implant 3D Printed Nose

Yong Leng Chuan*, and *Elliot* Andrews

Faculty of Built Environment, Engineering, Technology & Design, Taylor's University Lakeside Campus, Malaysia.

Abstract. Non-surgical rhinoplasty procedures which involves the use of injectable derma fillers are highly risky as patients are susceptible to side effects and complications that may cause unwanted changes in their appearance. This research explores an alternative method of non-surgical rhinoplasty for patients seeking augmentation of the nose with the use of three-dimensional (3D) printing. Most rhinoplasty procedures are conducted with the intention of enhancing the aesthetical features of the nose, a 3D model nose was designed based on the combination of the average and the ideal aesthetic parameters of the Northern European Caucasians and South Asia Chinese nose. The modelling of nose is done using the SolidWorks CAD software. An initial design was sketched in a polygon mesh form and further improved on. Different printing materials and infill densities were compared to determine the suitable printing technique. The final nose model is then printed using the Ultimaker 3D printer using Polylactic acid (PLA) with an infill density of 100% at a thickness of 1.4 mm. An inner layer to the 3D printed nose was developed for comfortable attachment of the nose model to human skin. The inner layer was fabricated using agar gelatine. Experiments were carried out to increase the strength and adhesiveness of the gelatine so that it could adhere to the human skin and the PLA surface. Tensile and adhesive strength tests were carried out to determine the suitable gel composition for the attachment of the nose to the user's face. The key outcome from the experiments using natural gelatine was capability of gel to act as an inner layer for the temporary attachment of the 3D nose model to the human skin

1 Introduction

Rhinoplasty can be defined as the surgical procedure of the reconstructing or reshaping of the human nose. This procedure is usually performed on individuals who have suffered nasal trauma and are in need of reconstruction of the nose. However, recent statistics suggest that rhinoplasty has been the procedure of choice for individuals seeking to aesthetically enhance their facial features by obtaining the ideal proportions of the facial structure. With the advancement of medical sciences, patients are given the option of a nonsurgical rhinoplasty procedure using injectable fillers for the augmentation of the nose. With a variety of types of fillers to choose from, cosmetic surgeons select the required material based on their properties and desired outcome of the patient.

* Corresponding author: LengChuan.Yong@taylors.edu.my

Some of the more common fillers used by cosmetic surgeons include hyaluronic acid, polyacrylamide gel and calcium hydroxyapatite which is a temporary filler capable of lasting for 10 to 14 month [1], [2]. These fillers are selected based its elasticity, hydrophilicity, viscosity, durability and its ability to mould according to the desired shape after injection. The selection process of the filler by a cosmetic surgeon is highly risky as patients are susceptible to side effects and complications such as bleeding, bruising, asymmetry of the nose shape, and also serious complications that lead to blindness [2]. It is important that surgeons make the selection of the fillers and injection techniques based on the individual patient and that the procedure is performed slowly and carefully to minimise the risks of the occurrence of these complications. With such complications arising from this nonsurgical procedure, people that want to aesthetically enhance their facial features are left with a single choice of surgically reshaping of the nose. This indicates that a need for an alternative method of nonsurgical rhinoplasty is to be presented.

The use of three-dimensional (3D) printing, a form of rapid prototyping technology, has been growing in many industries over the past decade. With capabilities of rapidly producing objects with complex geometrical designs, the use of this rapid prototyping technique has overtaken many conventional manufacturing processes [3]. 3D printing has also been making advances in the medical field with medical practitioners using bio-models to provide feedback and simulate complex anatomical movements that are otherwise difficult to replicate using 3D modelling software [4]. The ability of rapid prototyping technology to produce customised organ and implants to suit the needs of individuals opens a pathway for advancement in nonsurgical cosmetic procedures.

Ciocca et al. [5] have successfully fabricated an eyeglasses-supported nasal prosthesis and expected to provide a temporary prosthesis to a gunshot victim who lost his entire nose to the injury. One of the methods used to achieve this outcome was to obtain a 3D model of the patient's facial structure using a laser scanner. A CAD model of the prosthetic nose was generated based the digital model of the patient's face that was scanned. The 3D model nose was digitally placed over the digital model of the patient's facial structure. Adjustments were then made so that the prosthesis would fit comfortably over the defected area. A two-part mould was then fabricated using a 3D printer. The material used for the 3D printing of the mould was acrylonitrile butadiene styrene (ABS). The mould was then used to fabricate the silicone prosthesis using the silicone process procedures with 12 different silicone colours used to obtain a colour similar to the patient's skin colour [5]. The final step of the fabrication of the prosthesis was to connect the nose model to the eyeglasses using a bolt.

Similarly, Y. He et al. [6] studied the use of rapid prototyping in the manufacturing of soft tissue prostheses. The use of a 3D scanner was used to obtain data for the generation of a digital 3D model for design of a negative mould of the required body part. The prostheses that was fabricated by the author was a prostheses of a human ear. The mould was then printed using a 3D printer using ABS as the material of choice. Silicone is then poured into the mould to form the desire soft tissue prosthesis. The silicone prostheses were then removed from the mould to complete the fabrication process.

Unlike the research, this research aims to use 3D printing as the primary fabrication process for an alternative nonsurgical rhinoplasty procedure. Study of the relationship between the average nose and the perspective of the ideal nose for various ethnic groups was conducted. The result of this research can eventually contribute to the designing of the nose model using a Computer-Aided Design (CAD) software. Due to budget constraints the nose designs were limited to average dimensions taken from sample groups instead of individual patients. The designing of the nose model was based on a combination of the average nasal diameters of an ethnic group and the aesthetical ideal of the given ethnic group. The ethnic groups that were chosen in this research paper was the South Asian Chinese and the Northern European Caucasians. These specific groups were chosen to proof that this research is applicable to not only the local population but also the international population. The CAD design was then fabricated into a physical model using a 3D printer using the suitable printing techniques. The model is attached to the user through an adhesive inner layer that was made of synthesised natural gelatine instead of using a mechanical support. The inner layer was made using suitable ratios of agar, glycerol as plasticisers and crushed egg shell for an increase in adhesiveness and strength. The gel compositions underwent tensile and adhesive strength testing to determine its practicality.

2 Research Methodology

The research of the fabrication of a non-implant three-dimensional (3D) printed nose can be separated into two main parts which are the fabrication of the 3D model nose and the synthesis of natural gelatine as an inner layer for a comfortable attachment of the nose. The fabrication of the 3D model nose can be further broken down in to three subcategories which are the identification and selection of nasal dimensions, 3D modelling using computer-aided design (CAD) software and the printing techniques used to fabricate the nose model.

2.1 Identification and Selection of Nasal Dimensions

The shape and parameters of the human nose varies according to an individual's ethnicity, age and gender. The differences in nose shape, facial features and proportions causes the perspective of the ideal nasal proportions required to achieve an aesthetically enhanced look to differ according to the individual. Therefore, in order to conduct this research within the proposed budget without the use of a 3D scanner, the fabrication of the 3D printed nose was to be modelled based on ethnicity and gender of a sample age group. The sample age group chosen ranges from 21 - 29 years of age, set as a boundary for this research. Studies on the average nasal parameters of different ethnicity groups were conducted to identify the key points and dimensions that affect the shape of the nose based on each ethnic group. Some of the key parameters of a nose that was used in the CAD modelling can be found in figure 1. These key parameters include the nasal height, nasal width and nasal tip protrusion. A literature review was then conducted on the international perspective of the ideal proportions for the nose. The ideal nose was then compared to the average nose of two ethnic groups, the Caucasian nose and the Asian nose. Findings from the literature review showed that the average nose from neither ethnic groups conformed to the international perspective to the ideal nose. It was also found that the Asian perspective of the ideal nose differed from the international perspective which was based on the neoclassical canon of facial aesthetics. The review concluded that the success in achieving an enhancement of facial aesthetic value through rhinoplasty depended on the more on the ethnicity of the person than the ideal proportions. Upon obtaining the stated information, it was appropriate to model the 3D printed nose based on the European Caucasian nose and the Chinese Asian nose from each gender for this research.

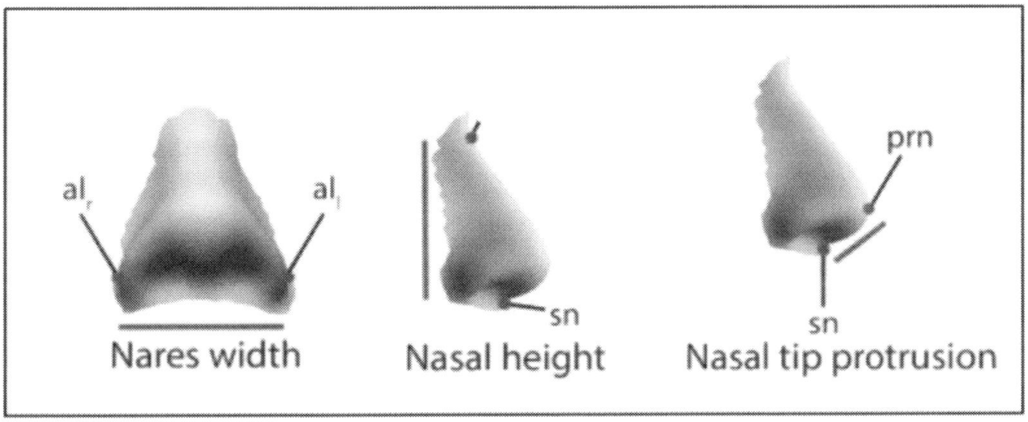

Fig. 1. Key parameters of the nose used in CAD modelling [7].

2.2 Geometrical Modelling

The 3D models of the human nose were modelled utilising the CAD software Solidworks 2016 Student Edition. Solidworks was selected as the preferred CAD software as it is available to students as a university resource. An initial design was modelled by sketching a low poly or polygon mesh version of the nose to achieve the desired shape of the nose. Fillets are then applied to the edges of the polygons to obtain the roundness of the nose. This initial prototype was designed as a base model that was to be further refined throughout the research duration. It was also modelled to determine the suitable printing material and techniques. This initial design was modelled based on the average Caucasian male nose.

A more refined version of the nose was then designed according to the appropriate dimensions of the average European Caucasian male nose obtained from the journal paper by Zaidi et al. [7]. These dimensions were combined with facial angles from the international perspective of the ideal nasal proportions. Once the desired model was sketch using the CAD software. A shell feature was used to obtain an outer shell of desired shape with the appropriate shell thickness. A tolerance value of 2 mm was set to allow space for the inner layer required for a comfortable attachment. Similarly, 3D nose models were fabricated for the Caucasian female nose and the Chinese Asian nose.

2.3 3D Printing Techniques

The completed 3D model was printed using the Ultimaker 3 printer that was available in the laboratory. There were two types of material available for the printing of the 3D model, which were Polylactic Acid (PLA) and Acrylonitrile Butadiene Styrene (ABS). PLA is considered to be the stronger of the two materials as it has a larger tensile strength compared to the ABS [8]. PLA is also has a higher elastic modulus compare to ABS [8]. On the other hand, ABS is more heat resistant of the two materials. The samples of each type of material was obtained so that a comparison could be made. The comparison made shows that due to the surrounding environment of the 3D printer, the print quality of the ABS material was not as good as the PLA material. The finishing around the edges of the ABS material was rough and cracks could be seen forming on the surface. The PLA material showed a significantly better print quality as no cracks on the surface could be found. The initial design of the nose model was printed using PLA material at a thickness of 2 mm with a 75% density. The print quality of the initial model was compared to the refined model of a Caucasian male nose which was printed using a thickness of 1.4 mm with a 100% density. The results of the comparison can be found in the results and discussion section. Once the desired 3D nose model was printed, a coat of skin colour paint was applied to give the 3D printed nose a more natural look.

2.4 Inner Layer using Synthesis of Natural Gelatine

The objective of synthesising natural gelatine as an inner layer was to provide a comfortable attachment of the 3D nose model to the user's face. This meant the gel was required to be adhesive towards the human skin. A type of natural gelatine that can be commonly found was agar which is a type of algae based natural gelatine widely used as a culinary ingredient. The gelation of agar exhibits good physical and mechanical properties such as flexibility, transparency, homogenous and is a hydrocolloid [9,10]. These characteristics make agar a suitable type of gelatine to be used as a coating material for food and edible products [9]. The combination of glycerol plasticisers improves the tensile strength and gives it a film like texture. However, increasing the concentration of the glycerol plasticiser decreases it tensile strength [9].

An experiment was conducted to test the physical characteristics of the agar gel when combined with glycerol plasticisers. The materials used in this experiment consist of commercial agar powder and food grade glycerol. Different concentrations of glycerol solution was added to the agar solution and compared. The gel samples were prepared with a thickness at 2 mm with a surface area of 2 cm x 2 cm. A qualitative analysis of these various compositions of agar and glycerol gel was carried out. The texture of the gel and an initial test for the adhesive properties of the gel was conducted on both the PLA surface and the human skin. This initial qualitative analysis provided a benchmark for the future experiments using the agar gel. One of the limitations of testing using human skin is that the adhesion properties of the gel could not be tested over time. Therefore, the following experiments were conducted using only the PLA material that was used to print the 3D nose model.

2.5 Strength and Adhesion Testing of Gel

The results from the initial test showed that the pure agar glycerol composition lacked in rigidity. One of the solutions that were provided was the addition of eggshells in the form of impurities to increase the strength of the agar gel, a theory similar to the addition of impurities to improve the mechanical strength of metals. A tensile test was conducted to determine the relationship between the tensile strength of the agar gel and the amount of egg shell added to the composition. Due to the limitations of using a tensile testing machine, the test was conducted by fixing one end of test sample agar gel strip and applying force to the other end of the gel strips using weights. The amount of weights that was required for the failure of the gel strips were used to calculate the tensile strength of the gel. The test samples of the agar gels that was prepared for the tensile testing can be seen in the figure 2.

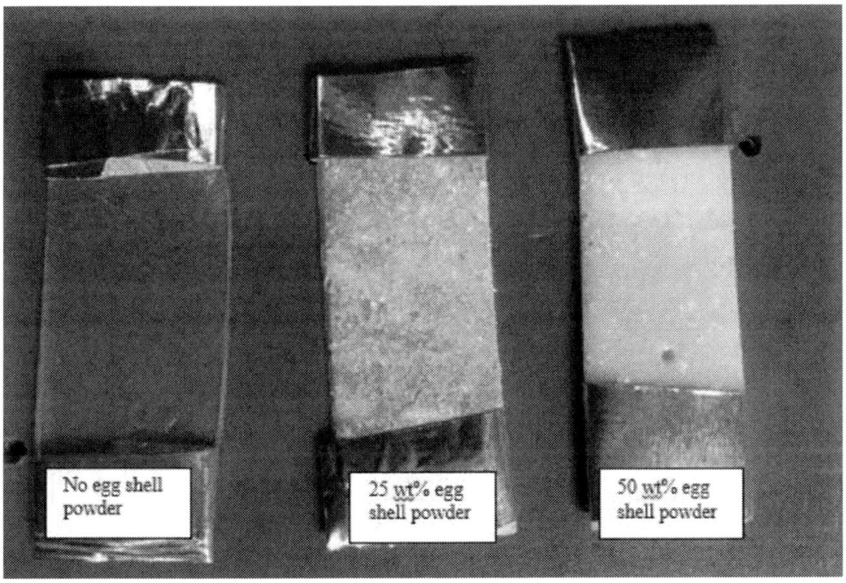

Fig. 2. Test samples for tensile testing.

The adhesion strength of the agar gel was tested using two flat PLA printed surfaces to simulate the surface of the 3D nose model. The agar glycerol gel of a 1 mm thickness was placed in between the PLA printed surfaces. One of the surfaces was fixed while force was applied using weights that were tied to the other surface. Weights were added until failure in the adhesion of the agar gel to the free PLA surface was detected. The adhesion test was conducted for three different samples of the agar gel and at three separate timings at intervals of 3 hours to determine the change in adhesion properties of the gel over time. The test samples of the agar gel had a surface area of 5 cm x 5 cm. The experimental set up for the adhesive testing of the agar gel can be seen in figure 3. The maximum adhesive strength of the agar gel was compared to the total weight of the 3D printed nose model which was approximately 4.2 g, to determine its feasibility.

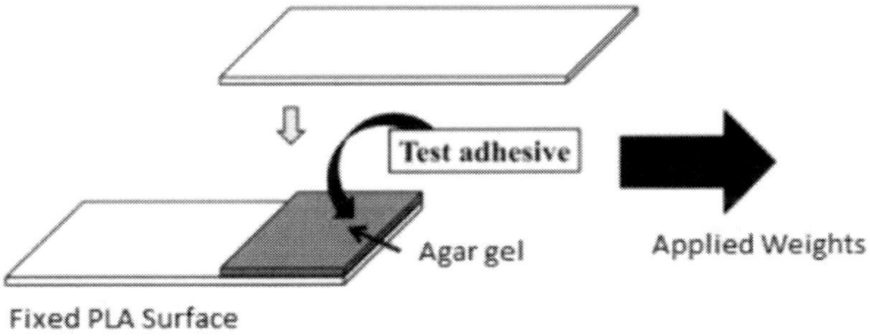

Fig. 3. Adhesive strength measurement.

3.0 Results and Discussion

3.1 3D Nose Model

The 3D model of the nose was designed using Solidworks with dimensions of the average Caucasian male nose. The average nose dimension were combined with the ideal aesthetic angles according to the neoclassical canons. Parameters of the ideal naso-facial angle by S. C. L. Leong et al. was used in the modelling of the nose [11]. The key dimensions that were used in the in modelling of the 3D nose can be found in table 1 in which a tolerance of 2 mm was added to the dimensions. The shell feature that can be found in the CAD software was used to create the outer shell of the nose shape with a thickness of 1 mm. The completed CAD model of the 3D Caucasian male nose can be found in figure 4. The Caucasian female nose and a Malaysian male nose was also designed using similar techniques using the measurements data of A. A. Zaidi et al. [7] for the Northern European Caucasian nose and Y. Dong et al. [12] Chinese Asian nose respective.

Table 1. Dimensions used for the modelling of the 3D nose [7,11,12].

Population	N. European		Chinese	
Gender	**Male**	**Female**	**Male**	**Female**
Nasal Height (mm)	51.59	48.62	54.31	52.44
Nasal Width (mm)	34.29	30.91	42.88	47.48
Nasal Tip Protrusion (mm)	17.62	16.43	20.48	19.38
Alar Base Width (mm)	35.66	33.00	34.46	31.56
Naso-facial Angle(°)	36	36	36	36

Fig. 4. The completed 3D model of the Northern European male nose using Solidworks.

3.2 Inner Layer using Synthesis of Natural Gelatine

Some of the benefits of using agar as the natural gelatine of choice was that it has no toxic effect on the human skin, it is also easily available to consumers and it displays good mechanical and hydrocolloid properties. Some of these characteristics include its high shear stability, high gel formation effect and high gel strength and stability. The addition of high concentration sugar solutions such as glycerol increases the strength of the gel. Therefore, the following experiment was conducted to determine the use of agar as a natural gelatine as the inner layer for the attachment of the 3D nose model.

A qualitative analysis was carried out to determine the physical properties of different compositions of agar gel and glycerol solution. In the initial experiment, glycerol solutions with different concentrations were prepared. Three qualitative tests were conducted on the gel properties which are the gel rigidity, adhesiveness to PLA printed surface and adhesiveness to human skin.

The analysis showed that compared to pure agar gel, the agar gel with 1% glycerol showed adhesive characteristics as it was capable of being placed on a PLA printed surface at 90° upright position. However, when rotated at 180° or at an upside down position, the 1% glycerol gel failed to stick to the PLA surface. With 2% glycerol added to the gel, the adhesiveness of the gel increased as it was able to stick to the PLA surface at 180° or upside down position. The gel was then tested against the human skin at both 90° and 180°. The gel with 3 wt% glycerol was found to be sticky but was easily crumbled when lifted from the drying pan. Observations from this experiment showed that the adhesiveness of the agar gel increased with the concentration of glycerol added to the gel. However, at a 2 mm thickness, the agar gel did not have the required strength and easily crumbled.

4. Conclusion

The use of rapid prototyping techniques for the fabrication of a 3D model nose presents an alternative method of nonsurgical rhinoplasty for patients seeking an aesthetic enhancement of the nose. This alternative method of nonsurgical rhinoplasty eliminates the risk of complications involving the injection of derma fillers that are commonly used in current nonsurgical rhinoplasty procedures.

The 3D noses that were fabricated using CAD and 3D printing was designed based on the average dimensions of an ethnic group instead of individuals due to budget limitations. The concept of using 3D printing as the primary fabrication method of the noses was proven. Accuracy on the desired nose shapes may be improved in future work.

References

1. M. E. Jasin, Facial Plast. Surg. Clin. North Am **21**, 2, pp. 241–252 (2013)
2. O. N. Johnson and T. C. Kontis, Facial plast Surg **32**, 5, pp. 500–506 (2016)
3. G. N. Levy, R. Schindel, and J. P. Kruth, CIRP Ann. - Manuf. Technol **52**, 2, pp. 589–609 (2003)
4. Y. Chen, F. Niu, B. Yu, J. Liu, M. Wang, and L. Gui, J. Craniofac. Surg **25**, 1, pp. 184–188 (2014)
5. L. Ciocca, M. Fantini, F. De Crescenzio, F. Persiani, and R. Scotti, J. Rehabil. Res. Dev **47**, 7, pp. 595–604 (2010)

6. Y. He, G. Xue, and J. Fu, Sci. Rep **4**, 1, p. 6973 (2015)
7. A. A. Zaidi, B. C. Mattern, P. Claes, B. Mcecoy, C. Hughes, and D. Shriver, *Investigating the case of human nose shape and climate adaptation*. 2017.
8. B. M. Tymrak, M. Kreiger, and J. M. Pearce, "Mechanical properties of components fabricated with open-source 3-D printers under realistic environmental conditions," Mater. Des **58**, pp. 242–246, 2014.
9. S. Agricultural and P. Pangkep **23**, 4, pp. 1669–1675 (2016).
10. D. Phan The, F. Debeaufort, A. Voilley, and D. Luu, Food Hydrocoll **23**, 3, pp. 691–699 (2009)
11. S. C. L. Leong and P. S. White, J Plast Reconstr Aesthet Surg **59**, 3, pp. 248–252, (2006)
12. Y. Dong, Y. Zhao, S. Bai, G. Wu, and B. Wang, Br. J. Plast. Surg **63**, 11, pp. 1832–1839 (2010)

Assessing Bioinspired Topographies for their Antifouling Potential Control Using Computational Fluid Dynamics (CFD)

Jacky Ling[1] and *Felicia* Wong Yen Myan[1,*]

[1]Department of Mechanical Engineering, School of Engineering, Taylor's University, 1, Jalan Taylors, 47500 Subang Jaya, Selangor, Malaysia.

Abstract. Biofouling is the accumulation of unwanted material on surfaces submerged or semi submerged over an extended period. This study investigates the antifouling performance of a new bioinspired topography design. A shark riblets inspired topography was designed with Solidworks and CFD simulations were antifouling performance. The study focuses on the fluid flow velocity, the wall shear stress and the appearance of vortices are to be noted to determine the possible locations biofouling would most probably occur. The inlet mass flow rate is 0.01 kgs^{-1} and a no-slip boundary condition was applied to the walls of the fluid domain. Simulations indicate that Velocity around the topography averaged at 7.213×10^{-3} ms^{-1}. However, vortices were observed between the gaps. High wall shear stress is observed at the peak of each topography. In contrast, wall shear stress is significantly low at the bed of the topography. This suggests the potential location for the accumulation of biofouling. Results show that bioinspired antifouling topography can be improved by reducing the frequency of gaps between features. Linear surfaces on the topography should also be minimized. This increases the avenues of flow for the fluid, thus potentially increasing shear stresses with surrounding fluid leading to better antifouling performance.

1 Introduction

The accumulation of unwelcomed biological organisms on a surface due to prolonged fluid flow is called biofouling. There are many industries that still face challenges because of this natural phenomenon [1]. Methods to prevent biofouling thus far have shown less than satisfactory effects. The nature and effects of biofouling come in different forms: biofouling in medical devices presents in layers of biofilms only; biofouling in marine and industrial applications contains layers of biofilm together with macrofouling and inorganic fouling. Macrofouling is similar as biofilm, however it is caused by the accumulation of macro sized organisms. Inorganic fouling on the other hand is caused by the accumulation of non-

* Corresponding author: feliciayenmyan.wong@taylors.edu.my

biological substances, such as deposits from corrosion particles, crystallization, suspended particles, fuel, oil or ice [2].

Biofouling leads to many negative repercussions in the maritime industry and also could cause fatalities in the medical field. Biofouling greatly affects medical field from many common operations such as implants (prosthetic & dental), blood transfusion, body organ sensors and any other medical operation involving interaction between a patient's body with the equipment. Medical equipment contaminated with biofouling material can be accidentally transferred to a patient leading to further illness or in more serious cases, death. In Malaysia, approximately 5000 deaths can be attributed to infections on medical equipment that were contaminated with biofouling material [3, 4].

Early efforts relied on toxic chemicals to deter biofouling. A well-known example is Tributylin (TBT) which has since been banned because of its harmful side effect towards marine life [5] Current research focuses on non-toxic approaches to deter biofouling and one such approach involves applies surface topographies. Field tests on various types of surface topographies should be supplemented by numerical studies to provide better evidences that surface topographies could be a sustainable antifouling approach.

Existing research on surface topographies focus on applying biomimicry to develop potential antifouling solutions. The inspiration for this research comes from observations of Self-cleaning abilities on some organisms allowing them to be free of the micro-organisms accumulation, lotus leaf and shark skin are both the common example of self-cleaning ability. Although the incorporation of biomimicry in antifouling purpose has only limited studies, but it has shown tremendous result in other relevant engineering applications. The most direct inspiration of nature towards antifouling would be from marine living organisms, although other source of inspiration should be viable too. Million years of evolution allowed some organisms to grow the ability to defend against harmful colonization on the hosts [6].

In numerical studies, computational fluid dynamics (CFD) is the sole approach to simulate the behavior of topographies regarding their abilities in antifouling. Through CFD, the fluid flow pattern and magnitude, together with other parameters such as vorticity and wall shear strain at specific location. These parameters are essential to determine the optimum design for better antifouling properties. This paper discusses studies on 3 different surface topographies inspired from the nature. The said parameters in each bio-inspired topography will be determined. Wall shear stress would be the main concern in deciding the antifouling performance because higher stress deny the movement of microbials, directly pushed them along the direction and prevent their localized settlements [7]. Input parameter such as fluid flow rate, viscosity, local environment condition plays the key role in the accuracy of results too and each of them will be discussed further.

2 Research Methodology

This research is conducted using CAD and CFD software due time and cost constraint, and thus field test will not be carried out.

2.1 Selection of Bioinspired Topographies

For this study, the inspiration for the bioinspired topography design came from combining features on several natural surfaces. This includes the textures found on shark's skin, lotus leaves and the pilot whale skin. All these surfaces possess natural self-cleaning abilities that prevents biofouling material from adhering to their surface so easily.

2.2 CAD 3D Model and CFD Setup

After determining how the surface topographies should be like, a CAD 3D model for each topography is drawn on SolidWorks. The detailed dimension for each of the geometry in a model is justified to suit the size of simulation that is going to be carried out. CFD meshing is another crucial part in determining accurate result, a good mesh usually is complicated and time consuming, however it is much needed for a research like this, thus a mesh convergence analysis would be required after each simulation. This is to determine the best meshing quality and the accuracy of results generated.

2.2.1 CAD Modelling

Three dimensional models of the topographies were generated with SolidWorks. The geometry dimension of topography is as shown in Fig. 1.

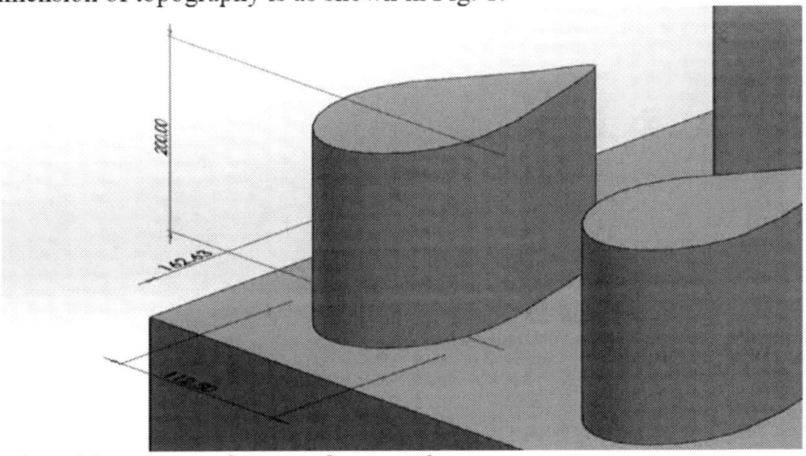

Fig. 1. Dimension of the geometry features of topography.

The geometry feature has width of 112.5 μm, 162.63 μm in length and 200 μm in height. The gaps between each feature is 150 μm. The topography has 10 rows and column each on a 2mm x 3mm plate that will be used later to create the fluid domain model for use in CFD. The fluid domain was represented by a solid block of 8mmx8mmx80mm as shown in Fig. 2.

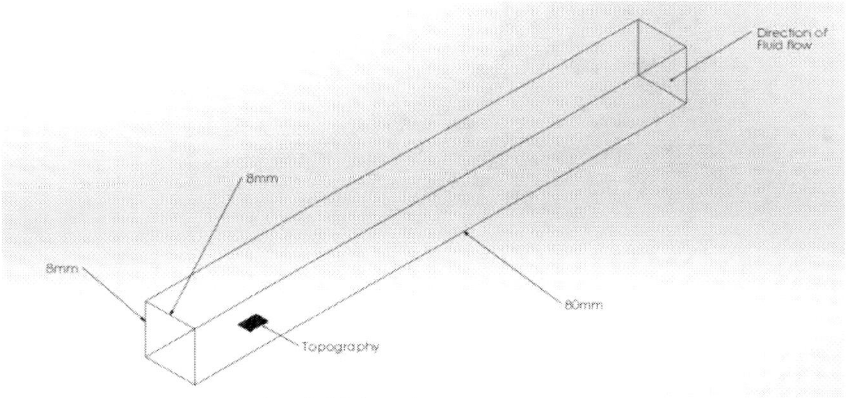

Fig. 2. Dimension of block representing fluid domain.

The entire solid part will be detected as fluid domain automatically in CFD. The topography solid created earlier on will used to create cavity on the fluid domain block. The topography is placed at the bottom middle plane and 65 mm away from the inlet plane.

2.2.2 Meshing Method

ANSYS Workbench is used to create the meshing for the model. Firstly, the entire model was meshed with tetrahedral shape elements. This shape provides finer mesh around curvature and edges. Edge and face sizing was used at the edges and bottom plane respectively.

Wall shear stress is important in this simulation, thus inflation around the walls are created with 2 layers inflation of elements size 0.1 mm. The resulting mesh are as shown in Fig. 5. Fig. 6 shows the close ups look at the meshing around the topography. To ease the setup of boundary conditions in CFD, the inlet, outlet and bottom wall topography was named respectively. This reduce the confusion when setting up boundary conditions, and makes previewing results easier.

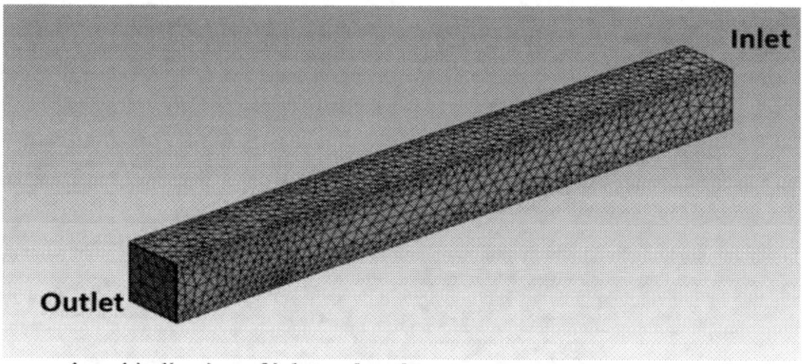

Fig. 5. Resulting mesh and indication of inlet and outlet.

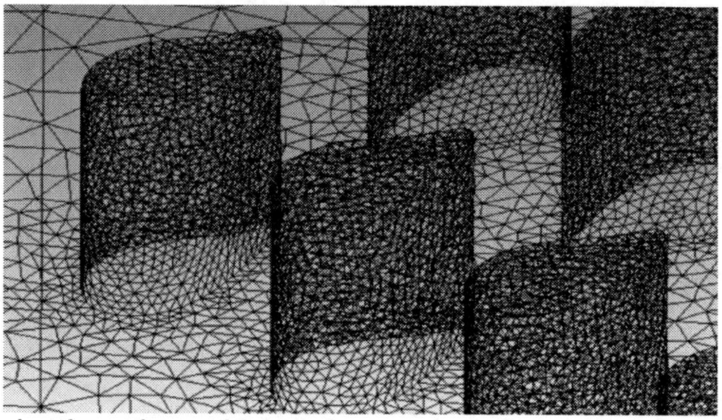

Fig. 6. Close-up look at the mesh around topography.

2.2.3 Numerical Setup

The use of Navier-Stokes equations (Equation 1 to 3) with the governing continuity equation (Equation 4) the mass and momentum to be conserved.

$$\frac{\partial \rho \mathbf{u}}{\partial t} + \nabla \cdot \rho \mathbf{u} V = -\frac{\partial p}{\partial x} + \nabla \cdot (\mu \nabla \mathbf{u}) + S_{m,x} \quad (1)$$

$$\frac{\partial \rho \mathbf{v}}{\partial t} + \nabla \cdot \rho \mathbf{v} V = -\frac{\partial p}{\partial y} + \nabla \cdot (\mu \nabla \mathbf{v}) + S_{m,y} \tag{2}$$

$$\frac{\partial \rho \mathbf{w}}{\partial t} + \nabla \cdot \rho \mathbf{w} V = -\frac{\partial p}{\partial z} + \nabla \cdot (\mu \nabla \mathbf{w}) + S_{m,z} \tag{3}$$

$$\frac{\partial \rho \mathbf{u}}{\partial t} + \nabla \cdot (\rho V) = 0 \tag{4}$$

The flow simulation required low Reynolds number formulation, transferred from a normal wall function, thus k-omega turbulence model was selected to ensure the accuracy of results. The density and dynamic viscosity of the fluid follows the default, which is 1000 $kgms^{-3}$ and 0.001 $kgm^{-1}s^{-1}$ respectively. Fig. 7 shows the boundary condition applied on the model. No-slip boundary conditions were applied on the wall of fluid domain except inlet and outlet. Inlet mass flow rate was set at 0.001 kgs^{-1} while the outlet was set as outflow.

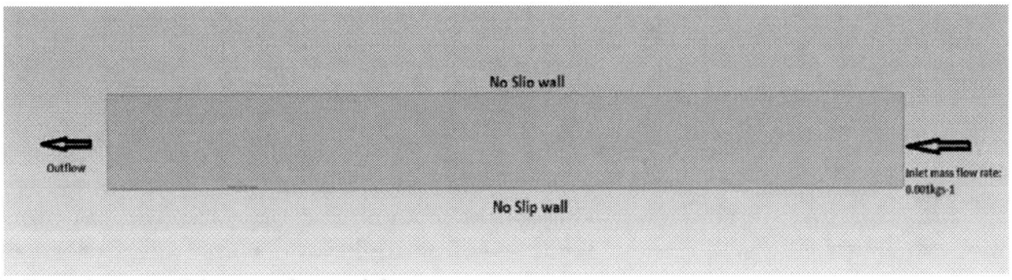

Fig. 7. Boundary conditions on the model.

3 Results and Discussion

3.1 Mesh Convergence Analysis

Mesh convergence analysis was carried out to determine the optimum number of mesh elements and assess the mesh quality. A few mesh refinements with different number of elements will be simulated and the difference in maximum wall shear stress will be compared to each other as shown in Table 1. The optimum number of elements is met when the difference in wall shear stress is less than 5%.

Table 1. Mesh convergence at the model, mass flow rate at inlet 0.01kgs-1.

			Optimum number of elements
Mesh refinements	1.4 million to 5.3 million	5.3 million to 6.1 million	5.3 million
Difference in wall shear stress	7.82%	2.34%	

For this model, when the number of mesh element increased from 1.4 million to 5.3 million, the difference in wall shear stress exceeds 5%, indicating the result are not stable and accurate enough. When the mesh refinement increased from 5.3 million to 6.1 million, the difference is only 2.34%, this suggest that the results are nearly converged and further increase in number of elements would not give any big differences in the accuracy of results.

3.2. Velocity Streamlines and Vector Plot

Velocity streamline and vector plot shows the fluid flow pattern around the topographies. Figure 8 presents the velocity streamline of fluid moving in between the topographies in the forward direction. The contour colours of the streamlines show that the velocity is increasing as the distance from the topographies increase in the Y direction. (Y=4mm). The results indicate that fluid flow is slow around the topographies. The velocity above the topography (Y=0.2mm) is an average of only 7.213×10^{-3} ms^{-1}.

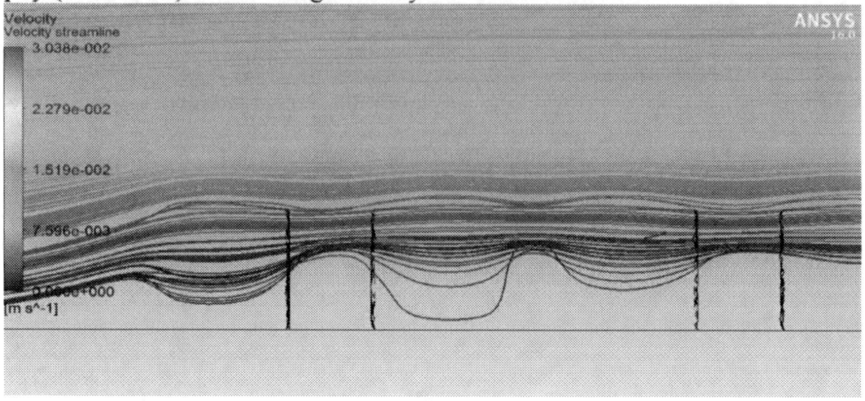

Fig. 8. Velocity streamline over topography.

Figure 9 presents the velocity vector profile between the gaps of Vortices were found in the flow direction, in between of subsequent topographies gap. The vortices formed and move in anti-clockwise direction (Fluid flow from right to left, in Z direction), preventing fluid from escaping the gap. This could also potentially trap the microorganisms in the gaps, allowing biofouling to occur.

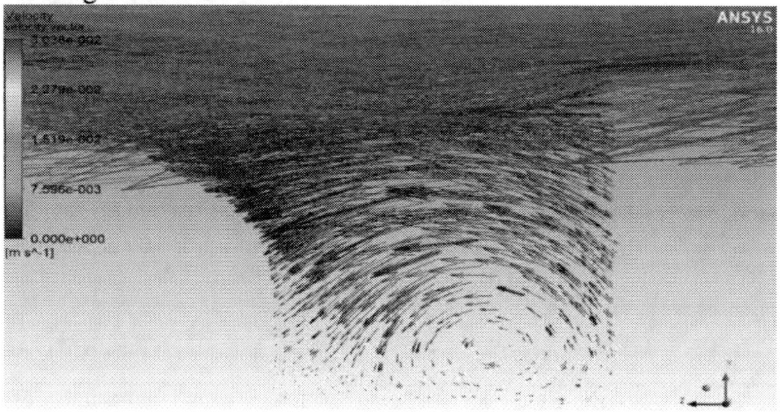

Fig. 91. Velocity vector plot showing vortices between the gaps of two adjacent feature in forward flow direction (right to left).

3.3. Wall Shear Stress

Several studies on biofouling have suggested that microorganisms have the ability to choose a suitable location for their settlements by adhesion. This indicates that surfaces that undergo higher wall shear would be more difficult for microorganisms to adhere to [8]. This indicates that high wall shear stress directly prevent biofouling. Fig. 10 presents the wall shear stress contour on one of the topography, viewing from side wall.

Fig. 10. Wall shear stress contour on the topography feature.

Fig. 11 presents the wall shear stress contour plot around the topographies at height of 100 um. Generally, the topographies have higher wall shear stress concentrated at the top, but significant low wall shear at the bed of topography. Highest wall shear stress appears at the tip of each topography, averaging at 0.231Pa. While the bed of topographies recorded an average 4.812 x 10^{-6} Pa. These values suggest that biofouling has the best chance to occur at the bed of topographies since the wall shear stress at the location are very weak.

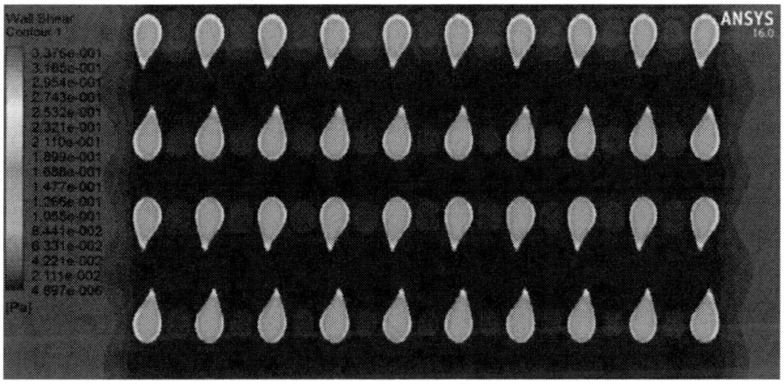

Fig. 11. Wall shear stress contour around the topography (100um above the bottom plane).

4 Conclusion

The numerical studies carried out determined the performance of bioinspired topographies against biofouling. The shark riblets inspired topography in this study exhibits average performance in antifouling based on the results collected in the form of velocity pattern and wall shear stress. The key problems to be noted from this study are as:

- The vortices formed in the gap of topographies
- Low wall shear stress at the bed of topographies

Further modifications on the topographies design aiming to eliminate the problem stated above should generate better results in term of antifouling performance.

References

1. Bixler, G. D. & Bhushan, B. Biofouling: lessons from nature. Philos. Trans. R. Soc.

370, 2381–2417 (2012).
2. A. I. Railkin, Ganf, T. A., O. G. Manylov, Marine biofouling colonization processes and defenses. Biofouling **20,** 129–131 (2004).
3. J. Chan, S. Wong, Biofouling types, impact and anti-fouling. Biofouling types, impact and anti-fouling 1–70 (2010).
4. M. E. Shirtliff, J. G. Leid, in (eds. Shirtliff, M. & Leid, J. G.) **3,** 272 (Springer Berlin Heidelberg, 2009).
5. F. W. Y. Myan, J. Walker, O. Paramor, The interaction of marine fouling organisms with topography of varied scale and geometry: a review. Biointerphases **8,** 27 (2013).
6. M. L. Carman *et al.* Engineered antifouling microtopographies - correlating wettability with cell attachment. Biofouling **22,** 11–21 (2006).
7. P. Halder *et al.* A novel approach to determine the efficacy of patterned surfaces for biofouling control in relation to its microfluidic environment. Biofouling **29,** 697–713 (2013).
8. Y. K. Lee, Y. J. Won, J. H.Yoo, K. H. Ahn, C. H. Lee, Flow analysis and fouling on the patterned membrane surface. J. Memb. Sci. **427,** 320–325 (2013).

Design and Optimisation of Home Scale Greywater Recycling Package

Kalaichelvan a/l Arugam[1], *Azadeh* Ghadimi[1,*], and *LH* Chang [2]

[1]Department of Mechanical Engineering, Taylor's University, Malaysia
[2]DD Techniche Sdn. Bhd., No. 11-3, Jalan USJ 21/1, UEP Subang Jaya, Selangor, 47630 Subang Jaya, Malaysia

Abstract. Water crisis is becoming one of the biggest challenges in some countries due to over population and drought. Therefore, government and non-government organisations in well-developed countries are encouraging industries to install water recycling package to reduce water usage and consume greywater instead. However the home scale is not yet available in the market. This research focuses on design a home scale greywater treatment system for toilet flushing within 3,000 to 5,000 USD as a collaboration with DD Techniche Sdn Bhd. Greywater is the waste water produced from laundry, kitchen and bathroom except the toilet waste and contains suspended solids, microorganisms, oil, nitrates and phosphorus which needs to be treated using suitable types of treatment for safe reuse. The treatments include physical, chemical and biological water treatment in which in this study the proposed system is a combination of all three. The design is evaluated using experimental methods to meet the standard of toilet flushing quality. The greywater system consists of the mechanical parts to control the flow of greywater and store the treated water including pumps, sand filtration tank, wire mesh chassis and collection tank. The final design has dimension of 6 m3 and optimised price within the budget.

1. Introduction

Water crisis is becoming one of the biggest threat in some countries. Especially, countries in Middle East are facing severe water crisis due to drought. Countries with high population also face water crisis due to high water usage [1]. Water crisis can be reduced by controlling the water usage. One of the successful method to control water usage is by recycling rain water and greywater. Government and non-government organisations are encouraging industries to have water recycling system to reduce water crisis. DD Techniche Sdn Bhd is a water recycling package Installer Company has seen the potential to market a home scale greywater treatment for toilet flushing within the budget ranging between 3,000 USD to 5,000 USD.

[1] Corresponding author: Azadeh.Ghadimi@taylors.edu.my

The company has come out with this initiative due the attention gained by the industrial scale greywater treatment package and the absence of home scale greywater treatment system. Greywater is the water produced from the laundry, kitchen and toilet except the toilet wastes [2]. Greywater contains phosphorus and nitrates due to the usage of detergent, shampoo and dishwasher liquid [3]. Greywater also contains bacteria, total suspended solids, oil and grease. The total suspended solids (TSS) are hair from the bathroom, dirt from laundry and food particles from kitchen. These contents in the greywater causes infection to the user if not treated. Therefore, the greywater must be treated before reusing it for toilet flushing.

There are three types of treatment, which are the physical treatment, chemical treatment and biological treatment. Physical treatment involves filtering out particles from water based on the size using wire mesh or membrane [4]. This treatment is applicable to filter the suspended solid and reduce the turbidity. Chemical treatment is process of adding chemicals into the water to neutralise the bacteria through salt addition [5]. Chemical treatment also involves coagulation process. Coagulation is the process of accumulating the bacteria together to increase the size and filter out using the physical treatment [2]. Biological treatment is microbial reaction using microorganism to convert bacteria into gas, water and biomass [6].

Greywater treatment must consist at least two types of treatment because chemical or biological treatment can treat bacteria but need to be supported by the physical treatment to filter the solid particles. The design of greywater treatment package for this project requires to have dimension of 2 m x 2 m x 2 m and budget within 3,000 USD and 5,000 USD. The design for this project is be inspired from the existing industrial scale water recycling system. The existing industrial designs for greywater recycling system are slow sand filtration, membrane bioreactor and water treatment with clarifiers.

The final design of greywater treatment package for this project is made up of pre-treatment stage with wire mesh and followed by slow sand filtration. This design consists all three types of water treatment stated above. The pre-treatment with wire mesh involves physical treatment to filter out food particles, hair and other suspended solids. The slow sand filtration involves physical treatment, biological treatment and chemical treatment. The water is accumulated at the top of sand to form platform for biological treatment. Then, the water passes through the sand and the waste from the biological treatment is filtered out by the sand layer. The nitrates in the water are be treated through denitrification by the sand layer.

The greywater treatment package consists of mechanical parts to control the flow and store the greywater for toilet flushing. The mechanical parts are pumps, sand filtration tank, wire mesh chassis and collection tank. The mechanical parts are designed based on the flow rate of the greywater. The pumps are used to pump the water from the collection tank to storage tank at the roof top for distribution. The sand filtration tank is designed based on the flow rate entering the sand filtration and requirement of sand height by the environmental organisations. Wire mesh are designed based on the size of particles to be filtered out and budget.

The design is evaluated by comparing the treated water with standard water quality for toilet flushing. The system must be able to meet the water quality for toilet flushing. The water quality is compared by measuring the total suspended solid (TSS), chemical oxygen demand (COD) and turbidity. These three parameters must meet the standard water quality for toilet flushing. The final design is represented in 3D model with all the mechanical parts.

The objectives of this project are to design a home scale greywater recycling package and optimise the design by having low cost compare to existing designs in the market. The final design is able to treat the greywater for toilet flushing. The design is optimised by having lower cost than existing home scale greywater package.

This project creates the chance for the public to use the greywater treatment system for household. These reduces the water usage for the public, which will reduce the cost if the system is used for long term. Reducing water usage results in reduction of water crisis. This system able to gradually bring advantage to the economy and environment. Since, public are not using any water recycling system currently because water recycling systems are only manufactured in industrial scale, this project able to fulfil the gap of designing a home scale water recycling system.

2. Methodology

This project involves designing, 3D modelling, prototyping and experimental evaluation. The designing involves the designing of pre-treatment stage with wire mesh, slow sand filtration and pump. The design is modelled into 3D prototype using SolidWorks Ver. 2016. The performance of the design is evaluated using the prototype.

2.1 Designing Stage

The greywater treatment system consists of pre-treatment filter, slow sand filtration, pump, and collection tank. The greywater enters to the pre-treatment filter and follow by the slow sand filtration. The treated water is collected in the collection tank. Then, the treated water is pumped to the storage tank using the auto water pump. The treated water is distributed to the toilets from storage tank.

The pre-treatment filter consists coarse and fine wire mesh. The pore size of coarse wire mesh is 1 mm and the pore size for the fine wire mesh is 0.5 mm. The coarse mesh is used to filter out hairs and particles that are larger than 1 mm. The fine mesh is used to filter out dirt and particles larger than 0.5 mm. The pre-treatment stage is designed vertically to use the gravitational force during the filtration. The gravitational force provides extra support to push the water through the pores of the wire mesh. The wire mesh is designed to have hemispherical surface to provide larger filtration area. The hemispherical shape also provides the slippery slope for the movement of filtered particles through the drainage pipe.

The slow sand filtration consists sand layer and gravel. The slow water flows on the sand causes the water to accumulate at the top to form water bed for the biological treatment [7]. The biological treatment transforms bacteria into gas, water and biomass. The biomass is filtered by the pores between the sand. The sand also involves chemical treatment. The sand neutralises thc nitrates through denitrifications [8].

The slow sand filtration is designed based on the volumetric flow rate equations and requirement by the environmental organisation [8]. Volumetric flow rate equation is applied to calculate the filtration area of the sand. Volumetric flow rate means the volume of water passes through medium per unit time. The volumetric flow rate equation is shown below.

$$A = \frac{Q}{v} \quad (1)$$

Where, A = surface area of the sand filter (m^2)
Q = flow rate onto the filter surface (L/h)
v = loading rate of sand (L/hm^2)

The flow rate of greywater in household are obtained from the journal published by the Texas A&M University from United States [9]. The data are shown in Table 2. The loading rate is set at 400 L/hm2 because it is the optimum loading rate for sand filtration based on the environmental organisations from United States and United Kingdom. Therefore, the surface area of the sand filter can be calculated by substituting the parameters.

The height of sand filtration and gravel are designed based on the standard dimension set by the environmental organisation from well-developed country. This dimensions are set by the organisations based on the feedbacks gathered from the existing slow sand filtration. The dimensions are listed in Table 3.

Table 1. Parameters for sand filtrations system [7,10]

Organisations	Parameters	Dimensions
Cisco Foundations	Sand depth	≥50 cm
	Loading rate	400 L/hm^2
	Gravel Depth	≥5 cm
American Water Works Association	Sand Depth	≥50 cm
	Loading rate	100 L/hm^2-400 L/hm^2

Based on requirement in Table 1, the height of sand is set to 50 cm and the height is validated by applying Darcy's law. Darcy's law is used to calculate the flow rate of liquid that passes through a porous medium. Darcy's law states that the water flows from high potential area to low potential area. Therefore, the water in the sand filter obeys this law where, the water on top flows to the bottom due to the potential difference. Darcy's equation (2) is given as shown below. The sand depth recommended by environmental body is validated using the Darcy's law.

$$Q_{max}=k.A.\frac{h+d}{d} \qquad (2)$$

Where, Q_{max}= Maximum flow rate (m^3/s)
k= Hydraulic conductivity of sand (m/s)
h= Depth of clearance above the sand (m)
d=Depth of sand (m)

The designing of pump includes the process of transferring the water from collection tank at the slow sand filtration to the storage tank. Auto water pump is used in this project because the slow sand filtration required longer time to fill in the collection tank. The pump should operate once the collection tank is full. Auto water pump activates based on the water level. Therefore, auto water pump is the ideal choice for this project.

The suitable pump can be chosen based on the water horsepower. The parameters that are related in the horsepower pump designing is the head loss and water flow rate. The equation relating these parameters are shown below. Head loss is the distance to pump the water from one point to another point. The pump for this project must be able to transfer the water from the collection tank to storage tank at the rooftop. Therefore, the head loss is the height of the house, which is generally 5.7 m.

$$WHP=\frac{Q\times H}{3960} \qquad (3)$$

Where, WHP= Water horsepower (hp)
Q= flow rate of water from the pump (gallon/minute)
H= head loss or distance to pump the water (foot)

2.2 3D Modelling

The design of the pre-treatment stage and slow sand filtration are modelled as software prototype using SolidWorks Ver. 2017. The pre-treatment stage consists of three parts, which are the mesh holder, filtration chassis and chassis connector. The mesh holder is placed in the filtration chassis and the coarse filtration chassis is connected to the fine filtration chassis using the chassis connector. Since, 4 inches PVC pipe commonly used as greywater outlet, the pre-treatment stage is designed to have same dimension as 4 inches PVC pipe for ease of installation. The drawing of mesh holder is shown in Figure 1.

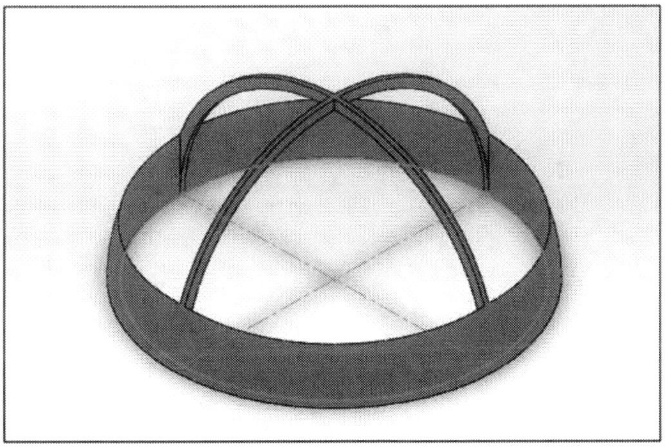

Fig. 1. Mesh holder

The slow sand filtration chassis was designed based on the calculations carried out using the equations and requirements as discussed in Section 2.1. The slow sand filtration consists two parts which are the chassis and connector. The chassis is used to fill the sand and gravel. The connecter is used to connect the sand filtration to the pre-treatment stage. The complete 3D model is shown in Figure 2.

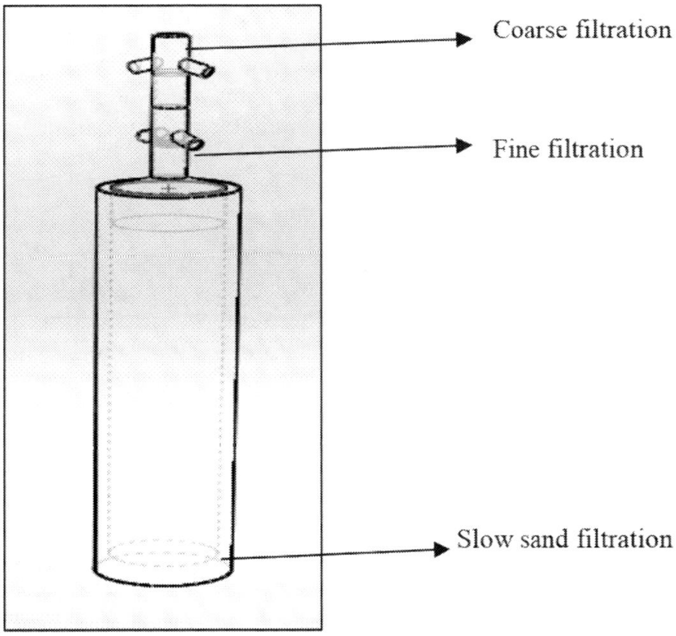

Fig. 2. Drawing of the whole filtration system

2.3 Prototyping

The prototyping stage involves the manufacturing of the pre-treatment filter and the sand filtration to evaluate the design. The pre-treatment filter is built using the 110 mm PVC pipe with PVC pipe connector. The wire mesh for fine screening and coarse screening are glued in the pipe using adhesive bonding. The complete prototype model for pre-treatment filter is shown Figure 3.

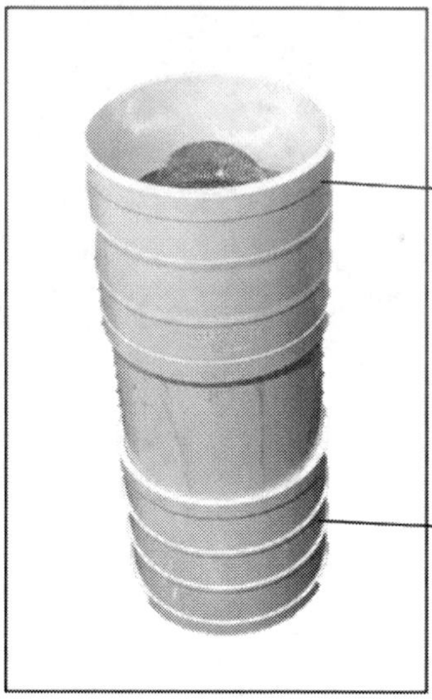

Fig. 3. Pre-treatment filter

The slow sand filtration is manufactured using plastic tanks. The slow sand filtration is manufactured in scaled down dimension because the prototype is only needed to evaluate whether the system can treat the greywater to meet standard quality of water for toilet flushing. The prototypes are manufactured for use of experiments to evaluate the design.

2.4 Experiments

The project involves the experiments to measure the flow rate entering the sand filter from the pre-treatment stage, total suspended solid (TSS), chemical oxygen demand (COD) and turbidity. The TSS, COD and turbidity is measured to evaluate whether the treated water meets the standard water quality for toilet flushing. The standard water quality for toilet flushing is gathered from various water quality department as shown in Table 2. The performance of the design is evaluated by comparing the treated water with parameters in Table 2.

Table 2. Water quality for toilet flushing based on American standards

Organisation	Parameter	Value
United States Environmental Protection Agency (US EPA) [11]	BOD	≤10 mg/L
	COD	-
	Turbidity	≤ 2 NTU
	TSS	-

World Health Organization [12]	BOD	≤10 mg/L
	COD	-
	Turbidity	-
	TSS	≤10 mg/L
Canada Health Department [13]	BOD	≤10 mg/L
	COD	-
	Turbidity	≤ 2 NTU
	TSS	≤10 mg/L
The International Water Quality published book titled Milestone in Water Reuse [14]	BOD	-
	COD	≤30 mg/L
	Turbidity	-
	TSS	-

2.4.1 Flow Rate Test

The objective of this experiment is to determine the flow out rate of the greywater from the pre-treatment stage. Flow out rate is needed to calculate the surface area for the slow sand filtration. The experiment is conducted to obtain the relationship between the flow in and flow out of the pre-treatment system. The relationship can be used to approximately measure the flow out rate at any particular flow in rate for the pre-treatment stage.

The experiment is conducted by measuring the flow in rate from tap water by adjusting the tap at different angle. The time taken to fill 2 L of water at each angle is recorded. The procedures are repeated by passing the tap water through the pre-treatment filter. The time taken to fill 2 L of water with the presence of pre-treatment filter represents the flow out rate and the time taken to fill 2 L of water with the absence of pre-treatment filter represents the flow in rate. The graph of flow out rate against flow in rate is plotted and the relationship is obtained. Based on the relationship, the flow out rate is substituted into equation 1 to calculate the surface area for sand filtration.

2.4.2 Turbidity Experiment

The turbidity of the treated water is measured using the turbidity tube. Turbidity tube is measuring cylinder with a reference disk at the bottom of the tube as an indicator. The treated water is filled up until the maximum level, where the reference plane is still visible as shown in Figure 4. Then, the height of the water level in the cylinder is measured. The height is converted into turbidity scale using the table obtained from lab manual prepared by Chicago River Schools Network [15].

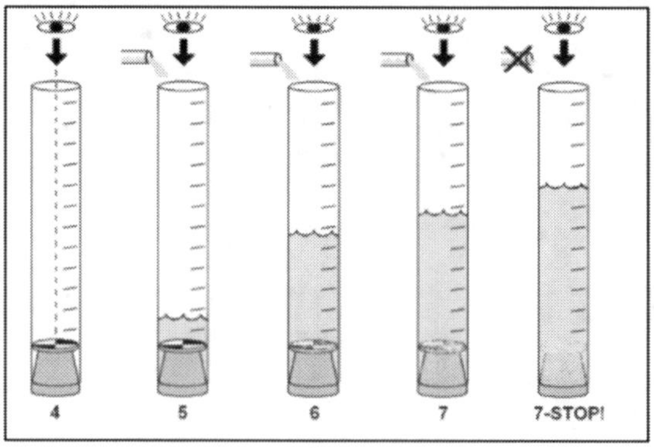

Fig. 4. Filling the treated water until the maximum level, where the disk is visible

2.4.3 Chemical Oxygen Demand Experiment

Chemical oxygen demand, COD is the total oxygen consumption by all the organic substances in the water [16]. High values of COD indicate the water has high level of microorganisms and pollution [17]. The COD experiments is carried out by adding nitric acid into the treated water. Then, the mixture is heated until the oxidation process completed and the remaining unreacted nitric acid is weighed. The mass of reacted nitric acid is calculated by subtracting the remaining unreacted nitric acid added to the treated water. The mass of dissolved nitric acid represents the COD of the treated water in g/L.

2.4.4 Total Suspended Solid Experiment

Total suspended solid (TSS) is the quantity of solid particles in the water. The TSS experiment involves the evaporation of the treated water. 1 L of treated water is evaporated under the sun and leftover residue after evaporation is weighed. The mass of leftover residue represents the TSS value in g/L.

2.5 Limitation

Limitation in this project is the usage of instruments for the measurement of COD, TSS and turbidity of the water. The COD, TSS and turbidity of the water are measured using manual experimental method. The instruments are not used because budgets or facility to borrow the instruments are not allocated by the industry supervisor. Therefore, the manual methods of determining this parameter involve limitation such as the scale and sensitivity. The turbidity test only able to give minimum value of 5 NTU.

3. Results and Discussion

The data of time taken to fill 2 L of water with the absence of pre-treatment filter is tabulated as shown in Table 3. Table 3 shows the flow in rate when the tap is at different angle. Table 4 represents the time taken to fill 2 L of water with the presence of pre-treatment filter. The graph of flow out rate against flow in rate is plotted as shown in Figure 7.

Table 3. Results of flow in rate

Tap angle (°)	Time Taken to fill 2 L (s)			Flow in rate (L/s)
	1st	2nd	Average	
45	0	0	0	0
90	101.42	105.2	103.31	0.0194
135	26.13	26.22	26.18	0.0764
270	25.35	25.53	25.44	0.0786
360	25.12	24.37	24.75	0.0808
405	25.24	23.51	24.375	0.0821
450	23.16	24.25	23.71	0.0844
495	21.68	23.88	22.78	0.0878

Table 4. Results of flow out rate

Tap angle (°)	Time Taken to fill 2 L with Filter (s)			Flow out rate (L/s)
	1st	2nd	Average	
45	0	0	0	0
90	107.37	105.92	106.645	0.019
135	29.18	28.87	29.03	0.0689
270	27.08	25.54	26.31	0.0760
360	27.68	23.33	25.51	0.0784
405	25.63	25.48	25.555	0.078
450	27.33	23.55	25.44	0.0786
495	24.7	23.16	23.93	0.0836

Graph in Figure 5 represents the relationship between the flow in rate and flow out rate. The relationship equation is obtained to calculate the flow entering the slow sand filtration. The flow out rate is represented as y and the flow in rate is represented as x. The flow out rate represents the flow rate entering the slow sand filtration. Then, the surface area required for the slow sand filtration is calculated using equation 1 and tabulated in Table 5. Therefore, the slow sand filtration is designed based on the calculated surface area as shown in Table 5. The sand filtration surface has 21 cm of radius. Therefore, the design can be used up to maximum of 4-bedroom house. Based on Table 5, the maximum flow rate entering the slow sand filtration is 54.76 L/h. Therefore, the maximum flow rate can be handled by the slow sand filtration can be calculated using the Darcy's equation as shown below. The sand depth, d is recommended to be at least 50 cm by the environmental bodies from United States. The hydraulic conductivity of the fine sand is given 2×10-4 m/s. The depth of clearance above the sand layer is 10 cm.

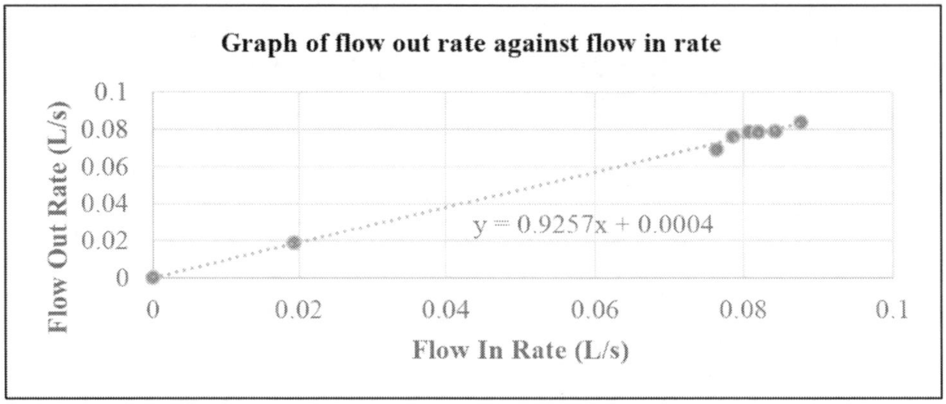

Fig. 5. Graph of flow out rate against flow in rate

$$
\begin{aligned}
Qmax &= k.A.(h+d)/d \\
&= (2\times 10\text{-}4)\ (\pi 0.212)\ [(0.1+0.5)/0.5] \\
&= 3.325\times 10\text{-}5 \text{ m3/s} \\
&= 119.70 \text{ L/h}
\end{aligned}
$$

Based on the calculated Qmax, the sand depth is designed to be 50 cm. This is because the flow of water entering the sand filtration is lower than the maximum flow rate can be handled by the slow sand filtration system with 50 cm sand depth. Based on the experiments and calculation the final design of the system with collection tank is shown in Figure 6 and Figure 7. The collection tank is design to collect 1000 L of capacity.

Table 5. Flow rate entering the slow sand filtration and surface area of sand

No of bedrooms	**Daily water use (gal)**	**Greywater flow rate (L/h)**	**Flow rate entering the sand filtration (L/h)**	**Surface Area of Sand (m^2)**	**Radius sand surface (cm)**
1 or 2	225	35.49	32.85	0.08	16.17
3	300	47.32	43.80	0.11	18.67
4	375	59.15	54.76	0.14	20.87

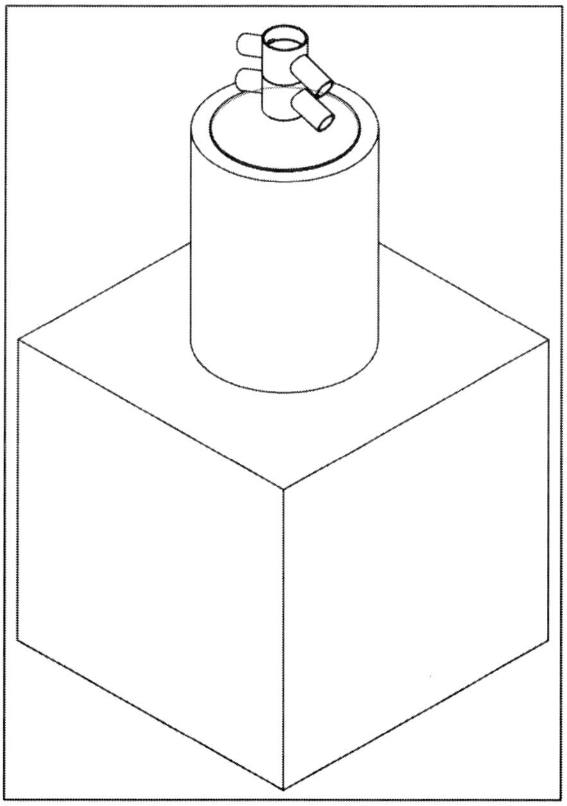

Fig. 6. 3D design of the system with the collection tank

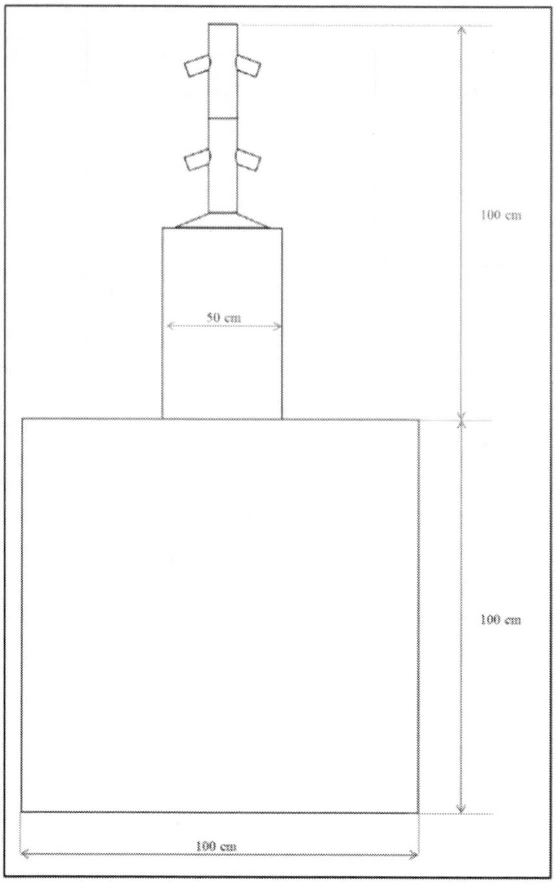

Fig. 7. Detailed design of the system with the collection tank

The design is evaluated by carrying out experiments to determine the COD, TSS and turbidity value of the treated water using the small-scaled prototype. The data collected are shown in Figure 8.

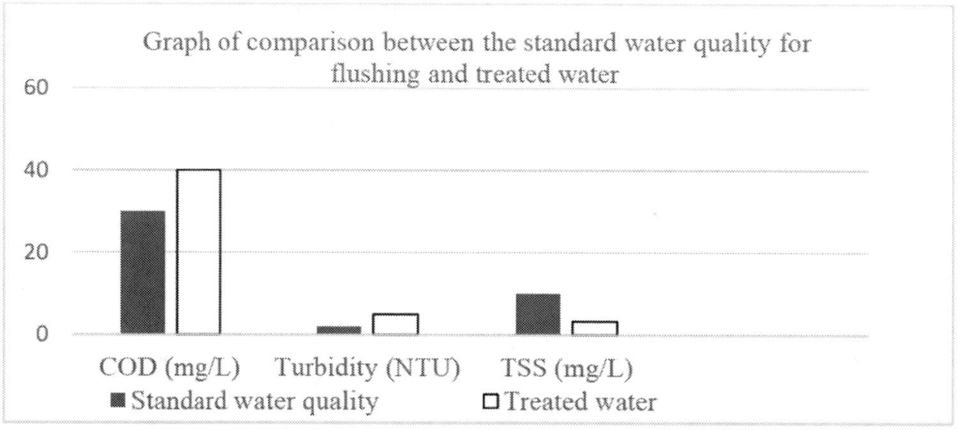

Fig. 8. Bar chart comparison for standard flushing water quality and treated water

The turbidity of the treated water in this experiments is less than 5 NTU. The total suspended solid in the treated water is 3.33 mg/L. The TSS value required for the toilet flushing water is less than 10 mg/L. Therefore, the design fulfils the requirement for the TSS value. The COD of the treated water is 40 mg/L. The required COD for the toilet flushing water is less than 30 mg/L. The COD requirement was not able to fulfilled because of the small-scale prototype.

Based on the finalised design, the price to manufactured one complete filter system is estimated. The complete system includes the plastic filter chassis for the pre-treatment filter and slow sand filtration, filter sand, wire mesh, auto water pump and collection tank. The price for each product is obtained and the total cost is calculated as shown in Table 6. The experiments to evaluate the performance of system by measuring the TSS, COD, and turbidity of the treated water are conducted using the prototype.

As can be seen in Table 6, the budget to manufacture one complete system using the design in this project is lower than cost of the existing greywater recycling package. The existing home scale greywater system in the market is a down-scaled of industry greywater recycling design. Therefore, the cost of this existing home scale greywater recycling package design is 8,100 USD. This design has high cost due to the technology that are used in this system. This system consists physical treatment using different size of mesh and membrane. Then, the biological treatment is carried out using the UV ray. The UV ray neutralises the microorganism in the water. The parts that cause the system to be expensive are the membrane, UV ray and pump. This design is relatively cheaper because this designs does not uses any membrane as the physical filtration and high performance pump for the water movement through the membrane. The membrane is replaced by the sand filtration which able to meet the same turbidity and pump is replaced by the gravitational force in the sand filtration. The UV ray is replaced by the sand filtration because the water and residue accumulated at the top of the sand layer forms biological reaction bed layer. The microorganisms are neutralised at top of the sand through biological treatment.

Table 6. List of price for the filtration system

Product	Price	Quantity	Total (USD)
Filter chassis	25.80 USD/kg	10.073	259.8834
Auto water pump	133.43 USD	1	133.43
Filter fine sand	2.41 USD/kg	117	281.97
Gravel	2.31 USD/kg	80	184.8
1000 L water tank	425.55 USD	1	425.55
1 mm wire mesh	60 USD/sheet	1	60
0.5 mm wire mesh	80 USD/sheet	1	80
Total Cost (USD)			**1425.63**

Therefore, the optimisation of the design involves the cost of the system. The recycling package is designed at relatively lower cost with same function and requirement. This low cost provides more chances for the customer to buy this product. This can increase the attention and business value of greywater recycling package.

4. Conclusion

The greywater treatment package is successfully designed within proposed budget The design of the system is within 6 m^3 including the collection tank. The greywater treatment package contains the pre-treatment filter, slow sand filtration and collection tank. The effect of the pre-treatment stage on the flow entering the slow sand filtration is studied to design the slow sand filtration. Based on the experiments, the slow sand filtration is designed based on the volumetric flow rate and Darcy's law. The material budget of a complete system is lower than the existing home scale greywater recycling system. This design is able to treat greywater to meet the standard water quality of toilet flushing.

References

[1] K. Bani-melhem and E. Smith, Chem. Eng. Journal., **198-199**, 201–210, (2012).
[2] M. S. Fountoulakis, N. Markakis, I. Petousi, and T. Manios, Sci. Total Environ. Journal, **551–552**, 706–711, (2016).
[3] J. A. Peeples, Wastewater Use Agric., **II**, 119–131, (2006).
[4] B. Je, A. Laine, S. Parsons, T. Stephenson, and S. Judd, Journal of Elsevier, **1**, 1999, (2000).
[5] F. Li, K. Wichmann, and R. Otterpohl, Sci. Total Environ. Journal, **407**, 11,. 3439–3449, (2009).
[6] B. A. Mittal, "Biological Wastewater Treatment," *FullTide Corporation*, 2–9, (2011).
[7] Y. Dunets, Siobhan; Zheng, "Slow sand filtration," *Greenh. Nurs. Watery*, (2011).
[8] W. T. Course, Sci. Total Environ. Journal, 1–2, (2012).
[9] B. Lesikar, R. Melton, D. Smith, and C. O. Neill, "*AgriLife Communications and marketing*, 3–08, (2003).
[10] Centre for Affordable Water TEchnilogy, "What is the minimum depth of filtration sand, underdrain gravel and separating gravel? (particularly." Cisco Foundation, 1, (2017).
[11] Members of WQA, "Residential Water Reuse Fact Sheet," *Water Quality Association*, 1–5, (2013).
[12] T. Bruursema, Plumbing Systems & Design Magazine, 15–22, (2011).
[13] American Water Standard, "NSF/ANSI Standard 350 For Water Reuse Treatment Systems," (2011).
[14] V. Lazarova, *Milestones in Water Reuse*. The International Water Association, (2013).
[15] Chicago School Network, "Turbidity," *Turbid. Test Procedure.*, 1–2, (2013).
[16] Y.-C. Kim, S. Sasaki, K. Yano, K. Ikebukuro, K. Hashimoto, and I. Karube, Anal. Chim. Acta, 432, **1**, 59–66, (2001).
[17] J. Chen, J. Zhang, Y. Xian, X. Ying, M. Liu, and L. Jin, Water Res.Journal, **39**, 7, 1340–1346, (2005).

Mechanical Characterization of Zirconia Ceramic Composite

Kelvin Chew Wai Jin[1,*], *Kam Dickson*[1] , and *Ramesh Singh*[2]

[1]Department of Mechanical Engineering, School of Engineering, Faculty of Built Environment, Engineering, Technology and Design, Taylor's University, 47500 Subang Jaya, Selangor, Malaysia.
[2] Centre of Advanced Manufacturing & Material Processing (AMMP), Department of Mechanical Engineering, Faculty of Engineering, University of Malaya, 50603 Kuala Lumpur, Malaysia.

Abstract. In this day and age, zirconia ceramics are used widely in the medical field as biomaterials for the replacement of damaged body parts. This is because zirconia is one of the closest replacements for bone tissue. However, there were a few cases regarding the failure of zirconia ceramic hip transplants. To overcome this issue, composite materials are being studied as they are able to combine different properties which are not present in a material. This study of Y-TZP/stainless steel 316 composite is carried out with the idea of providing a solution for failure of zirconia implants and also an improvement in biomaterials which will benefit the biomedical world. The study aims to determine the effects on how the increasing of stainless-steel content in the composite will affect the relative density, Vickers hardness, fracture toughness and ageing resistance of the Y-TZP/stainless steel 316 composite. To carry out the research, the composite samples were prepared by mixing the powder of each samples according to their determined content with ethanol. After the powders were mixed, the powders were then pressed, followed by undergoing a Cold Isostatic Press process (CIP) and then it underwent sintering at its determined temperature. After sintering, the samples underwent grinding and polishing before being ready for testing. The outcomes of the research showed that as the sintering temperature and the content of stainless-steel increase in the composite, the fracture toughness and ageing resistance improved while the Vickers hardness and relative density decreased.

1 Introduction

Composite materials are being used more frequently in this modern era as it tends to enable the combination of different properties which is not present in a material. In the medical field, different materials have been used for biomaterials study as a replacement for damaged body parts. Ceramics in particular has been given close attention as it is the closest replacement for bone tissue [1]. However, ceramics does not have a high toughness when it is compared to metals. Therefore, ceramic filled metal and metal filled ceramic composites are being studied to increase the overall toughness of the replacement implants.

* Corresponding author: KelvinWaiJin.Chew@taylors.edu.my

In this modern era, prostheses that are made of purely ceramics are slowly being replaced by composite materials [2]. This is due to the improvement of the major factors, which is the strength and durability of the prostheses. Thus, studies regarding composite materials as biomaterials are being developed rapidly.

Zirconia is a ceramic which is readily used in many different sectors, especially the medical field. It has different properties when it exists at different phases, and is usually added with stabilising agents so that the phases and characteristics do not change at different temperatures. It has a high tolerance towards corrosion, high strength, high density and also high resistance towards impact. Y-TZP (Yttria-stabilized Tetragonal Zirconia Polycrystalline) in particular is more compatible to the human body. Thus, it has been used for many years as prosthetics and has not shown negative responses. Besides that, there are many different types of stainless steel which is graded based on its mechanical properties, grade compositions and also its production specifications. Generally, compared to ceramics, stainless steel has a high corrosion resistance, high toughness, high ductility and high strength. To produce implants which should be more durable, composite materials can be used to replace ceramics.

Based on the study done by Kurtz et al.[3], composite materials such as zirconia toughened alumina is better compared to materials such as zirconia and alumina as it is able to combine both the advantageous properties of alumina and zirconia. The addition of both alumina and zirconia results in having a higher fracture toughness and strength which reduces the probability of dislocation and impingement which leads to failure of implants. It also reduces the hardness which exist in ceramics. The stress-assisted degradation will also be reduced drastically while the wear characteristics improves. The study showed that the use of composite materials would allow the implants to last for a longer duration compared to ceramics without the need to be revised.

Agudelo et al.[4] did a study to determine the relative density of composite materials, zirconia and 316 stainless steel with varied compositions and heating rates. The relative density of the samples was tested with different heating rates during sintering and the results showed that, when the content of stainless steel in the samples are below 50%, the relative density of the samples decreased as the heating rates increased. On the other hand, when the content of stainless steel in the samples were above 50%, the relative density of the samples increased as the heating rates increased.

Esteban et al. [5] conducted a study to determine the Vickers hardness and the fracture toughness of zirconia/stainless steel composites. The zirconia/stainless steel composites were made into samples with stainless steel concentrations ranging from 15 to 30%. The results of Esteban et al. [5] study showed that the Vickers hardness decreased and the fracture toughness increased as the stainless-steel content in the zirconia/stainless steel composites increased which follows the rule of mixture [6-8]. The results also showed that there was no plastic deformation when fractures occur in the zirconia/stainless steel composite samples. Besides that, the study also showed that the zirconia/stainless steel is not brittle compared to zirconia ceramic. These characteristics are important for the enhancement of materials used in biomedical implants, especially zirconia.

In the past, the medical field has reported a few cases regarding the failure of zirconia ceramic hip transplants. To overcome the issue, more and more studies have been carried out regarding the sintering of zirconia composites at low temperatures regarding its mechanical characteristics [9]. Regarding hip transplants, most of the studies carried out involve around the zirconia (Y-TZP) ceramic. In this study, the zirconia ceramic (Y-TZP) was mixed with stainless steel (316) to produce composite materials. The present paper aims to determine the relative density, Vickers hardness, fracture toughness and ageing resistance of the composite at different sintering temperatures. This study is carried out with the idea to enhance and aid the current medical field with an alternative solution regarding the production of transplants using different composite materials.

2 Methodology

2.1 Powder preparation

During the first stage of study, stainless steel 316 and Y-TZP powder were used based on Table 1. The stainless steel 316 powder and the Y-TZP powder were mixed and measured according to the content of stainless-steel for the preparation of the samples. Firstly, the digital ultrasonic bath (Powersonic UB-410, Thermoline Scientific, Australia) was filled with water. Then, 100ml of ethanol was poured into a beaker and both the stainless steel 316 powder and the Y-TZP powder were added into the beaker to undergo mixing. The weight percentage of each samples were fixed with the content of stainless steel ranging from 0 wt%, 10 wt%, 20 wt% and 30 wt% before both the powders were mixed. The beaker was placed in the Digital Ultrasonic Bath for 15 minutes and then taken out after the mixing was completed. The mixture then underwent attrition milling using the Laboratory Attritor Mill (01-HD, Union Press, United States) for 45 minutes at 500 rpm. Both the stainless steel 316 and Y-TZP powder mixture were reduced to a slurry. After mixing, the mixture was placed in a universal oven (UN 55, Memmert, Germany) overnight at 60°C and was left to dry. A sieve of 212μm was then used to sieve the mixture through to ensure that the powder mixture was fine.

2.2 Green sample preparation

The composite powders were then pressed using a hydraulic bench press (HP10-150, RMS Hydro, Malaysia) to form round tablets that have a diameter of 20 mm. Pressing was required to form a green body and ensure that the powder mixture was fine before cold isostatic pressing. Cold isostatic pressing was carried out using a cold isostatic press (KJYU150, Shanxi Golden Kaiyuan Co., Ltd, China) and was done by placing the green body round tablets into containers of the desired shape and under room temperature and a high pressure. This was done to ensure that the green body samples were not affected by heat or pressure when they were being pressed together. If heat or pressure was applied to the green body samples as it was being pressed together, the green body samples might have cracks after they were pressed together.

Table 1. Characteristics of Starting Powders.

Materials	Composition (wt%)	Theoretical density (g/cm^3)	Tap density (g/cm^3)	D_{50} (μm)
Stainless Steel 316	17 Cr-12 Ni-2.5 Mo-1 Si-2 Mn	8.00	4.4	21.9 ± 0.1
Y-TZP	5.2 Y_2O_3 (yttria)	6.10	6.0	0.6 ± 0.1

2.3 Sintering, grinding and polishing

The next stage of the study was where the samples underwent sintering to become specimens which were ready for testing. The samples underwent sintering by being placed into an electric box furnace (LT01, LT, Malaysia) and was heated with different temperatures which are 1250°C, 1300°C, 1400°C and 1500 °C. The sintering rate was 10° per minute and the sintering process had a holding time of 2 hours. Grinding paper was then used to polish and grind the samples after sintering. The type of grinding paper used was silicon carbide (SiC) paper. 5 different grades of grinding papers, namely 120, 240, 600, 800 and 1200 were used to ensure that the samples were grinded properly. After grinding, the samples underwent polishing. There were 2 different sizes of diamond bits from the diamond paste which were used, namely 6μm and 1μm. The samples were polished using the diamond paste on a polishing clothe. Both the grinding and polishing process were carried out using a polishing and grinding machine (GP2V, LT, Malaysia).

2.4 Measurement and Analysis

2.4.1 Relative density

After the samples were prepared, a series of different test were conducted on the samples to determine different mechanical properties. To determine the relative density of the composite samples, the water immersion technique using an analytical balance with density kit (AY220, Shimadzu, Japan) was used while the immersion medium was distilled water. The technique was done based on Archimedes principle. The theoretical density of Y-TZP was determined as 6.1 Mg.m^{-3} (or g/cm^3). The formula for the calculation of relative density is based on (eq 1).

$$\rho = \frac{W_a}{W_a - W_w}\rho_w \tag{1}$$

where

ρ	=	Relative density of sample (g/cm^3)
w_a	=	Weight of sample in air (g)
w_w	=	Weight of sample in water (g)
ρ_w	=	Density of distilled water (g/cm^3)

2.4.2 Vickers hardness

The Vickers hardness of the samples was tested using a Vickers hardness tester (MV-01, Matsuzawa, Japan) which had a pyramidal diamond indenter and an applied load of 10kg. The load was applied on to the samples slowly without impact and was held down in place on the sampled for a duration of 10 seconds. The physical quality if the indenter and the accuracy of the applied load was defined based on ASTM E284-99 [10] and ISO 14705 [11] and shown in Figure 1. After the indentation was done, the indentation was measured using a filar micrometer which was built in the attached microscope on the Vickers hardness machine. The microscope had an accuracy nearest to 0.1μm. The formula for the calculation of the Vickers's hardness is presented in (eq 2).

$$H_v = \frac{1.854P}{(D)^2} \tag{2}$$

where

P	=	Applied load (kg)
D	=	Average of two diagonals D_1 and D_2 (m)

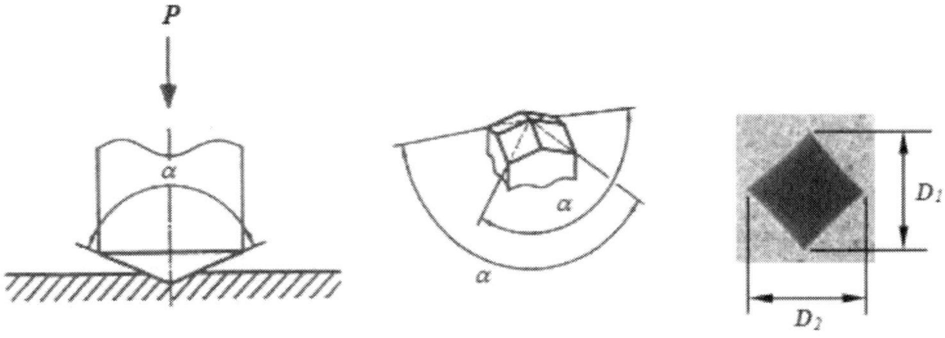

(a) (b)

Fig. 1. Schematic diagram showing the shape of the Vickers indenter (α =136° ±0.5°, angle between the opposite faces at the vertex of the pyramidal indenter) and (b) an indentation that is formed [11].

2.4.3 Fracture toughness

After the indentation was done on the sample, the fracture toughness was also measured based on the same indentations that were used to determine the Vickers hardness. The equation to calculate the fracture toughness is presented in (eq 3) which was proposed by Niihara et al. [12]. Selected parameters from Vickers indentation are depicted in Figure 2.

$$\left(\frac{K_{IC}\Phi}{H_V a^{\frac{1}{2}}}\right)\left(\left(\frac{H_V}{E\Phi}\right)^{\frac{2}{5}}\right) = 0.035\left(\left(\frac{L}{a}\right)^{\frac{-1}{2}}\right) \qquad (3)$$

where

K_{IC}	=	Fracture toughness (MPa. $m^{1/2}$)
H_V	=	Vickers hardness (GPa)
E	=	Young's Modulus (GPa), taken as 200 GPa for Y-TZP ceramic
a	=	Half-length of the average diagonal (m)
L	=	Average crack length (m)
Φ	=	Constraint factor taken as 3 [13]

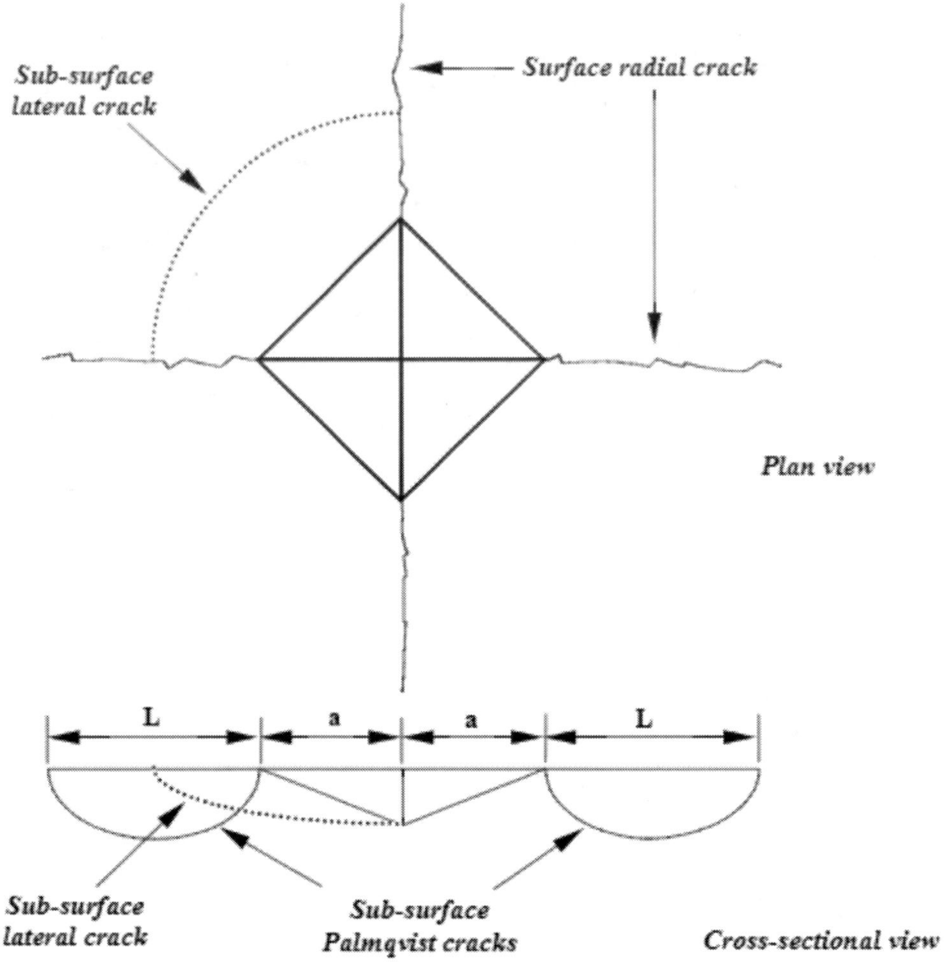

Fig. 2. Schematic diagram of a typical Palmqvist crack system observed after Vickers indentation. "*L*" refers to the crack length which was measured to compute the fracture toughness of the sample while the length of "*2a*" is equivalent to a diagonal of the Vickers indent (D_1 or D_2, as indicated in Fig. 1).

3 Results and Discussion

3.1 Relative Density

The relation between relative density and the sintering temperature of the composite material is shown in Figure 3. It is shown that, for the samples which contain 10 and 20 weight percentage of stainless steel 316/Y-TZP, 1300 °C is the optimum sintering temperature as it has its highest relative density. As for the sample which contained 30 weight percentage of stainless steel 316/Y-TZP, the relative density increases as the sintering temperature increases from 1250 °C to 1500 °C. For the sample which contains pure Y-TZP, it has the highest relative density when the sintering temperature was 1400 °C. The results showed that pure Y-TZP had a higher relative density compared to composite materials of ss316/Y-TZP at higher sintering temperatures. Besides that, it also showed that as the weight percentage of stainless steel in the composite material increased, the relative density of the composite material decreased. The results obtained also showed that it was in line with the findings based on the study of Agudelo et al. [4] to a certain extent.

Based on Agudelo et al. [4] study, as the weight percentage of stainless steel in ss316/Y-TZP samples increase, the relative density of the samples increased over sintering temperatures. This is true theoretically as the heating rates and temperatures increase, the density if the samples with increasing weight percentage of stainless steel increases. However the trend of the study suggested otherwise. When the percentage of stainless steel 316 increased past 10 wt%, the relative density decreased. This can be explained due to a few factors. Firstly, the method of processing for the samples in this study and the study carried out by Agudelo et al. [4] was different; thus, the processing method could have an effect on the results obtained. Another possibility could be attributed to the type of powder used in the study. To have a better understanding regarding the difference in trends, further study regarding microstructure examination via scanning electron microscope (SEM) is required.

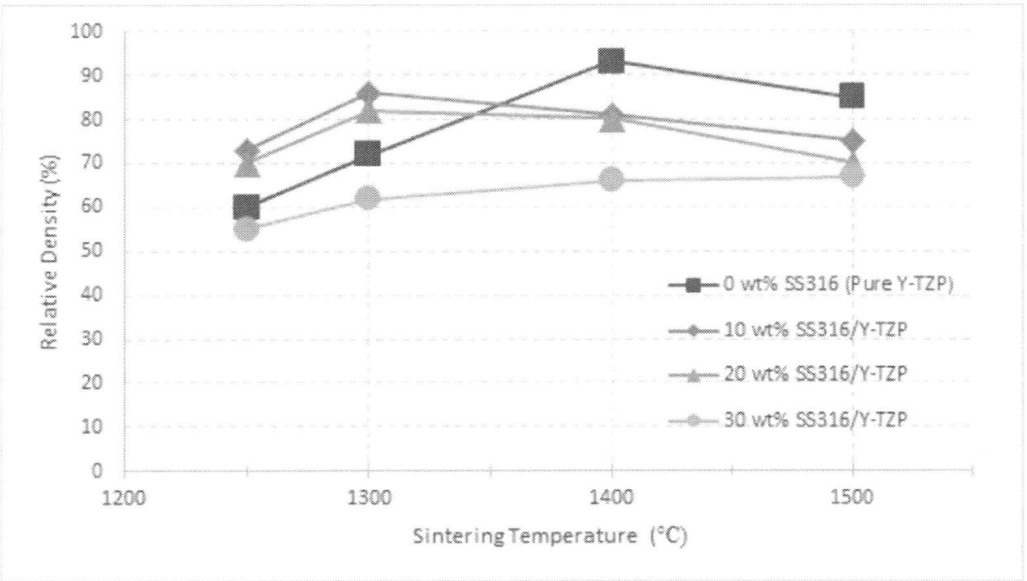

Fig. 3. Relative density of ss316/Y-TZP composites with different stainless steel ss316 contents sintered at various temperatures.

3.2 Vickers Hardness

Figure 4 shows the relation between sintering temperature and the Vickers hardness of the composite material. Based on Fig. 4, pure Y-TZP had the highest Vickers hardness which is 13.5 GPa when its sintering temperature is 1400 °C. However, all the composite materials with different weight percentage of stainless steel had the highest Vickers hardness at the sintering temperature of 1300 °C. This showed that the ideal sintering temperature for the stainless steel 316/Y-TZP composite was 1300 °C. It is also shown based on the results that when the weight percentage of stainless steel 316 in the stainless steel 316/Y-TZP composite increases, the Vickers hardness of the composite material will reduce. This is also true based on the study of Esteban et al. [5] which showed the same results compared to Figure 3. This can be explained due to the hardness of materials as stainless steel 316 is softer than Y-TZP, which also obeys the rule of mixture [6-8].

The trend in this study is similar to the trend of Esteban et al. [5] study. The reduction in Vickers hardness can be explained based on the hardness of ss316 and Y-TZP. Stainless steel 316 is softer compared to Y-TZP; thus, when the content of the ss316 in the ss316/Y-TZP composite increases, the Vickers hardness will reduce as the hardness of ss316/Y-TZP composite is lower as the composition has higher amounts of stainless steel which will also cause the composite to be softer.

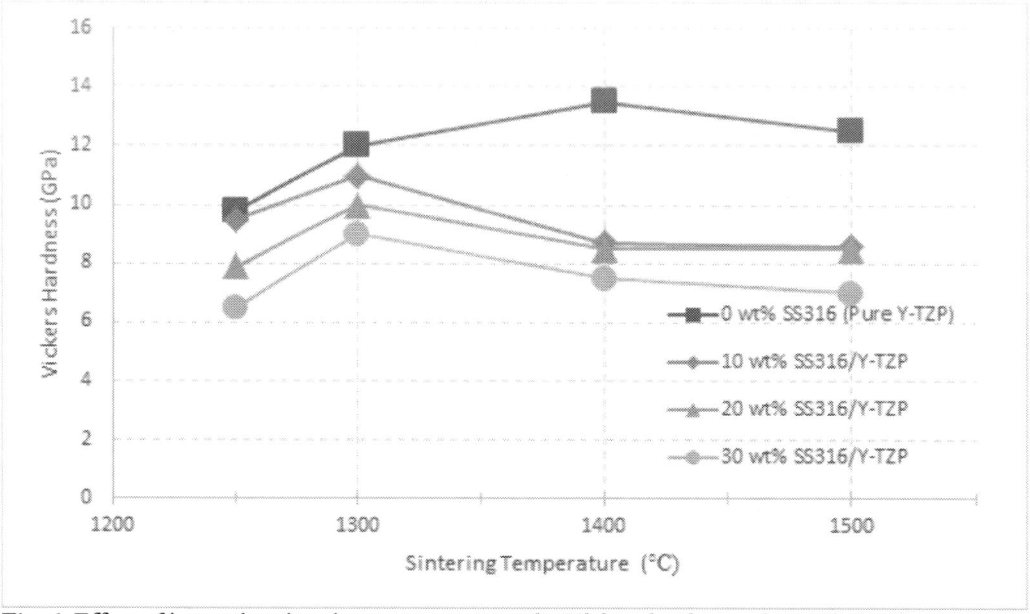

Fig. 4. Effect of increasing sintering temperature on the Vickers hardness of ss316/Y-TZP composites containing 0 to 30 wt% stainless steel ss316 content.

3.3 Fracture Toughness

Figure 5 exhibits the results of the fracture toughness of different weight percentage of ss316/Y-TZP at different sintering temperatures. Based on the results from Figures. 3, 4 and 5, it showed that once again the ideal temperature for sintering of 10 and 20 weight percentage ss316/Y-TZP is 1300°C. At 1300°C, both the 10 and 20 weight percentage ss316/Y-TZP had the highest fracture toughness. The results also showed that the composite materials which contained stainless steel had lower fracture toughness as the sintering temperature increased after 1300 °C. The results also showed that when the weight percentage of stainless steel in the composite increased, the fracture toughness also improved. However, this is only from 0 wt% to 20 wt% ss316/Y-TZP.

The trend of the results appeared somewhat similar to the results obtained by studies done by Esteban et al. [5]. For the 10 and 20 wt% ss316/Y-TZP samples, when the number of stainless-steel 316 particles increases in the composite, mismatched thermal expansions will occur between both the zirconia and stainless-steel particles and result in radial tensile stresses to be developed internally when cooling occurs. This in turn causes debonding to happen due to the lack of interfacial bonds between the particles. When cracks happen, the cracks will propagate smoothly between the weak interface of the zirconia and stainless-steel particles which in turn produce crack deflection processes that improves that fracture toughness [5]. Figure 6 shows the crack propagation across the Y-TZP and metal. However, for the 30 wt% ss316/Y-TZP, the number of stainless steel 316 particles in the composite is too large which causes mismatched thermal expansions that form thermal stresses in the composite material which is too large, causing the fracture toughness to reduce.

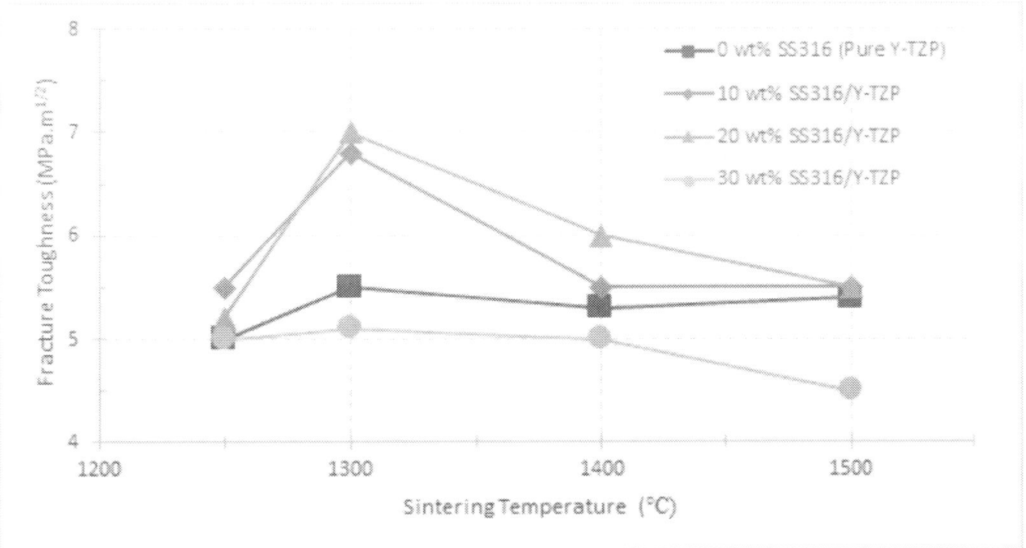

Fig. 5. Fracture toughness of ss316/Y-TZP composites sintered at various temperatures from 1250°C to 1500°C.

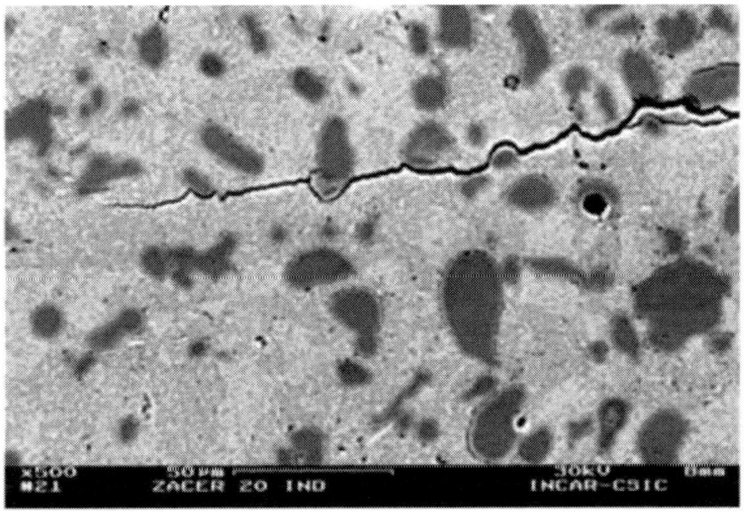

Fig. 6. SEM micrograph showing the propagation of a crack, induced by Vickers indentation, across the microstructure of a stainless steel/zirconia composite. Areas of lighter shade refer to zirconia while the darker phase refers to metal [5].

4 Conclusion

The mechanical characteristics of stainless steel 316/Y-TZP composite were studied through relative density, Vickers hardness and fracture toughness testing. Based on the results, it is shown that the ideal weight percentage of stainless steel 316 in the stainless steel 316/Y-TZP composite is 20 wt%. The results also showed that 30 wt% of stainless steel would result in the reduction of mechanical characteristics. Besides that, the ideal sintering temperature for the stainless steel 316/Y-TZP composite is 1300 °C as any temperature above that would result in deterioration of its mechanical characteristics. It is also found that the increase in stainless steel 316 weight percentage in the ss316/Y-TZP composite will result in areduction in relative density and Vickers hardness. However, the Vickers hardness of the composite will still be lower compared to pure Y-TZP.In addition to that, an increase in stainless steel 316 weight percentage in the ss316/Y-TZP composite will also result in an improvement in fracture toughness.

References

1. M.M. Stevens, Mater. Today, **11**, 18–25 (2008).
2. S.D. Heintze, A. Cavalleri, G. Zellweger, A. Büchler, G. Zappini, Dent. Mater., **24**, 1352–1361 (2008).
3. S.M. Kurtz, C. Arnholt, R. Huet, M. Ueno, W.L. Walter, J. Mech. Behav. Biomed. Mater., **31**, 107–116 (2014).
4. L. Grajales Agudelo, V.L. Arantes, P.L. di Lorenzo, Adv. Powder Technol., **802**, 120–124 (2014).
5. S. Lopez-Esteban, J.F. Bartolome, C. Pecharroman, S.R.H. Mello Castanho, J.S. Moya, Mater. Res., **4**, 217–222 (2001).
6. F.F. Lange, J. Mater. Sci., **17**, 247–254 (1982).
7. V.D. Krstic, W.H. Erickson, J. Mater. Sci., **23**, 4097–4102 (1988).
8. A.H. De Aza, J. Chevalier, G. Fantozzi, M. Schehl, R. Torrecillas, Biomater., **23**, 937–945 (2002).
9. G. Schierano, F. Mussano, M.G. Faga, G. Menicucci, C. Manzella, C. Sabione, T. Genova, M.M. Von Degerfeld, B. Peirone, A. Cassenti, P. Cassoni, S. Carossa, Biomed. Res. Int., **2015**, Article ID 157360 (2015).
10. ASTM E384-99, "Standard Test Method for Microindentation Hardness of Materials," *Annual Book of ASTM Standards* (1999).
11. ISO 14705, "Fine ceramics (advanced ceramics, advanced technical ceramics) – Test method for hardness of monolithic ceramics at room temperature" (2000).
12. K. Niihara, R. Morena, D.P.H. Hasselman, J. Mater. Sci. Letts., **1**, 13–16 (1982).
13. R.N. Katz, Mater. Sci. Eng., **71**, 227–249 (1985).

Preparation and Characterization of Durian Husk Fiber Filled Polylactic Acid Biocomposites

Man Chee Lee[1] , *Seong Chun* Koay[1*] , *Ming Yeng* Chan[2] , *Ming Meng* Pang[1] , *Pui May* Chou[1] , and *Kim Yeow* Tsai[3]

[1]School of Engineering, Faculty of Built Environment, Engineering, Technology and Design, Taylor's University Lakeside Campus, 47500 Subang Jaya, Malaysia
[2]Centre for Engineering Programmes, HELP College of Arts and Technology, 55200 Kuala Lumpur, Malaysia
[3]Faculty of Engineering, The University of Nottingham, 43500 Semenyih, Malaysia.

Abstract. Polylactic acid (PLA) is biodegradable thermoplastic that made from renewable raw material, but its high cost limited the application. Thus, addition of natural fiber can be effectively reduced the cost of PLA. This research is utilised natural fiber extracted from durian husk to made PLA biocomposites. This paper is focus on the effect of fiber content on tensile and thermal properties of PLA/durian husk fiber (DHF) biocomposites. The results found that the tensile strength and modulus of this biocomposites increased with increase of fiber content, but the strength still lower compared to neat PLA. Then, the elongation at break of biocomposites was expected decreased at higher fiber content. The PLA/DHF biocomposites with 60 phr fiber content exhibited tensile strength of 11 MPa, but it is too brittle yet for any application. The addition of DHF caused an early thermal degradation on PLA biocomposites. Then, the thermal stability of PLA biocomposites was decreased with more fiber content.

1 Introduction

In 2015, researchers found 8.3 billion metric tons of plastics waste were produced globally and 6.3 billion tons of this plastic products are turning into waste. As well known, not all the plastic can be recycle, thus only 9% of that waste total was recycled, 12% was used for energy recover and left were sent to landfills. The amount of plastics waste for land fill are expected to achieve 12 billion metric tons, if the current trends is keep going on [1]. Plastic material made from petroleum resources are chemically stable which hardly to biodegrade in environment. In present, part of the plastic product has replace by bioplastic that able to biodegradable after disposal. Polylactic acid (PLA) is a type of bioplastic that synthesized from natural material, such as corn starch, tapioca root and sugarcane [2-3]. Besides, PLA offers a biodegradable properties and also good mechanical strength, non-toxic and sustainable since it made from renewable resources [4-5]. Nowadays, the production of bioplastic over the world is about 750,000 tons per year. Unfortunately, this

* Corresponding author: SeongChun.Koay@taylors.edu.my

amount still very small compare to conventional plastic that produced 200 million tons per year. The main reason why PLA is seldom choose by industrial as their main choice of material, because the price of PLA resin is more expensive than conventional plastic [5-6].

In order to promote the usage of PLA material, natural filler can be introduced to PLA and make it into a biocomposites. The natural filler can be easily obtained from any agricultural crop and waste, for instance, corn cob [7], cocoa pod [8], coconut shell [3] and palm kernel shell [9]. The main benefit of natural filler very low cost, biodegradable, less abrasive to machine, and inexhaustible resources [10-11]. Thus, by making the PLA in to biocomposite, the addition of natural filler can be partially replace the PLA without influences its biodegradability and highly reduced the cost of final product. *Durio zibethinus Murray*, also called as durian, the famous fruit among the Southest Asia countries especially in Malaysia, Thailand, Indonesia and Philippine [12]. The durian fruit is made up of 40% of flesh and 60% of husk. Usually, the durian husk was discarded and end up on landfills or burnt which poses an environmental issues. In Malaysia, about 3,800 metric tons of durian fruits are produced in year 2016. Assuming every 60% of total amount produced durian fruits is husk, it estimated the durian husks waste produced in year 2016 was achieved 2,280 metric tons. The production of durian fruits are predicted to achieve 22,000 metric tons at year 2020 [13]. This meaning the amount durian husks waste will be increased. The durian husk consists 60.45% of cellulose, 15.45% of lignin and 13.09% of hemicellulose, and the content are similar to wood fiber [14]. For this reason, this research is underway to utilize the fiber obtained from durian husk and combined with PLA to produce biocomposites.

This present research is focus on the effect of fiber content on the tensile and thermal properties of durian husk fiber filled polylactic acid biocomposites.

2 Methodology

2.1 Research Materials

PLA used in this experiment was supplied by TT Biotechnologies Sdn. Bhd. (Malaysia) and durian husk was collected from durian fruit stall at SS2, Petaling Jaya (Selangor).

2.2 Preparation of Durian Husk Fiber

The durian husks obtained from durian fruit stall were washed with tap water and cut into small pieces. Then, the durian husks were dried using circulated air oven at 70°C until completely dried. The dried durian husks were ground into short fiber using mechanical grinder. Next, the short durian husk fiber (DHF) was passed through a sieve with mesh size of 600 micron to get the homogenous size of fibers. The DHF further dried using oven before compounding process.

2.3 Preparation of PLA/DHF Biocomposites

The PLA/DHF biocomposites were compounded using Haake Rheomix 600p at SIRIM Malaysia (Shah Alam). The biocomposites were prepared with fiber various from 15, 30, 45, and 60 phr (part per hundred resin). The processing temperature was fixed at 180°C and rotor speed of 80 rpm. The compounding procedures included: i) addition of PLA resin in to mixing chamfer for 1 minute to fully melt the PLA resin; ii) Then, the DHF was added into melted PLA and compounded for 5 minutes; iii) Last, the compound was removed from mixing chamber.

The PLA/DHF biocomposites were further molded into thin sheet with thickness of 1 mm. The hotpress machine (model Moore) was used in this molding process and temperature was set at 180°C. The molding procedures included: i) preheated to soften the compound for 2 min; ii) fully compressed the softened compound under 100 MPa pressure for 1 min; and iii) Last, cooled specimen under same pressure for 20 minutes. All the biocomposite sheets were cut into tensile specimen and dimension following ASTM D638.

2.4 Testing and Characterization

Tensile test was performance using Instron Universal Testing Machine (Model 5569). The tensile test was referring to ASTM D638. The cross-head speed of machine was set at 5 mm/min and a 15kN load cell was used. The tensile strength, modulus and elongation at break of the samples were recorded by Bluehill 3 software. A minimum of 7 specimens were tested for every formulated biocomposites.

Thermogravimetric analysis (TGA) was carried out using Pyris Diamond TGA (Perkin-Elmer). The specimen was prepared in small size with weight from 6 to 8 mg and placed in a ceramic pan. The specimen was heated from 30°C to 600°C at hearing rate of 10°C/min. The TGA analysis was run under nitrogen atmosphere with gas flow rate of 20 ml/mm.

3 Results and Discussion

3.1 Tensile Properties

Figure 1 shows the tensile strength of PLA/DHF biocomposites with different DHF content. The tensile strength of PLA/DHF biocomposite exhibits an increasing trend as the DHF content increased. Usually, natural fiber displays a higher mechanical strength compared to most plastic material [15]. Thus, the addition of DHF fiber might share part of the tensile load which subjected PLA matrix and increased the tensile strength of biocomposite. From the previous study found that the tensile strength of neat PLA was about 54 MPa [3]. The biocomposites exhibited lower strength compared to neat PLA. This is because the DHF used in this experiment was short fiber and the reinforcing ability of the fiber is highly depending on the fiber orientation. The short fibers usually is randomly orientated. The presence of fibers that orientated perpendicular to direct of applied load might unable to share the load which caused the strength of biocomposite to decrease. However, at higher fiber content, the presence of more fibers that parallel to direction of load and it will share load that subjected to biocomposite. Thus, the strength of biocomposite was raised when more fiber content was added. Similar findings have also been reported by Yu et al. [16].

The effect of different fiber content on PLA/DHF biocomposites is illustrated in Figure 2. As expected, the tensile modulus of the biocomposites gradually increases with increasing amount of DHF. It is because the modulus of DHF is usually higher than plastic [12]. Therefore, the presence of higher amount of DHF contributed to higher modulus of the PLA/DHF biocomposite. For this reason, the tensile modulus of PLA/DHF biocomposite increases due to addition of DHF. Moreover, Gunti et al. [17] also agreed that the modulus of biocomposite is highly depending on the amount of filler. Hence, the biocomposite with higher fiber content showed higher tensile modulus.

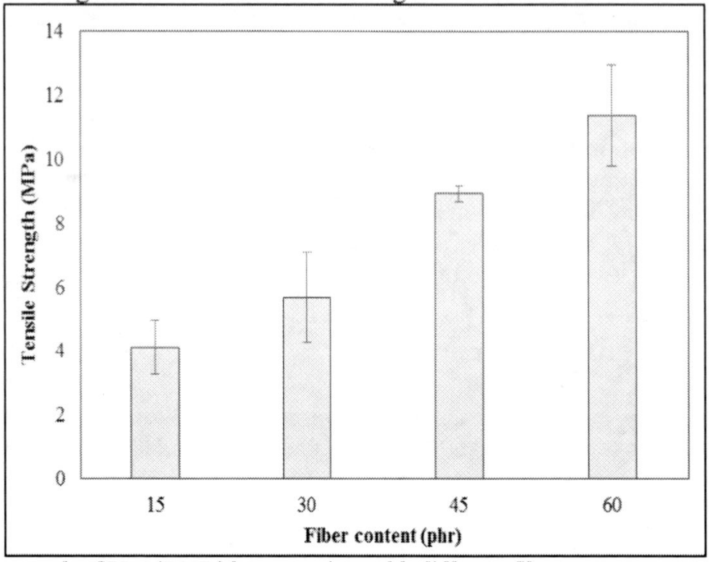

Fig. 1. Tensile strength of PLA/DHF biocomposites with different fiber content.

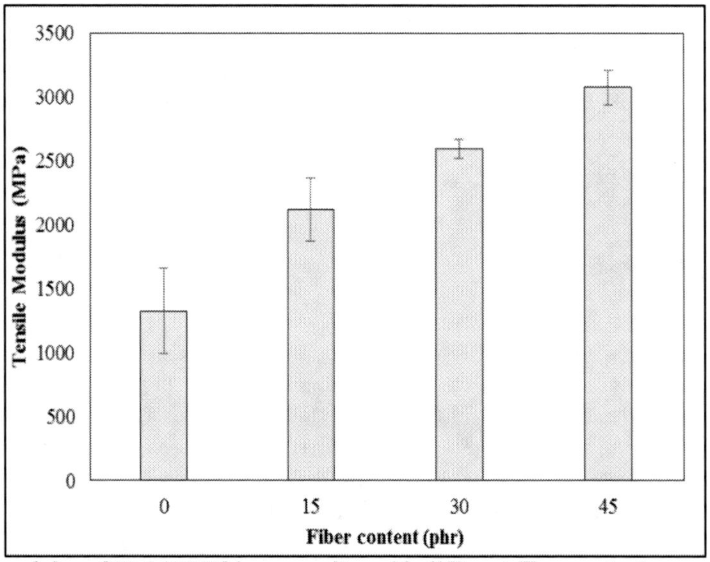

Fig. 2. Tensile modulus of PLA/DHF biocomposites with different fiber content.

Figure 3 displays the elongation at break of the PLA/DHF biocomposite with different DHF content. The result trend of tensile modulus of biocomposite typically opposite with elongation at break. The neat PLA is well known brittle material and its elongation at break was only 4.2% [3]. The addition of DHF had dramatically reduced the elongation at break of PLA. The result found the fiber content increased, the elongation at break of PLA/DHF biocomposite decreased. In general, natural fiber is rigid material that not elongate much [18]. Thus, the addition of DHF would not contribute on the elongation of PLA matrix. In addition, the friction presents between PLA matrix and DHF fiber will significantly reduce the mobility of the polymer chains. For this reason, PLA/DHF biocomposite with more fiber content became more rigid and brittle. This observation is in agreement with findings of many others researcher [19].

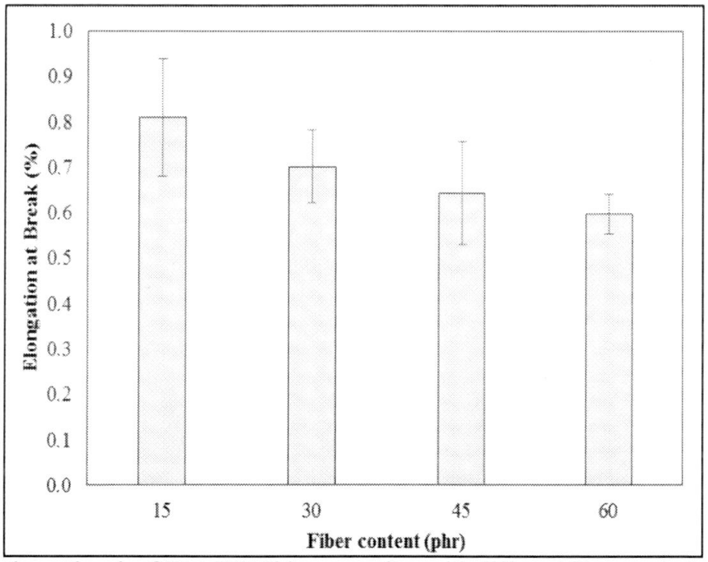

Fig. 3. Elongation at break of PLA/DHF biocomposites with different fiber content.

3.2 Thermal Properties

The TGA curves of DHF, neat PLA, and PLA/DHF biocomposites at selected fiber content is shown in Figure 4. All the data obtained from TGA curves are tabulated in Table 1. The TGA result shows the DHF was thermal degraded earlier than neat PLA and yielded high char residue at 600°C. The earlier thermal degradation of DHF was due to the removing of moisture and volatile compounds that found in DHF. Then, the thermal degradation of DHF was following by degradation of hemicellulose at temperature above 250°C. Furthermore, the thermal degradation of lignin and cellulose from DHF were found beyond 350°C [20]. The thermal degradation of lignin and cellulose were contributed to char formation that caused high amount of char residue. From Table 1, the temperature at 5% ($T_{d5\%}$) and 50 % ($T_{d50\%}$) weight loss of PLA/DHF biocomposites were shifted to lower temperature when fiber content increased from 30 to 60 phr. The PLA/DHF biocomposites also exhibited lower $T_{d5\%}$ and $T_{d50\%}$ compared to neat PLA. This indicated the PLA/DHF biocomposites have lower thermal stability compared to neat PLA. The biocomposite had lower thermal stability was due to components in DHF that decomposed when subjected to high temperature. The char residue of PLA/DHF biocomposites also increased with the increases of fiber content. As mentioned early, the formation of more char residue was due to thermal degradation of lignin and cellulose from DHF. The similar observation also found by other researchers [21-22].

4 Conclusion

The increase of fiber content increased the tensile strength and modulus of PLA/DHF biocomposites, but decrease in elongation at break. The PLA/DHF biocomposites exhibited lower strength compared to neat PLA, but the average tensile strength of PLA/DHF biocomposites with 60 phr of fiber still have 11 MPa. The strength is comparable to low density polyethylene, but its to brittle to make any application. Besides, the addition of DHF had caused an early thermal degradation on PLA/DHF biocomposites and also lower in thermal stability. The TGA results also found the biocomposite with more fiber content exhibited high char residue after decomposed at temperature 600°C.

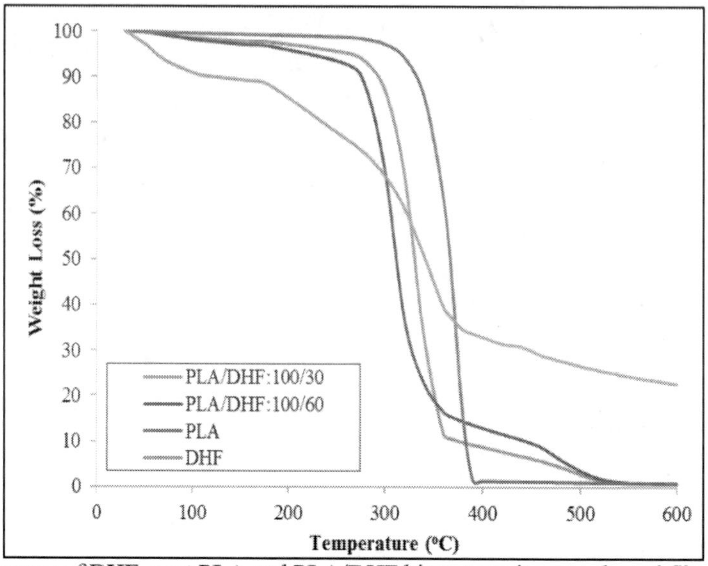

Fig. 4. TGA curves of DHF, neat PLA and PLA/DHF biocomposites at selected fiber content.

Table 1. Formatting sections, subsections and subsubsections.

Samples	Temperature at 5% weight loss (°C)	Temperature at 50% weight loss (°C)	Char residue at 600°C (%)
Neat PLA	313.4	365.1	0.7
DHF	55.5	339.3	22.7
PLA/DHF: 100/30	260.9	331.5	0.8
PLA/DHF: 100/60	232.6	307.7	0.9

5 Acknowledgement

The authors gratefully acknowledge use of services and facilities at Taylor's University, funded by Taylor's Research Grant Scheme (Project code: TRGS/ERFS/1/2017/SOE/033).

References

1. R. Geyer, J.R. Jambeck, K.L. Law, Sci. Adv. **3**, 1 (2017).
2. A.A. Yussuf, I. Massoumi, A. Hassan, J. Polym. Envir. **18**, 442 (2010)
3. K.S. Chun, S. Husseinsyah, H. Osman, Polym. Eng. Sci. **53**, 1109 (2013)
4. K. S. Chun, S. Husseinsyah, J. Thermoplast. Compos. Mater. **27**, 1667 (2014)
5. R.E. Drumright, P.R. Gruber, D.E. Henton, Adv. Mater. **12**, 1841 (2000)
6. P.A. Fowler, J.M. Hughes, R.M. Elias, J. Sci. Food Agricul. **86**, 1781 (2006)

7. C.M. Yeng, S. Husseinsyah, S.S. Ting, Polym. Plast. Technol. Eng. **52**, 1496 (2013)
8. K.S. Chun, S. Hussiensyah, C.M. Yeng, Polym. Bull. **73**, 3465 (2016)
9. S. Hussiensyah, K.S. Chun, A. Hadi, R. Ahmad, J. Vinyl Add. Technol. **22**, 200 (2016)
10. D. Mulinari, H. Voorwald, M. Cioffi, M. Dasilva, T. Dacruz, C. Saron, Compos. Sci. Technol. **69**, 214 (2009)
11. M.M. Pang, M.Y. Pun, Z.A.M. Ishak, J. Appl. Polym. Sci. **129**, 3656 (2013)
12. Y.M.W. Jason, M. Y. Chan, Adv. Polym. Technol. (to be published). doi:10.1002/adv.21921.
13. Department of Agricultural (Malaysia). Strategic Plan 2016 to 2020. (Md: Department of Agricultural, Malaysia, 2016)
14. R. Manshor, H. Anuar, W. Wan Nazri, M. Fitrie, Adv. Mater. Res. **576**, 212 (2012)
15. W.L. Ngo, M.M. Pang, L.C.Yong, Adv. Envir. Bio. **8**, 2742 (2014)
16. T. Yu, J. Ren, S. Ji, H. Yuan, Y. Li, Compos. Part A, **41**, 488 (2010)
17. R. Gunti, A. Ratna Prasad, A. Gupta. Polym. Compos. **37**, 1 (2016)
18. H. Salmah, S.C. Koay, O. Hakimah, J. Thermplast. Compos. Mater. **26**, 809 (2013)
19. K.S. Chun, C.M. Yeng, S. Husseinsyah, M.M. Pang, A. Ismail, J. Eng. Sci. Technol. **12**, 1165 (2017).
20. M.R. Manshor, H. Anuar, M.N. Nur Aimi, M.I. Fitrie, W.B. Wan Nazri, S.M. Sapuan, Y.A. El-shekeil, M.U. Wahit, Mater. Des. **59**, 279 (2014)
21. Y. Lei, Q. Wu, F. Yao, Y. Xu, Compos. Part A. **38**, 1664 (2007)
22. K.S. Chun, S. Husseinsyah, C.M. Yeng, J. Thermoplast. Compos. Mater. **29**, 1517 (2016)

Operational Research on Design and Process Optimization of Ozone Water Application in Kitchen

Zhun Jing Lee[1], *Pui May* Chou[1,*], *Choon Lih* Hoo[1], *Daniel* Lu[1]

[1]School of Engineering, Taylor's University, 47500 Subang Jaya, Selangor, Malaysia

Abstract. Food safety is a very important focus in the kitchen industry today, as bacteria such as E.Coli and Salmonella are very difficult to tackle. The objective of the present study was to optimize nozzle designs that use ozone technology to bring out the best results in cleaning and sterilizing the kitchen utensils in Taylor's University School of Hospitality kitchen area. This includes customization of the Medklinn International Sdn Bhd ozone machine and nozzle profiles that improve the effectiveness of ozone generated. Reduction or elimination of chemicals and water usage would be a part of the study. This will bring a huge impact on cost effectiveness, time saving and safety of the users. Return on investment (ROI) using ozone technology is calculated at the end of the research. To compare between the traditional way of cleaning and using ozone technology, the volume of water and dishwashing liquid used, and the Relative Light Units (RLU) before and after washing were recorded. The RLU numbers are found using the 3M Clean Trace measuring equipment. RLU was recorded to determine the cleanliness of the kitchen utensils before and after washing. It has been proved that ozone water with the accompaniment of the selected nozzle prototype is as efficient as the traditional way of cleaning.

1 Introduction

The cleanliness of kitchen area, utensils and stoves is essential in the food processing industry. Diarrhea is a concerning issue for travelers as 1 out 3 people will be affected in this health issue. Hikers and people who go on adventure trips are very prone to this problem[1]. In 2015, World Health Organization (WHO) has estimated that 1 in 10 people are prone to foodborne diseases which also killed 420,000 people[2]. It was also reported that Africa and South East Asia are the two regions that are in the high risk zone[2]. There were 125,000 of children under five years old are killed due to illness from foodborne diseases every year [2]. It was found out that 16% of the foodborne diseases outbreaks from England and Wales are linked with the food preparation in the house [3, 4].

* Corresponding author: PuiMay.Chou@taylors.edu.my

Cross contamination is one of the reason to the foodborne disease and it caused 28 out of 101 outbreaks to happen [3]. Cross contamination can happen either by hand to the plates, from the plates to the food, the utensils to the food, or even in the washing process with the usage of sponge and water [3, 5, 6, 7]. In order to reduce and eliminate this issue, different washing up systems have been created. Researchers believe that rather than tackling the cleanliness of water, the cleanliness of kitchen utensils and hand hygiene are two very important factors to prevent foodborne diseases[1, 8]. Cross contaminations on food happens very often in the process of cutting, trimming, rinsing and washing [8]. Bacteria infect almost all the kitchen area, as study shown in appendix 1 and 2, swap tests were taken in a hospital kitchen to show the amount of infected areas. These results were also picked up from pre-processing zones for food, storage area and cooking area.

Biofilm is the form of 99% of bacteria in the kitchen. Biofilm formation makes the bacteria to be able to live in either biotic or abiotic surfaces. Therefore, another study on bacteria infecting kitchen surfaces is done as well [9]. Table 1 shows the total colony forming unit count on a cafeteria kitchen. It can be seen that the bacteria count is quite alarming even in dishwashers, sink, countertop areas, food waste container and pots. These are some areas that show that even bacteria would live in cleaning environment and cooking pots which would promote cross contamination. This also shows that the existing steps of cleaning could not disinfect microorganisms efficiently, and potential biofilm formation is not effectively prevented as well [9].

Table 1. Total colony forming unit (CFU) count on a cafeteria kitchen [9]

Surface No.	Surface Area	Total Aerobic Count (log $CFU/100cm^2$)	Bacteria
1	Cold Room	5.90	Brachybacterium, Brevundimonas, Kocuria, Pantoea, Paracoccus, Roseomonas, Sphingobacterium
2	Pretreatment Countertops	3.83	Bacilus, Curtobacterium, Gordonia, Kocuria, Pantoea, Pseudomonas, Sphingomonas, Staphylococcus
3	Sink	4.14	Bacillus, Enterobacter, Raoultella
4	Faucet in the sink	5.35	Bacillus, Enhydrobacter, Escherichia, Kocuria
5	Roasting/frying countertop	3.63	Bacillus, Kocuria, Lysinibacillus
6	Spice rack	4.84	Bacillus, Microbacterium, Pantoea
7	Soup pot	3.59	Acinetobacter, Bacillus
8	Frying pot	2.16	Bacillus
9	Countertop for completed menu	3.16	Acidovorax, Acinetobacter, Bacillus, Chryseobacterium, Deinococcus, Enterobacter, Sphingomonas
10	Rice cooker	2.73	Acinetobacter, Bacillus, Chryseobacterium, Kocuria, Sphingomonas

11	Food waste container	3.21	Bacillus, Staphylococcus
12	Dishwasher	2.53	Bacillus
13	Sink for dishwashing	2.86	Bacillus
14	Stainless steel trays	1.60	Bacillus
15	Plastic wicker tray	4.07	Acinetobacter, Chryseobacterium, Dermacoccus, Exiguobacterium, Kocuria, Pantoea
16	Iron roasting pan	2.00	Acinetobacter, Kocuria
17	Knives	2.00	Acinetobacter, Bacillus
18	Cutting board (Completed Menu)	2.45	Arthrobacter, Bacillus, Curtobacterium, Kocuria, Microbacterium, Staphylococcus
19	Cutting board (vegetables)	2.97	Acinetobacter, Dermacoccus, Enterobacter, Kocuria, Leclercia, Methylobacterium, Roseomonas
20	Gloves	3.97	Acinetobacter, Bacillus, Kocuria
21	Apron	6.85	Bacillus, Kocura, Pseudoxanthomonas, Rhizobium
22	Fan	4.15	Bacillus
23	Floor	4.65	Achromobacter, Aeromonas, Bacillus, Cellulosimicrobium, Chryseobacterium, Cloacibacterium, Diaphorobacter, Kocuria, Pseudomonas

Hence, in order to prevent cross contamination that will lead to foodborne diseases, studies have been done to optimize the cleaning processes in the kitchen and also trying out with different disinfectants and cleaning agents. Sterilizing of kitchen surfaces using a variety of disinfectants is one of the ways to minimize microorganisms living on surfaces and also preventing cross contamination. In this case, Taylor's Hospitality and Culinary Arts School (TCHT) is used as the model of research to find out how chemical and disinfectant washing can help in the cleaning of kitchen utensils. A table (Table 2) shows the volume of water and detergent needed to clean 6 kitchen utensils and 2 kitchen stoves.

Table 2. Volume of water used to clean 6 kitchen utensils and 2 kitchen stoves

Items washed	Amount of dishwashing liquid used, ml	Amount of water used for cleaning, L
6 Kitchen utensils	30	143.7
2 Kitchen stoves	15	5.9L

TCHT has been searching for a way to reduce the cost of cleaning and also reducing the steps to do the cleaning so that cleaning of the kitchen would be faster and requires less man power. There are a few ways to optimize the current cleaning method to achieve TCHT's objective:

1. In order to reduce chemical usage, TCHT has the possibility to use higher concentration of dishwashing liquid so that less chemical will be used to dilute in water but producing the same effect. However, more concentrated dishwashing liquid will also be higher cost.

2. In order to achieve less man power, the steps of cleaning have to be reduced. An alternative which can bypass the use of degreasing and too much scrubbing can reduce man power. In this case, the alternative way has to be good in sterilizing, degreasing and cleaning off dirt in one shot. Hence, reducing cleaning steps.

Cleaning using ozone technology is one way that can reduce the use of chemicals as ozone is well known to be a good disinfectant because of its highly reactive properties[10, 11]. Ozone is also proven by many ozone technology companies to be a good degreaser. Hence, ozone is widely used in the western part of the world in industry cleaning, sterilizing and degreasing. However, ozone is not as well known in this tropical region as in the western world. Ozone technology is not unknown in kitchen area as ozone is also used in cleaning vegetables and fruits before cooking. One of the examples is in the cleaning of spinach [12].

Hence, this research is to introduce the usage of ozone technology in order to enhance and improve the ways to clean the kitchen utensils which takes up most of the cleaning time in Taylor's Hospitality and Culinary Arts School. The objective is to customize and develop designs to bring out effective applications of ozone technology in cleaning kitchen utensils. Furthermore, the research is carried out also to prove that the usage of ozone technology reduces the usage of chemicals, man powers and time consumption in cleaning process.

2 Methodology

This research is done by cleaning stainless steel kitchen utensils in Taylor's Hospitality and Culinary Arts (TCHT). The research done on different methods is shown in Table 3.

Table 3. Methods of testing

With Nozzle	Without Nozzle
Without ozone	Without ozone
With ozone	With ozone

2.1 Calculation of water and chemical usage in cleaning process

A water flow meter shown in Fig. 1 is attached to the outlet of the water pipe where water flows out for kitchen utensil washing. This is to measure the amount of water used during the whole cleaning process. The amount of water used is compared between the cleaning with and without ozone, and with and without nozzles. The volume of water was controlled during the research, not letting it to overrun before and after washing. The water was only let run when needed.

Fig. 1. Water flow meter with readings in m^3

A 10 ml syringe shown in Fig. 2 is used to measure the volume of dishwashing liquid used to clean the kitchen utensils. The dishwashing liquid is already diluted with the ratio of 1 L of detergent to 20 L of water. It is tested with the same methods as in Table 2. The detergent is added on by an increment of 1ml onto the dishwashing sponge until the worker feels that it is sufficient to clean the pot. This is done to ease the calculation for the efficiency of ozone water cleaning vs non-ozone water cleaning.

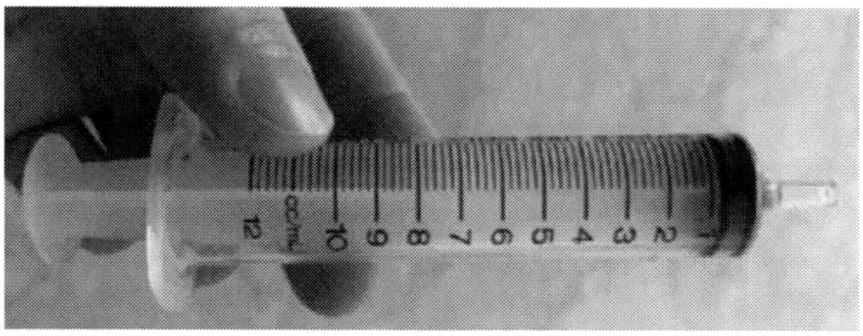

Fig. 2. 10 ml syringe for dishwashing liquid measurement

2.2 Customization of ozone machine

The ozone machine in Fig. 3 is provided by Medklinn International Sdn Bhd. The ozone capacity used was 500mg/hr. Ozone is also produced with the presence of oxygen as stated in the literature review. In this case, a consistent provided of oxygen is needed so that the ozone produced will be consistent as well. Therefore, an oxygen concentrator is used to provide the consistent amount of oxygen. A 1 LPM oxygen concentrator is used for this research. A venturi shown in Fig. 4 is used for suction of ozone gas to be diluted in water is MK584 (¾ inch inlet and outlet size). This will result in a 1.8 particles per million (PPM) (without nozzle) and 1.25 PPM (with nozzle) concentration of ozone.

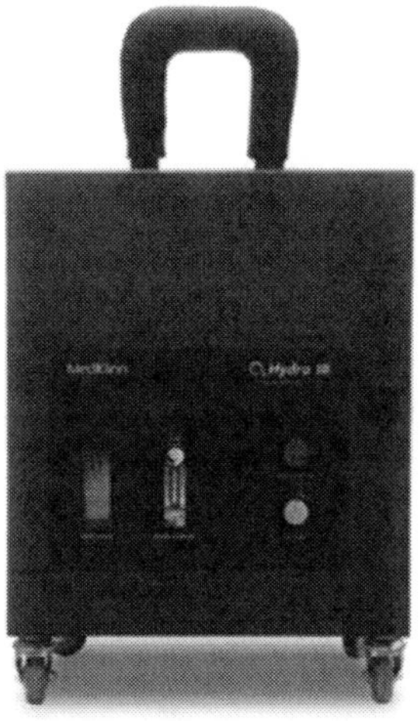

Fig. 3. Ozone Machine

Inlet (higher pressure)

Outlet (lower pressure)

Suction of gas or liquid

Fig. 4. MK584 Venturi [13]

2.3 Nozzle profile

A nozzle is used to have a higher water impact in the cleaning process to remove oil and stubborn dirt on the kitchen utensils. The nozzle design has 16 linear 1.50 mm diameter circular holes with a 3 mm linear distance between each hole. Fig. 5 shows the design of the nozzle used to clean the kitchen utensils. The design has the same number of holes compared to the nozzle used in the traditional way of cleaning in Taylors School of Hospitality and Culinary Arts (TCHT). However, the nozzle profile in TCHT is a shower pattern nozzle. A shower pattern nozzle shows a lower impact force compared to the one in linear. This is because most of the water flow is not blocked at the outlet, hence there is lesser force that pushes the water out of the orifices. However, in linear, the water is constrained to exit only at the middle of the outlet. Hence, pressure accumulates in the nozzle and more force is present to push the water out of the orifices.

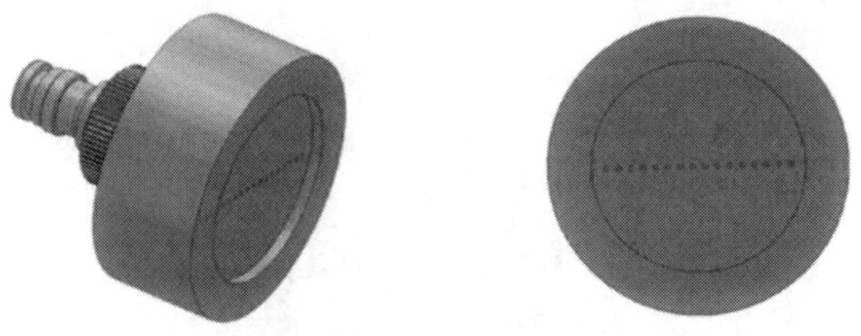

Fig. 5. Linear Circular Pattern Nozzle

ANSYS Fluent CFD is running on the nozzle profile to identify the velocity profile and also the pressure at the outlet. A multi-zone, mapped mesh type hexa/prism and free mesh type of tetra/pyramid meshing is done [Fig. 6]. Multi-zone, hex dominant is done because the outlets of the nozzles are having circular patterns and multi-zone work best for these conditions. It has a total of 108202 numbers of elements, average skewness 0.36274 and an average orthogonal quality of 0.79.

For the boundary conditions, Reynold's number is calculated for the flow to see whether the flow is a laminar or turbulent flow across the nozzle. The equation for Reynold's number is as follow:

$$\textit{Reynold's number} = \frac{\rho v D}{\mu} \text{ (Eq. 1)}$$

Where,
ρ: Density of water
v: Velocity of water
μ: Viscocity of water at 28°C

According to [14], viscosity and the density of water at 28°C is 0.8324 mPa.s and 0.9962 g/cm^3 respectively.

$$\textit{Reynold's number} = \frac{996.2 \frac{kg}{m^3} \times 3 \frac{m}{s} \times 0.0092m}{0.8324 \times 0.001 \frac{kg}{ms}} = 33{,}031.14$$

The flow is turbulent, Realizable k-ε is used as the viscous modelling during all the CFD analysis. The boundary conditions are set to 3 m/s inlet velocity with. This is due to the volumetric flow rate of 12 LPM measured from the pipes of the kitchen in Taylor's School of Hospitality and Culinary Arts.

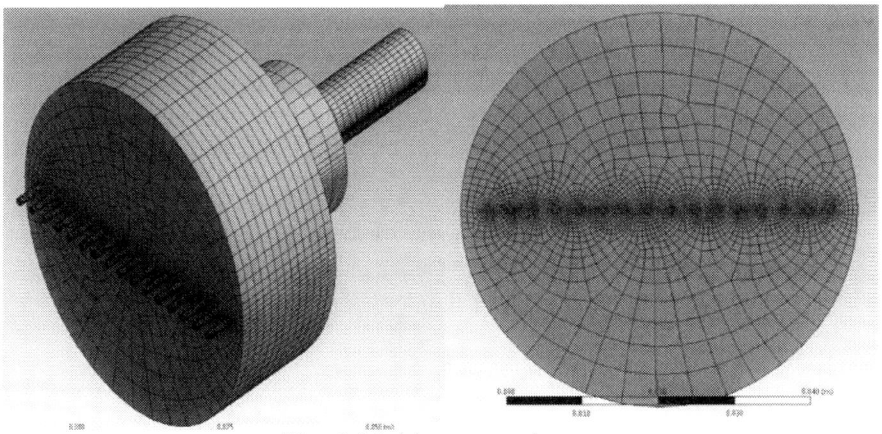

Fig. 6. Multi-zone mesh type

2.4 Cleanliness of kitchen utensil

3M Clean Trace shown in Fig. 7 is used as the equipment to test on the cleanliness of the kitchen utensil. Swabs are taken before and after cleaning of the kitchen utensil using the traditional way (without ozone water) and with ozone water. A cotton bud like stick shown in Fig. 7 are used to swab on the surface (10cm × 10cm) of the kitchen pot before cleaning, and done it again after cleaning. Same applies to with and without nozzle. The measurement is taken as Relative Light Units (RLU). RLU is the measurement of adenosine triphosphate (ATP) present on the kitchen utensil. The higher the RLU, the dirtier the kitchen utensil is. In normal practice, RLU higher than 11 is a caution and higher than 30 is a fail. Hence, RLU lower than 10 is in the safe region [15].

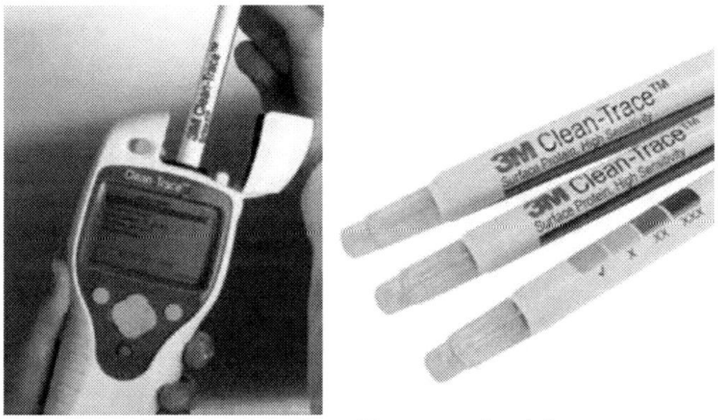

Fig. 7. 3M Clean Trace swab sticks

Kitchen towel and an oil control blotting paper are used to have a qualitative result on the dirt and oil left on the kitchen utensil after cleaning. Results are shown in the next section of discussion.

3 Results and discussion

3.1 Washing of stainless steel pot

Table 4 and 5 show the amount of water and detergent is used for the different methods of washing (as discussed in Table 2). The tables also show the cleanliness (RLU) of the pots before and after washing. Comparing the results with the background reading in Table 2, averagely one utensil needed 23.95 L of water to clean. However, during the research, the volume of water was controlled well hence, there is a big reduction in the volume of water.

Table 4. Cleaning of a stainless steel kitchen pot without nozzle

Variables	Traditional way	Ozone water
Volume of water, L	10.4	11.1
Volume of detergent, ml	4	6
Ozone level, PPM	-	1.8
Pot before washing	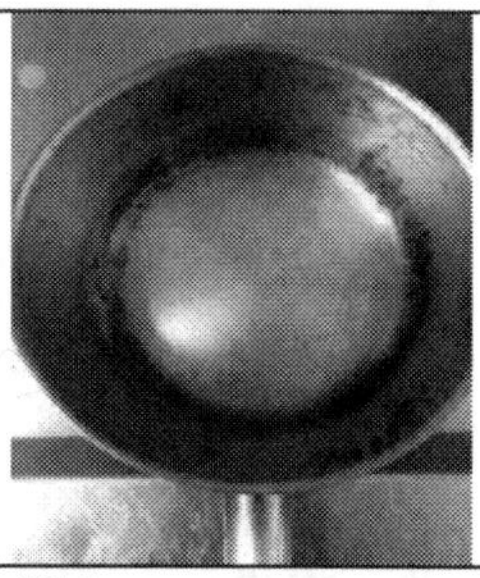	
Cleanliness before washing, RLU	674	312
Pot after washing		
Oil and dirt detection after washing	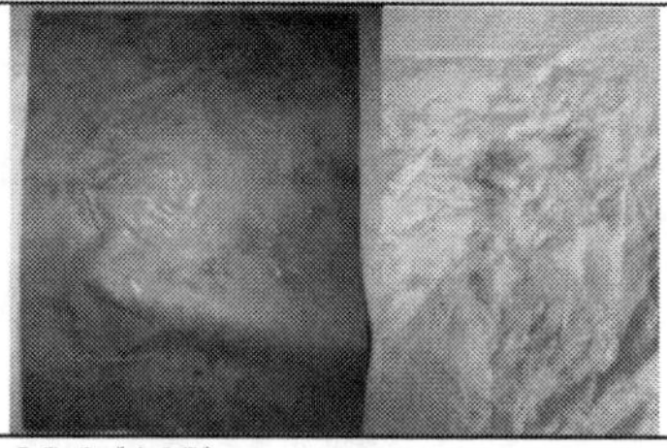	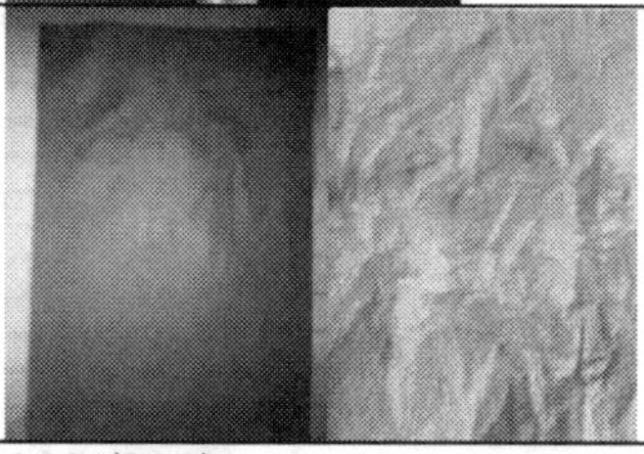
Cleanliness after washing, RLU (TOTAL REDUCTION OF RLU)	226 (448)	103 (209)
Temperature of water	29.6°C	29.1°C

Table 5. Cleaning of a stainless steel kitchen pot with nozzle

Variables	Traditional way	Ozone water
Volume of water, L	8	8
Volume of detergent, ml	5	5
Ozone level, PPM	-	1.25ppm
Pot before washing		
Cleanliness before washing, RLU	1106	679
Pot after washing		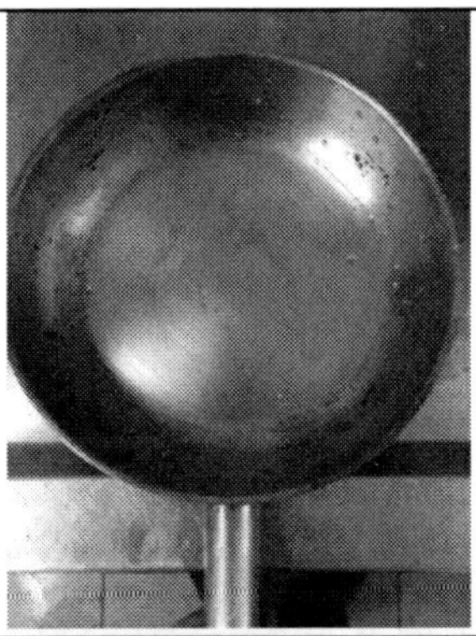
Oil and dirt detection after washing	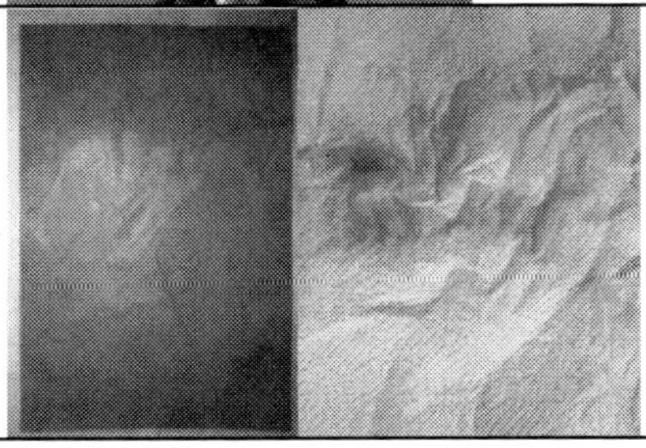	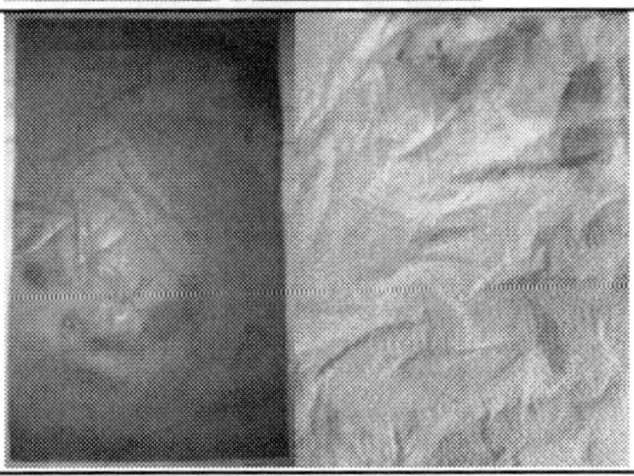
Cleanliness after washing, RLU (TOTAL REDUCTION OF RLU)	156 (950)	6 (673)
Temperature of water	29.5°C	29.1°C

From Table 4, the RLU reduction for traditional way of cleaning (using just detergent and normal 29.6°C of water) is 66.47% while for ozone water is 67%. Traditional way uses 700ml of water and 2ml of detergent lesser compared to ozone water to achieve the RLU. The difference in water volume and detergent, and also the RLU reduction are not significant in this experiment. Table 5 shows a significant difference in RLU reduction comparing the two methods of cleaning with the same amount of water and detergent used. There was a 85.9% reduction using the traditional way with nozzle and a 99.12% RLU reduction using ozone with nozzle. This has significantly shown that with the use of nozzle in the cleaning process, it increases the RLU reduction by about 20% ~ 30% compared to without the nozzle. Ozone water with the use of nozzle also shown the cleaning process to be very efficient.

The readings are as such because it is due to the properties of nozzle. Although with nozzle, the PPM level is lower than the one without nozzle, however nozzle plays a part in cleaning and also sterilizing. Ozone needs exposure time to the surface in order to kill bacteria, hence when ozone without nozzle is applied, the reduction rate is slower. When nozzle is applied in this case, the nozzle will give impact force to the utensil surface and clearing most of the dirt and bacteria, adding on with ozone, the sterilizing effect is much higher compared to without the nozzle.

So, comparing with nozzle with ozone and without ozone, although both shows more than 80% of RLU reduction, the one without ozone has always show RLU more than 10 which is unsafe to use according to [41]. When comparing without nozzle with ozone and without ozone, both have also failed to the requirement stated in [41]. Hence, ozone and nozzle is the best combination as the RLU reduction for all the cases has shown RLU less than 10, which is safe for consumers. This has proven that even though chemicals are used to wash utensils, it might not be safe for consumption purposes. Ozone has proven to be a strong disinfectant and the utensils that are cleansed by ozone will be definitely safe to use.

3.2 Nozzle and ozone suction in water

The inlet velocity is at 3 m/s and the outlet velocity as the result of the nozzle design is at about 12-14 m/s. Fig. 9 shows the velocity path lines of the flow of water through the nozzle. It can also be seen that there is back flow occurring as it hits the head of the nozzle. The backflow also causes the high pressure of 1×10^5 Pa to be accumulated in the nozzle itself as shown in Fig. 10. As the pressure accumulates in the nozzle, the velocity decreases, the air pressure in the venturi will find it difficult to push the ozone gas to dilute in the water through the venturi. This causes the decrease in ozone suction in the venturi which was from 1.8 PPM to 1.25 PPM (from Table 4 and 5).

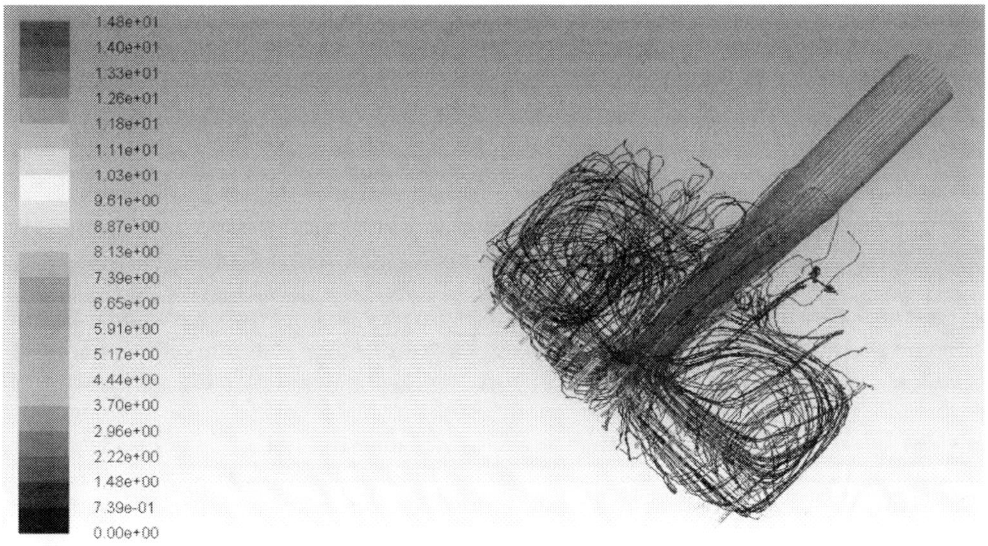

Fig. 9. Velocity path lines

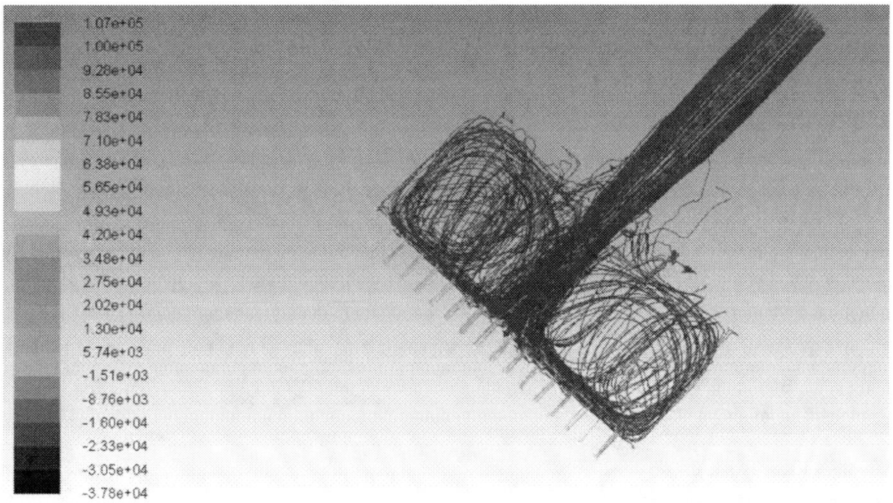

Fig. 10. Pressure path lines

According to [16], the impact force is based on the angle of water distribution from the nozzle. The bigger the angle of water distribution, the lower the impact force. This is because of the kinetic energy losses when it is needed to distribute the water at the big angles. Since the impact force equation is given by,

$$Fs = \frac{1}{2}mv^2 \text{ (Eq. 2)}$$

where the velocity at the outlet will play a part in determining the impact force. When the velocity is higher at the outlet, the impact force will be higher.

The design of the conventional nozzle is based on the nozzle used in TCHT kitchen shown in Appendix 3 and 4. According to literature review [16, 17], a nozzle that produces solid stream has the highest impact force compared to water distributed at an angle. Single hole will give a solid stream flow. However, a solid stream will also lose momentum quicker compared to others, even though all nozzles will lose momentum after the water exits from the orifice [17]. Hence, the distance of nozzle to the surface intended to be washed have to be nearer so that the water can hit the surface with minimal losses of momentum. A single hole is also not very user friendly to wash kitchen utensil as the focus area is small and usually a kitchen utensil is at sizes of 10 cm – 20 cm. So in order to optimize the washing, a linear pattern is done with 3 mm apart so that the water streams could wash the kitchen utensil at a bigger surface area. It can mimic a fan pattern nozzle but having solid streams as well to increase impact force.

4. Conclusion

In conclusion, the usage of ozone water in cleaning kitchen utensils has shown significant difference compared to using the traditional way of cleaning with the aid of a nozzle. The nozzle prototype has given the cleaning process a huge boost with a big increment of RLU reduction. The kitchen utensil is tested to be 99% clean after using ozone with the same amount of water and dishwashing liquid as to the traditional way of cleaning. The controlled research has definitely reduced the volume of water used and proven to be much more efficient in cleaning. In terms of manpower, there is no reduction because the time of cleaning is almost the same. The nozzle could help to ease the cleaning of stubborn dirt however, the time of cleaning is almost the same with and without nozzle. Ozone level in the water is also based on the pressure of water, as the backpressure is high, the ozone level dissolved in the water is lesser.

Acknowledgement

Throughout this research, I would like to acknowledge a few people that have guided me. Firstly, I would like to thank Medklinn International Sdn Bhd, for providing me the equipment to carry out the experiments. My extended acknowledgement to my industrial supervisor Mr. Lim Boon Han and Mr. Daniel Lu for giving me his advice based on his knowledge and experience in the ozone industry. I would also like to thank Taylor's School of Hospitality and Culinary Arts for the opportunity to allow me to carry out my research in the kitchen, alongside the supervision of Mr. Hisammudin and Mr. Kumar, the people in charge of the kitchen.

References

[1] J. S. Hargreaves and B. Cantab, "Laboratory Evaluation of the 3-Bowl System Used for Washing-Up Eating Utensils in the Field," Wilderness Environ. Med., vol. 17, no. 2, pp. 94–102, 2006.

[2] F. Chaib and Olivia Lawe-Davies, "WHO's first ever global estimates of foodborne diseases find children under 5 account for almost one third of deaths," 2017. [Online]. Available: http://www.who.int/mediacentre/news/releases/2015/foodborne-disease-estimates/en/. [Accessed: 17-Jun-2017].

[3] K. Mattick et al., "The survival of foodborne pathogens during domestic washing-up and subsequent transfer onto washing-up sponges , kitchen surfaces and food," vol. 85, pp. 213–226, 2003.

[4] H. Protection, A. Communicable, D. Surveillance, H. Protection, A. Specialist, and R. M. Division, "Foodborne general outbreaks of Salmonella Enteritidis phage type 4 infection, England and Wales , 1992 – 2002 : where are the risks ?," pp. 795–801, 2005.

[5] A. Valero, J. Carlos, G. Fongaro, M. Hern, and D. Rodríguez-l, "De fi nition of sampling procedures for collective-eating establishments based on the distribution of environmental microbiological contamination on food handlers , utensils and surfaces," vol. 77, no. 852, pp. 8–16, 2017.

[6] H. Sciences and Y. Sakyo-ku, "A study on Campylobacter jejuni cross-contamination during chilled broiler preparation," vol. 115, pp. 107–115, 2010.

[7] G. Stellato, A. La Storia, T. Cirillo, and D. Ercolini, "International Journal of Food Microbiology Bacterial biogeographical patterns in a cooking center for hospital foodservice," Int. J. Food Microbiol., vol. 193, pp. 99–108, 2015.

[8] F. Timothy, "Risk of Giardiasis from Consumption of Wilderness Water in North America : A Systematic Review of Epidemiologic Data," pp. 100–103.

[9] E. Seob, J. Eun, J. Kim, and O. Kyung, "LWT - Food Science and Technology Isolation of indigenous bacteria from a cafeteria kitchen and their bio fi lm formation and disinfectant susceptibility," LWT - Food Sci. Technol., vol. 77, pp. 376–382, 2017.

[10] G. H. R. Silva, L. A. Daniel, H. Bruning, and W. H. Rulkens, "Bioresource Technology Anaerobic effluent disinfection using ozone : Byproducts formation," Bioresour. Technol., vol. 101, no. 18, pp. 6981–6986, 2010.

[11] S. Tripathi, V. Pathak, D. M. Tripathi, and B. D. Tripathi, "Bioresource Technology Application of ozone based treatments of secondary effluents," Bioresour. Technol., vol. 102, no. 3, pp. 2481–2486, 2011.

[12] M. V Shynkaryk, T. I. Pyatkovskyy, A. E. Yousef, and S. K. Sastry, "Gaseous ozone treatment of baby spinach within the existing production chain for inactivation of Escherichia coli O157 : H7," J. Food Eng., vol. 191, pp. 10–18, 2016.

[13] "Venturi Fertilizer Applications - Wassertec. Cape Town", Wassertec.co.za, 2017. [Online]. Available: https://www.wassertec.co.za/venturi-fertilizer-applications/. [Accessed: 17- Jun- 2017].

[14] "Water – viscosity table and viscosity chart", Viscopedia.com, 2017. [Online]. Available: http://www.viscopedia.com/viscosity-tables/substances/water/. [Accessed: 17- Nov- 2017].

[15] "Food & Beverage - Hygiena Monitoring Systems - Setting ATP Limits - RLU Limits | Food and Beverage Products | Hygiena | Hygiena - Rapid Solutions for Hygiene Monitoring", Hygiena.com, 2017. [Online]. Available: https://www.hygiena.com/rlulimits-food.html. [Accessed: 26- Oct- 2017].

[16] "Cleaning", Spray-nozzle.co.uk, 2017. [Online]. Available: http://www.spray-nozzle.co.uk/spray-nozzles-by-industry/food-processing-new/cleaning. [Accessed: 20- Nov- 2017].

[17] "Solid Stream Nozzles for cleaning applications", Spray-nozzle.co.uk, 2017. [Online]. Available: http://www.spray-nozzle.co.uk/resource-links/applications/cleaning/solid-stream-nozzles. [Accessed: 17- Nov- 2017].

APPENDIX

Abundance of bacterial taxa in environmental swabs from the hospital cooking center in sampling 1. Only OTUs with >0.1% abundance in more than 20% of the samples are represented.

Taxon	S_BlastChiller_Wall	S_ColdStorage_Shelf	S_ColdStorage_Wall	K_Balance	K_Sink	K_Workbench	PPF_Sink	PPF_Workbench	PPR_Cutter	PPR_Sink	PPR_Workbench	PPV_CuttingBoard
Acinetobacter	0.003	0.007	0.001	0.004	0.009	0.001	0.000	0.012	0.001	0.017	0.001	0.000
Acinetobacter johnsonii	0.004	0.071	0.007	0.005	0.096	0.001	0.001	0.009	0.001	0.079	0.001	0.000
Alicyclobacillus	0.266	0.029	0.127	0.000	0.020	0.041	0.030	0.107	0.000	0.021	0.049	0.001
Caulobacter	0.001	0.002	0.001	0.000	0.001	0.000	0.002	0.008	0.000	0.001	0.000	0.046
Chryseobacterium	0.000	0.002	0.000	0.001	0.030	0.000	0.000	0.010	0.000	0.014	0.000	0.003
Corynebacterium	0.003	0.001	0.007	0.000	0.009	0.000	0.002	0.000	0.000	0.002	0.000	0.001
Enterococcus lactis	0.000	0.006	0.002	0.001	0.017	0.000	0.000	0.000	0.022	0.040	0.000	0.000
Kocuria	0.011	0.058	0.042	0.000	0.024	0.001	0.116	0.040	0.000	0.070	0.031	0.006
Kocuria marina	0.000	0.002	0.000	0.000	0.000	0.000	0.002	0.002	0.000	0.001	0.002	0.000
Kocuria palustris	0.006	0.010	0.006	0.000	0.012	0.000	0.006	0.000	0.000	0.028	0.002	0.000
Lactobacillus	0.004	0.000	0.007	0.000	0.002	0.001	0.015	0.004	0.000	0.001	0.000	0.000
Massilia timonae	0.004	0.003	0.001	0.000	0.007	0.000	0.100	0.004	0.000	0.000	0.000	0.091
Moraxellaceae	0.000	0.002	0.000	0.000	0.005	0.000	0.006	0.003	0.000	0.002	0.000	0.000
Paracoccus	0.017	0.007	0.009	0.000	0.018	0.000	0.084	0.004	0.003	0.003	0.001	0.104
Propionibacterium	0.025	0.004	0.009	0.000	0.001	0.006	0.004	0.015	0.000	0.001	0.002	0.000
Pseudomonas	0.028	0.024	0.011	0.241	0.258	0.273	0.001	0.002	0.426	0.010	0.603	0.001
Psychrobacter	0.110	0.190	0.005	0.316	0.023	0.236	0.093	0.001	0.110	0.192	0.000	0.000
Ralstonia	0.005	0.000	0.006	0.000	0.001	0.004	0.003	0.007	0.001	0.001	0.001	0.000
Serratia	0.007	0.000	0.019	0.000	0.001	0.002	0.002	0.004	0.000	0.003	0.001	0.000
Sphingomonas	0.006	0.001	0.006	0.000	0.002	0.000	0.015	0.000	0.000	0.001	0.001	0.057
Staphylococcus	0.005	0.008	0.003	0.000	0.007	0.002	0.002	0.000	0.000	0.004	0.000	0.000
Streptococcus	0.002	0.003	0.012	0.000	0.002	0.001	0.004	0.013	0.000	0.002	0.000	0.000
Streptococcus sanguinis	0.051	0.018	0.140	0.000	0.030	0.011	0.018	0.035	0.006	0.070	0.007	0.001
Streptococcus thermophilus	0.005	0.002	0.005	0.000	0.003	0.002	0.002	0.004	0.001	0.004	0.000	0.000
Other	0.018	0.047	0.022	0.011	0.023	0.016	0.023	0.046	0.010	0.014	0.009	0.016

Taxon	PPV_Grinder	PPV_Sink	PPV_Workbench	PPW_Sink	PPW_Workbench	T_Colander	T_CuttingBoard	T_Knife	T_Pan	T_SteelDish	T_SteelPincers	T_SteelShovel	T_SteelTray
Acinetobacter	0.015	0.082	0.010	0.018	0.026	0.007	0.001	0.000	0.001	0.000	0.003	0.392	0.001
Acinetobacter johnsonii	0.001	0.031	0.033	0.033	0.100	0.793	0.000	0.001	0.012	0.001	0.004	0.004	0.003
Alicyclobacillus	0.001	0.000	0.166	0.092	0.210	0.036	0.000	0.005	0.313	0.002	0.121	0.020	0.058
Caulobacter	0.001	0.022	0.002	0.005	0.006	0.000	0.000	0.000	0.004	0.000	0.001	0.001	0.003
Chryseobacterium	0.017	0.029	0.005	0.006	0.027	0.005	0.007	0.000	0.000	0.002	0.001	0.006	0.000
Corynebacterium	0.009	0.000	0.002	0.010	0.002	0.001	0.003	0.005	0.002	0.000	0.001	0.001	0.003
Enterococcus lactis	0.000	0.000	0.000	0.001	0.000	0.000	0.000	0.002	0.001	0.000	0.018	0.007	0.000
Kocuria	0.076	0.000	0.008	0.086	0.049	0.000	0.019	0.322	0.002	0.014	0.010	0.001	0.002
Kocuria marina	0.002	0.000	0.000	0.001	0.002	0.001	0.001	0.005	0.000	0.000	0.001	0.000	0.000
Kocuria palustris	0.198	0.000	0.001	0.022	0.005	0.000	0.000	0.005	0.000	0.002	0.001	0.000	0.001
Lactobacillus	0.000	0.000	0.011	0.033	0.004	0.001	0.002	0.001	0.023	0.000	0.003	0.001	0.007
Massilia timonae	0.049	0.137	0.002	0.061	0.006	0.001	0.000	0.000	0.000	0.000	0.001	0.001	0.000
Moraxellaceae	0.001	0.002	0.002	0.043	0.017	0.001	0.101	0.024	0.000	0.000	0.004	0.001	0.001
Paracoccus	0.023	0.088	0.009	0.004	0.013	0.000	0.000	0.000	0.002	0.079	0.001	0.012	0.002
Propionibacterium	0.010	0.000	0.034	0.004	0.008	0.003	0.000	0.000	0.016	0.000	0.010	0.003	0.034
Pseudomonas	0.001	0.040	0.038	0.007	0.007	0.019	0.092	0.029	0.006	0.000	0.003	0.004	0.002
Psychrobacter	0.094	0.000	0.008	0.005	0.001	0.004	0.276	0.035	0.000	0.001	0.011	0.001	0.002
Ralstonia	0.000	0.000	0.008	0.001	0.009	0.000	0.000	0.000	0.011	0.000	0.007	0.001	0.032
Serratia	0.000	0.000	0.016	0.001	0.007	0.001	0.000	0.000	0.043	0.000	0.006	0.000	0.003
Sphingomonas	0.000	0.026	0.003	0.007	0.005	0.001	0.000	0.000	0.004	0.033	0.000	0.001	0.001
Staphylococcus	0.000	0.000	0.010	0.014	0.004	0.001	0.000	0.000	0.004	0.000	0.001	0.000	0.004
Streptococcus	0.000	0.000	0.015	0.005	0.003	0.002	0.000	0.000	0.011	0.000	0.028	0.002	0.032
Streptococcus sanguinis	0.001	0.000	0.088	0.026	0.058	0.032	0.100	0.017	0.267	0.001	0.052	0.032	0.076
Streptococcus thermophilus	0.000	0.000	0.008	0.003	0.005	0.001	0.003	0.000	0.013	0.000	0.007	0.005	0.006
Other	0.031	0.019	0.028	0.025	0.020	0.011	0.014	0.014	0.006	0.049	0.030	0.012	0.017

Appendix 1: Bacteria count in hospital food service sample 1.

Abundance of bacterial taxa in environmental swabs from the hospital cooking center in sampling 2. Only OTUs with >0.1% abundance in more than 20% of the samples are represented.

Taxon	S_BlastChiller_Wall	S_ColdStorage_Shelf	S_ColdStorage_Wall	K_Balance	K_Sink	K_Workbench	PPF_Sink	PPF_Workbench	PPR_Cutter	PPR_Sink	PPR_Workbench	PPV_CuttingBoard
Acinetobacter	0.000	0.000	0.007	0.000	0.056	0.005	0.017	0.032	0.031	0.006	0.159	0.133
Acinetobacter johnsonii	0.001	0.001	0.024	0.000	0.032	0.002	0.005	0.014	0.020	0.023	0.319	0.656
Alicyclobacillus	0.001	0.004	0.067	0.000	0.002	0.076	0.000	0.015	0.000	0.005	0.001	0.000
Brochothrix	0.002	0.285	0.013	0.024	0.000	0.005	0.000	0.003	0.004	0.030	0.000	0.000
Caulobacteraceae	0.000	0.000	0.001	0.000	0.007	0.004	0.000	0.002	0.000	0.039	0.001	0.000
Chryseobacterium	0.000	0.000	0.004	0.000	0.053	0.000	0.036	0.034	0.002	0.039	0.020	0.000
Corynebacterium	0.000	0.001	0.012	0.000	0.000	0.016	0.000	0.001	0.023	0.000	0.007	0.000
Kocuria	0.009	0.002	0.158	0.000	0.004	0.015	0.488	0.006	0.068	0.230	0.003	0.000
Kocuria palustris	0.001	0.001	0.051	0.000	0.023	0.004	0.012	0.000	0.007	0.056	0.000	0.000
Massilia timonae	0.001	0.000	0.002	0.000	0.008	0.033	0.061	0.032	0.001	0.097	0.008	0.009
Microbacterium	0.000	0.000	0.006	0.000	0.005	0.001	0.000	0.029	0.000	0.022	0.001	0.000
Moraxellaceae	0.000	0.000	0.003	0.000	0.036	0.007	0.001	0.000	0.035	0.039	0.081	0.009
Paracoccus	0.000	0.000	0.057	0.000	0.007	0.042	0.042	0.209	0.018	0.054	0.023	0.000
Propionibacterium	0.003	0.003	0.046	0.000	0.001	0.033	0.000	0.008	0.000	0.001	0.000	0.000
Pseudomonas	0.794	0.020	0.020	0.951	0.005	0.005	0.038	0.006	0.012	0.016	0.002	0.022
Psychrobacter	0.000	0.069	0.004	0.000	0.161	0.145	0.041	0.026	0.224	0.028	0.003	0.002
Rhodobacteraceae	0.000	0.000	0.006	0.000	0.002	0.000	0.000	0.015	0.000	0.016	0.023	0.001
Sphingobacterium faecium	0.000	0.000	0.000	0.000	0.008	0.000	0.058	0.037	0.002	0.000	0.000	0.000
Sphingomonas	0.003	0.000	0.002	0.000	0.001	0.000	0.001	0.056	0.000	0.004	0.001	0.000
Stenotrophomonas	0.000	0.000	0.000	0.000	0.002	0.000	0.020	0.009	0.006	0.004	0.003	0.000
Other	0.035	0.007	0.048	0.005	0.044	0.028	0.026	0.023	0.040	0.034	0.035	0.015

Taxon	PPV_Grinder	PPV_Sink	PPV_Workbench	PPW_Sink	PPW_Workbench	T_Colander	T_CuttingBoard	T_Knife	T_Pan	T_SteelDish	T_SteelPincers	T_SteelShovel	T_SteelTray
Acinetobacter	0.000	0.011	0.031	0.045	0.031	0.051	0.003	0.017	0.001	0.087	0.003	0.044	0.004
Acinetobacter johnsonii	0.000	0.074	0.100	0.011	0.407	0.516	0.001	0.001	0.007	0.535	0.009	0.545	0.058
Alicyclobacillus	0.008	0.000	0.000	0.000	0.001	0.072	0.000	0.000	0.015	0.004	0.091	0.008	0.006
Brochothrix	0.000	0.000	0.000	0.002	0.000	0.000	0.000	0.000	0.000	0.000	0.000	0.001	0.004
Caulobacteraceae	0.000	0.016	0.021	0.026	0.000	0.003	0.010	0.000	0.001	0.000	0.000	0.000	0.000
Chryseobacterium	0.002	0.007	0.094	0.083	0.002	0.003	0.303	0.081	0.000	0.163	0.001	0.002	0.003
Corynebacterium	0.001	0.000	0.001	0.002	0.010	0.000	0.008	0.001	0.001	0.000	0.028	0.002	0.001
Kocuria	0.042	0.018	0.031	0.032	0.013	0.001	0.040	0.008	0.010	0.002	0.070	0.001	0.012
Kocuria palustris	0.007	0.002	0.042	0.009	0.000	0.000	0.001	0.001	0.002	0.000	0.006	0.000	0.000
Massilia timonae	0.000	0.003	0.018	0.025	0.022	0.003	0.025	0.299	0.000	0.000	0.003	0.001	0.010
Microbacterium	0.001	0.042	0.001	0.020	0.001	0.001	0.003	0.000	0.000	0.000	0.002	0.000	0.000
Moraxellaceae	0.000	0.017	0.001	0.045	0.015	0.011	0.003	0.001	0.005	0.002	0.011	0.007	0.003
Paracoccus	0.008	0.086	0.098	0.048	0.000	0.003	0.001	0.050	0.008	0.000	0.007	0.001	0.000
Propionibacterium	0.001	0.000	0.000	0.000	0.001	0.004	0.000	0.000	0.203	0.001	0.010	0.002	0.001
Pseudomonas	0.001	0.221	0.104	0.036	0.387	0.007	0.117	0.024	0.004	0.072	0.010	0.002	0.001
Psychrobacter	0.000	0.002	0.001	0.150	0.002	0.009	0.073	0.190	0.295	0.000	0.004	0.006	0.391
Rhodobacteraceae	0.013	0.039	0.003	0.002	0.000	0.001	0.000	0.001	0.001	0.000	0.002	0.001	0.000
Sphingobacterium faecium	0.000	0.000	0.000	0.019	0.000	0.001	0.080	0.000	0.000	0.000	0.000	0.001	0.000
Sphingomonas	0.003	0.020	0.012	0.012	0.000	0.000	0.001	0.000	0.000	0.000	0.006	0.000	0.000
Stenotrophomonas	0.001	0.010	0.050	0.010	0.001	0.001	0.045	0.004	0.000	0.000	0.000	0.001	0.000
Other	0.187	0.047	0.027	0.037	0.013	0.037	0.037	0.034	0.044	0.011	0.133	0.047	0.025

Appendix 2: Bacteria count in hospital service sample 2.

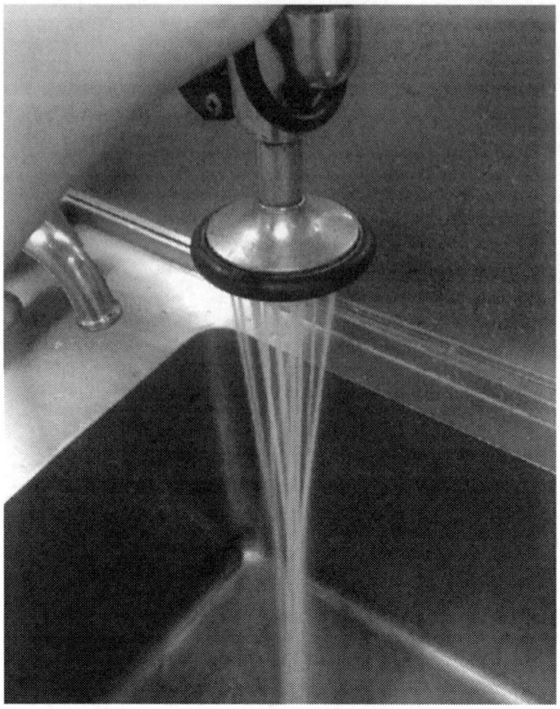

Appendix 3: Focus point of the water spray used in Taylor's School of Hospitality and Culinary Arts kitchen.

Appendix 4: Water spray profile.

Stabilising an Inverted Pendulum with PID Controller

Yon Yaw Lim[1], *Choon Lih* Hoo[2,*], and *Yen Myan* Felicia Wong[1]

[1]School of Engineering, Mechanical Engineering, Taylor's University, Malaysia
[2]School of Engineering and Physics Sciences, Mechanical Engineering, Heriot-Watt University, Malaysia

Abstract. Inverted pendulum is a system in which the centre of the mass is above the pivot point, where the mass can freely rotate. The inverted pendulum has a unique trait; it is unpredictable, non-linear and consists of multiple variables. Balancing by PID controller is a continuous process where it corrects the feedback system error from the difference between the measured value and the desired value. This research mainly focusses on balancing an inverted pendulum with reaction wheel. The research objectives are to construct a self-balanced inverted pendulum and using PID controller to control the stability of the pendulum. The PID configuration is then evaluated based on the response of the system. The idea is to use the reaction torque generated by the motor to counter balance the inverted pendulum. The factor which governs the amount of torque generated is the height of the pendulum and the mass of the wheel. To balance the pendulum, tuning the PID gain is essential. Proportional gain is tuned first to get oscillation, next is to tune the integral and derivative gain to get a smoother and quicker response. Idea is to get short settling time, and minimum overshoot percentage. Hypothesis is that higher proportional gain will give a faster response rate and the acceleration of the motor is the key on generating torque. A simulation of the pendulum falling is simulated and the results are recorded in term of the response of the pendulum against time. At initial point, proportional gain, integral gain and derivative gain are set to zero to validate the simulation. The finding in this research is that torque is generated by the acceleration of the reaction wheel. Higher acceleration gives a high torque. Others findings is the PID parameter; Proportional gain increases the response rate; Integral gain is used to eliminate steady state error; Derivative gain is used to lessen the overshoot.

1 Introduction

Inverted pendulum is a system in which the centre of mass is above the pivot point, where the mass can freely rotate. For an example, balancing a long stick on a palm is considered as inverted pendulum. The inverted pendulum has a unique trait where it is unpredictable, non-linear and consists of multiple variables. Which is why, inverted pendulum is usually

* Corresponding author: c.hoo@hw.ac.uk

used to benchmark a control algorithm such as PID controllers, state space representation, neural networks, fuzzy control and genetic algorithms. To balance an inverted pendulum, torque must be applied on the pivot to counter balance the falling pendulum. This can done by moving the pivot point horizontally and vertically by using a cart or a reaction wheel which generate a reaction torque from spinning in opposite direction [1].

The concept of inverted pendulum is very common in the world either in man-made structure or natural things. One of the great examples of an inverted pendulum that can be seen in natural things is human being. To stand up right, human body must be constantly balance itself. Putting an unconscious person upright will lead to the person falling. A humanoid robot is also one of the example where the concept of inverted pendulum can be applied in [2]. Inverted pendulum can also be seen in advance technology such as rockets and missile where the centre of gravity is placed behind the centre of drag, results in aerodynamic instability [3]. As for the latest technology, it would be the Segway [4].

As mentioned above, there are many ways to balance an inverted pendulum. The most common way is to use a moving cart to move the pivot point to create counter moment. Another uncommon way of balancing is by rotating clockwise and anticlockwise to generate moment. Out of the many variations out there, the focus here is to balance an inverted pendulum by using reaction wheel. Unlike other type of inverted pendulum such as rotary or cart they require a runway or platform to balance itself because it is balanced by motion of the whole pendulum. However, reaction wheel inverted pendulum is balance by torque generated by the motor. This requires no platform or runway making it compact and rigid.

Reaction wheel is mainly used in the satellite to provide precise rotation. Newton 3rd law states that for every action there is an equal and opposite reaction, thus the name of reaction wheel. To rotate the satellite in space, satellite rotated by spinning the reaction in the opposite direction. This allows the satellite to have rotational motion through its rotational axis. However, reaction does not generate translation motion, it cannot move from one point to another [5]. Because the wheel has mass, and it is moving in circular motion, the wheel itself has angular momentum. The angular momentum is conserved as long as there is no external torque applies to the system. Other than that, object with mass also subjected to moment of inertia. By relating the relation above, theoretically a reaction wheel inverted pendulum can be model out [6].

Control is one of the most essential elements nowadays, such as control in a power plant. A power plant without control is disastrous. A common control system consists of the controller, sensor and an actuator. Actuator is the component where it converts electrical signal to mechanical energy to change the physical environment. Sensor reads the change in physical environment and sends the electrical signal to the controller. Controller acts as the processor, which processes the value from the feedback sensor. It calculates the difference of the environmental value and target value, which also known as response error. This value is then processed by the controller and the correct signal is sent to the actuator. A target value is the value which the user desire to achieve. Below in Fig. 1 is a basic control diagram [7].

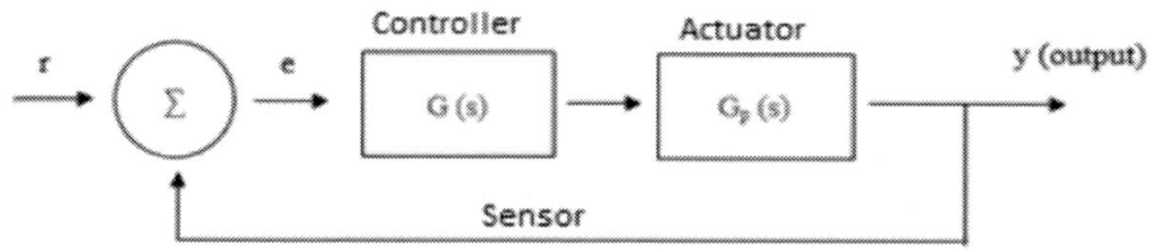

Fig. 1. Control diagram.

Where,
r = input
e = response error
G(s) = Controller
$G_p(s)$ = actuator

1.1 Evaluating the stability of system in general

There is numerous way to test the stability of a system. One of the most common way is to test the linear system of the transfer function [8].
Consider the system has two main block-diagrams $G_p(s)$ and G(s).
The transfer function in loop:
The output can be expressed as, $y = G_p G$
The response error can be expressed as, e = r - y
The ratio of the output and response error is the closed loop function,

$$\frac{y}{r} = \frac{G_p\, G}{e+y} = G_{cl}(s) \qquad (1)$$

$G_{cl}(s)$ = a closed loop function

The transfer function above is in complex form. Once the function is converted into Laplace form, the data can be presented graphically on a complex s – plane. A closed loop function is always in the fraction form as shown in equation (1). The numerator in a function is also known as zero whereas the denominator is known as pole. The 'circle' mark on the s-plane shown in Fig. 2 represents the zero, whereas the 'cross' mark represents the pole. Both of these factors govern the stability of the system. If a pole exists on the right plane, it means that there is an output component which gradually increases as time progress, which result in instability in the system. In other words, to achieve stability, the pole needed to be on the left half plane. Fig. 3 shows a clearer response of the system depending which point the function located. In the centre of Fig. 3 shows the general response of the system, with either gradually decrease, increase or remain constant. Y – axis represents the frequency of the response, where the higher the magnitude of the imaginary number indicates higher frequency.

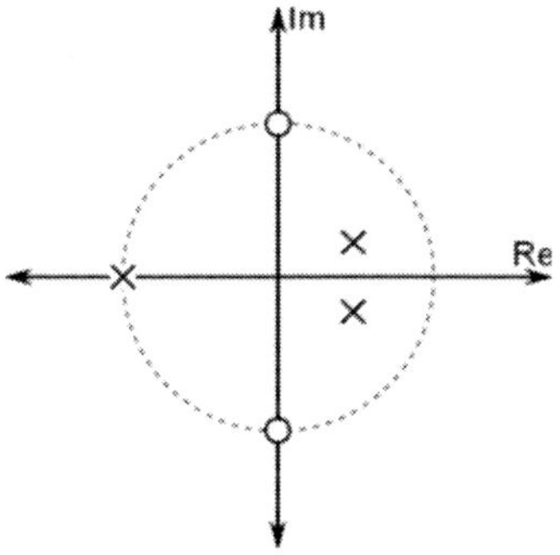

Fig. 2. Data representation in a complex s-plane.

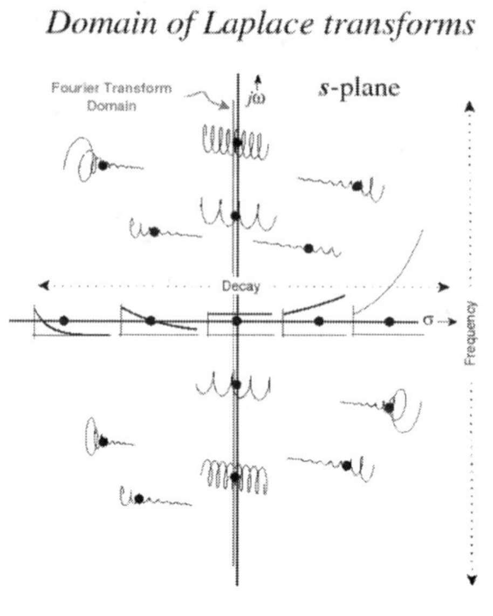

Fig. 3. Complex s-plane [8].

1.2 PID controller

PID is a very well-known feedback controller, due to its efficiency and reliability. PID consists of three parameters, proportional gain, integrated and derivative. Like most of the controller, it calculates the differences between the measured value and the desired value as shown in Fig. 4. One of the examples of PID in daily life is air conditioning. First air conditioning is set to a desire temperature by using the remote. Most of the air conditioning has a built-in temperature sensor to feedback the ambient temperature. The purpose of the PID is to constantly receive feedback from the ambient temperature and send the suitable voltage to the system to lower down the temperature. Once the desire temperature is achieved, the PID controller will then try to maintain the temperature within a range. This range is known as the steady state error. A good controller has a low range of steady state error, also achieve the desire value in a shorter time with low overshooting. The proportional gain acts as multiplier which multiplies the error response. It is also one the factor to make the response quicker, however too much will result in oscillation and eventually unstable. The function of an integral gain is sum up all the errors produce to reduce the steady state error, whereas the derivative gain is product of rate of change of the error. The function is to increase the reaction magnitude when there is a change in error. However, it is sensitive to noises such as disturbance in the electrical signal.

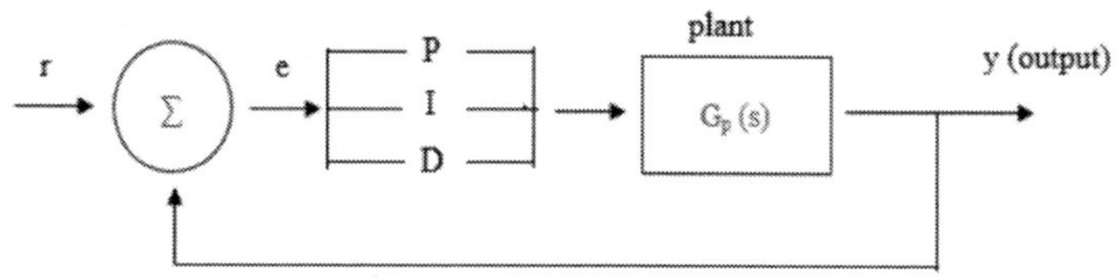

Fig. 4. Flow process with PID.

The objectives of this research are to build an auto-balanced inverted pendulum and validate the performance and stability of the inverted pendulum by comparing various set of PID setup based on other research. The hypothesis is that any object that moves in circular motion produces angular momentum and it is conserved unless it is subjected to external torque. By using this theory, it is possible to balance an inverted pendulum by placing the reaction wheel at the end of the pendulum. Balancing is done by rotating and stopping the reaction wheel against the falling direction of the pendulum. Stopping the reaction wheel in a short duration, result in high deceleration, hence producing a counter force to balance the inverted pendulum. The best configuration of the PID controller in this research can be obtained by comparing the result in terms of performance and stability [8].

2 Research Methodology

2.1 Design of the inverted pendulum

The inverted pendulum on this research is very similar to reaction wheel discussed earlier. Instead of balancing on one axis, the purpose of this research is to construct a two-axis reaction wheel inverted pendulum. The main components of the design consist of a pendulum body, two reaction wheels, two motors, gyroscope and a microcontroller. The overall design is as shown in Fig. 5.

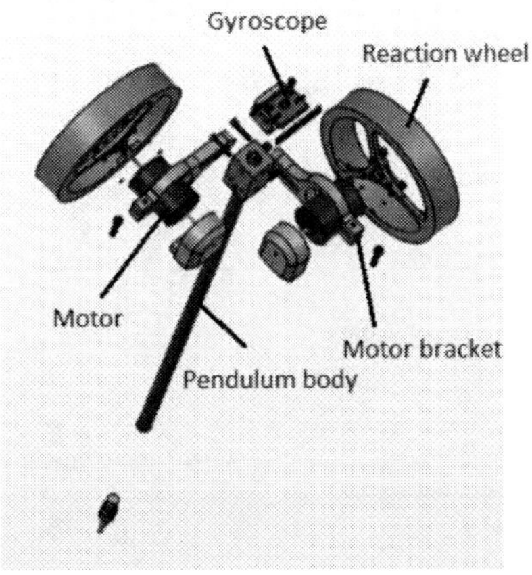

Fig. 5. The overall design of the pendulum.

The pendulum body must be made from light material such as aluminium. Lighter body require lesser force to counter the falling thus, weaker motor can be used. Stronger motor is expensive and big in size. Furthermore, it adds on weight into the system.

Due to budget limitation, smaller motor is used to fit into the budget. The limitation of small motor is that, the torque generated by the motor is small. Thus, a lighter and smaller reaction wheel is used. The wheel has a dimension of 6 cm diameter and 2 cm depth. It is 3D printed with PLA (polylactic acid) filament. The density is set to 30% because of light weight requirement. The height of the pendulum is designed to be approximately 30 cm. This height is decided based on the amount of counter torque generated by the motor.

The torque generated by the motor is 14.7×10^{-3} Nm, this value is based on the specification of the motor. This number is bigger than the torque generated by the weight which is 5.11×10^{-3} Nm. The weight is calculated by multiplying the weight of the pendulum and distance of the pivot point to centre of gravity. The whole pendulum including motor has a mass of 150 grams. The distance of the centre of gravity to the pivot point is approximately 20 cm. Assumption here is the maximum angle the pendulum can be tilted before falling is 10 degrees. Thus, after resolve the weight in vertical component the maximum torque generated by weight is 5.11×10^{-3} Nm. A bigger value is needed because overshoot better than having insufficient enough torque to counter the balance.

Furthermore, the speed of the wheel can be controlled to reduce the amount of overshooting.

As mentioned above the height is designed to be 30 cm because shorter height only generates small amount of torque to counter balance the pendulum. Although greater height generates more torque, however the torque generated by the weight is also increases as the height increases. Thus, 30 cm is the optimum height for the inverted pendulum.

2.2 Electronic components of the inverted pendulum

Electrical components are the vital of the inverted pendulum. Table 1 shows the description of each component.

Direct current (DC) motor is used due to the ability to provide high RPM. The importance of reaction wheel is to create torque, and this is proportional to the acceleration of the wheel. High acceleration can be obtained by changing the velocity in a short time. This can be achieved by DC motor; hence DC motor is used. Since this is the dual axis system, a gyroscope is a better sensor to measure the angular position of the pendulum. Potentiometer is not a suitable selection because it can only read one axis and having two in the system, only result in complex. As for microcontroller, Arduino Uno is selected due to its reliability and low cost. Other alternative such as raspberry pi does not have the same quality as Uno. Furthermore, the software used to tune the PID is Arduino as well, thus more compatible. Mega Arduino is not chosen because of the high cost. With the number of sensors and motors, Uno is sufficient to do the job. In Fig. 6, shows the connections of all the electrical components to the microcontroller.

Table 1. Electrical components in the system.

Electrical components	Description
Actuator	6V DC motor
Sensor	Gyroscope IMU MPU-6050 up to 6 degrees of freedom
Microcontroller	Arduino Uno

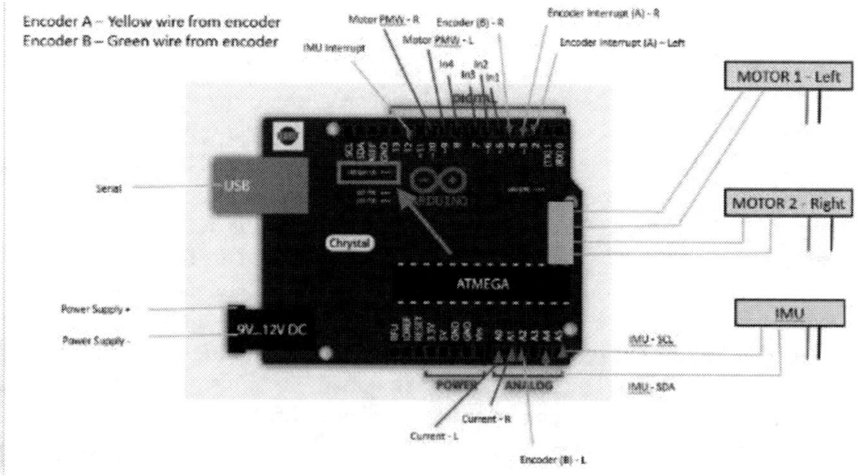

Fig. 6. Electrical connection of the system.

2.3 Mathematical modelling

2.3.1 Modelling using vectorial dynamic

The structure of the system can breakdown into forces to further understand the system. Following Fig. 7 is the schematic of the system. Fig. 8 is the free body diagram of the system showing the point where the force is acting. Here is a list of abbreviation used below,

G_p = pendulum weight
L_ω = length of pivot to wheel
m_ω = wheel mass
L_p = length of pivot to centre of gravity of pendulum
T_m = motor torque
M_p = pendulum mass
G_ω = weight of wheel

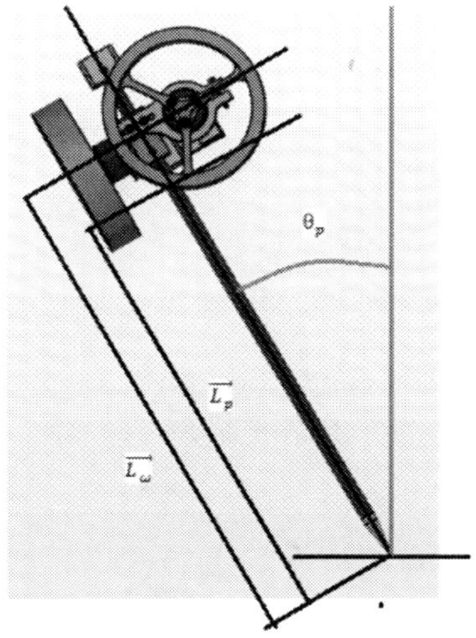

Fig. 7. The schematic of reaction wheel pendulum.

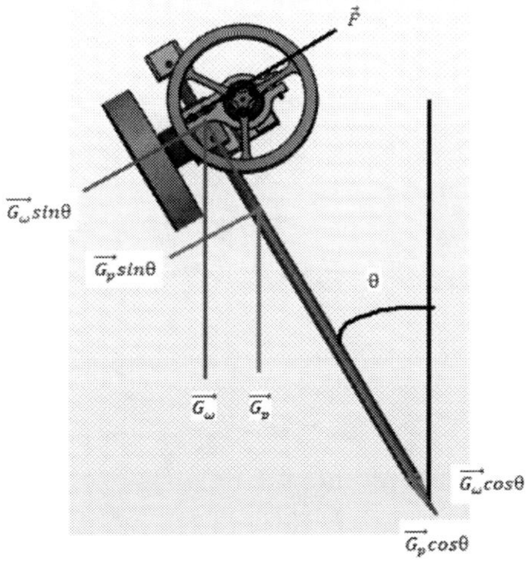

Fig. 8. Forces on the system.

The moment of the system is stated as

$$\overrightarrow{M_F} + \overrightarrow{M_{G_\omega}} + \overrightarrow{M_{G_p}} = I\vec{\epsilon} \tag{2}$$

Where,

ϵ = the acceleration of whole system

I = the moment of inertia

$\overrightarrow{M_F}$ = Moment caused by Force

$\overrightarrow{M_{G_\omega}}$ = Moment caused by wheel

$\overrightarrow{M_{G_p}}$ = Moment caused by pendulum

$$\overrightarrow{M_F} = \vec{R} \times \vec{F} \tag{3}$$

$$\overrightarrow{M_{G_\omega}} = \overrightarrow{L_\omega} \times \overrightarrow{G_\omega} \tag{4}$$

$$\overrightarrow{M_{G_p}} = \overrightarrow{L_p} \times \overrightarrow{G_p} \tag{5}$$

Where R is the radius of the reaction wheel. In this setup, the momentum force, F can be assumed to be approximately same as the torque generated by the motor due to negligible mechanical loses from the motor to the reaction wheel [8]. Thus, Eq. (2) can be written as

$$-T_m + G_\omega L_\omega \sin\theta + G_p L_p \sin\theta = \ddot{\theta} I \tag{6}$$

2.3.2 Modelling using analytical dynamics

To obtain the equation of the system, alternative way such as analytical dynamics is reliable.

The total energy of the system can be break down into the kinetic and potential energy. The difference in kinetic (T) and potential energy (V) is also known as Lagrangian equation.

$$L = T - V \tag{7}$$

Equation of motion expressed in Lagrangian form. This is a general formula,

$$\frac{d}{dt}\left(\frac{\partial L}{\partial \dot{\theta}}\right) - \left(\frac{\partial L}{\partial \theta}\right) = Q \tag{8}$$

Where Q is dissipative force. The force in the system is conservative, thus

$$\frac{d}{dt}\left(\frac{\partial L}{\partial \dot{\theta}}\right) - \left(\frac{\partial L}{\partial \theta}\right) = 0 \tag{9}$$

Moment of inertia is additive,

$$m_t = m_p + m_\omega \tag{10}$$

$$m_t L_t = m_p L_p + m_\omega L_\omega \tag{11}$$

$$I_{total} = I_p + I_\omega = m_p L_p{}^2 + m_\omega L_\omega{}^2 \tag{12}$$

The potential and kinetic energy are expressed as, with the assumption V = 0 at ground level

$$V = m_t gh = m_t g L_t \cos\theta_p = g(m_p L_p + m_\omega L_\omega)\cos\theta_p \tag{13}$$

$$T = \frac{1}{2} I_{total} \dot{\theta}^2{}_p + \frac{1}{2} I_\omega \dot{\theta}^2{}_\omega \tag{14}$$

Substitute the previous two equations into the Eq. (7)

$$L = \frac{1}{2} I_{total} \dot{\theta}^2{}_p + \frac{1}{2} I_\omega \dot{\theta}^2{}_\omega - m_t g L_t \cos\theta_p \tag{15}$$

Partial derivative is calculated with the first generalized coordinate, θ_p

$$\frac{\partial L}{\partial \dot{\theta}_p} = I_{total}\dot{\theta}_p \tag{16}$$

$$\frac{d}{dt}\left(\frac{\partial L}{\partial \dot{\theta}_p}\right) = I_{total}\ddot{\theta}_p = \left(I_p + m_\omega L_\omega{}^2\right)\ddot{\theta}_p \tag{17}$$

On the other hand, $\left(\frac{\partial L}{\partial q_j}\right)$ is express as the moment of inertia

$$\frac{\partial L}{\partial \theta_p} = m_t g L_t \sin\theta_p \tag{18}$$

By substituting equation (17) and (18) to (8), the result is

$$\left(I_p + m_\omega L_\omega{}^2\right)\ddot{\theta}_p - m_t g L_t \sin\theta_p = Q \tag{19}$$

Replace Q with the torque generated by the motor into Eq. (19),

$$\left(I_p + m_\omega L_\omega{}^2\right)\ddot{\theta}_p - m_t g L_t \sin\theta_p = T_m \tag{20}$$

Notice in Eq. (13), the potential energy is in term of cosine. This is because when dealing with the potential energy, only the height is considered, which is the vertical axis.

3 Results and Discussions

The purpose of this research is to identify the effect of PID configuration on the response rate of inverted pendulum. The expected outcome of this research is fluctuation should be considerably low, but the settling time will be high. This is most likely due to the charge up time for the reaction wheel to deliver a counter moment. Unlike the other system where the moment is instantaneous generated by the motor, reaction wheel need to take up speed to generate counter moment.

A simulation of the inverted pendulum is designed by using Simulink. The purpose of the simulation is to theoretically simulate the outcome of the pendulum. The simulated result is then compared with the practical result to evaluate the difference between the theory and practical performance. The Simulink design is shown below in Fig. 9.

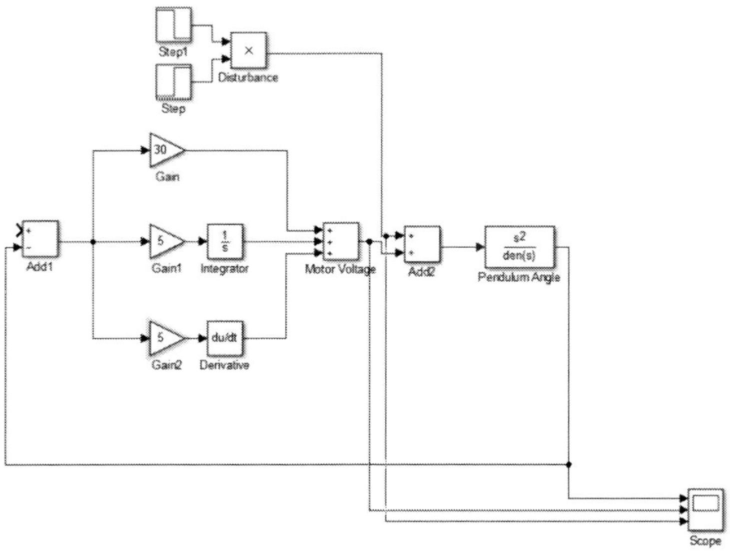

Fig. 9. Design of the Simulink.

A basic PID control system consists of PID, plant, actuator and sensor to feedback the output. In Fig. 10, the loop first starts by having a constant with 0 magnitude going through the PID controller. The PID controller is then produce an output as motor voltage to give specific amount of torque to balance the pendulum. After that the change in angle position is then measured by the sensor and feedback to the PID controller. 'Step' shown in Fig. 9 is the source of the disturbance which made the pendulum to be unbalance. The scope shown in the Simulink design is the result of the response rate and there are three entries to the scope which produces three variables in graph as shown in Fig. 10.

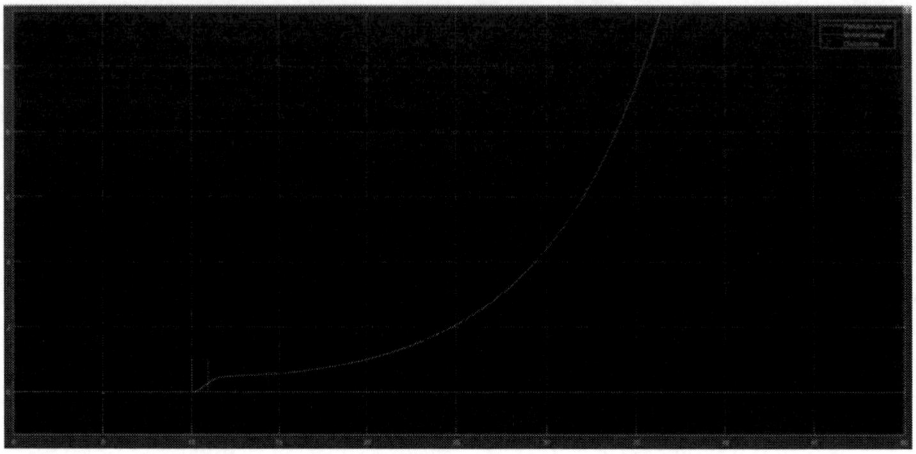

Fig. 10. System response when $K_p = 0$, $K_i = 0$ and $K_d = 0$.

3.1 Ziegler-Nichols tuning method

One of the way to tune an unstable system using PID controller is the Ziegler-Nichols turning method. This method is very common and reliable because tuning can be done without the mathematical models of the system. This method can be used on system with known mathematical modelling as well. Ziegler-Nichols tuning method tune a system by judging the response of the system. To use this method, first, only the K_p (Proportional gain) is tuned to certain amount until the response start oscillating. Second is to tune either K_i (Integral gain) or K_d (Derivative gain) to stabilize the system. Finally, the last step is to tune the last parameter to improve the stability of the system either by reducing the fluctuation or reducing the steady state error [10].

When PID gain is zero. The response shows that the system is unstable as it grows exponentially away from stable point as the disturbance introduced into the system. Yellow line indicates the angle position of the pendulum, red line indicates the disturbance and blue line indicates the response of the motor voltage. With the zero gain, the motor is deactivated thus the voltage response is stagnant as shown in Fig. 10.

Fig. 11 shows when the K_p has a value of 15. The response is similar as the previous response but with a few fluctuations at the beginning.

Fig. 11. System response when $K_p = 15$, $K_i = 0$ and $K_d = 0$.

To start balancing an unbalance system, K_p is commonly the priority to tune. In this case, the proportional gain is tuned to 15 whereas the integral gain and derivative gain remain zero. With the K_p at 15, the angle of the pendulum decreases a slight amount before growing exponentially. This shows that there is a slight improvement to the stability of the system, however the voltage response of the motor has a very steep gradient towards infinity. To counter the pendulum from falling, the motor must generate torque in opposite direction, which is why the voltage response has a mirror pattern against the falling of pendulum. As the voltage response going towards infinity, which means the system require infinite amount of energy to balance the system, practically this is unachievable. Thus, higher K_p is required.

Higher K_p value is input to the simulation and the result is shown in Fig. 12. More fluctuations occurred before system went unstable.

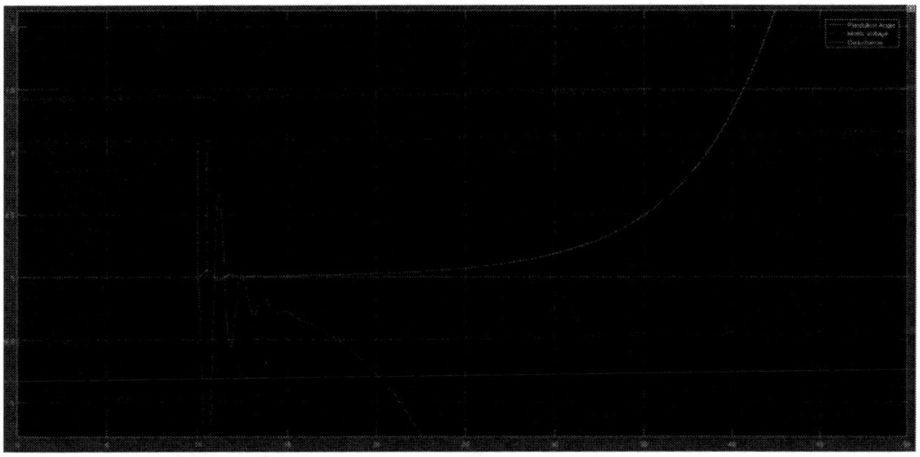

Fig. 12. System response when $K_p = 30$, $K_i = 0$ and $K_d = 0$.

As higher K_p is added, the fluctuation occurrences increase at the beginning then gradually went unstable. The rate which the system became unstable is significantly lower with twice amount of K_p. Voltage motor also share the similar traits. This shows that the system is more stable when higher proportional gain is input into the system. One of the reason the stable time is longer is because motor voltage has more attempts to balance the pendulum. Fluctuations shows in the motor voltage is the result of motor trying to balance the pendulum. However, the lack of other parameter gains it is still unable to balance the pendulum in a longer duration. Thus, other gains are needed for the system to be balanced.

Fig. 13. System response when $K_p = 30$, $K_i = 5$ and $K_d = 0$.

With $K_p = 30$ and $K_i = 5$, the system is no longer unstable. As compared to Fig. 12, this response stays stable after a little fluctuation. The amplitude of the fluctuation is significantly lower as well. On the other hand, the motor voltage has more visible fluctuations and stay constant as the system is stable. The response of the pendulum has a biggest fluctuation when an external force is acted on it. The amplitude of the fluctuation is then reduced gradually until reaches zero, which is the stable point. This indicates that when the external force acted on the pendulum, it swings away from equilibrium. As it is away from equilibrium the motor generates opposite torque to bring the pendulum back to equilibrium, which is why the blue line has a peak when the response of the pendulum has a peak. After the pendulum reaches stability voltage is still supplying the motor to maintain the balance of the system, which explains the motor voltage is not in zero.

Fig. 14. System response when $K_p = 30$, $K_i = 5$ and $K_d = 5$.

All the parameters are input into the PID controller, the system became more stable as shown in Fig. 14. Only one visible fluctuation and the system is balanced without overshooting that may cause another oscillation. With addition derivative component of the PID control, the controller is now more sensitive in change in error. The moment change in error is read by the controller, it reacted quickly to neutral the change. Integral is important to make this happen as well. Only with the help of K_i, the system managed to stabilise with low stead state error else the pendulum will still swing left and right at equilibrium.

After the simulation, theoretically it is proven it is possible to balance the inverted pendulum by using PID controller. Next step will be testing on an actual inverted pendulum. The actual inverted pendulum has finish manufactured as shown in Fig. 15. Same procedure will be applied on the actual inverted pendulum in future test.

Fig. 15. The actual inverted pendulum.

4 Conclusions

The simulation done by Simulink shows concrete proof that an inverted pendulum can be balanced PID controller. With Ziegler-Nichols tuning method, even if the mathematical modelling unknown it is possible to balance it by evaluating the response of the system and tune each gain until a positive result is achieve. In PID, proportional gain is responsible for amplifying the error; integral gain is to reduce the steady state error; derivative is sensitive in change in error. Based on the simulation done in Simulink, high K_p is require to balance the pendulum and the value K_i and K_d can be as low as 5 and it is sufficient for the system to be balanced.

References

1. M.V. Subbotin, Proj. Report, Univ. California, St. Barbar., (2004)

2. C. Walck, Graduate Theses and Dissertations, (2013)
3. M. Olivares, P. Albertos, ISA Trans., **53**, 5 (2014)
4. A. Özbey, E. Uzal, H. Yildiz, Appl. Mech. Mater., **464**, (2014)
5. A. Rahimi, K.D. Kumar, H. Alighanbari, Acta Astronaut., **134**, (2017)
6. C.N. Me-, Dissertation, (2016)
7. S. Kumar, D. Sain, S. Kumar, S. Ghosh, AEUE - Int. J. Electron. Commun., **78**, (2017)
8. A. Tepljakov, E.A. Gonzalez, E. Petlenkov, J. Belikov, C.A. Monje, I. Petr, ISA Trans., **60,** (2016).
9. C. Liu, Bachelor ' s Thesis, Significance, no. 6704071, (2011)
10. T. Yucelen, O. Kayma, S. Kur, IFAC, **39**, 14 (2006)

Evaluation of SIPIC01 and SIPIC02 on Motor Speed Control

Kah Kit Wong[1], *Choon Lih Hoo*[2], *Mohd Hardie Hidayat Mohyi*[1,*]
[1]School of Engineering, Taylor's University, Malaysia
[2]School of Engineering and Physical Sciences, Heriot-Watt University, Malaysia

Abstract. Due to its simplicity, Proportional-Integral (PI) controller still remains as the widely used controller for motor speed control system. However, PI controller exhibits windup phenomenon when the motor operates in a saturated state, which may cause degradation to the control system. In order to overcome the windup phenomenon, many researches have introduced various types of anti-windup methods such as the Conditioning Technique (CI), Tracking Back Calculation (TBC), Integral State Prediction (ISP), Steady-state Integral Proportional Integral Controller-01 (SIPIC01) and Steady-state Integral Proportional Integral Controller-02 (SIPIC02). These are anti-windup techniques with integral control switching mechanism, coupling of proportional gain, k_p, and integral gain, k_i. Due to the coupled k_p and k_i, tuning motor performance is a difficult task with short settling time without experiencing overshoot. SIPIC01 and SIPIC02 are robust anti-windup methods without a switching mechanism and exhibit decoupling feature. SIPIC01 and SIPIC02 have shown better dynamic performance compared to CI, TBC and ISP. However, SIPIC01 has not been compared to SIPIC02 in terms of their decoupling effect flexibility and dynamic performance. The decoupling effect was verified using MATLAB simulation, while the performance analysis was verified through hardware simulation and testing by using Scilab. The results obtained from the simulation showed that both SIPIC01 and SIPIC02 consist of decoupling features that allow a perfomance with coexistence of zero or minimum overshoot with short settling time. However, SIPIC02 consists of longer rise and settling time as compared to SIPIC01. Therefore, it can be concluded that SIPIC01 is better than SIPIC02 in term of dynamic performance.

1. Introduction

A control system is one of the important components in a system which is responsible in producing desired output. Control system comprises of open-loop and closed-loop systems.

Open-loop control system is a simple control system without any feedback loop in which the input is totally independent from the output of a system [1]. Other than its

[1] Corresponding author: hardie.hidayat@taylors.edu.my

simplicity, open-loop control system is generally stable and economical. However, the output of the open-loop control system is inaccurate and unreliable. This is due to the system disability to auto-correct any error signals since there is no feedback loop. In contrast, a closed-loop control system is a control system with feedback loop ability [1]. This type of control system contains the error auto-correction operation that gives a more accurate and reliable output. However, the design of a closed-loop control system is complicated.

In modern control, Artificial Intelligence (AI) is a type of closed-loop control system that utilises the complex computation mechanism to imitate human thinking and thought process. Fuzzy logic controller (FLC) [2], artificial neural network (ANN) [3], and genetic algorithm (GA) [4] are some of the commonly discussed AI controls which are robust with good dynamic performance in motor speed control. Due to the classical Proportional Integral (PI) controller ease of use and simplicity in applications, it is still remain as the current widespread controller choice for motor speed control system.

However, PI controller experiences windup which normally happens when the output of the controller exceed the input limit of the plant which caused the system to operates in a saturated state [5]. This phenomenon will lead to poor control response such as increased overshoot, long settling time and system instability.

In order to improve the performance of the PI controller, the windup phenomenon should be eliminated. There are many anti-windup techniques developed to avoid windup phenomenon such as the Conditional Integration (CI), Tracking Back Calculation (TBC), Integral State Prediction (ISP), Steady-state Integral Pi Controller-01 (SIPIC01) and Steady-state Integral Pi Controller-02 (SIPIC02) [5]. Currently, both SIPIC01 and SIPIC02 have been compared with the existing anti-windup PI controllers and the result shows that both of them perform better in term of dynamic performance on motor speed control [6]. However, the evaluation between the two has yet to be studied. Hence, the dynamic performance between SIPIC01 and SIPIC02 is investigated in this study. This work serves to expand the knowledge of the field of anti-windup PI controller.

2. Types of Anti-Windup PI Controller

In order to overcome the windup phenomenon, many researches introduced different types of anti-windup schemes. Out of the schemes available; CI, TBC and ISP appeared to be the choice among researchers. Though they operate differently by altering their integral control component in the saturated state, they have the similarity that resume to the conventional PI controller under unsaturated state. Consequently, they experience difficulty in tuning when operating as PI controller due to their coupled tuning gains [5].

CI, TBC and ISP have similar integral control under unsaturated state but consist of different integral control under saturated state. Table 1 shows the integral control of the respective anti-windup method under different states. Table 2 shows the advantages and disadvantages of these anti-windup methods [5]. The e, k_a, u, v, ωi, q_{ss} and q shown in Table 1 denote the input error, anti-windup gain, controller output, plant input, positive parameter of the low-pass filter, steady state integral component and integral state respectively.

Table 1. Comparison of integral control of anti-windup schemes under different states

Controllers	Integral Control Under Unsaturated State, u = v	Integral Control Under Saturated State, u ≠ v	Switching Mechanism
Conditional Integration (CI)	e	0	Yes

Tracking Back Calculation (TBC)	e	$e - k_a (u - v)$	Yes
Integral State Prediction (ISP)	e	$\omega_i (q_{ss} - q)$	Yes

Table 2. The advantages and disadvantages of anti-windup schemes

Controllers	Advantages	Disadvantages
Conditional Integration (CI)	• No overshoot	• Integral value is large when the controller output differ from the plant input. • Difficult to choose value of gain • Consists of coupling effect
Tracking Back Calculation (TBC)	• Large value of anti-windup gain, k_a is allowed to be chosen • Improve overshoot	• Error will occur if the gain is too big. • Consists of coupling effect
Integral State Prediction (ISP)	• Steady state value is predicted to prevent state change.	• Integrator value is limited. • Consists of coupling effect

3. Generalisation of PI Controllers

The main purpose of generalisation of PI controllers is to develop a generic expression which can be used to represent all the PI-related control methods. According to [7], the function $k_i k_t(q_{SS} - q)$ can generalise any PI-related controller and the general equation is shown in Equation (1) where l, u, v and w are natural numbers while A_l, B_u and C are integers.

$$k_i k_t(q_{SS} - q) = \sum_{l=0}^{l=v} A_l q^{(l)} + \sum_{u=0}^{u=w} B_u e^{(u)} + C \quad (1)$$

The k_i, k_t, q_{ss}, q and e represent the integral gain, external load constant, steady state integral component, integral state and input error respectively while the l and u denote the number of coefficient of A and B for q and e respectively. Both l and u also denote the number of order for q and e respectively while C represent any constant. Each term in the function of $k_i k_t(q_{SS} - q)$ carries certain characteristic of the control system. For example, the existence of the constant C will give a pole at the origin which will result in steady state error [7]. In order to obtain a zero steady state error control performance, Equation (2) must be met.

$$\lim_{s \to 0} k_i k_t s \left(\frac{q_{SS}}{s} - Q(s)\right) = 0 \quad (2)$$

Following this condition (2), the expression of $\frac{q_{SS}}{s} - Q(s)$ needs to be reduced into a function of s which consists of no pole at the origin [7]. However, with the presence of constant C in the function of $k_i k_t(q_{SS} - q)$, Equation (2) will never be satisfied and the steady state error cannot be eliminated.

Besides, the existence of error term, e will lead to the coupling of k_p and k_i tuning parameters which increase the difficulty to obtain a performance of zero overshoot and short settling time. According to [7], the absence of the error term will decouple the k_p and k_i which allows a performance with coexistence of zero overshoot and short settling time.

4. Steady-state Integral Proportional Integral Controller 01 (SIPIC01)

According to the characteristic of the term in the general Equation (1) stated in section 3, Hoo et al. proposed a structure of the controller as shown in Equation (3) where A_1 can be any constant [8], [9].

$$k_i k_t (q_{ss} - q) = A_1 \dot{q} \tag{3}$$

In order to develop a controller with no steady state error and decoupling effect, A_1 can be equal to k_t as A_1 can be any constant. By substituting $A_1 = k_t$ into Equation (3), Equation (4) which describes the integral control of SIPIC01 will be formed. The Laplace transformation of Equation (4), with *Q(s)* the subject is shown in Equation (5).

$$k_i (q_{ss} - q) = \dot{q} \tag{4}$$

$$Q(s) = \frac{q(0) + k_i \frac{q_{ss}}{s}}{s + k_i} \tag{5}$$

Equation (6) is a second order error dynamic equation (EDE) formed from a common first order system plant, *P*(s) controlled by a PI controller as shown in Figure 1, where *V*(s) and T_L(s) are the control output and external load respectively. The error equation of SIPIC01 (7) can be derived through the substitution of Equation (5) into Equation (6). Equation (8) corresponds to the damping ratio of SIPIC01, which is formed from the denominator of Equation (7) by comparing with the general characteristic equation, $s^2 + 2\zeta\omega s + \omega^2$, where ζ and ω are the damping ratio and natural frequency respectively.

$$E(s) = \frac{e(0)\left(\frac{b}{f} + k_d k_t\right) + k_i k_t \left(\frac{q_{ss}}{s} - Q(s)\right)}{\left(\frac{b}{f} + k_d k_t\right)s + \frac{a}{f} + k_p k_t} \tag{6}$$

$$E(s) = \frac{e(0)\frac{b}{f}}{\left(\frac{b}{f}s + \frac{a}{f} + k_p k_t\right)} + \frac{k_i k_t (q_{ss} - q(0))}{(s + k_i)\left(\frac{b}{f}s + \frac{a}{f} + k_p k_t\right)} \tag{7}$$

$$\zeta = \frac{a + f k_p k_t + b k_i}{2\sqrt{b}\sqrt{a k_i + f k_p k_t k_i}} \tag{8}$$

As shown in Equation (7), there is no pole at the origin and this indicates that there is no steady state error. Besides, only k_p controls the $e(0)$ while k_p and k_i control the $k_i k_t (q_{ss} - q(0))$ in two separate terms. Hence, it can be concluded that SIPIC01 consists of decoupling effect which allows the controller to obtain short settling and rise time without experiencing overshoot.

5. Steady-state Integral Proportional Integral Controller 02 (SIPIC02)

Chiah at al. [6] proposed another anti wind-up PI controller based on general Equation (1) with the structure of the controller shown in Equation (9) where A_1 and A_2 can be any constant.

$$k_i k_t (q_{ss} - q) = A_1 \dot{q} + A_2 e \tag{9}$$

One of the possible option for A_1 and A_2 is k_t. By substituting $A_1 = A_2 = k_t$ into Equation (9), the integral control of the SIPIC02 can be obtained (10). The Laplace transformation of Equation (10), with *Q(s)* the subject is shown in Equation (11).

$$k_i (q_{ss} - q) = \dot{q} + e \tag{10}$$

$$Q(s) = \frac{q(0) + k_i \frac{q_{ss}}{s} - E(s)}{s + k_i} \tag{11}$$

From Equation (11) and Equation (6) with zero derivative gain, the error equation of SIPIC02 is shown in Equation (12). Equation (13) is the damping ratio of SIPIC02 obtained from the denominator of Equation (12) by comparing with the characteristic equation.

$$E(s) = \frac{e(0)\frac{b}{f}(s + k_i) + k_i k_t (q_{ss} - q(0))}{\left(\frac{b}{f}s + \frac{a}{f} + k_p k_t\right)(s + k_i) - k_i k_t} \tag{12}$$

$$\zeta = \frac{a + f k_p k_t + b k_i}{2\sqrt{b}\sqrt{a k_i + f k_p k_t k_i - f k_i k_t}} \tag{13}$$

As shown in Equation (12), there is no pole at the origin and this indicates that there is no steady state error. Theoretically, SIPIC02 consists of coupling effect which will prevent the controller from achieving a performance of zero overshoot and short settling and rise time. This is because the $k_i k_t$ in Equation (12) does not allow distinctive separable factors in the denominator. The coexistence of k_p and k_i in the pole hinder the decoupling. However, according to Chiah et al. [6], SIPIC02 shows a promising performance with short settling and rise time while maintaining low or zero overshoot.

6. Simulation and Experimental Testing

The theoretical analysis of SIPIC01 and SIPIC02 done so far will be verified in this section by conducting simulation and hardware testing. The simulation for decoupling effect will be described in section 6.1 while the hardware simulation and experiment testing for speed control performance will be described in section 6.2.

6.1 Simulation for Decoupling Effect

MATLAB/Simulink software was used in to verify the decoupling effect of SIPIC01 and SIPIC02 through simulation for a second order EDE system. The block diagram of the simulation for decoupling mode is illustrated in Figure 2. For simplicity, the other parameters are set to unity, the input is set as 1000 rpm and the system plant is set as $P(s) = \frac{1}{s+1}$. The simulation is conducted under two conditions which are with and without load. For simulation with load, the load, L is set as 200. The simulation for both conditions are repeated for input equal to 2000 rpm.

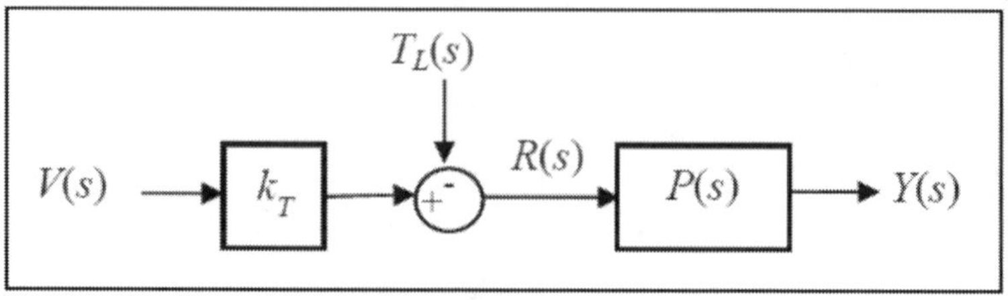

Fig 1. The block diagram for general control system

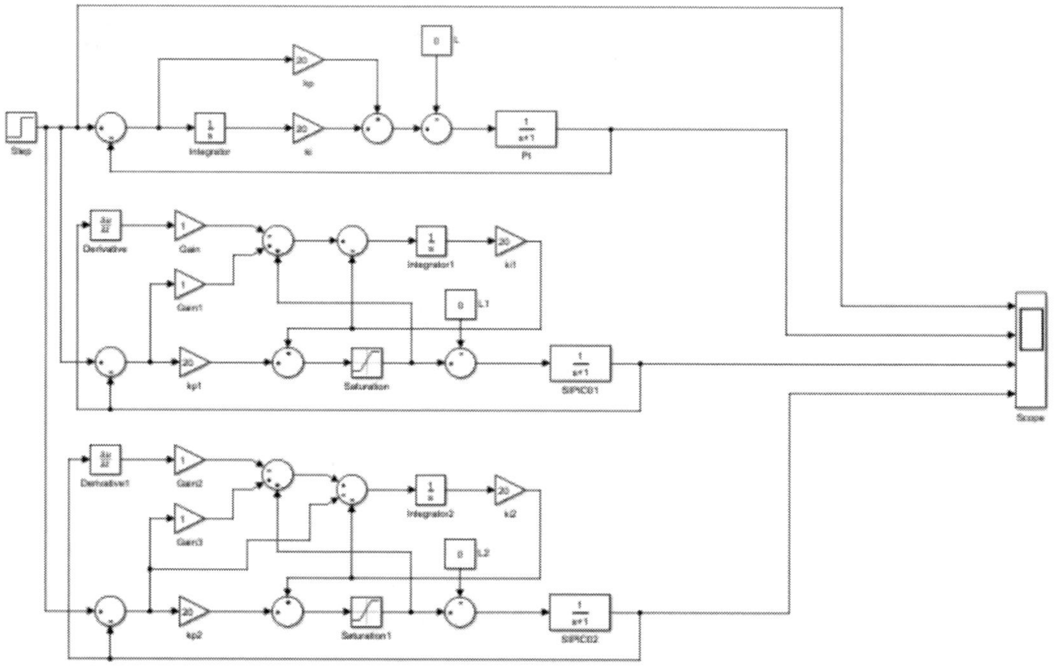

Fig. 2. The block diagram of the simulation for decoupling mode

6.2 Simulation and Experiment Testing for Control Performance

The simulation testing of SIPIC01 and SIPIC02 for control performance is done through Scilab/Scicoslab. They are tested under three cases which are no load, load 1 and load 2. Load 1 is a mild steel black plating with moment of inertia of 8.63×10^{-5} kgm^2 whereas the material for load 2 is an aluminium plating with moment of inertia of 2.83×10^{-5} kgm^2. Figure 3 shows the block diagram used for the simulation testing of SIPIC01 and SIPIC02. In order to ensure that the model is reliable, system identification is performed by using Scilab/Scicoslab. The simulation result is later compared with the real hardware testing result.

For the experimental testing, the setup for the DC servo motor speed control is shown in Figure 4. The setup includes Scilab/Scicoslab, DC servo motor, host controller computer, load tuner, motor encoder, power amplifier, terminal system unit and real time interfacing platform data acquisition. Similar to the simulation, the experimental testing for SIPIC01 and SIPIC02 are also tested under the three cases. For each case, the testing was done for two speeds input, which are 50 rad/s and 100 rad/s. The specification of the DC motor is shown in Table 3.

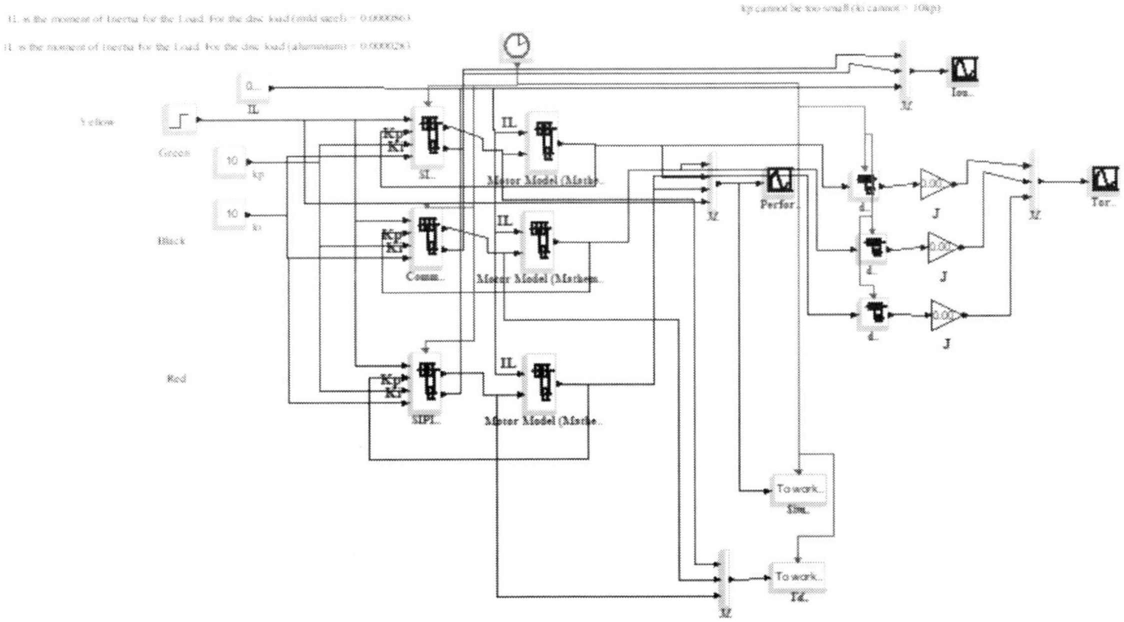

Fig. 3. The block diagram for the simulation testing

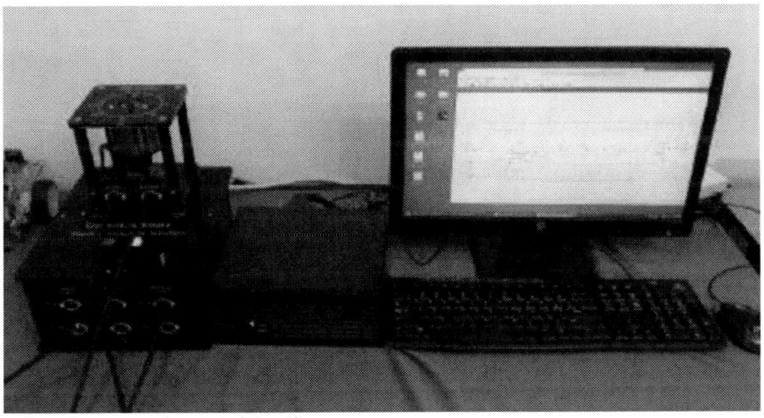

Fig. 4. Setup for the experimental testing

Table 3. Specification of the DC motor

Characteristics	Values
Maximum supply voltage	40 Vdc
Maximum continuous torque	14 Ncm
Maximum peak torque	36 Ncm
Motor voltage constant	10.3 V at 1000rpm

Motor torque constant	9.0 Ncm/A
Mechanical time constant	20ms
Rotor inertia	0.214 kgcm
Terminal resistance	7.8 Ohms
Rated speed	1600 rpm
No load speed	2600 rpm @ 24 Vdc
Rated torque	12 Ncm
Peak torque	27 Ncm

7. Simulation Result

7.1 Decoupling Effect

Figures 5 and 6 show the result of simulation for decoupling effect for $k_p = 1$, $k_i = 20$ at 1000 rpm under no load and loading condition respectively. PI has the shortest rise time and greatest overshoot, while both SIPIC01 and SIPIC02 have no overshoot. The response of simulation for decoupling effect for PI, SIPIC01 and SIPIC02 at 1000 rpm and 2000 rpm under no load condition is shown in Table 4. The response of simulation for decoupling effect under loaded condition is not shown but will be discussed in this section.

As shown in Table 4, the rise time and settling time for PI reduced with increasing k_p or k_i gains. However, the overshoot percentage for PI increased when k_i is increased. This observation can be explained based on the damping ratio equation (14). By referring to Equation (14), the damping ratio of PI is directly proportional to k_p but inversely proportional to k_i. The damping ratio of PI increases with increasing k_p but conversely with increasing k_i.

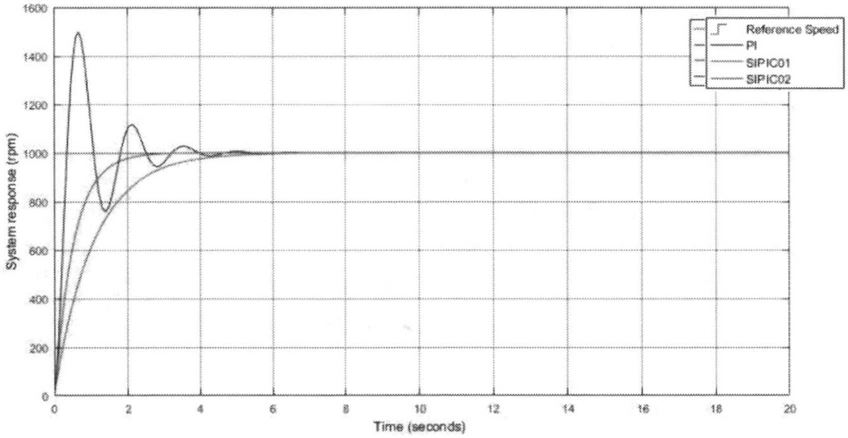

Fig. 5. Simulation for decoupling effect for $k_p = 1$, $k_i = 20$ at 1000 rpm under no load condition.

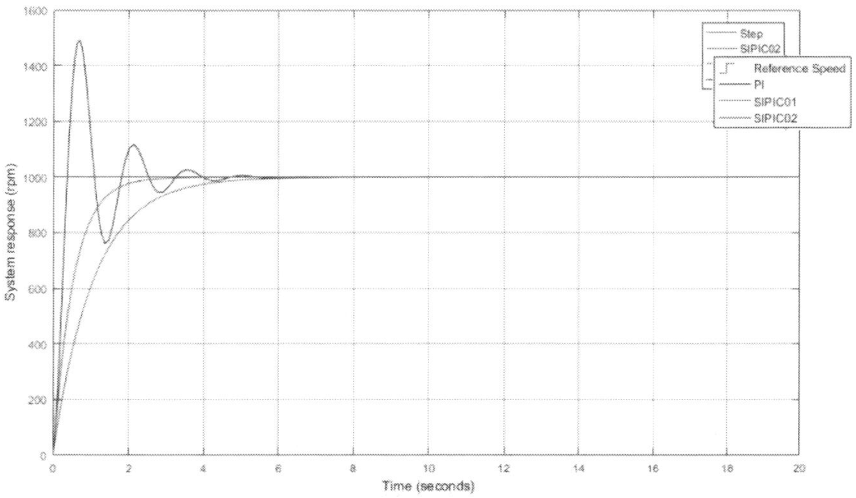

Fig. 6. Simulation for decoupling effect for $k_p = 1$, $k_i = 20$ at 1000 rpm under loaded condition.

The system is said to experience overshoot when the damping ratio is < 1. The lower the damping ratio, the higher the percentage of overshoot.

$$\zeta = \frac{a+fk_pk_t}{2b\sqrt{\frac{f}{b}k_ik_t}} \qquad (14)$$

As for SIPIC01 and SIPIC02, their rise time and settling time also reduced with increasing k_p or k_i gains. However, both of them do not experience overshoot. This phenomenon can be explained based on Equations (8) and (13). Since the tuning parameters, k_p and k_i exist in the numerator and denominator, both of them increase or decrease together with the damping ratio. As mentioned above, increasing k_p or k_i gains will reduce the rise and settling time of SIPIC01 and SIPIC02. However, increasing k_p or k_i gains will only have slight effect on the damping ratio due to the existence of k_p and k_i gains in the numerator and denominator of Equations (8) and (13). Hence, it can be said that both SIPIC01 and SIPIC02 consist of decoupling features that allows a performance with coexistence of zero overshoot and short settling time. By comparing SIPIC01 and SIPIC02, SIPIC02 consists of longer rise and settling time. This is due to the $-fk_ik_t$ term in the denominator of Equation (13). With the $-fk_ik_t$ term, the denominator of Equation (13) will always be smaller as compared to the denominator of Equation (8). This will cause the SIPIC02 to have a greater damping ratio as compared to SIPIC01. The over damped response with greater damping ratio will lead to longer rise and settling time. By increasing k_p or k_i gains, the rise and settling time of SIPIC02 have a greater amount of reduction as compared to SIPIC01. The performance under loaded condition is similar to the result under no load condition.

7.2 Control Performance

The simulation for speed control at 100 rad/s for k_p = 5, k_i=10 under no load condition is shown in Figure 7 while the experimental testing for speed control at 100 rad/s for k_p = 5, k_i=10 under no load condition is shown in Figure 8. Based on Figure 7, PI consists of the shortest rise and settling time, whereas SIPIC02 has the longest rise and settling time. PI experienced overshoot while both SIPIC01 and SIPIC02 did not experience overshoot. Tables 5 and 6 summarise the rise time, settling time and percentage overshoot of PI, SIPIC01 and SIPIC02 under no load condition. According to Table 5, increasing k_p or k_i gains will reduce the rise and settling time for PI, SIPIC01 and SIPIC02. When the k_p is 1, PI has the shortest rise and settling time while SIPIC02 has the longest rise and settling time. The controllers are said to have reached their lowest boundary of rise and settling time when the k_p is ≥ 5. Therefore, all controllers consist of similar rise and settling time. According to Table 6, there is no overshoot for all controllers at 50 rad/s under no load condition. The same performance can be observed when the controllers are tested at 100 rad/s under the same condition. However, PI experienced overshoot when k_i is equal to 10. When k_p increases, the percentage overshoot decreases. This can be explained by Equation (14) where the k_p is directly proportional to the damping ratio. When the damping ratio increases, the percentage overshoot decreases. The rise time and settling time for both SIPIC01 and SIPIC02 reduced by increasing k_p or k_i gains without experiencing overshoot. This is due to the decoupling effect of k_p or k_i gains in SIPIC01 and SIPIC02. By comparing SIPIC01 and SIPIC02, SIPIC01 has a much shorter rise and settling time while both of them do not experience overshoot. Hence, it can be said that SIPIC01 is better in dynamic performance as compared to SIPIC02.

Table 4. The response of simulation for decoupling effect for PI, SIPIC01 and SIPIC02 under no load condition

k_p	k_i	1000 rpm									2000 rpm								
		Rise Time (s)			Settling Time (s)			Overshoot (%)			Rise Time (s)			Settling Time (s)			Overshoot (%)		
		PI	SIPIC01	SIPIC02	PI	SIPIC01	SIPIC02	PI	SIPIC01	SIPIC02	PI	SIPIC01	SIPIC02	PI	SIPIC01	SIPIC02	PI	SIPIC01	SIPIC02
1	1	2.196	2.196	5.069	2.996	2.997	6.997	0	0	0	2.196	2.196	5.069	2.996	2.997	6.997	0	0	0
1	5	0.595	1.202	2.696	2.138	1.642	3.667	23.444	0	0	0.595	1.202	2.696	2.138	1.642	3.667	23.444	0	0
1	10	0.392	1.128	2.435	2.42	1.558	3.316	37.060	0	0	0.392	1.128	2.435	2.420	1.558	3.316	37.060	0	0
1	15	0.309	1.111	2.352	2.712	1.536	3.205	44.739	0	0	0.309	1.111	2.352	2.712	1.536	3.205	49.926	0	0
1	20	0.263	1.105	2.312	2.952	1.526	3.152	49.925	0	0	0.263	1.105	2.312	2.952	1.526	3.152	49.926	0	0
5	1	2.201	0.752	0.983	6.264	1.392	1.846	0	0	0	2.201	0.752	0.983	6.264	1.392	1.846	0	0	0
5	5	0.438	0.439	0.555	0.600	0.601	0.786	0	0	0	0.438	0.439	0.555	0.600	0.601	0.786	0	0	0
5	10	0.318	0.394	0.496	1.088	0.534	0.681	6.702	0	0	0.318	0.394	0.495	1.088	0.534	0.681	6.702	0	0
5	15	0.265	0.380	0.475	1.079	0.518	0.650	11.925	0	0	0.265	0.380	0.475	1.079	0.518	0.650	11.925	0	0
5	20	0.233	0.374	0.465	0.977	0.512	0.636	16.048	0	0	0.233	0.374	0.465	0.977	0.512	0.636	16.049	0	0
10	1	0.345	0.307	0.329	5.656	0.704	0.857	0	0	0	0.345	0.324	0.350	5.656	0.707	0.883	0	0	0
10	5	0.261	0.240	0.268	0.455	0.342	0.400	0	0	0	0.261	0.259	0.289	0.455	0.366	0.430	0	0	0
10	10	0.219	0.219	0.248	0.300	0.301	0.348	0	0	0	0.219	0.242	0.270	0.300	0.331	0.379	0	0	0
10	15	0.195	0.211	0.239	0.252	0.288	0.330	2.564	0	0	0.195	0.236	0.261	0.252	0.323	0.363	2.564	0	0
10	20	0.178	0.207	0.234	0.225	0.283	0.321	4.884	0	0	0.178	0.234	0.257	0.225	0.32	0.355	4.884	0	0
15	1	0.187	0.184	0.189	2.650	0.360	0.409	0	0	0	0.187	0.231	0.237	2.65	0.395	0.456	0	0	0
15	5	0.170	0.164	0.017	0.294	0.240	0.266	0	0	0	0.170	0.210	0.222	0.294	0.289	0.320	0	0	0
15	10	0.156	0.153	0.166	0.228	0.216	0.239	0	0	0	0.156	0.202	0.214	0.228	0.271	0.294	0	0	0
15	15	0.145	0.149	0.161	0.200	0.207	0.228	0	0	0	0.145	0.200	0.210	0.200	0.267	0.286	0	0	0
15	20	0.137	0.146	0.158	0.183	0.203	0.222	1.353	0	0	0.137	0.199	0.208	0.183	0.265	0.282	1.353	0	0
20	1	0.130	0.139	0.140	0.251	0.232	0.244	0	0	0	0.130	0.200	0.203	0.251	0.296	0.312	0	0	0
20	5	0.124	0.129	0.135	0.202	0.189	0.203	0	0	0	0.124	0.191	0.197	0.202	0.255	0.272	0	0	0
20	10	0.118	0.124	0.131	0.176	0.175	0.189	0	0	0	0.118	0.188	0.193	0.176	0.245	0.259	0	0	0
20	15	0.113	0.121	0.128	0.161	0.169	0.182	0	0	0	0.113	0.187	0.191	0.161	0.243	0.254	0	0	0
20	20	0.109	0.120	0.127	0.150	0.166	0.178	0	0	0	0.109	0.186	0.190	0.150	0.242	0.251	0	0	0

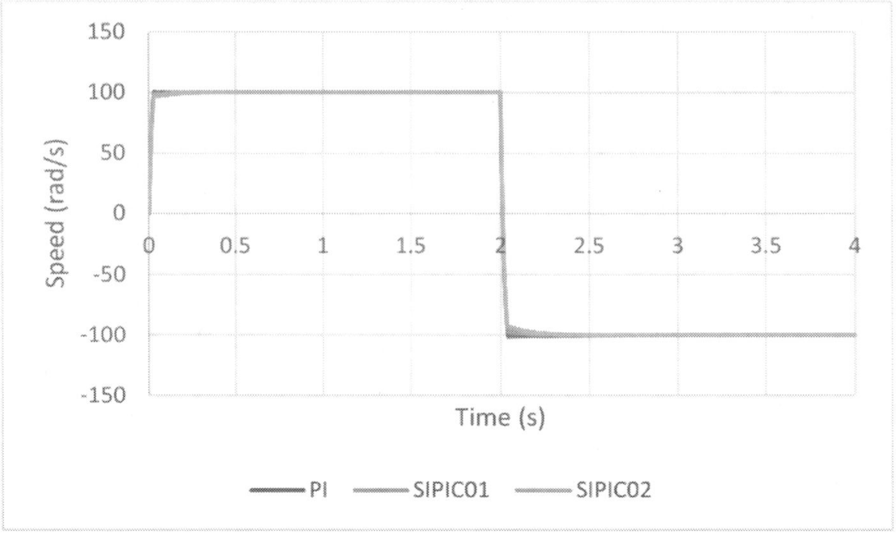

Fig. 7. Simulation for speed control at 100 rad/s for $k_p = 5$, $k_i = 10$ under no load condition.

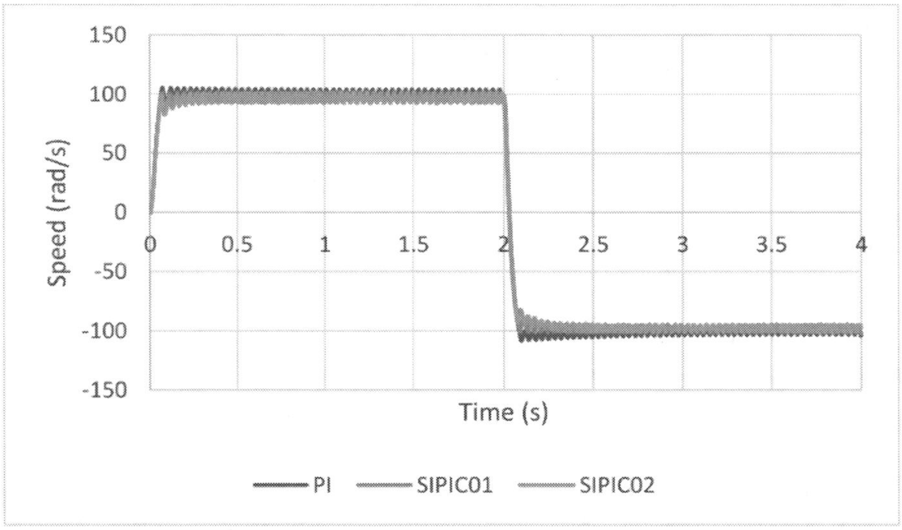

Fig. 8. Experimental testing for speed control at 100 rad/s for $k_p = 5$, $k_i = 10$ under no load condition.

Tables 7 and 8 summarise the rise time, settling time and percentage overshoot of PI, SIPIC01 and SIPIC02 under load 1 (mild steel) condition. According Table 7, the average rise and settling time of PI, SIPIC01 and SIPIC02 are longer compared to no load condition which is due to the higher moment of inertia. The PI experienced inconsistent settling time with increasing k_p or k_i. This is due to the coupling effect of k_p or k_i as both of them affect each other in the response. According to Table 8, SIPIC01 experience insignificant amount of overshoot which is less than 1.2 % when k_i is equal to 10. Tables 9 and 10 summarise the rise time, settling time and percentage overshoot of PI, SIPIC01 and SIPIC02 under load 2 (aluminium) condition. As shown in Table 9 and 10, the controllers have the similar performance to the load 1 condition. The only difference is that all controllers have shorter average rise and settling time which is due to the lower moment of inertia. Besides, SIPIC02 experienced overshoot under load 2 condition. However, the amount is insignificant.

Table 5. The rise time and settling time of PI, SIPIC01 and SIPIC02 under no load condition

Speed (rad/s)	k_p	k_i	Rise Time (s)			Settling Time (s)		
			PI	SIPIC01	SIPIC02	PI	SIPIC01	SIPIC02
50	1	1	0.009	0.011	0.518	0.742	0.708	2.427
	1	5	0.008	0.009	0.377	0.106	0.143	1.758
	1	10	0.007	0.008	0.334	0.016	0.072	1.032
	5	1	0.006	0.006	0.006	0.010	0.010	0.010
	5	5	0.006	0.006	0.006	0.010	0.010	0.010
	5	10	0.006	0.006	0.006	0.010	0.010	0.010
	10	1	0.006	0.006	0.006	0.010	0.010	0.010
	10	5	0.006	0.006	0.006	0.010	0.010	0.010
	10	10	0.006	0.006	0.006	0.010	0.010	0.010
100	1	1	0.021	0.023	0.885	0.690	0.709	2.341
	1	5	0.020	0.021	0.678	0.029	0.143	2.021
	1	10	0.020	0.020	0.564	0.027	0.073	1.263
	5	1	0.020	0.020	0.020	0.027	0.027	0.027
	5	5	0.020	0.020	0.020	0.027	0.027	0.027
	5	10	0.020	0.020	0.020	0.027	0.027	0.027
	10	1	0.020	0.020	0.020	0.027	0.027	0.027
	10	5	0.020	0.020	0.020	0.027	0.027	0.027
	10	10	0.020	0.020	0.020	0.027	0.027	0.027

Table 6. The overshoot percentage of PI, SIPIC01 and SIPIC02 under no load condition

Speed (rad/s)	k_p	k_i	Overshoot (%)		
			PI	SIPIC01	SIPIC02
50	1	1	0	0	0
	1	5	0	0	0
	1	10	0	0	0
	5	1	0	0	0
	5	5	0	0	0
	5	10	0	0	0
	10	1	0	0	0
	10	5	0	0	0
	10	10	0	0	0
100	1	1	0	0	0
	1	5	0	0	0
	1	10	0.693	0	0
	5	1	0	0	0
	5	5	0	0	0
	5	10	0.462	0	0
	10	1	0	0	0
	10	5	0	0	0
	10	10	0.221	0	0

Table 7. The rise time and settling time of PI, SIPIC01 and SIPIC02 under load 1 condition

Speed (rad/s)	k_p	k_i	Rise Time (s)			Settling Time (s)		
			PI	SIPIC01	SIPIC02	PI	SIPIC01	SIPIC02
50	1	1	0.048	0.054	1.084	0.565	0.632	2.226
	1	5	0.038	0.042	0.792	0.051	0.067	2.001
	1	10	0.037	0.038	0.640	0.155	0.052	1.336
	5	1	0.037	0.037	0.037	0.047	0.047	0.047
	5	5	0.037	0.037	0.037	0.047	0.047	0.047
	5	10	0.037	0.037	0.037	0.047	0.047	0.047
	10	1	0.037	0.037	0.037	0.047	0.047	0.047
	10	5	0.037	0.037	0.037	0.047	0.047	0.047
	10	10	0.037	0.037	0.037	0.047	0.047	0.047
100	1	1	0.107	0.109	2.001	0.160	0.630	1.830
	1	5	0.107	0.107	1.619	0.424	0.131	2.142
	1	10	0.107	0.107	1.130	0.445	0.130	1.830
	5	1	0.107	0.107	0.107	0.130	0.130	0.130
	5	5	0.107	0.107	0.107	0.130	0.130	0.130
	5	10	0.107	0.107	0.107	0.460	0.130	0.130
	10	1	0.107	0.107	0.107	0.130	0.130	0.130
	10	5	0.107	0.107	0.107	0.130	0.130	0.130
	10	10	0.107	0.107	0.107	0.130	0.130	0.130

Table 8. The overshoot percentage of PI, SIPIC01 and SIPIC02 under load 1 condition

Speed (rad/s)	k_p	k_i	Overshoot (%)		
			PI	SIPIC01	SIPIC02
50	1	1	0	0	0
	1	5	0.822	0	0
	1	10	9.934	0.339	0
	5	1	0	0	0
	5	5	0.362	0	0
	5	10	2.694	0.296	0
	10	1	0	0	0
	10	5	0.829	0	0
	10	10	1.974	0.804	0
100	1	1	0	0	0
	1	5	13.417	0	0
	1	10	23.940	1.173	0
	5	1	0	0	0
	5	5	3.406	0	0
	5	10	8.743	0.353	0
	10	1	0	0	0
	10	5	1.834	0	0
	10	10	4.608	0.302	0

Table 9. The rise time and settling time of PI, SIPIC01 and SIPIC02 under load 2 condition

Speed (rad/s)	k_p	k_i	Rise Time (s)			Settling Time (s)		
			PI	SIPIC01	SIPIC02	PI	SIPIC01	SIPIC02
50	1	1	0.022	0.025	0.728	0.685	0.684	2.398
	1	5	0.017	0.020	0.544	0.031	0.116	1.925
	1	10	0.016	0.018	0.465	0.024	0.041	1.164
	5	1	0.016	0.016	0.016	0.022	0.022	0.022
	5	5	0.016	0.016	0.016	0.022	0.022	0.022
	5	10	0.016	0.016	0.016	0.022	0.022	0.022
	10	1	0.016	0.016	0.016	0.022	0.022	0.022
	10	5	0.016	0.016	0.016	0.022	0.022	0.022
	10	10	0.016	0.016	0.016	0.022	0.022	0.022
100	1	1	0.049	0.052	1.572	0.547	0.685	2.174
	1	5	0.049	0.049	1.088	0.060	0.115	2.103
	1	10	0.049	0.049	0.834	0.194	0.061	1.533
	5	1	0.049	0.049	0.049	0.060	0.060	0.060
	5	5	0.049	0.049	0.049	0.060	0.060	0.060
	5	10	0.049	0.049	0.049	0.060	0.060	0.060
	10	1	0.049	0.049	0.049	0.060	0.060	0.060
	10	5	0.049	0.049	0.049	0.060	0.060	0.060
	10	10	0.049	0.049	0.049	0.060	0.060	0.060

Table 10. The overshoot percentage of PI, SIPIC01 and SIPIC02 under load 2 condition

Speed (rad/s)	k_p	k_i	Overshoot (%)		
			PI	SIPIC01	SIPIC02
50	1	1	0	0	0
	1	5	0	0	0
	1	10	0.697	0	0
	5	1	0	0	0
	5	5	0.635	0.081	0
	5	10	1.723	0.485	0
	10	1	1.325	1.263	1.132
	10	5	1.772	1.496	0.888
	10	10	2.144	1.697	0.588
100	1	1	0	0	0
	1	5	1.924	0	0
	1	10	12.639	0	0
	5	1	0	0	0
	5	5	0.746	0	0
	5	10	3.351	0	0
	10	1	0	0	0
	10	5	0.650	0	0
	10	10	1.959	0.095	0

Based on the rise time, settling time and overshoot percentage under three load cases, it can be said that SIPIC01 and SIPIC02 are better in dynamic performance as compared to PI. By comparing SIPIC01 and SIPIC02, SIPIC01 has a much faster rise and settling time. Although the percentage overshoot of SIPIC01 is slightly greater as compared to SIPIC02, the amount of overshoot experienced by SIPIC01 and SIPIC02 are insignificant. Therefore, it can be said that SIPIC01 is better in dynamic performance as compared to SIPIC02.

8. Conclusions

In conclusion, both SIPIC01 and SIPIC02 consist of decoupling feature which allow them to produce a performance with coexistence of zero or minimum overshoot and short settling time. However, by comparing SIPIC01 and SIPIC02, SIPIC02 consists of longer rise and settling time. Hence, it can be concluded that SIPIC01 is better than SIPIC02 in term of dynamic performance. As shown in Tables 6, 8 and 10, SIPIC02 does not experience overshoot except under loading 2 condition at 50 rad/s. The zero overshoot performance of SIPIC02 may be due to its longer rise and settling time and cannot reach its maximum speed with the simulated period. In future, work will be conducted to verify the stability of SIPIC02. From this research, increasing k_p or k_i gains gives a reduction trend for the rise and settling time of SIPIC02. A more in-depth study of the dynamic performance and mathematical model of SIPIC02 by determining the integral output and torque will be performed to investigate this observation.

References

1. R. C. Dorf and R. H. Bishop, "Introduction to Control Systems," *Mod. Control Syst.*, 2–23 (2000).
2. W. Yan, D. Wang, P. Jia, and W. Li, "The PWM speed regulation of DC motor based on intelligent control," *Syst. Eng. Procedia*, **3**, 259–267 (2012).
3. P. M. Menghal and A. J. Laxmi, "Dynamic analysis & artificial intelligent control of induction motor drives," *Annu. IEEE India Conf.*, **December**, 1–6 (2014).
4. S. K. Suman and V. K. Giri, "Performance Analysis of optimization techniques and Intelligence Techniques for speed control of DC motor," *African J. Basic Appl. Sci.*, **9**, **no. 2**, 97–104 (2017).
5. X. L. Li, J. G. Park, and H. B. Shin, "Comparison and evaluation of anti-windup PI controllers," *J. Power Electron.*, **11, no. 1**, 45–50 (2011).
6. T. L. Chiah, C. L. I. H. Hoo, E. Chin, and Y. A. U. Chung, "Hardware Simulation of a New Anti-Windup PI Control For Motor Speed Application," *J. Eng. Sci. Technol.*, 43–57 (2017).
7. C. L. Hoo, S. M. Haris, E. C. Y. Chung, and N. A. N. Mohamed, "The generalisation and decoupling mode of PI-based control: theoretical approach," *ICIC Express Lett.*, **4**, **no. 5**, 1–6 (2010).
8. C. L. Hoo, E. C. Y. Chung, S. M. Haris, and N. A. N. Mohamed, "Steady-state integral proportional integral controller for PI motor speed controllers," *J. Power Electron.*, **9**, **no. 6**, 1777–1782 (2015).
9. C. L. Hoo, S. M. Haris, C. Y. E. Chung, and N. A. N. Mohamed, "Steady-state integral proportional integral controller for PI motor speed controllers: a theoretical approach," *ICIC Express Lett.*, **4**, **no. 5**, 1–6 (2010).

Analysis of a Hybrid Mechanical Regenerative Braking System

Matthew Toh Xiang Wen[1,] and *Douglas* Tong Kum Tien[2,1]

[1]School of Engineering, Taylor's University, Malaysia
[2]School of Engineering, Taylor's University, Malaysia

Abstract. Regenerative braking systems for conventional vehicles are gaining attention as fossil fuels continue to be depleted. The major forms of regenerative braking systems include electrical and mechanical systems, with the former being more widely adopted at present. However mechanical systems are still feasible, including the possible hybrid systems of two mechanical energy recovery systems. A literature study was made to compare the various mechanical energy recovery systems. These systems were compared based on their advantages and disadvantages with regards to energy storage, usage, and maintenance. Based on the comparison, the most promising concept appeared to be one that combined the flywheel and the pneumatic energy recovery systems. A CAD model of this hybrid system was produced to better visualise the design. This was followed by analytical modelling of the energy recovery systems. The analysis indicated that the angular velocity had an extremely significant impact on the power loss and energy efficiency. The results showed that the hybrid system can provide better efficiency but only when operating within certain parameters. Future work is required to further improve the efficiency of this hybrid system.

1 Introduction

The slowing down or stopping of a moving automotive vehicle or machine is known as braking. Braking is commonly performed through mechanical means such as the use of friction brakes. Stationary brake pads are forced against the rotating brake discs mounted on the driveshaft to slow it down or to bring it to a stop. Kinetic energy from the rotating driveshaft is transformed into heat as a result of the friction between the brake pads and discs. This heat dissipation could cause energy losses of up to 30 percent [1] leading to an increase in fuel consumption. To minimize the loss, part of this energy could be captured and reused through a regenerative braking system (RBS).

Majority of road vehicles depend on fossil fuel to power an internal combustion engine (ICE). Efforts are being made to reduce fuel consumption. Research into RBS being one of these efforts. RBS are typically mechanical, electrical, or chemical systems that allow the stored energy to be reused to accelerate the vehicle. High-tech hybrid systems have been produced which combine the conventional brakes with a RBS. A RBS usually combines

[1] Corresponding author: DouglasKumTien.Tong@taylors.edu.my

one method of energy storage with one method of energy conversion [2]. At present, the electrical storage hybrid vehicles appeal to the consumer market with the environment-friendly concept and reduction in fuel costs for the end user [3].

A large amount of energy is usually lost during transformations of energy. Since energy is lost due to various factors such as friction, air resistance, and electrical resistance, the increase in the number of energy transformations will only cause a decrease in energy efficiency. A battery storage system shows this effect as it contains energy conversions of mechanical to electrical and to chemical to store the energy from braking. Besides the losses of energy, batteries are made up of rare-earth elements, increasing maintenance costs due to replacement of batteries after several years of use [4]. Furthermore the batteries are only operated in 55% to 65% of the total state of charge due to discharging efficiencies and to prolong battery life [5].

Currently, the public is becoming more aware about the kinetic energy recovery systems (KERS) that use batteries to store energy as these are being integrated into the latest models of everyday road vehicles. The electrical KERS however have limitations as they are less effective at lower speeds and require conventional brakes to provide sufficient braking. This is crucial when driving conditions vary, such as low-friction road surfaces or when driving in the rain.

Another form of energy recovery systems (ERS) also takes advantage of the inertia of the vehicle, converting kinetic energy of the moving wheels and driveshaft into stored mechanical energy and then releasing this energy back to the driveshaft for acceleration. Examples of these ERS include flywheel systems, pressurised accumulator systems, elastomeric systems, and regenerative shock absorber systems [4].

As the wheels of a vehicle are rotating, it follows that storage of energy in a rotational form would be able to reduce energy losses due to reducing the number of energy transformations. Flywheels can store energy in the form of rotational kinetic energy. Flywheels are popular mechanical energy recovery systems, as they have high power densities and are able to compete with batteries in power storage. Flywheel KERS have been used before and have shown up to 30% fuel reduction [4]. The flywheel KERS is simple in design with a clutch that connects the driveshaft to the flywheel during braking and thus helps with the braking of the vehicle while storing energy in the flywheel. A 65 kg flywheel KERS can store up to 60 kW of power and 580 kJ of energy for a 1800 kg vehicle [4].

Pressurised accumulators such as pneumatic or hydraulic systems use pressurised fluids to store energy. Gear pumps connect to the driveshaft that are driven to push air into the accumulators to be stored. A pneumatic accumulator can store high amounts of energy. Storage of up to 400 kJ or energy, with hydraulic accumulators storing about 90 kJ of energy have been shown [4]. Energy storage is however limited to the size of the accumulator and efficiencies of the accumulators at variable speeds. At low speeds, fixed mechanical losses cause drops in efficiency [6]. Elastomeric systems are able to store energy in the form of torsional energy if connected to the driveshaft and a fixed end, but store low energy. Storage of energy of about only 30 kJ is available, albeit for a long time [4]. The elastomer will however require frequent replacements due to decomposition caused by hysteresis. Regenerative shock absorbers recover energy from vertical oscillations of the vehicle, but only show at most a 10% improvement in efficiency [4]. Furthermore, smooth surfaces may not provide the maximum efficiency increase and may not be worth implementing due to the additional mass of the system.

Based on the review of literature, this study focused on the simulation of a mechanically-exclusive system that combines two mechanical KERS. The intention was to minimize the number of energy transformations and therefore maximize the amount of power delivered to the wheels which would reduce fuel consumption.

2 Research Methodology

2.1 Concept Comparisons

Various theoretical designs of hybrid mechanical RBS were conceptualised based on a critical literature review of mechanical systems. The designs were then compared amongst each other to determine the most suitable concept for further analysis. A theoretical design of a non-hybrid mechanical RBS was also produced and analyzed for comparison against the hybrid concept.

2.1.1 Advantages and Disadvantages of Energy Recovery Systems

The previously mentioned ERS were compared with regards to their advantages and disadvantages concerning energy storage, energy losses, longevity of life and other relevant parameters. Table 2.1 summarises these advantages and disadvantages for each ERS.

Table 2.1. Advantages and disadvantages of various energy recovery systems

Energy Recovery System	Advantages	Disadvantages
Electric	Proven technology Readily available in the market	Costly due to replacements Losses of energy due to energy transformations Long recharging times
Flywheel	Reduced losses during energy capture High energy storage capabilities Technology well proven	Loss of rotational energy over time due to friction and air resistance
Hydraulic/Pneumatic	Very low losses during energy storage Not as costly as other ERS	Limited energy storage
Elastomeric	Ejection characteristic for high acceleration	Low life span Limited energy storage

Extensive research is currently conducted into hybrid electrical-mechanical RBS and it is as considered one of the more attractive fields of research for energy regeneration. However based on the literature reviewed on the ERS, a fully mechanical hybrid RBS can also be promising. Mechanical systems such as flywheels have the ability to capture and store high amounts of energy, while hydraulic or pneumatic systems are not costly and have low energy storage losses. The accumulator system is also able to charge and discharge more periodically than batteries that require lengthy recharging durations. By combining two of these mechanical ERS based on their advantages in energy storage, losses, or lifespan, it was anticipated that a more energy efficient systems could be produced.

2.1.2 Selection of Design Concept for Analysis

These hybrid system design concepts include various combinations of the flywheel, hydraulic, and elastomeric ERS. As hybrid RBS consists of two main steps of energy harvesting and energy storage, the hybrid mechanical system will implement the same concept with two mechanical systems to collect and store energy from braking respectively.

As energy harvesting or capture systems require the ability to store a high initial amount of energy from the high angular velocity of the wheels of a vehicle, high storage capacity systems are preferred. Flywheel systems fit this category as they are able to store high amounts of energy while keeping the energy in the form of rotational energy, minimising the number of energy transformations and thus reducing the energy losses incurred. A hollow cylinder flywheel is also able to minimise the chance of disintegration at too high speeds [7].

Hydraulic, pneumatic and elastomeric systems are not as suitable for the initial energy capture and storage compared to the flywheel due to their limited energy storage. Hydraulic and pneumatic accumulators are limited to the amount of energy stored based on the size of the accumulators, while elastomerics under high usage conditions undergo hysteresis, requiring frequent replacements.

With flywheels identified as the most suitable system for initial energy capture and storage, a secondary energy storage system in the form of either the pressurised accumulator or elastomer can be considered for storing a portion of the initial energy. The accumulator has low energy losses which allows efficient energy storage, while the elastomer has an ejection characteristic that allows for acceleration during energy transfer back to the flywheel. As the flywheel will be storing energy as rotational energy, the transfer to the secondary energy storage in the form of rotation is preferred, once again to reduce losses. Both the accumulator and elastomer are able to utilise rotational motion, with accumulators running on gear pumps and elastomers storing energy through torsion. However, the decomposition of the elastomers in elastomeric systems is unavoidable. Thus, the pneumatic or hydraulic accumulators would be more effective in terms of lifespan and maintenance. As pneumatic accumulators use air or gas as opposed to liquids such as oil that is used by hydraulics, they are preferred for high speed applications due to the lower viscosity of gases. With potentially high speeds of the wheels and thus the flywheel, the pneumatic accumulator would be the more viable option.

A direct comparison was made between a non-hybrid mechanical system and the hybrid mechanical RBS was done by modeling them analytically. As the hybrid mechanical RBS selected for analysis was the flywheel-pneumatic system, the conventional mechanical RBS that was simulated for comparison was be the flywheel RBS. This comparison would show any changes in efficiency due to the addition of the pneumatic accumulator to the flywheel.

2.2 Design of Regenerative Braking Systems

The concept of combining two mechanical ERS to form a hybrid mechanical RBS is shown in Figure 2.1. The initial energy capture and storage system directly collects energy from the driveshaft connected during braking, and then transfers the energy captured to the secondary energy storage system. The stored energy is transferred back to the initial energy storage system and thus to the driveshaft during the acceleration of the vehicle.

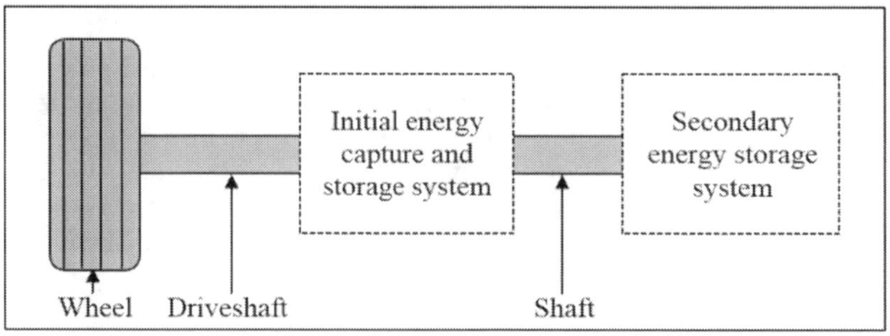

Fig. 2.1. General combination of systems for conceptualised hybrid mechanical RBS

2.2.1 Flywheel Regenerative Braking System

Flywheels by themselves are able to store large amounts of energy but lose energy over time due to air resistance and friction losses of the bearing. Figure 2.2 shows the flywheel that connects to the driveshaft through a conventional clutch system, such as those found in manual transmission vehicles. As the system is not a hybrid system, only one ERS is used to both capture and store energy.

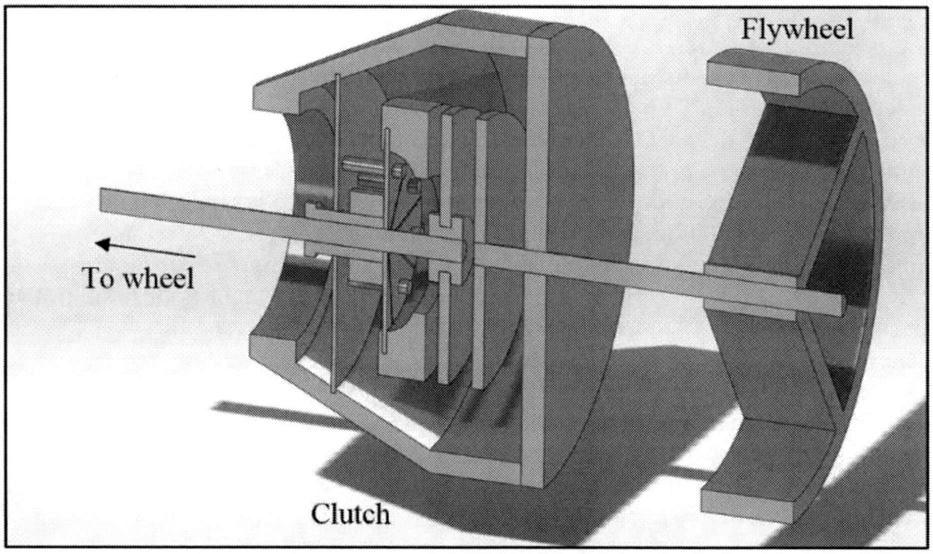

Fig. 2.2. Design of the flywheel RBS

2.2.2 Flywheel-Pneumatic Regenerative Braking System

As flywheels lose energy over time due to drag and friction between bearings, the addition of an accumulator to store a portion of the energy would help reduce the energy losses of the flywheel, as accumulators lose much less energy over time. Releasing this energy back to the flywheel provides higher energy that can be transferred back to the driveshaft for acceleration. This is the rationale behind the hybrid concept.

Figures 2.3 and 2.4 show the conceptual design of the flywheel-pneumatic hybrid RBS. The wheel mounted on the driveshaft will transfer the rotational energy to the flywheel through the activation of a conventional clutch system upon braking. A similar clutch is used in between the flywheel and pneumatic accumulator to transfer energy between the two systems. The accumulator is connected to the clutch by a shaft to the gear pump that draws air into the accumulator. A secondary gear is used to transfer the energy back to the flywheel in the same rotational direction

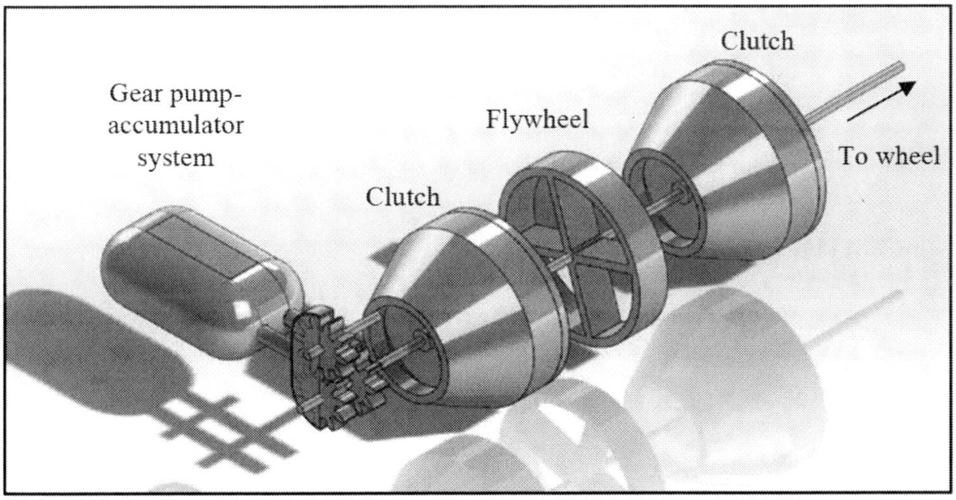

Fig. 2.3. Theoretical Design of the flywheel-pneumatic hybrid RBS

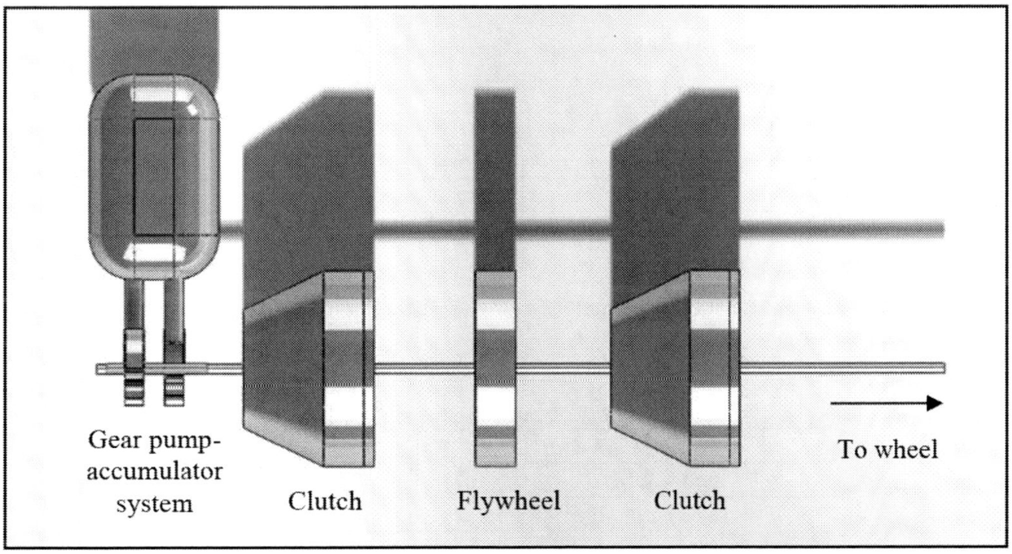

Fig. 2.4. Top view of the flywheel-pneumatic hybrid RBS

2.3 Analysis

The analysis was performed by running simulations of analytical models using Microsoft Excel version 2016. This computer software was used to process various parameters and energy equations for the flywheel RBS and the flywheel-pneumatic hybrid RBS.

2.3.1 Parameters

The flywheel was modeled with a set of variable parameters shown in Table 2.2. These variables determined the energy storage and losses of the ERS of the systems. The values of the variables shown in the table were applied to the analysis of the system and then varied to determine the effect of these parameters on the energy efficiency of the system. The flywheel was chosen to be a rim type with spokes to reduce chance of disintegration of material [7]. The angular velocity of the flywheel was chosen based on low speed applications below 3000 rpm and high-speed applications between 3000 rpm and 60000 rpm. The flywheel was chosen to be of S-glass epoxy material as it has more than double the energy density value of common cast iron flywheels [8]. The mass was varied around the average flywheel mass of around 65 kg [4]. This volume change was done by reducing thickness of the flywheel while the diameter values were kept constant. This was done by determining the volume of the cylinder based on the varying mass to thus alter the thickness value as shown in Eq. (1):

Volume of Flywheel, V_f (m^3):

$$V_f = \pi t \left[\left(\frac{d_o}{2}\right)^2 - \left(\frac{d_i}{2}\right)^2 \right] \tag{1}$$

Where

t = Flywheel thickness (m)
d_o= Outer diameter of the flywheel (m)
d_i= Inner diameter of the flywheel (m)

Table 2.2. Initial variable flywheel parameters for analysis

Flywheel parameter	Value
Mass, m (kg)	70.00
Outer diameter, d_o (m)	0.50
Inner diameter, d_i (m)	0.25
Thickness, T_f (m)	0.05

Table 2.3 shows constant parameters such as the density of air, dynamic viscousity of air, and gravitational acceleration.

Table 2.3. Parameters for simulations

Constant	Value
Density of air, ρ_α (kg/m^3)	1.225
Dynamic viscousity of air, β_α (Ns/m^2)	1.983×10^{-5}
Gravitational acceleration, g (m/s^2)	9.810

Parameters for the flywheel ball bearing are shown in Table 2.4. The ball bearing is used to connect the flywheel to the shaft and is a deep groove ball bearing. They include the constant friction coefficient of ball bearings, radius of the shaft that connects to the ball bearings, and the corrective force factor for unbalanced and gyroscopic forces [9].

Table 2.4. Ball bearing parameters for simulations

Ball bearing parameter	Value
Friction coefficient, μ	1.50×10^{-3}

Radius of shaft, r_s (m)	0.05
Corrective force factor, γ	1.00

The clutch discs were modeled with parameters based on technical specifications of clutch disc manufacturers [10]. These values are shown in Table 2.5.

Table 2.5. Clutch discs parameters

Clutch discs parameter	Value
Clutch mass, M_C (kg)	5.44311
Clutch diameter, D_C (m)	0.18415
Clutch radius, r_C (m)	0.09208

2.3.2 Energy equations

The equations applied to simulate the two RBS included mass moment of inertia, energy storage, power losses, energy losses, and final energy efficiency equations. These equations are shown in Eq. (1) previously mentioned and Eqs. (2) to (13) as follows:

Mass moment of inertia of a flywheel, *I* (kg.m²) [11]:

$$I = \frac{1}{2}.m.(r_o^2 + r_i^2) \quad (2)$$

Where

m = Mass of flywheel (kg)
r_o = Outer radius of flywheel (m)
r_i = Inner radius of flywheel (m)

Initial kinetic energy of a flywheel, KE_i (J) [7]:

$$KE_i = \frac{1}{2}.I.\omega^2 \quad (3)$$

Where

ω = Angular velocity of the flywheel (rad/s)

Power loss due to air resistance of a flywheel, $P_{a,l}$ (W) [9]:

$$P_{a,l} = 0.04.\rho_a^{0.8}.\beta_a^{0.2}.(\omega.r)^{2.8}.(2.r)^{1.8}.(\alpha + 0.33) \quad (4)$$

Where

ρ_α = Density of air (kg/m³)
ρ_α = Dynamic viscosity of air (Ns/m²)
ω = Angular velocity of the cylinder (rad/s)
r = Outer radius of the cylinder (m)
Geometric ratio, $\alpha = \frac{h}{2.r}$

Where
h = Thickness of the cylinder (m)

Power loss due to friction of bearings of a flywheel, $P_{b,l}$ (W) [9]:

$$P_{b,l} = \mu . \gamma . r_s . m . g . \omega_b \tag{5}$$

Where

μ = Friction coefficient
γ = Corrective force factor considering unbalanced and gyroscopic forces
r_s = Radius of the shaft (m)
m = Mass of the flywheel (kg)
g = Gravitational acceleration (m/s^2)
ω_b = Angular velocity of the bearing (rad/s)

Total power loss of a flywheel, P_T (W) [9]:

$$P_T = P_{\alpha,l} + P_{b,l} \tag{6}$$

Total energy lost due to power losses, E_l (J):

$$E_l = P_T \times t \tag{7}$$

Where

t = Time of flywheel rotation (s)

Final energy of the flywheel system, E_f (J):

$$KE_f = KE_i - E_l \tag{8}$$

Final angular velocity of the flywheel, ω_f (rad/s):

$$\omega_f = \sqrt{\frac{2 \times KE_f}{I}} \tag{9}$$

Final energy efficiency of the system, η_E (%):

$$\eta_E = \frac{KE_f}{KE_i} \times 100\% \tag{10}$$

Moment of inertia of clutch discs 1 and 2, I_1 and I_2 (kg.m^2) [12]:

$$I_1 = I_2 = \frac{1}{2} \times m_C \times r_C^2 \tag{11}$$

Where

m_C = Mass of clutch disc (kg)
r_C = Radius of clutch disc (m)

Energy loss due to clutch system, P_C (W) [13]:

$$E_C = \frac{I_1 I_2 \times (\omega_1 - \omega_2)^2}{2 I_1 I_2} \tag{12}$$

Where

ω_1 = Angular velocity of clutch disc 1 (rad/s)
ω_1 = Angular velocity of clutch disc 2 (rad/s)

Energy efficiency of gear pump-accumulator system, η_A (%) [14]:

$$\eta_S = \eta_{GP} \times \eta_A \times \eta_{GP} \tag{13}$$

Where

η_S = Efficiency of gear pump-accumulator system (%)
η_{GP} = Efficiency of gear pump (%)
η_A = Efficiency of pneumatic accumulator (%)

2.3.3 Assumptions

The analytical modeling was based on several assumptions as follow:

- Mass of the spokes of the flywheel were considered to be negligible
- The operation of the clutch system happens instantaneously
- Corrective force factor was assumed to be 1 based on assumption that gyroscopic forces do not highly affect the power losses due to bearing friction
- Dynamic viscousity of air based on assumption that air is at room temperature

3 Results and Discussion

3.1 Validation of the Analytical Model

The analytical model was validated with existing literature [15]. The parameters that were set are as shown in Table 3.1, while the results are shown in Figure 3.1:

Table 3.1. Parameters provided by literature source used in simulation [15].

Parameter	Value
Outer diameter of flywheel, d_o (m)	0.330
Inner diameter of flywheel, d_i (m)	0

Average radius of ball bearing, r_b (m)	0.0385
Density of air, ρ_a (kg/m^3)	1.225
Dynamic viscousity of air, β_a (Ns/m^2)	1.983×10^{-5}

The flywheel from literature was that of a solid flywheel without an inner diameter usually due to the ring flywheel design. The radius of the ball bearing was averaged and assumed as the radius of shaft, r_s from Table 2.4.

Figure 3.1 shows that the power loss differences between the analytical model and literature was small, with the gradient difference of about only 7%. Power losses itself by about 5 to 7 W between the model and literature.

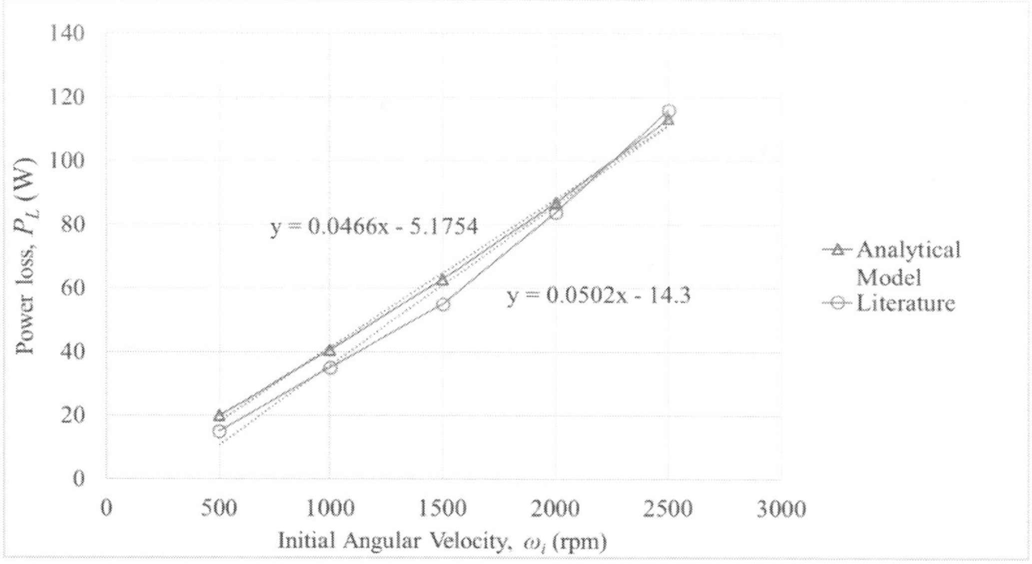

Fig. 3.1. Comparison of power loss between analytical model and literature.

Further comparisons with other literature sources mentioned that flywheels have about 70% to 90% efficiency for a short amount of time after the flywheel is left to rotate freely [16,17]. From the calculated efficiency of the modeled flywheel system using Eqs. (1) to (10), the results showed that at many varying parameters tested a 90% to 100% energy efficiency was available for the first few seconds of the simulation. The values then drop quickly to below 90%, as in line with the information gained from the literature.

3.1 Comparison of Energy Efficiencies of Simulated Systems

3.1.1 Effect of Mass of Flywheel on Energy Efficiency

A heavier flywheel of 100 kg resulted the best performance in terms of energy efficiency values, while lower mass flywheels of 8.5 kg and 25 kg showed higher improvements in the hybrid RBS over the Flywheel RBS counterparts, as shown in Figure 3.2. Initial increase in energy efficiency occurs as the ratio of amount of energy recovered from braking relative to the losses due to the inefficiency of the gear system increases. Eventually, the power losses due to increasing angular velocity outweigh the energy stored in the accumulator and thus cause the efficiency to drop.

The overall improvements due to mass begin to decrease at a faster rate than the heavier flywheels, causing a limitation to the required total time for positive energy efficiency gains. Heavier flywheels will also require power from the driveshaft to accelerate a vehicle carrying the heavier flywheel. This trade-off between weight and power needs to be carefully considered in view of overall fuel efficiency.

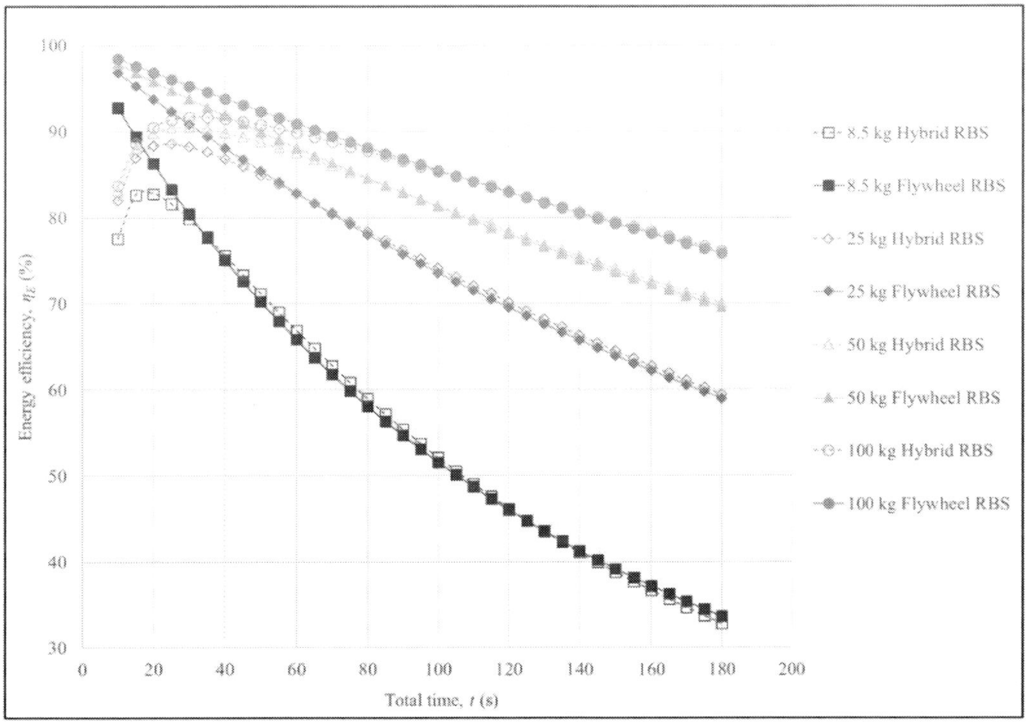

Fig 3.2. Effect of various flywheel masses on energy efficiency

3.1.2 Effect of Initial Angular Velocity on Energy Efficiency

Low angular velocity of 750 rpm is the most energy efficient but also has the lowest maximum energy storage available as shown in Figure 3.3. Thus, a low angular velocity may not be able to power a vehicle for acceleration as compared with a faster flywheel system that is less energy efficient but has more energy available to be used by the driver of the vehicle.

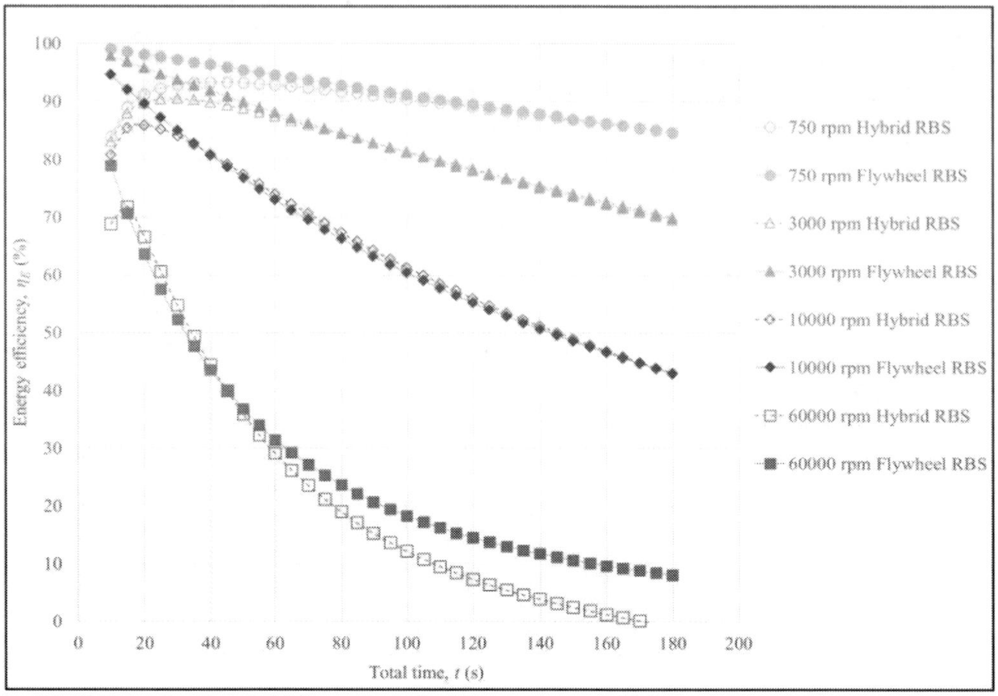

Fig. 3.3. Effect of various initial angular velocities on energy efficiency

3.1.3 Effect of Energy Transfer and Storage Time on Energy Efficiency

Low transfer times resulted in the highest energy efficiency as shown in Figure 3.4. However, the inconsistency of vehicle idle time during daily driving may not always call for total energy storage times of 180 seconds or less. As the energy efficiency of shorter transfer times close up on the energy efficiency provided by larger transfer times, an expected breakeven point and negative difference may cause the longer transfer times to be m

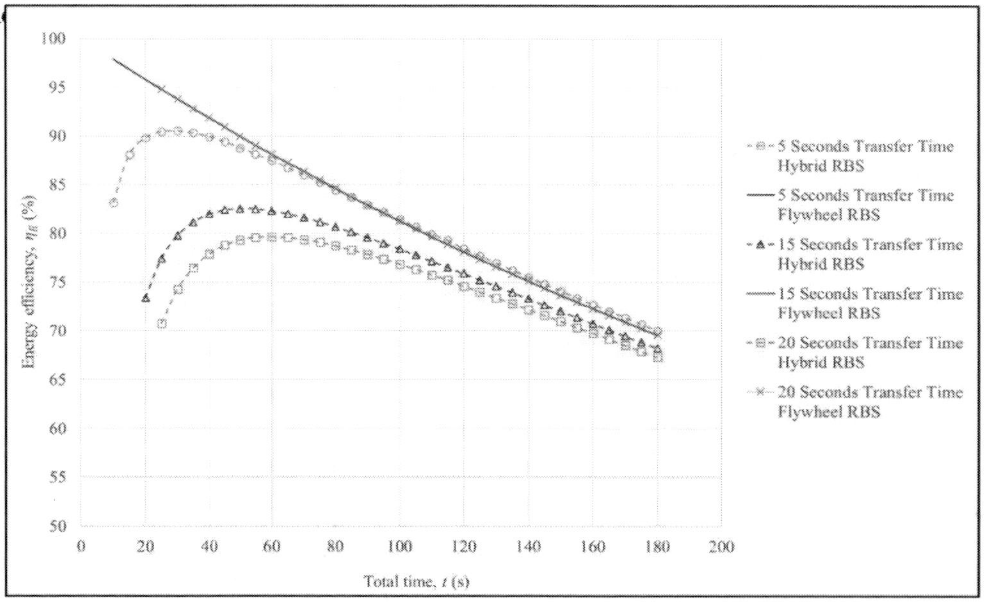

Fig. 3.4. Effect of various transfer and storage times on energy efficiency

3.2 Comparison with Existing *Flybrid* Flywheel System

Figure 3.5 shows the highest energy efficiency values for the flywheel and HRBS compared with the current *Flybrid* Flywheel system. The hybrid system was able to improve on the *Flybrid* Flywheel system, but was not able to surpass the flywheel system in terms of energy efficiency. This comparison between the modeled and current systems is however not a direct comparison due to the very different parameters and also that the *Flybrid* Flywheel efficiency was not given as a function of time in the literature. Specifically, the increase in mass of the flywheel will cause an increase in power needed by the vehicle to accelerate, while the lower mass *Flybrid* Flywheel system does not add as much to the total vehicle mass.

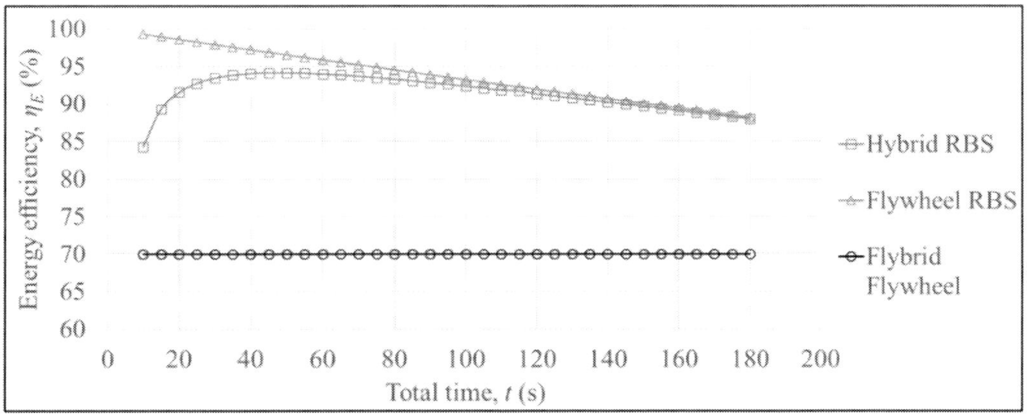

Fig. 3.5. Comparison of simulated systems with existing *Flybrid* Flywheel system

4 Conclusions and Future Work

The energy losses can be seen as factors that greatly affect the energy efficiencies of the regenerative braking systems. This is due to the high angular velocities that cause very high power losses to air resistance and bearing friction. This analysis has shown that the flywheel-pneumatic hybrid system is at times able to provide an improvement in energy efficiency to current systems within a range of suitable parameters only such as a high mass value of 50 kg to 100 kg, low angular velocity of 750 rpm to 3000 rpm, and low or short energy transfer times of 5 seconds or less between the flywheel and the pneumatic accumulator.

However, overall energy values that are available for usage are not enough to take advantage of the energy efficiency due to the low angular velocity, which the *Flybrid* Flywheel system avoids by mitigating the losses. Furthermore, as the mass of the flywheel increases, power required to accelerate the vehicle increases. Research could be done to determine if the energy efficiency provided by the increase in mass is able to overcome the additional energy required to power the heavier vehicle.

It is hoped that these findings can contribute to further the field of knowledge of hybrid mechanical RBS regarding energy storage and efficiency. Future work could focus on reducing energy losses and producing working prototypes for experimental analysis.

References

1. P. Gupta, A. Kumar, and S. Deb, International Journal of Mechanical And Production Engineering, **2,** 75–78 (2014)
2. R. Ramakrishnan, S. S. Hiremath, and M. Singaperumal, J. Mech. Sci. Technol., **26**, 1321–1331 (2012)
3. U. Diego-Ayala, et al., Proc. Inst. Mech. Eng. Part D J. Automob. Eng., **222**, 2087–2101 (2008)
4. Gabriel-Buenaventura, Alejandro, and B. Azzopardi, Renew. Sustain. Energy Rev., **41**, 955–964 (2015)
5. J. B. M. William, C. David, J Automobile Engineering, **1,** 1-14 (2012)
6. S. J. Clegg, Institute of Transport Studies, University of Leeds, **471** (1996)
7. H. Liu and J. Jiang, Energy Build, **39**, 599–604 (2007)
8. A. Shinde, et al., Global Journal of Enterprise Information System, **9**, 95-99 (2017)
9. X. Zhang, C. Mi, *Vehicle Power Management,* 86-89 (2011)
10. Tilton Engineering. "Ford Clutch-Flywheel Assemblies". Internet: http://tiltonracing.com/product/ford-clutch-flywheel-assemblies/ (2017)
11. B. Bolund, et al., Renewable and Sustainable Energy Reviews, **11**, 235-258 (2007)
12. C. R. Nave. "Thick hoop and hollow cylinders". Internet: http://hyperphysics.phy-astr.gsu.edu/hbase/ihoop.html (2016)
13. B. I. Silviu, et al., Recent Researches in Automatic Control, **May**, 252-256 (2011)
14. R. J. V. Mark, J. W. Steven, J. R. Donald, J. M. Timothy, US Patent 8,622,860 (2014)
15. J. Abrahamsson, et al., energies, **5**, 2794-2817 (2012)
16. I. Hadjipaschalis, et al., Renewable and Sustainable Energy Reviews, **13**, 1513-1522 (2009)
17. A. Boretti, International Journal of Hydrogen Energy, **35**, 8417-8424 (2010)

Sintered Properties of Stainless Steel-doped Y-TZP Ceramics

Kelvin Chew Wai Jin[1,*], *Mervyn Marlon Anak Mathew Minggat*[1] , and *Ramesh Singh*[2]

[1]Department of Mechanical Engineering, School of Engineering, Faculty of Built Environment, Engineering, Technology and Design, Taylor's University, 47500 Subang Jaya, Selangor, Malaysia.
[2] Centre of Advanced Manufacturing & Material Processing (AMMP), Department of Mechanical Engineering, Faculty of Engineering, University of Malaya, 50603 Kuala Lumpur, Malaysia.

Abstract. Yttria stabalised Tetragonal Zirconia Polycrystalline (Y-TZP) has been extensively renowned for its vast outstanding contribution in medical field such as primary function for implant biomaterial which acts as restoration of damaged body parts. Nonetheless in the past decades, there have been various reported cases that point the failure of hip transplants involving material of zirconia ceramic. This eventually leads to the birth of this research project where it studied the properties effects of Y-TZP as it was added or doped with stainless steel 316 (SS316). In this research project, the parameter of weight percentage for stainless steel 316 (SS316) to be doped with Y-TZP were as follows; 0 wt%, 0.1 wt%, 0.5 wt% and 1 wt%. The outcome of this experiment then were measured and analysed as according to properties of relative density, Vickers hardness and fracture toughness. The experiment was conducted by firstly mixed the Y-TZP and SS316 powders together as according to the weight percentage of SS316 as well as addition of ethanol solution that act for medium of enhancement. The mixture then underwent compacting pressure bench press as well as Cold Isostatic Pressure (CIP) in order to produce the green sample. After that, the green sample underwent sintering process via pressureless sintering method at mentioned ranges temperature of 1250°C, 1300°C, 1400°C and up to 1500°C. Subsequently, the sample then underwent grinding and polishing process before proceeding to the measurement and analysis process. The outcome of the measurement and analysis process revealed that as the content of SS316 doped with Y-TZP increases as well as the increased in sintering temperature, the relative density and fracture toughness profound to be increased while Vickers hardness indicated as decreased.

1 Introduction

In this robust technological era, experimentation involving combination of two distinct materials which consist of different properties respectively often been performed via doping method in order to attained successful outcome of material with superior properties. In the past for medical field, there were numerous of different materials have been studied for the

* Corresponding author: KelvinWaiJin.Chew@taylors.edu.my

objective to produce a durable implant biomaterial which function as alternative substitution to repair damaged body parts. Zirconia ceramic in particular has gain enormous popularity and attention from surgeon and scientist across the world due to its advantage as the closest restoration for bone tissue [1]. Nevertheless, zirconia ceramic alone has been identified in which it does not exhibit a high toughness as comparison made with metals. Therefore, metals do gain consideration in which it can be added to zirconia ceramic through doping and creating metal doped ceramic or ceramic doped metal. This method and process aim to enhance the overall replacement implants toughness. Moreover, this doped material progressively replaced the prostheses that comprised of purely ceramic as taking concerned to improve the prostheses major aspects of strength and durability [2, 3]. Hence, the studies regarding doped material to produce advance biomaterial has developed rapidly.

In addition, zirconia ceramic has been described to exhibit different properties when taking into account that it exists at distinct phases. In this case, zirconia ceramic often added with dopant or sintering additive in order for the characteristics and phases do not alter or transformed at different temperatures. Nonetheless, the advantages of zirconia ceramic were such that include high density and strength, high tolerance towards corrosion as well as high resistance towards impact. A specific zirconia ceramic type known as Yttria-stabilised Tetragonal Zirconia Polycrystals (Y-TZP) often been used as implant material in human body due to its high compatibility. Therefore, it has been applied as prosthetics for numerous years and yet displayed negative feedback. Moreover, stainless steel usually comes with different types and graded as according to its production specifications, grade composition as well as mechanical properties. Customarily when comparing zirconia ceramic with stainless steel, stainless steel in this case has displayed to have high strength, ductility, toughness as well as high corrosion resistance. Thus, the doped material can be utilised as a substitution for pure ceramic in order to create a more durable implant material.

A study was conducted to examine the doped material relative density such that considering the factors of varied zirconia and stainless steel compositions as well as heating rates [4]. In this case, the samples relative density was determined based on different sintering heating rates. The results then revealed that the samples relative density decreased when the stainless steel weight percentage (wt%) falls below 50% as well as during the increased in sintering heating rates. Vice versa, samples relative density increased when the stainless steel weight percentage (wt%) exceeding 50% as well as during the increased in sintering heating rates. A study was performed to find out the fracture toughness and Vickers hardness of zirconia stainless steel composite [5]. In this investigation, the samples were made as according to stainless steel concentration with ranges from 15% to 30%.

The results displayed to obey the mixture law such that when stainless steel content inside the zirconia/stainless steel composite increased, the fracture toughness indicated as increased while Vickers hardness described as decreased [6, 7]. In addition when subjected to fractures, there was no observation of plastic deformation. As comparison to zirconia ceramic, zirconia/stainless steel composite do not exhibit brittle condition [8]. In terms of biomedical implants, these types of characteristics often denoted as vital for material enhancement. In the previous years, there were various cases had been reported that address the serious issue of zirconia ceramic hip transplant failure. In order to provide solution to this dilemma, various thorough studies have been conducted that primarily emphasize on sintering of zirconia doped material at low temperature as well as mechanical properties [9]. In this research project, it fundamental objective was to study the effects of properties when adding dopant stainless steel 316 to Y-TZP. This report presented the experimental flow process as well as results for the tested physical and mechanical properties.

2 Methodology

2.1 Powder preparation

The starting materials for this research project methodology begins with acquired and gathered the two essential material substances that may include Y-TZP and stainless steel 316 powders. Table 1 shows the Y-TZP and stainless steel 316 summaries of typical characteristics accordingly. The powder preparation process was initiated with powder synthesis process in which both Y-TZP and stainless steel 316 powders were been mixed together as well as measured based on the stainless steel 316 content. Before both of the powders to be mixed together, each of the samples weight percentage were fixed as according to the stainless steel 316 content which ranges from 0 wt%, 0.1 wt%, 0.5 wt% and 1 wt%. Ethanol with measurement of 100 ml was firstly poured into the beaker and only then followed by poured both of the stainless steel 316 and Y-TZP together into the beaker in order to undergo the mixing process.

Table 1. Characteristics of Zirconia and Stainless Steel powders.

Material	Grade /Type	Composition (wt%)	Powder Particle Size (μm)	Particle Shape
Zirconia	Stabilized tetragonal zirconia, Y-TZP	5.2 Y_2O_3, HfO_2 <5, Al_2O_3 0.1-0.4, ZrO_2 (remaining)	0.04	Spherical
Stainless Steel	316	Cr 16-18, Ni 10-14, Mo 2-3, C 0.08, Fe (remaining)	5-20	Spherical

The equipment of Digital Ultrasonic Bath (Powersonic UB-410, Thermoline Scientific, Australia) was utilised by filled up the equipment with distilled water. Next, the beaker then placed inside the Digital Ultrasonic Bath equipment for 15 minutes of thorough mixing process with continuous stirring and beaker later was removed from equipment itself once achieved completion of mixing. After that, the machine of Laboratory Attritor Mill (01-HD, Union Press, United States) was utilised by pouring the mixture inside it with the purpose for the mixture to undergo the attrition milling process where further break down of powder mixture for duration of 45 minutes with running at 500 rpm. Furthermore, 5 mm size of zirconia milling ball which account quarter of the mixture amount were added to the Laboratory Attritor Mill machine also with aim to enhance the entire milling process completion. A sieve utility of measurement mesh size 212 μm was used for the first time to sieve the mixture where zirconia ball were trapped and minimised the mixture to slurry conditions. The remaining mixture then created into a cake form and undergoes baking process where the mixture was stored inside the Universal Oven (UN 55, Memmert, Germany) which heated for 60°C for overnight duration and subsequently left dry. The cake mixture after the heating process was crushed into smaller form using hand tools. After that, a sieve utility of measurement mesh size 212 μm was utilised for second time with objective to produce the final powder composition of definite fine. Table 2 shows the summaries of typical properties of Y-TZP and stainless steel 316, respectively, at room temperature.

Table 2. Typical properties of Y-TZP and Stainless Steel 316 at room temperature.

Materials	Density (g/cm^3)	Young's Modulus (GPa)	Vickers Hardness (GPa)	Fracture Toughness ($MPa.m^{1/2}$)
Zirconia (Y-TZP)	6.02	200	12.5	13
Stainless Steel 316	8.00	193	1.47	112-278

2.2 Green sample preparation

The machine of Hydraulic Bench Press (HP10-150, RMS Hydro, Malaysia) was utilised for the compacting process of the powder mixture in which the powder mixture was poured into a round shaped die with diameter measurement of 20 mm. The subjected high pressure bench press was 1000 bar for 10 seconds. After that, the Cold Isostatic Press (KJYU150, Shanxi Golden Kaiyuan Co., Ltd, China) machine was used for the CIP process where the green body round tablets were been placed inside a glove as well as placed it again inside desired square shaped container. On the other hand, the container in this case then to be stored under room temperature which in this case aim for neither heat nor pressure to penetrate and ultimately affect the green body round tablets during the pressing process. In overall for this current process stage, the green body round tablets indeed can likewise be referred as green samples.

2.3 Sintering

The type of applied sintering process method was indeed pressureless sintering method and to be performed after the formation of green samples. In this process, it begins by placing the green samples inside the crucible which to be stored inside the Electric Box Furnace (LT10, LT, Malaysia). The applied heating for sintering temperatures were distinct such that ranging from 1250°C, 1300°C, 1400°C and up to 1500°C where only one sintering temperature was performed in one day. The process heating rate was also conserved at 10°C per minute. In addition, the furnace holding time was kept for 2 hours and after the process, the heating was stopped. The samples then let to be cooled down at rate of at 10°C per minute where it explained that the samples cooled through natural cooling method until approaching ambient.

2.4 Grinding and Polishing

2.4.1 Grinding

After the completion of sintering process, the samples were vitally required to be grinded and followed by polished. The equipment of Polishing & Grinding Machine (GP2V, LT, Malaysia) was utilised for both of grinding and polishing processes. For grinding process, the Polishing & Grinding Machine was installed with grinding papers and operates at 50 rpm in which two minutes for each sample. The applied grinding papers range with five grit sizes in total such that namely 120, 240, 600, 800 and 1200 that consists with surface material of silicon carbide (SiC). The samples were grinded only on one side and the other side used as samples labelling for clear identification purposes. Furthermore, the grinding results for each grade were displayed as visible one direction line and vitally covering the entire grinded side of samples surface. Each visible direction lines also represent each grade of grinding papers respectively.

2.4.2 Polishing

Polishing process was performed subsequently after the grinding process in which polishing clothes were utilised and injected with diamond paste as well as lubricant via syringe device. The polishing clothes were installed to the equipment of Polishing & Grinding Machine (GP2V, LT, Malaysia) and operated at 20 rpm. In addition, the diamond paste in this case consists of diamond bit with grades measurement of 1 μm and 6 μm. Moreover, the lubricant likewise been injected to the polishing clothes as well as was to function by ease the spreading of diamond paste evenly across the entire surface of the samples. The polishing process was performed for duration of two minute only for each sample.

2.5 Measurement and Analysis

2.5.1 Relative density

The density of samples was determined in which the testing was conducted by adhering strictly according to Archimedes principle and water immersion technique. ASTM C-29 was used as reference of standard to relate with the determined relative density. In this process, an analytical balance with density kit (AY220, Shimadzu, Japan) was employed by filled up the equipment with distilled water that function as immersion medium. The value for density of water was declared as 0.99654 kg/m^3.The stated Y-TZP theoretical density was 6.10 g/cm^3 and the following formula (eq 1) was used to calculate the relative density of samples.

$$\rho = \frac{W_a}{W_a - W_w} \rho_w \quad (1)$$

where

ρ = Relative density of sample (g/cm^3)
w_a = Weight of sample in air (g)
w_w = Weight of sample in water (g)
ρ_w = Density of distilled water (g/cm^3)

2.5.2 Vickers hardness

The samples Vickers hardness were experimented by utilising the equipment of Vickers hardness tester (MV-01, Matsuzawa, Japan) that consist of pyramidal diamond indenter as well as subjected load application of 10 kg. Only minimum duration of 10 seconds for held down implementation. ASTM E284-99 and ISO 14705 can likewise be used as a standard reference for all experiment parameter. In broad, the impression was indicated as square such that presence with two diagonals that has identical length. The indentation in this process was measured by using the Vickers hardness tester machine microscope that has been installed together with filar micrometer which consist advantage accuracy of 0.1 μm. The equation to calculate the samples Vickers hardness was presented below as (eq 2). Figure 1 displays the schematic diagram of a Vickers hardness indenter profile which described as angle between the opposite faces at the pyramidal indenter vertex and consequent resulting indentation on the sample.

$$H_v = \frac{1.854P}{(D)^2} \tag{2}$$

where

P	=	Applied load (kg)
D	=	Average of two diagonals D_1 and D_2 (m)

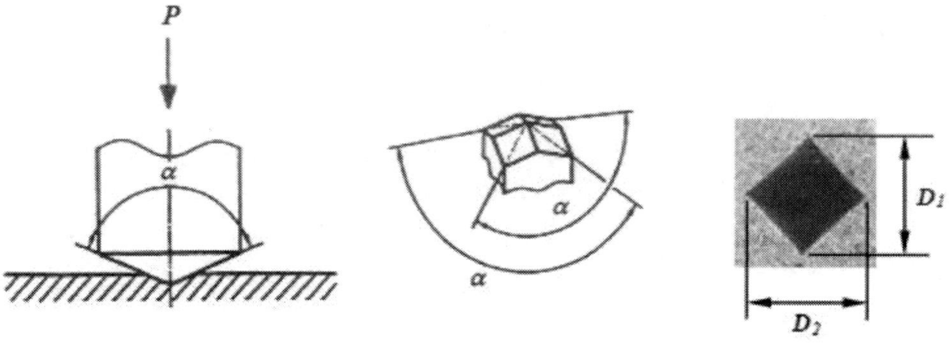

(a) (b)

Fig. 1. (a) Schematic diagram of Vickers hardness indenter with profile angles α =136° ±0.5° (angle between the opposite faces at the vertex of the pyramidal indenter) and (b) an indentation that is formed on the sample.

2.5.3 Fracture toughness

The samples fracture toughness were determined with an identical method of testing as the one of Vickers hardness due to it used similar indentation. The microscope filar micrometer was used to determine the crack length (L1, L2, L3, and L4) and indentation length (D) after the formation of indentation. The equation for fracture toughness was indicated below as (eq 3). Figure 2 displays the schematic diagram regarding typical Palmqvist crack system that emerged after Vickers indentation. In this case also, "L" was described as the crack measured length which used to calculate the samples fracture toughness. The Vickers indent diagonals (D1 and D2) were suggested to be identical with the length of 2a by referring to Fig. 1.

$$\left(\frac{K_{IC}\Phi}{H_V a^{\frac{1}{2}}}\right)\left(\left(\frac{H_V}{E\Phi}\right)^{\frac{2}{5}}\right) = 0.035\left(\left(\frac{L}{a}\right)^{\frac{-1}{2}}\right) \tag{3}$$

where

K_{IC}	=	Fracture toughness (MPa. $m^{1/2}$)
H_V	=	Vickers hardness (GPa)
E	=	Young's Modulus (GPa), taken as 200 GPa for Y-TZP ceramic
a	=	Half-length of the average diagonal (m)
L	=	Average crack length (m)
Φ	=	Constraint factor taken as 3

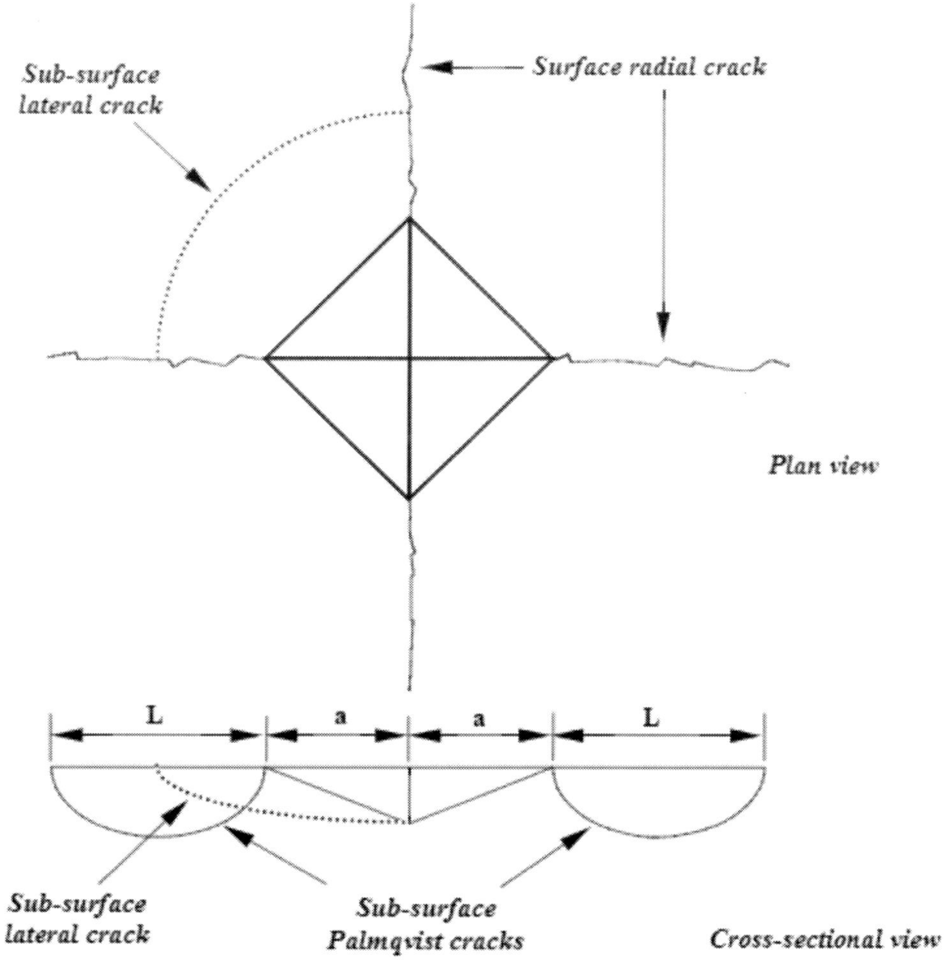

Fig. 2. Schematic diagram regarding the typical Palmqvist crack system that emerged after Vickers indentation. Parameter "L" refers to the measured crack length used for the computation of both hardness and fracture toughness. The Vickers indent diagonals (D_1 and D_2) are equivalent to the length of "2a" as indicated in Fig. 1.

2.5.4 Accelerated Ageing Test

The first step for ageing test was obligatory required the samples to be stored inside the acid digestion vessel whilst being subjected to hydrothermal conditions for increasing time intervals. Consecutively, the samples were then be immersed inside distilled water with 20 ml in measurement. After the immersion process, it later required the samples to be transferred to an oven for heating and baking process as well as provided with temperature below 180°C for duration of five hours. Segregations of samples were later to be done as according to the corresponding ageing time intervals and subsequently examined using X-ray diffraction (XRD) to monitor the rate of zirconia phase transformation.

3 Results and Discussion

3.1 Relative Density

By referring to the plotted graph in Figure 3, the lowest identified relative density was 55% which consist of 1 wt% of SS316 as well as sintered at 1250°C. The relative density for 1 wt% of SS316 then continue to increased when sintering temperature increased from 1250°C to 1400°C and remained static for sintering temperature between 1400°C to 1500°C. For 0 wt% SS316 and 0.5 wt% SS316, the results showed that the maximum sintering temperature was 1400°C due to obtained the highest relative densities which were 93% and 92% respectively. In overall, considering sintering temperature between 1250°C and 1300°C, the relative densities increased when weight percentage of stainless steel increased. It can be observed that adding dopant SS316 by 0.5 wt% to Y-TZP was beneficial to enhance the densification but provided that it must be sintered at low temperature of 1300°C.

Nonetheless taking into account of sintering temperature between 1400°C and 1500°C, all of the samples exhibit no improvement or decreased in relative densities. This can be supported that grain coarsening could possible cause the decreased in densities as higher sintering temperature was applied. In addition, when a higher sintering temperature was applied such that exceeding 1500°C, zirconia underwent partial transformation such that transform into monoclinic or cubic phases [28]. The samples doped with 0.5 wt% SS316 indicated to exhibit the optimum benefit in terms of densification where it attained the highest relative density of 90% and 89% when sintered at temperature 1300°C and 1500°C. For the case 0 wt% SS316 which denoted as undoped or pure Y-TZP, the achieved density was 72% when sintered at 1300°C. Considering that sintering temperature was identical such that at 1300°C but different weight percentage of SS316 such that 0 wt% SS316 and 0.5 wt% SS316, the samples of doped 0.5 wt% SS316 displayed a higher relative density as compared to 0 wt% SS316. This in turn proved that adding SS316 into Y-TZP indeed contributed an enhancement for consolidation process but must in line with condition of lower sintering temperature such that below 1500°C.

Taking into account of sintering temperature between 1250°C and 1300°C which indicated nearly all samples showed an increased, as weight percentage of the dopant increased, the samples relative density indicated as increased. This was considered as truthful as compared to theory such that it mentioned stainless steel density increased when heating rate increased and therefore, promote the increased in samples densities [10]. This also means that samples with more weight percentage of stainless steel exhibits a higher density due to the total accumulated of an increased in densities based on the amount. However, considering the opposite case such that for 0.5 wt% SS316 which showed a decreased in relative density when sintered from 1250°C to 1300°C, the trend line was the opposite way. The reason for this may be due to method of processing for the samples such as shorter sintering holding time. A shorter holding time may cause the reverse of the material density value such that it became decreased [11]. Another possible reason for the decreased in relative density was due to uneven powder composition. It considered as extremely difficult to obtain the most ideal powder composition.

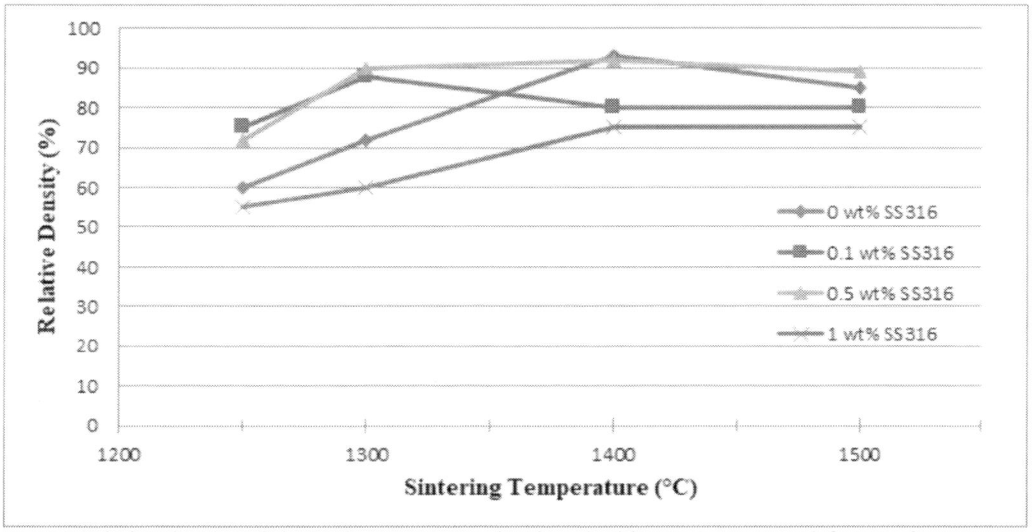

Fig. 3. Relative density of sintered Y-TZP ceramics doped with 0.1 to 1 wt% stainless steel SS316.

3.2 Vickers Hardness

As according to the plotted graph in Figure 4, all samples exposed to have an increased in Vickers hardness when considering sintering temperature was increased from 1250°C to 1300°C. As sintering temperature continued to increase from 1300°C to 1400°C, all doped samples showed a rapid declined in Vickers hardness but for undoped sample on the other hand indicated as increased. When sintering temperature increased higher from 1400°C to 1500°C, all samples for both doped and undoped presented as decreased. Considering sintering temperature at 1300°C, sample of 0.5 wt% SS316 revealed to achieved the highest among all samples such that 13.3 GPa. In comparison with 0 wt% SS316 which denoted as pure Y-TZP or undoped sample, 0.5 wt% SS316 showed a higher Vickers hardness than 0 wt% SS316 with values 13.3 GPa and 12 GPa respectively. For all doped samples, the optimum sintering temperature to obtain the highest sintering temperature was 1300°C.

At sintering temperature of 1300°C, the Vickers hardness values for samples of 0.1 wt% SS316, 0.5 wt% SS316 and 1 wt% SS316 were 13 GPa, 13.3 GPa and 11.7 GPa. For undoped sample of 0 wt% SS316, the optimum sintering temperature was 1400°C in which it attained it highest Vickers hardness value of 13.5 GPa. Considering an increased sintering temperature between 1250°C to 1300°C, all samples exhibit improvement of Vickers hardness due to sintered body strengthens densification. Nonetheless, for the case of decreased in Vickers hardness as sintering temperature continued to increase from 1300°C to 1500°C, this issue occurred due to grain coarsening that related to cubic structure [12]. By referring at sintering temperature of 1500°C, Vickers hardness decreased as weight percentage of SS316 increased. This likewise follows the trend line as well as obeys the mixture law such that explained Y-TZP was harder than SS316.

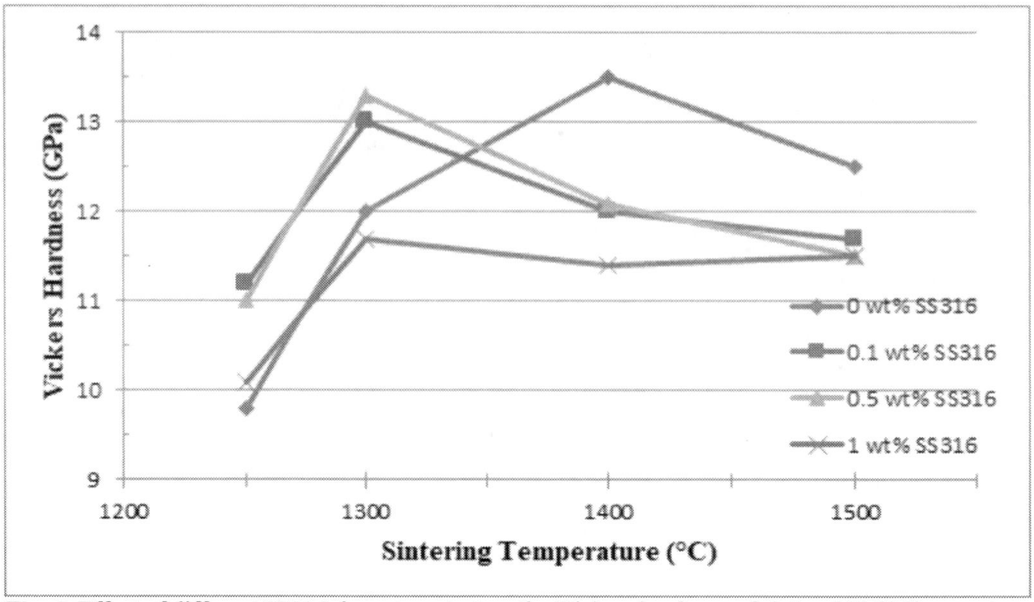

Fig. 4. Effect of different SS316 dopant content on the Vickers hardness of sintered Y-TZP ceramics.

3.3 Fracture Toughness

Based on the plotted graph in Figure 5, samples of 0 wt% SS316, 0.1 wt% SS316 and 1 wt% SS316 profound to exhibit an increased in fracture toughness when sintering temperature increased from 1250°C to 1300°C. On the contrast side, sample of 0.5 wt% SS316 displayed a decreased in fracture toughness when sintering temperature increased from 1250°C to 1400°C. At sintering temperature of 1300°C, it indicated as the optimum sintering temperature for samples of 0 wt% SS316, 0.1 wt% SS316 and 1 wt% SS316 as all samples showed their highest fracture toughness with values 5.5 MPa.m$^{1/2}$, 5.5 MPa.m$^{1/2}$ and 6 MPa.m$^{1/2}$ respectively. As sintering temperature increased from 1400°C to 1500°C, samples of 0 wt% SS316, 0.1 wt% SS316 and 0.5 wt% SS316 identified as increased in fracture toughness whereas sample of 1 wt% SS316 showed a decreased in fracture toughness. At sintering temperature of 1500°C, sample of 0.5 wt% SS316 has the highest fracture toughness with value of 5.7 MPa.m$^{1/2}$ and 1500°C was declared as its optimum sintering temperature.

Mismatched of thermal expansions occurred at the section between stainless steel and zirconia particles provided when there was an increased in stainless steel particles [13]. Hence, radial tensile stress was developed internally during cooling process and consequently contributed towards particles with inadequate interfacial bonds which cause debonding. In addition, crack also then formed in a method that it smoothly inseminates between the unsteady interface of zirconia and stainless steel particles [33]. This in turn leads to the enhancement of fracture toughness via the process of crack deflection. Sintering temperature on the other hand suggested that it marginally affected the fracture toughness and this also considering that the all wt% SS316 exhibit fracture toughness with values varied from 5.2 MPa.m$^{1/2}$ to 5.7 MPa.m$^{1/2}$ when sintered at sintering temperature of 1500°C. In this case, homogenization developed over stabilizer which also existed within the samples ZrO_2 grains and thus, produce an improved stability of tetragonal grains [14].

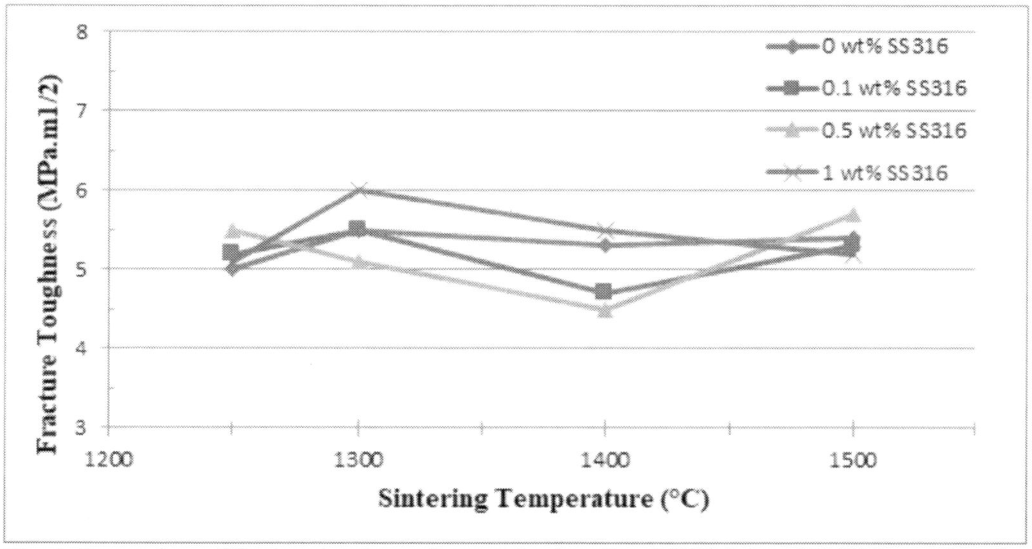

Fig. 5. Comparison of fracture toughness for pure (undoped) and SS316-doped Y-TZP ceramics after sintering at various temperatures between 1250°C and 1500°C.

3.4 Ageing Resistance

The accelerated ageing test was subjected to selected samples with aim to determine the samples ageing resistance due to low temperature degradation (LTD) when exposed towards humid environments. Generally, samples were subjected to superheated steam at temperature of 180°C as well as 10 bar of pressure with increasing duration time. In this case, XRD analysis was utilised with the objective to monitor the samples phase content as well as identified the monoclinic zirconia (m-ZrO_2) phase content. Figure 6 shows the monoclinic phase content for samples of 0 wt% SS316 and 0.1 wt% SS316 accordingly with increasing ageing duration up to 24 hours. The monoclinic phase content was monitored for ageing duration of 0, 1, 6, 12 and 24 hours respectively. Sample of 0.1 wt% SS316 with sintering temperature of 1300°C was chosen as reference due to optimum composition and consist of the highest mechanical properties. By observing the trend line on the plotted graph of monoclinic phase content against ageing duration in figure 6 above, it has been identified that pure Y-TZP or known as 0 wt% SS316 exhibit a rapidly continuous increasing trend line. The sample for 0 wt% SS316 then breaks during the ageing treatment for the next 12 hours which tells that data was absence for analysis.

After an hour of exposure, 0 wt% SS316 displayed 11 vol% of monoclinic phase content which was higher and drastic in comparison to 0.1 wt% SS316 that presented as 6.5 vol%. As approaching 24 hours of exposure, 0 wt% SS316 closed the monoclinic phase content at 91 vol%. This essentially means that the longer the ageing duration, the higher the development of monoclinic phase content [15]. In addition, this was considered as not ideal due to monoclinic promote failure of material. The more the monoclinic phase, the worst the material properties become. Moreover, development of crack and surface spalling were observed at the sample via visual inspection as results of progressive ageing [15]. On the other hand, sample of 0.1 wt% SS316 had presented a better ageing resistance but sample then disintegrated at 12 hours which closed the monoclinic phase content with 20 vol%. This can be proved based on the trend line for 0.1 wt% SS316 stopped at 12 hours.

The disintegrated of samples will be consideration for further study. By comparing both samples of 0 wt% SS316 and 0.1 wt% SS316 at similar 12 hours of exposure, 0.1 wt%

SS316 was considered ideal as compared to 0 wt% SS316 due to it showed significant low of monoclinic phase content. This proved that 0.1 wt% SS316 exhibit better ageing resistance and phenomena possibly due to finer grain size obtained through sintering process. Hence, sample of 0 wt% SS316 can be further upgrade to prevent the disintegration in the future works. Pure Y-TZP (0 wt% SS316) was considered as instable which leads to lower ageing resistance. Hence, it changes phases easily from tetragonal to monoclinic supposed that polar water molecules penetrate the sample particles. Phase transformation occurs in such a way that if one of the grain experienced stress, this will interfere the neighbouring grain and hence, phase changes take placed and promote low temperature degradation. This project also aims to achieve an ideal stable matrix in order to prevent phase transformation. By doing so, ageing rate likewise can be minimised via the reduction of deterioration rate, microcracking and surface roughening.

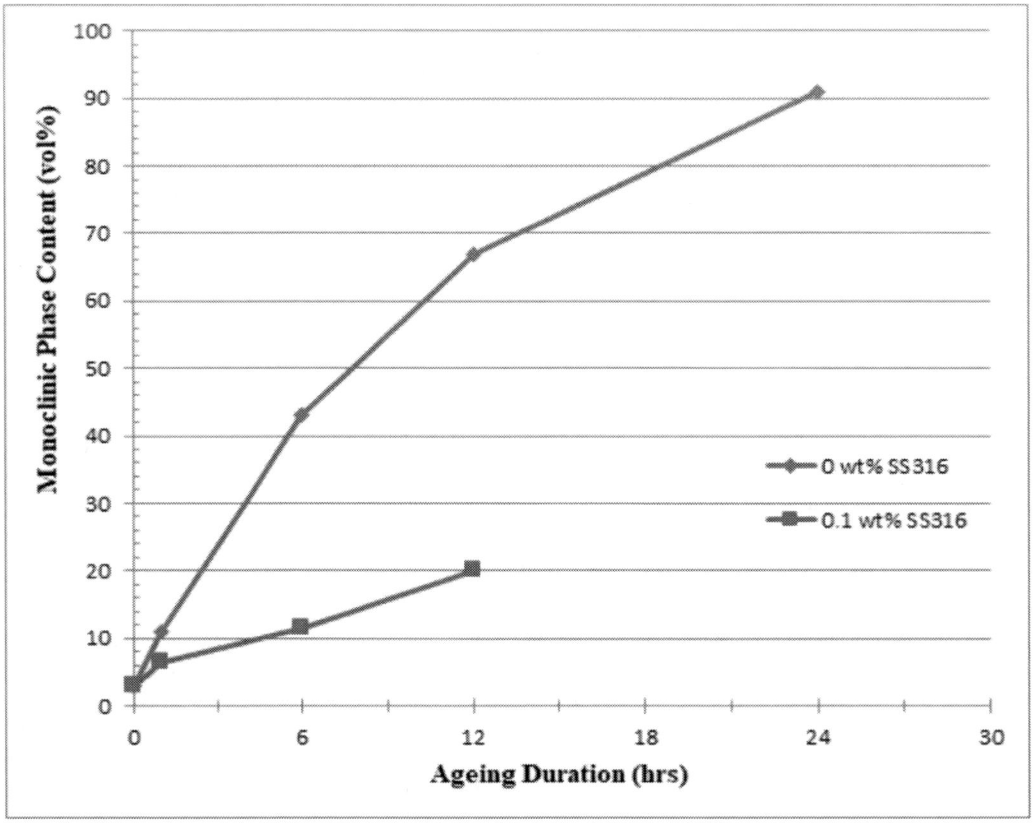

Fig. 6. Monoclinic zirconia phase content of undoped and 0.1 wt% SS316-doped Y-TZP samples at various accelerated ageing time intervals of exposure at hydrothermal conditions of 180°C and 10 bar.

4 Conclusion

The experiment has successfully been conducted to investigate the advantageous and effects of adding sintering additive stainless steel 316 to be doped with Y-TZP ceramics. It has been identified that the highest density would be 93% for the case of 0 wt% SS316 with sintered at sintering temperature of 1400°C. For Vickers hardness, the lowest value profound to be 9.8 GPa when taking into account of 0 wt% SS316. In addition, the optimum value for fracture toughness would be 6 $MPa.m^{1/2}$ which considering for the 1 wt% SS316. By referencing to the all the plotted graph for all wt% of SS316 for sintering temperature between 1400°C to 1500°C, the effects of adding stainless steel 316 to be doped with Y-TZP were such that the relative density nearly static, Vickers hardness decrease and fracture toughness increase. A low monoclinic phase content for samples was also the targeted outcome.

References

1. K.J. Burg, S. Porter, J.F. Kellam, Biomater., **21**, 2347–2359 (2000).
2. E.M. Maynard, Annu. Rev. Biomed. Eng., **3**, 145–68 (2001).
3. S.M. Kurtz, C. Arnholt, R. Huet, M. Ueno, W.L. Walter, J. Mech. Behav. Biomed. Mater., **31**, 107–116 (2014).
4. S. Lopez-Esteban, J.F. Bartolome, C. Pecharroman, S.R.H. Mello Castanho, J.S. Moya, Mater. Res., **4**, 217–222 (2001).
5. K.L. Lin, M. Singh, R. Asthana, J. Eur. Ceram. Soc., **34**, 355–372 (2014).
6. I. Espitia-Cabrera, H. Orozco-Hernández, R. Torres-Sánchez, M. E. Contreras-García, P. Bartolo-Pérez, L. Martínez, Mater. Lett., **58**, 191–195 (2004).
7. A. Balamurugan, S. Kannan, S. Rajeswari, Mater. Lett., **57**, 4202–4205 (2003).
8. V.K. Balla, P.P. Bandyopadhyay, S. Bose, A. Bandyopadhyay, Scr. Mater., **57**, 861–864 (2007).
9. Lenntech Com, "Stainless steel mechanical properties," Stainl. steel Mech. Prop. (2012).
10. C. Lorenzo-Martin, J.A. Flores-Vazquez, D. Gomez-Garcia, A. Monuz-Bernabe, A. Domiguez-Rodriguez, D. XueMing, J. Gomez-Herrero, J. Eur. Ceram. Soc., **22**, 2603–2607 (2002).
11. S. Butković, M. Oruč, E. Šarić, M. Mehmedović, Mater. Tehnol., **46**, 185–190 (2012).
12. S. Du, L. Xu, X. Zhang, P. Hu, W. Han, Mater. Chem. Phys., **116**, 76–80 (2009).
13. R.M. German, Crit. Rev. Solid State Mater. Sci., **35**, 263–305 (2010).
14. G. Wang, S. Zhang, C. Bian, H. Kong, J. Prosthet. Dent., **112**, 576–583 (2014).
15. E. Zapata-Solvas, D. Gómez-García, C. García-Gañán, and A. Domínguez-Rodríguez, J. Eur. Ceram. Soc., **27**, 3325–3329 (2007).

Design and Analysis of Collapsible Scissor Bridge

Mohamad Nabil Aklif Biro*, and *Noor Zafirah* Abu Bakar

Department of Mechanical Engineering, School of Engineering, Taylor's University, Malaysia

Abstract. Collapsible scissor bridge is a portable bridge that can be deployed during emergency state to access remote areas that are affected by disaster such as flood. The objective of this research is to design a collapsible scissor bridge which is able to be transported by a 4x4 vehicle and to be deployed to connect remote areas. The design is done by using Solidworks and numerical analysis for structural strength is conducted via ANSYS. The research starts with parameters setting and modelling. Finite element analysis is conducted to analyze the strength by determining the safety factor of the bridge. Kutzbach equation is also analyzed to ensure that the mechanism is able to meet the targeted degree of motion. There are five major components of the scissor structure; pin, deck, cross shaft and deck shaft. The structure is controlled by hydraulic pump driven by a motor for the motions. Material used in simulation is A36 structural steel due to limited library in ANSYS. However, the proposed material is Fiber Reinforced Polymer (FRP) composites as they have a high strength to weight ratio. FRP also tends to be corrosion resistance and this characteristic is useful in flooded area.

1 Introduction

Flood is one of the common disasters in tropical countries. Malaysia East Coasts Environmental Research Institute of University Sultan Zainal Abiden (USZA) claimed that 29,800 km^2 of vulnerable area to this disaster sourcing from a few numbers of river basins in peninsular Malaysia and almost 4.82 million residents are at risk especially during monsoon seasons [1]. As a result, certain areas that are heavily affected have high potential to loss access to the outside world during flood.

According to the Annual Report 2014 by Malaysia Rescue Team or BOMBA, it is suggested that they revise their existing disaster Standard Operating Procedure [2]. It is also mentioned in the report that they would have to increase logistics capabilities such as adding more equipment and have more boats and trucks to standby for this disastrous phenomenon. However, this suggestion might not be the best solution to this problem in accessing remote areas. In conjunction to that, Research team at Hiroshima University had successfully developed and tested a folding bridge which is inspired by Japanese traditional origami [3]. The bridge is intended to be deployed during chaotic situations. Nevertheless,

* Corresponding author: nabil.aklif@yahoo.com

the structure made is heavy and complex and took about one hour to assemble the components [4].

Adapting from the Nuremberg scissor concept, a collapsible bridge is one the solutions to provide passage for evacuation or bring in supplies. Nuremberg scissors is a planar mechanism consisting several links and pins forming a scissor-like structure which has the ability to collapse and expand with a given ratio. This bridge is able to operate with a minimum amount of force to expand and retract. With a force in the horizontal direction of the link, it provides a collateral force to the other link and expands the bridge. The motion of the structure must satisfy Kutzbach equation or criterion by determining the degree of freedom of the mechanism. This is done by obtaining the number of links and number and types of joints that is included in the system as a whole [5].

The project objectives are to design a collapsible light weight scissor-like bridge, to simulate its motion and to assess its structural strength. The mechanism of the bridge must satisfy Kutzbach equation which produces two degree of motion with specific number of links and joints.

The design scope does not consider the ground elevation or ground damping during flood. Soil damping naturally occur to areas where rivers are near and also during heavy rainfall that causes flood. Upon operation of the bridge at the area, ground condition might cause the bridge to operate at a certain angle which is not specifically analyzed in this project. Regarding the material, the proposed material for future development is FRP. Due to limited library function in ANSYS, A36 structural steel is chosen for numerical simulation.

2 Research Methodology

2.1 Design Parameters

The bridge is designed to fit a typical 4x4 pick-up truck which has average dimensions of width, W_V: 1860 mm and height, H_V: 1815 mm. Thus, the outer dimensions of the structure are set to be slightly bigger with width, W_S: 2200 mm and height, H_S: 2000 mm. The expansion ratio is adapted from an industrial scissor lift which has a common ratio of 1:5. Thus, the minimum length for the expansion mode of the bridge is 14000 mm. The condition of expansion and retraction modes are shown in Fig. 1.

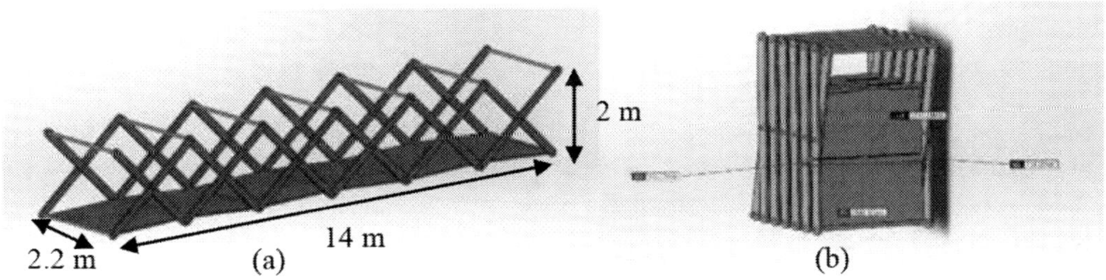

Fig. 1. Scissor structure (a) Expansion mode and (b) Retraction mode

The structure is designed to withstand maximum allowable load which are the summation of the total weight of all bridge components and the gross weight of the vehicle which is 1950 kg. The gross weight of the vehicle consists of the weight of the vehicle plus the passengers and cargo weight.

2.2 Components Design

The components of the bridge are categorized into structural components and auxiliary components. Fig. 2 shows the components as a complete structure which consists of six sets of scissor link cross, auxiliary components (pump and hydraulic cylinder) and structural support to mount the auxiliary components and the first set of scissor link cross.

2.2.1 Structural components

The set of scissor link cross consists of four links, a deck, a cross shaft, a deck shaft and a pin as shown in Fig. 3. There are 56 parts in total for the structure. SolidWorks is used to create the design and Motion Study function is used to visualize the motion of the structure.

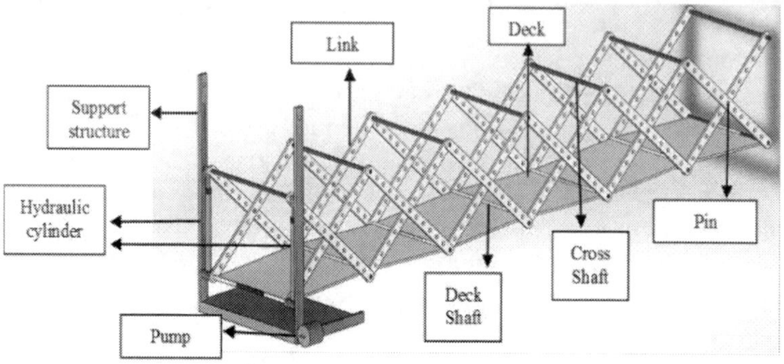

Fig. 2. Complete model of the bridge

Fig. 3. Structural components : (a) Link, (b) deck, (c) cross shaft, and (d) deck shaft

The dimension of the link is selected according to the optimum strength and weight of the link. These links hold up 40% of the total weight of the bridge and are considered as a major component to the structure. The deck serves as the main unit to allow traffic on the scissor bridge to pass through. The design of the deck is made with the interlocking method. With multiple arrangement of the deck, it can be combined to expand the length. Another feature of using this interlocking design is it allows the folding mechanism for the bridge to retract to the folded state. The cross shaft holds the links together and provides angular movement to the links. This cross shaft is designed in the shape of a normal cylindrical rod. Deck shaft serves similar purpose to the cross shaft except that in the middle part it has a flat surface to allow the deck to rest on it when the structure is opened at horizontal state. The pin serves a simple purpose to the structure which is to hold the links in position and to provide angular movement or rotation to the links. In other words, the pin acts similar to a bearing. However, in the real scale scissor bridge, this pin must accompanied with sets of bearing as it is an integrated component with the links.

2.2.1 Auxiliary components

The auxiliary components consisted of a hydraulic cylinder, single-acting type and a pump. Single-acting cylinder as shown in Fig. 4 is a type of cylinder that is pressurized at only one end and the other end is vented or exposed to the atmosphere [6]. The motion of the scissor bridge is governed by the principle of work and energy equation as shown in Eq.(1).

$$T_1 + \sum U_{1\text{-}2} = T_2 \tag{1}$$

$$T_i = \tfrac{1}{2} m v_i^2 \tag{2}$$

$$\sum U_{1\text{-}2} = F\Delta d + mgh \tag{3}$$

T_1 represents the kinetic energy at initial state and T_2 is the kinetic energy at final state as shown in Eq.(2). Velocity v_1 at initial state is always equal to zero and velocity v_2 at final state is a dependent variable. $\sum U_{1\text{-}2}$ consists of the work done by the piston inside the hydraulic cylinder and work done by gravitational force as shown in Eq.(3). The hydraulic cylinder force F is caused by hydraulic pressure and Δd is the piston displacement. The variable m is the total mass of the structure and g is the gravitational acceleration which is assume to be 9.81 m/s^2 and both are constant. The vertical distance travelled by the bridge during expansion and retraction is given by h and can be found during the motion simulation. By solving Eq.(1-3), the hydraulic cylinder force F can be determined. By applying Eq.(4), the power of the pump P required to move the structure can be calculated. The pressure inside the cylinder p can also be determined by using Eq.(5) where A is the cylinder's constant cross sectional area.

$$P = Fv_2 \tag{4}$$

$$p = F/A \tag{5}$$

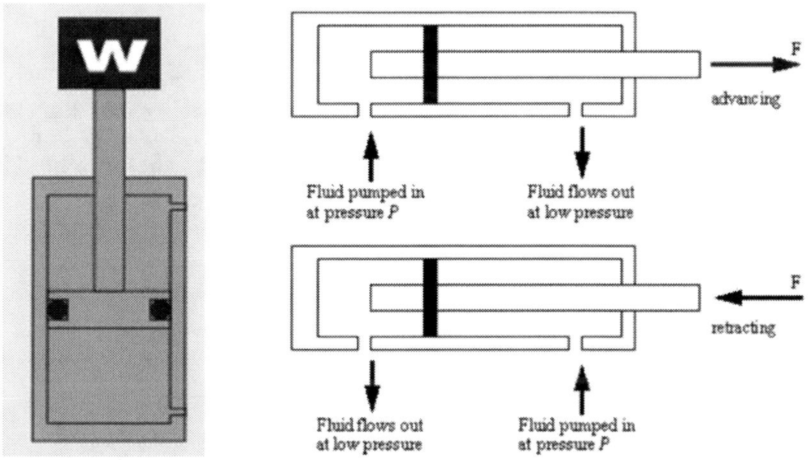

Fig.4. Single-acting cylinder [6]

2.3 Structural Analysis

2.3.1 Stress analysis by ANSYS Workbench (Static Structural)

Stress analysis is simulated by utilizing ANSYS Workbench with the feature of Static Structural. In this analysis, the main objective is to validate whether the structure assembly is able to withstand the targeted load by showing a major deformation. In ANSYS Workbench, the simulation is done with a few fixed parameters. Meshing is done by using coarse span angle center for a faster simulation and meshing processes. Element size is set to default which the software decides the best element size for the model. The nodes obtained from this meshing is 302927 and generated 131688 elements, which shows sufficient numbers of element to proceed the analysis.

Safety factor is determined from the numerical result by using Eq.(6).

$$\textit{Safety factor} = \textit{Yield stress/Allowable Stress} \tag{6}$$

2.4 Motion analysis

2.4.1 SolidWorks Motion Study

Motion Study is done by Solidworks. In this function, force is applied to the piston to provide movement to the structure and the animation of the structure is successfully developed. The motion is vice versa as the force to the piston is directed to the opposite side.

2.4.2 Kutzbach equation (Degree of Freedom)

It is important to determine the degree of freedom of a design or mechanism to ensure the movability. This is done by determining the number of links and joints included in the system. In this case, taking one set of the scissor structure should suffice to prove that the degree of freedom is two; in horizontal and vertical motion of the structure. Kutzbach equation is shown in Eq.(7).

$$n = 3(l-1) - 2j - h \tag{7}$$

Where, n = number of degree of freedom, l = total number of links, j = total number of primary joints and h = total number of higher-order joints.

2.5 Material Selection

As suggested by studies, Fibre Reinforced Polymer composites has some advantages when compared to conventional construction materials [7]. FRP is very strong as it dominates the high specific strength and stiffness. It also has a high strength to weight ratio which means the future scissor bridge can be lighter than expected if FRP is chosen to be its main material [8]. Other than that, it is a corrosion resistance material and this could reduce the maintenance cost of the bridge [9]. As claimed by studies, deck with FRP are usually multilayer which is also known popularly as sandwiches. The weight of this FRP is approximately lower by 20% compared to the conventional concrete slab [7]. However, this section is not covered in detail as it is not in the scope of the project and this preliminary study should suffice for the project up to this stage. Throughout this project, A36 structural steel is selected for simulation due to the limitations experienced as ANSYS library is not able to support mixture of material properties.

3 Results and Discussion

3.1 Dimension Table

Several design iterations are needed to ensure that the structure is at its optimum strength and meet the scope of the project. Table 1 below shows the summary of the dimensions of the components for the scissor structure.

Table 1. Components list and design specifications

Component	Function	Dimension	Material
Link	• Structural support	• Length: 3312 mm • Hole diameter: 80 mm	Steel (A36)
Cross Shaft	• Structural support • Provide angular movement	• Length: 2440 mm • Diameter: 78 mm	Steel (A36)

Deck	• Main feature as platform for usage	• Length: 2340 mm • Width: 2200 mm • Hole diameter: 78 mm	Steel (A36)
Pin	• Minor structural support • Provide angular movement	• Length: 120 mm • Diameter: 78 mm	Steel (A36)
Deck Shaft	• Structural support • Provide angular movement • Support deck placement	• Length: 2440 mm • Diameter: 78 mm	Steel (A36)

3.2 Stress analysis

Fig. 5 shows the forces applied to the structure which in total of 210070 N. These forces are assumed to be the forces that will be exerted by the 2.6 unit of 4x4 pick-up truck during its operation. In this simulation, both of the end of the structure is set to be the fixed support.

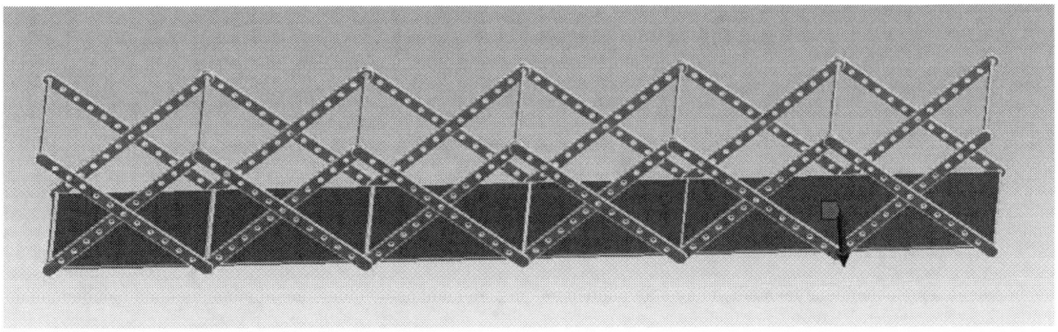

Fig.5. Uniform load applied to the structure

Fig. 6 shows the total deformation of the scissor structure after load of 210070 N is applied to the middle section (red contour). The maximum deformation is 15.741 mm. This shows that when the targeted load which is the typical 4x4 pick-up truck passes by, the structure is able to withstand the force exerted. Take note that the load applied is the gross weight of the truck which means it is inclusive of the weight of the vehicle with its cargo on the back side. With remarks from a study made by a group of researchers, the similar design of structure passed the experimental test to carry two set of loads; 13.6 kN and 10 kN and also it is claimed that the analytical value had less than 10% of error [10].

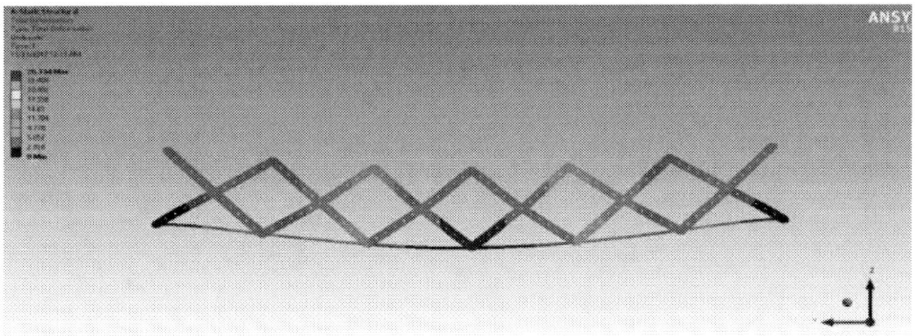

Fig.6. Total Deformation contour of scissor structure

There are two type of stresses that can be obtained from ANSYS Static Structural analysis, Von-Mises stress (equivalent stress) and also maximum principle stress. Based on Fig. 7, the highest value of maximum principle stress is 138.64 MPa and from Fig. 8 the equivalent stress has the value of 149.48 MPa at maximum.

Fig.7. Maximum Principle Stress on structure

Fig.8. Von-Mises stress

These values are used to obtain the safety factor of the scissor bridge by using the Eq.(6). In this case, yield stress is provided by the properties of A36 structural steel which is 250 MPa. By using the value from maximum principle stress and equivalent stress (von-mises) which is 138.64 MPa and 149.48 MPa respectively, the factor of safety is obtained.

From maximum principle stress, the safety factor is 1.803. From equivalent stress, the safety factor is 1.672.

These values support that the structure is able to withstand the load of 210070 N and to clarify this value, the load value used is a summation of the total weight of the structure and the maximum load of 2.6 units of the trucks on the bridge by taking one length of the truck, 5277 mm and fitting as many truck as possible onto the 14000 mm scissor bridge.

However, this value changes when only one 4x4 truck is allowed to pass at one time. When this condition is used, the load applied will be only 19100 N and the maximum deflection is 13.451 mm with 118.47 MPa of maximum principle stress and 127.73 MPa of equivalent stress (von-mises). These values can be found in Fig. 9, Fig. 10 and Fig. 11.

Fig.9. Total deflection for one truck loading

Fig.10. Maximum principle stress for one truck loading

Fig.11. Von-mises stress for one truck loading

3.3 Motion Analysis

3.3.1 Force Needed to Operate Scissor Bridge by Work-Energy Principle

The force required for the structure to expand is governed by Eq.(1- 3) with an input of velocity in kinetic energy equation.

Table 2 below shows the variation of velocity input and force required. From Table 2, the optimum velocity of the structure is 4.43 m/s as it requires minimum amount of force to expand, which is 40.06184 N and force to retract is 324393.8 N. Higher force is needed to retract because of the extra work needed to counter the gravitational force downwards as the piston needs to move upwards.

Table 2. Calculation summary for Varied Velocity

Velocity of Structure (m/s)	Force to expand (N)	Force to retract (kN)
4	-29.5967	294.4308
4.3	-9.23875	315.0129
4.43	0.040062	324.3938
5	43.98626	368.8238
6	133.921	459.7485
7	240.2075	567.205

Fixing the cross section area of the cylindrical hydraulic A with design of the piston radius of 0.125 m, the pressure inside is P = 6608.48 kPa by applying Eq.(5). For the selection of motor specification by power, the force and the velocity are multiplied as shown in Eq.(4) and the values can be compared with readily available motor in the market.

3.3.2 Kutzbach equation

By extracting one set of the scissor structure, the number of links is l = 5, the number of primary joints is j = 5, and the number of higher-order joints is h = 0. By applying Eq.(7), the number of motion or degree of freedom is n = 2. Thus, it proved that the design mechanism is workable and able to produce two kind motions which are vertically and horizontally as shown in Fig. 12.

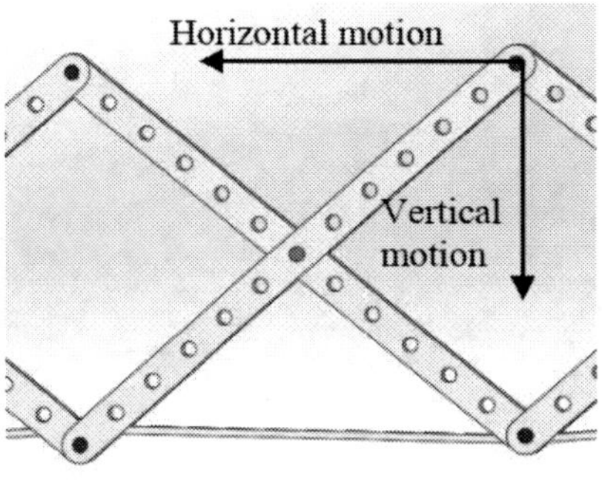

Fig.12 One set of scissor structure motions

4 Conclusion and Recommendation

This paper presents a thorough theoretical and numerical analysis of the scissor structure accompanied with the auxiliary components. Design of the structure are based on the literature reviews and cited studies from other research papers supported with structure analysis in results and discussion section. The structure able to satisfy Kutzbach equation and has satisfactory safety factor. Hence, the objectives of the project are also met Following from that, this project has numerical analysis on the collapsible bridge and also

This paper also reveals the basic design of the collapsible scissor bridge by parts and each function of the parts are justified.

Moving forward, numerical analysis using FRP as main material should be further investigated to design the bridge to be more lightweight and portable. The fluid analysis on the structure caused by wind can also be a potential area to analyze to ensure that the structure design is more robust.

5 References

1. M.B. Gasim, M.E. Toriman, M.G. Abdullahi, FLOODS IN MALAYSIA Historical Reviews, Causes, Effects and Mitigations Approach, IJIR **2**(4) 59-65 (2014).
2. JBPM, Annual Report 2014, Fire and Rescue Department of Malaysia (2014).
3. Y. Chikahiro, I. Ario, M. Nakazawa, S. Ono, J. Holnicki-Szulc, P. Pawlowski, C. Graczykowski, An experimental study on the design method of a real-sized Mobile Bridge for a moving vehicle, MRAS IV **136** 93 (2014).
4. N. Bronner, MB 4.0: The Future Of Bridge Technology. http://www.pitt.edu/~nbb8/eti.html. (Accessed 18-Nov-2017 2017).
5. R. Khurmi, J. Gupta, *Theory of machines*, Eurasia (1976).
6. M. Cylinder, Design Engineering Guide, Actuant Corp (2012).
7. P.B. Potyrala, Use of Fibre Reinforced Polymeromposites in Bridge Construction. State of the Art in Hybrid and All-composite Structures, Universitat Politècnica de Catalunya. Escola Tècnica Superior d'Enginyers de Camins, Canals i Ports de Barcelona. Departament d'Enginyeria de la Construcció (2011).
8. T.A. Hoffard, L.J. Malvar, Fiber-reinforced polymer composites in bridges: a state-of-the-art report, Naval Facilities Engineering command Port Hueneme ca Engineering Service Center (2005).
9. T. Keller, Use of Fibre Reinforted Polvmers in Bridge Coni truc tion, (2003).
10. Y. Chikahiro, I. Ario, M. Nakazawa, S. Ono, J. Holnicki-Szulc, P. Pawlowski, C. Graczykowski, Structural Analysis and Experimental Study for realsized MB Travelable Vehicles, *Proc. of the 8th European Nonlinear Dynamics Conference* pp. 1-2. (2014)

Comparisons of Flow Patterns over a Hierarchical and a Non-hierarchical Surface in Relation to Biofouling Control

Mohammed Ridha Bin Ahmad Fawzan[1] and *Felicia* Wong Yen Myan[1,*]

1Department of Mechanical Engineering, School of Engineering, Taylor's University, 1, Jalan Taylors, 47500 Subang Jaya, Selangor, Malaysia.

Abstract. Biofouling can be defined as unwanted deposition and development of organisms on submerged surfaces. It is a major problem as it causes water contamination, infrastructures damage and increase in maintenance and operational cost especially in the shipping industry. There are a few methods that can prevent this problem. One of the most effective methods which is using chemicals particularly Tributyltin has been banned due to adverse effects on the environment. One of the non-toxic methods found to be effective is surface modification which involves altering the surface topography so that it becomes a low-fouling or a non-stick surface to biofouling organisms. Current literature suggested that non-hierarchical topographies has lower antifouling performance compared to hierarchical topographies. It is still unclear if the effects of the flow on these topographies could have aided in their antifouling properties. This research will use Computational Fluid Dynamics (CFD) simulations to study the flow on these two topographies which also involves comparison study of the topographies used. According to the results obtained, it is shown that hierarchical topography has higher antifouling performance compared to non-hierarchical topography. This is because the fluid characteristics at the hierarchical topography is more favorable in controlling biofouling. In addition, hierarchical topography has higher wall shear stress distribution compared to non-hierarchical topography

1 Introduction

Biofouling is any unwanted deposition and development of organisms on submerged surfaces [1]. It is a major problem as it contaminates water and medical instruments inflicting health risk. Biofouling also causes unnecessary drag leading to low fuel efficiency of marine industry and reduces operational range. Other than that, biofouling promotes physical damage to infrastructures resulting in increased maintenance cost [2]. US$2.3 million were spent annually on a vessel due to biofouling of the vessel hull which causes increase in fuel consumption by 20.4% due to drag [3]. In oil and gas sector, the cost to

* Corresponding author: feliciayenmyan.wong@taylors.edu.my

repair equipment damaged by biofouling is approximately US$ 960 to US$ 1280 million [4]. These examples show that it is important for researchers to develop antifouling measures to control this problem.

T. Vladkova [2] explained in her study that there are three principles to control biofouling. The first is to use the mechanical approach to detach the biofoulers from the surface. The second approach is the application of chemical agents to exterminate or inactivate the biofouling organisms. The last approach is surface modification which involves altering either surface chemical composition or surface topography so that it becomes a low-fouling or non-stick surface to biofouling organisms [5]. Rosenhahn et al. [6] also recommended similar approach which include toxic antifouling coatings, fouling release coating and foul release via surface modification as shown in Figure 1.

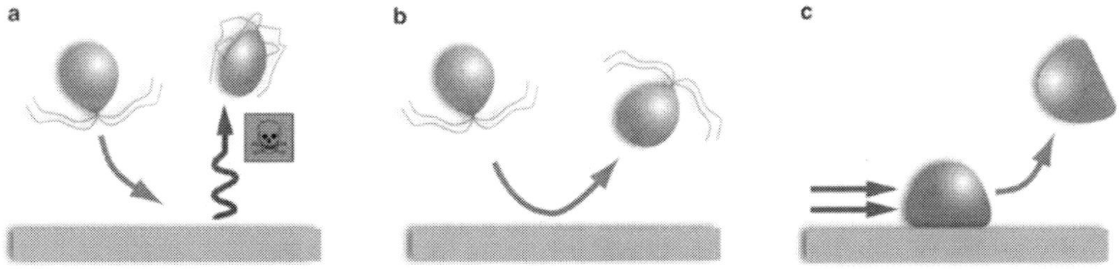

Fig. 1. Approaches to prevent biofouling: (a) toxic antifouling, (b) foul release, (c) foul inhibition [6]

A variety of chemical approaches were used to prevent biofouling of surfaces and some of it has been banned because of its adverse effects on the environment. Decades ago, a type of biocide named Tributyltin (TBT) was used as a coating on vessel hulls to prevent biofouling and by far this is the most effective method to mitigate biofouling in the history of antifouling technology to date [7]. However, this method was banned by the International Maritime Organisation (IMO) in November 2001. This is because IMO discovered that TBT causes detrimental effects on the environment like abnormal growth of the oysters, imposex (female gastropods developing male characteristics) and others [8–10]. K. A. Dafforn [7] also mentioned about the usage of copper as a co-biocide to increase the efficacy of TBT in paints for vessel hulls [11]. Copper is essential for metabolic functioning but it becomes toxic when the bioavailability level of copper exceeds the tolerance level of an organism [12]. The adverse effects of copper on the environment also eventually lead to restrictions of copper usage in paints for vessel hulls in countries like Canada, Denmark and the United States [13]. These were two popular chemical approaches that were used to control biofouling.

As mentioned above, surface modification is another method that can be used for antifouling applications. This method is preferable over other chemical approaches because it is environmental friendly. Previous studies have investigated the effects of surface topographies modification and the results are promising. The surface topographies' modification includes modifying the size and geometry. In recent years, these studies apply Computational Fluid Dynamics (CFD) to study the flow of the fluid on top of these modified surfaces [14]. By studying the flow, researchers can better ascertain the optimum size and geometry of surface topographies that can control biofouling effectively. More studies using this method is needed because it is relatively new compared to other methods and also to measure its efficacy for antifouling applications.

2 Research Methodology

2.1 Geometry Modeling

Solidworks 2013 was used in creating the fluid domain. The fluid domain consists of a rectangular block as shown in Figure 2 with a length of 80 mm and a square cross section of width and height of 8 mm. The distance between the inlet of the fluid domain and the topography array is 65 mm. The creation and placement of the array nearby outlet follows the Hagen-Poiseuille's flow theory [15]. This theory explains that in a fluid will develop a laminar flow in any fluid domain with a very significant difference in length compared to the size of its cross section. The development of the laminar flow is induced by the pressure drop along the flow. Two models were created for this research.

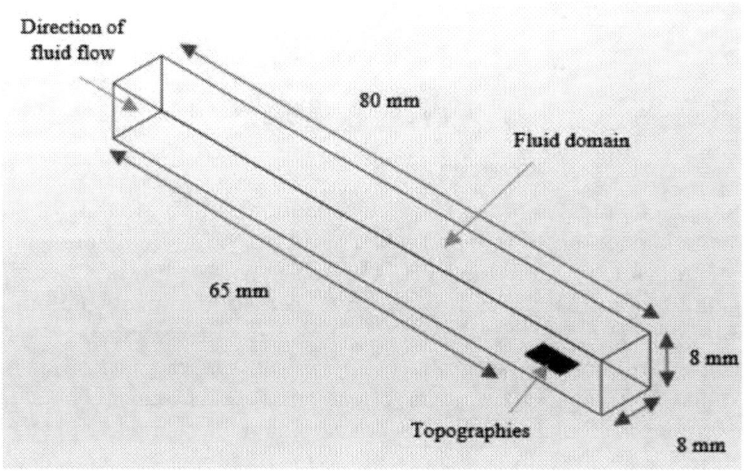

Fig 1. Dimensions of the fluid domain and the location of the topography array in the fluid domain.

Figure 3 shows a close up view of non-hierarchical topography while Figure 4 shows close up view of hierarchical topography with circular pillars. The selection of the geometry and size for these topographies are based on the literature.

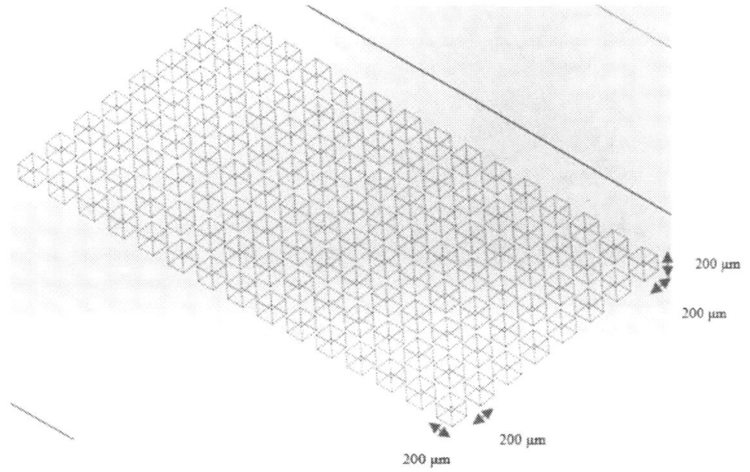

Fig. 3. Dimensions of the non-hierarchical topography

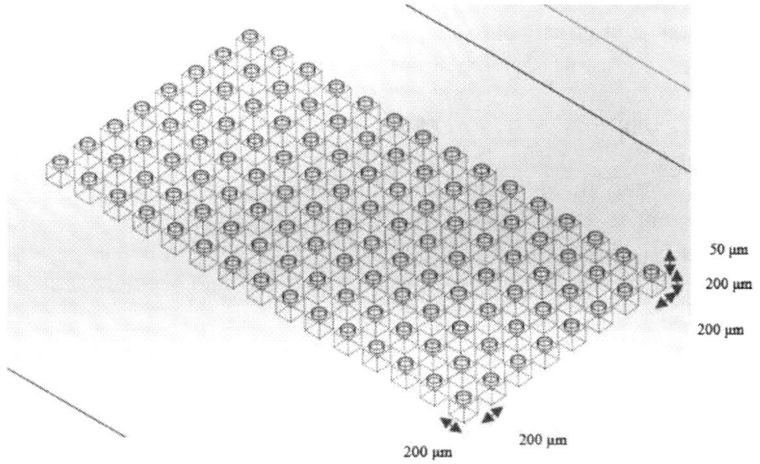

Fig. 2. Dimensions of hierarchical topography

Halder et al. [14] , the author found that well circumscribed geometries like circle, rectangle, hexagon and others develops higher shear bounded isolated areas compared to continuous geometries like ridges. Hence, cube was chosen as the first layer for non-hierarchical topography. Circular pillar was chosen as another well circumscribed geometry for the second layer of the topography to develop hierarchical topography. According to Nick Aldred et al. [16], the optimum size of topography to control biofouling is between 64-256 µm. Most of the biofouling organisms failed to attach to the surface at this range of size [16]. Therefore, all dimensions used in this research are within this range as shown in Figure 3 and 4

2.2 Meshing Method

The quality of the meshes on all models must be assessed to preserve the accuracy of the simulations. Models created using Solidworks were exported as STEP format files into ANSYS. Then, the meshing module was opened to start the meshing process. A fine mesh was created with tetrahedral elements and a sample is shown in Figures 5.

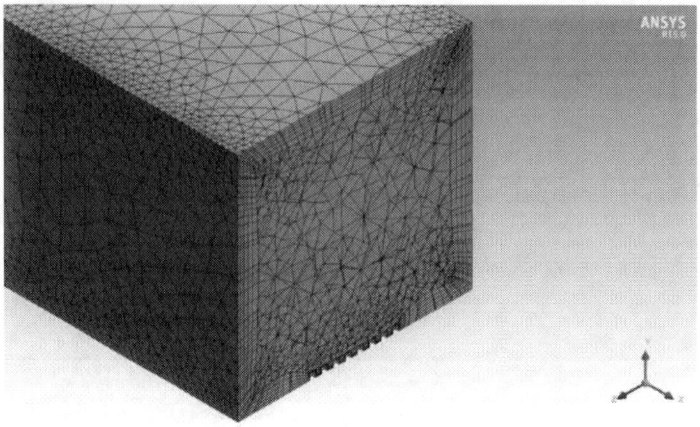

Fig. 5. Close up view of the meshing around the topography.

2.3 Boundary and Numerical Setup

The boundary and numerical setup for the research was done using CFD software package ANSYS FLUENT, version 15. The fluid used for this simulation is water with density ρ, of 1000 kgm^{-3} and dynamic viscosity μ, of 0.00001 $kgm^{-1}s^{-1}$. The k-ω turbulence model was used to simulate the flow as this model is able to provide higher accuracy of wall shear results [14]. The wall of the fluid domain was set with no-slip boundary condition. According to the study done by Halder et al. [14], the mass flow rate was set at 0.001 kgs^{-1} at the inlet and outlet. It is important to note that the mass flow rate was set faster than Halder's study as the model used in this research is longer and has larger cross section [14].

The Navier-Stokes equations (Equation 1 to 3) and the continuity equation (Equation 4) was used in this flow study. The continuity equation ensures the conservation of mass and momentum [17].

$$\frac{\partial \rho \mathbf{u}}{\partial t} + \nabla \cdot \rho \mathbf{u} V = -\frac{\partial p}{\partial x} + \nabla \cdot (\mu \nabla \mathbf{u}) + S_{m,x} \quad (1)$$

$$\frac{\partial \rho \mathbf{v}}{\partial t} + \nabla \cdot \rho \mathbf{v} V = -\frac{\partial p}{\partial y} + \nabla \cdot (\mu \nabla \mathbf{v}) + S_{m,y} \quad (2)$$

$$\frac{\partial \rho \mathbf{w}}{\partial t} + \nabla \cdot \rho \mathbf{w} V = -\frac{\partial p}{\partial z} + \nabla \cdot (\mu \nabla \mathbf{w}) + S_{m,z} \quad (3)$$

$$\frac{\partial \rho \mathbf{u}}{\partial t} + \nabla \cdot (\rho V) = 0 \quad (4)$$

Where u, v and w represents the velocity in x, y and z direction, ρ and μ are the fluid density and dynamic viscosity, V is the velocity and $S_{m,x}$, $S_{m,y}$, and $S_{m,z}$ are the source terms for x, y and z component.

3 Results and Discussion

3.1 Grid Independence Analysis

A grid independence analysis is an important part of this research in order to find an optimum number of mesh elements. This is required to ensure that the computational time taken are minimized while getting accurate results.

These analyses followed the grid independence analysis used by Halder et al. [14] to study flow over a few topographies. In Halder's study, the number of mesh elements was increased in successive models and simulation was carried out for each successive models at inlet velocities of 0.1 ms^{-1}. Wall shear stress was taken as the key parameter to determine the optimum number of mesh elements. The percentage difference the key parameter for each mesh refinement is then calculated to get the relative error. Relative error of less than 5% is acceptable for each successive models used in mesh refinements. Table 1 shows the grid independence analysis done in this research.

Table 1. Grid independence analysis of the models

3D model				Optimum number of mesh elements
Non-hierarchical	Mesh refinements	522171 → 2233727	2233727 → 6481004	2233727
	Relative error (%)	45.83	3.33	
Hierarchical	Mesh refinements	817734 → 1941795	1941795 → 2169321	1941795
	Relative error (%)	13.04	4.17	

3.2 CFD Results

3.2.1 Hagen-Poiseuille's Theory

The development of the flow follows the Hagen-Poiseuille's Theory. Figure 6 shows the development of the flow from the inlet to become a parabolic flow before it reaches the topography at YZ plane (z = 0). The velocity contour plot in Figure 6 shows that the flow has developed a fully parabolic flow before it reaching the topography array.

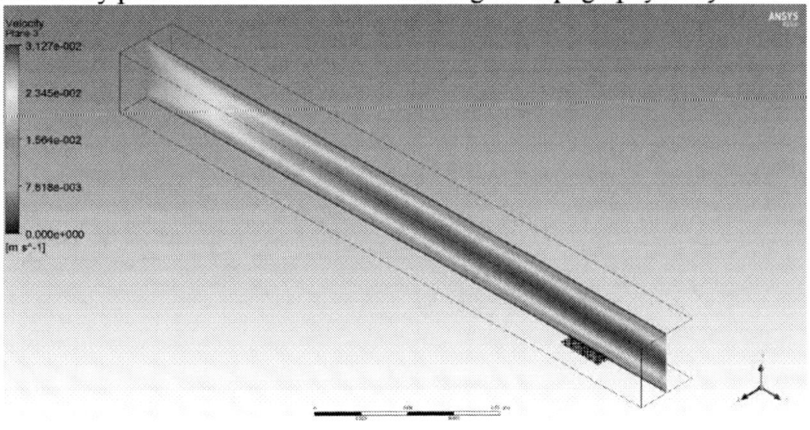

Fig. 6. The velocity contour plot at XY plane ($z = 0$)

This is further confirmed by the measured velocity profile of the fluid domain shown in Figure 7. A fully developed parabolic flow is proven where the maximum fluid velocity is located at the central height of the fluid domain and zero velocity at the walls of the domain. Both topography models in this research exhibits similar results.

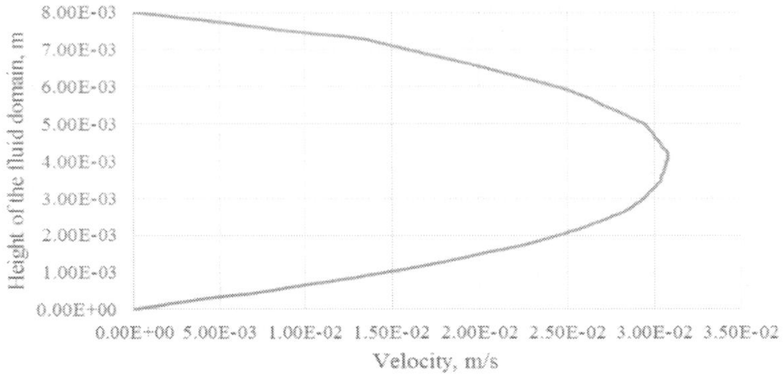

Fig. 7. Velocity profile before it reaches the topography

3.2.2 Velocity Streamlines and Vector Plot

Figures 8 shows streamlines for fluid flow over the non-hierarchical topography. The streamlines show that fluid flowing above the surface of the cube is moving at a faster speed. Fluid then flows into the gaps in between the cubes and exits the topography at a lower velocity. The average velocity above the gap is 4.41×10^{-4} ms^{-1}.

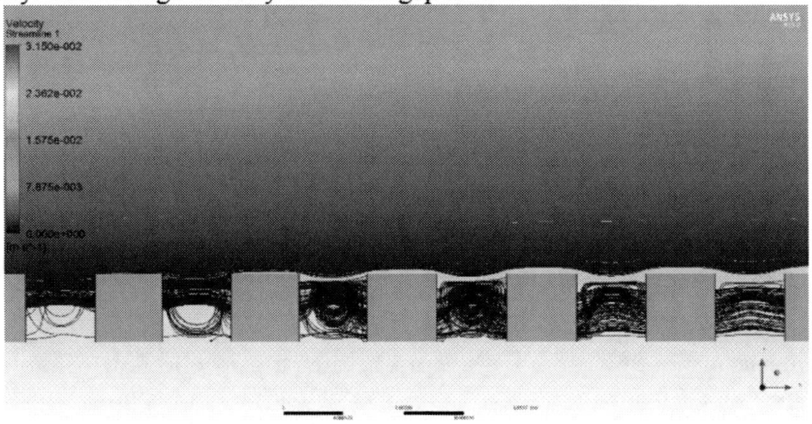

Fig. 8. Streamlines over non-hierarchical topographies

The vector plot profile shown in Figure 9 the fluid flow into the gaps at its lowest velocity and there is formation of vortices and slips. These vortices were rotating in clockwise direction in the gaps.

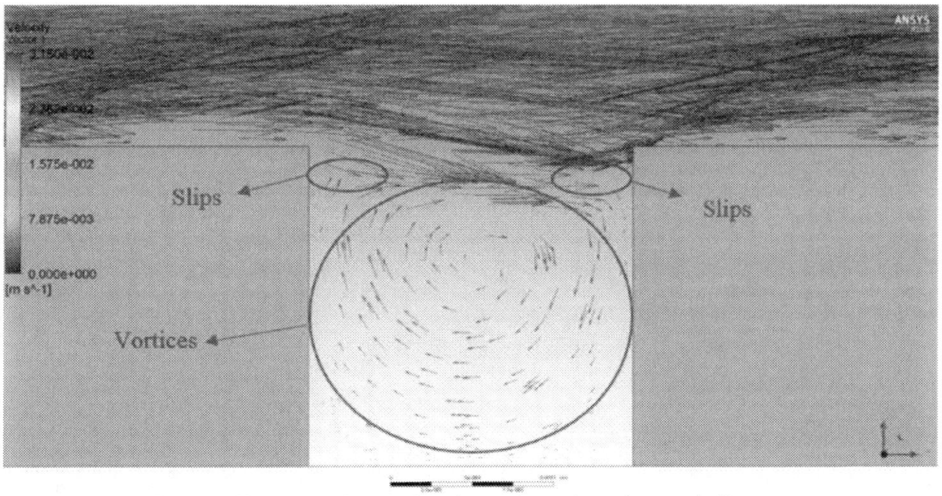

Fig. 9. Vector plot for non-hierarchical showing formation of vortices and slips

Figure 10 shows a streamlines plot for fluid flow through the hierarchical topography. The streamlines show that fluid flowing above the surface of the cube has higher speed and exits the topography array at a lower velocity. The characteristics of the flow is similar to non-hierarchical topography. However, the average velocity above the gap is 8.474×10^{-4} ms^{-1} which is higher than the average velocity of flow above the gaps of the non-hierarchical topography. This indicates the possibility of the hierarchical topography exhibiting better antifouling potential because flows at higher velocity would reduce the chances for prolonged exploration by biofouling organisms[18].

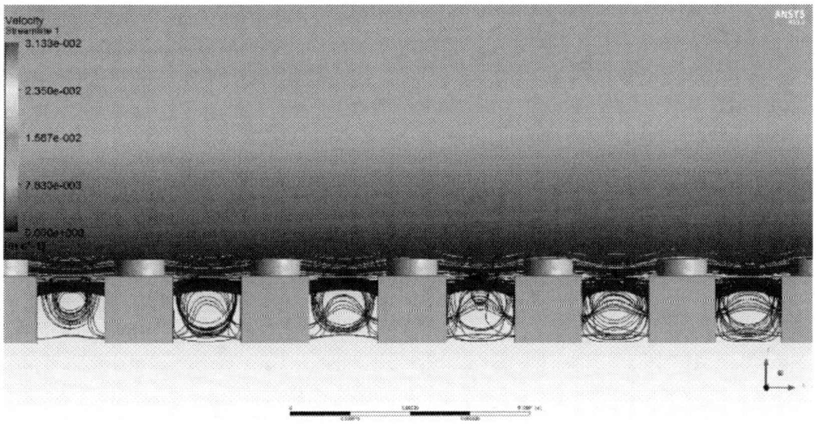

Fig. 10. Streamlines over hierarchical topographies

The vector profile in Figure 11 shows that fluid flow in gaps between topographical features is at the lowest velocity. There is also formations of vortices and slips. The vortices were rotating in clockwise direction in the gaps.

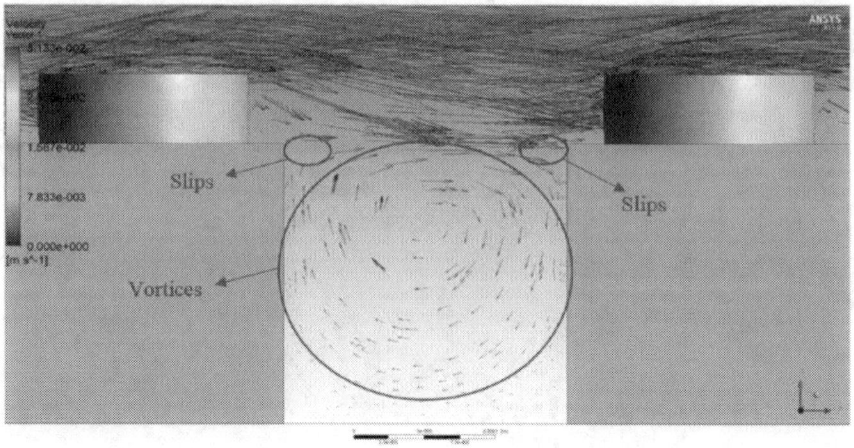

Fig. 11. Vector plot for non-hierarchical showing formation of vortices and slips

In a study conducted by Friedmann et. al [19], similar findings were found where there is formation of vortices in between the riblets. The redirection of the flow of the fluid into and out of the gaps between riblets had significantly decreased fluid velocity. The author also explained that the velocity of the fluid at the peaks of the riblets is high and causes it to slip over the fluid in between the riblets.

In this study, the findings are slightly similar to Friedmann's study. When the fluid flows through the topography, the fluid is redirected into the gaps in between these topographies. As a result, it causes interruption of the flow direction and there will be significant reduction of velocity in this area. Hence, there will be slip in between different layer of fluid flow with different velocities. Then, the vortices were formed as the slips force the fluid in the gaps to rotate [19].

Slips and vortices could possibly help microorganisms to settle in the gaps. This is because at this region, the velocity of the fluid is low which makes it favorable for the microorganisms or other biofouling material to settle and accumulate in this region. Since these gaps are not enclosed and there is fluid flowing around these spaces, there is a possibility for the biofouling materials to be brought out along with the flow that exists in the area.

3.2.3 Wall Shear Stress

Studies have shown substrates that are favorable for settlement and attachment tend to be explored at length by microorganisms. For instance, in a study conducted by Rosenhahn et al. [6], spinning action by algae spores at a place determines the strength of adhesion for permanent attachment. Therefore, biofouling can be prevented by reducing the contact between the microorganisms with the surface.

The possibility of the microorganisms to be detached from the surface can be measure by the wall shear stress. This is one of the crucial parameters in determining the antifouling potential of a surface [14], [20]. The shear stress of the adjacent fluid layer to the wall can be calculated by wall shear stress and it is induced by the velocity of the fluid flow over the topographies.

Figure 12 shows the wall shear stress distribution on the non-hierarchical topography. The highest wall shear stress can only be seen slightly at the edges on the first column from the right which is the first column of the topography being hit by the flow of fluid. There is less or minimal wall shear stress at the spaces in between the topographies. The middle and left most region shows similar wall shear stress which is low.

Fig. 32. Wall shear contour plot for non-hierarchical topography

Figure 13 shows the wall shear stress distribution on the hierarchical topography. The pattern of wall shear distribution is different from the non-hierarchical topography. It is observed that almost all edges of the circular pillars are experiencing higher degrees of shear stresses. There is also less or minimal wall shear stress at the spaces in between the topographies. It is important to notice that only the first column from the right shows the most edges with high wall shear stress as it is the first column to be hit by the flow of the fluid. Nevertheless, the wall shear stress distribution in Figures 12 and 13 shows that hierarchical topography has higher wall shear stress distribution compared to the non-hierarchical topography.

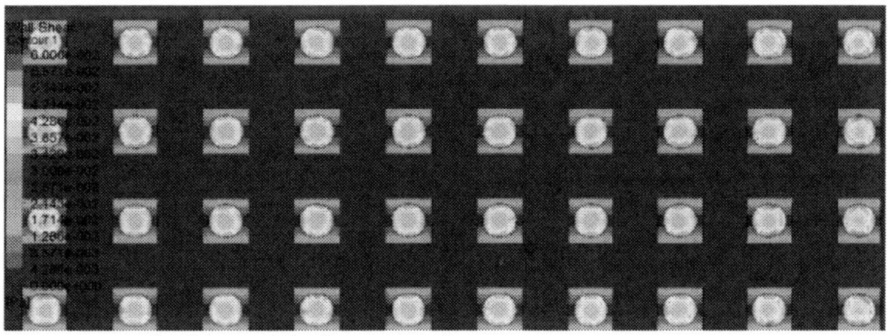

Fig. 43. Wall shear contour for hierarchical topography

Since hierarchical topography induce higher flow stresses, it is likely that this topography would exhibit better antifouling performance as compared to the non-hierarchical topography. High shear stresses is an important parameter in shearing off any microorganisms that is trying to attach to the substrates. Similar results were obtained in a study conducted by Lee et al. [21] where the antifouling performance of micro-sized prisms was investigated. In this study, the peaks of the prisms were less fouled in lab experiments as it exerts higher shear stress in CFD simulations [21].

Figure 14 shows a comparison of wall shear at the top of both hierarchical and non-hierarchical topographies. The graph shows that fluctuations of wall shear stress for the hierarchical surface is higher as compared to the non-hierarchical topography. This fluctuation of wall shear will help in preventing the settlement of biofouling material. The range for wall shear of hierarchical topography is between 0.00281 Pa and 0.00452 Pa while non-hierarchical topography is between 0.00165 Pa and 0.00301 Pa. The data clearly shows that the hierarchical topography has higher potential for antifouling applications.

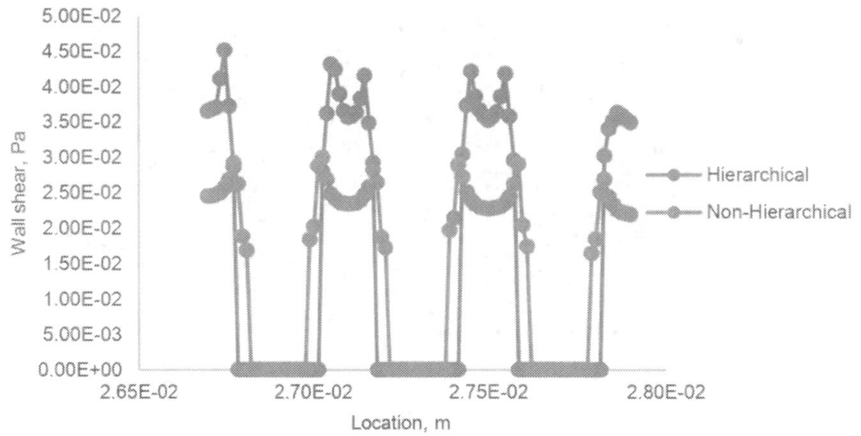

Fig. 14. Graph of wall shear for both topographies

4 Conclusion

The methodology of analysing the effect of hierarchical and non-hierarchical topography in controlling biofouling using CFD (ANSYS-CFX version 15, Solidworks 2013) has been demonstrated. The numerical method and boundary settings used in this study is sufficient to analyse the flow characteristics. This proved that numerical analysis is one of the best methods in studying the flow patterns over these topographies. It is shown that hierarchical topography has higher antifouling performance compared to non-hierarchical topography. This is because hierarchical topography has higher wall shear stress distribution compared to non-hierarchical topography. In addition, the flow characteristics of hierarchical topography indicates that it is superior in controlling biofouling compared to non-hierarchical topography.

References

1. H.-C. Flemming, "Biofouling in water systems--cases, causes and countermeasures." Appl. Microbiol. Biotechnol., **9**, no. 6, 629–40, (2002).
2. T. Vladkova, "Surface modification approach to control biofouling," in Marine and Industrial Biofouling, **7,** 135–163, (2009).
3. M. P. Schultz, J. a. Bendick, E. R. Holm, and W. M. Hertel, "Economic impact of biofouling on a naval surface ship," Biofouling, 27, **1**, 87–98, (2011).
4. R. Steinhagen, H. Müller-Steinhagen, and K. Maani, "Problems and Costs due to Heat Exchanger Fouling in New Zealand Industries," Heat Transf. Eng., **14**, (1993).
5. T. Vladkova, "Surface Engineering for non-toxic Biofouling Control," J. Univ. Chem. Technol. Metall., **42**, 239–256, (2007).
6. A. Rosenhahn and G. H. Sendra, "Surface sensing and settlement strategies of marine biofouling organisms," Biointerphases, **7**, 1–13, (2012).
7. K. A. Dafforn, J. A. Lewis, and E. L. Johnston, "Antifouling strategies: History and regulation, ecological impacts and mitigation," Mar. Pollut. Bull., **62**, 453–465, (2011).
8. Imo, "International Convention on the Control of Harmful Anti-Fouling Systems on Ships," Afs/Conf/26, **1,** 1–28, (2001).

9. P. E. Gibbs and G. W. Bryan, "Reproductive Failure in Populations of the Dog-Whelk, Nucella Lapillus, Caused by Imposex Induced by Tributyltin from Antifouling Paints," J. Mar. Biol. Assoc. United Kingdom, **66**, 767–777, (1986).
10. C. Alzieu, J. Sanjuan, P. Michel, M. Borel, and J. P. Dreno, "Monitoring and assessment of Butyltins in Atlantic coastal waters," *Mar. Pollut. Bull.*, **20**, 22–26, (1989).
11. J. A. Nichols, "Antifouling paints: Use on boats in San Diego Bay and a way to minimize adverse impacts," *Environ. Manage.*, **12**, 243–247, (1988).
12. Z. C. Xie, C. W. Nga, P. Y. Qian, and J. W. Qiu, "Responses of polychaete Hydroides elegans life stages to copper stress," Mar. Ecol. Prog. Ser., **285**, 89–96, (2005).
13. K. Schiff, D. Diehl, and A. Valkirs, "Copper emissions from antifouling paint on recreational vessels," Mar. Pollut. Bull., **48**, 371–377, (2004).
14. P. Halder *et al.*, "A novel approach to determine the efficacy of patterned surfaces for biofouling control in relation to its microfluidic environment," Biofouling, **29**, 697–713, (2013).
15. N. A. Mortensen, F. Okkels, and H. Bruus, "Reexamination of Hagen-Poiseuille flow: Shape dependence of the hydraulic resistance in microchannels," Phys. Rev. E - Stat. Nonlinear, Soft Matter Phys., **71**, (2005).
16. N. Aldred, A. Scardino, A. Cavaco, R. de Nys, and A. S. Clare, "Attachment strength is a key factor in the selection of surfaces by barnacle cyprids (Balanus amphitrite) during settlement.," Biofouling, **26**, 287–299, (2010).
17. P. M. Palaskar, "Effect of Side Taper on Aerodynamics Drag of a Simple Body Shape with Diffuser and without Diffuser MIRA Reference Car and Geometry," SAE Tech. Pap., **15**, 786-795, (2016).
18. J. F. Schumacher *et al.*, "Engineered antifouling microtopographies - effect of feature size, geometry, and roughness on settlement of zoospores of the green alga Ulva.," Biofouling, **23**, 55–62, (2007).
19. E. Friedmann, J. Portl, and T. Richter, "A study of shark skin and its drag reducing mechanism," in Advances in Mathematical Fluid Mechanics - Dedicated to Giovanni Paolo Galdi on the Occasion of His 60th Birthday, **50**, 271–285, (2010).
20. P. Halder *et al.*, "An assessment of the dynamic stability of microorganisms on patterned surfaces in relation to biofouling control.," Biofouling, **30**, 695–707, (2014).
21. Y. K. Lee, Y. J. Won, J. H. Yoo, K. H. Ahn, and C. H. Lee, "Flow analysis and fouling on the patterned membrane surface," J. Memb. Sci., **427**, 320–325, (2013).

The Enhancement of Pre-Storage Filtration Efficiency for the Rainwater Harvesting System in Malaysia

Yoong Sion Ong[1], *Ken* Sim Ong[2], *Y.K.* Tan[3], *Azadeh* Ghadimi[1,*]

[1]School of Engineering, Taylor's University, Malaysia
[2]Bina Plastic Industries Sdn. Bhd., Malaysia
[3]Besstem Plastic (M) Sdn. Bhd., Malaysia

Abstract. A conventional design of rainwater harvesting system collects and directs the rainwater through water piping from roof of building to the water storage. The filtration system which locates before the water tank storage and first flush bypass system is the main focus of the research. A filtration system consists of a control volume of filter compartment, filter screen (stainless steel mesh) and water piping that direct the water flow. The filtration efficiency of an existing filter "3P Volume Filter VF1" by industrial company is enhanced. A full scale filter design prototype with filter screen of 1000 μm stainless steel metal mesh is tested to compare with the original filter system design. Three types of water inlet setups are tested. Among the proposed water inlet setups, the 90° inlet setup with extension provides the best filtration rate per unit time, following by the 45° inlet setup. The 45° and 90° inlet setup has similar filtration efficiency at low to medium flow rate while 45° inlet setup has better efficiency at high flow rate. The filtration efficiency with the 90° inlet setup with extension is observed to maintain at highest value at medium to high flow rate. The overall filtration performance achieved by the 90° inlet setup with extension at low to high flow rate is between 34.1 to 35.7%.

1. Introduction

Rainwater harvesting practice can be traced since thousands of years using different types of techniques to collect and store the rainwater from rooftops and land surfaces. A conventional design of rainwater harvesting system will collect and direct the rainwater through water piping to the water storage [1].

The dimensions of the rainwater harvesting system are highly dependent on the size of rainwater catchment area. For industrial scale usage, a simple rainwater filtration process is carried out before it is stored in storage.

[1] Corresponding author: Azadeh57ghadimi@gmail.com

Particles like leaves, rocks, debris and dusts are removed at the filtration process to avoid the risk of clogging up in the water storage.

A conventional rainwater harvesting system consists of four major components which are shown as Fig. 1. The rainwater will first be collected by the catchment area at the rooftop. Filration system which locates before the water tank storage and first flush bypass system is the main focus of the research. A filtration system consists of a control volume of filter compartment, filter screen (stainless steel mesh) and water piping that direct the water flow.

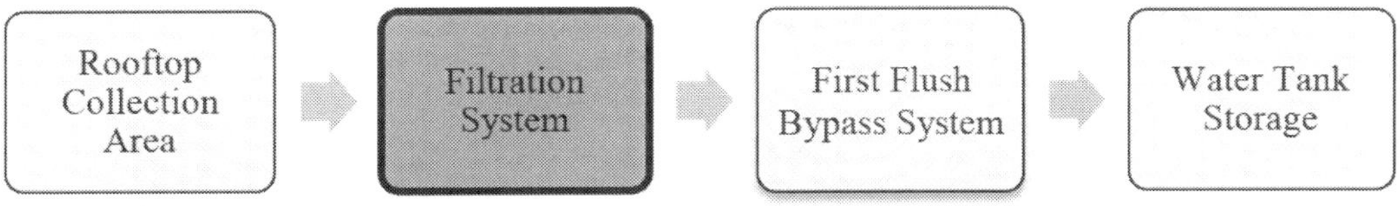

Fig 1. Components of rainwater harvesting system.

Large industrial scale filtration system which commonly used in the market is based on a parallel setup concept with a few of horizontal positioned filter compartments as shown in Fig. 2. More filter compartments can be added by increasing the number of the inlet.

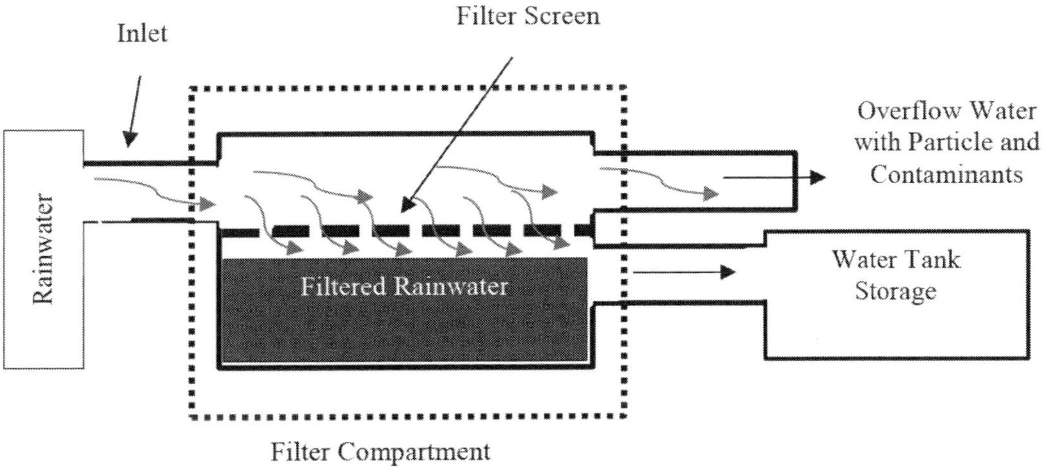

Fig.2. Schematic diagram of single filtration compartment in filtration system.

The inlet will direct the rainwater into the filter compartment which consists of a metal mesh filter screen for filtration process. The large particles and contaminants will be pushed out of the filter compartment by rainwater flow. Filtered rainwater that has passed through the horizontal positioned filter screen due to gravitational force will be stored into water tank storage. However, the performance of the design will decrease significantly during high rainwater flow rate operation. The small mesh size of the filter screen not only resist the particles from passing through, it also decreased the permeability of rainwater. The high rainwater flow rate which has filled up the whole water piping will only stay a short period in the filter compartment. There is just a small portion of water near the filter screen that can be directed to the water tank storage while large portion of water have overflowed and left the filter compartment [2].

1.1 Existing Filter "3P Volume Filter VF1"

This research was an industrial project in collaboration with the industrial partners, Bina Plastic Industries Sdn. Bhd. and Besstem Plastic Sdn. Bhd, which is a class leading plastic pipe and rainwater filtration system supplier in Malaysia respectively. Both companies have provided technical support to build the prototype and product testing. Fig. 3 shows the filter system 3P volume filter VF1 and its filter design, which is a product from Germany currently used by DD Techniche Sdn. Bhd. Company specialises in installation of rainwater harvesting system for both residential and commercial buildings in Malaysia. The rainfalls in Malaysia are commonly heavy but in a short period of time. The high filtration rate from the filtration system is desired to take advantage in harvesting more filtered rainwater in a limited duration. The projects that the company had done previously were to be capable of sustain a high rainwater flow rate range from 50 L/s to 200 L/s.

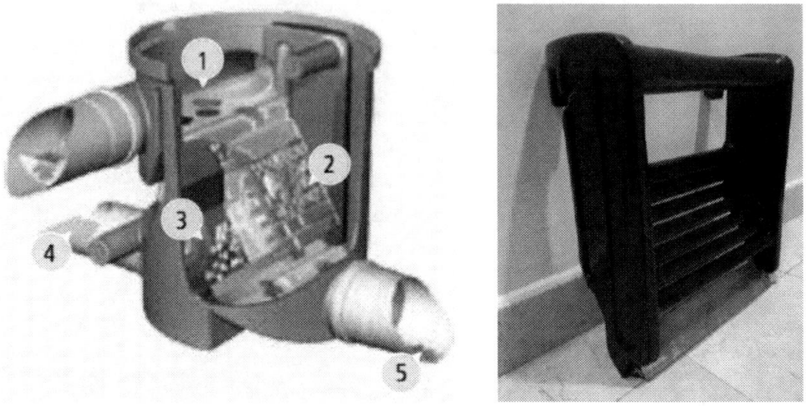

(a) (b)

Fig 3. (a) 3P volume filter VF1 by 3P Technik from Germany [3]; (b) Original filter design.

The design and functionality of the filter VF1 is similar to the parallel setup concept as discussed in Fig. 2. The filter VF1 is installed by the company whenever the installation location is limited, which will be discussed in this paper. The most significant difference is that the filter screen is placed at an incline instead of horizontal position. Refer to Fig 3(a), the rainwater flow into the filter compartment and accumulate at top section of the filter compartment at stage 1. At stage 2, the rainwater will be filtered while flowing along the slanted metal mesh filter screen (Fig. 3(b)). Next, the clean water that has passed through the filter screen is collected at the bottom section of the filter compartment at stage 3. While the clean water is directed to the water tank storage at stage 4, the suspended particle will be removed togetehr with the overflow water through a drain outlet at stage 5. The filter design does provide a self-cleaning process to increase the maintenance interval. However, the filter has the filter process characteristic same as the horizontal filtration method as large portion of rainwater flows towards the drain outlet directly during high flow rate.

Fig. 4 shows the official data by the company regarding on the graph of hydraulic efficiency against inflow of rainwater for filter VF1. Based on Fig. 4, the filter achieve 100% efficiency at flow rate of 0.01 L/s to 0.2 L/s but started to drop after that. It can be seen that the filter compartment has started to fail to capture all the inflow rainwater even before reaching flow rate of 1 L/s. Furthermore, the efficiency is just around 50% at flow rate of 3.5 L/s, which means only 1.75 L of rainwater is being filtered and collected into water tank storage every second. At this point, the filter has reached its peak filter performance as shown in the sudden drop of efficiency in the graph. Further increment of inflow of rainwater will only result in more overflow rainwater due to failure to capture. Hydraulic efficiency is the filtration rate efficiency, η is equal to the fraction of total filtered water, $Q_{out,filtered}$ devided by total inflow water, Q_{in} which is shown as Eq. (1).

$$\eta = \frac{Q_{out,filtered}}{Q_{in}} \quad (1)$$

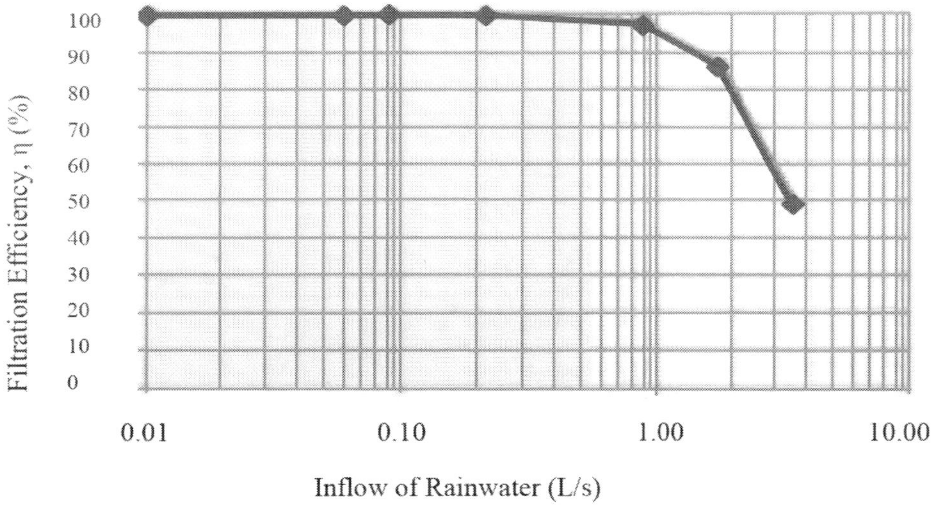

Fig 4. Graph of hydraulic efficiency vs inflow of rainwater for filter VF1 [3].

This paper aims to identify suitable filtration method to achieve higher filtration efficiency under high flow rate to coordinate with the heavy rainfall in Malaysia. The research focuses on developing a sustainable mechanical operated rainwater filtration system for home and small industrial scale usage. The desired solution is to attain higher hydraulic efficiency at higher inflow rate for this specific filter VF1 by applying suitable modifications without modifying the shape of filter compartment. The study of existing filtration system methods and designs available in the market globally were conducted and their specifications were compared. A prototype filter screen was built by integrating features from different types of filtration system. The filter screen must be able to fit inside the original compartment of filter VF1. The rainwater filtration efficiency, in term of filtration rate difference between the original and prototype filter screen design is to be obtained throughout the testing.

2. Specifications Comparisons of Different Types of Filter System

2.1 Cyclonic Rotation Filter

A cyclone filter is a design of rotational separator which leads the water flow to be circular in the filtration compartment. Fig. 5 shows a schematic diagram of rotational separator which direct the water from inlet to rotate around the compartment. The centrifugal forces produced by the rotating water will acts on large particles and contaminants pushing them outwards leaving the centre vertical axis-z. With the assist from gravitational forces, the suspended particles and contaminants will move towards the bottom of the compartment due to no drainage outlet while leaving the cleaned water to be directed into the storage.

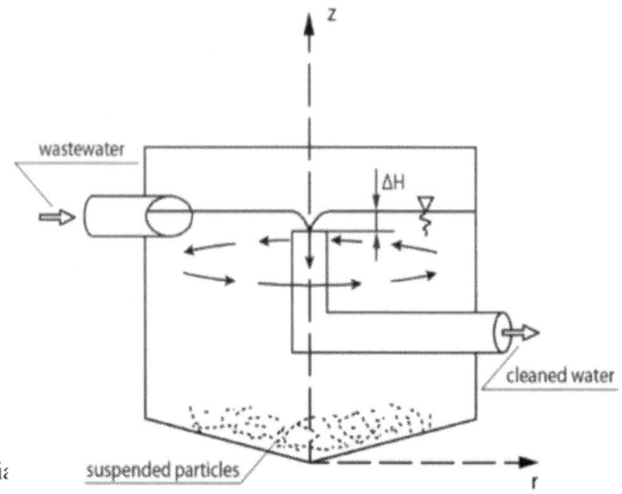

Fig. 5. Schematic di

The cyclone filtration method performs well under high flow rates of water. However, the suspended particles are not removed from the cyclone filter compartment throughout the operation [4]. Fig. 6 shows an improved design of cyclone filtration method product by Rainharvesting Systems Company from UK. This design overcomes the challenges by having a self-cleaning process which will remove the suspended particle out of the filter compartment. The rainwater flow into the compartment from the inlet and perform cyclonic rotation. When the rainwater is approaching the cylindrical shape metal mesh filter, the centrifugal force will pull the water pass through the metal screen. The collected filtered water will be directed to the water tank storage, leaving the filtered particle falling through centre of the filter and are washed off by the overflow water [5]. However, it only work on high flow rate, else, the rainwater will not perform rotation and flow down the drain directly.

Fig 6. Wisy vortex filter by Rainharvesting Systems company from UK [5].

2.2 Gravity Downpipe Filter

The conventional method of installing the downpipe in any building is vertically from top to bottom. The rainwater catchment area is normally placed on the roof, which the collected rainwater will flow down in the pipe vertically at high speed. By utilising the gravitational force with the speed of rainwater flow, the rainwater can pass through the filter screen right below the inlet easily due to the extra downward force. Fig. 7 shows a filter product by Rainharvesting Company from Australia which utilises the gravitational force. The rainwater is directed from top of the filter and the cleaned water is directed to the water tank storage through the outlet at below. During this process, all the particles like leaves and debris are deflected by the filter screen and washed off out of the filter along the slanted filter screen. This design enhances the filtration efficiency as the particles are quickly removed without contaminate on the filter screen [6].

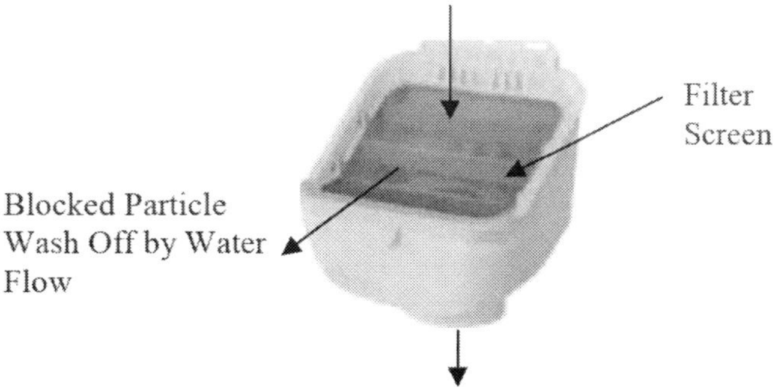

Fig 7. Leaf eater advanced filter by Rainharvesting company from Australia [6].

2.3 Justifications on Selecting the Suitable Filtration Features for the Prototype

Table 1 shows the specifications of different filter product in the market. By referring to Table 1, the general relationship between filter mesh sizes, filter rate and filtration method can be determined. First of all, product (1) using the largest mesh size among the other competitors. The filter has the horizontal filtration method design which only achieve a filter rate of 90 L/min. Secondly, a medium mesh size range from 0.8 – 0.955 mm is used for filter product (2) to (4) which utilise the gravitational force to speed up the rainwater along the downpipe. Even though a small mesh size is used, the filter manages to achieve a filter flow rate range from 113 – 120 L/min. The last group of products (5) to (7) have the feature of cyclonic rotation to filter the rainwater. It is found that they have the smallest mesh size among the competitors, however their filtration performances are better than both horizontal and gravity filtration method.

Table 1. Specifications of different filter product in the market.

Ref.	No.	Company (Country)	Metal Mesh Size	Roof Area	Rainwater Filter Rate	Filtration Method
[3]	(1)	3P Technik (Germany)	1.4 mm	350 m^2	90 L/min	Horizontal
[7]	(2)	SuperHead (Australia)	0.8mm	-	120 L/min	Gravity

[6]	(3)	RainHarvesting (Australia)	0.8 mm	-	113 L/min	Gravity
[8]	(4)	BRAE (US)	0.955	-	-	Gravity
[9]	(5)	Rainharvestmo (US)	0.28 mm	232 m^2	90% Recovery	Cyclonic Rotation
[5]	(6)	Rainharvesting (UK)	0.28 mm	200 m^2	66 L/min (not tested on maximum condition)	Cyclonic Rotation
[10]	(7)	Allegiance India (India)	0.25 mm	112 - 500 m^2	105-340 L/min (depends on the volume of collector)	Cyclonic Rotation

Based on real life application experiences, the space of the room that is prepared for the installation of rainwater system in the building and factory is very limited, especially the height of the room. During the installation, the height measured from ceiling to the ground is taken into consideration, conventionally 3.5 m per floor. The size of the water tank storage is normally installed with the largest size to store more filtered clean rainwater. Besides that, a stand is placed below the water tank storage for easier maintenance and cleaning purposes which result in a total of around 3 m in height. By taking into the consideration of the atmospheric pressure acting on water, the water tank storage inlet must be positioned as high as possible. Otherwise, the cleaned rainwater will not able to be pushed into water tank storage at higher level without the support from water pump if the inlet is positioned at lower level. Fig. 8 shows the schematic diagram of the installation of rainwater filtration system in a room for industrial scale usage.

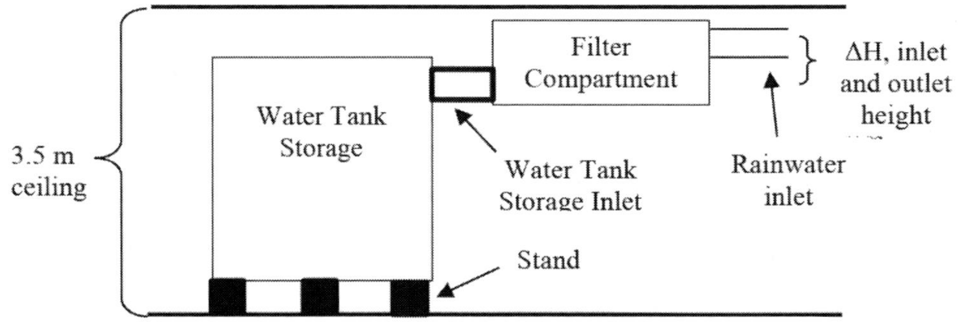

Fig. 8. Rainwater filtration system in a room for industrial scale usage.

Due to the space constraint of room, the overall height of the rainwater filter compartment must be maintained at smallest scale. Therefore, the height difference between rainwater inlet and water tank storage inlet, ΔH need to be taken into consideration at design stage. The compact sizing of the VF1 filter with dimension of 470 mm x 490 mm x 451 mm is very capable and has advantage during the installation in real life application.

The cyclonic rotation filtration method require a big and long in vertical dimension filter compartment to make it functional. Hence, it is more suitable for the installation area without space constraint. It is observed that the horizontal filtration method shows the least efficiency in filtering rainwater at high flow rate as discussed previously. However, by integrating the features from gravity downpipe filtration method may obtain significant improvement in filtration rate. Therefore, this feature is selected.

3. Development of New Filter Screen Design and Fabrication of Prototype

There were two preliminary designs for the filter screen before the final approved design was developed. To construct the new filter screen design to accurately fit into the filter compartment, the filter compartment VF1 was modelled into 3D geometry using SolidWorks (2015) as shown in Fig. 9. It had made the task easier to sketch and modify the new filter screen design within a round shape compartment with complicated structural design within it.

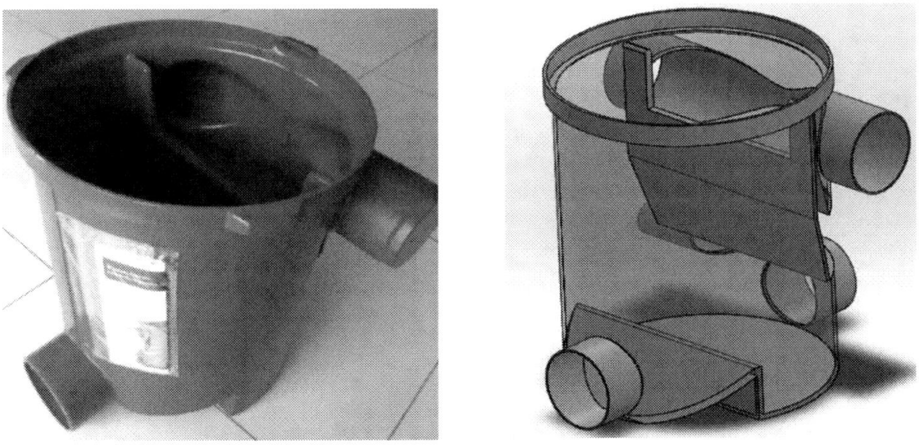

(a) (b)

Fig. 9. Existing filter compartment VF1: (a) Real life product; (b) 3D geometry model.

The new filter screen holder was designed to cover as much filtration area as it can within the limited space within the filter compartment. The model was then inserted into the original filter compartment VF1 to check any error in dimensions to achieve high fit and finish quality as shown in Fig. 10.

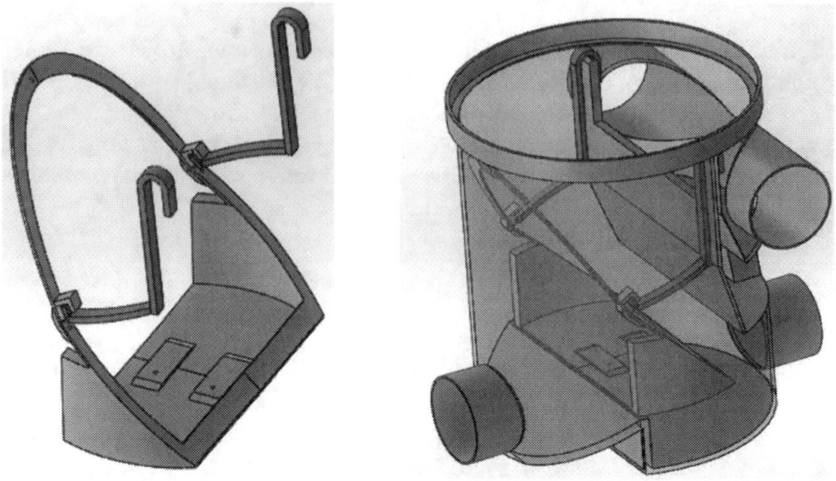

(a) (b)

Fig. 10. (a) 3D geometry model of new filter screen holder; (b) Filter screen holder inserted into original filter compartment VF1 model.

The approved 3D geometry model was then fabricated into full scale prototype using 3D printer with plastic PLA material. The filter screen holder was then attached a layer of 1000 μm stainless steel metal mesh, the same metal mesh size used for the original filter. The completed full scale prototype was fitted to the original filter compartment VF1 as shown in Fig. 11.

(a) (b)

Fig. 10. (a) Full scale new filter screen holder; (b) Filter screen holder inserted into original filter compartment VF1.

4. Experimental Testing of Filtration Rate for Original and New Design

The experimental equipment is set up as shown in Fig. 11. The setup is based on real life installation application where the water source is collected from the downpipe that linked to the roof. The horizontal water pipe will direct the water to the rainwater harvesting system room where the water flow into the filter compartment VF1 through the water inlet.

 https://doi.org/10.1051/matecconf/201815202015

Fi

Tap water is used for the testing of filtration rate because the lack of rainwater harvesting system in testing location. The water source is fed in through water tap and water pump to mimic the rainwater flow. Before starting the experiment, the water flow rate is measured. The filtration rate for the original filter system design is tested using the original method. Due to the horizontal positioned water piping that direct the rainwater flow, the positioning of water inlet is manipulated to test the differences in filtration rate as shown in Fig. 12. The step is taken to integrate the features of gravity downpipe filtration system onto the new filter system design. For the original filter system design, the 90° inlet will fill up the water at top section of the filter compartment and flow down to the filter screen. For the new filter system design, the 90° and 45° inlet allow the water to flow towards the filter screen directly.

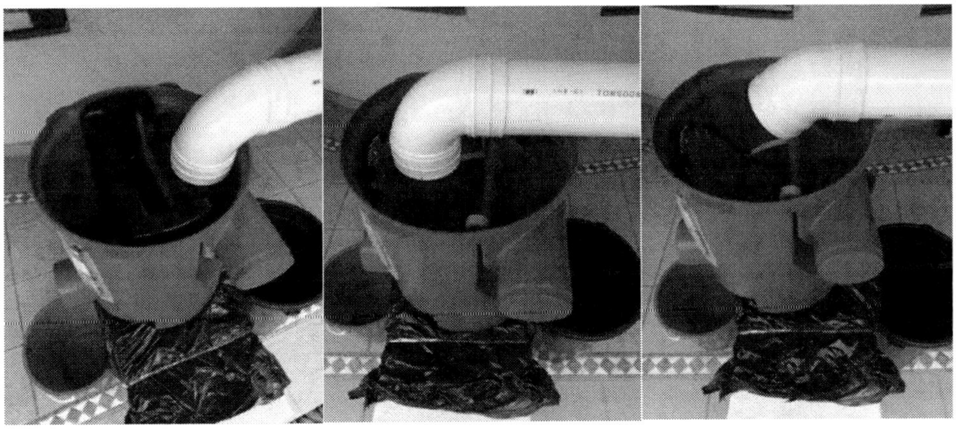

(a) (b) (c)

Fig. 12. (a) Original filter system design with 90° inlet; (b) New filter system design with 90° inlet; (c) New filter system design with 45° inlet.

The filtration rate testing for both the original and new filter system design are carried out and the data are tabulated. Each round of testing is run for 15 seconds to collect the filtered clean water and 3 sets of data are recorded to obtain the average value. Low, medium and high flow rate for water inlet are used in testing. The low flow rate is obtained through normal water tap while medium and high flow rate are obtained by using a water pump with a reservoir.

Before starting the experiment, the water flow rate is measured. The low flow rate water is feed into the experimental equipment through the water source inlet and allow to run for 5 second to reach steady flow within the system. Next, an empty pail is used to collect the water from the water inlet for 15 seconds and then weighted on the scale. It is measured that the mass of empty pail is 1.5 kg, the total water mass flow rate at the inlet, $Q_{Total\ inlet}$ of the equipment can be obtain using the Eq. (2).

$$Q_{Total\ inlet} = \frac{m_T - m_{pail}}{t} \quad (2)$$

Where m_T is the total mass of the water with the pail and m_{pail} is the mass of the pail and t is the duration to collect the water, in this case t is 15 seconds. The density of water is 1000 kgm-3, therefore, 1 L/s mass flow rate of water at the inlet can be converted as 1 kg/s volume flow rate.

5. Results and Discussions

The original filter design was attached into the filter compartment VF1 for filtration testing. Table 2 shows the data of filtration rate testing for the original filter system design with the original filtration method which consists of low flow rate (0.203 L/s and 0.301 L/s), medium flow rate (1.134 L/s and 1.598 L/s) and high flow rate (2.167 L/s).

Table 2. Data of filtration rate testing for the original filter system design.

Inlet flow rate (L/s)	Total filtered water per second by original filter system design (L/s)				Filtration Efficiency
	1	2	3	Average	
0.203	0.137	0.122	0.130	0.130	0.639
0.301	0.187	0.198	0.191	0.192	0.638
1.134	0.705	0.689	0.711	0.702	0.619
1.598	0.949	0.956	0.962	0.956	0.598
2.167	1.208	1.197	1.205	1.203	0.555

Based on Table 2, it is observed that the filtration efficiency has significant difference with the official data by the company as shown in Fig. 4. At 0.2 L/s of inlet flow rate, the tested filtration efficiency is just 63.9% instead of nearly 100% from the official data. As the inlet flow rate increases, the filtration rate of the original filter system design continues to decrease and hence results in just 55.5% filtration efficiency at inlet flow rate of 2.167 L/s. Fig. 13 shows the graph of hydraulic efficiency against inlet flow rate for the original filter system design.

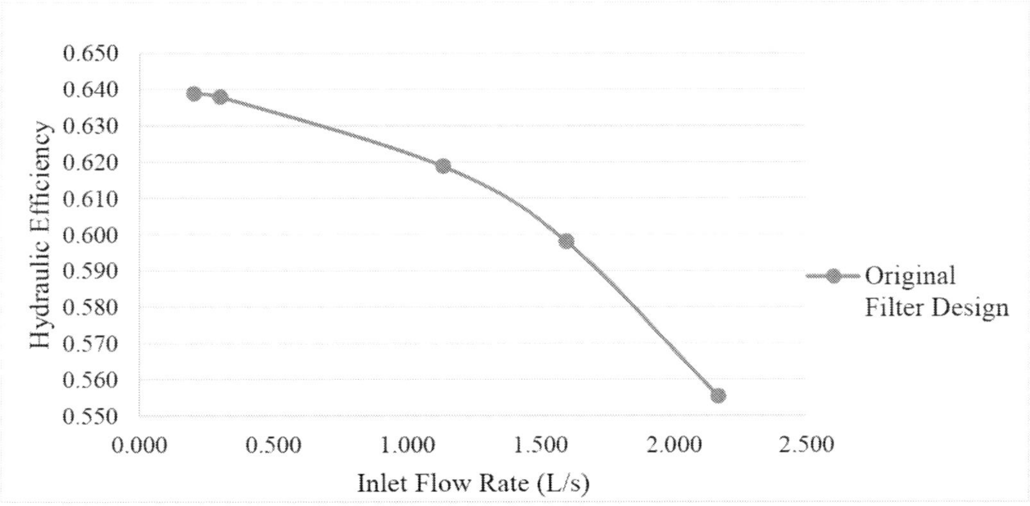

Fig. 13. Graph of hydraulic efficiency vs inlet flow rate.

By comparing Fig. 13 with Fig. 4, it is observed that the experimental data is nowhere near to the official data from the company. As a reference from both the graphs at inlet flow rate of 2 L/s, the tested hydraulic efficiency is just 56% while the official data shows an efficiency of 80%. Since the testing is done by the current equipment, the obtained experimental data is selected to be the reference for comparison with the new filter system design.

Throughout the filtration rate testing for the new filter system design, there is sign of water loss can be observed through overflow drain outlet of filter compartment. For the 45° inlet setup, the water is directly flowing towards the filter screen rather perpendicularly as shown in Fig. 14(a). This has caused some of the water being reflected away from the filter screen without being filtered and hence the loss of collectable water. As for the 90° inlet setup, due to the sudden change in direction of water flow, a portion of water do not flow downward vertically onto the filter screen as planned. Instead, the water is reflected by the wall of the 90° water pipe joint and flow towards the overflow drain outlet as shown in Fig. 14(b). The results of water loss from the 45° inlet setup is acceptable as it already meets the requirement of integrating the features from gravity downpipe filtration method. However, enhancement is needed to apply for the 90° inlet setup to determine the full potential of filtration rate. Therefore, a simple modification by adding a transparent plastic cover as extension around the 90° inlet joint to avoid the reflected water leaving the filter screen. Fig. 14 (c) shows the addition of transparent plastic cover extension around the 90° joint.

(a) (b) (c)

Fig. 14. (a) Water flow with 45° inlet; (b) Water flow with 90° inlet; (c) Addition of transparent plastic cover extension around the 90° joint

Table 3 shows the data of filtration rate testing for the new filter system design with integrated features of gravity downpipe filtration method. By referring to Table 3, there are no significant differences of filtration efficiency for 45° and 90° inlet setup between low and medium flow rate. The 45° inlet setup has slightly better filtration efficiency of 78.1% at high flow rate compare to 76% for the 90° inlet setup. Although there are water loss due to reflection at both inlet setup, there is still slightly advantage for the water to flow directly towards the filter screen with the 45° setup. The extra water flow speed act as external force to push more water pass through the filter screen while the 90° setup reduce the flow speed and reflect the water to opposite direction as the water hit the pipe wall. Hence, the 45° setup achieves better filtration rate by comparing to the 90° setup.

Table 3. Data of filtration rate testing for the new filter system design

Type of Inlet	**Inlet flow rate (L/s)**	**New Filter System Design**				**Filtration Efficiency**
		1	2	3	Average	
45°	0.203	0.200	0.198	0.199	0.199	0.980
	0.301	0.296	0.297	0.299	0.297	0.988
	1.134	0.997	0.995	1.006	0.999	0.881
	1.598	1.359	1.367	1.364	1.363	0.853
	2.167	1.684	1.701	1.692	1.692	0.781
90°	0.203	0.199	0.201	0.199	0.200	0.984
	0.301	0.297	0.299	0.296	0.297	0.988
	1.134	0.990	1.005	0.996	0.997	0.879
	1.598	1.368	1.357	1.362	1.362	0.853
	2.167	1.641	1.652	1.647	1.647	0.760
90° with Extension	0.203	0.198	0.199	0.200	0.199	0.980
	0.301	0.299	0.296	0.297	0.297	0.988
	1.134	1.071	1.076	1.079	1.075	0.948
	1.598	1.477	1.472	1.485	1.478	0.925

	2.167	1.984	1.975	1.970	1.976	0.912

For the 90° inlet setup with the transparent plastic cover extension, it is proven to be a good addition modification onto the filtration system based on Table 3. Water is now able to flow downwards vertically onto the filter screen with no wall reflection. The filtration efficiency is observed to maintain at higher value at medium to high flow rate. An overall increasing in filtration efficiency of 6.9% to 15.2% is achieved by comparing with the 90° inlet without the extension and 6.7% to 13.1% with the 45° inlet setup. Fig. 15 shows the graph of hydraulic efficiency against inlet flow rate for the new filter system design.

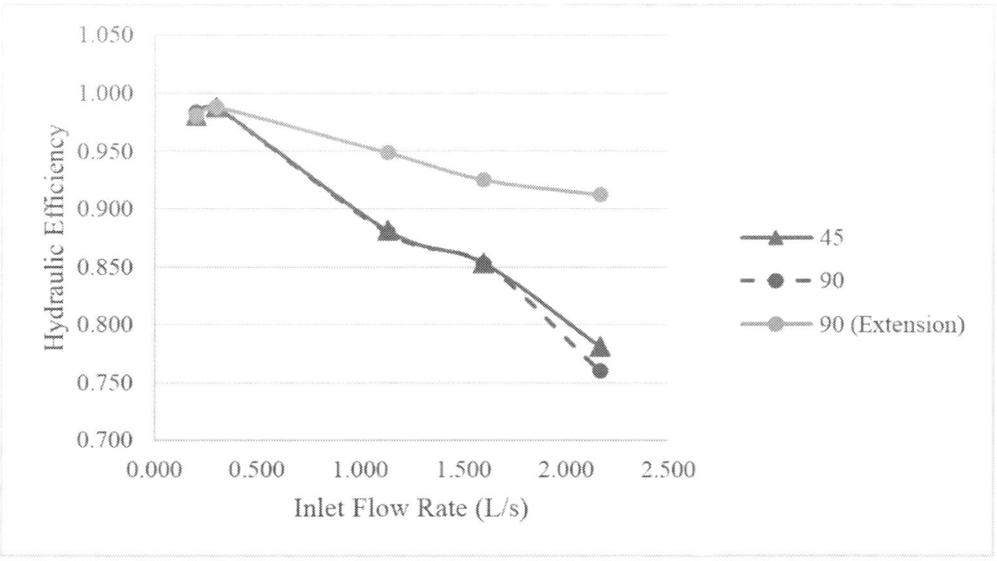

Fig. 15. Graph of hydraulic efficiency vs inlet flow rate with different inlet setup

Based on Fig. 15. It is observed that the 45° and 90° inlet setup has similar filtration efficiency at low to medium flow rate while 45° inlet setup has slightly better efficiency at high flow rate. The 90° inlet setup with extension has the highest efficiency among the three setups. By comparing the filtration efficiency with the original filter system design, all the new filter design manage to achieve higher filtration rate. The overall improved filtration performance achieved by the 90° inlet setup with extension from low to high flow rate is between 34.1to 35.7% compared to the original filter system design.

6. Conclusion

In conclusion, the objective to improve the pre-storage filtration efficiency of a rainwater harvesting system is achieved. Suitable filtration methods and modifications for the original filter system design to sustain high rainwater flow rate are proposed. The selections are justified with the researches done on the existing filter system in the market globally. The proposed prototype meets the requirements of fitting accurately inside the filter compartment VF1 and is operated mechanically.

Besides that, the rainwater filtration efficiency, in term of filtration rate difference between the original and prototype filter screen design is obtained throughout the testing. First, there is significant differences of filtration efficiency between the experimental and official data from the company for the original filter system design. As the testing is done by the current equipment, the obtained experimental data is selected to be the reference for comparison with the new filter system design.

Among the proposed water inlet setups for the new filter system design, the 90° inlet setup with extension provides the best filtration rate per unit time, following by the 45° inlet setup and the 90° inlet setup. The 45° and 90° inlet setup has similar filtration efficiency at low to medium flow rate while 45° inlet setup has slightly better efficiency at high flow rate. The filtration efficiency with the 90° inlet setup with extension is observed to maintain at highest value at medium to high flow rate. An increment in filtration efficiency of 6.9% to 15.2% is achieved by comparing with the 90° inlet without the extension and 6.7% to 13.1% with the 45° inlet setup. In comparison with the original filter system design, the overall improved filtration performance achieved by the 90° inlet setup with extension at low to high flow rate is between 34.1to 35.7%.

In this paper, the experimental results only cover one high flow rate of 2.167 L/s at the water inlet. The water pump that is used to represent the real life rainwater harvesting operation has reached it maximum output in the experiment equipment setup. Therefore, a water pump with higher water pumping rate can be used in future to further test the peak filtration rate of original and new filter system design. Moreover, more layers of filter screen with 1000 μm stainless steel metal mesh can be attached onto the new filter design to test on the filtration rate as future works. The smaller pore size of the filter screen will provide better filtration quality. The relationship between filtration efficiency against reduced pore size of filter screen can be determined.

References

[1] A. I. Che-Ani, N. Shaari, A. Sairi, M. F. M. Zain, and M. M. Tahir, Eur. J. Sci. Res., **34**, 1, 132–140, (2009).

[2] R. H. Kim, S. Lee, and J. O. Kim, Desalination, **177**, 121–132, (2005).

[3] 3P Technik, “3P Technik Filtersysteme GmbH - Volume Filter VF1”, (2017).

[4] M. A. Gronowska, Arch. Hydro-Engineering Environ. Mech., **59**, 49–62, (2012).

[5] Rainharvesting Systems UK, “WISY® Vortex Filters - Rainharvesting Systems”, (2016).

[6] Rainharvesting Australia, “Leaf Eater Advanced | Rain Harvesting”, (2010).

[7] Superwall Systems, “Superhead Rainwater Tanks Filter - First Flush Diverters Australia”, (2015).

[8] BRAE, “Commercial Rain Heads”. (2012).

[9] Rainharvestmo, “Rainwater Harvesting Supply Company, LLC | Products1 Vortex Filter”, (2016).

[10] Allegiance Enterprises, “Rooftop Rainwater Harvesting Filters Products - Allegiance Enterprises”, (2017).

Fabrication of Non-Implant 3D Printed Skin

Yong Leng Chuan*, and *Shivani* Amish Kumar Pandya

Faculty of Built Environment, Engineering, Technology & Design, Taylor's University Lakeside Campus, Malaysia.

Abstract. Many bandages tend to be harmful when being removed from the human skin. This is a crucial issue, especially faced by burn victims. When bandages are removed from the burn wound, they tend to be harmful by peeling off the newly formed layer of skin over the burn wound. Such nature causes the patient to endure a longer recovery time with additional pain. The objective of this project is to 3D print artificial skin for the victims of burn wounds by using natural gelation. The main aim for creating the artificial skin will be used in place of the current burn wound treatment techniques of dressing the wounds in bandages. The inner layer of this skin was lined with a natural adhesive, a thin layer of agar-agar, which has been reinforced with crushed eggshells to increase its adhesive strength and durability. The synthesized gel contained non adhesive behavior, yet aids in wound healing abilities. Applying hydrocolloids ensures that the wound is kept cool and the gel also ensures efficient heat transfer. This was done so that less sweating occurs on the patient. Based on the experiments that were conducted, the results conclude that the best ratio of artificial skin layer would be 2:1 of agar gel: crushed eggshells. This golden ratio of gel: crushed eggshells for the longest period of time for attachment on the skin without sweating, is achieved. The skin will be printed using Acrylonitrile-Butadiene-Styrene (ABS). The colour of the skin and the shape of the skin was individually designed for each specific patient. The inner gel has the capabilities of reducing the rehabilitation time, without compromising the comfort of the patient. This approach has the potential to be used as a new method to treat burn wounds..

1 Introduction

The main purpose of creating the 3D printed skin is to aid burn victims. According to the World Health Organization (WHO), there are an estimated of 265,000 fatalities every year caused by burns alone [1]. There are countless more people with burn injuries that are admitted into hospitals every year. Treatment for burn patients depends on the severity of the burn and how many layers of skin have been damaged from the burn. That will also determine the number of days the treatment will take and how many days the patient needs to be hospitalized.

* Corresponding author: LengChuan.Yong@taylors.edu.my

When victims have severe burn wounds, it will be a long and painful process to treat them all properly. Not all the skin layers can be treated at once; hence multiple surgeries have to be performed. The human skin has to be given some time to heal on its own before surgery can be performed again [2]. Whilst waiting for the patient to heal, the wound cannot be exposed to air and must be wrapped with gauze and bandages to prevent bacterial infections [3]. Despite this being a commonly practiced method that is used by hospitals worldwide, there are many accidents that are caused by the bandages themselves. When the bandages are removed from the area of the burn, it may cause tearing and stripping of the skin, which causes the burned skin to be further irritated, hence prolonging the recovery time. Based on a study of 155 residents in long-term care facilities who were at least 65 years of age, the incidence of new tape-related skin injuries was 15.5% over an 8-week period with skin injuries developing at 34 sites in 24 patients [4].

For burn victims, plastic surgery is a luxury and is not affordable for everyone. It is an expensive procedure that may contain plenty of risks. Especially, in the case of a deep skin injury, multiple procedures are required to fix the wound and get the skin surface looking normal again by placing multiple layers of skin grafts. Non-implant skin can be an alternative to solve this issue. This can be used by people of all financial means and has significantly less risks compared to undergoing surgery. Non implant skin means that there is no surgery required as the skin will not be embedded into your body. This is a cheaper alternative and relatively safer than surgery. The non-implant skin can be reused and the gel created with natural substances is compatible with the human skin, which will also promote recovery [5]. Fortunately, medical technology constantly advances, creating better healthcare and opportunities for patients. New discoveries have been made in the field of 3D printing, where a 3D printer is used to print tissues samples and skin layers that can be implanted into the human body using stem cells. However, this too is still an experimental procedure and costs a lot of money [6]. There are two main reasons for creating 3D printed skin for burn victims. Firstly, there is for the purpose of covering up their scars after their burn wounds have been treated. Secondly, it can be used for the healing process of the burn victims. The second reason is what will be discussed in this report based on the healing of the wounds of the burn victims. This is because the gel layer synthesized has natural organic substances, all of which promote skin growth, which aids in the recovery time of the burn victims.

A 3D printer is used so that the skin can be easily customized to the different types of structures that each person has. 3D printing is an efficient option as it can cater to each patient's specific applications. It is also a fast prototyping method with high accuracy and low fabrication cost. The filament to be used for the 3D printer is the acrylonitrile butadiene styrene (ABS). ABS is used as the filament because it is a material that can be easily sterilized and that is easy to mould. With medical applications, sterilization is a very important factor [7]. Another crucial element for the non-implant 3D skin will be the gelatinous substance that will be used to attach the 3D printed skin to the patient's burnt skin. The gel will be synthesized using hydrocolloids which have excellent healing properties for the skin and have adhesive strength that will be able to replace the current adhesives in bandages.

This research is done to have an alternative method to apply to the current care of burn victims. Instead of using the bandages and any medical adhesives, a 3D printed skin mask is to be used to promote healing. Furthermore, the gel produced will be nontoxic and hence will not be an irritant to the skin.

2.0 Research Methodology

2.1 Raw materials for gel formulation

The raw material that has been used to synthesize the inner layer of the artificial skin is agar, crushed eggshells and glycerol. The eggshells were roughly crushed using a pestle and mortar and were mixed with the agar. The reason for using roughly crushed eggshells compared to using eggshell powder is that the hand crushed eggshells contain the permeable membranes of an eggshell. The permeability of these membranes ensure that the entire solution remains "breathable". Agar has excellent rheological properties and hence have many applications in the biomedical industry. Agar is from a plant origin, along the walls of certain types of algae. The agar in the seaweed provides the flexible motion for the seaweed to be able to follow the current of the ocean [8]. The flexibility of agar can be useful in the project as the contours of the human body are not easy to adapt to and hence the flexibility will ensure the easy adhesion of the gel to the skin. The agar was then reinforced with crushed eggshells and glycerol.

The eggshell of a chicken egg contains 95% of calcium carbonate and 5% of organic materials such as collagen. Collagen is the main component in the extracellular matrix of all tissue types. It is biocompatible and is non-toxic. This collagen will be helpful in tissue regeneration [9]. It is useful for burn victims as it also would help speed up the recovery of the patient.

Another substance that was tested with was glycerol. A preliminary experiment conducted with plain agar and glycerol was also not particularly effective in showing good spreading.

During the formulation, the crushed eggshell powder was added to the boiled agar solution and blended homogeneously. This was tested in different concentrations to see which one was the optimum solution. Secondly, the eggshell powder was added to the solution along with glycerol. This was then compared with the solution of agar and eggshells. The best synthesized gel was chosen from these experiments.

2.2 Breathability

Breathability is one of the crucial factors of this project. The gel that is created should be "breathable". This is to ensure that no sweating occurs between the gel layer and the human skin. If there is sweat that pools up between the two layers, it will cause slippage of the artificial skin and hence would reduce the duration that the skin can remain attached to the human skin. Agar is a permeable substance that allows diffusion to occur. This will ensure that the sweating of the skin will not pool. Furthermore, all the additives that are tested with to create the gel, are permeable. The eggshells are used in crushed form instead of powder form because when the eggshells are crushed, the inner and the outer membrane is also included in the mixture. Both the inner and out membrane are permeable. This is added to the mixture when creating the gel. Furthermore, glycerol is also a substance that permeable. Hence, when the gel mixture is created, it is ensured that there is low risk of slippage of the artificial skin. Moreover, there ABS that is used to print the artificial skin also has the properties of being water permeable. This means that the sweat that would occur would have room to escape and would not affect the gel layer.

2.3 Determination of gel : crushed eggshells ratio

The aim of this step is to determine the best spreading and adhesion by adding crushed eggshell powder and glycerol to the agar gel. The factors that will be considered in the synthesizing of the gel is the concentration of the crushed eggshell powder that is put in the agar gel. The experiment was conducted by testing the gel and 3D printed skin on a test subject. The time taken for the gel to dry and the time that the printed skin remained stuck to the human skin were both measured.

The aim of this experiment is eventually to determine the best combination of the three raw materials – agar, crushed eggshells and glycerol, to be able to produce a gel that is quick to dry on the skin and has the ability to stick to the human skin for longer periods of time. Figure 1 below shows the raw materials used while Figure 2 shows the mixture of the substances.

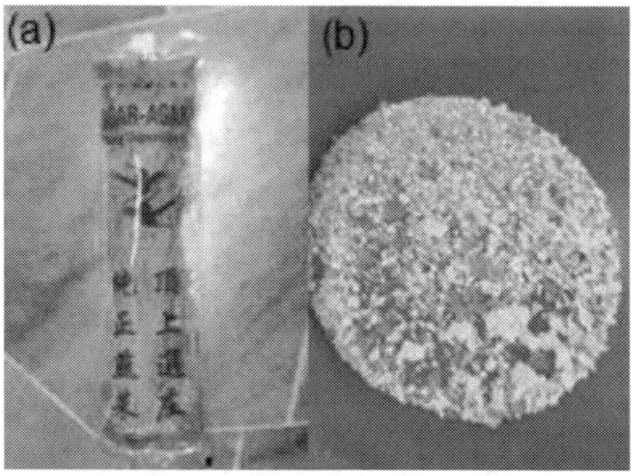

Fig. 1. (a) Dry Agar. (b) Crushed Eggshells

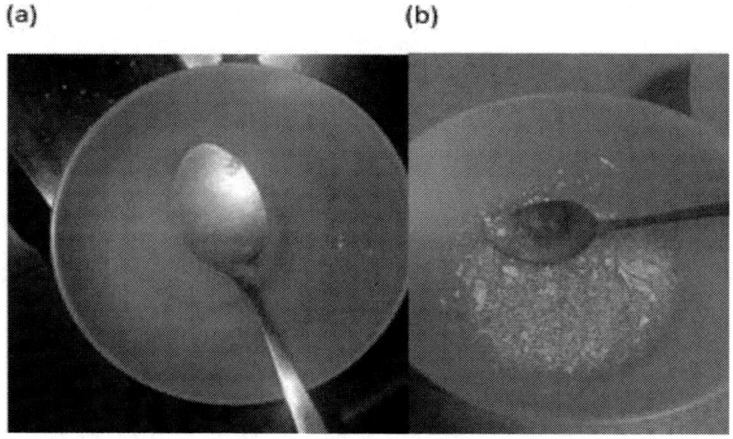

Fig. 2. (a) Combination of agar gel and glycerol. (b) Combination of agar gel, glycerol and eggshells.

The results for this experiment were found by applying a layer of the gel synthesized on the 3D printed skin. This 3D printed skin is held in place until the solution is dry on the human skin. The time taken for the solution to dry on the human skin is measured and then the time the skin remains on the human skin is also measured and recorded. The thin layer of the gel spread on the 3D printed skin and how the 3D printed skin was tested on the human skin is shown in the Figure 3 below. For this research, there was only a single test subject as the concept was trying to be proven where this new research could be a viable alternative in place of bandages. To be able to incorporate this into the medical field, further testing is required on more skin types.

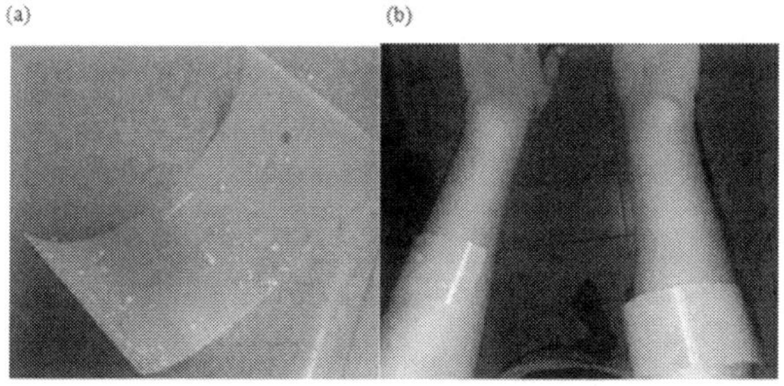

Fig. 3. (a) A layer of synthesized gel (agar, crushed eggshells and glycerol) applied on the 3D printed skin. (b) The adhesion testing of the 3D printed skin on the human skin

2.4 3D Printed Artificial Skin Outer Layer

To design the 3D printed skin, measurements were taken of an actual test subject and the skin was printed to fit the user's arm perfectly. Once the measurements of the subject's arm were taken, design was modelled using Solidworks. The material selected to print the skin was ABS. Figure 4 below shows the customized arm design of the printed skin in Solidworks.

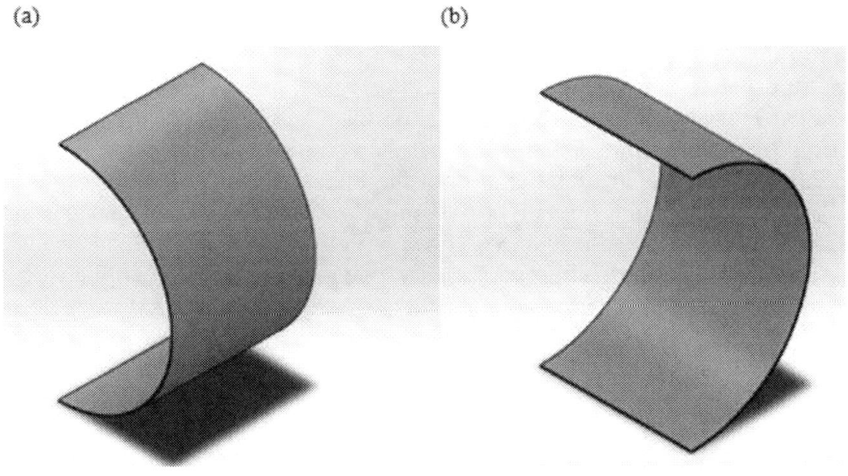

Fig. 4. The arm design of the 3D printed skin

The measurements of the artificial skin had the height of 8cm and the width of 5cm, with the radius of curvature of 4cm. The thickness selected for the skin was 0.5 mm. This is selected as the thickness of the skin as the actual thickness of human skin is 3 mm and the aim was to ensure the printed skin was light enough to not be able to be felt by the user and flexible enough to fit the contours of the human body, without being too fragile to break. Figure 5 below shows the actual printed skin.

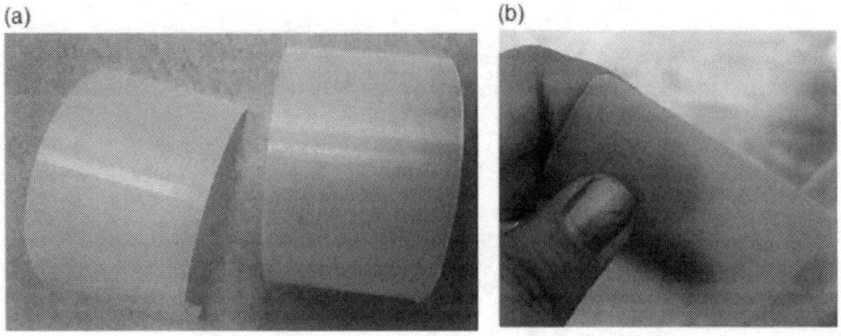

Fig. 5. (a) The 3D printed skin for the arm, (b) The 3D printed skin for the arm with rounded edges

The issue of safety always arises when creating something for human usage. This is why, special consideration was made when creating the skin. It was initially printed as a rectangular piece with sharp edges to ensure the dimensions taken from the test subject were correct. Once the printed piece was fitted to check if the skin fits the wound, the edges of the skin were filed to ensure that the sharp edges do not do further damage to the skin.

3.0 Results and Discussion

There are two experiments that were conducted. The first experiment tests which mixture is optimum for the gel synthesis. The second experiment tests the different concentrations of eggshells to see how the varying concentrations of eggshells affect the adhesive strength of the gel. A preliminary experiment was conducted to minimize the mixtures to be tested. The initial experiment tested the time taken for the mixture to dry on the human skin along with the duration it adhered to the human skin. The agar gel with the glycerol takes the shortest time on average to completely stick the paper on the human skin. Whilst plain agar had a lower time to dry completely, the adhesiveness of agar was weaker than that of the rest of the mixtures. This means that while doing routine work with the arm, the paper detached itself from the human skin. This means that the adhesive strength of plain agar is too weak to be able to allow the artificial skin onto the human skin if it cannot support a piece of paper to stay stuck to the human skin. Similarly, for the glycerol and the agar mixture, the mixture has very low adhesiveness. Since both the substances are water based, it has low stickiness and the paper slips very easily when some force is applied on it. Hence, for further experiments, the plain agar and the agar with glycerol mixture were not considered and only the other two mixtures were tested.

The reason for the agar and the glycerol mixture to dry faster compared to the neat agar is because both glycerol and agar are water based solutions, which has lower viscosity compared to the neat agar. The addition of glycerol to the neat agar makes it a more watery solution and hence has less adhesiveness and the ability to dry quicker.

3.1 Gel formation theory

The first experiment conducted was to compare the two different mixtures that are tested to synthesize the inner gel. The first experiment measures the time taken for the mixture to dry on the human skin and the duration for which it remains attached on the skin. The results showed when the two different mixtures are applied to the 3D printed surface and given time to allow them to continue adhering to the human skin. Figure 7 below represents the duration for which the printed skin remained adhered to the human skin.

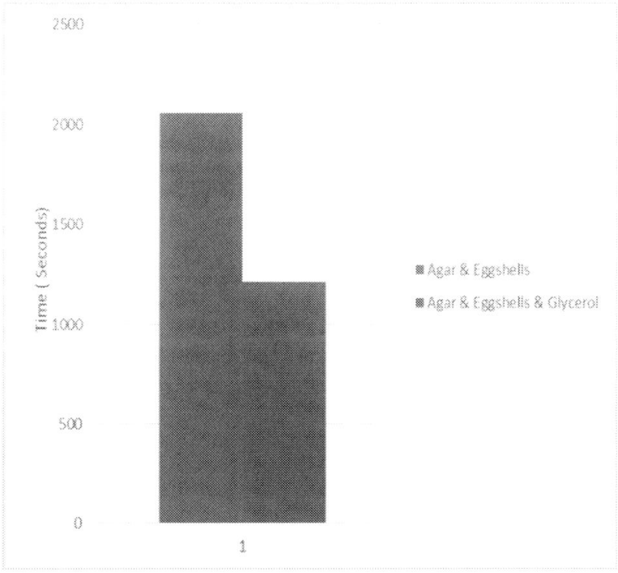

Fig. 7. Graph representing Experiment 1

From the results obtained in the first experiment that the more effective solution to be used as the gel is the agar and the eggshells when compared to the agar, eggshells and the glycerol. The agar and eggshells solution takes a lesser time to dry and has the capabilities of remaining on the human skin for a significantly longer period of time. This results were obtained after conducting the experiment twice and taking the average reading for more accurate results. The experiment is also conducted at the same time period and in the same area so there are no environmental factors that affect the results of the experiment.

3.2 Gel Adhesion Testing based on different crushed eggshells composition

Another factor that can affect the duration for which the synthesized gel can remain on the human skin is the percentage of eggshells that are added into each mixture. Hence, the gels are tested again with different concentrations of eggshells to be able to see which concentration is optimum to be used for the gel synthesis. The results below shows that the experiment conducted when different concentrations of eggshells were used for both the separate mixtures. Figure 8 shows the graphical representation of the results obtained.

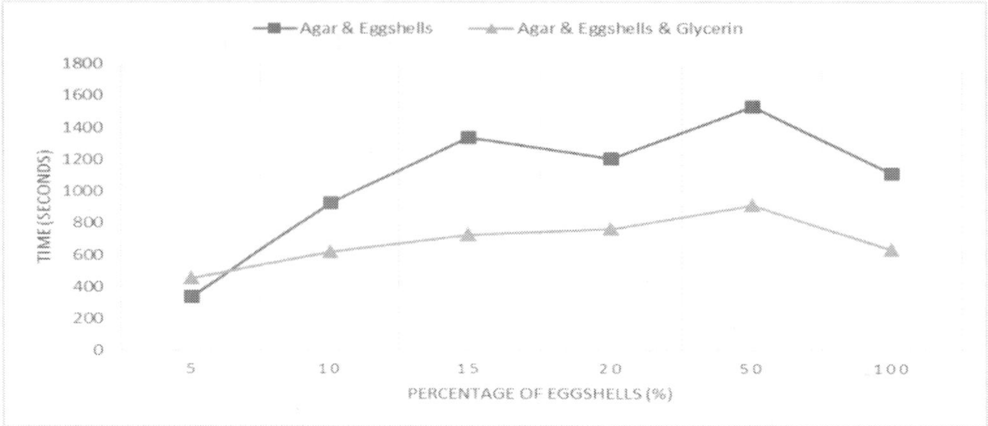

Fig. 8. Graph representing Experiment 2

It can be concluded from the results obtained in the second experiment that the optimum concentration of eggshells that should be used for the gel is 50%. The agar with the 50% concentration of eggshells has the longest duration of time that the printed skin was stuck to the human skin. When compared between the two mixtures, the longest duration that the mixture with glycerol is 15 minutes, whereas the longest duration that the mixture without glycerol is about 1.6 times more than the mixture with glycerol at 25 minutes. Hence, it can be said that the mixture that should be used to synthesize the gel should be agar gel with 50% eggshells.

It can be observed that at 20% of eggshells concentration, there is an anomaly in the results. It can be seen that at the 20% concentration of crushed eggshells, the duration of it sticking to the human skin decreases. This is due to some inconsistencies in the experimental results. This is why it does not follow the trend of the overall graph.

At 100% of eggshells, the boiled agar solution is already saturated with the crushed eggshells and hence does not produce more efficient results in terms of the duration of time it is allowed to stick on the human skin. The trend of the graph is observed closely, hence it can be deduced that the optimal percentage of eggshells to be added to the boiled agar solution is 50% of eggshells, without the additional glycerol.

The results show that the better solution to use as the gel is the agar and crushed eggshell solution without the usage of glycerol. This is because the usage of glycerol causes slippage of the artificial skin. The addition of the glycerol makes the solution reduce in adhesive strength. The adhesive strength is measured on the basis of the duration the artificial skin can adhere to the human skin without any external applied force. For the purposes of this experiment, the adhesiveness is not tested based on the amount of force each mixture can handle. This is why, the selected solution for spreading on the artificial skin is agar and 50% of crushed eggshells.

4.0 Conclusion

The optimum ratio of gel: crushed eggshells is 2:1. This ratio showed that the ability of the 3D printed outer layer to adhere to the human skin has been achieved. Using the optimum ratio, there was a short drying time along with it being able to adhere to the human skin for a long duration of time. Furthermore, the artificial skin showed no adhesive stains or marks after the skin was peeled off the human skin. This is beneficial for burn victims as there would be no additional disinfecting of the wound required once the artificial skin is removed.

The curvatures of the human body are also another consideration when designing the exact area for the 3D printed skin. It is important to consider the tolerances and the limitations of the 3D printer when the measurements are taken. Each printed skin has to be designed specifically for that user. The edges of the skin have designed to be rounded and not sharp so that it is safer for the user to use.

References

1. M. W. Michell, H. M. Oliveira, M. P. Kinsky, S. U. Vaid, D. N. Herndon, and G. C. Kramer, J. Burn Care Res. **27**, 6, pp. 819–825 (2006)
2. M. Bezuhly and J. S. Fish, Plast. Reconstr. Surg. **130**, 2, pp. 349e–358e (2012)
3. D. Church, S. Elsayed, O. Reid, B. Winston, and R. Lindsay, Clinical Microbiology Reviews **19**, 2, pp. 403–434 (2006)
4. Konya C, Sanada H, Sugama J, et al. J Clin Nurs **9**, pp. 1236-42 (2010)
5. T. Lefrancois, K. Mehta, V. Sullivan, S. Lin, M. Glazebrook, Foot Ankle Surg **23**, 4, pp. 215-224 (2016)
6. A. N. I. D. Gantry, K. Johnson, A. Dubey, C. Harris, and D. Sasaki, Radiother. Oncol **120**, pp. S33–S34 (2016)
7. R. Ricotti et al., Phys. Medica **32**, 12, pp. 1698–1706 (2016)
8. W.-K. Lee, Y.-Y. Lim, A. T.-C. Leow, P. Namasivayam, J. Ong Abdullah, and C.-L. Ho, Carbohydr. Polym **164**, pp. 23–30 (2017)
9. S. Hsieh, H. H. Chou, C. W. Hsieh, D. C. Wub, C. H. Kuo, and F. H. Lin, Food Chem **141**, 3, pp. 2117–2121 (2013)

Lift and Drag of Non-conventional Wings at Subsonic Speeds and Zero Angle of Attack - An Experimental Investigation

Abdulkareem Shafiq Mahdi Al-Obaidi[1,*] and *Ting Chern* Wei[1]

[1]School of Engineering, Faculty of Built Environment, Engineering, Technology & Design, Taylor's University, Subang Jaya, Selangor DE, Malaysia

Abstract. Various non-conventional wing development shows potential in increasing the aerodynamic performance of airplanes. If the non-conventional wing only improves the aerodynamic performance by a small margin, conventional wing is still a better option for airline operators. This provides opportunity to continue research on non-conventional configurations that can greatly saves the fuel consumption. This research was conducted to examine the lift and drag of non-conventional wings at low subsonic speed and low angle of attack. Analytical method based on DATCOM was used to calculate the lift and drag coefficients of non-conventional cranked wing for comparison with experimental results obtained experimentally using Taylor's wind tunnel (TWT). Experimental lift coefficient shows similar values with the analytical results but experimental drag coefficient had an average difference of 44%. The experimental setup and calibration of TWT were verified and further case studies on nonconventional wing model featuring trailing edge notches were carried out. Analysis of the results from case studies shows that generally the effect of varying the number of notches only had significant effect on drag reduction if the notch depth was higher. For flight condition that does not exceed 4° angle of attack, lower number of notches at higher notch depth had the best aerodynamic performance. On the other hand, for flight condition that requires cruise angle of attack that exceeds 4°, higher number of notches at higher notch depth had the best aerodynamic performance.

1 Introduction

Airline operators are aware of the high demand for air travelling as the number of passengers increased exponentially over the past two decades. These numbers are expected to further increase in the future, providing two main challenges for airline operators. The first challenge is to lower down operating cost of airplanes in order to increase profit and secondly to uphold the company's corporate social responsibilities in putting effort to reduce air pollutions caused by combustion emission [1]. A viable solution is to improve aerodynamic performance of airplanes with the aim of reducing fuel consumptions which will decrease operating cost and also minimize negative environmental impact caused by

* Corresponding author: abdulkareem.mahdi@taylors.edu.my

emissions from airplanes. One key component in improving the aerodynamic performance of an airplane is to improve the design of wings because it is the main source of lift force. Referring to Fig. 1, when lift force, L is greater than the weight, W, the airplane will have more upward force and when drag force, D is lesser than thrust force, T, the airplane will have lesser resistance moving forward. This indicates that having a better lift to drag ratio can improve the aerodynamic performance because for the same amount of fuel consumption, the airplane can now travel for a longer distance. The potential of applying non-conventional wing design to improve aerodynamic performance of an airplane can be considered as an option for airline operators [1].

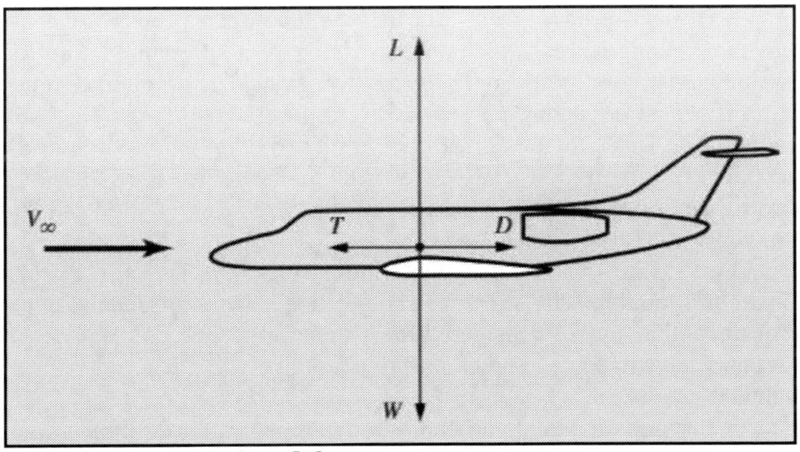

Fig. 1. Forces acting on an airplane [2].

The development of new technologies in the field of non-conventional wings is also part of the NASA's subsonic fixed wing project that aims to improve aerodynamic performance of subsonic transport airplane in the future [3]. Currently, C-wing, blended wing body (BWB) and split wing are the more common non-conventional wing configurations under development. These non-conventional configurations have the same aim of improving aerodynamic performance by reducing fuel consumptions.

Kaushik and Anemaat [4] conducted research on the split wing which has a non-conventional configuration where the left and right wing of the airplane are not aligned. The purpose of this design is to remove the horizontal tail from the low tail which is a common feature on conventional airplanes. Conventional airplanes having horizontal tail experiences downwash from the wings onto the horizontal tail which requires high power control that will leads to additional drag force. If the flaps on the wings are deployed, even more drag force is produced. The removal of horizontal tail on the split wing decreases drag force which will lead to a better fuel consumption. However, in terms of safety aspect, the split wing has asymmetric lift that will results in high tendency for the airplane to roll.

Reist and Zingg [5] studied the blended wing body (BWB) which has a non-conventional configuration of combining the airplane body and wings as one component. This configuration was proven to have high potential in improving aerodynamic performance on larger airplanes. The BWB was designed to have lesser wetted surface area that aims to reduce skin drag friction that contributes to total drag force. Reducing the skin drag friction improves the lift to drag ratio of the BWB. Additional advantage of the BWB is to have a lower operating noise due to the acoustic shielding provided by the aerodynamic design. However, Lyu and Martins [6] stated that the BWB configuration has a challenge in maintaining trim due to uneven pressure distribution over the body and wings. Furthermore, the BWB features thick airfoil shape that increases the difficulty in achieving good lift to drag ratio at a reasonable deck angle which is also a concern.

Skinner and Zare-Betash [7] conducted research on the non-conventional C-wing configuration that features a non-planer c-layout that provides high stability and control through the reduction of wing tip vortices. The decrease in wing tip vortices also improves lift to drag ratio through reducing vortex drag. The limitation of the C-wing is the challenge in manufacturing the uncommon wing architecture but this is more minor compared to the disadvantages of the split wing and BWB. Based on the advantages and disadvantages, C-wing is more suitable to be applied on commercial airplanes when compared to split wing and BWB.

The development of non-conventional wings have high potential in improving aerodynamic performance of airplanes but the manufacturing process of non-conventional designs are more complicated which increases the cost of manufacturing. Due to this reason, airline operators will only seriously consider applying non-conventional configuration if the aerodynamic performance can be greatly improved to cover the high manufacturing cost. This provides great opportunity to conduct research and develop new non-conventional configuration that is worth to be considered for application on commercial airplanes. One of nature's creations, the bird wing is good candidate to inspire a novel non-conventional design.

Different bird species have different wing shapes but are all considered as non-conventional configurations. Besides the compositions of the bird wing such as muscles, bones and feathers that are not considered from an aerodynamic point of view, the different wing shapes may have impacted and contributed positively to the aerodynamic performance of bird flight.

Dvorak [8] performs research on the aerodynamics of bird flight which consists of different functional studies from different parts of the wings. A bird wing can be mainly divided into two parts which are the primaries and secondaries. From the studies, lift force is mainly generated from the secondaries that resemble the airfoil shape for the bird. Primaries also contribute by forming tip slots that reduces wing tip vortex that produces induced drag. A bird varies the tail area located just behind the wing to maintain stability and balance. Overall, main studies on the lift can be performed on the secondaries section to act as a fixed wing for airplanes.

This research was conducted to compare the lift and drag of a non-conventional cranked wing by using analytical method from Data Compendium (DATCOM) and also experimental method using Taylor's wind tunnel (TWT). Further studies were performed to determine the lift and drag of non-conventional wings with the effects of notches on the trailing edge. The non-conventional configuration of having notches shape was inspired from the secondaries section of a bird wing which is the main section to generate lift. Most articles available are researching on the application of the flapping mechanism of bird wing and there are limited studies on the effect of shape, especially notches on the trailing edge of bird wing. This provides opportunity to research on the aerodynamic characteristics, lift and drag of fixed wing with different numbers and different sizes of trailing edge notches.

To determine the wing design with better lift to drag ratio, the non-conventional configuration featuring notches will be compared with conventional wing configuration through experimental investigation using TWT. The scope of this experimental investigation is at conditions of low subsonic speeds and low angle of attacks close to zero.

2 Research Methodology

This research examines the lift and drag using analytical method and mainly experimental method which is practicing a quantitative study approach. The research methodology includes analytical method for cranked wing, three dimension (3D) modelling of test model, low subsonic speed wind tunnel and case studies on trailing edge notches.

2.1 Analytical method

Hoak [9] published a document that consists of four volumes with the title of DATCOM. The DATCOM discussed about analytical prediction methods for static aerodynamic coefficients and dynamic derivatives. Analytical methods in DATCOM allow calculation of individual flying component and summed up for the total result of the whole flying body. This research applies DATCOM analytical method which is in fact a semi-empirical approach, to calculate lift and drag of cranked wing at subsonic speeds. Semi-empirical approach consists of calculated values summed up with values obtained from experimental graphs.

2.1.1 Calculation of lift force for cranked wing

Based on DATCOM method, the rate of change of lift coefficient with angle of attack for non-conventional cranked wing at subsonic speed can be determined by applying Eq. (1):

$$C_{L_\alpha} = \left(C_{L_\alpha}\right)_{pred} \frac{\left(C_{L_\alpha}\right)_{test}}{\left(C_{L_\alpha}\right)_{pred}} \tag{1}$$

where $\left(C_{L_\alpha}\right)_{pred}$ is the predicted lift-curve slope of the cranked wing and $\frac{\left(C_{L_\alpha}\right)_{test}}{\left(C_{L_\alpha}\right)_{pred}}$ is the empirical correlation factor of subsonic lift-curve slope for cranked wing. $\frac{\left(C_{L_\alpha}\right)_{test}}{\left(C_{L_\alpha}\right)_{pred}}$ can be obtained from Fig. 4.1.3.2-52 of DATCOM [9] which is presented as a function of aspect ratio and Mach number. $\left(C_{L_\alpha}\right)_{pred}$ can be obtained from Fig. 4.1.3.2-49 of DATCOM [9] which is presented as a function of aspect ratio, surface-roughness height, Mach number parameter and sweepback angles of mid-chord line of exposed wing ,where all parameters can be calculated using Eqs. (2) to (4):

$$\left(\cos\Lambda_{c/2}\right)_{eff} = \frac{1}{S_W}\left[\left(\cos\Lambda_{c/2}\right)_i S_i + \left(\cos\Lambda_{c/2}\right)_o S_o\right] \tag{2}$$

$$c_{l_{\alpha M}} = c_{l_\alpha} / \beta \tag{3}$$

$$k = c_{l_{\alpha M}} / (2\pi / \beta) \tag{4}$$

where $\left(\cos\Lambda_{c/2}\right)_{eff}$ is the effective sweepback angle of mid-chord line of exposed wing, $\left(\cos\Lambda_{c/2}\right)_i$ is the sweepback angles of mid-chord line of wing inboard panels, $\left(\cos\Lambda_{c/2}\right)_o$ is the sweepback angles of mid-chord line of wing outboard panels, S_W is the wing area , S_i is the total area of inboard panels of wing and S_o is the total area of outboard panels of wing, $c_{l_{\alpha M}}$ is the lift-curve slope corrected for compressibility effect, c_{l_α} is the section lift-curve slope with angle of attack at constant flap deflection, and β is the Mach number parameter and k is the surface-roughness height.

2.1.2 Calculation of drag force for cranked wing

For the case of drag force, the total drag can be divided into C_{D_o}, zero-lift drag and C_{DL}, drag due to lift which can be express as Eq. (5):

$$C_D = C_{D_o} + C_{DL} \tag{5}$$

Zero-lift drag consists of friction drag and form drag which are mainly affected by surface roughness, wetted area and also airfoil shape. Zero-lift drag of non-conventional cranked wing at subsonic speed can be calculated using Eq. (6):

$$C_{D_o} = C_f \left[1 + L\left(\frac{t}{c}\right) + 100\left(\frac{t}{c}\right)^4 \right] R_{LS} \frac{S_{wet}}{S_{ref}} \tag{6}$$

where C_f is the turbulent flat-plate skin friction coefficient that can be obtained from Fig. 4.1.5.1-26 of DATCOM [9] which is presented as a function of Mach number and Reynolds number based on the reference length, L is the airfoil thickness location parameter and is equal to 1.2 for $(t/c)_{max}$ located at $x_t \geq 0.3c$, t/c is the average streamwise thickness ratio of the wing, R_{LS} is the lifting-surface correction factor obtained from Fig. 4.1.5.1-28b of DATCOM [9] which is presented as a function of the Mach number and the cosine of the sweep angle of the airfoil maximum thickness line, $(\Lambda_{t/2})_{max}$, S_{wet} is the wetted area of the wing and S_{ref} is the reference area which is equal to the S_{wet}. Since the cranked wing features inboard and outboard panels, C_{Do} for both panels has to be calculated and summed up for total zero-lift drag.

To calculate drag due to lift of non-conventional cranked wing at subsonic speed, apply Eq. (7):

$$C_{DL} = \frac{C_L^2}{\pi A e} + \Delta C_{DL} = \frac{C_{DL}}{C_L^2} C_L^2 + \Delta C_{DL} \tag{7}$$

where C_L is the wing lift coefficient, A is the aspect ratio, e is the span-efficiency factor over the parabolic-drag region and ΔC_{DL} is the portion of the drag due to lift resulting from a breakdown in the leading-edge suction at lift coefficients above the parabolic-drag-polar region. An empirical correlation of ΔC_{DL} as a function of lift coefficient and aspect ratio is available in Fig. 4.1.5.2-54 of DATCOM [9]. Additional equations required are Eqs. (8) to (10):

$$\frac{C_{DL}}{C_L^2} = \frac{1}{\pi A e} \tag{8}$$

$$e = \frac{C_{L\alpha} / A}{R'(C_{L\alpha} / A) + \pi(1 - R')} \tag{9}$$

$$R' = R_i(n_B - d/b) + R_o(1 - n_B) \tag{10}$$

where R_i is inboard panel leading-edge-suction parameters, R_o is outboard panel leading-edge-suction parameters. R_i and R_o can be obtained from Fig. 4.1.5.2-53 of DATCOM [9] which is presented as a function of the Reynolds number based on the leading edge angle, Mach number and leading edge radius which is given as 0.00229 of mean chord length. n_B

is the dimensionless distance from plane of symmetry to break span station, d/b is the ratio of maximum body width to wing span, and e is the span-efficiency factor over the parabolic-drag region.

2.2 Test model

To compare the experimental results of lift and drag with the calculated lift and drag using DATCOM analytical method, an identical wing model has to be 3D modelled and manufactured. The wing model is having a non-conventional cranked wing configuration and using the NACA 0012 airfoil.

2.2.1 3D modelling

The cranked wing was modelled using 3D computer aided design (CAD) software of SolidWorks 2016 version. The method used for 3D model was to import x and y coordinates of the NACA 0012, available in Abbott ad Von Doenhoff [10] and generate the airfoil shape using the imported coordinates. These coordinates are in ratio to chord length. Lofted boss function is used to generate solid between each airfoil to form a cranked wing. Figure 2 shows the top view of the cranked wing with dimensions that are same with the dimensions used for the analytical calculation.

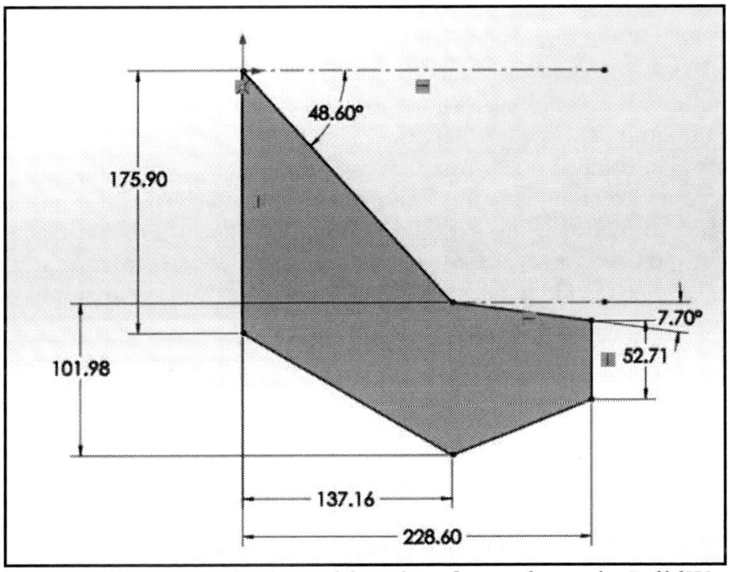

Fig. 2. Dimensions of the cranked wing with units of mm shown in SolidWorks.

2.2.2 Manufacturing

The cranked wing SolidWorks drawing was sent to 3D printer branded as Raise 3D for 3D printing using PolyLactic Acid (PLA) as the printing filament. The print density is set to 100%. Figure 3 shows the 3D printed cranked wing with NACA 0012 airfoil as the test model.

Fig. 3. 3D printed cranked wing.

2.3 Low subsonic speed wind tunnel

The experiments of this research were conducted using a wind tunnel located at Taylor's University, named as Taylor's Wind Tunnel (TWT). Figure 4 shows the TWT that consists of different sections which are the settling camber, contraction section, test section, diffuser and fan. The TWT is an open circuit subsonic wind tunnel model of TDA_M630/12AA/7/28 with a contraction ratio of 3:4:1 [11]. Open circuit wind tunnel, requires air following through an essentially straight path from the entrance to the exhaust section after the fan [12]. The test section of the TWT is rectangular in shape and consists of a square cross section of 0.303 m x 0.303 m with a length of 0.885 m. The TWT is able to collect force and moment readings through a Gamma DAQ F/T transducer. Readings obtained will be displaced using a PC1 DAQ card model of NI PCI-6220 installed in the computer. The 0.63 m diameter fan is powered by a 3 horsepower motor with an output torque of 4145 V / 50 Hz [11].

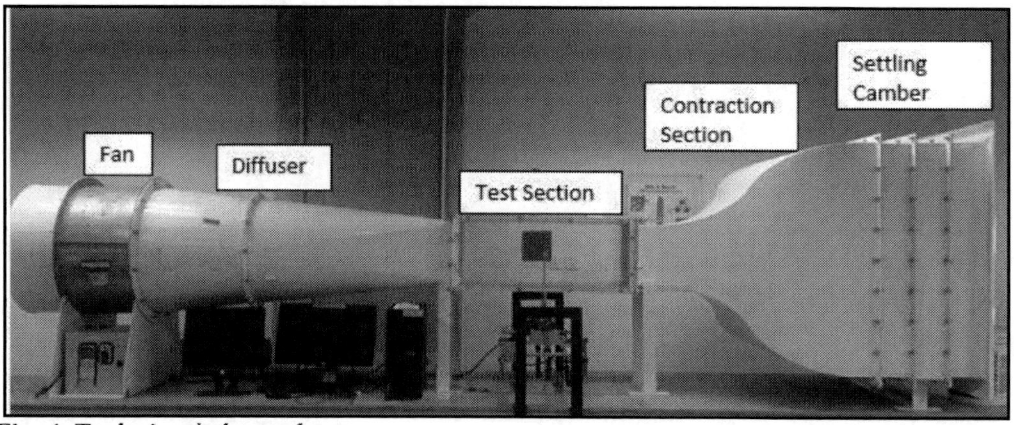

Fig. 4. Taylor's wind tunnel.

2.3.1 Experimental setup and conditions

The cranked wing model was mounted on a test rig and placed in the test section as shown in Fig. 5. The location of test rig was mounted 0.1266 m measured from the leading edge of root chord, which is the aerodynamic centre of the test model. Aerodynamic centre of non-conventional cranked wing can be calculated by using methods presented by Alexander [13].

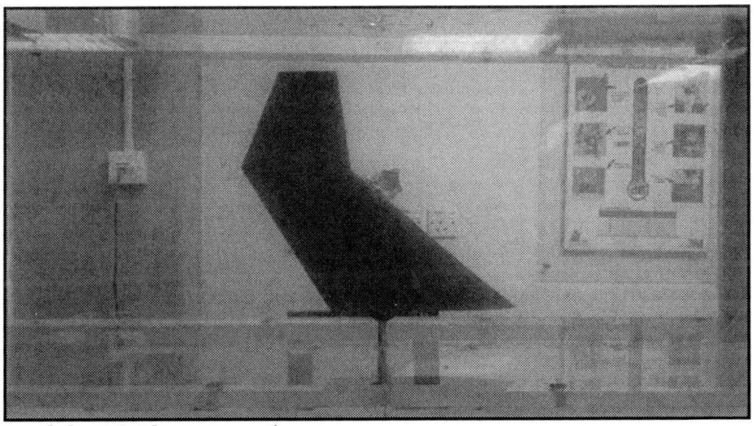

Fig. 5. Test model setup in test section.

The experimental was conducted at Reynolds number, Re = 1×10^5. To achieve this Reynolds number, the velocity of air flow was calculated to be 11.79 m/s. The variable that can be vary on the TWT was the frequency of the fan, and the linear equation between frequency and velocity is express as Eq. (11), which was derived from conducting experiment using hot wire and data logger.

$$f = 1.437V + 0.4359 \tag{11}$$

where f is the frequency of the fan and V is the velocity of air.

2.3.2 Calculating lift coefficient and drag coefficient

The reading obtained from the TWT computers is lift and drag force instead of lift and drag coefficient which can be calculated using Eqs. (12) and (13):

$$\text{Lift Force, } L = C_L \frac{1}{2}\rho V^2 S \tag{12}$$

$$\text{Drag Force, } D = C_D \frac{1}{2}\rho V^2 S \tag{13}$$

where C_L is the lift coefficient, C_D is the drag coefficient, ρ is the density of air, and S is the wing area. Lift and drag force were obtained by averaging 30 respective force readings from the raw data presented in TWT computer for different angle of attacks. Angle of attack of the test model is adjusted based on a protractor located on top of the test section.

2.3.3 Effect of test rig on drag coefficient

One important factor that has to be taken in consideration is the effect of test rig. The test rig that was used to mount the test wing model will affect the readings taken from TWT.

However the presence of the test rig only affects the drag force and will not affect the lift force due to the shape of the test rig. Since it is a cylindrical rod, it was assume to have symmetrical shape and no difference in angle of attack which will not produce lift force as the air flow passes the test rig. However, the test rig will induce drag by contributing to the friction drag and form drag. Because of that the drag force obtained from TWT is the combination of wing model and test rig while the lift force reading was not affected. To obtain the drag coefficient only for the test model, an equation as shown in Eq. (14) [14] was applied. The effect of test rig was considered in all experiments because it caused large difference in the drag coefficient.

$$C_{D_{Model}} = C_{D_{Total}} - C_{D_{Rig}} \frac{S_{Rig}}{S_{Model}} \tag{14}$$

where $C_{D_{Model}}$ is the drag coefficient of test wing model, $C_{D_{Total}}$, is the total drag coefficient, $C_{D_{Rig}}$, is drag coefficient of test rig, S_{Rig}, is the area of test rig, and S_{Model}, is the wing area of test model.

2.4 Case studies on the trailing edge notches

As mention in the introduction, a new non-conventional wing design can be development through inspiration from nature's creation, the bird wing. However, not all types of bird wings are suitable to be applied on airplanes because most bird species applied flapping flight which requires constant movement of wings and some bird species features wings that are unable to fly. In fact only wings from soaring bird species can have potential to be applied on airplanes. Soaring bird species have the ability to maintain flying without flapping its wing which also saves energy through the reduction of metabolic caused by flapping wing movement [15]. This situation is similar to the fixed wing of airplanes. Besides having good aerodynamic characteristic of high aspect ratio, the shape of soaring bird wing may also contribute to the increase in lift to drag ratio.

This research aims to examine the lift and drag of wing model featuring trailing edge notches. Figure 6 shows the notches from the secondaries of a golden eagle which consists of rounded edge. Notches on the primaries have shaper edge and have the ability to bend upwards to act as a wing tip which reduces size of vortices around the wing tip [16]. Experiment will be conducted only on the rounded edge notches because the main wing section that generates lift for the bird is the secondaries wing which has similar shape to airfoil. Experimental investigation was conducted by determining the lift and drag of a NACA 0012 wing with a chord length of 0.15 m and a span length of 0.15 m which gives an aspect ratio of 1. Keeping the aspect ratio and wing area as constant, add notches on the trailing edge for different case studies as shown in Table 1 to study the effects of trailing edge notches on the lift and drag force. Figure 7 shows the top view and isometric view of the 3D test model for trailing edge notches.

Table 1. Case studies on trailing edge notches.

Experiments	Number of Notches	Depth of Notches (mm)	Wing Area (mm^2)	Aspect Ratio
Case 1	-	-	22500	1
Case 2	3	10 mm	22500	1
Case 3	5	10 mm	22500	1
Case 4	3	15 mm	22500	1
Case 5	5	15 mm	22500	1

Fig. 6. Rounded edge notches of a golden eagle [17].

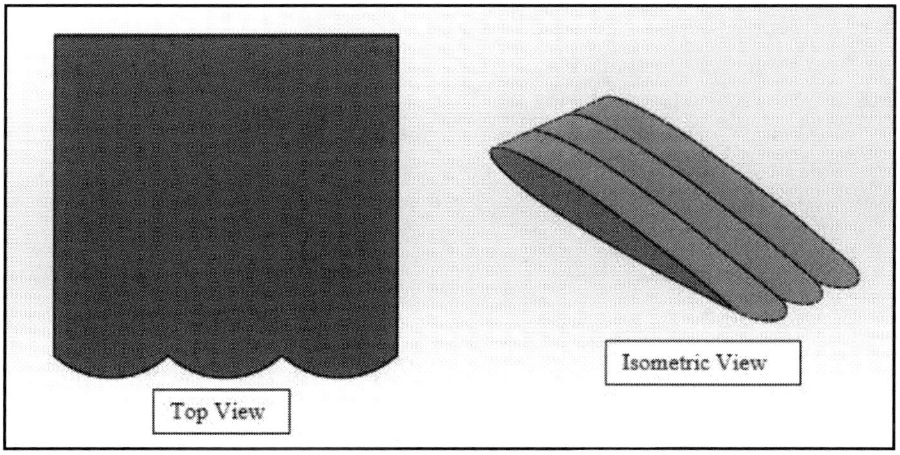

Fig. 7. Top view and isometric view of test model with tailing edge notches in solidworks (not according to scale).

3 Results and discussion

Results obtained according to the research methodology will be divided into analytical results and experimental results for more detail discussion. Table 2 shows the list of parameters and values used to perform calculation for both analytical and experimental methods.

3.1 Analytical results

Analytical results of the lift coefficient, C_L and drag coefficient, C_D of non-conventional cranked wing was calculated using DATCOM method and the results were plotted in Figs. 8 to 10. Based on Fig. 8, the lift coefficient of the test model is directly proportional to the angle of attack with a gradient of 0.0605. At zero angle of attack the lift coefficient is zero because the test model is using NACA 0012, which is a symmetrical airfoil. In the case of drag coefficient, an increase in the angle of attack regardless of positive or negative angle will still result in an increasing value which is shown in Fig. 9. However, unlike lift coefficient, drag coefficient is presence even at zero angle of attack. This is because total drag coefficient is the summation of zero-lift drag, C_{Do} and drag due to lift C_{DL}. At zero angle of attack, even there is no drag due to lift, friction drag and form drag are still contributing to the zero-lift drag with the value of 0.011. For further analyzation, a polar curve as shown in Fig. 10 was plotted, to determine the maximum lift to drag ratio of the test model. The point of maximum lift to drag ratio is actually the tangent line to the C_L vs C_D curve from zero y-intercept. For the non-conventional cranked wing test model, the maximum lift to drag ratio was found to have a value of ten.

Table 2. Parameters and values used for calculation.

Parameters	Values
Mach number, Ma	0.4
Mach number parameter, β	0.91652
Aspect ratio, A	4
Wing area, S_W	0.02612 m^2
Total area of inboard panels of wing, S_i	0.01906 m^2
Total area of outboard panels of wing, S_o	0.00707 m^2
Sweepback angles of mid-chord line of wing inboard panels, $(\Lambda_{c/2})_i$	40.9°
Sweepback angles of mid-chord line of wing outboard panels, $(\Lambda_{c/2})_o$	-7.7°
Section lift-curve slope with angle of attack at constant flap deflection, $c_{l\alpha}$	5.7869 per rad
Reynolds number, R_l	5.97×10^6 per m
Mean chord length of inboard panel, $\overline{C}_i$	0.14219 m
Mean chord length of outboard panel, $\overline{C}_o$	0.08001 m
Leading edge angles of wing inboard panels, $(\Lambda_{LE})_i$	48.6°
Leading edge angles of wing outboard panels, $(\Lambda_{LE})_o$	7.7°
Aspect ratio of outboard panels, A_o	2.37
Taper ratio of outboard panels, λ_o	0.517
Dimensionless distance from plane of symmetry to break span station, n_B	0.6
Ratio of maximum body width to wing span, d/b	0.139
Sweep angles of the airfoil maximum thickness line of wing inboard panels, $(\Lambda_{t/2})_{max,i}$	44.2°
Sweep angles of the airfoil maximum thickness line of wing outboard panels, $(\Lambda_{t/2})_{max,o}$	-2.01°
Wetted area of inboard panels of wing, $(S_{Wet})_{max,i}$	0.03812 m^2
Wetted area of outboard panels of wing, $(S_{Wet})_{max,o}$	0.01414 m^2
Average streamwise thickness ratio of the wing, t/c	0.12
Density of air, ρ	1.225 kg/m^3
Viscosity of air, μ	1.81×10^{-5}
Mean chord length, $\overline{C}$	0.12526 m
Airfoil	NACA 0012

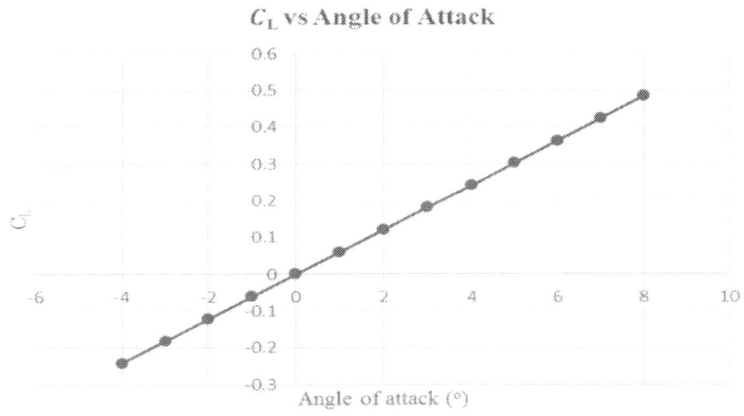

Fig. 8. Lift coefficient against angle of attack.

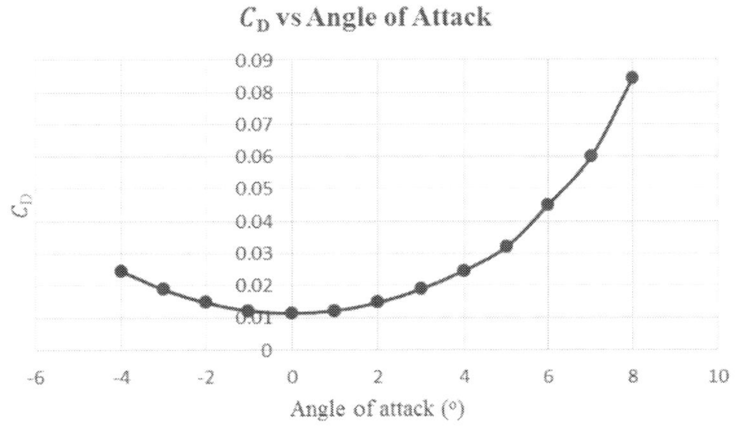

Fig. 9. Drag coefficient against angle of attack.

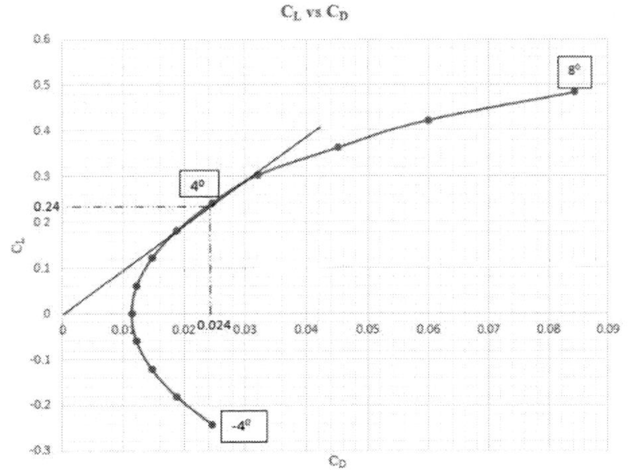

Fig. 10. Polar curve graph of lift coefficient against drag coefficient.

3.2 Experimental results

Experimental data was obtained by using Taylors' wind tunnel and the results were plotted in Figs. 11 and 12 for comparison with the analytical results to verify the experimental setup before conducting experimental investigation on the effect of trailing edge notches. Referring to Fig. 11, the experimental lift coefficient was approximate to be linearly increasing with angle of attack at a gradient of 0.0608 which is very near to the 0.0605 gradient value found using the analytical results. However, as shown in Fig. 12, experimental drag coefficient is not as accurate as experimental lift coefficient. Experimental drag coefficient has an average difference of 44% when compared to analytical drag coefficient. This error is caused by the drag induced by the test rig holding the test model. The presence of the test rig does not greatly affect the lift coefficient because of its symmetrical shape but it increases the friction drag and form drag that contributes to the total drag force. Because of that, the experimental drag coefficient curve is located higher than the analytical drag coefficient curve but still showing similar curve trend. An experiment was conducted to determine the drag force induced by the test rig and by applying Eq. (14), the drag coefficient for only the test model was calculated. Figure 13 indicates that after subtracting the drag coefficient caused by test rig, the accuracy of experimental results had improved greatly. Based on the comparison, it is found that TWT gives accurate drag coefficient results but less precise than lift coefficient which are both accurate and precise.

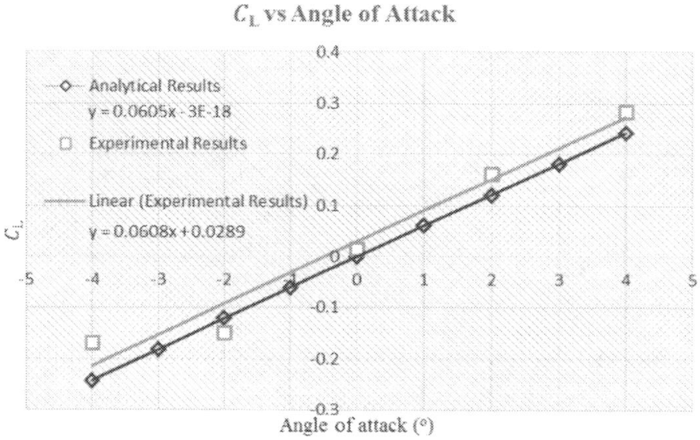

Fig. 11. Lift coefficient against angle of attack.

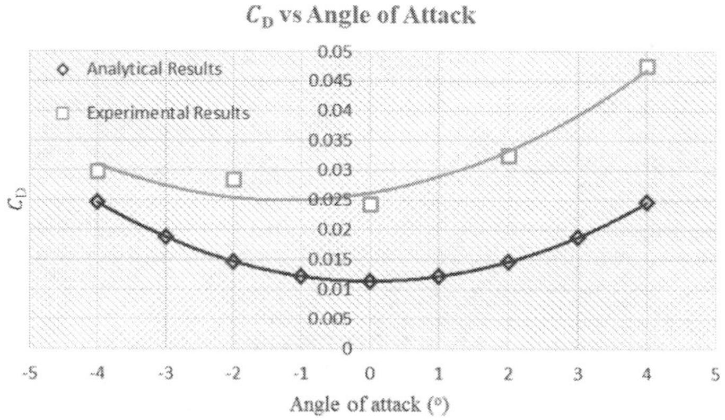

Fig. 12. Drag coefficient against angle of attack.

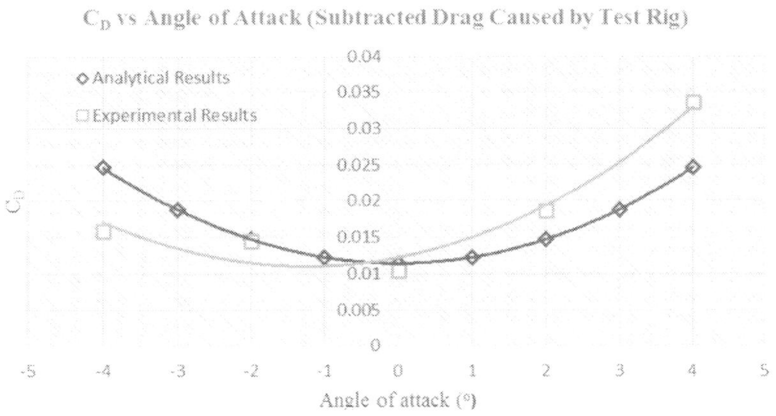

Fig. 13. Drag coefficient (only wing model) against angle of attack.

The experimental setup was verified and further experiments was carried out to study the effects on lift and drag of non-conventional test wing models featuring trailing edge notches and comparing the results with conventional wing model with the same wing area and aspect ratio. Based on the experimental results, it was found that lift force was not greatly affected because all wing models are having the same wing area. However, applying notch shape on the trailing edge may have reduced the vortices at the wake area behind the trailing edge which may result in reduction of drag force. Further analysis from case studies shows that at notch depth of 10 mm, there was no significant reduction in terms of drag even by varying the number of notches from three notches to five notches. However the effect of drag reduction can be clearly seen when the notch depth was increases to 15 mm. With comparison to conventional wing model, non-conventional wing model featuring three notches with 15 mm notch depth was chosen as the best model for flight condition that does not exceed 4° angle of attack while non-conventional wing model featuring five notches with 15 mm notch depth was the better option for flight condition that will exceed 4° angle of attack.

4 Conclusions

To examine the lift and drag of non-conventional wings at low subsonic speed and low angle of attack, analytical and experimental approach was practiced. By using DATCOM analytical method, the lift and drag coefficient of a cranked wing was calculated and plotted in graphs as a function of angle of attack. When compared with experimental results, it was found that the lift and drag obtained from TWT are accurate, provided that the effect of the extra drag caused by the test rig is subtracted. The experimental setup was verify using the comparison results from cranked wing model before conducting further case studies on the non-conventional wing model featuring various trailing edge notches. By analysing the experimental results, it can be concluded that the effects of varying the number of notches only had significant effect on drag reduction if the notch depth is higher. For flight condition that does not exceed 4° angle of attack, lower number of notches at higher notch depth has the best aerodynamic performance, while higher number of notches at higher notch depth has better aerodynamic performance for flight condition that requires cruising angle of attack that exceed 4° angle.

List of Abbreviations

TWT	Taylor's Wind Tunnel
BWB	Blended Wing Body
DATCOM	Data Compendium
3D	Three Dimension
NACA	National Advisory Committee for Aeronautics

References

1. Z.W. Teo, T.H. New, B. Nagel, V. Gollnick, *AIAA SciTech Forum Expo.*, **1**, pp. 1-10 (2016)
2. J.D. Anderson, *Fundamentals of aerodynamics*, (McGraw-Hill Education, 2010)
3. B.K. Stanford, C.V. Jutte, C.D. Wieseman, *AIAA J.*, **54**(1) pp. 293-309 (2016)
4. B. Kaushik, W. Anemaat, *28th Int. Congr. Aeronaut. Sci.*, pp. 1-13 (2012)
5. T.A. Reist, D.W. Zingg, *AIAA*, **1**, pp. 1-13 (2014)
6. P.Z. Lyu, J.R.R.A. Martins, J. Aircr., pp. 1-129 (2014)
7. S. Skinner, H. Zare-Betash, *2015 Aerospace Symposium - Aerospace Ecosystem*. p. 1 (2015)
8. R. Dvorak, Exp. Fluid Mech. **114**, pp. 1-8 (2016)
9. D.E. Hoak, *USAF Stability and Control DATCOM,* (Flight Control Div. Air Force Flight Dyn. Lab. Rept. 0410, 1972)
10. I.H. Abbott, A.E. Von Doenhoff, *Theory of wing sections: including a summary of airfoil data* (2nd Ed. New York: Dover Publication, 1959)
11. A.S.M. Al-Obaidi, S. Eftekhari, *3rd eureca conference*, Dec 2014 Taylor's University, (2014)
12. A.P. Jewel, B. Barlow, W.H. Rae, *Low-speed wind tunnel testing* (3rd Ed. Wiley-Interscience, 1999)
13. M.G. Alexander, *Subsonic Wind Tunnel Testing Handbook*, **5**, pp. 1-291 (1991)
14. A.S.M. Al-Obaidi, H. Wibowo, *8th eureca conference*, July 2017, Taylor's University (2017)
15. J. M. Mallon, K.L. Bildstein, T.E. Katzner, Auk, **133**(1) pp. 1-8 (2016)
16. M.A. Aldheeb, W. Asrar, E. Sulaeman, A.A. Omar, J. Aerosp. Technol. Manag. **8**(1) pp. 7-17 (2016)
17. P.W. Trail, Identification guides for wildlife law enforcement No. **15** pp. 1-26 (2014)

In-situ Study of Seating Static Comfort in Passenger Vehicles

Umi Salmah Seraj, *Mohd Farid* Aladdin [1,*]

School of Engineering, Faculty of Built Environment, Engineering, Technology and Design, Taylor's University, Malaysia.

Abstract. In today's automotive market, comfort is huge selling point of a vehicle. Priority is given by buyers to how comfortable a seat feels during purchase decisions. The measure of comfort is harmonious mix of many aspects such as human ergonomics and physiological factors. However, a gap still exists between objective and subjective measures due to lack of emphasis by past researchers. This is particularly obvious in the lumbar support feature that has still not been able to address the health problems related to driving. This project focuses on bridging the gap by giving users the ability to define true preferred posture in realistic settings. This is done by the creation of a apparatus that allows users to individually manipulate the seat contour for optimum support in more segments than just lumbar area. The experiment is performed in 3 parts, where in each part different segments of the apparatus are manipulatable (lumbar segment, sacral & thoracic segment, and all segments). Sixty human subjects' statistics are recorded (gender, age, BMI and height) and the subjects are palpated to locate internal joints. These joints are marked and postural angles between them are measured using a goniometer. In each seat configuration, the angles are measured and a comfort rating is taken to be compared. It was found that the posture angles are different among the 3 experiments, and there is a change in comfort felt. Some human factors have also been proven to contribute heavily to angles chosen by occupants.

1 Introduction

The comfort facility of a seat is now as equally important as the functional and aesthetic aspects of automobiles for consumers to have an enjoyable ride [1]. The feeling of comfort has been collectively agreed on as [2]: a rise of a subjectively-defined personal character; exaggerated by physical, physiological, psychological characters; and a response to the environment. Normally the term comfort is more appropriately defined and studied as the absence of discomfort as it is easier to quantify discomfort than to measure comfort [2]. There is a large body of literature devoted to the study of automotive seating comfort, likely due to the economic costs associated with discomfort and injury in the motor-vehicle workplace. Studies have shown that low-back pain and serious conditions like spinal disc herniation worsen with increased driving hours [3], thus requiring ergonomic interventions in the early designing stages.

[1] Corresponding author: MohdFarid.Aladdin@taylors.edu.my

The two categories in ergonomic criteria assessment studies are the physiological and anthropometric factors [4] . Due in large part to Akerblom's work, anthropometric studies have long been considered a key aspect of comfortable seating, repeatedly verified by Parkinson and Reed [5], Ziolek et al., [6] and Seitz et al., [7]. Designers aim to ensure that a wide range of people can fit and utilize the seat, commonly adapting the constraining anthropometric values of 5th percentile female and 95th percentile male for corresponding seat dimensions. However, the identification of posture based on anthropometry is difficult because people with similar body dimensions have proven to assume completely different positions [8]. Basic demographic information (e.g. BMI, age, etc.) is usually included to understand its effects on and classify the varying assumed position.

Examples of physiological factors on the other hand are: objective measurement using pressure distribution [9,10], vibration analysis [11,12], surface electromyography [13,14] and temperature at seat (On the other hand, the subjective measurement methods often use a rating scale and a questionnaire that elicits a rating for each of the defined seat regions in the experiments. Most research involves the changing of a critical aspect of the seat to form various iterations around which the research can be conducted and subjective comfort relationships studied. Some previous experiments define the subjective regions based on the part of the body that are interfaced with the cushion (upper back, middle back, lower back, etc.) [2] while other researchers tend to identify regions on the surface area of the seat (e.g. seat back - upper/mid/lower).

However, ergonomics criteria related to physiology have also recently come under scrutiny. Reed et al. described the automobile seat designer's dilemma of prescribing a physiologically suitable seated posture while accommodating a driver's preference, reasoning that prescribed postures sometimes compromise actual comfort. An important finding by Reed et al. vitally highlighted the incompatibility between the common practice of designing seatbacks to induce a large degree of lumbar lordosis and satisfying occupant-selected spinal configurations (which are more kyphotic for some). Lordosis is associated with lower pressure in the lumbar intervertebral discs compared with more flexed (flat or kyphotic) lumbar spine postures, therefore regarded as more comfortable. Schneider verified this incompatibility in a follow-up study. Kolich et al. came to a similar conclusion.

Due to the varying preferences, comfortable accommodation in the lumbar region is thought to be best achieved through adjustability. However, evaluation of market products confirm that existing lumbar supports do not function as well as intended even with a wide range of adjustments. Yun Choi et al. found that even the recommended maximum prominence from past studies produced insignificant restoration of lordosis, while causing additional discomfort due to increased pressure concentrations and peaks at the lumbar area from firm padding. Warrow et al. described the conflict between the recommended heights of the support verses the large variability in occupant lumbar spine configuration, highlighting the placement issue of the support. More importantly, Eby discovered that the task of driving constrains postures to a great extent and prevent occupants from selecting lordotic postures that may cause them difficulty in vision and manoeuvrability, particularly those with extreme body proportions. This conclusion was reached by many other researchers.

The discrepancy in recommendations has proved to be difficult task for designers to interpret and integrate into current seats. A more recent discovery in seating is the appropriateness of sacral + thoracic support over lumbar support for seating comfort. Nairn,

B.C found that sacral support will stabilize the pelvis in the proper neutral position, and lower thoracic support will apply an extension force that will result in elevation of the sternum and optimal curvature of the spine, restoring the proper axial/neutral relationship of the thorax and pelvis. This finding has been incorporated into many industrial products (e.g.: office chairs, massage equipment, wheelchairs, etc.) Latest seat configurations in luxury vehicle brands have also somewhat developed supports in these regions, although no standardization has been set due to lack of experimentation and data.

As stated by Hanson et al., it is not sustainable to rely on old recommendations when competing on a changing market. Therefore, it is necessary to study how much improvement the sacral and thoracic supports bring to the comfort of users in a passenger vehicle. With no current method to do so, an apparatus must be constructed to allow the study of different supports and the corresponding postures, and evaluate the differences in comfort rating felt by the passengers. Joint postural angles can be used to compare preferred postures as they are quantifiable and easily compared to one another, and verifiable with existing data where comfort feelings are not. It is also necessary to study the human factors that affect the chosen posture, which can be done by classification of the passengers' physical characteristics. Understanding this is important for future prediction, modelling and experimentation.

Therefore, the objectives of this study aligned to the theoretical framework and arising problems are: 1) To create an apparatus that allows drivers to define their own preferred posture in realistic settings. 2) To identify the potential differences between existing and preferred posture in terms of postural angles, as well as the resulting difference in comfort. 3) To evaluate the human factors that affect the outcoming preferred posture the most in terms of postural angles.

2 Methodology

2.1 Design of Support Apparatus

There are two important considerations for this artefact, which are:

a) adaptability to the adjustment preferences of a large variety of subjects; and
b) adherence to current seating contour constraints (lateral support).

The apparatus is only capable of forward extension (support prominence). By having separate adjustable segments at different fixed heights, the intended height of the prominence can also be somewhat varied. It has a maximum adjustment of 45 mm to accommodate any extreme postures. The anthropometric values used to ensure a wide fit was taken between the 5th and 95th percentile of Malaysian drivers. The sitting shoulder height for 95th percentile male (645.55 mm) is chosen to allow subjects in that range to make adjustments in the upper back area. The width should also be according to the 95th percentile male (480.00 mm), however due to current seats already having lateral supports, the flatback was set to be 280 mm which is the average width of the flat surface. The curvature was measured at 80 mm radius for most vehicles, which will be the minimum distance of the supports from the flatback to not be affected by them.

The apparatus has 3 main parts, which are the flatback, the jacks and the supports. To keep within the project's budget, most materials are obtained free or at a lower cost from associates. The jacks are made of aluminium links, steel left-hand/right-hand male threaded rods, steel left-hand (LH) and right-hand (RH) female nuts, steel left-hand and right-hand female hex coupling nuts and steel flat plates. The flatback is made of a wooden board, with fabric to cover the wood and avoid damage to car seats. The supports utilized armrest cushion for wheelchairs. An issue was the sizing, as the longest armrests available were 14 inches (355.6 mm) which is 125 mm shorter than the intended length. This might affect the comfort felt by larger occupants. The width of the armrest was 2.25 inches (57.15 mm), making the vertical distance between the centres of each cushion when stacked lengthwise also at least 57.15 mm. This value is close enough to the common 50mm height interval in many researches.

The tests performed to ensure the prototype met the following requirements were:

a) Constraint fitting – the product needs to fit to the seat constraints.
b) Comfort feel – the product should not cause unnecessary discomfort to the subject as it diminishes the purpose of the study.
c) Versatility – the product's adjustability should be able to conform to the preference of a variety of subjects.
d) Stability – the product need to be sturdy and stable under the weight of subjects.

The most challenging aspect during testing was stabilizing the prototype. The weight caused a resistance force on the prototype that halted its expansion capability because the links were not strong enough on their own to support the occupant. The expansion of the prototype was not uniform on both sides and a bending moment was being produced that made the jack deflect slightly downwards when expanded. Two alternatives were finalized to address these issues: scissor mechanism or support block. The support block solution was chosen as the solution as the alternative was too time-consuming and its outcome unpredictable. The block idea involves including an expandable hollow block in the centre of the jack to direct most of the weight exerted on the cushion onto it. The strong opposing forces of the block will be able to balance out the weight of the occupant, and it's structure will ensure both sides expand/contract simultaneously. Deflection will also no longer occur as the block ensures only linear movement as according to its path. Figure 1 shows the completed prototype and apparatus

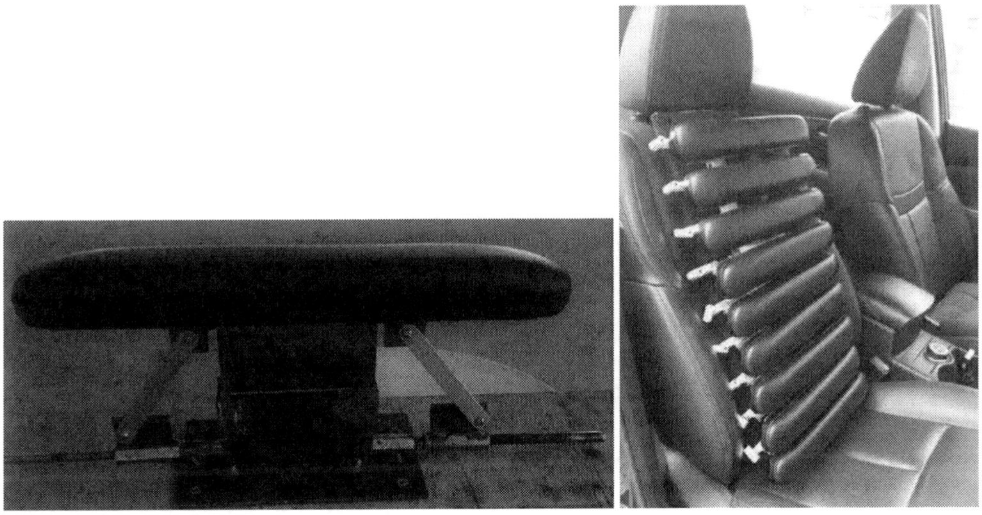

Figure 1. First successful prototype (left); completed apparatus tested in vehicle (right)

2.2 Postural Study

There are many ways to represent a vehicle occupant's posture including body landmarks, external body contours and segment linkage models. The chosen method is in terms of orientations (joint angles) of body segments. In the torso, the lumbar, thorax and cervical regions of the spine are each represented by a single segment instead of a singular torso segment to clearly represent the intra-subject posture changes corresponding to change in backrest supports as shown in Figure 2. Additional segments would provide better illustration, however since most studies have utilized these three-segment torso division, it would be easier to compare the data with verified value. Palpated surface landmarks shown in Figure 3 of the spinous process will be used to identify the joint locations that define the linkage.

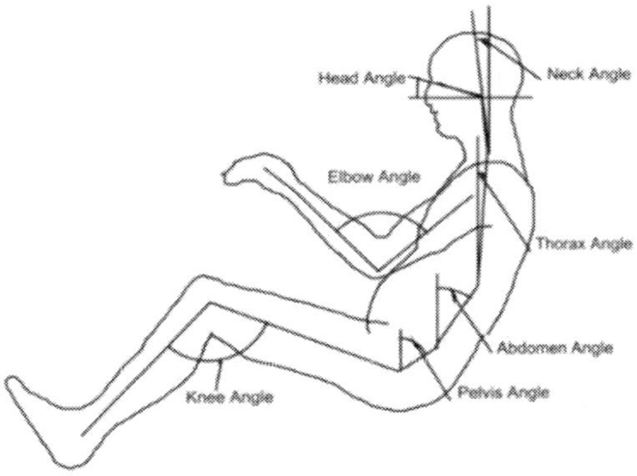

Fig. 2. Segment joint angles [15]

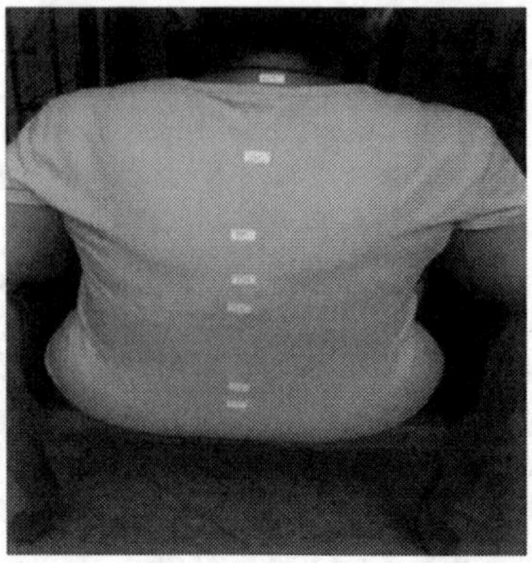

Fig. 3. Identification of palpation points

2.3 Landmark-to-Joint Transformations

The landmark-to-joint transformation methods used here are mostly based on the data analysis and summary presented by Reed, who referred his work from several sources, including a three-volume publication by Robbins. Robbins used external landmark data to estimate the internal joint locations of three categories of humans (large male, mid-sized male and small female) from experimentation and previous sources. The locations of joints relative to the surface landmarks are calculated using simple linear scaling relationships. Ultimately, the scaling approach for all subjects uses the midsize-male data as the reference geometry because the underlying data on which the landmark-to-joint transformations are based relied exclusively on male subjects, and because there are very minor differences between each category.

Joint	Landmarks
Upper Neck (atlanto-occipital)	Infraorbitale, Tragion
Lower Neck (C7/T1)	C7, Suprasternale
Upper Lumbar (T12/L1)	T8, T12, C7, Suprasternale
Lower Lumbar (L5/S1)	ASIS, PS, PSIS
Hip	ASIS, PS, PSIS
Shoulder	Acromion, C7, Suprasternale
Knee	Lateral Femoral Condyle (plus Hip Joint and Lateral Malleolus)
Elbow	Lateral Humeral Condyle (plus Shoulder Joint and Wrist land-mark)
Wrist	Wrist
Ankle	Lateral Malleolus (plus Lateral Femoral Condyle and Hip Joint)

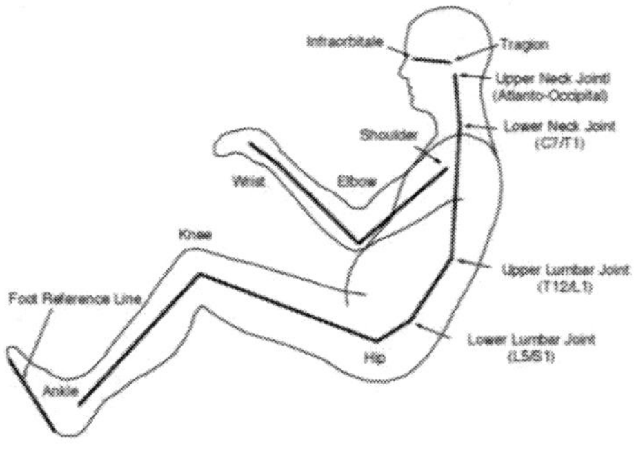

Fig. 4. Internal joints of human body with description [15]

Since the definition of drivers' seat positions are almost always made in relation to their hip locations, the determination of the pelvis geometry is one the main parts of this study. Due to the occurrence of laterally tilted pelvises, the hip joint locations are calculated in three dimensions, then averaged in the XZ plane to obtain a mean hip joint location. The hip joint and lower lumbar joint locations are calculated using the anterior-superior iliac spine (ASIS) landmarks (right and left), the pubic symphysis (PS) landmark, and the posterior-superior iliac spine landmark (PSIS). Seidel [16] defined three pelvis dimensions to further increase accuracy of the location: pelvis width (PW) and pelvis height (PH) and pelvis depth (PD). Notably, Seidel also found no important differences between male and female pelvis in these scaling relationships.

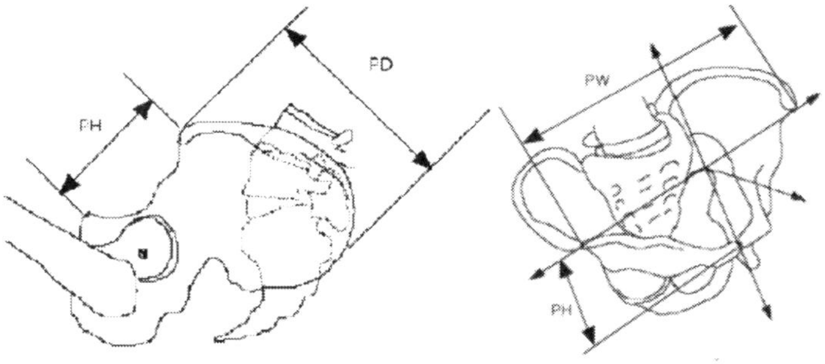

Fig. 5. PW and PH dimensions (left) [15]; PW and PH dimensions (right) [15]

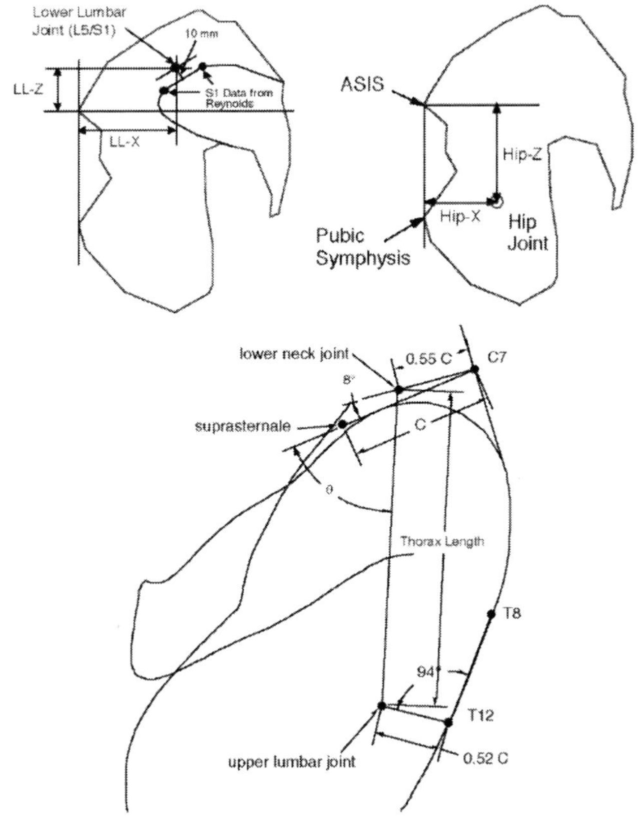

Fig. 6. Lower lumbar joint scaling (left) [15]; hip joint scaling (middle) [15]; upper lumbar and lower neck joint scaling (right) [15]

2.4 Flesh Margin with BMI and Age Effects

Recent papers published corrected hip joint estimates based on adiposity of high-BMI-value subjects as well as old age [17], [18]. The adjusted calculations shown below were based on the fact that skeletal dimensions are mostly independent of body mass index (unless when referring to those with unusual muscle development, e.g: athletes and professional body builders) as well as the degeneration of bones with age.

The pelvic depth (PD) was adjusted by subtracting the effect of BMI shown in Equation 1. The adjusted depth was calculated as:

$$PDadj = PDmin + PDmeas - PDpred \tag{1}$$

where PDmin would be the mean expected PD for the subject with the lowest BMI based on the regression given in Equation 2. PDpred is the PD predicted by Equation 2 for the subject accounting for BMI, and PDmeas is the measured PD for the subject.

$$PDpred = 65.6 + 4.38 * BMI, R2 = 0.43 \tag{2}$$

The bispinous (bi-ASIS) breadth (BB) was adjusted for BMI by subtracting the effect of BMI as shown in Equation 3. (BBpred) was subtracted from the measured bispinous breadth (BBmeas) and the result added to BBmin per Equation 4.

$$BBadj = BBmin + BBmeas - BBpred \tag{3}$$

$$BBpred = 165.3 + 2.64 * BMI, R2 = 0.23 \tag{4}$$

The PSIS flesh margin was set to a vector along the pelvic x-axis with a length of 0.0006*BMI^3 mm based on the adjustment in Reed's [17] study of the ASIS flesh margin. Due to the lack of study at the intimate region of the pubic symphysis, flesh margin at that area will be kept at the 25mm estimate.

The hip joint was also adjusted for age as shown by equation 5. LL is length leg measured in millimetres and the origin from which the equation is measured is the midpoint of the biponous (bi-ASIS or PW).

$$Hip - Z = -18 - 0.063LL - 5.403age \tag{5}$$

3 Experimental Set-up

The car chosen to be used as the test vehicle is a 2015 Nissan X-Trail due to long seat tracks and wide opening doors for all subjects to use the apparatus without constriction.

4 main factors used in experiments: Stature (Height), BMI, Age and Gender (Sex) as recommended from literature review. Sample size of 50 subjects were calculated with a 0.75 power and 0.12 error margin. The subjects are firstly given a survey to fill in.

The subjects are then categorized by the four factors of BMI, age, stature and gender. For stature, an online calculator is used to calculate the percentile of each occupant based on their height. Occupants below 25th percentile are considered short while those above 75th percentile are considered tall, and the remaining in between are average. For BMI, the occupants are classified according to the World Health Organization (WHO)'s descriptions of body mass index. However, the 'overweight' and 'obese' categories, the were merged due to the lack of participants falling in the obese category, with most of them being relatively close to the overweight limit. For age, occupants are divided to being above and below 50.

The study is carried out in a few stages. Firstly, the participants are palpated and marked according to the methods previously described. Then, the joint locations of the occupants are marked using the discussed transformations and apparatus like rulers, multi-angle rulers, Vernier callipers and spirit level gauges. Then the experiments are conducted. Participants are asked to assume their most comfortable posture with hands on the steering wheel and right leg on the accelerator (Figure 7). There are 3 parts of the experiment. In the first one, occupants are only allowed to adjust the lumbar supports. Second part is adjustment of the thoracic and sacral supports. Third is adjustment of any and all supports. At the end of each adjustment, the joint angles are measured using a digitalized angle ruler for better accuracy (Figure 8), and the occupants fill in the comfort ratings. Finally, the last parts of the survey are filled in.

Ahead of conducting the experiment, the participants adjusted the seat recline, seat distance, seat height, steering wheel distance and height. The seat recline angle and seat and steering wheel height were fixed after this point. Seat distance and steering wheel distance was allowed to be varied because occupants may feel too constrained after increasing the prominence of the seat apparatus.

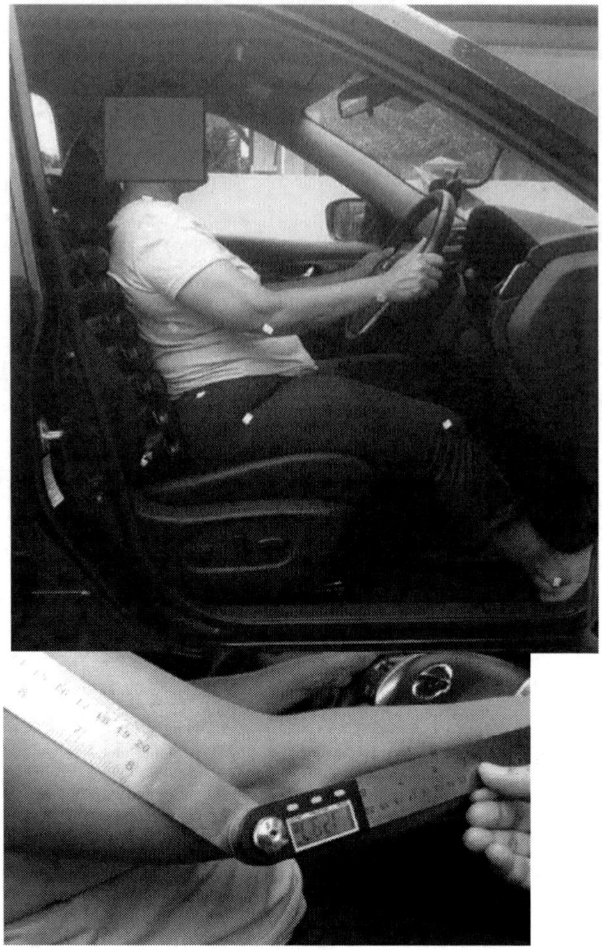

Fig. 7. Subject during experiment with angle measurement method

4 Result and Discussion

Of the 50 participants, there were 28 females and 22 males. 12 were short, 22 were average and 16 were tall. 29 are below 50 and 21 are above 50. 10 were underweight, 23 were normal weight and 17 were overweight/obese. Angle values of this study are compared to 12 other studies conducted between this year and 1998 in Figures 10-14 below. Aside from previous studies, the measured values are also compared to predetermined range of motion values. The comparison is necessary to ensure the angles measured from this study are reasonable within human capability and preference, and sustainable in future researches. More importantly, comparison will ensure the method and apparatus used is somewhat accurate.

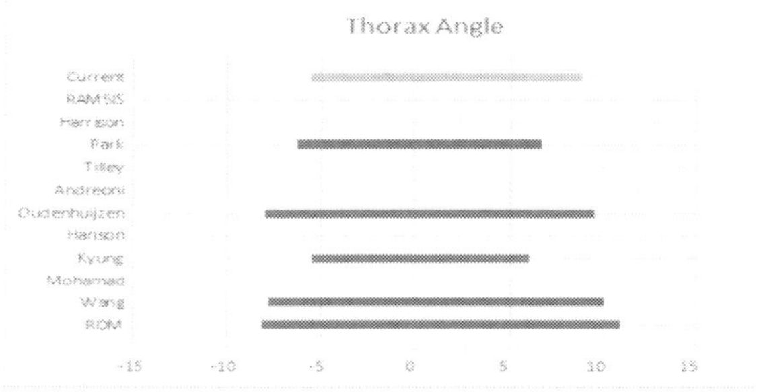

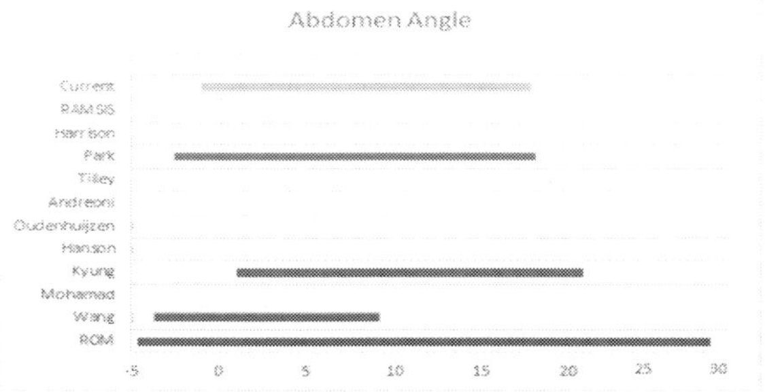

a) Thorax comparison b) Abdomen comparison

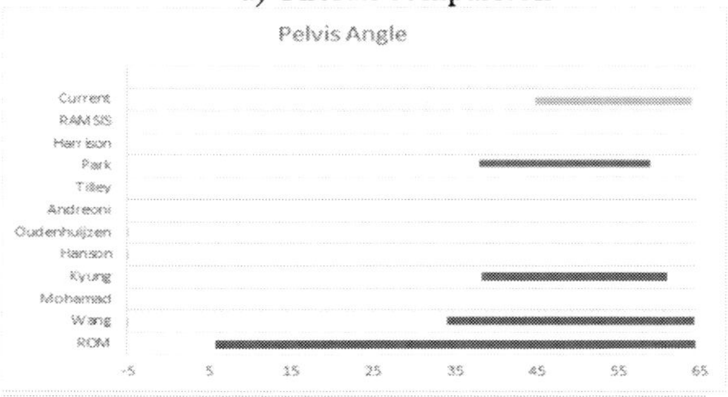

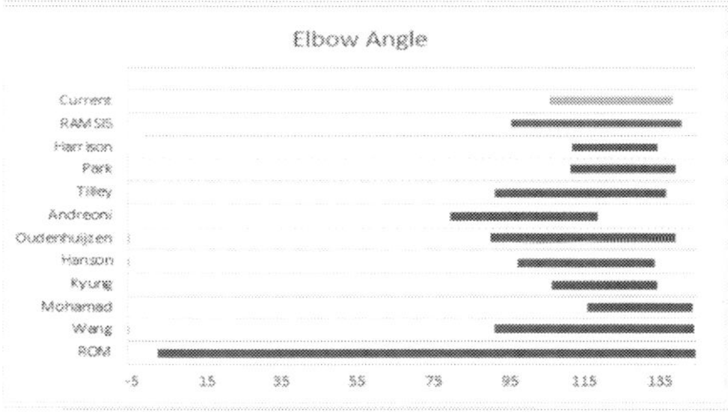

c) Pelvis comparison d) Elbow comparison

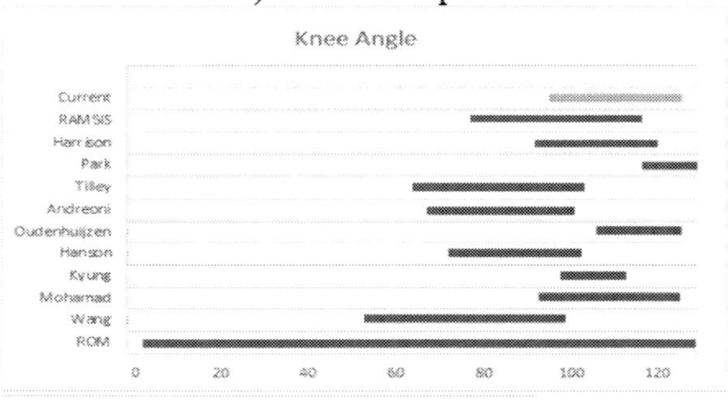

e) Knee comparison

Fig. 8. Range of angles for a) Thorax, b) abdomen, c) Pelvis, d) elbow and e) knee preferred and comparison with literatures

Firstly, all angles are verified to be within acceptable ROM, so the apparatus was confirmed to be reliable. Moving on from that, the measured data was compared to previous studies, and some differences were noted. Due to the complexity of measuring posture and the interdependence of joint angles, the main problem in comparing different studies is that the results are influenced by the heterogeneity of experimental settings. Factors like data type (2D or 3D), number of participants, number of joints, recording time, symmetry, experimental method (field of laboratory), car type (sedan, SUV, etc.), study region, optimum vs. preferred, questionnaires, measurement method and reference point (H-point or Seat Reference Point).

One-Way Repeated Measures ANOVA was performed to test for difference in the 5 angles between each experiment. The null hypotheses of this ANOVA test states that there are no differences in the means of angles between the three seat configurations. The alternate hypotheses state that there are differences in the means of angles between the three seat configurations. The different seat configurations did not lead to any statistically significant changes in the elbow angle, F (2, 48) = 1.503, p = .235. or in the knee angle F (2, 48) = 1.218, p = .307. The null hypothesis is accepted for these two cases. This would mean that regardless of the seat configuration, participants did not alter their knee and elbow angles, possibly due to constricts from the task of driving. However, there were statistically significant differences in the thorax angle F (1.545, 39.353) = 5.138, p = .045, as well as in the abdomen angle F (1.505, 38.591) = 17.096, p = .000 and the pelvis angle F (1.693, 42.170) = 3.466, p = .042. For these cases, the null hypothesis is rejected and the alternate hypothesis is accepted.

To identify the human factors and its interactions' effects on each joint, Factorial ANOVA was carried out. It is important to understand how Factorial ANOVA works in SPSS. The analysis essentially tests for the variance between means. In this software, the analysis chooses one standard mean and compares all the rest of the means to it. The default settings are to choose the combination of the highest numerical coded factors. Categories are coded as seen in Table 3:

Table 1. Coding for each category

Stature	3 - Average	2 - Short	1 - Tall
BMI	3 - Normal	2 - Underweight	1 - Overweight/Obese
Age		2 - Below 50	1 - Above 50
Gender		2 - Male	1 - Female

This is to set the standard as a normal BMI, male, below 50 with average stature. Every mean angle of all other categories in any combination is compared to this standard. The ANOVA analysis is then performed with the null hypothesis that the mean of all other combination of categories are equal to the mean of the standard. The parameter estimates in Table 4 tell us which specific category combinations are most influential on the angles. The intercept below is the standard mean angle of a normal BMI, male, below 50 with average stature

Table 2. Combination of Parameter estimates on different body parts

Thorax

Parameter	B	Std. Error	Sig. level	Observed Power
Intercept	11.0	0.222	0.000	0.738
(BMI = 1)	+14.0	4.184	0.020	0.514
(Gender = 2) * (BMI = 1)	+16.0	3.254	0.018	0.582
(Age = 1) * (Gender = 1)	+9.0	0.360	0.031	0.640
(Age = 1) * (Stature = 2)	-10.0	0.252	0.028	0.715
(Stature = 1) * (BMI = 1)	+13.0	3.895	0.023	0.534
(Stature = 1) * (Gender = 1) * (Age =1)	-9.0	0.428	0.031	0.685

Abdomen

Parameter	B	Std. Error	Sig. level	Observed Power
Intercept	23.0	0.141	0.000	0.749
(Age = 1)	-19.0	0.190	0.038	0.725
(Gender = 2) * (BMI = 1)	-17.0	5.125	0.032	0.599
(Age = 1) * (Gender = 1)	-21.0	0.235	0.040	0.692
(Stature =1) * (BMI = 1)	-24.0	4.605	0.046	0.612
(Stature =1) * (Gender =1) * (Age = 1)	-25.0	0.360	0.008	0.677

Pelvis

Parameter	B	Std.	Sig. level	Observed

		Error		Power
Intercept	41.0	0.192	0.000	0.741
(BMI = 1)	+12.0	6.420	0.028	0.489
(Gender = 2) * (BMI = 1)	+18.0	8.214	0.009	0.475
(Gender = 1) * (BMI = 1)	+9.0	6.104	0.031	0.501

Elbow

Parameter	B	Std. Error	Sig. level	Observed Power
Intercept	114.0	0.184	0.000	0.748
(Age = 1)	-11.0	0.345	0.022	0.692
(Stature = 1)	-9.0	0.241	0.034	0.710
(Stature = 2)	+16.0	0.219	0.018	0.721

Knee

Parameter	B	Std. Error	Sig. level	Observed Power
Intercept	129.0	0.202	0.000	0.742
(BMI = 1)	-23.0	0.451	0.013	0.652
(Gender = 1) * (BMI = 1)	+10.0	0.520	0.024	0.640

By studying the perimeter estimates, we can tell which interactions or main factors caused the biggest deviation from the standard mean. In the thorax angle, the interaction of male and high BMI caused the biggest change in angle, at an estimated mean of 26.0°. In the abdomen, the biggest effect was from a tall, older female making the angle -2.0° In the pelvis, it was the interaction of male with high BMI, causing a 59.0° angle. In the elbow, it was the from the short occupants causing a 130.0° angle. Finally, in the occupants with high BMI, causing an angle of 106.0°. There are larger errors in some of the mean angles, like the pelvis and abdomen, where the joints used are covered with a lot of fat and harder to palpation and predict.

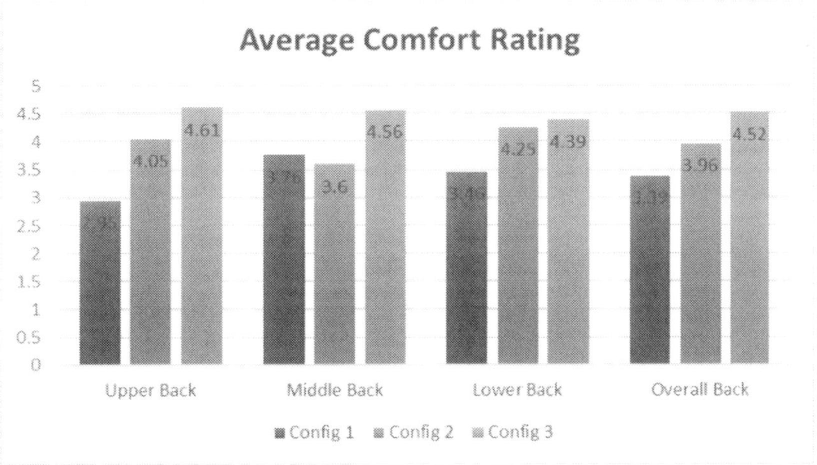

Fig. 9. Graph of Comfort Ratings

The comfort rating survey divided the back into 3 segments: upper, middle and lower. Ratings were on a scale of 1 – 5, 1 being 'Uncomfortable' and 5 being 'Comfortable'. The overall back comfort is calculated as the average comfort of all three areas. From the graph in Figure 15, there was a definitely increase in all three segments and the overall back in the last experiment of the seat that had all supports utilized. In the first experiments, occupants felt very little comfort in the upper back and lower back, though the middle was relatively higher. This could be due to the lumbar support only allowing occupants to have support in that region and not in the upper and lower. It could be possible that occupants had to compromise one for the other. In the second experiments, with sacral and thoracic support, most occupants had better overall back comfort, though they were some who felt the middle part was now lacking. Finally, in the last experiments, all three segments were the highest with the highest overall back rating. It is clear that there is a definite different in the comfort felt by occupants in their true preferred posture compared to ones that they had less adjustability in.

5 Conclusion

All the objectives of the experiment were met. The apparatus was able to allow occupants to define their preferred posture in a realistic setting with accuracy. The incorporation of the apparatus was proven to cause changes in postural angles that positively affected the static comfort of occupants. It was also identified that human factors affect the joint angles in a variety of ways, and have different influences on different joints. A limitation that may have potentially influenced the accuracy of the results is the palpation and determination of joints that may not have been very accurate. For future work, a more advanced apparatus could be created using similar concept that will allow more adjustments (support tilt, height variation, etc.) and have lesser influence over the subject's comfort perception. The angles recorded in this study can be used to verify predicted posture from Digital Human Models in the future, as well as benchmark data for similar experimental set-ups. More objective measures like EMG, pressure mapping and vibration analysis can be done as extensions to further validate the efficiency of the sacral + thoracic support.

References

1 D. Mohamad, B.M.Deros, D.A. Wahab, D.D.I. Daruis and A.R. Ismail, "Integration of Comfort into a Driver ' s Car Seat Design Using Image Analysis Darliana Mohamad",*American Journal of Applied Science*, vol. 7, no. 7, pp. 937–942, 2010.

2 M. P. De Looze, L. F. M. Kuijt-Evers, and J. Van Dieën, "Sitting comfort and discomfort and the relationships with objective measures," *Ergonomics*, vol. 46, no. 10, pp. 985–997, 2003.

3 M. P. Reed, L. W. Schneider, and L. L. Ricci, "Survey of Auto Seat Design Recommendations for Improved Comfort," no. April, pp. 1–96, 1994.

4 B. M. Deros *et al.*, "ScienceDirect Incorporating Malaysian's Population Anthropometry Data in the Design of an Ergonomic Driver's Seat," *Procedia - Soc. Behav. Sci.*, vol. 195, no. 195, pp. 2753–2760, 2015.

5 M. P. Reed, M. B. Parkinson, and D. B. Chaffin, "A New Approach to Modeling Driver Reach," *Sae Tech. Pap. Ser.*, no. 724, 2003.

6 S. A. Ziolek, "Beyond Percentiles : An Examination of Occupant Anthropometry and Seat Design," *Hum. Factors*, no. 724, 2004.

7 T. Seitz and H. Bubb, "SAE TECHNICAL Measuring of Human Anthropometry , Posture and Motion," no. 724, 1999.

8 B. Deros *et al.*, "Advances in Natural and Applied Sciences Incorporating Anthropometrics Data in Designing Driver ' s Seat for Malaysian Made Compact Cars," vol. 8, no. June, pp. 915–919, 2014.

9 M. D. Stinson, A. Porter-Armstrong, and P. Eakin, "Seat-interface pressure: A pilot study of the relationship to gender, body mass index, and seating position," *Arch. Phys. Med. Rehabil.*, vol. 84, no. 3 SUPPL. 1, pp. 405–409, 2003.

10 G. Andreoni, G. C. Santambrogio, M. Rabuffetti, and A. Pedotti, "Method for the analysis of posture and interface pressure of car drivers," *Appl. Ergon.*, vol. 33, no. 6, pp. 511–522, 2002.

11 M. J. Griffin, "Whole-Body Vibration," *Encycl. Vib.*, pp. 1570–1578, 2001.

12 H. Nahvi, M. H. Fouladi, and M. J. Mohd Nor, "Evaluation of whole-body vibration and ride comfort in a passenger car," *Int. J. Acoust. Vib.*, vol. 14, no. 3, pp. 143–149, 2009.

13 T. R. Bush, F. T. Mills, K. Thakurta, R. P. Hubbard, and J. Vorro, "The Use of Electromyography for Seat Assessment and Comfort Evaluation," *Soc. Automot. Eng.*, no. 412, pp. 143–1995, 1995.

14 S. J. Park and B. Min, "Development of the Evaluating System for Ride Comfort and Fatigue in Vehicle," *SAE Tech. Pap.*, no. 724, 2013.

15 M. P. Reed, M. A. Manary, and L. W. Schneider, "Methods for Measuring and Representing Automotive Occupant Posture," *Soc. Automot. Eng.*, no. 724, 1999.

16 G. K. Seidel, D. M. Marchinda, M. Dijkers, and R. W. Soutas-Little, "Hip joint center location from palpable bony landmarks-A cadaver study," *J. Biomech.*, vol. 28, no. 8, pp. 995–998, 1995.

17 M. P. Reed, S. M. Ebert-Hamilton, and J. D. Rupp, "Effects of Obesity on Seat Belt Fit," *Traffic Inj. Prev.*, vol. 13, no. 4, pp. 364–372, 2012.
18 R. Hara, J. McGinley, C. Briggs, R. Baker, and M. Sangeux, "Predicting the location of the hip joint centres, impact of age group and sex," *Sci. Rep.*, vol. 6, no. 1, p. 37707, 2016.

Preparation and Characterization of Wood Plastic Composite Made Up of Durian Husk Fiber and Recycled Polystyrene Foam.

Seong Chun Koay[1,*] , *Varnesh* Subramanian[1] , *Ming Yeng* Chan[2] , *Ming Meng* Pang[1] , *Kim Yeow* Tsai[3] , and *Kean How* Cheah[4]

[1]School of Engineering, Faculty of Built Environment, Engineering, Technology and Design, Taylor's University Lakeside Campus, 47500 Subang Jaya, Malaysia
[2]Centre for Engineering Programmes, HELP College of Arts and Technology, 55200 Kuala Lumpur, Malaysia
[3]Faculty of Engineering, The University of Nottingham, 43500 Semenyih, Malaysia.
[4]School of Engineering and Physical Science, Heriot-Watt University Malaysia, 62200 Putrajaya, Malaysia.

Abstract. Polystyrene foam is one of the major plastic waste that hardly to recycle. The present research is aims to recycle polystyrene foam as raw material to produce wood plastic composites (WPC). The WPC was produced from recycled polystyrene (rPS) and durian husk fiber (DHF) using melt compound and compression moulding processes. This paper is focus on effect of fiber content on tensile and thermal properties of rPS/DHF composite. The results found the tensile strength modulus of this WPC increased at higher fiber content, but elongation at break was reduced. However, this composites exhibited an early thermal degradation when subjected to high temperature and this was commonly found among WPC. The thermal degradation of rPS/DHF composites yielded high percentage of char residue due to char formation of DHF. Overall, the rPS/DHF composites with 60 phr fiber content able to achieved strength slight above 16 MPa without any chemical treatment additives. This indicates the rPS/DHF composites can be a potential WPC if further modify with to improve its strength.

1 Introduction

Wood Plastic Composites (WPC) is composite material made from wood sawdust and thermoplastic material [1]. WPC also considers as an eco-friendly material that is potential to replace natural wood. Thus, the market demand of WPC around the world is getting increases every year. In 2015, the global WPC production had reached 2,695 kilo tons and the market is expected to grow. WPC is widely found in replacing natural wood products like window fitting, flooring panel, door frame, and furniture. IKEA's ODGER chair is one of the example of WPC furniture that made from recycled plastic and wood fiber [2-3]. The soft wood fiber is the main choice of wood fiber that used to produce WPC. Nowadays, WPC also made from rice husk and recycled plastic, especially in China [2]. The WPC made from recycling plastic and agricultural waste materials able to ensure the sustainability [4-5].

Natural fiber is the main components in WPC, which up to 80% of fiber can used in extrusion process and 30 to 40% of fiber can be used in injection moulding [2]. The natural fiber usually obtained from wood and also agriculture crop or waste, such as palm oil empty fruit bunch, rice husk, and corn stalk [6-8]. Durian is a tropical fruit that widely planted across Malaysia. The production of durian in Malaysia alone had achieved 3,800 metric tonnes in year 2016 and amount of durian fruits is estimated to reach 22,000 metric tonnes in year 2020 [9]. Figure 1 shows the durian husk shares up to 70% of a durian fruit and it usually discarded as waste after consumed the flesh [10]. This means it is about 7 tonnes of estimated wet durian husks will be generated for every 10 tonnes of produced durian fruit. The increase of durian produce will generates more durian husk waste. The discarded durian husks usually end up in landfills or burnt which contributed to environmental pollution. From the literature found durian husk actually contains huge amount of lignocellulose fiber and those compositions are similar to wood fiber [11-12]. In regards, this research is underway to utilise durian husk fiber to WPC.

Polystyrene foam is an expanded polystyrene that commonly called as Styrofoam. The polystyrene foam are widely used many industries for packaging and storage purposes, because it is very cheap in production cost, light in weight, high impact absorption and good heat insulation [13]. The usage of polystyrene foam was very high over the world, but the disposal of polystyrene foam into landfills also very high. The main reason is because the recycle rate of polystyrene foam is relatively low due to it is bulky as it made up of 98% of air and only 2% of polystyrene [14]. The recycling process of polystyrene foam is not economic, because the transportation expenses for sending polystyrene from to recycling plant is very high. Thus, industry is less interest on recycling polystyrene foam. For this reason, the present research is working on developing new value added WPC using recycled polystyrene foam and durian husk fiber.

This present paper is focuses on the effect of fiber content on the tensile and thermal properties of WPC made from durian husk fiber and recycled polystyrene foam.

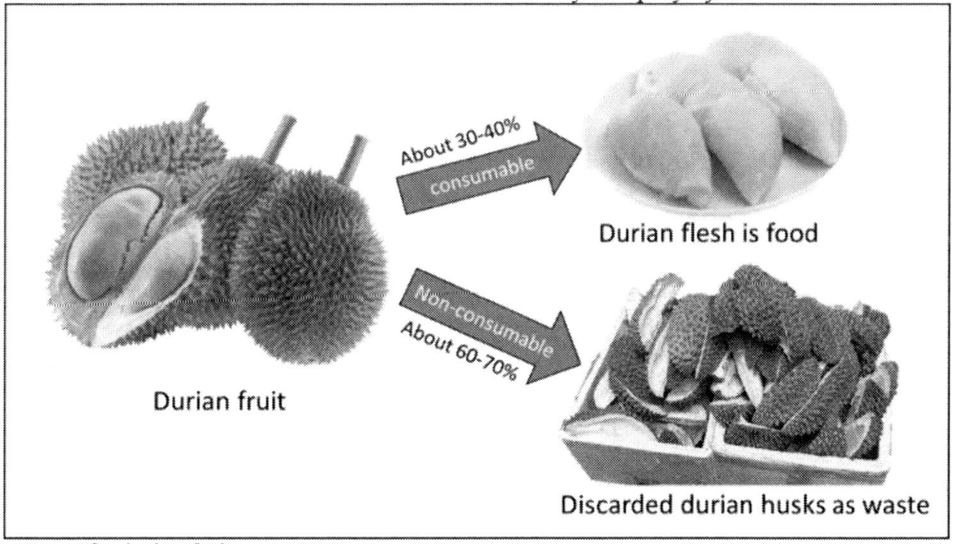

Fig. 1. Parts of a durian fruit.

2 Methodology

2.1 Research Materials

The durian husk was obtained from durian fruit stall at SS2, Petaling Jaya (Selangor). The polystyrene foam used in this experiment was the packaging Styrofoam used for electric appliance and it usually discarded as waste at electric appliance shop. The acetone used in this experiment was supplied by Evergreen Engineering & Resources (Selangor, Malaysia).

2.2 Preparation of Durian Husk Fiber

First, the collected durian husks (DHs) were cleaned with tap water to remove any dirt and cut into small pieces. Next, the small piece of DHs were dried using oven at 70°C. After that, the dried DHs were ground into short fiber using mechanical grinder. Then, the short durian husk fiber (DHF) was sieved into homogenous size of fibers (mesh size of 600 micron). The DHF be dried again before compounding process in order to avoid high moisture content.

2.3 Preparation of Recycled Polystyrene

The collected polystyrene foam was cut into small piece and dissolved in acetone solvent in order to remove air from foam. Then, the dissolute polystyrene foam was filtered and dried using an oven at 70°C to remove acetone. The recycled polystyrene (rPS) was cut into smaller pieces and stored for compounding process.

2.4 Preparation of WPC

The WPC was prepared from rPS and DHF using Haake Rheomix 600p (Brand: Thermo Fisher Scientific Inc). The processing temperature was set at 190°C and rotor speed of 80 rpm. The WPCs were prepared with fiber content of 15, 30, 45, and 60 part per hundred resin (phr). The WPC compound was molded into 1 mm thickness sheet using hot press machine (model: Moore). The molding temperature was similar to compounding process and the pressure was set at 100 kPa. The operating procedures are: i) preheat the compound for 4 minutes, ii) fully compress the compound for 1 minute, iii) cool the specimen to temperature below 50°C, iv) remove the composite sheet from mold. The composite sheet was cut into tensile specimen and the dimensions of tensile specimen were following ASTM D638 standards.

2.5 Testing and Characterization

The tensile testing for the rPS/DHF composite was carried out based on the ASTM D638 standards using an Instron universal testing machine (model 5596). The crosshead speed was set at 5 mm/min and a 15 kN loading cell was used to perform the test. Tensile properties of the composites were automatically obtained through the computerized system. A minimum of 7 specimens were tested for each formulation. Thermogravimetric analysis (TGA) was carried out using Pyris Diamond TGA (brand: Perkin-Elmer). The specimen was prepared in small size with weight from 6 to 8 mg and placed in a ceramic pan. Then, the specimen was undergone thermal scan from 30°C to 600°C and the heating rate was set at 10°C/min. The TGA analysis was run under nitrogen atmosphere with gas flow rate of 20 ml/mm.

3 Results and Discussion

3.1 Tensile Properties

The effect of fiber content on tensile strength of rPS/DHF composites is illustrated in Figure 2. The average tensile strength of neat rPS was recorded at 26 MPa. The addition of fiber caused a drastic decreases of tensile strength on rPS matrix. Generally, the strength of composite materials are highly depending on fiber orientation, length/diameter ratio, and interfacial adhesion [15]. In this case, the DHF was a short fiber and it have random orientation when dispersed in matrix. The short fiber that orientated parallel to direct of applied stress would contributed to improve of strength as part of the stress would share by the fiber. From literature, natural fibers were typically exhibited higher tensile strength and modulus compared conventional plastic [16]. Thus, if the fiber in the composite able to share part of the load, meaning it will increase the strength of the composite. On the other hand, the fiber might also orientated perpendicular to the direct of applied stress and the fiber did not able to share the stress but it might initial the cracking along the fiber-matrix interface. Hence, the strength of the composite will dramatically decreases. The result found the increases of fiber content had increased the tensile strength of rPS/DHF composites. At low fiber content, the amount of fiber orientated parallel to direct of applied stress is less. Thus, the strength of the composite has drastically reduced. However, the increases of fiber content might increases the amount of fiber that able to share the stress. For this reason, rPS/DHF composites shown higher strength at 60 phr of fiber content. The similar observation also reported by many other researchers [17-18].

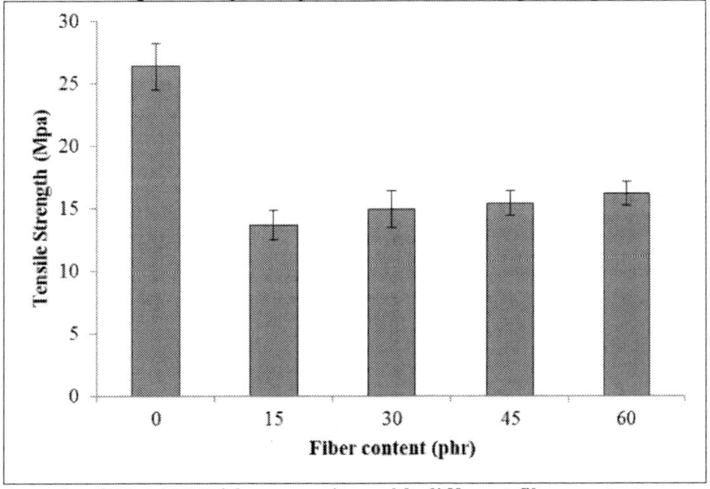

Fig. 2. Tensile strength of PLA/DHF biocomposites with different fiber content.

Figure 3 shows the increases of fiber content increased the tensile modulus of rPS/DHF composites. As mentioned above, the modulus of natural fiber usually higher than plastic material, thus the obtained results trend was expected. The rPS/DHF composites had 32% in average higher tensile modulus compared to neat rPS. The presence of DHF have share part of the stress applied to rPS matrix. Thus, the tensile modulus of composites is increased at more fiber content. Besides, the present of friction between fiber-matrix also reduced the chain mobility of matrix. For this reason, the flexibility of rPS was further decreased. Then, the composite become more rigid and brittle. Furthermore, Manshor et al. [12] also agreed that presence of more fiber was tend to increase the modulus of composites.

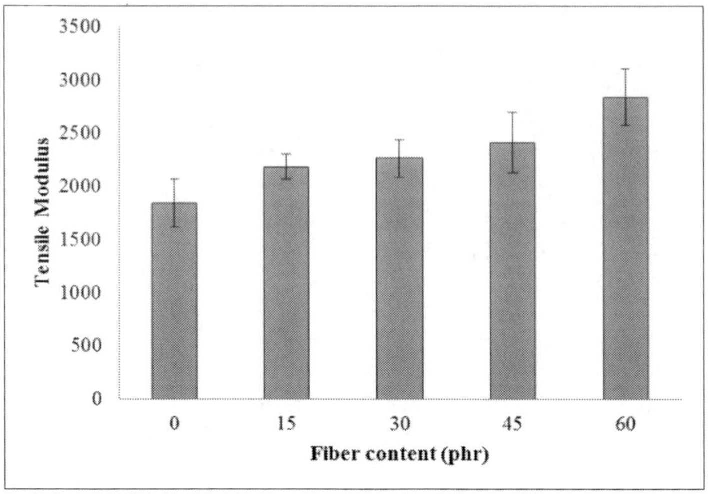

Fig. 3. Tensile modulus of PLA/DHF biocomposites with different fiber content.

The elongation at break of neat rPS and rPS/DHF composites with different fiber content are shown in Figure 4. As expected, the result trend of elongation at break and tensile modulus was in opposite. The rPS was rigid and it exhibited only average of 2.8% of elongation at break. The increases of fiber content was significantly reduced the elongation at break of composites. The increases of fiber content have caused the elongation at break of composites dropped 50% in average. As discussed earlier, the addition of fiber have friction with rPS matrix at interface region. The fiber-matrix friction would reduce the chain mobility of the matrix. Thus, the composites usually more rigid and brittle if fiber content was more. This observation is in agreement with findings of many others researcher [19-20].

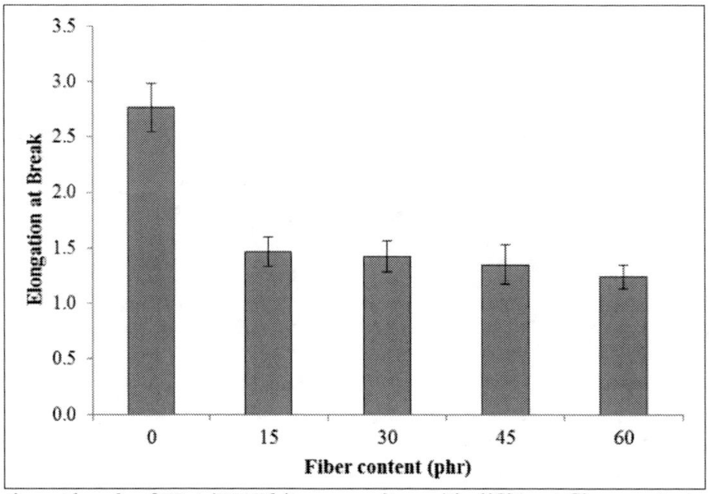

Fig. 4. Elongation at break of PLA/DHF biocomposites with different fiber content.

3.2 Thermal Properties

Figure 5 illustrates the TGA curves of DHF, neat rPS and rPS/DHF composites at selected fiber content. The data calculated from TGA curves are tabulated in Table 1. The neat rPS exhibited a single step thermal degradation above temperature 300°C. However, DHF was initially showing weight loss above temperature 50°C. This is due to the weight loss of removing moisture and volatile compounds from DHF. Then, the weight loss of DHF was following by thermal degradation of hemicellulose above temperature 150°C. Last, the thermal degradation of DHF was involved cellulose and lignin were started at temperature of 300°C. The cellulose and lignin were decomposed and yields high amount of char residue. For Table 1, the thermal degradation temperature at 5% weight loss ($T_{d5\%}$) of rPS/DHF composites were decreased as the increase of fiber content. This indicates early thermal degradation occurs on rPS/DHF composites due to the weight loss from moisture, volatile compound and hemicellulose of DHF. In contrast, the thermal degradation temperature at 50% weight loss ($T_{d50\%}$) of rPS/DHF composites have shifted to higher temperature when added with more DHF. This was due to the char formation effect that caused the increased of thermal stability of composite at high temperature. As mentioned earlier, the char formation was due to the thermal decomposition of cellulose and lignin. Then, char usually is higher thermal stability and the presence of char in composites material provides a shielding effect which caused the thermal degradation of composite delayed. Hence, the rPS/DHF composites exhibited a better thermal stability at higher temperature. The char formation also contributed to high percentage of residue content of rPS/DHF composites. The similar findings regarding to effect of char formation influenced the thermal properties of composites also reported by others researcher [21-22].

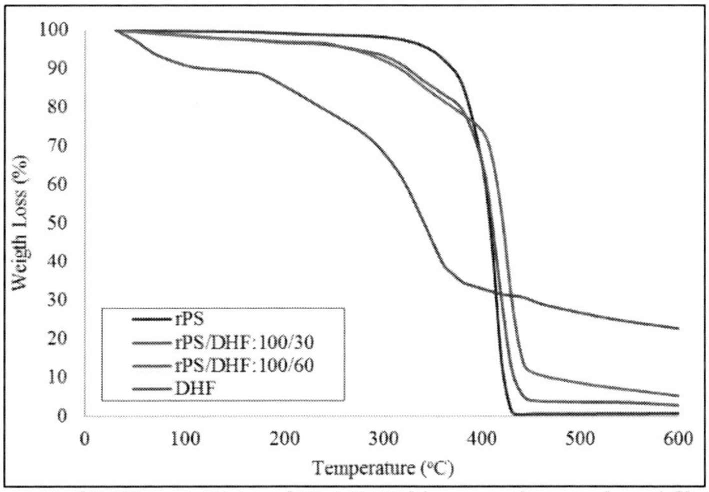

Fig. 5. TGA curves of DHF, neat PLA and PLA/DHF biocomposites at selected fiber content.

Table 1. Formatting sections, subsections and subsubsections.

Samples	Temperature at 5% weight loss (°C)	Temperature at 50% weight loss (°C)	Char residue at 600°C (%)
Neat rPS	348.9	408.1	0.2
DHF	55.5	339.3	22.7
PLA/DHF: 100/30	248.5	409.1	2.8
PLA/DHF: 100/60	245.0	420.4	5.4

4 Conclusion

In conclusion, addition of more fiber content increased the tensile strength and modulus of rPS/DHF composites increased, but decreased the elongation at break. The rPS/DHF composites with 60 phr of fiber content exhibited strength about 16 MPa and modulus of 2842 MPa without any modification. This means rPS/DHF composites can be potential new WPC material. The TGA results shows the composites had early thermal degradation and it was command to WPC. The thermal stability of rPS/DHF composites also increased at high temperature due to char formation.

References

1. S. Butylina, O. Martikka and T. Kärki, Polym. Plast. Technol. Eng. **50**, 1040 (2011).
2. A, Eder, Bioplast. Mag. **8**, 16 (2016)
3. K.S. Chun, S. Husseinsyah, H. Osman, Polym. Eng. Sci. **53**, 1109 (2013)
4. N. Ayrilmis and S. Jarusombuti, J. Compos. Mater. **45**, 103 (2011)
5. K. S. Chun, S. Husseinsyah, J. Thermoplast. Compos. Mater. **27**, 1667 (2014)
6.
7. K.S. Chun, C.M. Yeng, S. Husseinsyah, M.M. Pang, A. Ismail, J. Eng. Sci. Technol. **12**, 1165 (2017).
8. W.L. Ngo, M.M. Pang, L.C.Yong, Adv. Envir. Bio. **8**, 2742 (2014)
9. A.A. Yussuf, I. Massoumi and A. Hassan, J. Polym. Envir. **18**, 422 (2010)
10. Department of Agricultural (Malaysia). Strategic Plan 2016 to 2020. (Md: Department of Agricultural, Malaysia, 2016)
11. Y.M.W. Jason, M. Y. Chan, Adv. Polym. Technol. (to be published). doi:10.1002/adv.21921.
12. M.R. Manshor, H. Anuar, M.N. Nur Aimi, M.I. Fitrie, W.B. Wan Nazri, S.M. Sapuan, Y.A. El-shekeil, M.U. Wahit, Mater. Des. **59**, 279 (2014)
13. R. Manshor, H. Anuar, W. Wan Nazri, M. Fitrie, Adv. Mater. Res. **576**, 212 (2012)
14. J. Schellenberg and M. Wallis, J. Cell. Plast. **46**, 209 (2010)
15. Kan and R. Demirboga, J. Mater. Process. Technol. **209**, 2994 (2009)
16. X. Hu, J. Fang, F. Xu, B. Dong, Y. Xiao and L. Wang, Sci. Rep. **6**, 1 (2016)
17. K.L. Pickering, M.G. Aruan Efendy, T.M. Le, Compos. Part A. **83**, 98 (2016)
18. L. Danyadi, J. Moczo, and B. Pukanszky, Compos. **41**, 199 (2010)
19. K.S. Chun, S. Hussiensyah, C.M. Yeng, Polym. Bull. **73**, 3465 (2016)
20. K.S. Chun, S. Husseinsyah, C.M. Yeng, J. Thermoplast. Compos. Mater. **29**, 1517 (2016)
21. C.M. Yeng, S. Husseinsyah, S.S. Ting, Polym. Plast. Technol. Eng. **52**, 1496 (2013)
22. S. Hussiensyah, K.S. Chun, A. Hadi, R. Ahmad, J. Vinyl Add. Technol. **22**, 200 (2016)

Development of a PID Controlled Arduino-Based Stabiliser

Yee Lyn Wah[1], *Choon Lih* Hoo[2,*], *Yen Myan* Felicia Wong[1] and *Jun Jet* Tai[1]

[1]School of Engineering, Mechanical Engineering, Taylor's University, Malaysia
[2]School of Engineering and Physics Sciences, Mechanical Engineering, Heriot-Watt University, Malaysia

Abstract. Inverted pendulum remained as the most popular topic for control theory researches because of its characteristic of being non-linear, unstable and under-actuated system. It is ideal for verification, validation and enhancement of control theory by stabilizing the inverted pendulum in an upright position using various controller and stabilizer mechanism. For this project, Proportional-Integral-Derivative (PID) controller is used to stabilize the inverted pendulum by tuning the respective gains (k_P, k_I, and k_D) to control the parameters of inverted pendulum which includes the rise time, settling time, overshoot and steady-state error in cooperation with of Arduino microcontroller. The objective of this project is to design and build a stabilizer mechanism with the integration of mechanical and electrical components to stabilize two Directional (2D) inverted pendulum similar to 3D printer mechanism. Besides that, PID controller will be tuned in Arduino microcontroller and control the output of stabilizer mechanism. The stabilizer mechanism is designed in SolidWorks software and built using various manufacturing techniques, raw materials and 3D printing, while the electronics components such as gyroscope and Direct Current (DC) motors are controlled using Arduino Due in C++ language. The gyroscope determines the tilting angle of the pendulum as a feedback in the control loop, and the gains of PID are used to control the speed and direction of DC motor to provide sufficient force/torque to keep the inverted pendulum in an upright position. The stabilizer mechanism with inverted pendulum has been built and the gains of PID have been tuned using "trial and error" method as friction is now taken into consideration. The inverted pendulum is successfully stabilized in an upright position (0o measure at z-axis) using control theory.

1 Introduction

Inverted pendulum remained as one of the most popular problems existing in control theory and has been studied in many researches for the control community [1]. The unstable, nonlinear and under-actuated characteristics of an inverted pendulum system raise a challenging task for testing different control algorithms due to its large uncertainties attached with the system [2]. There are many methods to stabilize the inverted pendulum

* Corresponding author: c.hoo@hw.ac.uk

and one of the methods is by using a Proportional-Integral-Derivative (PID) controller. PID controller is a control instrument that is commonly used in the industrial control applications to regulate processes as part of a control loop [1].

Each controller in a PID contains of their respective tuning gain, k_P, k_I, and k_D. Tuning any of the gain in PID controller affects the control system parameters: rise time, settling time, overshoot and steady-state error. Rise time is the time taken for the waveform to go from 0.1 to 0.9 of the final value. Settling time is the time taken for the transient's damped oscillations to reach and stay within ±2% of the steady-state value. Overshoot is the amount of waveform overshoot from the final value or the steady-state value. Steady-state error is the difference between output value and desired output value [3].

A one-dimensional (1D) inverted pendulum is commonly used as study material. The inverted pendulum is placed on a cart and stabilized using linearization method. The distance travelled by the cart produces force to keep the inverted pendulum upright. The distance travelled by the cart alters based on the difference of pendulum angle relative to y-axis [1]. The pendulum angle is measured and the difference of the pendulum angle is calculated and fed into PID controller for stabilizing the inverted pendulum by producing an ideal system response with optimal gain value to drive the cart for a certain distance. The limitation of one axis inverted pendulum is that the stability of the pendulum can only be studied for one axis.

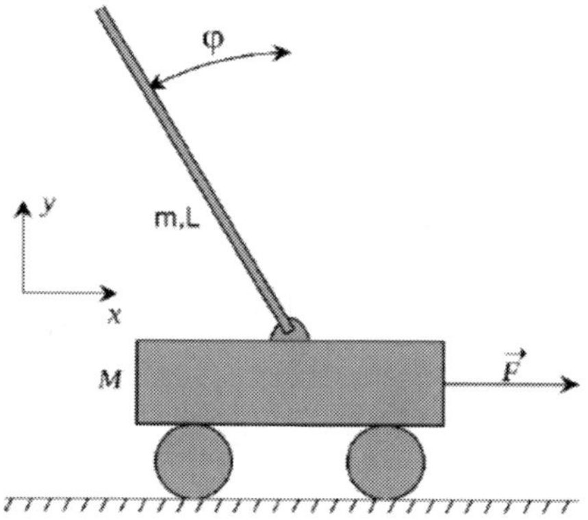

Fig. 1. Cart pendulum system. [1]

To further study on the inverted pendulum, a two-dimensional (2D) inverted pendulum will be chosen. Increasing one axis will increase the number of uncertainties and nonlinearity of the pendulum which introduces a wider range for testing control algorithms. A mechanism with the integration of mechanical components and electronics components will be built to stabilize the 2D inverted pendulum using pulley system and sliding mechanism. The difference of the pendulum angle relative to x-axis and y-axis is measured using a gyroscope sensor.

A gyroscope sensor determines the position, orientation and angular velocity using Coriolis force applied to a vibrating object [4].When the gyroscope is rotated, the Coriolis force acts upon the drive arms and produces vertical vibration. The vertical vibration then produces a sensing motion in the sensing arms. The motion on the sensing arms creates a potential difference where the angular velocity is measured and converted into small current signals. The signals are then amplified and fed into a controller. Gyroscope sensor is placed onto the 2D inverted pendulum. The orientation and position of inverted pendulum are measured, converted into signals and compared with the desired output value in the control system and the error is calculated [5]. The error is then fed into PID controller and each parameter of the gain is tuned to produce and output angle to drive the mechanism to stabilize the 2D inverted pendulum.

2 Design of the Stabiliser

The stabiliser design consists of two parts, which are the mechanical and electrical parts. Mechanical part contains the inverted pendulum, pulley system and sliding mechanism. The PID control and communication between the microcontroller and the actuators (motors) are discussed under electrical part.

2.1 Mechanical system

Linearisation method is one of the easiest methods to stabilise an inverted pendulum. A cart moves in x-axis and y-axis linearly to a desired position where the pendulum is stabilized when the pendulum tilts but does not exceed the tilting maximum angle (30o relative to z-axis), if tilting angle is exceed the pendulum falls to maximum position. Using this method, the first stage of this project is to build a 2D inverted pendulum on a moving cart and the stabiliser mechanism to stabilize or balance the inverted pendulum to an upright position.

The ideology of the stabiliser mechanism is a mock of a 3D printer mechanism shown in Fig. 2 because of the accuracy and precision exhibited by a 3D printer. The extruder of 3D printer moves on along the Cartesian plane in x-axis and y-axis with four directions on the gantry. The extruder receives xy coordinates from the software, where each part of the object needs to be printed and the heater cartridge on extruder heats and melt the filament to be deposit onto the bed of 3D printer [6]. Therefore, 3D printer has high accuracy and precision in coordination to print the objects as they are designed to be. To balance an inverted pendulum requires the accuracy and precision exhibited by the 3D printer, hence the stabiliser mechanism is built similarly to the mechanism of a 3D printer.

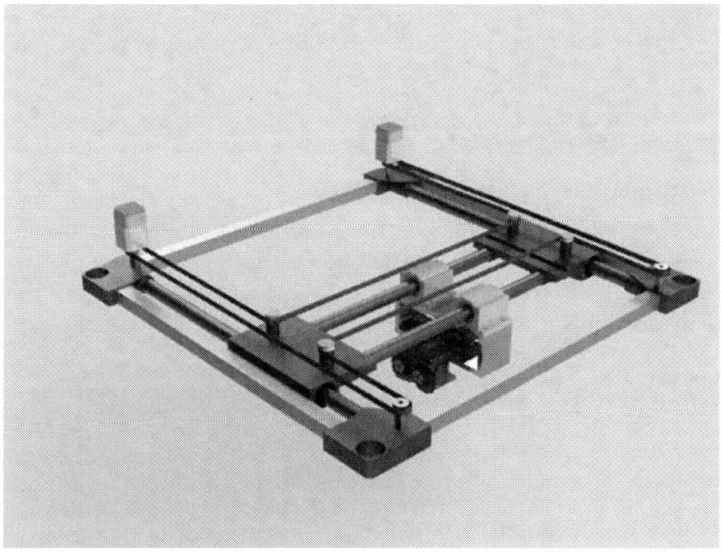

Fig. 2. 3D printer mechanism.

From the 3D printer mechanism, a pulley system and four sliding mechanism is implemented to allow the 3D printer to move at x-axis and y-axis. The similar pulley system and sliding mechanism is implemented on the cart to form a stabiliser mechanism and moves the cart of inverted pendulum along the x-axis and y-axis.

2.1.1 Pulley system

The pulley system is implemented to convert rotational motion produced by motors into translational motion for the inverted pendulum cart. The components in this system include timing pulleys, timing belts, flanged bearings and solid rods. A 60T timing pulley is mounted on the shaft of Direct-Current (DC) motor and connected to 20T timing pulley [7], with a closed-loop timing belt [8], as shown in Fig. 3.

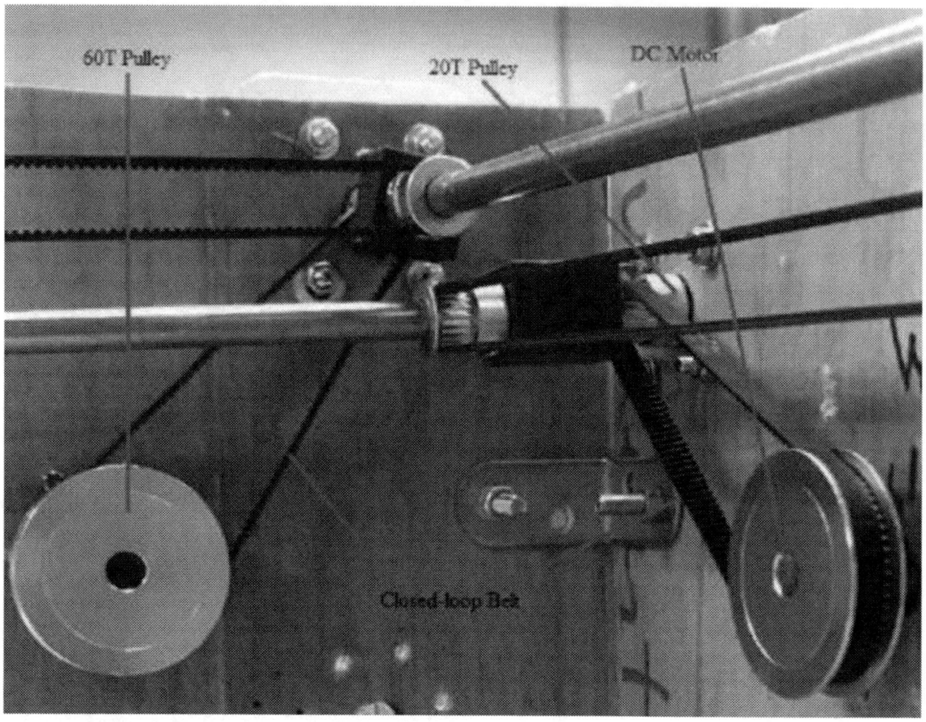

Fig. 3. Connection of 60T and 20T with DC motor.

Due to the friction and requirement of high acceleration to hold the inverted pendulum upright, the DC motor rated speed was not sufficient to provide such acceleration. Therefore, the number of teeth on pulley attached to motor is triple the number of teeth on pulley attached to solid rod to increase the translational speed by three times using pulley system.

$$\frac{N_1}{N_2} = \frac{T_2}{T_1} \tag{1}$$

Where;
N_1 = Speed of gear on DC motor
N_2 = Speed of gear on Solid Rod
T_1 = Number of teeth of gear on DC motor
T_2 = Number of teeth of gear on Solid Rod

2.1.2 Sliding mechanism

The sliding mechanism is integrated with the pulley system to provide guidance for inverted pendulum cart to move along X-axis and y-axis. Since there are two axes, the pulley system of x-axis overlaps with the pulley system of y-axis, therefore the design of inverted pendulum cart requires high accuracy of dimension where the solid rod of y-axis on cart is perpendicular to x-axis and does not collide or touches each other. Similarly, the design of sliding mechanism to hold the inverted pendulum cart by each end of stabiliser mechanism requires high accuracy for inverted pendulum cart to travel in straight line on x-axis and y-axis. The designs are shown in Fig. 4 and Fig. 5.

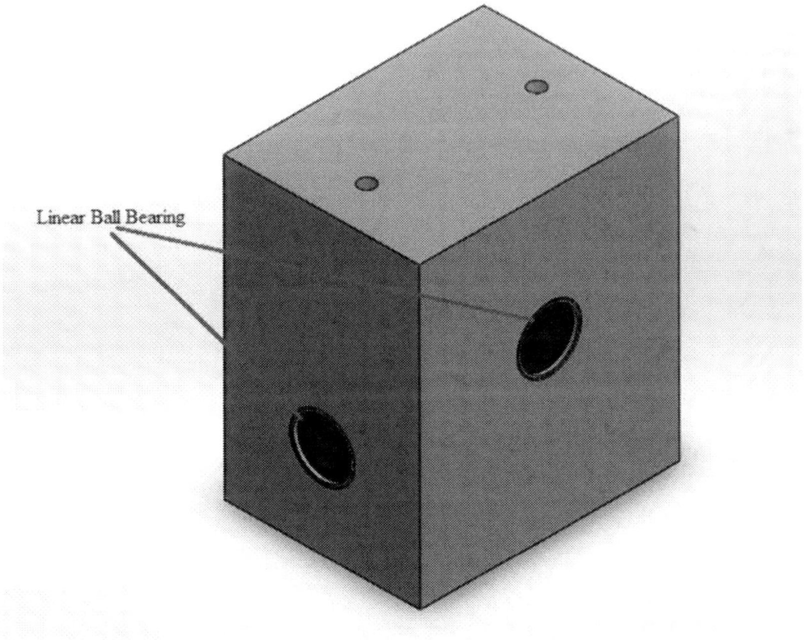

Fig. 4. Inverted pendulum cart.

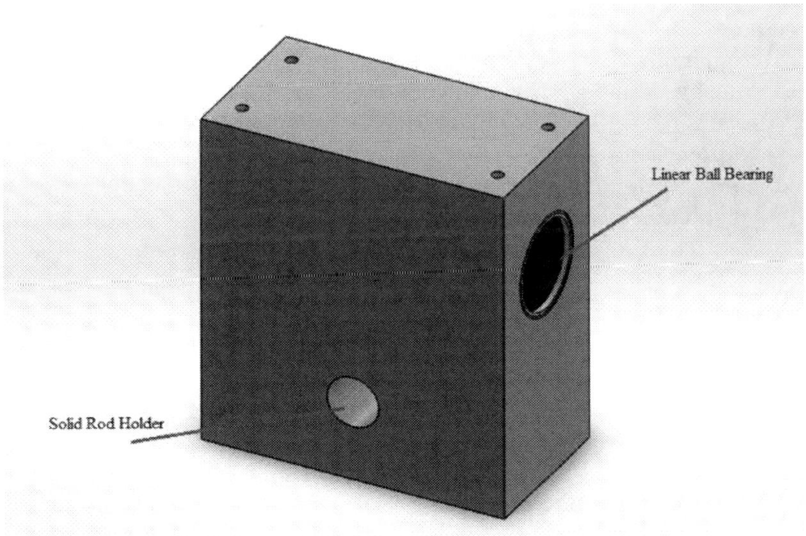

Fig. 5. Sliding mechanism.

The main component for sliding mechanism is the linear ball bearing. The linear ball bearings assist in reduction of friction by allowing solid rod of y-axis to rotate while the sliding mechanism of x-axis to have precise translational motion [9]. Due to high accuracy of dimension, the sliding mechanism parts are manufactured through 3D printing and linear ball bearing is inserted into each block of sliding mechanism.

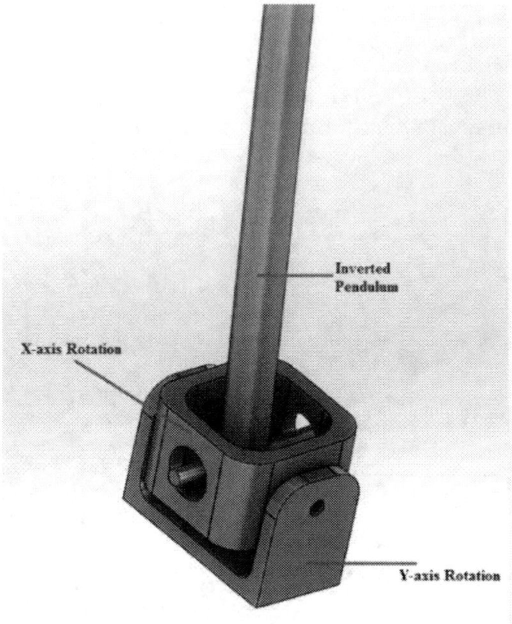

Fig. 6. Inverted pendulum holder.

The gyroscope is attached onto the inverted pendulum to determine the position and orientation of inverted pendulum which will be converted into electrical signals and fed into a microcontroller to control the direction of rotation and speed of both DC motors.

2.2 Electrical and electronics methodology

The translation motion of the cart is controlled by 2 DC Geared motor as each DC geared motor is assigned to control one axis of translational motion. Both motors are coupled with the motors through the pulley system and sliding mechanism. The DC geared motor is used to rotate a vertical gear and a horizontal gear for a transmission of torque from horizontal to vertical and rotates the timing belt pulley to pull the timing belt on the cart. In order for the cart to have sufficient force in balancing the inverted pendulum, the torque and speed of DC geared motor are significant in producing the amount of force. The speed of motor is measured in revolution per minute (rpm) and torque would be newton per meter (N.m). Both DC geared motor has the same specification where each motor has the speed of 380 rpm (0.12 m/s) and torque of 637.43 mN.m (6.5 kg.cm) [10], which is able to produce the force needed to keep the inverted pendulum upright.

The tilting angle of pendulum is measured using gyroscope and the current signals from gyroscope is fed into microcontroller to convert angle analog signals to digital signals to provide a feedback. If the angle of pendulum is more than 0° and less than 5°, the difference between desired angle of pendulum (0°) and measured angle of pendulum is calculated for an error signals to be converted into Pulse Width Modulation (PWM) signals. The error signals determine the amount of PMW signals to be fed into DC geared motor and rotate the motor in clockwise direction. If the angle of pendulum is less than 0° and

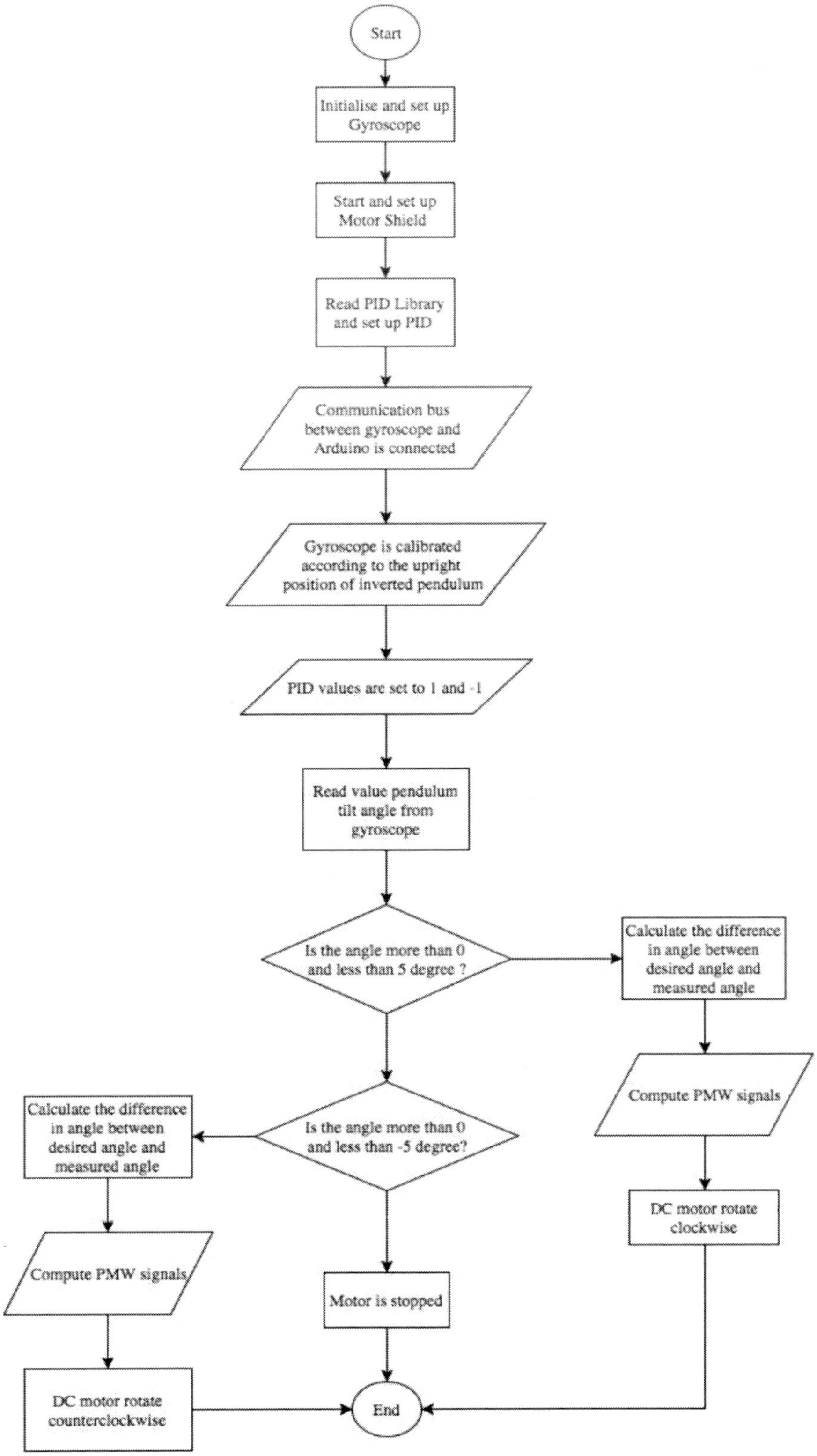

Fig. 7. Flowchart of electrical components.

less than -5°, the difference between desired angle of pendulum (0°) and measured angle of pendulum is calculated for an error signals to be converted into PWM signals. The PWM signals is fed into DC geared motor and rotate the motor in counter-clockwise direction. However, if the angle of pendulum remains at 0°, the PWM signals are LOW to stop the DC geared motor from rotating. Fig. 7 shows the flowchart for the DC Geared motor.

Methods from Research Methodology are executed in order to achieve the objectives of this research. Experimental testing on the stabilizer mechanism and electrical and electrical components are done such ensuring the stabilizer mechanism is able to move the cart in x-axis and y-axis, inverted pendulum tilt only along x-axis and y-axis, angle of tilt for inverted pendulum is read by gyroscope, microcontroller process the angle of tilt, tuning the PID gains and vary the speed and direction of DC motors.

3 Results and Discussion

3.1 Mechanical system

The 20T timing pulley is mounted on a solid rod which is connected from one end to another end of the frame of stabiliser mechanism with flanged bearing to allow smoother rotation of solid rod [11], shown in Fig. 8. When the DC motor rotates, it rotates the 60T timing pulley and with the closed-loop timing belt, the rotation is translated to the 20T timing pulley and solid rod.

Fig. 8. Rotational motion components.

The pulley system is also used to move the inverted pendulum and cart by having two sets of 20T timing pulley mounted on a solid rod that hinged on the frame of stabiliser mechanism as shown in Fig. 9. This is implemented on the right end and left end of the frame of stabiliser mechanism. Both of the solid rod are connected with timing belt to form a closed-loop and the ends of timing belt are attached onto the sliding mechanism of inverted pendulum. This allows the inverted pendulum cart to move along x-axis. The same configuration is implemented for the cart to move along y-axis.

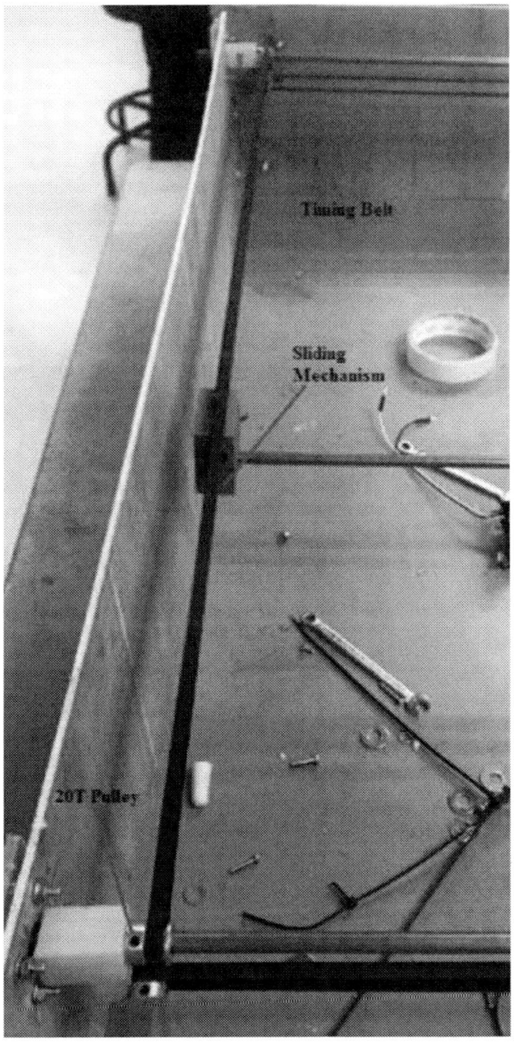

Fig. 9. Left end of stabiliser mechanism.

As the DC motor rotates clockwise, the 60T pulley translate the clockwise rotational motion to 20T pulleys and solid rod. The rotational motion is then converted to translational motion by rotating the timing belt attached on sliding mechanism allowing the cart to move at positive direction of x-axis. While if DC motor rotates counterclockwise, the cart moves at negative direction of x-axis. This applies in moving the cart for positive and negative direction of y-axis. This creates force through acceleration of motor to keep the inverted pendulum at an upright position, $\theta = 0°$.

3.2 Electrical and electronic system

The overall schematic diagram of electrical and electronic system is shown in Fig. 10 on how each components are related to each other and the process of the system.

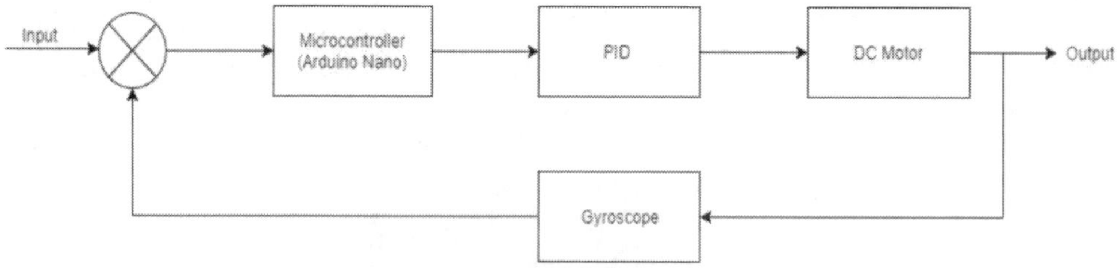

Fig. 10. Schematic diagram of electrical and electronic system.

3.2.1 Microcontroller (Arduino Nano)

Microcontroller forms a communication between input devices and output devices through programming languages. In this research, Arduino Nano communicates with gyroscope and DC motors through C++ programming to identify the angle of tilt from inverted pendulum as input and the speed required from DC motor to keep the inverted pendulum upright as output. Acceleration of DC motors are converted into linear force by pulling the timing belt in clockwise and counter-clockwise direction subsequently moving the cart in x-axis and y-axis. However, to keep the inverted pendulum upright, the acceleration is constantly changing and increases the difficulties in providing theoretical value of force required to balance the inverted pendulum. PID controller is used to assist the response of DC motor towards the angle of tilt from inverted pendulum. The ideal response characteristic includes low rise time, small percentage of overshoot and zero steady-state error of DC motor. This is done by tuning the gains of PID controller which can be alternated in C++ programming of Arduino Nano. The angle of tilt is monitored through Serial Monitor from Arduino Nano and the gains are tuned according to measured angle of tilt until desired value of angle of tilt is obtained. The current gains of PID for x-axis and y-axis are -1 to 1 which includes the friction arise from pulley system and sliding mechanism.

3.2.2 Gyroscope (MPU6050)

The gyroscope is attached perpendicular to the inverted pendulum facing downwards because the jumper wires to be connected from gyroscope to Arduino Nano cannot exceed 5cm. This is due to the disturbance of noise and interference on the bandwidth of gyroscope when transferring signals to Arduino Nano. The longer the jumper wire, the higher the resistance, noise and interference of bandwith. Therefore, shorter jumper wires are recommended. The gyroscope is tuned in such way that it only measures the x-axis and y-axis rotational motion. Z-axis and acceleration motion are tuned out to reduce the processing time of signals (faster signal transfer between gyroscope and Arduino Nano) as these two parameters are not in the scope of this research. Gyroscope has an internal voltage regulator to reduce 5V supplied from Arduino Nano to 3.3V as required to eliminate excess of voltage and causes damage to the gyroscope.

3 ports of gyroscope are connected to Arduino Nano namely, Signal Data (SDA), Signal Clock (SCL) and Interrupt. SDA processes the rotational motion (x-axis and y-axis) of inverted pendulum, stores the value as angle of tilt and send it to Arduino Nano. SCL is the frequency of transmission of data between gyroscope and a microcontroller and the built in frequency range about 400kHz which allows one data to be transferred at 2.6 nano second. The Interrupt port is to identify the status of gyroscope. When the gyroscope is ready to function, Interrupt port sends a signal to Arduino Nano and the program in Ardunio will run because the gyroscope requires initialization before running the program. Due to the limitation of bandwidth from gyroscope, the maximum tilting angle of inverted cannot exceed $\pm 5°$ on x-axis and y-axis.

3.2.3 Motor shield

The current and voltage draw with load of one DC motor are 1A and 12V. Both DC motors are connected in parallel which means the total voltage needed would be 12V, however the current would be 2A. The maximum voltage can be supplied by Arduino Nano is only up to 5V, insufficient to power even one DC motor. To assist and increase the voltage and ampere supplied to DC motors, a motor shield is used. The motor shield is able to supply up to 6A of current to each motor but only 5V can be drawn out from motor shield. However, the motor shield has input ports for external voltage supply to be connected to DC motor, hence an external power supply with 5A and 12V is connected to the motor shield to increase the voltage and current for both DC motors. Each motor connects to three ports of motor shield. Two ports are to control the direction of rotation (clockwise and counterclockwise) for DC motor. One port is to power the DC motor with voltage accumulated from power supply and motor shield. The motor shield receives processed signals from Arduino Nano and controls the speed and direction of rotation of DC motors and converts rotational motion to translational motion on pulley system.

Table 1. List of function achievable by stabiliser mechanism and inverted pendulum.

Tasks	Achievable?
Transmission of rotational motion to linear motion by DC motors and pulley system	√
Increases the speed of rotational motion of pulley system by 3 times of the speed of DC motors	√
Moving the cart along x-axis	√
Moving the cart along y-axis	√
Moving the cart along x-axis and y-axis at the same time	√
Tilting inverted pendulum along x-axis	√
Tilting inverted pendulum along y-axis	√
Tilting inverted pendulum along x-axis and y-axis	√
Eliminates rotational motion of inverted pendulum and moving along z-axis	√
Inverted pendulum balanced at x-axis	√
Inverted pendulum balanced at y-axis	√
Inverted pendulum balanced from negative x-axis	√
Inverted pendulum balanced from positive x-axis	√
Inverted pendulum balanced from negative y-axis	√
Inverted pendulum balanced from positive y-axis	√
Limited tilt angel can be recovered	√

As all components are assembled, communication of electrical components are programmed by C++ programming and PID gains tuned, experimental testing is done on the stabiliser mechanism to balance the inverted pendulum. Table 1 indicates the achievable tasks by stabiliser mechanism and inverted pendulum.

3 Conclusion

In conclusion, the stabiliser mechanism has been successfully designed and built. The inverted pendulum cart is able to travel along the x-axis and y-axis and produce force to keep the inverted pendulum in an upright position. The gyroscope is able to measure the angle of tilt of inverted pendulum and sends values to Arduino Nano to be processed. By analyzing the received signals, the PID gains are tuned through C++ programming in Arduino Nano and the response of inverted pendulum and DC motors are observed. The inverted pendulum is balanced in an upright position from negative and positive of x-axis only and y-axis only.

The future work includes balancing the inverted pendulum from various coordination on x-axis and y-axis as with increasing number of coordination, the PID gains will be tuned differently in order to receive same outcome as balancing inverted pendulum from negative and positive of x-axis only and y-axis only. The response curve of an inverted pendulum changes with number of coordination or number of uncertainties. The aim of this future work is to obtain characteristic includes low rise time, small percentage of overshoot and zero steady-state error of inverted pendulum and identifies the mathematical modelling to determine gains for PID controller that stabilizes 2D inverted pendulum.

References

1. K.H. Lundberg, J.K. Roberge, Proc. IEEE Conf. Decis. Control, **5** (2003)
2. "PID Theory Explained - National Instruments." [Online]. Available: http://www.ni.com/white-paper/3782/en/. [Accessed: 27-Apr-2017].
3. A. N. K. Nasir, R. M. T. R. Ismail, M. A. Ahmad, Intl. Conf. On Control, Automation, Robotics & Vision, (2010)
4. L. Blitzer, Am. J. Phys., **33**, 12 (1965)
5. "Accelerometer vs. Gyroscope: What's the Difference?" [Online]. Available: http://www.livescience.com/40103-accelerometer-vs-gyroscope.html. [Accessed: 27-Apr-2017]
6. "How exactly does 3D printing work? - Technology & Science - CBC News." [Online]. Available: http://www.cbc.ca/news/technology/how-exactly-does-3d-printing-work-1.1371800. [Accessed: 27-Oct-2017]
7. "GT2 Pulley 20 Teeth 8mm Bore – 6mm Belt Width - 3D Gadgets Malaysia." [Online]. Available: http://3dgadgets.my/belt-pulley/228-gt2-pulley-20-teeth-8mm-bore-6mm-belt-width.html. [Accessed: 27-Oct-2017]
8. "Gt2 Closed-loop Belt 200mm - 3D Gadgets Malaysia." [Online]. Available: http://3dgadgets.my/belt-pulley/244-gt2-closed-loop-belt-200mm.html. [Accessed: 27-Oct-2017]
9. "LM8LUU Linear Ball Bearing - 3D Gadgets Malaysia." [Online]. Available: http://3dgadgets.my/bearings/242-lm8luu-linear-ball-bearing.html. [Accessed: 27-Oct-2017]

10. "DC Geared Motor 12V High Power 380RP (end 9/14/2017 1:15 PM)." [Online]. Available: https://www.lelong.com.my/dc-geared-motor-12v-high-power-380rpm-stelectronics-182767385-2017-09-Sale-P.htm. [Accessed: 27-Oct-2017]
11. "F688ZZ Flanged Bearing - 3D Gadgets Malaysia." [Online]. Available: http://3dgadgets.my/bearings/240-f688zz-flanged-bearing.html. [Accessed: 27-Oct-2017]

Electrical Energy Harvesting from Cooker's Wasted Heat with Using Conduction Cooling

Wincent, *Ghafour* Amouzard Mahdiraji[1,*]

School of Engineering, Taylor's University, 47500, Subang Jaya, Selangor, Malaysia

Abstract. In order meet the demand of electricity in current era, the need for new sources of energy even in very minimal amount, could be done with proper research and technology advancement in order to convert as much wasted energy as possible. Collecting and analyses cooker's wasted heat as a main wasted energy source become the main interest for this research. This application can be installed either in household usage or commercial usage. Based on majority stove in household datasheet it shown that the efficiency of the stove is approximately 50%. With half of the efficiency turn into wasted heat, this application is suitable for thermoelectric generator (TEG) to harvest the heat. The objective of this research is to determine whether the thermoelectric generator (TEG) would able to power the 3V LED light as a small lighting system in household. Several designs with five TEGs in series circuit are tested to the application to analyses which method generated a better result. Since this research only focus in using a conduction cooling, aluminum heat sink will be utilized either for heat absorption or heat rejection. The maximum temperature differences between hot side and cold side is 209.83 °C with average power approximately 0.1 W.

1 Introduction

In this contemporary society, energy demand has become one of the crucial challenge. Based on Energy Information Administration, the demand of energy in the society is annually increase without any sight of peaking. Thus, this challenge has become one of the popular topic to discuss and to motivate the society to increase the energy supply. A lot of efforts is made by researchers to meet the demand of the energy by considering either other source of energy or renewable energy. An improvement has been done by other extensive research on renewable energy such as wind turbine, tidal energy, and solar power to supply the energy demand and, yet it still could not entirely cover the demand. Therefore, new methods to gain an extra energy is developed and one of the method is to absorb the wasted heat produce in daily household appliances.

Thermoelectric energy is known to be the usable source of electricity from the conservation of wasted heat and this energy is operate based on Peltier and Seebeck effecting. In year 1821, German physicist Thomas Johann Seebeck discover that a temperature differences is existed between two dissimilar metals, electricity will be

---

[1] Corresponding author: Ghafour.AmouzadMahdiraji@taylors.edu.my

generated since the electron level in each material shifted differently. Electron from the surface that is supplied with enough heat will eventually jump to the metal with colder surface. Thus, the movement of the electron is known to generate the thermoelectric energy [1].

Thermal energy is commonly being classify as a by-product from various forms of energy such as mechanical energy, chemical energy and electrical energy. It is known that, machinery such as washing machine, refrigerator, cooker is used daily by the people either in household or commercial. With the usage of the machinery, the electrical energy that supplied it will generated a lot of wasted heat due to the moving part of the machine or the heat loss due to the appliances product efficiency. Therefore, this wasted heat energy contain wasted efficiency percentage that can be absorb and covert to the usable electrical energy [2].

Due to the increasing of the energy demand, research is done to find the new method to discover renewable and available energy. As such, by absorb the wasted heat from the machinery heat loss and convert them to usable electrical energy. To perform this research, Thermoelectric generator (TEG) is become one of the solution. This technology is initially used to convert the geothermal energy to the electrical energy.

Although a thermocouple will be the first thing that comes to mind when thermal energy is mentioned, a TEG cannot be related to thermocouple. This is because both devices have a different usage purpose. Thermocouple is temperature gauge device which only measures the temperature in one space, while thermoelectric generator is used to covert the heat energy to usable electricity that depend on the temperature differences between two surfaces [3].

Thermoelectric technology is a rather new, developing technology that is inexpensive to build and is space saving due to their small size as shown in Figure 1. It is also easy to use. In addition, this technology is sustainable as well as it is eco-friendly and does not produce greenhouse gases or sound pollution. A downside to this technology is its low efficiency when generating electricity, which is of approximately 5–10% of the total heat energy [4][5].

Fig. 1. Thermoelectric Generator.

The concept behind kitchen appliances, such as a cooker is studied in order to understand the heat transfer from the cooker's flame and to discover the most optimum location to install the thermoelectric generator. Efficiency and total power generated from each design will be calculated together with the total cost. This is to determine whether this application would be a success.

In year 2005, researcher form Xiamen University has done and experiment with applying same TEG material Bismuth telluride with antimony doping to the usage on the plantation machine. During its experiment it found that the temperature difference that produce by the machine heat with water cooling technique as heat rejection is approximately 77 °C. this produce the ZT value as big as 1.15 and voltage generated is less than 0.5 V [6].

By reducing the thermal conductivity, it may enhance the ZT. Pei He Jiangqing, a Student from School of material science and engineering in Beihang University, found that Ca-doped BiCuSeO material exhibits low thermal conductivity and resulted in increasing the ZT by 0.9 at 923 K [7]. Meanwhile, Zhao et al. from Northwestern University reported that SnSe single crystal that measured along with room temperature orthorhombic unit cell b axis produced the ZT of 2.6 at 923 K [8].

2 Research methodology

The method in conducting this research is discusses in this section. The research that is conducted is a quantitative study based on the simplest and most effective method. TEG has been by so many researchers especially for energy experts. Several types of thermoelectric modules have been utilised and fabricated. However, in this study only Bi2Te3 thermoelectric module is used due to its reliability and capability to produce the higher ZT among other type in room temperature. Moreover, the availability of Bi2Te3 thermoelectric module is common in the market. Although thermoelectric generator is easy to operate, multiple experiments are required to find the most effective method [9].

FYP1 project will focus on studying literature review for both thermoelectric module and cookers from either journal or online to have a better understanding on what it will dealing with and what the device are. This includes executing initial experiment two of designs and find the best location to install the TEG, and lastly is purchase the suitable equipment. ANSYS software will be used to find the best location to install the TEG. For the first design (Design A) the TEG is attach to the acrylic and hanged near the heat source which is approximately 5cm. small heatsink is place at the cold side in order to release the heat from the hot side. This design has a convection heat transfer thus the temperature obtained was lower compare to the other design. For second design TEG (Design B) will be place to the aluminium plate which design to have a hole at the left, this is to allow the fire is able to go through to heat the pot/pan. To place the TEG, ANSYS computer software is utilized to find the best location, and at the purple line as show in Figure 2 which is approximately 10cm from the heat source was selected in order to prevent the TEG overheat easily conduction heat transfer was applied in this design. Both, the simulation and the two experiments will be done in implement stage where initial experiment is conducted. Result of the experiment will be analysed and both method will be compared.

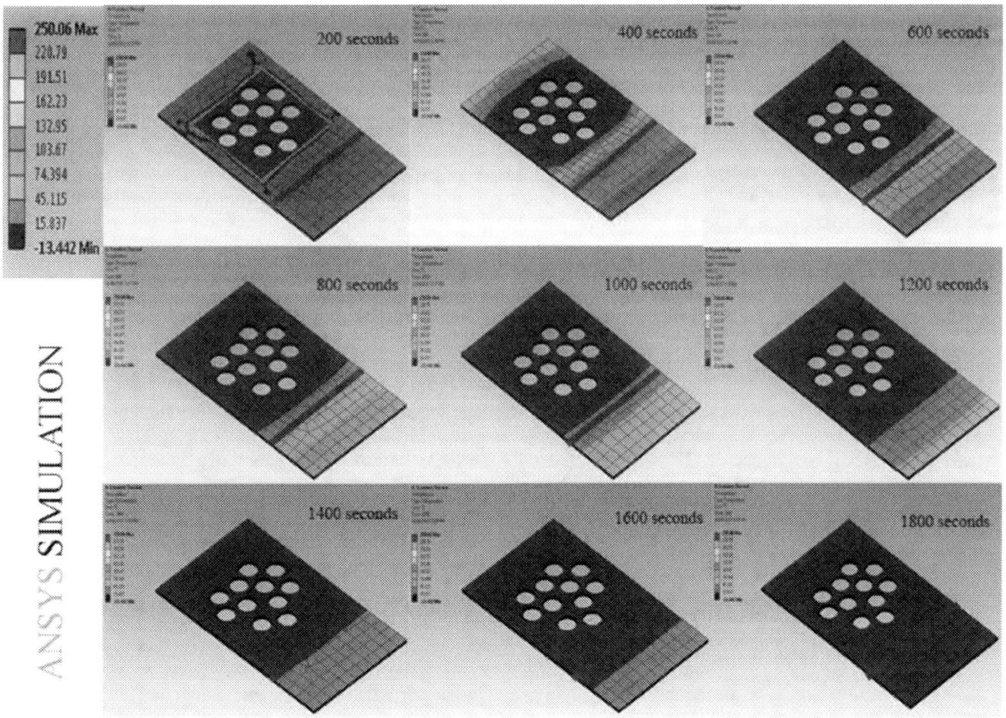

Fig. 2. ANSYS simulation for aluminium plate.

Meanwhile, to observe the differences of the result, another three Designs which are convection heat transfer and two calcium silicate with TEG facing aluminium plate (Design C), convection heat transfer and single calcium silicate with TEG facing heat source (Design D) and Design E, convection and conduction with TEG attach with L aluminium plate are tested for the experiment. Not only new designs are proposed, but high performance Bi2Te3 TEG, calcium silicate Ca2SiO4 and big aluminium heat sink also introduce to the experiment since the old TEG performance is consider low and inefficient, the calcium silicate is to prevent the heat to transfer to the cold side and heat sink is to reject more heat to generate large temperature differences. However, with using new high performance TEG, the experiment is only conducted with using single TEG instead of using multiple TEG that form into series circuit. This is done to check whether the TEG is efficient to use before purchase multiple of them. Figure 7 show that the visual experiment setup for design C, in this design the calcium silicate that place near to the heat source is aims to reduce the convection heat transfer to the heatsink so that the temperature differences is larger. While the calcium silicate in the bottom of TEG is aims to prevent the TEG to experience over heat since the maximum temperature of high performance TEG can withstand is 300 °C while the aluminium plate itself has temperature of 270 °C by using infrared thermometer gun to measure. Small heatsink is attach to the hot side which is inside the calcium silicate in order to absorb more heat and big heat sink is attached to the cold side to maximize the heat rejection. This design place near the edge of the aluminium edge which is 20cm from the heat source due to consideration of the convection heat transfer to the big heatsink. Since the hot side is cover by calcium silicate, temperature of the calcium silicate will become the temperature for the hot side.

While, Design D is designs to place the TEG facing the heat source with 4 differences distance which are 15cm, 10cm, 5cm and 2.5cm from the heat source. calcium silicate is used to cover and protect the TEG and to prevent the heat to transfer to the aluminium heatsink which will be affecting the result. Small heatsink is attach to the hot side to absorb maximum heat from the heat source while big aluminium is attach to the cold side to reject the heat. In order to obtain a perfect connection, thermal grease is applied on both hot side and cold side. In addition, aluminium plate is placed to produce a better heat transfer by convection. 4 experiments are conducted and compare to each other to analyse the changes of generating voltage.

Lastly, Design E as shown in Figure 3 is conducted by placing one end of the aluminium and steel L bracket to attach to hot side of the TEG and the other end to fire, this is done so that conduction heat transfer is occurred and to maximize the heat absorption. Meanwhile same as the previous design, big heat sink is placed at the cold side to reject the heat and aluminium plate is placed to ensure the convection heat transfer is maximized. The results of both L bracket's material are compare. Since aluminium have a better heat conductivity with 209.5 W/m.K, three differences size of aluminium L bracket 30cm, 20 cm and 15 cm (Figure 4) are tested and compared. Again, the best results among three differences size of L bracket is tested with four difference resistors which are 10 Ω, 7.5 Ω, 2.2 Ω and 1Ω. This is to find the power produce from the single high performance TEG. The reason several resistors are tested is to observe the behaviour of the TEG.

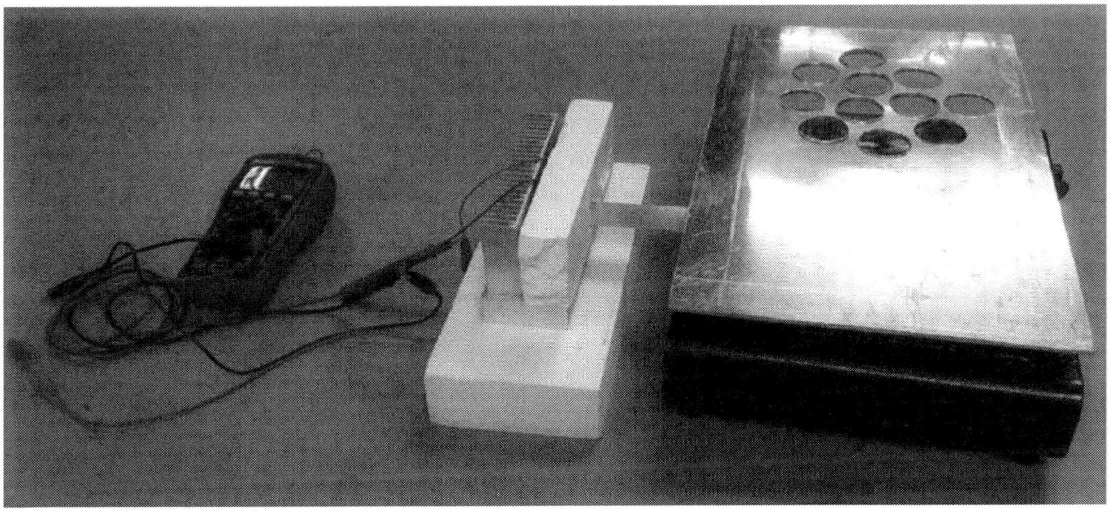

Fig. 3. 20 cm aluminium L bracket experiment setup.

Fig. 4: Steel and aluminium L bracket.

3 Result and discussion

Based on table 1 it shows the average result of each experiment. For experiments design C, D and E, the power are zero because no resistor is connected to the experiment. Resistor is only applied to experiment design E4 and not applied to the others is because based on the result, E4 experiment generate the highest voltage which is 822.71 mV. four different resistors are applied to E4 to observe the changes of the current by the TEG. Based on the observation it is shows that the current generated by the TEG is indirectly perpendicular to the resistant of the resistor. Table 1 also shown that the design E4 experiment with 1 Ω resistor produce the highest power with approximately 0.1 W. In order to increase the generated power, TEG can be connect series with another 4 TEGs.

Table 1. Experiment result based on each model.

Design	mV	mA	P(W)	Th	Tc	dt	ZT	Efficiency
A (5 TEG)	40.65	8.55	0.0003	148.4	134.5	13.9	0.0326	0.0008
B (5 TEG)	100.54	7.69	0.0008	271.3	242	29.3	0.0688	0.0028
C	28.64	0	0.0000	66.66	39.99	26.67	0.0626	0.0076
D (15cm)	7.23	0	0.0000	37.39	27.49	9.9	0.0232	0.0018
D (10cm)	12.49	0	0.0000	45.75	32.36	13.39	0.0314	0.0026
D (5cm)	17.02	0	0.0000	59.59	34.22	25.37	0.0595	0.0078
D (2.5cm)	79.55	0	0.0000	126.7	37.25	89.45	0.2099	0.0506
E1 (steel)	25.09	0	0.0000	80.33	34.47	45.86	0.1076	0.0202
E2 (30cm Al)	258.63	0	0.0000	100.98	35.51	65.47	0.1536	0.0337
E3 (20cm Al)	409.12	0	0.0000	113.12	37.59	75.53	0.1772	0.0400
E4 (15cm Al)	822.71	0	0.0000	228.52	46.59	181.93	0.4269	0.1108
E4 (10Ω)	891.81	53.091	0.0473	212.64	44.64	168	0.3942	0.1027
E4 (7.5Ω)	935.13	59.645	0.0558	196.9	39.95	156.95	0.3683	0.0986
E4 (2.2Ω)	916.71	63.954	0.0586	195.72	46.41	149.31	0.3504	0.0884
E4 (1Ω)	883.68	112.16	0.0991	254.76	44.93	209.83	0.4924	0.1306

For experiment design A it shows that the setup was inefficient since the voltage generated is relatively low with small temperature difference. During the experiment, the voltage of the TEG is keep fluctuating up and down rapidly. This show the connection is unstable and it only generate a very small current which is not worth the investment. Meanwhile, for design B experiment is more suitable to utilize for the superior quality TEG since the temperature in both side of the TEG is too high. During the experiment, the soldering flux on the TEG cables were melted and spoiled the device after the first experiment.

In experiment design C the result is also not satisfy since the average voltage result is low. This is because the TEG is located too far from the heat source and the convection heat transfer is tend to spread to environment instead of absorb by TEG. While, if the TEG placed nearer to the heat source this will result the big aluminium heat sink that supposed to reject heat from the TEG to act inversely. Meanwhile, experiment design D is done with testing 4 difference distances. In this experiment it shows that distance of 2.5 cm show the best result. However, the hot temperature of the TEG is still consider low since the average temperature is only half from the heat source temperature which is 126.7 °C.

The last design is created in order to allow the TEG hot side can reach approximately >200 °C. Initially 30 cm steel and aluminium L bracket are tested to compare which material provide a better result based on the voltage. Since aluminium has higher heat conductivity, 3 difference size of it is created. The longer the aluminium bracket, the greater heat loss will be occurred due to convection heat transfer. Therefore, shorter aluminium L bracket is preferable. Not only have higher conduction heat transfer, but shorter L bracket also allow the TEG to get closer to the heat source where convection and radiation heat transfers is higher. To prevent the heat to transfer to the big aluminium heatsink, calcium silicate is place as the TEG holder as shown if Figure 3. The shortest aluminium bracket is chosen as the final design since the result is quite satisfying and able to produce both average closed and close circuit voltage 0.8 V with 0.12 A in closed circuit current.

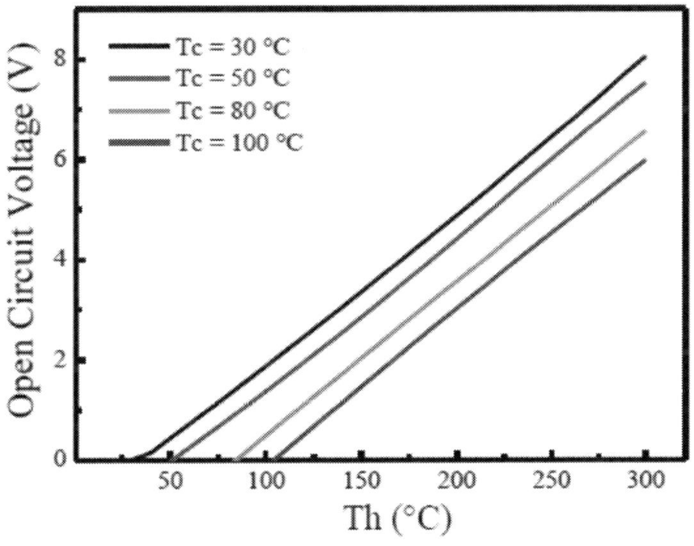

Fig. 5. Chart for open circuit voltage versus Th under various Tc.

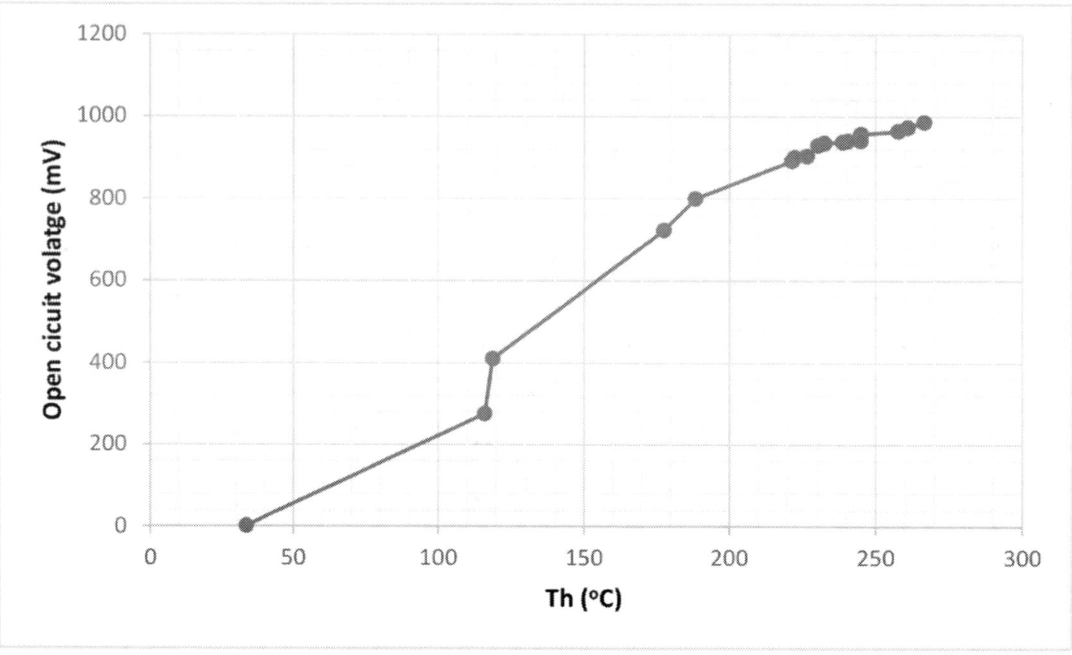

Fig. 6. Experiment chart for open circuit voltage versus Th under Tc 50 °C.

Figure 5 it shows the performance of the high performance TEG based on the provided device's data sheet. Since the average of the Tc of the experiment is approximately 45 °C, the result will be compare to the red colour line where the temperature of Tc is 50 °C. Meanwhile in Figure 6 shows the result of the Voltage generated with average Tc 45 °C. in the experiment result it shows that the voltage generated is increase rapidly when hot side start reaches temperature 100 °C to 220 °C this is because the temperature difference is also increase rapidly. While, the voltage only increases slightly when temperature reach 220 °C and above since the temperature of the heat supply is only 270 °C and with effect of external factor such as environment temperature, the temperature on the hot side is only fluctuating between 230 °C to 260 °C. Based on the comparison between two graphs, it shows that the voltage that generated in the experiment is relatively low even the temperature Th is follow the temperature from the data sheet. Several factors are stated as the factor that cause the changes such as inconstant Tc during the experiment, fluctuate Th during experiment and environment condition. In order to investigate another factor, more experiment will be conducted to the future. Table 1 shows that the maximum power that can reach by the TEG is only 0.1 W in design E4. The voltage of open circuit E4 and closed circuit 1 Ω E4 still remain the same, this mean that this circuit still able to undergo smaller resistant. 0.5 and 0.2 Ω resistant will be tested in the future to observe the changes and 5 TEGs in series closed circuit also will be tested to increase the power. It is expected by combining 5 TEGs into Series closed circuit with smaller resistant, the result will be better and worth the investment since less maintenance is require and it is long last technology.

4 Conclusion

In conclusion, TEG technology is consider as green technology since this technology have no moving part, long last and require less maintenance. The temperature differences between hot side and cold side of the TEG has become the main parameter in order to generate electricity. Generated energy from the TEG is directed perpendicular with the temperature difference. The higher the temperature differences so do the electric generated. To maximize the efficiency of the power generated, all type of heat transfer which are conduction, convection and radiation is apply to the latest design. While, only conduction cooling with aluminium heat sink was applied to reduce to ease the installation and usage. Design E4 produce the best result in open circuit with voltage 822.71 mV. Thus, this design is continued to close circuit with using four difference resistors. Based on the result, design E4 with 1 Ω resistor produce the best result with a power of around 0.1 W.

References

[1] P. Gokhale, B. Loganathan, J. Crowe, A. Date, and A. Date, Energy Procedia, **110** (2017)

[2] D. Li, J.C. Li, X.Y. Qin, J. Zhang, H.X. Xin, C.J. Song, L. Wang, Energy, **116** (2016)

[3] X. Zhang and L.-D. Zhao, J. Mater., **1**, 2 (2015)

[4] D. Champier, Energy Convers. Manag., **140** (2017)

[5] R. Y. Nuwayhid, A. Shihadeh, and N. Ghaddar, Energy Convers. Manag., **46** (2005)

[6] M. Otake, K. Sato, O. Sugiyama, and S. Kaneko, Solid State Ionics, **172** (2004)

[7] H. Pang, Y. Y. Piao, Y. Q. Tan, G. Y. Jiang, J. H. Wang, and Z. M. Li, Mater. Lett., **107** (2013)

[8] L.D. Zhao, C. Chang, G. Tan, M.G. Kanatzidis, Energy & Env. Sci. (2016)

[9] Y.-L. Pei et al., NPG Asia Mater., **5** (2013)

Anti-windup PI controller, SIPIC For Motor Position Control

Kah Kit Wong[1], *Choon Lih Hoo*[2], *Mohd Hardie Hidayat Mohyi* [1,*]

[1]School of Engineering, Taylor's University, Malaysia
[2]School of Engineering and Physical Sciences, Heriot-Watt University, Malaysia

Abstract. Control system plays a major role in the industry nowadays as it simplifies workload and reduce manpower. Among all the controlled applicable field, control system is heavily used in motor speed and motor position controls. Although there are various types of controllers available in the market, PID controller remains as one of the most used controller due to its simplicity. Unfortunately, PID controller experiences windup phenomenon which affects the controller's performance. This paper proposes a new type of anti-windup PI controller, SIPIC for motor position control application and aims to validate the performance of this controller as compared to conventional PI controller. To test the ability of the controllers, both controllers were experimented using hardware testing. The settings conditions of with and without loadings were used under two different inputs of 0° to 90° and 270° to 90°. The results obtained show that under without loadings, both controller showed favourable performances. Though, SIPIC controller slightly outperforms PI controller by having lower overshoot and shorter settling time for a wider range of gains. The rise time of both controllers are similar as it is the lowest possible rise time due to hardware limitations. Experiment results with loading condition, for both inputs and when Kp is 1 and Ki is 15, PI controller shows unstable performance by having large amount of oscillations and overshoots. The settling time was unable to be determined as the controller did not settle within the given step time. On the other hand, at the same gain, SIPIC controller still shows acceptable performance. This shows that SIPIC controller is more favourable by having better stable performance for a wider range of gains while PI controller needs to be finely tuned to a specific gain to obtain desired results..

1. Introduction

In the 21st century where manual manpower is increasingly being replaced by automatic robotic system, control system plays a major role in ensuring the quality of product and efficiency of the automated system.

* Corresponding author: hardie.hidayat@taylors.edu.my

The involvement of control system in the industry can mainly be found in speed control and position control. In position control system, mechanical position usually is the system output for an arm or a pointer. The controller performance highly affects the accuracy of arm/pointer system output. The position control system applications can be commonly found in automated systems such as solar tracking system and robotic system. Servomotor is usually used as the actuator for position control system. Servomotor is a type of direct current (DC) motor that is capable of precise control for velocity and acceleration as well as angular and linear positions. Many types of controllers have been used in the application of position control systems. These controllers usually include the use of fuzzy logic, fractional order, finite control with model predictive controllers as well as a proposed controller that combines the characteristics of μ-synthesis and neural network [1].

The proposed controller is a modified crisp fuzzy logic controller where modifications were done on both the input and the output of the controller. The input of the controller uses crisp rectangular adjacent non-overlapping membership function which is distributed along the discourse universe whereas modification for the output was done on the computation method [1]. The controller showed acceptable performance and shorter processing time than conventional fuzzy logic controller [2].

Another proposed controller is based on finite control set model predictive control for the application of permanent magnet synchronous motor (PMSM). This controller predicts the future behavior of the actuator and perform action that is chosen from a model set [3]. Other than that, many also applied fractional order calculus in motor position control which uses a fractional order PI controller and compared to integer order PI controller. The fractional order controller contains two extra variables which allow more flexible and precise tuning in the applications. Tt can be concluded that fractional order PI controller experience less overshoot and lower maximum percentage of error than conventional PI controller [4]. Further expanded application of fractional order added with differential evolution algorithm is also introduced in PI controllers. This method adds a self-tuning ability to the controller for the application of voice coil motor (VCM). Result shows that the proposed controller experience lower settling time, tracking error as well as lower maximum overshoot as compared to conventional PI controller and standalone fractional order PI controller [5].

In this paper, Proportional-integral-derivatives (PID) controller is chosen due to its simplicity in applications [6]. By tuning the controller's three individual components, overshoots, steady-state errors, rise times and settling times of the system can be altered easily. The limitation of a conventional PID controller is often caused by the limitation of the actuator being used instead of the controller itself. When the output of the controller exceeds the limit of the actuator, the controller begins to lose the ability to control the system. Example can be shown through the flushing mechanism. The system controls the opening of the valve to limit the flow rate of the water depending on the tank level. However, if the error continues to grow even after the valve is fully open, the flow rate of the water will not continue to increase due to the limitation of the valve. This condition is known as the integral windup phenomenon, where the integral component continues to offset the error even when the output remains the same.

Over the years, several anti-windup schemes had been proposed to eliminate the integral windup phenomenon and to increase the efficiency of PID controller. These schemes include the Conditional Integration (CI), Tracking Back Calculation (TBC), Integral State Prediction (ISP), pole placement method and gain scheduling method.

Conditional Integration is one of the most common schemes and it function by switching off the integral component during saturation state then turns it back on when the controller enters a linear state again. Tracking Back Calculation measures the difference between the output signal and the non-saturation state signal before feeding the error back to obtain higher accuracy. Another scheme is the Integral State Prediction (ISP). This scheme predicts the steady-state value and feed the value back to the integral state when the controller re-enters linear state [7].

Recently, both pole placement and gain scheduling methods were introduced in PID. The desired plant is first transformed into a type-1 plant as type-0 plant will always have error. Pole placement method is then used for the derivation of the controller gain [8]. Lastly, gain scheduling proposed the use of various gains and constants for different controller states for higher accuracy [9].

This paper uses a new type of controller known as Steady-state Integral Proportional Integral Controller (SIPIC) scheme for the application of position control system. Steady-state Integral Proportional Integral Controller (SIPIC) scheme is also another proposed controller that can overcome integral windup. It does that by having a separate closed loop feedback with the steady state integral value as an input. This constantly leads the integral control towards the steady state integral value[7]. It was mentioned earlier that different component of PID controller will have different effect on the performance and for PI controller, it will often have a short settling time with overshoot performance. SIPIC scheme made it possible to have a controller with short settling time with no overshoot performance due to its tuning decoupling ability. This ability allows the proportional gain (K_p) and integral gain (K_i) to be tune separately.

SIPIC has shown ideal performance with little overshoot while maintaining short settling time on a DC motor speed control application[7]. However, no work had been done on applying the SIPIC scheme on a DC motor position control application. Therefore, this research aims to validate the usage of proposed SIPIC controller in the application of motor position control. The objectives of this research are as below:

1. To develop a new anti-windup PI controller that is suitable for motor position control.
2. To validate the stability and performance of the new PI controller by studying the overshoot, steady-state error and settling time of the controller.
3. To investigate and determine the range of gain that will maximize the performance of the PI controller for motor position control.

2. RESEARCH METHODOLOGY

2.1 Experimental Setup

Hardware testing was done to verify the proposed SIPIC controller. Comparison was done between SIPIC controller and conventional PI controller. The performance of the controllers is evaluated based on theirs steady-state error, overshoot, rise time as well as settling time. SCILAB and SCICOSLAB software were used to carry out experiment and data analysis for both the controllers. Figure below shows the equipment used for experimental testing purpose.

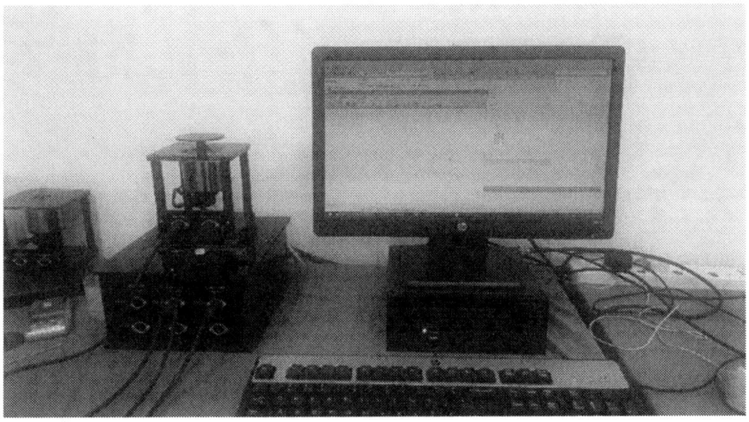

Fig 1. Experimental setup

2.2 SIPIC

This section focuses on the SIPIC controller and its characteristics. Based on previous work done by [7], Eq. 1 shows the generic template where SIPIC was developed from. $Q(S)$, q_{ss}, $f(s)$, s and n represents Laplace form of integral state, integral during steady state, function in Laplace form, Laplace domain and non-negative integer respectively. A and B are constants. Eq. 2 below shows integral component of SIPIC after various substitutions and derivations while Eq. 3 shows the error equation of SIPIC.

$$\frac{q_{ss}}{s} - Q(S) = As^n f(s) + B \tag{1}$$

$$k_i q_{ss} = v - \frac{b}{f k_T}\dot{y} + \frac{a}{f k_T} e \tag{2}$$

$$E(S) = \frac{e(0)\frac{b}{f}(s+k_i) + k_i k_T [q_{ss} - q(0)]}{\left(\frac{b}{f}s + \frac{a}{f} + k_p k_T\right)(s+k_i) - k_i k_T} \tag{3}$$

2.3 Input Reference

Based on the previous work done on SIPIC controller, the input reference was modified for the application of motor position control. Fig. 2 shows the block diagram for motor position control. From this, Eq. 4 was derived, and Eq. 5 shows the final input reference for motor position control for SIPIC controller. Fig. 3 shows the SIPIC controller block diagram for experimental setup whereas Table 1 shows the parameter used to obtain the input reference.

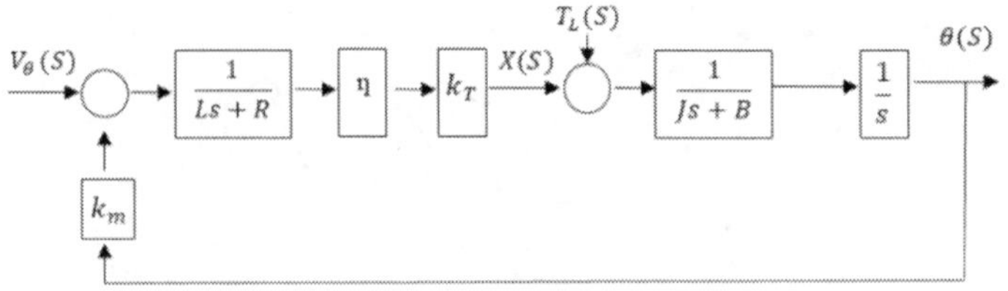

Fig. 2. Motor position control block diagram

$$X(S) = \frac{v_\theta(S) - \theta(s)k_m}{Ls+R} * \eta k_T \tag{4}$$

$$k_i q_{ss} = v_{ss} = v_\theta + \frac{LJ}{\eta k_T}\dddot{e} + \frac{B+RJ}{\eta k_T}\ddot{e} + \frac{BR}{\eta k_T}\dot{e} + \frac{\eta k_T k_m}{\eta k_T} e \tag{5}$$

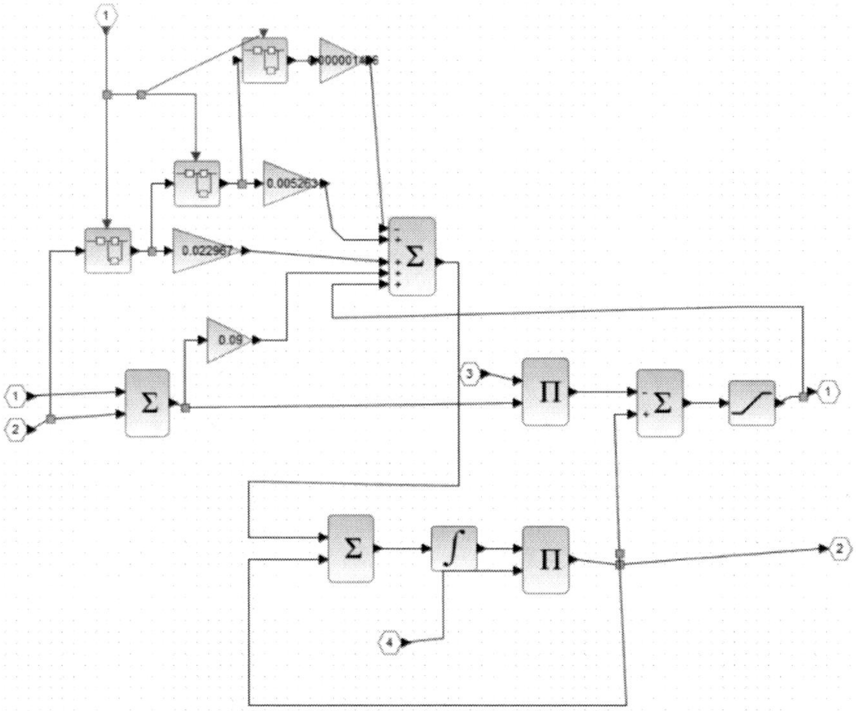

Fig. 3. SIPIC experimental setup block diagram

Table 1. Parameters for experiment [7]

Characteristics	Values
Viscous damping coefficient, *B*	2.12 x 10^{-4} kg m^2/s
Inductance, *L*	0.005 H
Moment of inertia of motor, *J*	2.14 x 10^{-5} kg m^2
Torque constant, K_T	0.09 Nm/A
Back-emf constant, K_m	0.09 Nm/A
Efficiency, η	0.8
Resistance, *R*	7.8 Ω
Moment of inertia of mild steel black plating	8.63 x 10^{-5} kgm^2

3. Results and Discussions

This section shows the results obtained from hardware testing. Necessary information was obtained from the extracted data and analysis was done to verify the performance of each controller. Comparison was done between the two controllers to verify the proposed SIPIC controller for motor position control under no load condition and under load condition.

Two different step inputs were used for testing. For the first step input, the controller travels from 0° to 90° and for the second input, the pointer travel from 270° to 90°. To put this into real life application, the first input is similar as to when changing steering direction of a vehicle while travelling in straight line while the second input is similar as to changing the direction of a vehicle from left directly to right. The stability in each case is different.

All the data obtained was tabulated below from table 2 to table 4. The table indicates

3.1 No Load

3.1.1 0 to 90

Table 2. Results for no load condition first input

		Overshoot		Rise Time		Settling Time	
K_p	K_i	PI	SIPIC	PI	SIPIC	PI	SIPIC
1	1	0.132	0.05	0.078	0.086	1.737	2.34
1	5	0.452	0.21	0.064	0.091	0.504	0.671
1	10	0.722	0.374	0.056	0.095	0.576	0.981
1	15	0.949	0.487	0.05	0.094	0.93	0.762
1	20	1.09	0.543	0.046	0.095	1.228	1.047
2	1	0.251	0.204	0.04	0.042	0.156	0.632
2	5	0.371	0.188	0.037	0.048	0.688	0.434
2	10	0.521	0.295	0.035	0.055	0.175	0.333
2	15	0.634	0.411	0.034	0.059	0.242	0.541
2	20	0.663	0.502	0.034	0.061	0.239	0.486
3	1	0.389	0.358	0.027	0.029	0.165	0.167
3	5	0.452	0.242	0.027	0.033	0.161	0.352
3	10	0.502	0.305	0.026	0.038	0.496	0.251
3	15	0.524	0.367	0.026	0.043	0.396	0.233

3	20	0.638	0.44	0.025	0.045	0.328	0.394
4	1	0.471	0.418	0.022	0.023	0.171	0.139
4	5	0.506	0.308	0.022	0.026	0.392	0.311
4	10	0.553	0.283	0.022	0.03	0.438	0.232
4	15	0.565	0.352	0.022	0.033	0.369	0.197
4	20	0.65	0.427	0.021	0.036	0.349	0.322
5	1	0.515	0.471	0.021	0.021	0.172	0.165
5	5	0.55	0.33	0.02	0.022	0.183	0.314
5	10	0.59	0.286	0.02	0.024	0.187	0.223
5	15	0.59	0.327	0.02	0.03	0.193	0.187
5	20	0.619	0.396	0.019	0.03	0.19	0.172
10	1	0.537	0.524	0.019	0.019	0.176	0.179
10	5	0.55	0.474	0.019	0.019	0.183	0.287
10	10	0.575	0.443	0.019	0.019	0.188	0.223
10	15	0.597	0.43	0.019	0.019	0.191	0.178
10	20	0.606	0.455	0.019	0.019	0.195	0.145
15	1	0.543	0.521	0.019	0.019	0.178	0.154
15	5	0.55	0.493	0.019	0.019	0.193	0.231
15	10	0.553	0.458	0.019	0.019	0.204	0.201
15	15	0.562	0.443	0.019	0.019	0.21	0.168
15	20	0.581	0.449	0.019	0.019	0.209	0.144
20	1	0.534	0.531	0.019	0.019	0.201	0.177
20	5	0.543	0.506	0.019	0.019	0.197	0.168
20	10	0.553	0.477	0.019	0.019	0.216	0.176
20	15	0.556	0.462	0.019	0.019	0.225	0.154
20	20	0.568	0.436	0.019	0.019	0.228	0.136

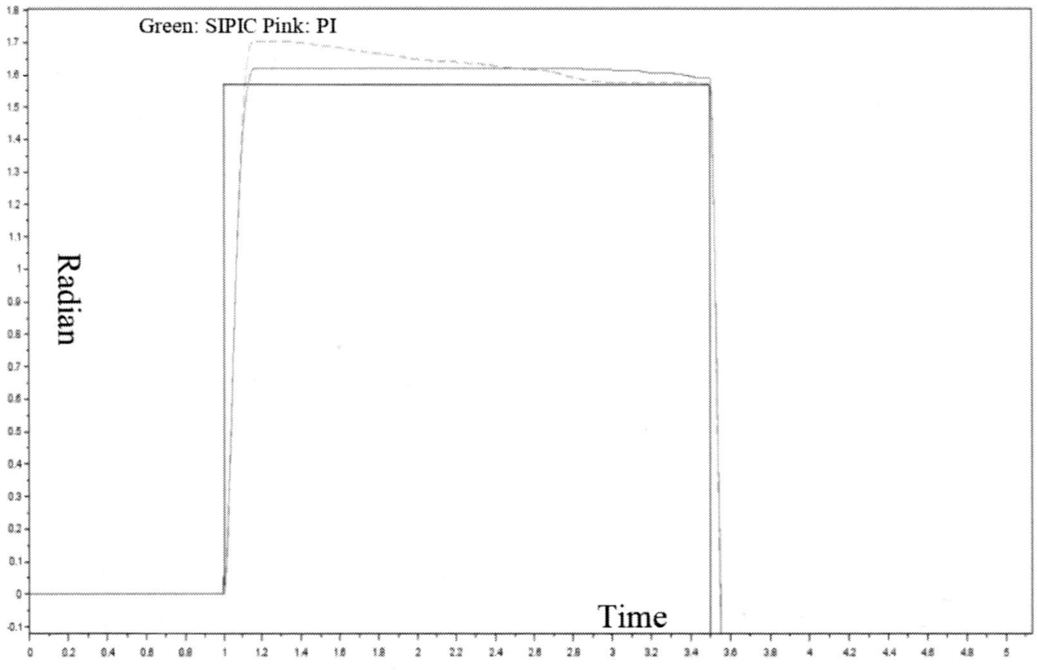

Fig. 4. Performance graph when $K_p = 1$, $K_i = 1$

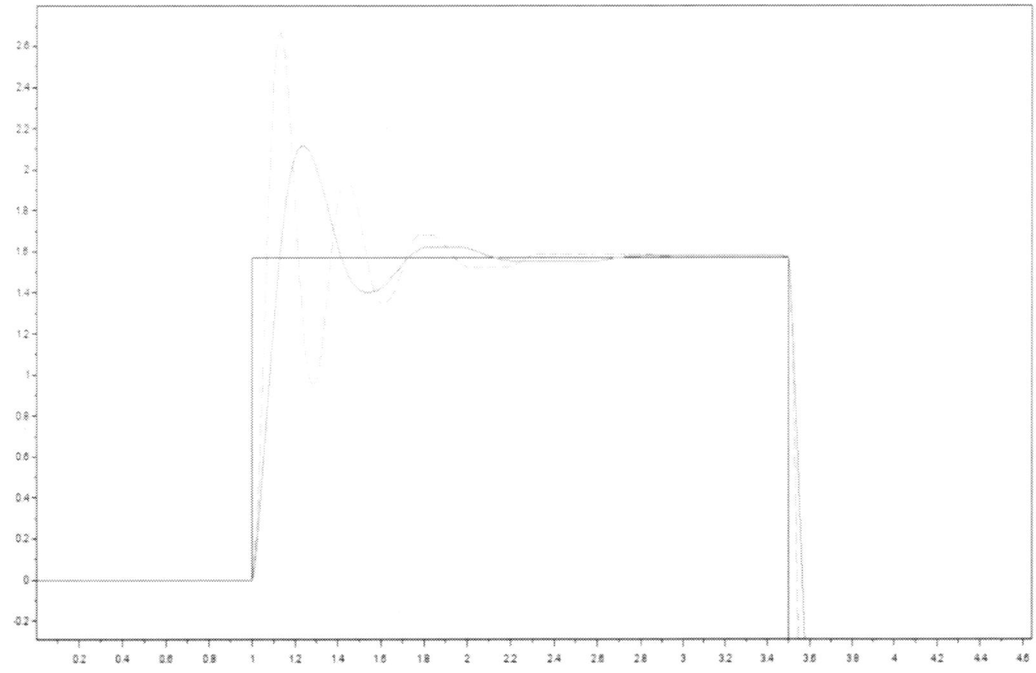

Fig 5. Performance graph when $K_p = 1$, $K_i = 20$

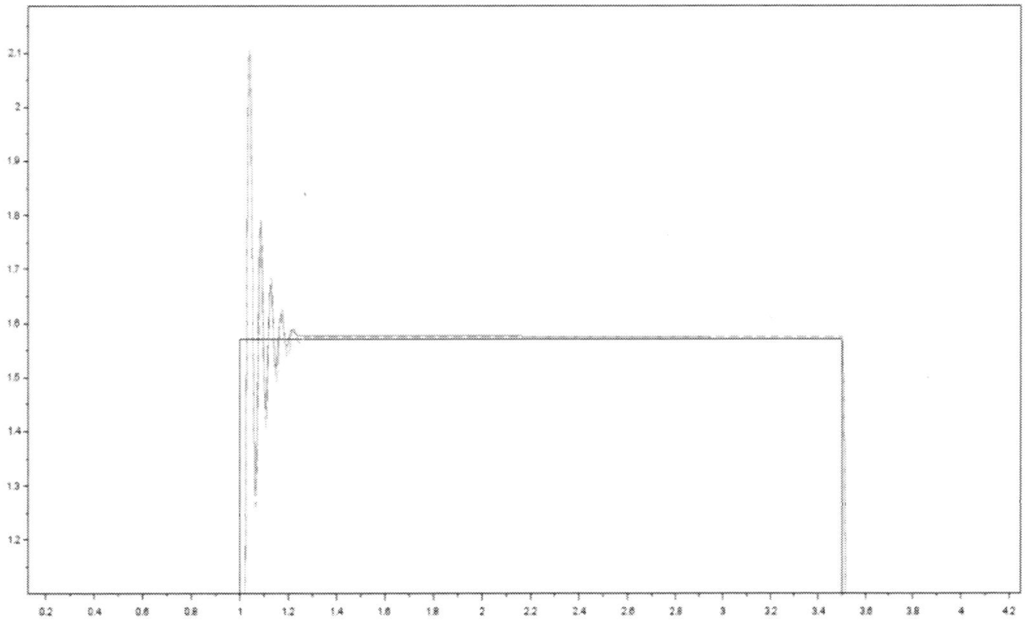

Fig. 6. Performance graph when $K_p = 20$, $K_i = 1$

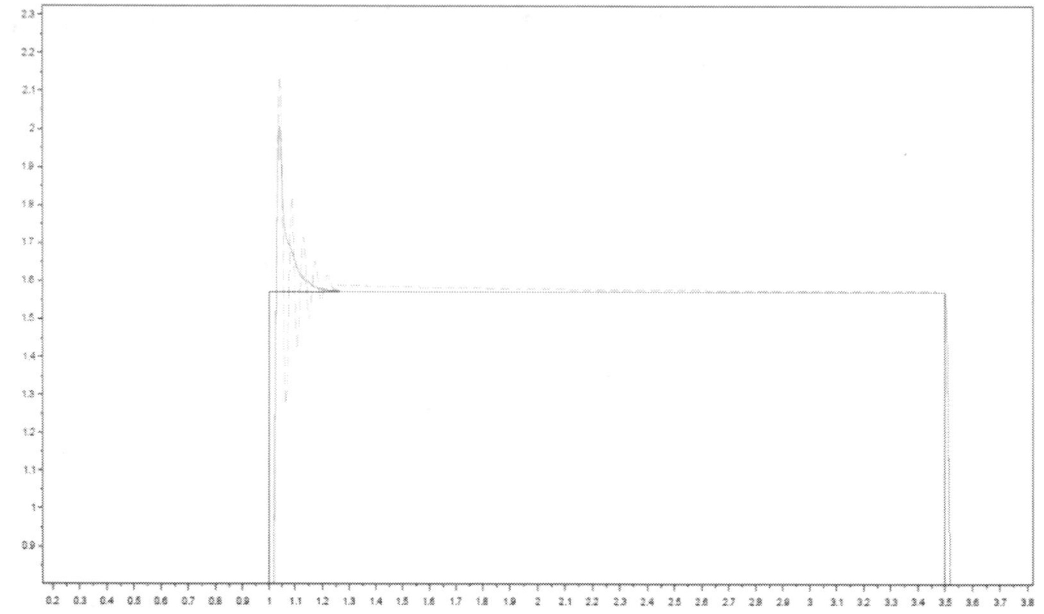

Fig. 7. Performance graph when $K_p = 20$, $K_i = 20$

Table 2 above shows the performance of both controllers when K_p is 1 and K_i is 1, 5, 10 and 15 respectively. As the integral gain increases, both controllers experience an increase in overshoot but SIPIC controller have a significantly lower overshoot especially at higher K_i value. Moving on to the rise time of the controllers, different pattern is seen in each controller. For PI controller, rise time decrease as the integral gain increase but for SIPIC, rise time increase alongside the integral gain. However, SIPIC shows higher rise time compared to PI controller. Lastly, the settling time of the controllers is compared. Under comparison, when both the gain is 1, SIPIC shows a higher settling time. When the integral gain is increased to 5, both controller experience a big drop in settling time but as the integral gain continue to increase, settling time gradually increase in PI controller but experience minor up and down changes in SIPIC controller.

Next, K_i is kept at 1 and K_p is increased from 1 to 5, it is seen from results that the overshoot for both controllers increased significantly. Rise time and settling time decreases for both controller. Both controllers now experienced similar overshoot, rise time as well as settling time. As the integral value increases, PI controller experience slight increase in overshoot but SIPIC controller experience slight decrease in overshoot. The rise time of PI controller remain constant whereas SIPIC controller has slight increase in rise time when the integral gain increases. For settling time, PI controller experience slight increase. For SIPIC controller, when the K_i value is increase from 1 to 5, there is an increase in settling time but setting time experience a descent when the K_i value is further increased.

The K_p value was then increased to 10. The increased of K_p value causes the same effect to both controllers, increase in overshoot and settling time but decrease in rise time. SIPIC controller still has a lower overshoot than PI controller, similar settling time and both controller now have the same rise time. As the integral gain increase, overshoot for PI controller has a slight increase but a slight decrease is experienced by SIPIC controller. There are no changes in rise time for both controller as this is the lowest possible rise time due to hardware limitation. PI controller continue to experience slight increase in settling time as K_i increases while settling time for SIPIC controller first increase then decreases, like where the K_p is 5.

Lastly, K_p was increased to15. The rise time for both controllers is still 0.019s, same as before. The increase in K_p causes a slight increase in overshoot and settling time of PI controller but a slight decrease in overshoot and settling time of SIPIC controller. Increasing K_i value continue to cause slight increase in overshoot of PI controller but decrease in overshoot for SIPIC controller. Same pattern is also seen in settling time of both controller. The increase in K_i value results in increase of settling time for PI controller but for SIPIC controller, the settling time first increases then start to decrease when the K_i value is further increased from 5.

Under no load condition first input, both controller shows very close performance. The proposed SIPIC controller have lower overshoot at all parameter as compared to conventional PI controller. SIPIC controller experiences higher rise time at low K_p value but both controller experience similar rise time when K_p value is further increased. Comparing the condition when both K_p and K_i is 15 as shown in Fig. 6, SIPIC controllers outperform PI controller by having a lower overshoot as well as lower settling time. This is because at higher K_i value, SIPIC controller experience decrease in both overshoot and settling time but PI controller experience increase in both components.

3.1.2 270 to 90

Table 3. Results for no load condition second input

		Overshoot		**Rise Time**		**Settling Time**	
K_p	K_i	**PI**	**SIPIC**	**PI**	**SIPIC**	**PI**	**SIPIC**
1	1	0.358	0.213	0.069	0.076	2.209	1.942
1	5	0.986	0.468	0.061	0.089	0.584	0.647
1	10	1.486	0.785	0.053	0.091	0.499	0.868
1	15	1.756	0.967	0.049	0.095	1.223	1.137
1	20	2.149	1.087	0.044	0.094	1.093	1.377
2	1	0.433	0.427	0.036	0.039	0.191	1.302
2	5	0.704	0.44	0.033	0.046	0.941	0.523
2	10	0.983	0.619	0.034	0.055	0.612	0.337
2	15	1.156	0.801	0.033	0.058	0.561	0.523
2	20	1.285	1.002	0.032	0.06	0.448	0.702
3	1	0.616	0.54	0.032	0.033	0.169	1.288
3	5	0.769	0.531	0.031	0.035	1.023	0.515
3	10	0.923	0.644	0.03	0.037	0.676	0.309
3	15	1.055	0.776	0.03	0.042	0.528	0.419
3	20	1.244	0.936	0.03	0.043	0.424	0.391
4	1	0.669	0.631	0.031	0.031	0.216	1.005
4	5	0.782	0.587	0.03	0.032	1.11	0.501
4	10	0.901	0.682	0.03	0.033	0.857	0.318
4	15	1.03	0.851	0.03	0.034	0.649	0.217
4	20	1.15	0.952	0.03	0.036	0.552	0.333

5	1	0.697	0.685	0.03	0.03	0.184	0.943
5	5	0.782	0.672	0.03	0.03	1.212	0.485
5	10	0.905	0.769	0.029	0.031	0.916	0.328
5	15	0.964	0.873	0.029	0.03	0.721	0.223
5	20	1.043	1.011	0.029	0.03	0.618	0.298
10	1	0.682	0.682	0.029	0.03	0.196	0.496
10	5	0.71	0.691	0.029	0.03	0.255	0.435
10	10	0.776	0.741	0.029	0.03	0.98	0.307
10	15	0.826	0.801	0.029	0.03	0.943	0.237
10	20	0.864	0.892	0.029	0.03	0.864	0.185
15	1	0.694	0.682	0.029	0.03	0.214	0.224
15	5	0.704	0.688	0.029	0.03	0.223	0.36
15	10	0.732	0.691	0.029	0.03	0.631	0.267
15	15	0.769	0.713	0.029	0.03	0.935	0.219
15	20	0.798	0.744	0.029	0.03	0.912	0.182
20	1	0.691	0.691	0.029	0.03	0.219	0.199
20	5	0.697	0.688	0.029	0.03	0.24	0.267
20	10	0.719	0.704	0.029	0.03	0.291	0.235
20	15	0.738	0.694	0.029	0.03	0.701	0.204
20	20	0.769	0.716	0.029	0.03	0.866	0.174

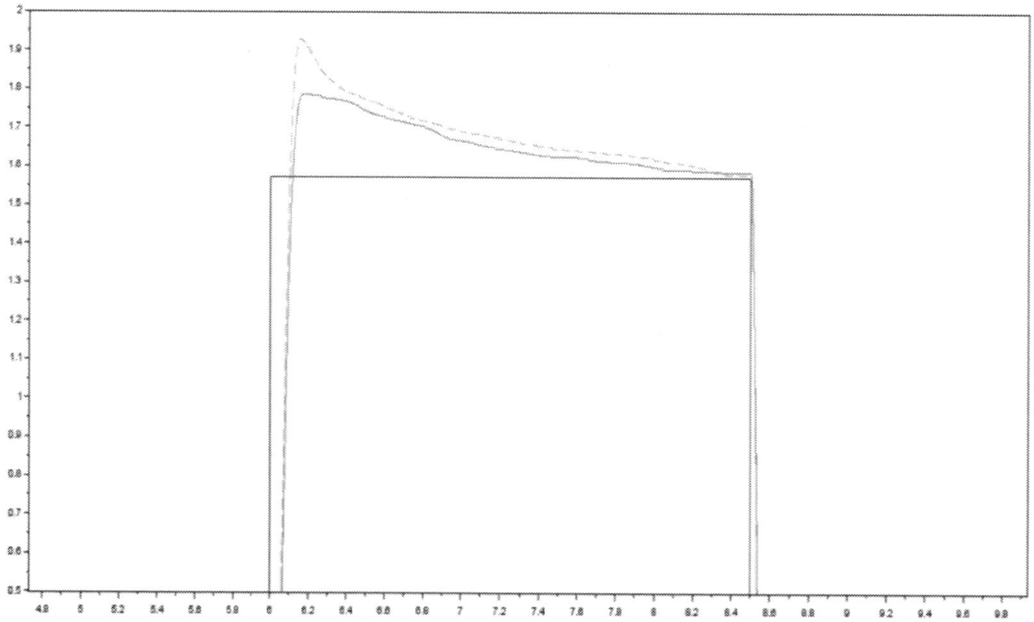

Fig. 8. Performance graph when $K_p = 1$, $K_i = 1$

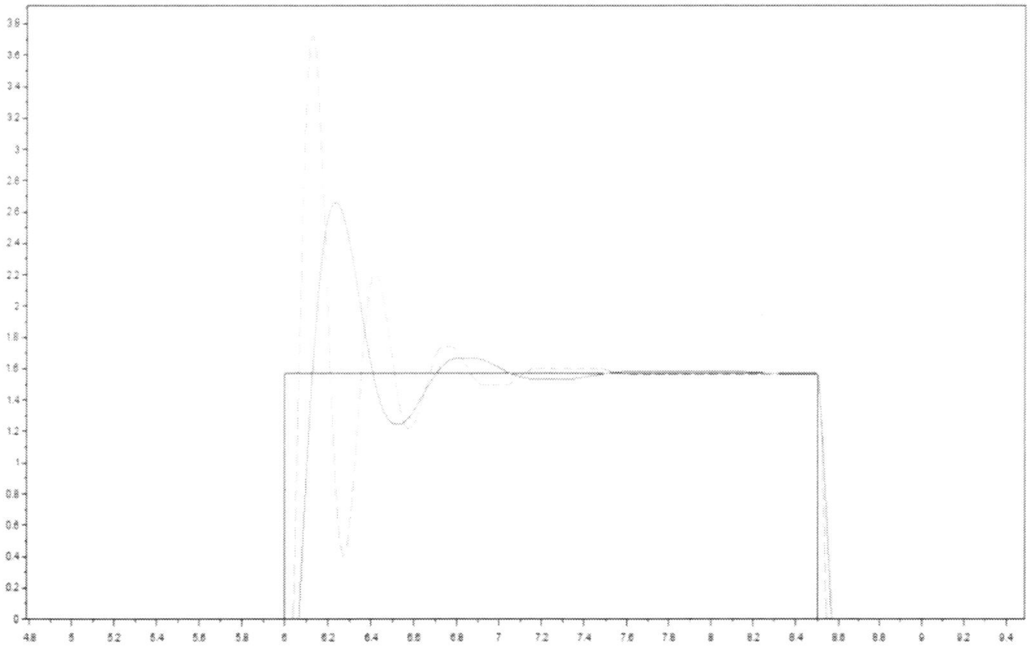

Fig. 9. Performance graph when $K_p = 1$, $K_i = 20$

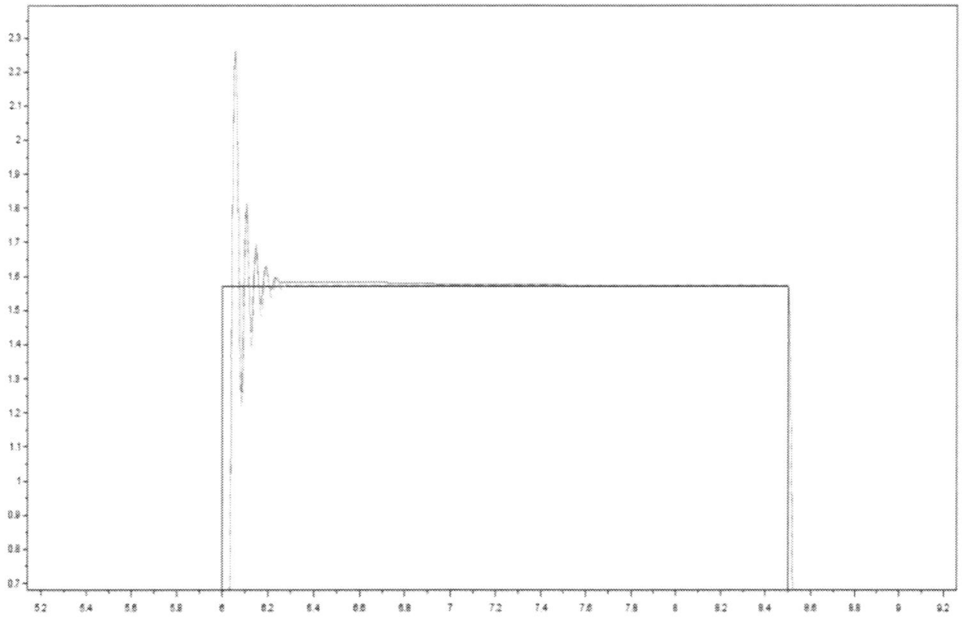

Fig. 10. Performance graph when $K_p = 20$, $K_i = 1$

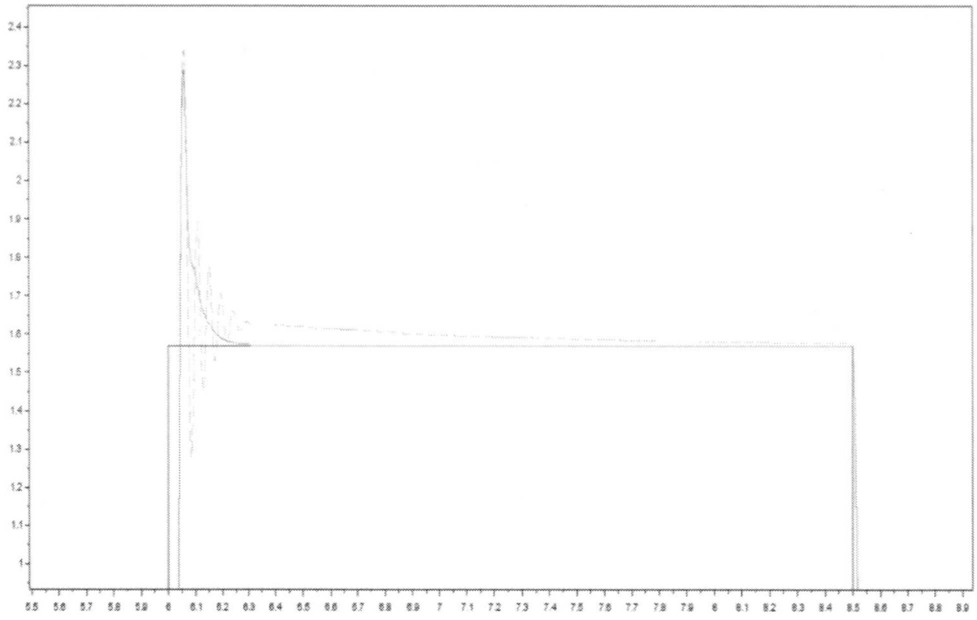

Fig. 11. Performance graph when $K_p = 20$, $K_i = 20$

As shown in Fig. 6, when both value is 1, SIPIC controller has lower overshoot and settling time as compared to PI controller, but higher rise time. When the K_i value is increased, both controller experience an increase in overshoot but the increase in PI controller are far more significant. Rise time for PI controller decreases as the value of Ki increases but SIPIC controller experience higher rise time at higher K_i value. For settling time, both controller first experience a decrease before increasing again as Ki value is further increased. When K_p is 1 and K_i is 15, SIPIC controller performs slightly better by having lower overshoot and settling time but higher rise time as compared to PI controller.

Moving on, K_p value was increased from 1 to 5. Compared to when K_p value is 1, both controller now have higher overshoot, lower rise time and lower settling time. As the value of K_i increases, both controller experience an increase in overshoot. The changes in rise time are insignificant in both controller. For settling time, SIPIC controller goes through a steady decrease in settling time but an increase then decrease in the PI controller. As the K_i value increases, the settling time of the SIPIC controller starts to be significantly lower than the PI controller.

When the K_p is further increased to 10 and 15, the rise time of both controllers remain the same as it is the lowest rise time possible. Both controller also experience minor changes in overshoot with SIPIC controller still having lower overshoot than PI controller. For settling time, SIPIC controller continue to show higher settling time than PI controller when the K_i value is small but SIPIC controller then outperform PI controller when the K_i value increases.

3.2 With Load

3.2.1 0 to 90

Table 4. Results for loading condition first input

		Overshoot		Rise Time		Settling Time	
K_p	K_i	PI	SIPIC	PI	SIPIC	PI	SIPIC
1	1	0.543	0.521	0.096	0.1	0.532	1.18
1	5	1.121	0.697	0.083	0.097	1.314	1.423
1	10	1.624	0.876	0.075	0.095	N/A	1.804
1	15	N/A	0.923	0.068	0.09	N/A	2.012
1	20	N/A	0.942	0.063	0.09	N/A	1.986
2	1	0.81	0.697	0.063	0.064	1.595	0.667
2	5	0.917	0.766	0.058	0.06	0.967	0.631
2	10	1.165	0.854	0.054	0.061	1.157	0.957
2	15	1.354	0.905	0.051	0.061	N/A	1.038
2	20	1.476	0.945	0.049	0.06	N/A	1.449
3	1	0.813	0.807	0.046	0.047	0.586	0.579
3	5	0.958	0.848	0.044	0.047	0.736	0.609
3	10	1.043	0.927	0.043	0.046	0.874	0.831
3	15	1.212	0.952	0.043	0.046	1.413	0.963
3	20	1.26	0.952	0.041	0.046	1.722	1.034
4	1	0.917	0.886	0.039	0.04	0.626	0.615
4	5	0.967	0.92	0.039	0.04	0.709	0.65
4	10	1.052	0.955	0.037	0.039	0.824	0.642
4	15	1.134	0.977	0.038	0.039	0.951	0.643
4	20	1.256	1.005	0.036	0.039	1.044	0.912
5	1	0.895	0.939	0.037	0.037	0.593	0.762
5	5	0.961	0.971	0.039	0.037	0.718	0.65
5	10	1.037	1.037	0.036	0.037	0.73	0.658
5	15	1.106	1.065	0.036	0.037	0.815	0.667
5	20	1.168	1.118	0.036	0.037	0.929	0.672
10	1	0.889	0.92	0.035	0.035	0.652	0.664
10	5	0.933	1.074	0.036	0.035	0.717	0.644
10	10	0.974	1.209	0.035	0.035	0.791	0.68
10	15	1.002	1.344	0.035	0.035	0.802	0.683
10	20	1.043	1.426	0.035	0.035	0.799	0.858
15	1	0.908	0.917	0.035	0.035	0.724	0.711
15	5	0.92	1.046	0.035	0.035	0.772	0.73
15	10	0.942	1.172	0.035	0.035	0.791	0.764
15	15	0.958	1.25	0.035	0.035	0.792	0.789

15	20	0.989	1.319	0.035	0.035	0.898	0.964
20	1	0.876	0.895	0.035	0.035	0.785	0.834
20	5	0.908	0.999	0.035	0.035	0.869	0.833
20	10	0.911	1.09	0.035	0.035	0.875	0.834
20	15	0.952	1.178	0.035	0.035	0.883	0.876
20	20	0.964	1.209	0.035	0.035	0.958	1.119

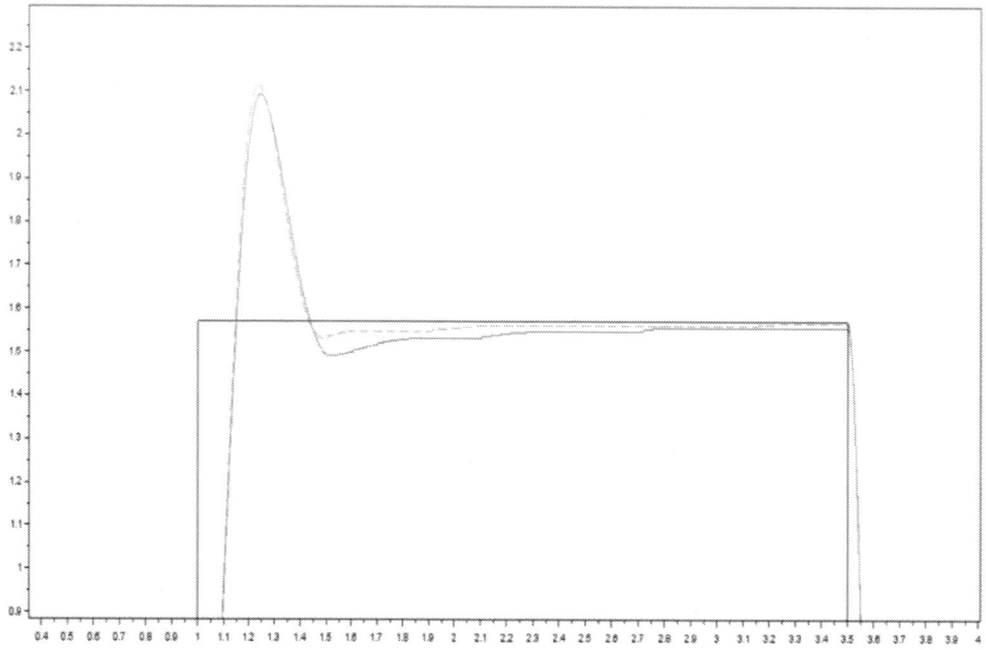

Fig. 12. Performance graph when $K_p = 1$, $K_i = 1$

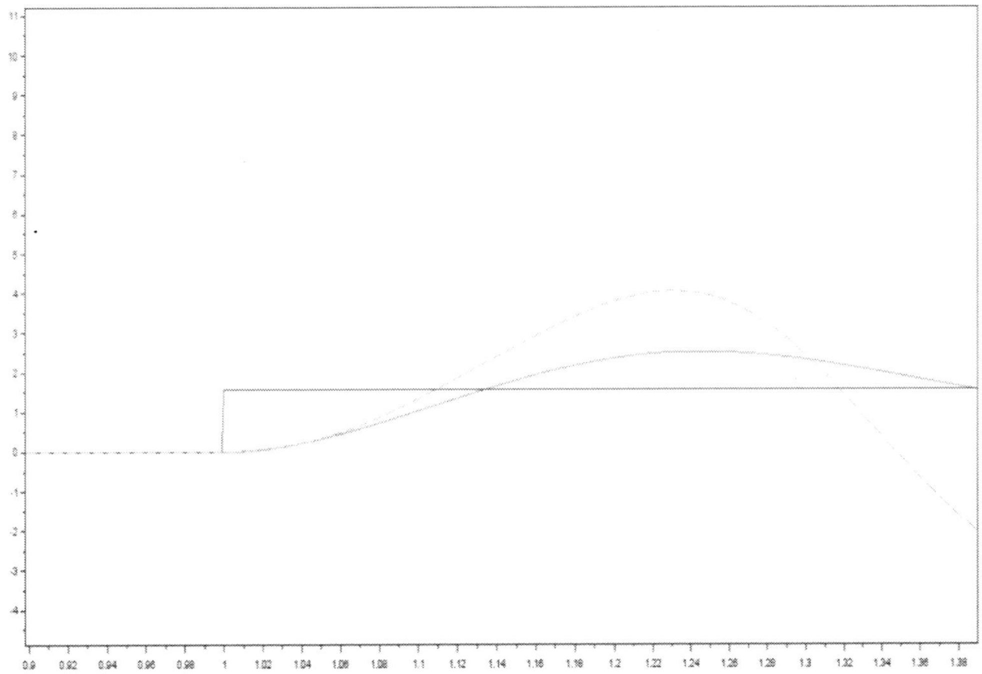

Fig. 13. Performance graph when $K_p = 1$, $K_i = 20$

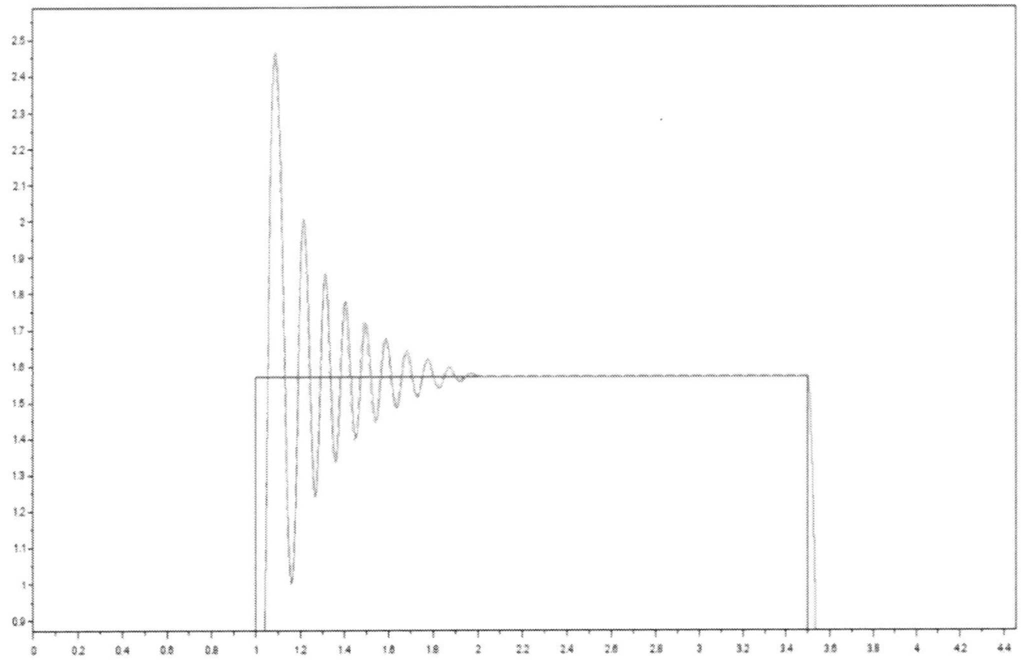

Fig. 14. Performance graph when $K_p = 20$, $K_i = 1$

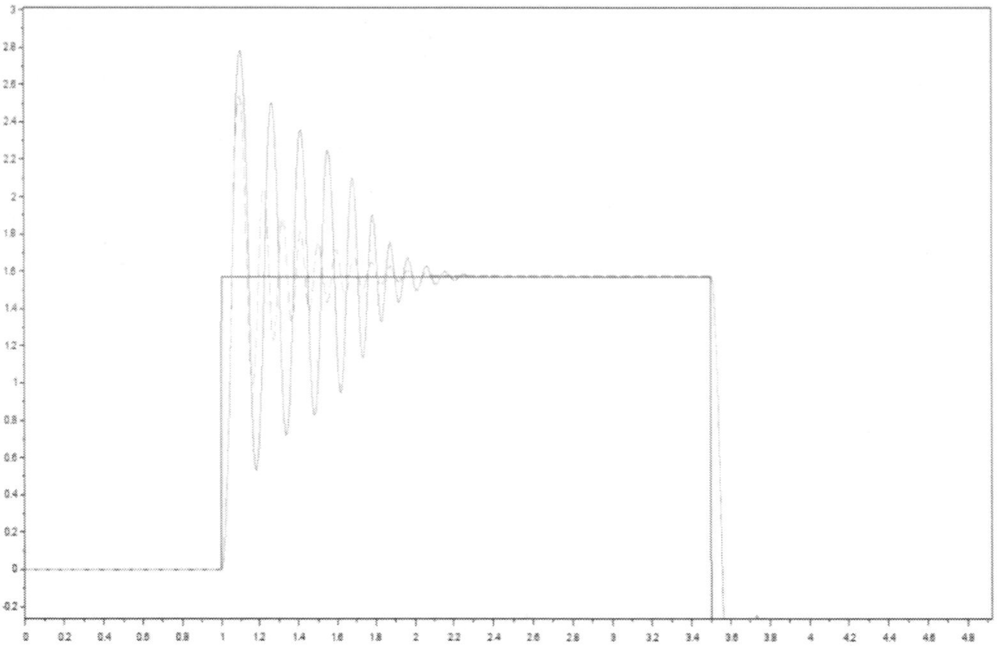

Fig. 15. Performance graph when $K_p = 20$, $K_i = 20$

Table 4 shows results of controller first input under loading condition. When both K_p and K_i is 1 as shown in Fig. 10, SIPIC controller showed lower overshoot, shorter rise time and settling time as compared to PI controller. As the K_i value increases, both controller experience an increase in overshoot and decrease in rise time. Both controller also experience an increase in the settling time of the controller but for PI, the settling time were unable to be determined because the controller have not settled in the given step time. When K_p is 1 and K_i is 15 as shown in Fig. 11, PI controller shows very high oscillation and the overshoot, rise time and settling time were all not able to be determined. SIPIC controller on the other hand, still showed a very stable performance. This proves that SIPIC controller is more stable than conventional PI controller.

As the K_p value is increases from 1 to 5, both controller experience higher overshoot but lower rise time. PI controller experienced slightly longer slightly time but SIPIC controller experienced shorter settling time. As the value of K_i increases, the overshoot for both controller increases and both controller showed similar overshoot. The difference in rise time of both controllers is insignificant while SIPIC controller shows shorter settling time as K_i increases.

The K_p value is increased to 10 with K_i value being 1, 5, 10 and 15 respectively. Compared to when K_p was 5, both controller experience slight decrease in overshoot and rise time. PI controller has a longer settling time than before while SIPIC controller now has shorter settling time. As K_i increases, overshoot for both controllers increases while PI controller outperform SIPIC controller by having lower overshoot. Both controller shows similar rise time as this is the lowest possible value. Both controllers also experience longer settling time as K_i increases but SIPIC performs slightly better by having slightly shorter settling time.

Lastly, K_p value was increased to 15. Rise time remains the same for both controllers. The increase in K_p value causes minor changes to overshoot in both controller while increasing the settling time for both controller. As K_i increases, similar pattern as before can be seen. Both overshoot and settling time increases alongside the K_i value and PI controller still have lower overshoot while settling time are similar for both controller.

3.2.2 270 to 90

Table 5. Results for loading condition second input

		Overshoot		Rise Time		Settling Time	
K_p	K_i	PI	SIPIC	PI	SIPIC	PI	SIPIC
1	1	1.209	1.033	0.09	0.098	N/A	0.647
1	5	2.18	1.429	0.101	0.092	1.507	1.137
1	10	3.251	1.586	0.052	0.094	N/A	1.679
1	15	N/A	1.778	0.052	0.091	N/A	1.756
1	20	N/A	1.765	0.052	0.091	N/A	2.07
2	1	1.47	1.316	0.059	0.06	N/A	0.682
2	5	1.765	1.586	0.057	0.058	0.938	0.789
2	10	2.224	1.674	0.054	0.059	1.525	1.046
2	15	2.673	1.759	0.054	0.059	N/A	1.238
2	20	3.349	1.85	0.052	0.058	N/A	1.638
3	1	1.467	1.495	0.053	0.053	0.757	1.529
3	5	1.75	1.743	0.053	0.053	1.395	0.758
3	10	1.988	1.982	0.053	0.053	1.016	0.978
3	15	2.321	2.139	0.052	0.052	1.42	1.102
3	20	2.557	2.287	0.052	0.052	1.97	1.269
4	1	1.436	1.514	0.052	0.052	0.723	1.436
4	5	1.674	1.825	0.052	0.053	1.541	0.756
4	10	1.875	2.152	0.052	0.053	0.889	0.88
4	15	2.067	2.447	0.052	0.052	1.123	0.916
4	20	2.347	2.598	0.051	0.052	1.185	1.147
5	1	1.414	1.527	0.052	0.052	0.73	0.824
5	5	1.599	1.9	0.051	0.052	0.788	0.791
5	10	1.803	2.281	0.052	0.051	1.081	0.907
5	15	1.985	2.516	0.052	0.052	0.968	0.94
5	20	2.18	2.651	0.052	0.051	1.062	1.064
10	1	1.388	1.495	0.052	0.051	0.769	0.72
10	5	1.508	1.74	0.052	0.051	0.858	0.774
10	10	1.624	1.951	0.052	0.051	1.171	0.834
10	15	1.712	2.108	0.052	0.051	1.034	0.93
10	20	1.806	2.19	0.052	0.051	1.006	1.144
15	1	1.42	1.442	0.052	0.051	0.837	0.83

15	5	1.467	1.655	0.052	0.051	0.952	0.855
15	10	1.527	1.803	0.052	0.051	0.978	0.895
15	15	1.58	1.916	0.052	0.051	1.112	0.999
15	20	1.687	1.966	0.052	0.051	1.094	1.198
20	1	1.379	1.414	0.052	0.051	0.893	0.946
20	5	1.436	1.52	0.052	0.051	0.948	0.954
20	10	1.464	1.69	0.052	0.051	1.045	0.966
20	15	1.539	1.794	0.052	0.051	1.127	1.059
20	20	1.596	1.809	0.052	0.051	1.153	1.373

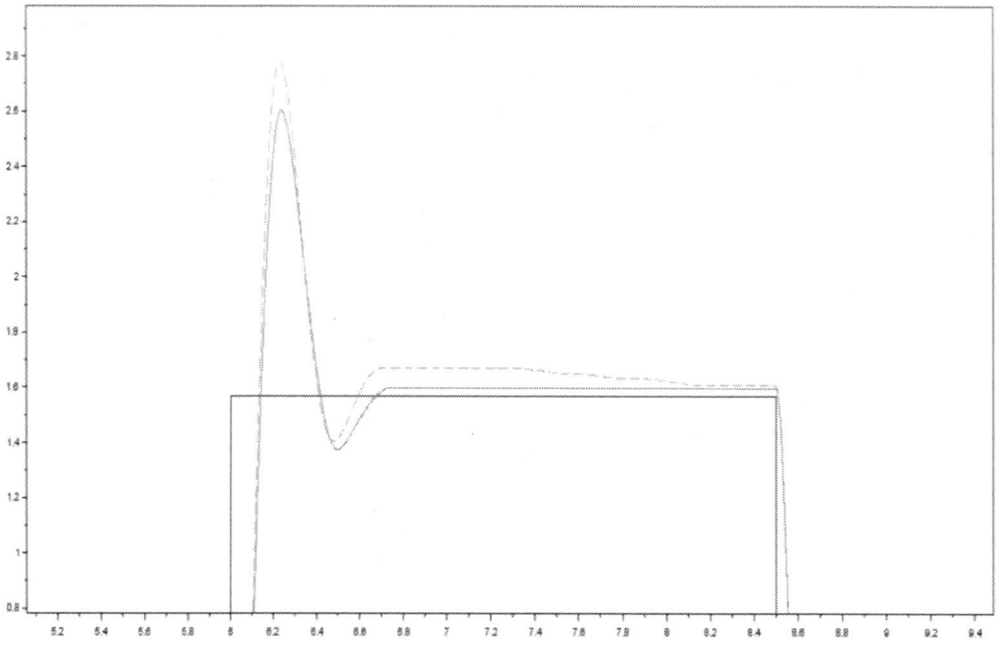

Fig. 16. Performance graph when $K_p = 1$, $K_i = 1$

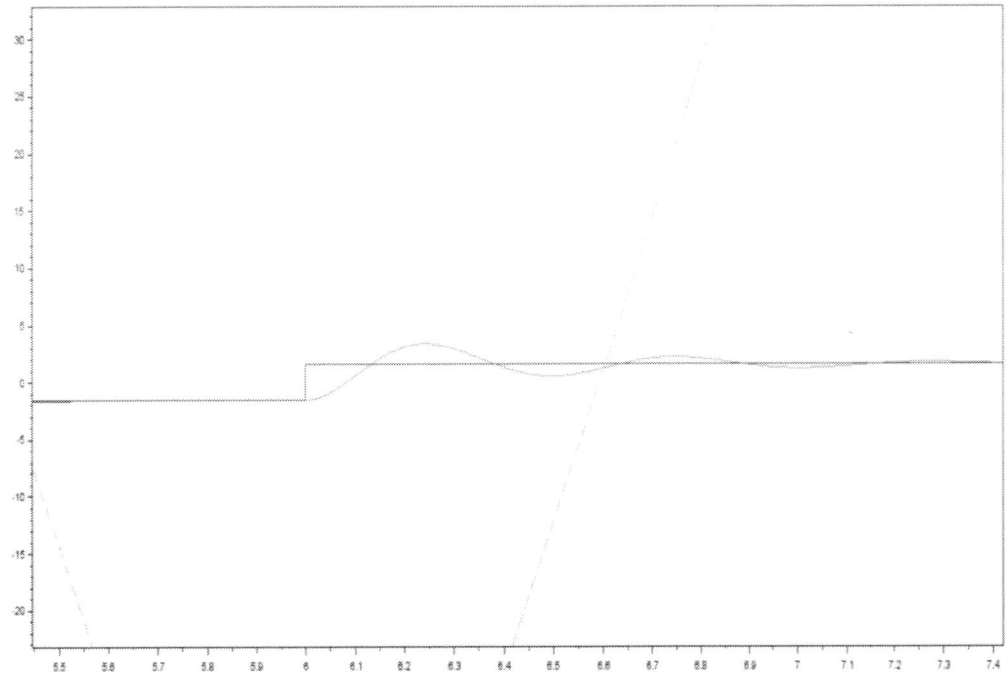

Fig. 17. Performance graph when $K_p = 1$, $K_i = 20$

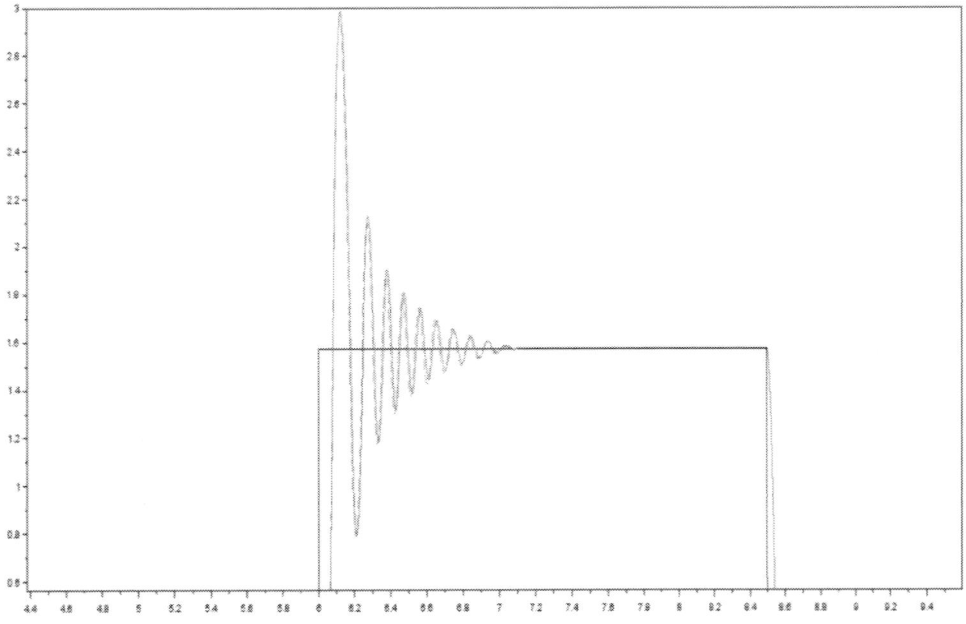

Fig. 18. Performance graph when $K_p = 20$, $K_i = 1$

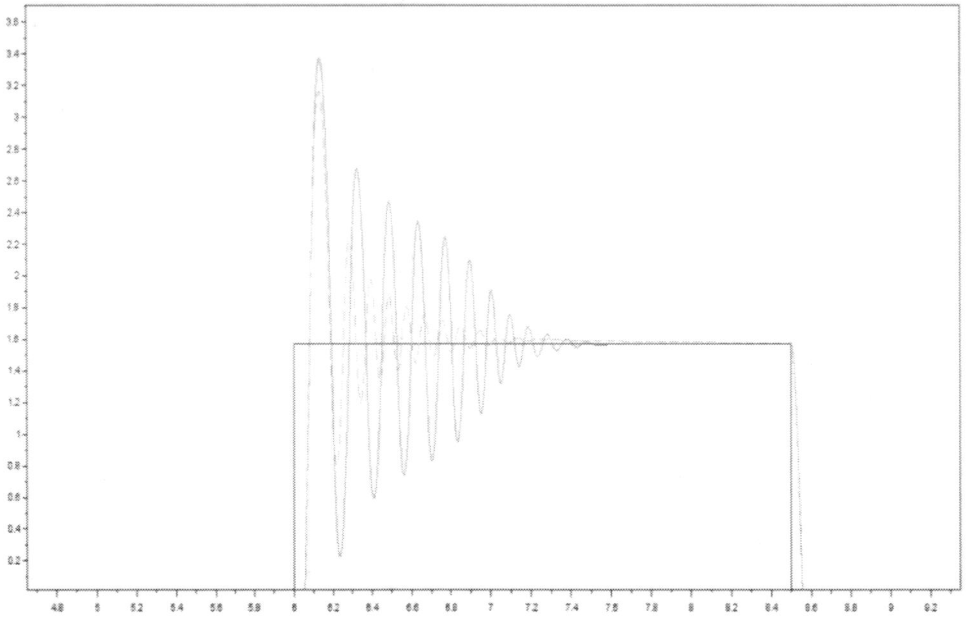

Fig. 19. Performance graph when $K_p = 20$, $K_i = 20$

Table 5 show the results obtained for second input under loading condition. When both K_p and K_i is 1 as shown in Fig. 13, PI controller shows higher overshoot, longer settling time but slightly shorter rise time as compared to SIPIC controller. As K_i value increases, a major increase in overshoot can be seen in the PI controller and while SIPIC controller also experience an increase in overshoot, the increase is subtle as compared to PI controller. A very minimal spike was seen in the PI controller when Ki value was increased from 1 to 5 but the rise time then decreases when the value was further increased to 10. For settling time, a steady increase can be seen in the SIPIC controller as the K_i value increases. However, for the PI controller, it was unable to settle inside the desired range within the step time given. Fig. 14 shows the controller performance when K_p is 1 and K_i is 15, the overshoot and settling time experienced by the PI controller is far higher than the SIPIC controller.

Moving on to $K_p = 5$, the increase in K_p value causes the overshoot of both controller to increase and an increase in rise time can be seen in both controller. A more stable performance is seen in the PI controller as the controller can settle in the given step time while the SIPIC controller has a slightly longer settling time due to the increase in K_p value. As the K_i value increases, overshoot starts to increase in both controller and PI controller has a lower overshoot as compared to SIPIC controller. Rise time of both controllers are similar while PI controller shows slightly shorter settling time as compared to SIPIC controller.

When the K_p value is increased to 10, both controller experienced a decrease in overshoot but slight decrease in the settling time. Rise time remains the same as it is the lowest possible rise time due to limitation of hardware. The pattern for overshoot is similar as before where overshoot increase alongside the K_i value with PI controller having lower overshoot than SIPIC controller. Similar trend can also be found in settling time of both controller. As K_i increases, both controller starts to have longer settling time and SIPIC controller perform slightly better in this aspect.

As the K_p value further increases to 15, PI controller continue to see an increase in overshoot while SIPIC controller experience decrease in overshoot. Rise time remains the same for both controller and settling time increased for both controller. The same trend is seen for both overshoot and settling time of both controllers as K_i value increases. When both K_p and K_i is 15 as seen in Fig. 15, PI controller performs better in terms of overshoot while SIPIC controller shows more ideal performance in terms of settling time.

4. Conclusion

To conclude this paper, a new type of anti-windup PI controller was used for the application of motor position control and the performance of the proposed controller was used to compare with the performance of a conventional PI controller. The controllers were tested under no load and under loading condition while two different inputs were used to test the controllers: 0° to 90° and 270° to 90°. Results show that under no load condition, SIPIC controller performs better than PI controller by having lower overshoot and lower settling time across various gains. Under loading condition, PI controller showed to be unstable under some of the gains as it experience large amount of oscillations and overshoot as well as very long settling time whereas under the same gain, the proposed SIPIC controller still shows acceptable performance. As a conclusion, the proposed SIPIC controller shows a more reliable performance as compared to PI controller when experiment was conducted. A wider range of gain can be used for the SIPIC controller whereas PI controller only performs well under a small range of gain.

As this paper only compares the performance between the proposed SIPIC controller and conventional PI controller, the future work includes further validating the performance of SIPIC controller in motor position control by comparing it with other anti-windup PI controller. SIPIC controller can also be further refined to achieve lower overshoot, shorter rise time and settling time. Wider range of gain can also be used to test the performance of SIPIC controller to further test the stability of the controller.

References

1. A. . Benmakhlouf, A. . Louchene, and D. . Djarah, "Fuzzy logic and modified crisp logic applied to a DC motor position control," *Control Intell. Syst.*, **38**, **no. 3**, 173–180 (2010).
2. N. Bianchi and M. Dai Pre, "Active power filter control using neural network technologies," *IEE Proceedings-Electric Power Appl.*, **150**, **no. 2**, 139–145, (2003).
3. V. Slapak, K. Kyslan, and F. Durovsky, "Position Controller for PMSM Based on Finite Control Set Model Predictive Control," **no. 1**, 17–22 (2016).
4. K. Orman, A. Basci, A. Derdiyok, E. Engineering, and E. Engineering, "Speed and Direction Angle Control of Four Wheel Drive Skid-Steered Mobile Robot by Using Fractional Order PI Controller," 14–20 (2016).
5. S.-Y. Chen and C.-S. Chia, "Precision Position Control of a Voice Coil Motor Using Self-Tuning Fractional Order Proportional-Integral-Derivative Control," *Micromachines*, **7**, **no. 11**, 207 (2016).
6. K. Kawamura and Y. Ishida, "Approach to New Model Recovery Anti-windup Scheme with PID Controller," *2015 3rd Int. Conf. Artif. Intell. Model. Simul.*, 208–212 (2015).
7. C. T. Ling, "A New Anti-windup PI control for Motor Speed Application," **no. June** (2016).
8. K. Sakai and Y. Ishida, "*A Design of an Improved Anti-Windup Control Using a PI*

Controller Based on a Pole Placement Method," **no. 2**, 1–7 (2016).
9. S. C. Pratama, E. Susanto, and A. S. Wibowo, "Design and implementation of water level control using gain scheduling PID back calculation integrator Anti Windup," *ICCEREC 2016 - Int. Conf. Control. Electron. Renew. Energy, Commun., Conf. Proc.*, 101–104 (2017).

Design and Fabrication of Small Vertical-Take-Off-Landing Unmanned Aerial Vehicle

Yasir Ashraf Abd Rahman[1], *Mohammad Taghi* Hajibeigy[1,*], *Abdulkareem* Shafiq Mahdi Al-Obaidi[1], and *Kean How* Cheah[2]

[1]School of Engineering, Taylor's University, Malaysia

[2]School of Engineering and Physical Sciences, Heriot-Watt University, Malaysia

Abstract. Modern UAVs available in the market have well-developed to cater to the countless field of application. UAVs have their own limitations in terms of flight range and manoeuvrability. The traditional fixed-wing UAVs can fly for long distance but require runways or wide-open spaces for take-off and landing. On the other hand, the more trending multirotor UAVs are extremely manoeuvrable but cannot be used for long-distance flights because of their slower speeds and relatively higher consumption of energy. This study proposed the implementation of hybrid VTOL UAV which has the manoeuvring advantage of a multirotor UAV while having the ability to travel fast to reach a further distance. The design methodology and fabrication method are discussed extensively which would be followed by a number of flight tests to prove the concept. The proposed UAV would be equipped with quadcopter motors and a horizontal thrust motor for vertical and horizontal flight modes respectively.

Keywords: UAV, VTOL, RC aircraft, drone

1 Introduction

Unmanned Aerial Vehicles (UAV) are aircrafts which are controlled remotely via direct radio frequency communication or are equipped with autonomous devices to performed autonomous mission flights. The active development in the field of UAV in recent years has led to its increase in versatility in terms of its functionality and range of application. Early UAVs development is generally divided into two separate types, Fixed Wing UAV (FWUAV), and rotorcraft UAV (RUAV). FWUAVs typically have longer endurance and thus can reach further than RUAV in a single flight, but require runways for take-offs and landings. RUAVs on the other hand, are more manoeuvrable and able to take-off and land vertically without any runway but has much lesser endurance due to their low efficiency in power consumption and low speed.

[1] Corresponding author: MohammadTaghi.Hajibeigy@Taylors.edu.my

Therefore, to bridge the performance gap between RUAVs and FWUAVs, the solution is to have a hybrid Vertical-Take-Off-Landing (VTOL) aircraft. This type of aircraft will have the advantages of both FWUAV and RUAV and to a substantial extent should be able to eliminate their shortcomings. In the aviation industry, major companies have already successfully mass-produced and commercialized numerous models of VTOL aircraft. A famous example of such aircraft is the Bell Boeing V-22 Osprey which combined the features of fixed wing aircraft and rotary wing aircraft via the concept of tilt rotor mechanism. The tilt-rotor mechanism is only one of the concepts that were proven for successful VTOL flights. Other researchers have also explored other concepts to incorporate into similar hybrid VTOL aircrafts. In the small to medium UAV applications, particularly in the Radio-Controlled (RC) drone world, only a few small-scale projects have been successful in flying VTOL aircrafts [1]. Also, those projects that are successful experienced great difficulty in optimizing the transition flight from the rotor to fixed-wing flight and vice versa. In addition, manufacturers are still developing their own VTOL UAV for aerial-photography drone market, but they are still yet to be commercialized.

The hybrid VTOL UAV design, however, has its own shortcomings compared to both FWUAV and RUAV. The addition of extra propulsion system to cater for horizontal and vertical flights resulted in significant extra weight carried by the vehicle. Normally, a conventional RUAV model would only need four brushless motors for its vertical propulsion and other manoeuvres. An addition of horizontal flight would require at least an extra motor for horizontal propulsion as well as a number of servo motors for different manoeuvres. There is also a notable increase in the complexity of the design and control mechanism. The common VTOL UAV design can be classified into three types. This include 1) TYPE1, which is a fixed wing aircraft with tilting rotor mechanism, 2) TYPE2, which has no wings and has fixed vertical rotor(s), 3) TYPE3, which has wings and vertical rotor(s) [2].

Each of these designs has its own complexity, with the TYPE1 being most complex than others. Moreover, due to the difficulty in optimizing the transition flight from vertical to horizontal flight and vice versa, the VTOL UAV may lose its balance during the transition process and can lose a lot of altitudes. Therefore, it is vital to design the VTOL UAV that is lightweight, able to perform transition flight safely and also relatively simple yet efficient propulsion system.

This study reviews previous attempts at developing a VTOL UAVs as well as current projects and combine suitable concepts into a better efficient prototype of a VTOL UAV. Particularly, this study would emphasize the control development on the transition flight mode for better stability and safety. This study proposed the implementation of a TYPE3 VTOL UAV without the tilting mechanism as the scope of this research. TYPE3 VTOL UAV is the least complex system and better stability and safety during the transition flight mode as compared to the VTOL UAV with the tilting mechanism. Another factor of the choice of VTOL type is limited time and cost factor.

The proposed aircraft motor configuration would consist of four motors for its vertical propulsion, a single motor for the horizontal propulsion and servo motors on the control surface for horizontal flight manoeuvring. Basically, the vertical propulsion motors would be attached to the fixed wing platform to produce a hybrid VTOL UAV system. Unlike the tilt-rotor VTOL UAV, the vertical propulsion motors would not tilt and would function to fulfil the vertical flight mode while the horizontal motor is used for horizontal flight. This would reduce the system's complexity as a result of different flight modes are controlled by separate actuators.

In this paper, theoretical concepts behind the flight dynamics of a VTOL UAV are established which would be the basis of this study. Later, the detailed design process is presented where the estimations, calculations, selection of materials, conceptual designs and 3D modellings are justified by the mission requirement stated. At the end of the design process, fabrication and assembly process of a VTOL UAV prototype are done which are followed by a series of software simulations and actual flight tests to support the concept of a non-tilting rotor VTOL UAV.

The concepts of how a fixed-wing aircraft and multirotor works are quite simple. However, developing a VTOL hybrid aircraft requires integrating the flight characteristics of both fixed-wing aircraft and multirotor platforms. Therefore, it is important to first have a separate understanding of the general theory of how a fixed-wing aircraft and multirotor aircraft works as well as the effective control technique to stabilize the whole system.

The primary aim of this project is to determine the potential performance of a small VTOL UAV. First, the VTOL UAV is to be designed and fabricated using scientific methodologies. Then, a number of flight tests are performed under different weather conditions to assess the stability performance during hover, transition and winged flight. The secondary aim is to identify the flight duration of a VTOL UAV compared to a common RUAV of similar specifications.

2 Theoretical Backgrounds

2.1 General Theory

Basically, a VTOL aircraft has the ability to fly like a fixed-wing aeroplane and at the same time hover like a multirotor. A body-fixed coordinate system is usually used in flight dynamics where the coordinate system is fixed relative to the aircraft.

For a multirotor aircraft in motion, it is appropriate to classify the three rotation axes that can alter the craft's attitude as 1) Roll: Rotation of x-axis, 2) Pitch: Rotation of y-axis, 3) Yaw, rotation of z-axis. These axes of rotation are shown in Figure 1 below.

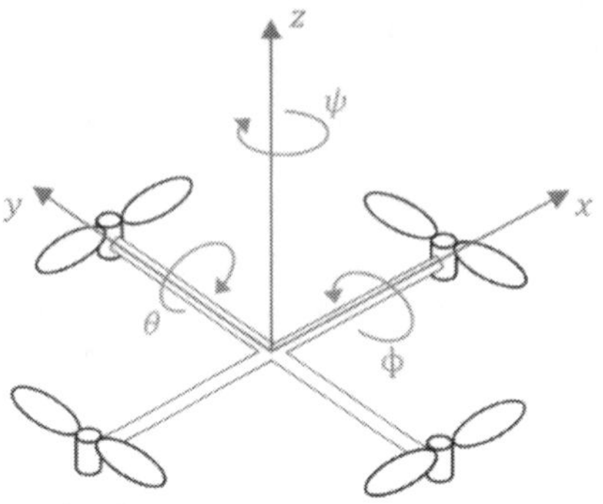

Fig. 1. The coordinate system of a multirotor aircraft.

A common multirotor propeller configuration usually consists of four propellers or better known as a quadcopter. Other known multirotor configurations include tricopter, hexacopter and octocopter. Different configurations are for the different area of application and have performance variation in terms of flight time, agility, stability and payload capacity [3][4]. The propeller's main task is to generate enough lift force to bring the aircraft up to the sky and also deliver adequate thrust to move its current position to forward, backward, left and right. The attitude of multirotors is actuated by the difference in rotation speed of each propeller. Figure 2 shows how the aircraft's attitude change and the corresponding body movement when the rotation speed of certain propellers is manipulated [5]. There is four basic movements for multirotors.

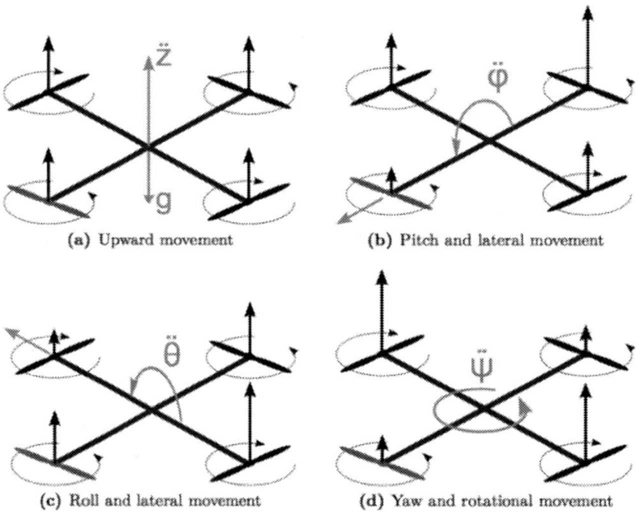

Fig. 2. Attitude and Body Movement Change due to rotation speed change in the individual propeller.

The multirotor must be able to achieve a reasonably stable flight during hover before attempting to manoeuvre it in its pitch, roll and yaw axes. Stable flight requires that the pitch and roll axes must be first stabilized. The multirotor will be at risk of tipping over and crash if these axes are not controlled adequately. The pitch and roll attitudes are determined using the gyroscope and accelerometer which is built onboard the Flight Controller Unit (FCU). As for the yaw axis, it is less critical to stabilizing the axis provided that the multirotor is still controllable. Unstable yaw axis would slightly drift the multirotor body in that axis but can be balanced by using the radio controller.

For a fixed-wing aeroplane in motion, the coordinate system is similar to that of the multirotor. A clear difference is that aeroplanes use propellants that are aligned to its fuselage for forwarding propulsion as depicted in Figure 3 below. Lift in this case is generated by the aerodynamic pressure difference between the upper and lower surface of the wing. The attitude of the aircraft is controlled using the control surfaces attached to the main wing, horizontal tail wing and vertical tail wing. The three control surfaces are called 1) Elevator: Control the pitch axis, 2) Aileron: Control the roll axis, 3) Rudder: Control the yaw axis.

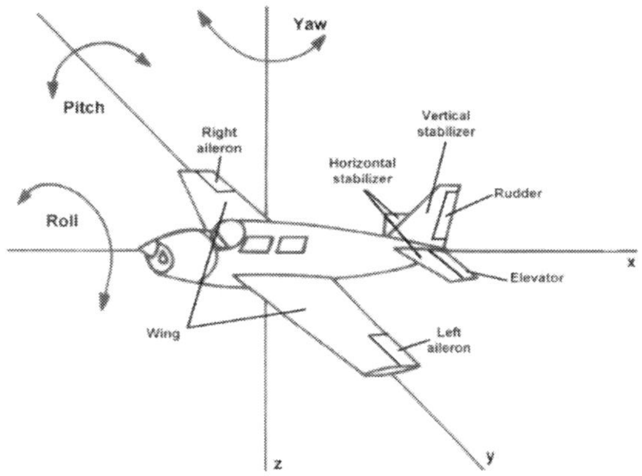

Fig. 3. Coordinate System of a fixed-wing aircraft [6].

Similar to multirotors, to fly fairly stable while in straight-and-level flight, the roll and pitch axes must be stabilized properly. In full-scale manned aircraft, this can be done by directly trimming the ailerons and elevators. In the case of small-scale unmanned fixed-wing aircraft, the roll and pitch axes can be stabilized by using the attitude sensors such as the accelerometer and gyroscope.

Designing and fabricating a hybrid VTOL UAV system involves considering the flight dynamics of both multirotor and fixed-wing systems. The dynamics would become more complex due to propulsion units for longitudinal and vertical movements. To simplify the hybrid VTOL UAV concept, during vertical flight mode, the UAV is assumed to follow the flight dynamics of a multirotor system, whereas, during horizontal flight mode, it is expected that the craft's flight dynamics follow fixed-wing flight dynamics. During the transition flight mode, the aircraft will be in a varying stated and the gravity can be overcome by both vertical and longitudinal propulsion units until the craft has complete the transition process. Optimizing the altitude requires the controlling motor outputs properly. In the transition to horizontal flight mode, lack of vertical motor outputs while not achieving minimum stall speed would severely decrease the altitude. Inversely, while transiting to vertical flight mode, reducing the longitudinal motor output below minimum stall speed before the vertical motors achieve adequate thrust would lead to reducing altitude. It should be noted that maximum airspeed is limited to the rated power of the longitudinal propulsion unit. Moreover, additional structure to support the vertical propulsion system may add to the drag during horizontal flight mode. However, these extra drags can be counteracted by having wings with high lift coefficient characteristics.

3.2 PID Control Technique

Proportional, integral and derivative (PID) control is one of the most attractive techniques used to stabilize as well as optimize the control of UAVs [7]. During the flight, it is preferred to have a stable flight where the aircraft maintained its desired attitude effortlessly without external inputs from the user. In cases where wind or other external disturbance detected, the FCU as processing unit will compute the errors between current and desired attitude and perform necessary corrections. PID control tries to balance out this disturbance by giving out more motor and/or servo outputs accordingly. Figure 4 represents a general block diagram for any PID control system.

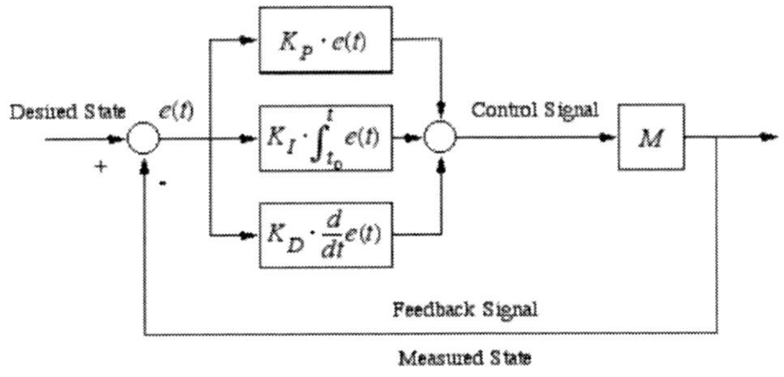

Fig. 4. General PID Block Diagram.

PID control can be easily understood by representing it in the block diagram shown in Figure 4 above. The steps for PID control are:

1. The desired state is first determined and the previously measured state is presented at the beginning of the loop to compute the state error *e(t) = Desired State – Measured State*
2. The proportional, integral and derivative terms are calculated as K_P*e(t), K_I*(integral of e(t)) and K_D*(derivative of e(t)) respectively.
3. The 3 terms are totaled up to obtain the motor/servo output, $u(t) = P + I + D$

In reality, it would be difficult to obtain a stable system by concurrently tuning the control gains, (K_P, K_I and K_D) of the PID controller on all three axes. Therefore, the more critical axes, which are the pitch and roll are first stabilized separately before the yaw axis. PID Control method will yield the desired stability performance characteristics on all axes when the control gains are tuned correctly.

PID control is the most widely used method in theoretical approach and engineering applications partly because exhaustive dynamics models are not required to implement this method. Obtaining the best control gain parameters can be done by modifying the gains live during flight tests according to the control effect of the UAV until the desired response is achieved [8]. Many researchers who have resorted to developing UAV based on PID control have achieved stable flights on their experimental platforms in different modes. Therefore, due to its reliability, PID control is adopted in this study.

3 Design Methodology

3.1 Flow Chart

The flow chart for the overall research design shown in Figure 5 was prepared at the beginning of this project. This flow chart will serve as the general guidance to steer this project in the right direction. The flowchart is divided into two parts, in which the first part concentrate on the design and the second part focus on the fabrication and flight tests. The outcome of this research is to study the performance of the proposed hybrid VTOL UAV by analysis of flight data recorded during test flights.

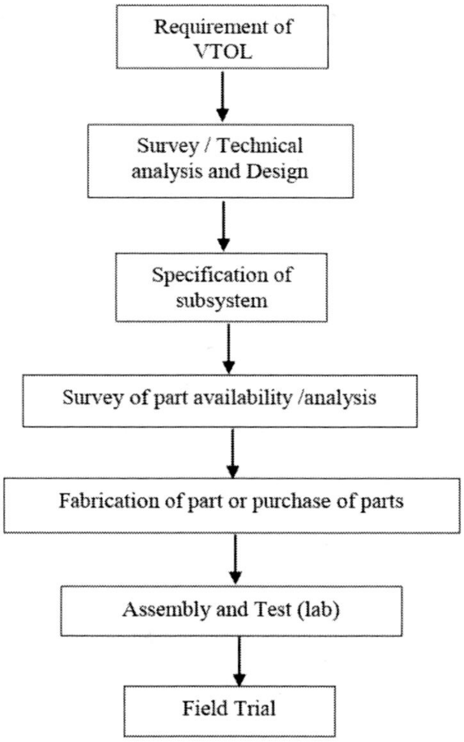

Fig. 5. Design Methodology Flowchart

3.2 Specifying Requirements

The requirements of the proposed TYPE3 VTOL UAV are established after considering the information obtained from a literature survey conducted. These requirements are set in a way that the UAV is practically feasible in term of design, fabrication and flight test.

There are a few basic UAV requirements established in this study. First, TYPE3 VTOL UAV, which consists of conventional fixed wing aircraft attached with fixed vertical rotors to be incorporated into the VTOL design. The UAV should have long flight range and ability to reach the desired location as fast as possible at the reasonable time. The UAV should be manoeuvrable at narrow and confined spaces like a helicopter. Maximum payload of 1 kg is to be allowed which may consist of imaging equipment, gimbals, etc. The UAV should be capable of performing VTOL operations with acceptable safety and stability characteristics. The transition flight mode from rotor flight to fixed-wing flight and vice versa is to be handled manually or autonomously. All-Up-Weight (AUW) of less than 3.5 kg. The wingspan of about 1.5 – 2.0 m. Five-rotor system (vertical quadrotors and single horizontal rotor) for stability and reduction in complexity. The expected cruise velocity range of approximately 15 – 20 m/s. The stall speed is thus should be around 70% of the minimum cruise velocity. Lastly, it would be equipped with versatile flight controller unit able to handle manual controls and execute autonomous missions. The final build of this proposed prototype is roughly shown in Figure 6 where in the future, minor modifications may be necessary.

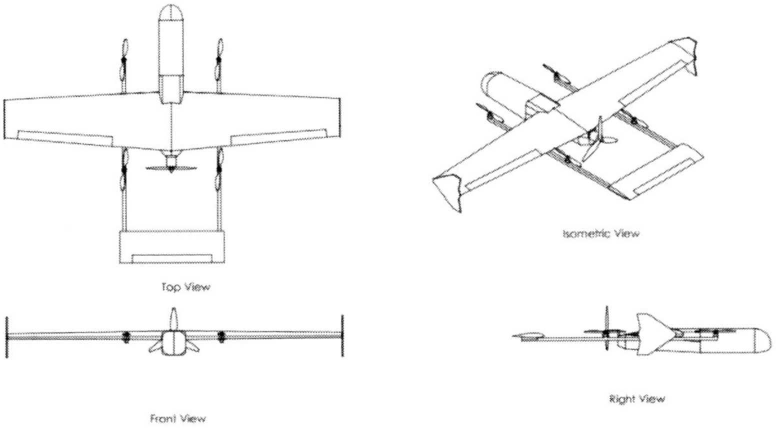

Fig. 6. Overview of the proposed VTOL UAV.

3.3 Test Protocol

The prototype would be conducted in both manual and autonomous flights. The UAV would take off vertically in vertical flying mode, then perform transition to horizontal flying mode, fly straight and level to a certain designated waypoint, then perform transition to vertical flying mode again and finally land the UAV safely. For a manual flight, a ground user would pilot the UAV via radio control transmission. This mission profile can be simply visualized as in Figure 7 below.

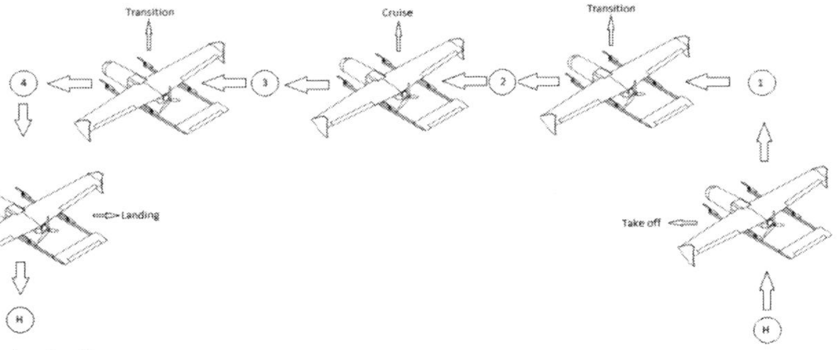

Fig. 7. Prototype Mission Profile.

From Figure 7, the biggest challenge is to perform the transition flight from horizontal flight to vertical flight mode and vice versa. During this portion of the flight, the UAV may lose some altitude due to the reduction in lift force vector while slowing down the vertical motors. Hence, the horizontal motors need to produce enough thrust to accelerate the UAV quickly for the wings to replace the vertical motors in generating lift. Another method to address this possible altitude lose is to develop an effective control system where the vertical motors would continue to generate enough lift while in transition until it passed the minimum stall speed.

3.4 Structural Design Parameters

To design a small-scale VTOL UAV, it is important to limit certain parameters of sizes to a practically reasonable figure so that other parameters concerning dimensions would be feasible for fabrication. The design parameters chosen should also obey known theoretical concepts and equations to confirm that the VTOL UAV will be able to accomplish stable and safety flights across all its flight modes. The general procedures to determine basic structural design parameters are:

1. Estimate the All-Up-Weight of the aircraft
2. Wing Design: Determine the size of the wing loading, wingspan, aspect ratio and chord
3. Fuselage Design: Determine the size of the fuselage
4. Tail Design: Determine the size of the tail wing and its distance from the main wing

These general procedures will be used to estimate the size of the prototype. After fabrication and flight tests, the actual dimensions will be adjusted accordingly.

3.5 Weight Estimation

Before fabricating the prototype, some components which are readily available was weighed while the weight of components that needed to be work is estimated. The total All-Up-Weight (AUW) would be designed slightly higher than the total estimated weights. This is because normally, the actual prototype weight will be higher than the initial weight calculations. Estimating the extra weight also increase the margin for the extra payload. It is important to determine the right AUW for the prototype as the subsequence design calculations are based entirely on this parameter see Table 1.

Table 1. Weight Estimate for proposed prototype

Item	Mass (g)	Quantity	Total Mass (g)
Motor	85	5	425
ESC and PDB	65	5	325
Battery 3s 6000mAh	422	1	422
FCU, gps, power module, receiver	114	1	114
Propeller	12	5	60
Servo	60	4	240
Aluminium frame	700	1	700
Polystyrene Board	600	1	600
Plywood base	200	1	200
Consumables (bolts, nuts, cable ties, glues, etc)	150	1	150
Total All-Up-Weight			3236

Hence, AUW would be estimated a bit higher than the calculated value to be 3.5 kg.

3.6 Wing Design

First, Wing Loading, W_l is to be calculated (Equation 1).

$$W_l = \frac{AUW}{S_W} \tag{1}$$

The wing loading needs to be as low as possible to ensure minimum stress exerted during flight can be sustained by the wing. Wing loading value is more important than the actual weight of the aircraft because it tells how much weight is carried by each unit of wing area, S_W. From literature survey, 10 kg/m2 is selected which would make S_W be about 0.32 m^2. Next, the wingspan, b can be calculated from the wing's Aspect Ratio formula given in Equation (2) below.

$$AR = b^2/S \tag{2}$$

Generally, for normal aircrafts that are designed for low-speed flights have typical Aspect Ratios around 5-10. Higher Aspect Ratio increases the gliding capability of the aircraft and can fly at lower speed. AR = 10 is chosen so that the proposed VTOL UAV can safely glide in case of motor failure. Therefore, b_W = 1.8 m.

to determine the wing chord, C_W, the wing is considered to have a rectangular shape, this is shown in Equation 3.

$$S = b * C \tag{3}$$

Hence, C_W = 0.18 m

For the Aileron sizing, the suitable Aileron area, S_a = 20% of S_W
Therefore,
$S_a = 0.32*0.2 = 0.064$ m^2
$b_a = (0.064*10)^{0.5} = 0.8$ m
$C_a = 0.064/0.8 = 0.08$ m

For the chosen wing size, Reynold's number, Re has to be confirmed to be in the laminar region. Reynold's number is given as in Equation 4.

$$Re = \frac{VC}{\nu} \tag{4}$$

At STP air condition, ρ = 1.225 kg/m^3, ν = 1.46e-5 m^2/s

For the estimated cruise speeds,
$Re_{@V=15m/s}$ = 1.85e-5
$Re_{@V=20m/s}$ = 2.46e-5

Therefore, the flow will be in the laminar region.

3.7 Tail Design

For the tail sizing, the calculations are similar to the wing sizing calculations. For this study, a flat horizontal tail consisting of only elevator control surface is chosen. The yaw axis of the prototype would not be controlled as it would add to the complexity and for the time being, is unnecessary to proof the concept of hybrid VTOL UAV. Instead, the yaw axis can still be controlled while the UAV is in vertical flight mode.

The initial step is to determine the location of the tail wing from the main wing. This location should produce enough moment to control the pitch axis. As a general rule, the distance from leading edge of the wing to the leading edge of the tail should be around thrice of C_W, in this case, it would be approximately 0.6 m.

The next step is to determine the surface area of the horizontal tail, S_{ht}, which is normally about 30% of S_W.

$S_{ht} = 0.3*0.32 = 0.096$ m^2

The Aspect Ratio of the horizontal tail is taken as 3. The subsequent calculations to find tail span and tail chord are the same as Equation (2) and (3) respectively. This gives b_{ht} = 0.53 m and C_{ht} = 0.18 m.

For the Elevator sizing, the the suitable Elevator area, S_e = 20% of S_{ht}
Therefore,
$S_e = 0.096*0.2 = 0.0192$ m^2
b_e = (0.0192*3)0.5 = 0.24 m
$C_e = 0.0192/0.24 = 0.08$ m

3.8 Fuselage Design

The fuselage does not have any specific calculations as the prototype will use the F450 quadcopter frame (Figure 8) as the base fuselage. This frame's aerodynamic shape can be improved to cater for the horizontal flight mode by covering the centre hub using light materials and incorporating shape that can improve lift and reduce drag. Covering the centre hub also protects the exposed main electronics part such as the flight controller unit, battery, GPS unit and power distribution board.

Fig. 8. F450 frame.

3.9 Material Selection

Key and critical parameters in the design of such UAV is choice of material. Every single part of the smallest item such as a screw to the largest part such as a wing, weight is the critical factor when coming to selection. Ideally, the lightest material and yet roust against mechanical (drop, shock) and environment (waterproof, salt spray compliant, altitude /low pressure, oil/chemical contamination/corrosion) shall be selected. Such UAV that normally uses in harsh environment shall comply with some international standard of complaint to ensure the product is robust and can be used in any kind of application/situation. High investment in design/fabrication and qualification test shall be anticipated.

However, the final choice of material and choice of manufacturing/fabrication techniques is also depending on cost affordability and quantity to be produced. For this study which is considered a prototype or PoF (Proof of Concept) model, the choice of material is not at all compliant with the above Mil Standard. What most important in this study is the VTOL is really working and demonstrable. Material for the wing should be light enough and easily available. In this study, wing material is made out of Depron foam which is very lightweight, easy to shape and low cost [9]. The frame support would consist of a combination of the ready-made F450 frame to house the electronics components and hollow 1" by ½" aluminium bar.

3.10 Electronics Design

No specific electronic design take place in this stud. Most electronic parts such as the motor, control module, radio frequency modules and battery/charger are COTFs (commercial on the shelf) item. Main activities in the electronics are on parts/systems integration activities.
Key and critical electronic part /subsystem of the UAV are:

- Radio Remote control
- Brushless Motor
- Battery capacity /voltage

3.10.1 Brushless Motor

Most important in the selection of this motor are a specification of RPM vs supply voltage and durability and also some spec of waterproof and temperature. Long flight duration could cause heat issue and to some extent, it could melt the insulation of the motor coils and causing a short circuit. It is advisable not to use the motor up to the maximum limit of the voltage supply. For real design for commercial use, waterproof or splash proof is a must for this motor. The IP65 standard which outlined protection against contact and infiltration of water and dirt (IP) standard recognized internationally and is available free download from the internet. For this study which is proof of concept project, a requirement such as IP65 is not really considered. What most important is the capability of the term are RPM and weight.

3.10.2 Battery

There are many battery technologies available commercially, named NiCd, Li-Ion, Li-Polymer, Lead Acid etc. For this mobile application, capacity density (how energy stored in a specific volume/weight) is critical. The smaller the size with the higher energy density is the preferred technology. As far as today is concern Li-Ion battery is the most common battery technology for the mobile application. It is a stable technology. Li-polymer is better in term of energy capacity and with a flexible shape but is more expensive. NiCd is obsolete technology and Lead Acid is for automotive application.

Rechargeable Li-Polymer battery is the best choice, cheap and good energy capacity and current rating value. In addition, this batter has a good life cycle (charge/discharge) 1000 cycle, long lasting usage.

3.10.3 Electronics Layout and Operation

Figure 9 below is the wiring diagram on the electric part of the VTOL-UAV. The heart or the core engine of the system is the Pihawx Flight Control Unit, a control processor unit. Its control the rotation or RPM of the motor during takeoff, landing or horizontal movement of the VTOL-UAV.

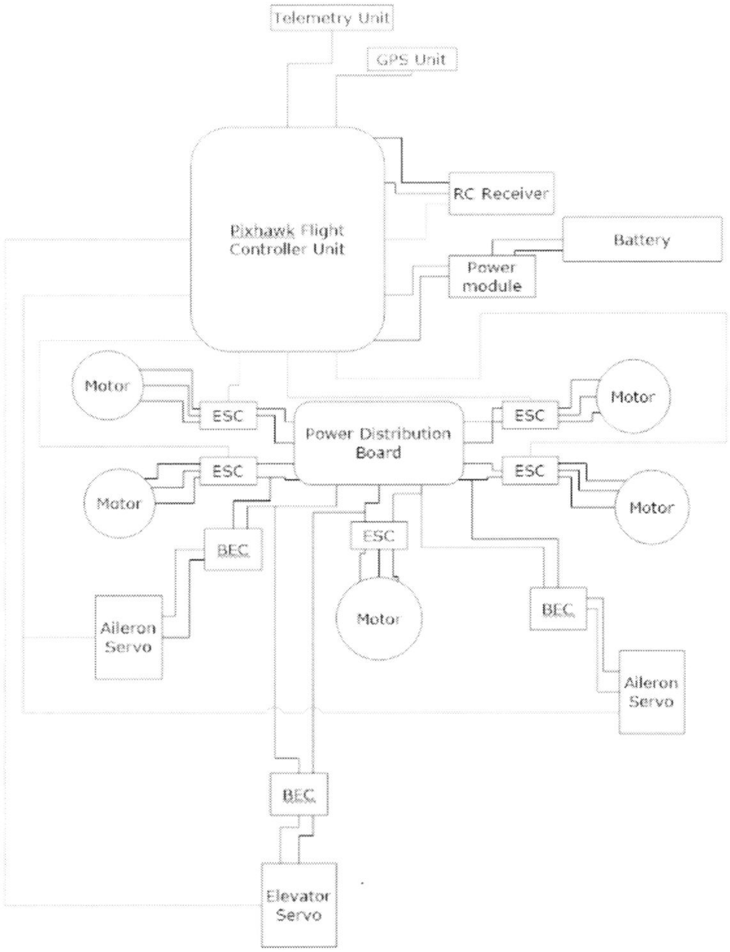

Fig. 9. Wiring Diagram for Onboard.

The input of speed (increase or decrease the motor RPM) come from the operator at the ground. Data was transmitted from the transmitter RC at the ground to receive remote control on the VTOL-UAV by microwave signal operating at 2.4Ghz (ISM unlicensed band for commercial application). Other data transmitter from the ground is the controlling /navigation info for the aircraft. The range (how far the UAV can fly from the operator) is determined by the transmitted power of the RC at the ground and the receiver sensitivity of the RC receiver on board the UAV. It must be a line of sight link, i.e. any blockage such as rain, cloud, trees will degrade the reachable range of the remote control.

Beyond the reachable range of the RC or due to blockage it will cause the VTOL-UAV be flying with control and the risk it falls/ drops to ground will be unavoidable.

On board of the VTOL-UAV there is GPS and telemetry unit. GPS device read the position, speed? and altitude of the UAV. These GPS data together with information on battery status and sensor (such as a camera) could be sent back to the ground centre the telemetry transmitter operating at 915Mhz. In this prototype, mode camera was not installed as it is not part of the scope of the project.

The supply for all the system comes from the reachable 11.1V LI-Ion battery. The battery could be later changed to 14.8V (will extra cell) to increase the RPM of the motor, in case the 11.1V does not adequate to lift the VTOL-UAV.

The duration of the flight depends on the capacity of the battery. The higher battery capacity the longer the VTOL-UAV can fly.

3.11 VTOL mechanism

A key element to ensure the VTOL capability of the four (4) motors with rotary blade stability to lift the load. All the four rotary blades will be at the same speed (Rotation per minute) and acceleration rate i.e. changing the RPM from one state to another state. Increasing the RPM means increasing the height or takeoff stage. Decreasing the speed or RPM mean reducing the altitude or to land. The same RPM for all the vertical blade and weight balance are the keys to ensure stability during takeoff landing and horizontal movement. The speed of the propeller and stability will be achieved by means of PID stability control techniques as described in the earlier section 3.2 above.

Horizontal movement (speed of the UAV) is by mean of controlling the RPM of then horizontal motor/propeller. Attitude control and stability are achieved by adjusting the control surface accordingly and tuning the PID control gains. Manual adjustments of speed/RPM of all motor/propeller are remotely done by mean of radio remote control while during autopilot missions, the UAV would follow the designated speed as programmed in the software.

3.12 3D Modelling

The 3D modelling is done in Solidworks software. Basically, the critical parts that need to be modelled are the main wing and the horizontal tail wing which would be made from Depron foam. Other parts such as the plastic and aluminium frames are less important to model because these parts are known for their reliability and strength during actual flight test and would not be easily be damaged by light crashes. The 3D modelling for wing section and tail section are shown in Figure 10 and Figure 11 respectively. The steps for the 3D modelling are:

1. Determine the airfoil shape of the wing
2. Project the airfoil cross-section into 3D model
3. Assemble in the wings with the frame

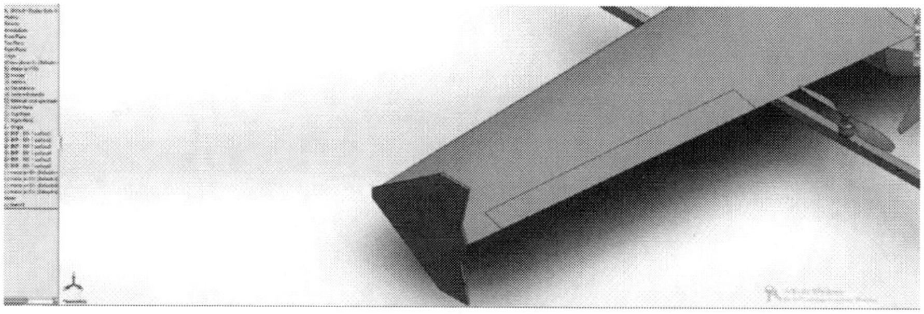

Fig. 10. 3D Modelling of Wing Section.

The cross-section (airfoil shape) of the wing is modelled using standard NACA coordinates which are available online. For the proposed prototype, the suitable airfoil would be the NACA 3414 airfoil family. It has low camber and the lower surface is quite flat and therefore would produce good Lift-Drag ratio for the estimated airspeeds.

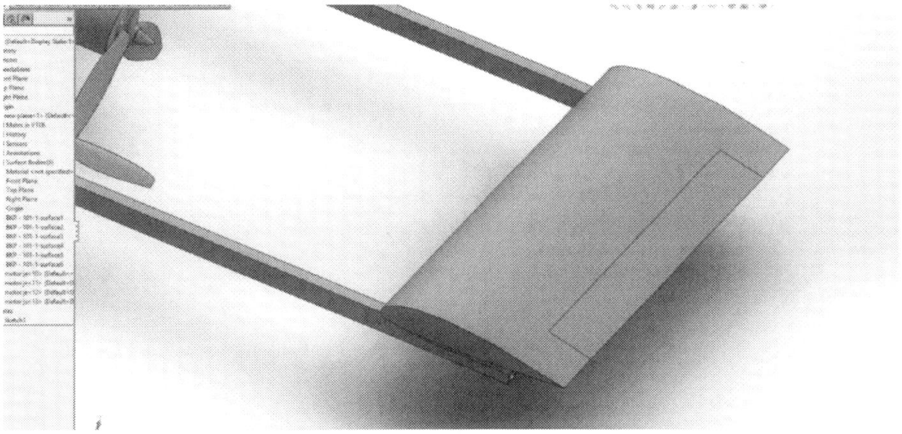

Fig. 11. 3D Modelling of Tail Section.

For the horizontal tail cross-section, NACA 0012 airfoil family is used which is thinner and result in lower drag. As calculated before, the area of the horizontal tail is quite small and therefore the tail is expected to be able to withstand pressure during flight. In addition, both the main and tail wing would be reinforced with aluminium angled beam if the actual flight tests fail to sustain it from crashes.

3.13 Software setups

Mission planner software is used to setup the Pixhawk flight controller unit. This software uploads the correct firmware to Pixhawk to match the prototype's vehicle type which is a hybrid VTOL UAV. This software is also used to tune the PID control gains to achieve optimum stability. This can be done during actual flight tests where the gains are adjusted accordingly by the pilot or can be set before flight tests and Trial-and-Error method is used to obtain the best result. Figure 12 shows a screenshot of tuning tab inside Mission Planner.

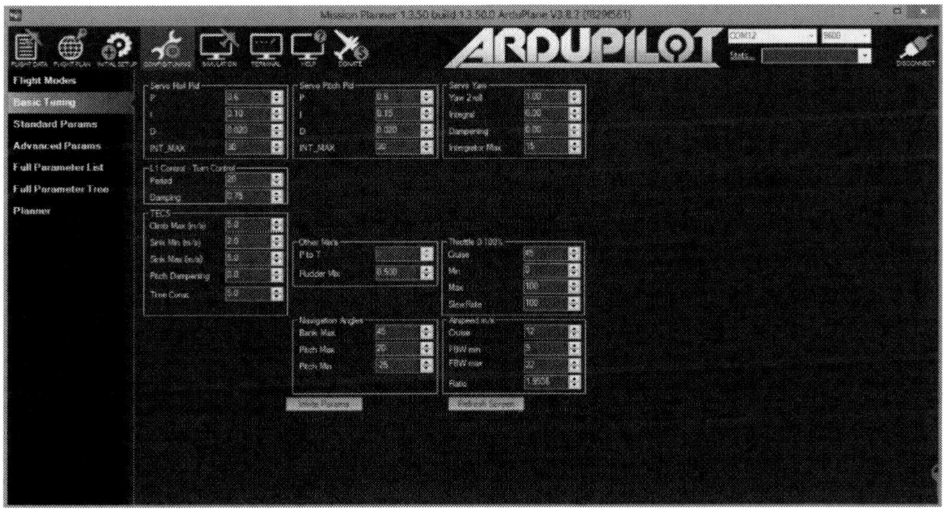

Fig. 12. Mission Planner PID Tuning Tab.

4 Results and Discussions

The fabrication was done successfully and the final prototype was tested to be fully functional. The built of the prototype is shown in Figure 13. After complete fabrication and assembly have been done, physical inspection along with subsystems and full system tests were performed.

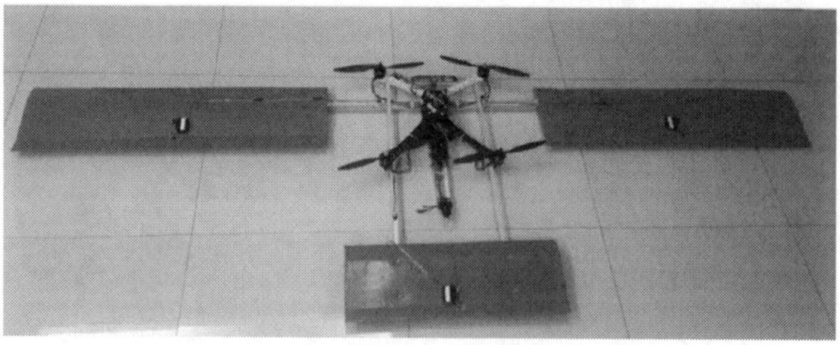

Fig. 13. Final Prototype After Complete Fabrication and Assembly.

Apart from this, the data flash logs from actual flight tests were successfully recorded for further analysis. Using the advanced FCU which is equipped onboard the system, various type of data can be recorded. However, the data that are of interest for this project includes the PID response performance within the 3 axes of rotation, the vibration level about the three axes and also the flight trajectory data.

The flight modes that were used when the data was recorded are stabilized, altitude hold and loiter mode. These three modes are the flight modes in vertical flights. As for the horizontal flight modes, the flight tests could not be done yet due to limitations in available open space that are safe for flight tests. In stabilize mode, the position and altitude control is performed manually. In altitude hold mode, the altitude is controlled autonomously while the UAV position can still be drifted due to the wind. In loiter mode, both altitude and position hold is controlled autonomously. In all these three modes, the accelerometer onboard the FCU is automatically stabilized where minimum user intervention is required to hold the UAV in a perfect hover.

The graph is shown in Figure 14 and Figure 15 highlight the vibration level (m/s/s) as well as the roll response (m/s/s) against time. This graph displays the flight time of about 1 minute to ease the visualization of the graphs. In summary, the stability performance in a roll and other axes was at acceptable levels where the response of the system, recorded in the unit of m/s/s, followed closely to the desired inputs either from the user or the FCU itself as it tried to auto-stabilize the axes continuously. Further tuning of the PID gain values may be required to obtain a more desired performance. As for the vibration levels, the values recorded also fell within an acceptable level of vibrations in all axes (below 60 m/s/s). This suggests that for future works, it would be suitable to install a camera onboard for in-flight monitoring and for other purposes.

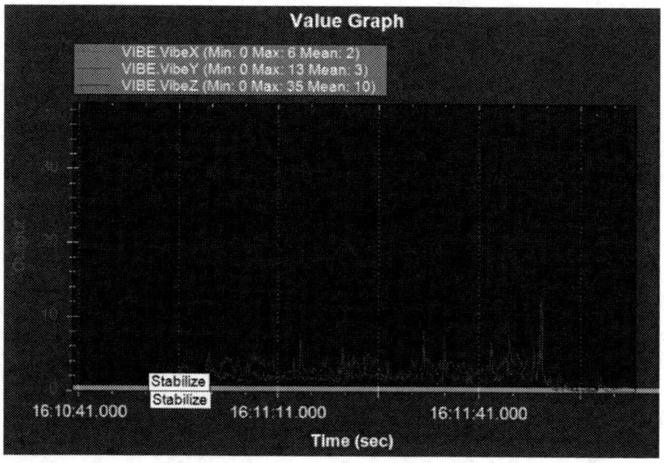

Fig. 14. Graph of Vibration levels in all three axes.

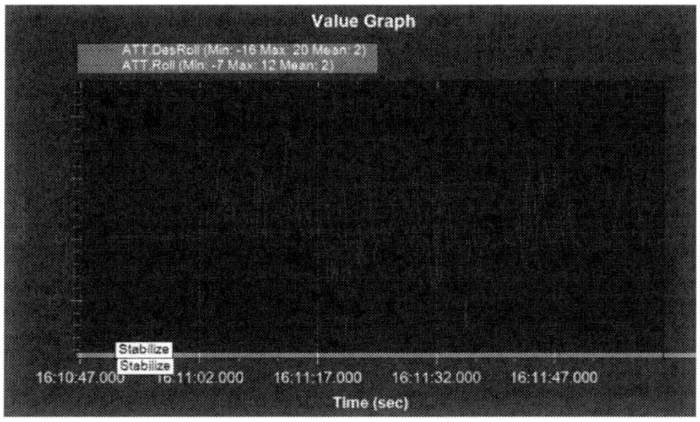

Fig. 15. Graph for desired roll and roll response axis

5 Conclusion

Based on the theoretical study and flight data analysis, the proposed hybrid VTOL UAV system would achieve the predicted flight performance in terms of stability and flight range. This study encountered some challenges where the fabrication of the prototype cannot be completed within the specified time due to delay in receiving new part (motors) due to defect and also sourcing the right material for the wing.

6 Future Work

The model VTOL-UAV assembly is still in progress. Main components such as the motors/ and RCs are functional well. Below show some of the pictures of parts or and the construction of the assembly which is still in progress.

The design and development of "VTOL" vertical takeoff and landing UAV prototype is really a preliminary stage in the design, and commercialization of this platform needs much more improvement. With the limitation of facilities, budget and expertise, it is difficult to release the prototype to fully meeting the targeted objectives. The VTOL is not only mechanical product that requires Only mechanical expertise, it does require competency in electronic and RF (Radio frequency) communication as well. Using COTF solution for the electronic and communication could be adequate as starting point in the study phase but customization does require in order to be able to meet the desire targets I,e. range, robustness, interference-free etc.

The proposed product is not a hobby kind of UAV, it is rather to be used for the professional application. Certification and validation by authorized agencies such as DCA and SIRIM are required. The following are recommendation and steps in order to proceed the project.

7 References

1. A. K. Patra, K. Patel, G. S. Bundela, and V. P, "Design and Development of Transition Autopilot For VTOL UAV," *Int. J. Innov. Eng. Res. Technol.*, **vol. 4**, no. 6, pp. 57–72 (2017)
2. D. Jo and Y. Kwon, "Analysis of VTOL UAV Propellant Technology," *J. Comput. Commun.*, **vol. 5**, pp. 76–82 (2017)
3. V. Kadamatt, "Tricopter vs Quadcopter vs Hexacopter: A Comparison," *Droney Bee*, 2017. [Online]. Available: http://www.droneybee.com/tricopter-vs-quadcopter-vs-hexacopter/. [Accessed: 24-Oct-2017].
4. F. Thomas, "What is a Tricopter, Quadcopter, Hexacopter and Octocopter," *Droners Guides*, 2017 [Online] Available: https://www.dronersguides.com/what-is-a-quadcopter/. [Accessed: 24-Oct-2017].
5. S. Said, "Quadcopter Mechanics," 2014. [Online]. Available: http://uav-society.blogspot.my/2014/06/quadcopter-mechanics.html. [Accessed: 24-Oct-2017]
6. J. Yoon, "Adverse Yaw and Aircraft Turns," 2001. [Online]. Available: http://www.aerospaceweb.org/question/dynamics/q0045.shtml. [Accessed: 24-Oct-2017]
7. J. Dryden and R. Barbaccia, "Quadcopter Design Project," (2014)
8. L. Zhong, H. Yuqing, Y. Liying, and H. Jianda, "Control techniques of tilt rotor unmanned aerial vehicle systems : A review," *Chinese J. Aeronaut.*, **vol. 30**, no. 1, pp. 135–148, (2017)
9. M. F. Saharudin, "Development of tilt-rotor unmanned aerial vehicle (UAV): material selection and structural analysis on wing design," in *IOP Conference Series: Materials Science and Engineering* (2017)

The Application of Image Processing to Solve Occlusion Issue in Object Tracking

Yun Zhe Cheong[1] and *Wei Jen* Chew[1*]

[1] School of Engineering, Taylor's University, 47500 Subang Jaya, Selangor, Malaysia.

Abstract. Object tracking is a computer vision field that involves identifying and tracking either a single or multiple objects in an environment. This is extremely useful to help observe the movements of the target object like people in the street or cars on the road. However, a common issue with tracking an object in an environment with many moving objects is occlusion. Occlusion can cause the system to lose track of the object being tracked or after overlapping, the wrong object will be tracked instead. In this paper, a system that is able to correctly track occluded objects is proposed. This system includes algorithms such as foreground object segmentation, colour tracking, object specification and occlusion handling. An input video is input to the system and every single frame of the video is analysed. The foreground objects are segmented with object segmentation algorithm and tracked with the colour tracking algorithm. An ID is assigned to each tracked object. Results obtained shows that the proposed system is able to continuously track an object and maintain the correct identity even after is has been occluded by another object.

1 Introduction

Multiple object tracking (MOT) is a process that commonly uses an image processing system to track objects in a busy environment [1]. Development of multiple object tracking has begun since 1988 [2]. Since then, many different tracking techniques have been proposed to improve and enhance the current object tracking techniques [3]. Typically, multiple object tracking recognizes and tracks each unique object with a specific ID [4]. However, there are several issues that affect the MOT process. One of the issue is the occlusion of objects [5].

Occlusion often occurs when two or more objects come too close and seemingly merge or combine with each other. Image processing system with object tracking often wrongly track the occluded objects [6]. Sometimes, after occlusion, the system will wrongly identify the initially tracked object as a new object [7].

Figure 1 depicts the occlusion issues in 3 images. As observed from the 3 images in Figure 1, the normal colour tracking system only track the exposed region and neglected the occluded region. As observed in the second image of Figure 1, the blue object was wrongly

* Corresponding author: weijen.chew@taylors.edu.my

tracked as 2 separate objects. The main objective in this paper is to solve the occlusion issue that occurs in object tracking with the application of image processing.

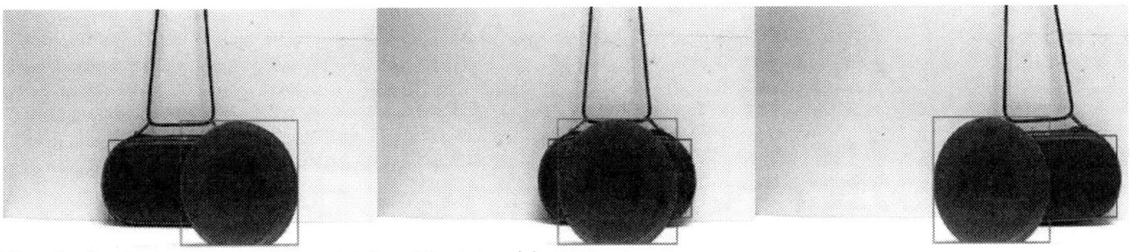

Fig. 1. Occlusion issues in multiple object tracking.

2 Research Methodology

Generally, the research methodology is separated into four different phases. These phases are the modelling phase, simulation phase, measuring phase and data collection phase, as shown in Figure 2.

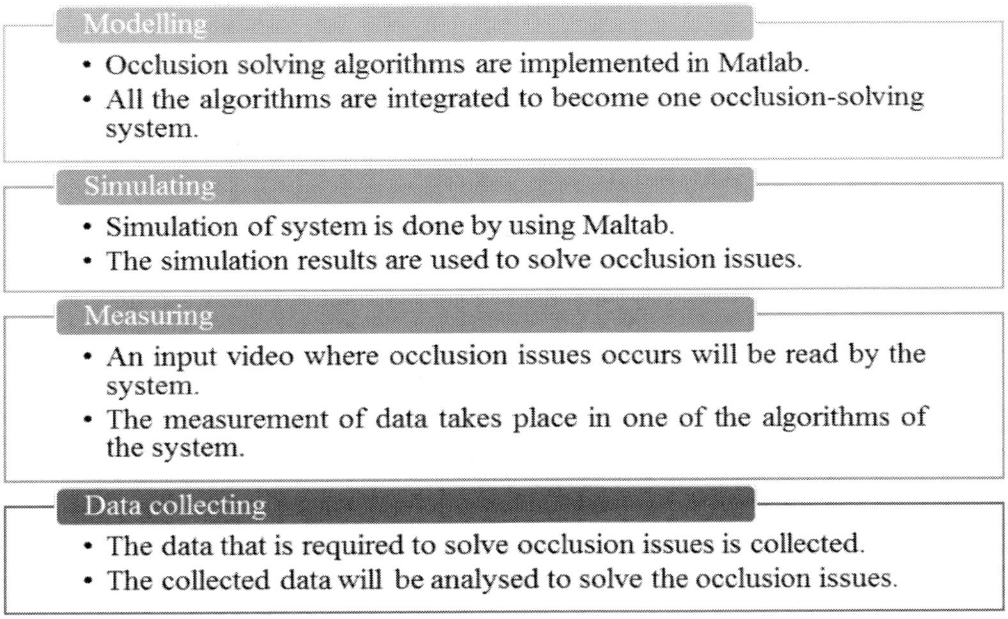

Fig. 2. Project overview chart.

Firstly, an input video will be read by the system. Next, the system extract and reads every frame from the input video with the frame extracting function. After all the frames from the video has been read, each frame will then be processed by foreground object segmentation algorithm. Mean colour values of the foreground objects will be detected with the colour detection algorithm and then the foreground objects will be tracked based on their own mean colours values with a given ID. In order to identify tracked foreground objects clearly, the coordinate, area, width, and height of each foreground object will be identified using the object specifying algorithm. Finally, the occlusion solving algorithm will be implemented to solve any occlusion issues the video had.

 https://doi.org/10.1051/matecconf/201815203001

2.1 Overall Configuration of the system

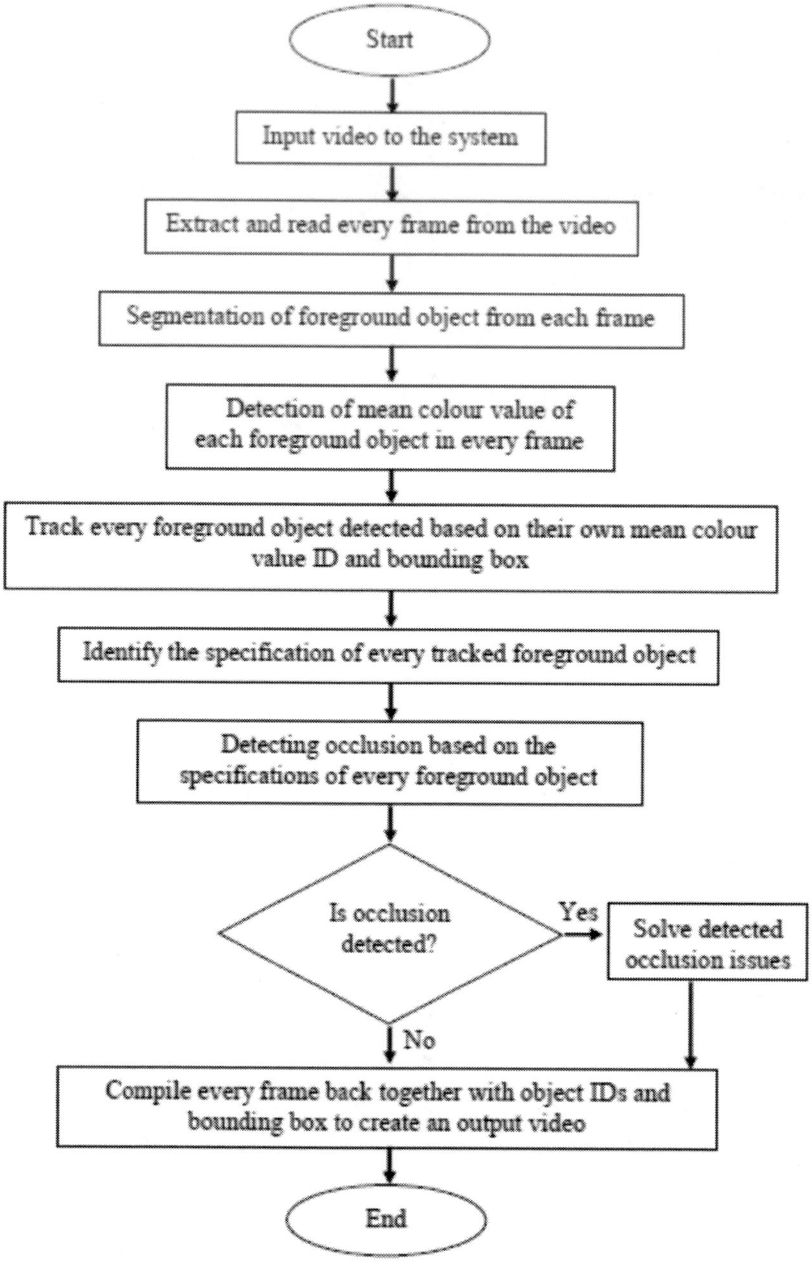

Fig. 3. Flowchart of the proposed system.

The flowchart of proposed system is shown in Figure 3. Generally, the system starts off by extracting all the frames of the input video and then read it frame by frame to segment the foreground objects. After the segmentation of foreground objects, the system will further analyse the mean colour value of each object and track it with an ID and a bounding box. The system identifies the tracked object's specifications such as height, width and area to detect the occurrence of an occlusion. The system will then proceed to handle the occlusion issue if it occurs. Finally, after handling the occlusion issue, the system will compile every post-processed frame into one output video.

2.2 Object segmentation and colour tracking algorithms

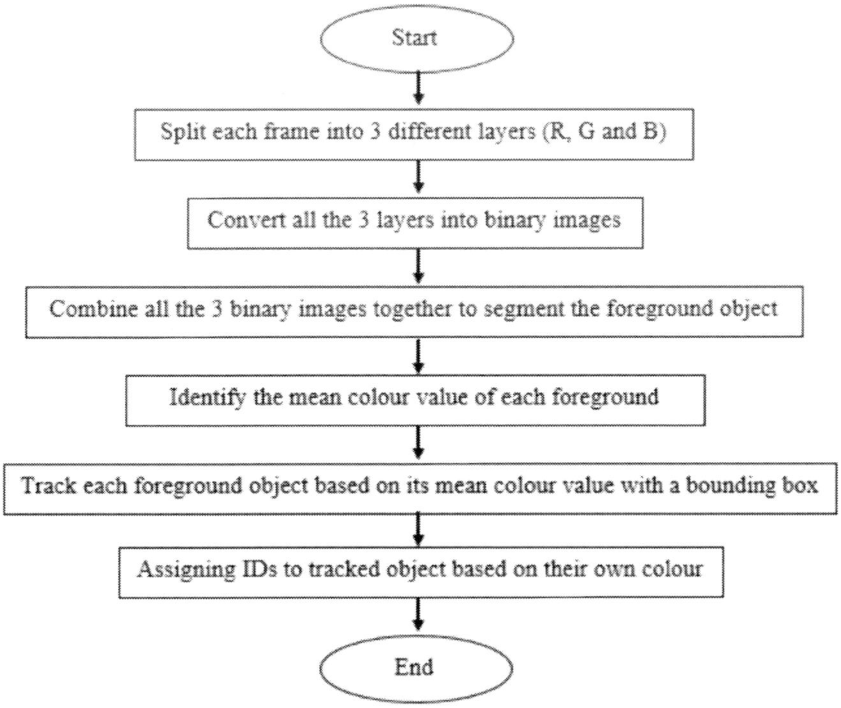

Fig. 4. Flowchart of object segmentation and colour tracking algorithms.

Figure 4 shows the flowchart of object segmentation and colour tracking algorithms. The segmentation of foreground objects is done by combining 3 R, G and B layers. After the object segmentation process, the mean colour value of each segmented object is identified. After the mean colour of each foreground object is identified, the algorithm proceeds by tracking the object with a bounding box and assigning an ID to the object in the frame.

2.3 Object specifying algorithm

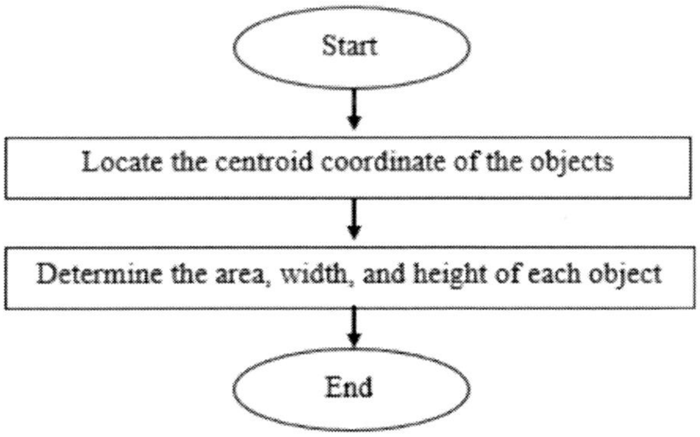

Fig. 5. Flowchart of object specifying algorithm.

Figure 5 shows the flowchart of the object specifying algorithm. Technically, this algorithm detects every tracked object specification in every frame. First, the algorithm locates the centroid coordinate of every tracked object. Next, it determines and saves the area, width and height of every tracked object.

2.4 Occlusion handling algorithm

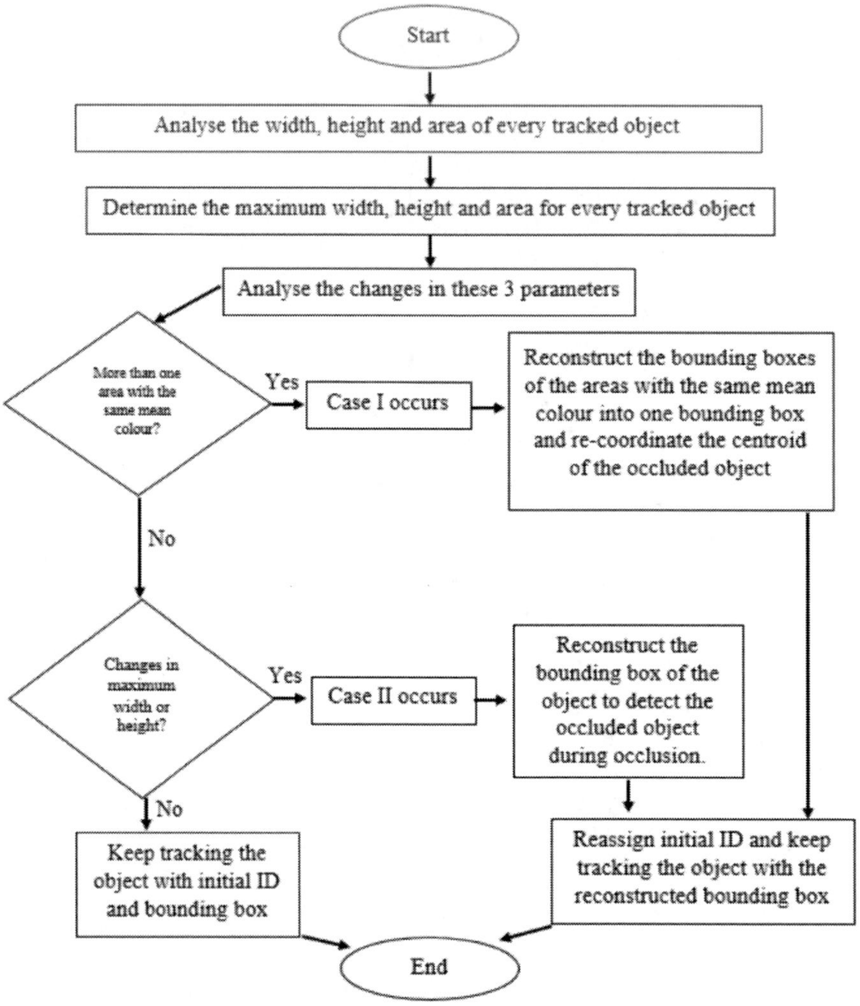

Fig. 6. Flowchart of occlusion handling algorithm.

Figure 6 shows the flowchart of the occlusion handling algorithm. In this algorithm, the saved data from the previous algorithm will be analysed to detect whether occlusion issues occur in a frame. Before deciding whether occlusion do occur, the maximum width, height and area of every tracked object is obtained from the previous algorithm. Case I occurs when 2 or more areas of with the same mean colour is detected. Technically, an object with a mean colour only has one area. If the number of areas of the same colour is 2 or more, then the object is being occluded in the middle while exposing its edges at the left and right sides. Therefore, it is considered that the object is moving horizontally from left to right and being blocked in the middle by another object. The Case I occlusion issue will be solved by replacing the 2 bounding boxes of two different areas with a reconstructed bounding box to track the middle-occluded object from behind. Case II occurs when the change in width or height of an object is detected. For example, considering the same motion of moving horizontally from left to right. When the width of an object changes significantly along with the area of the object, the object might be occluded either on its left side or right side. Case II occlusion is solved by reconstructing the smaller bounding box that only tracks the exposed part of an object, with a new bounding box that tracks the whole object including its occluded part

2.5 Limitation of the proposed system

The proposed system's scope is limited to solving occlusion issues in recorded video with multiple single-coloured objects and a white background. Also, the shapes of all the objects must remain unchanged throughout the video.

3 Simulation Results

For this paper, three different videos with different occlusion issues were tested with the proposed system. The results were recorded and discussed below.

3.1 Object segmentation and colour tracking

In the object segmentation and colour tracking algorithm, the foreground objects are segmented out from the background. Firstly, the RGB frames in a single frame is split into 3 different layers. These 3 different layers are R layer, G layer and B layer, each layer carries its own colour value. R layer carries red value, whereas G and B carry green value and blue value respectively. Next, every layer is converted to binary images. By combining all the 3 RGB binary layers together with certain binary thresholding value, the foreground objects can be segmented from the white background. Since white background has lower mean colour value, it does not consider as a foreground object. The combination of the 3 RGB binary layers are shown in II of Figure 7.

The average mean colour of an object, V_a, is determined by calculating the average value of R, G, B (R_a,G_a,B_a) in the pixels inside the bounding box. After obtaining the average mean colour, the range of mean colour is set as [$R_a\pm8\%$,$G_a\pm8\%$,$B_a\pm8\%$] and any pixel within the range of mean colour is treated as part of a foreground object's pixel. Once a foreground object with a mean colour is tracked, its pixel-area can be obtained as shown in IV and V of Figure 7. Finally, ID is given to every tracked foreground object according to its mean colour. For example, in Figure 7, ID:B is given to a blue object and ID:R is given to a red object.

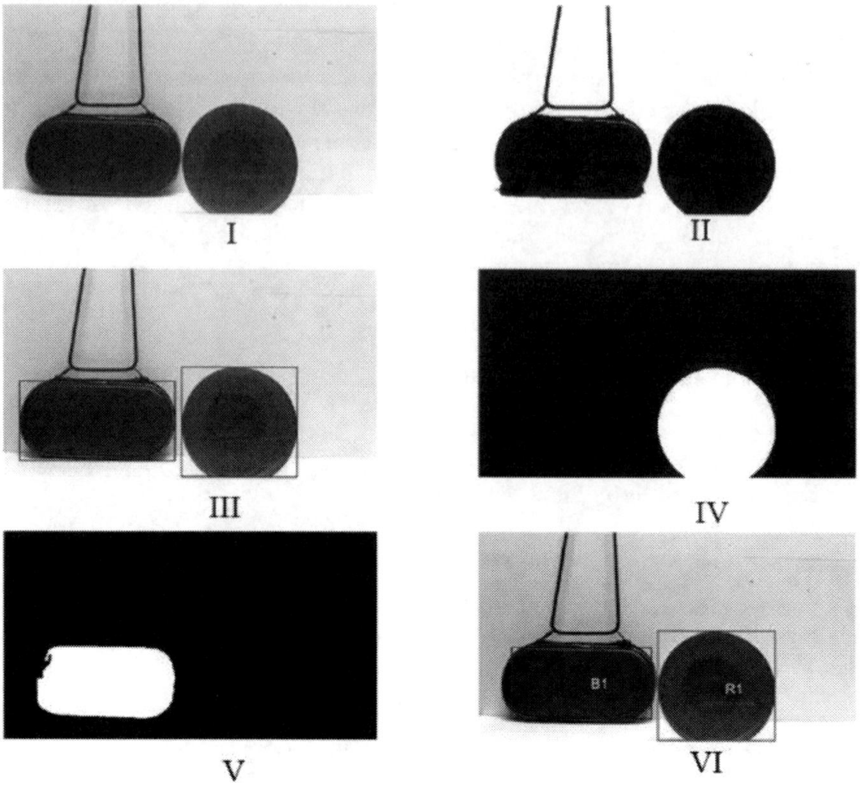

Fig. 7. Object segmentation and colour detection results.

Figure 7 shows the results of object segmentation and colour tracking algorithm. The process of the algorithm is explained below:

I) The original extracted image of a frame.
II) The foreground objects are segmented by processing R, G, and B layers.
III) The mean colour value inside the box is determined.
IV) Red colour object is detected, the white region is the area of the red object.
V) Same as the previous step, the blue object is detected and the area of blue object is represented in the white region.
VI) Both blue and red objects are tracked with different bounding boxes and IDs are assigned to both objects.

The criteria for constructing bounding boxes will be explained in the next section.

3.2 Object specification

Object specification algorithm is considered as a short algorithm as it only determines the specification of every tracked object. Technically, this algorithm determines the maximum areas of all foreground objects at the beginning of every input video. After the maximum area of an object is determined, the algorithm will proceed to locate the coordinates of centroid of the object. Next, the properties of the bounding box will also be determined with this algorithm.

Construction of bounding box requires four different properties. These properties are right (r_b), top (t_b), width (w_b) and height (h_b). r_b and t_b refers to the right and top starting points of the bounding box respectively. Whereas w_b and h_b determine the width and height of the bounding box.

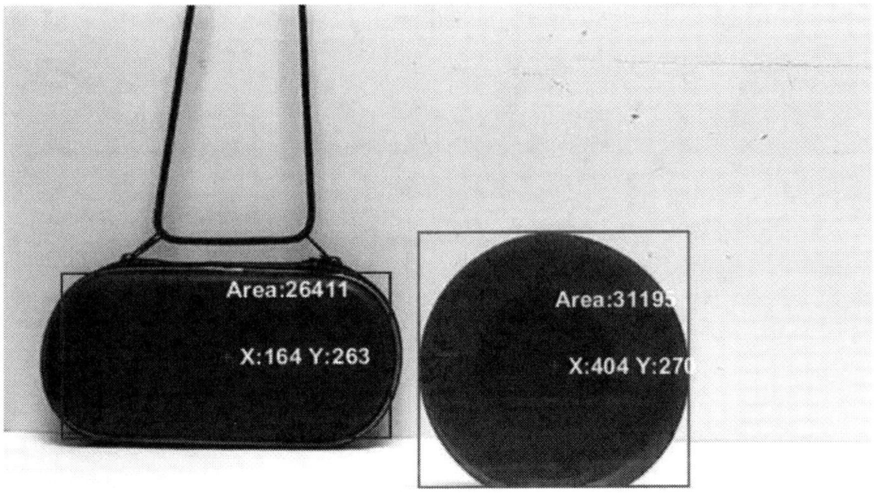

Fig. 8. Object specification result.

Figure 8 shows that each segmented object is tracked with a bounding box with its centroid coordinates identified. The centroid coordinates of the objects are important to justify which case of occlusion issue occurs.

3.3 Occlusion handling

The occlusion handling algorithm handles occlusion issues that occurs throughout the input video. In this algorithm, the bounding box of an occluded object is reshaped to accurately track occluded object with its true size. The bounding box is reshaped according to how the object is occluded. Following the project scope, there are two potential occlusion cases, Case I and Case II. Case I only occurs when an object is being occluded at the middle with its left and right sides exposed as shown in Figure 9. Case II occurs when an object is being partially occluded by another object on one side, either left or right. The different cases of occlusion issues are solved with the occlusion handling algorithm.

Case I

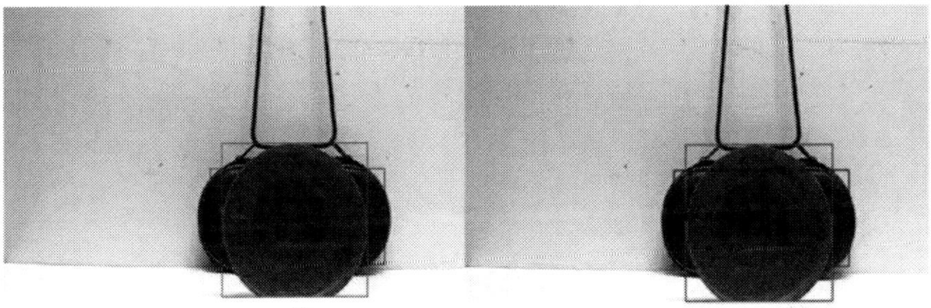

I II

Fig. 9. Object is being occluded in the middle.

When the blue object is occluded by the red object in the middle, two different bounding boxes wrongly track a single object as two different objects with the same colour as shown in I (left) of Figure 9. To rectify this occlusion issue, a new bounding box is constructed to replace the old ones as shown in II (right) of Figure 9. In this case, the new bounding box is constructed with maximum width data obtained from the previous frame.

Since the new bounding box is tracking an occluded object, the initial ID assigned for the object will be retained along with word "occluded" to indicate the occurrence of occlusion.

Case II

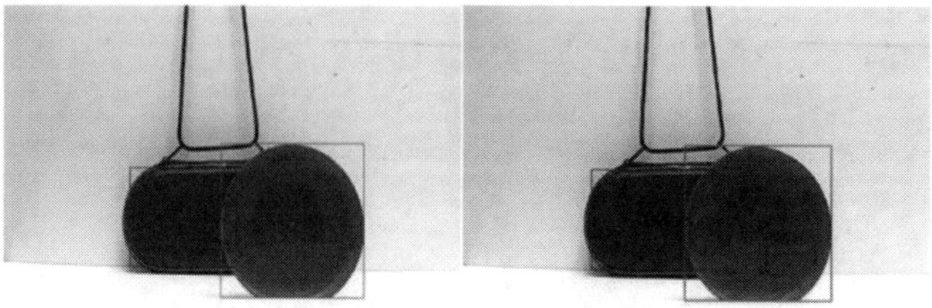

I II

Fig. 10. Object is being occluded on the right side.

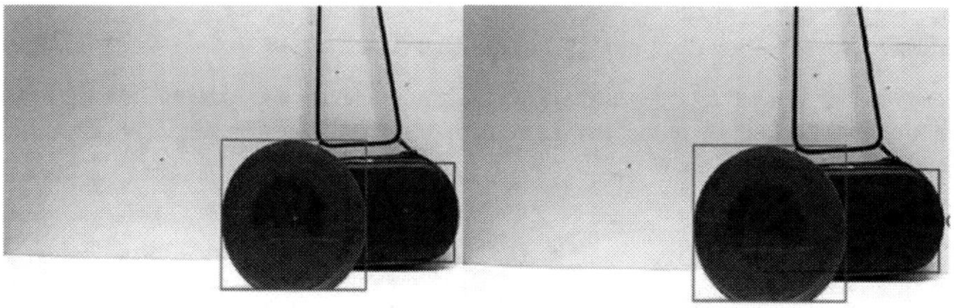

I II

Fig. 11. Object is being occluded on the left side.

When either side of the blue object is occluded as shown in I of both Figure 10 and Figure 11, the bounding box only tracked the exposed portion of the blue object. To rectify this occlusion issue, the bounding box is reconstructed to track the blue object even when it is occluded. In Figure 10, the right side of the blue object is occluded. The system analyses the centroid coordinate of both the blue and red object to select the appropriate expansion of the bounding box. In this case, the x-coordinate of the centroid of the blue object is less than the red object and the exposed area of the blue object is smaller than the maximum determined area of blue object. Thus, the bounding box is expanded to the right to track the occluded right portion of the blue object as shown in II of Figure 10. The same principle is used to solve the occlusion issue in Figure 11, where the x-coordinate of the centroid of blue object is greater than the red object. Figure 12 and 13 shows other examples of occlusion handling by the proposed system

Fig. 12. Object is being occluded on the left side.

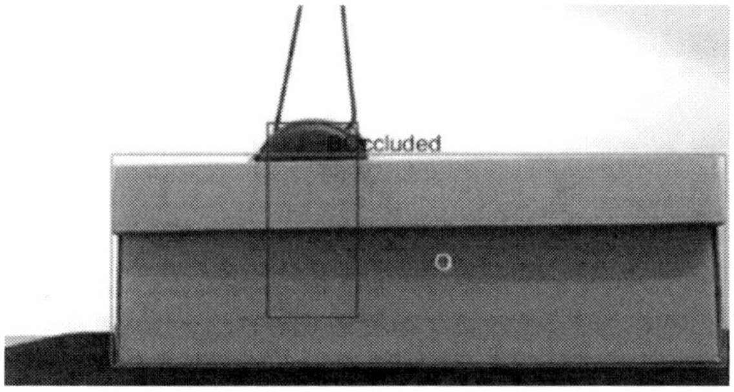

Fig. 13. Object being occluded at the bottom.

4 Conclusion

In conclusion, the occlusion issue that occurs in the input video is properly handled by using the system proposed in this paper. This proposed method was able to handle 2 objects overlapping each other and was able to identify the occluded object as a single object even though it was blocked in the middle. For future work, further improvements need to be done for handling occlusion in real-life situations with complex-coloured objects such as pedestrian or vehicles.

References

1. Y. Cai, H. Wang, X. Chen, and L. Chen, "Multilevel framework to handle object occlusions for real-time tracking," pp. 1–8, 2016.
2. Pylyshyn, Z. (2007). Multiple object tracking. [online] Scholarpedia. Available at: http://www.scholarpedia.org/article/Multiple_object_tracking. [Accessed 26 Oct. 2017]
3. P. Ding and Y. Song, "Robust Object Tracking using Color and Depth Images with a Depth based Occlusion Handling and Recovery," pp. 930–935, 2015.
4. I. Engineering and S. Province, "Variable Scale and Anti-occlusion Object Tracking Method with Multiple Feature Integration," pp. 426–433, 2016.
5. W. Luo, J. Xing, X. Zhang, X. Zhao, and T.-K. Kim, "Multiple Object Tracking: A Literature Review," arXiv, 1-39. Retrieved from http//arxiv.org/abs/1409.7618, vol. V, no. 212, 2014.
6. M. Intelligence, Q. Zhao, W. Tian, Q. Zhang, and J. Wei, "Robust object tracking method dealing with occlusion," 2016.
7. Z. M. et al. Shehzad, Muhammad Imran, Yasir A. Shah, "K -means based multiple objects tracking with long-term occlusion handling," 2016.

Comparative Study between Planar and Bent Antenna Characterization

Lau Zheng Tung[1], *Ghafour* Amouzed Mahdiraji[1,*] and Lee Chia Ping[2]

[1]School of Engineering, Taylor's University, 47500, Subang Jaya, Selangor, Malaysia
[2]Heriot-Watt University Malaysia, 62200 Putra Jaya Malaysia

Abstract. Smaller designs of antenna usually implemented on Wi-Fi modules or navigation system devices which requires smaller bandwidth such as Planar Antenna. There are some antenna application requires bent antenna such as wearable UWB antenna but the substrate used will have some defects after bending and crumpling from time to time. What if the planar antenna is designed that it can be bent and the bending affects the impedance matching and shift in resonant frequency? So, this research aims to achieve the limitation of the planar antenna to be bent at certain angle and given frequency range (1-8 GHz). The planar antenna is designed in the Computer Simulation Technology (CST) Software and the proposed antenna is then bend in the simulation software. The analysis of comparative evaluation for both planar and bent antenna is carried out on different characteristics such as evaluation between S-Parameters & Frequency and evaluation between Voltage Standing Wavelength Ratio (VSWR) & Frequency. The behaviour of radiation pattern for both antennas can also be obtained in the CST simulation.

1 Introduction

Antenna has been widely used in various form of communication in different applications from machine to machine. The application that requires the usage of antenna includes, radio broadcast, navigation systems, satellite communication, remote sensing in defence radar or medical scanning device as well. Planar antennas are commonly used in sensing device due to low cost and simple implementation with other system. The frequencies of the Planar antenna can reach as high as 77 GHz which can be performed in automotive sensing devices. Typical antennas that people had worked on usually control by the different bandwidth and return loss of the signal during the receiving of the signal. While extensive of studies have been related to planar antenna, the studies on non-planar antenna such as bent antenna are limited. The bent antennas mostly are the solutions to some application as they are suitable to be implemented onto structures or objects that are not in planar form. There are researches identified the draw backs for bent antenna which would lead to certain distortion of parameters which includes, the impedance bandwidth,

[1] Corresponding author: Ghafour.AmouzadMahdiraji@taylors.edu.my

input impedance, directivity and phase angle when antenna is exposed to bending [1], [2]. Some cases where the substrate material is applied on the antenna with better parameters, allows the antenna to operate properly with better efficiency even though the antenna is bent [2].

S parameters are defined in terms of incident and reflected travelling waves which are simple to be used in analysis. S parameters are usually easy to be measured by using network analyzers and it is popular in any Radio Frequency (RF)/Microwave circuit design and testing. S Parameters are complex matrix that shows Reflection/Transmission characteristics (Amplitude/Phase) in frequency domain.

Figure 1 shows a two-port device which has four S-parameters. The first number following the "S" is port where the signal emerges while the second number is where the signal is applied. So, S21 is a measure of signal coming out from port 2 relative to RF stimulus entering port 1. If the numbers are the same (S11), this indicates that the signal has a reflection measurement. Quantify of reflection and transmission characteristics of devices is available with amplitude and phase information. There are few measured terms are scalar in nature otherwise others are in vector. For example, return loss is a scalar measurement of reflection, while impedance results from a vector reflection measurement [3]. Meanwhile, like group delay, are referred to purely phase-related measurement. For Reflection, S11 /S22: the parameters that we can identify includes, Reflections (Return loss), Impedance, Admittance, VSWR. For Transmission, S21/S12: the parameters that we can identify includes, Gain/Loss (Insertion loss), Phase, Group delay (Delay time). To identify any loss of signals that was sent, a return signal of the transmission signal is needed which indicates the Reflection, S11 parameters. Return Loss is measured at the input to the coaxial cable connected to antenna and it is the difference between forward and reflected power in dB. The Return Loss calculation is given by Radiated Power (PR) which the power is reflected back, divided by Incident Power (PT) which the power transmitted by the source. The Return Loss must be as small as possible for maximum power transfer and the ratio, PR/PT should also be as small as possible and the return loss should also have the value which is as large a negative as possible.

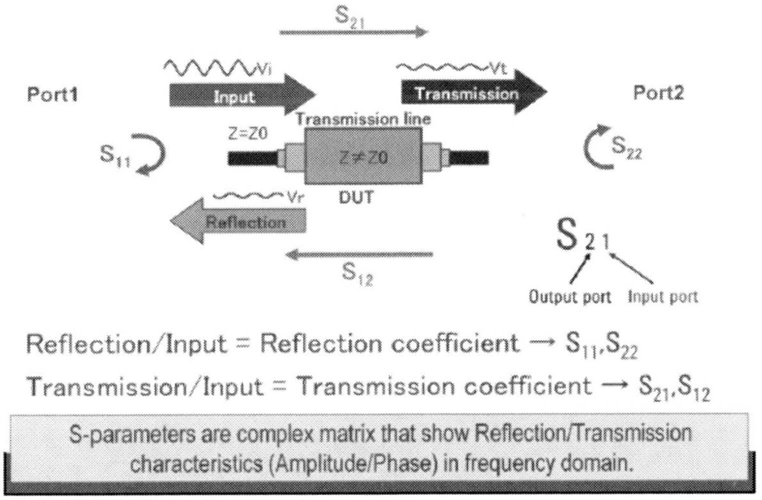

Fig. 1. Reflection/Transmission Characteristics (Amplitude/Phase) in Frequency domain [3].

2 Methodology

2.1 Antenna design

Figure 2 shows the antenna design with the dimensions. The antenna is designed on FR-4 substrate with the dimension given; dielectric constant, ε_r of 4.3, loss tangent of 0.025 and thickness of 0.16 mm. The copper tape with thickness of 0.035 mm is used as conductive material for patch and ground. Most research recommended transmission line model to design the width, W and length, *L* of the antenna [4], [5]. The implementation of transmission line model is fairly easy and antenna designs will acquire with reasonably good performance in terms of return loss and efficiency during the simulation of antenna. According to transmission line model, the length, *L* and width, W of the patch are calculated as [5]:

2.1.1 Calculation for patch width:

The dimension of the antenna depends on the operating/resonant frequency and value of dielectric constant.

$$W = \frac{c_o}{2f_o}\sqrt{\frac{2}{\varepsilon_r + 1}} \tag{1}$$

Where *W* is antenna width in mm, c_o is speed of light in free space with 3×10^8, f_o is operating frequency or resonant frequency, and ε_r is dielectric constant of substrate.

2.1.2 Calculation for effective permittivity

The effective permittivity is important in the designing process of the antenna. There is radiation travelling through from patch towards the ground, followed by passing through air then passing through the substrate and this occurrence is referred as 'Fringing'. As we know that the air and substrate have different dielectric constant, so we need to find the effective of the dielectric constant [6].

$$\varepsilon_{reff} = \frac{(\varepsilon_r + 1)}{2} + \frac{(\varepsilon_r - 1)}{2}\left(1 + 12\frac{h}{W}\right)^{-0.5} \tag{2}$$

Where ε_{eff} is the effective permittivity, *h* is the height of the substrate in mm, ε_r is the dielectric constant of substrate and *W* is the antenna with in mm.

2.1.3 Calculation for effective length:

$$L_{eff} = \frac{c_o}{2f_o\sqrt{\varepsilon_{reff}}} \tag{3}$$

Where L_{eff} is the effective length of patch, ε_{eff} is the effective permittivity, c_o is speed of light in free space with 3×10^8, and f_o is operating frequency or resonant frequency.

2.1.4 Calculation for length extension:

The size of the antenna is increased by an amount of length extension, ΔL due to fringing [6].

$$\Delta L = 0.412h \left[\frac{\left(\varepsilon_{reff} + 0.3\right)\left(\frac{W}{h} + 0.264\right)}{\left(\varepsilon_{reff} - 0.258\right)\left(\frac{W}{h} + 0.8\right)} \right] \quad (4)$$

Where ΔL is the length extension, ε_{eff} is the effective permittivity, h is the height of the substrate in mm, and W is the antenna width in mm.

2.1.5 Calculation for patch length:

$$L = L_{eff} - 2\Delta L \quad (5)$$

Where L is the length of patch in mm, L_{eff} is the effective length of patch and ΔL is the length extension.

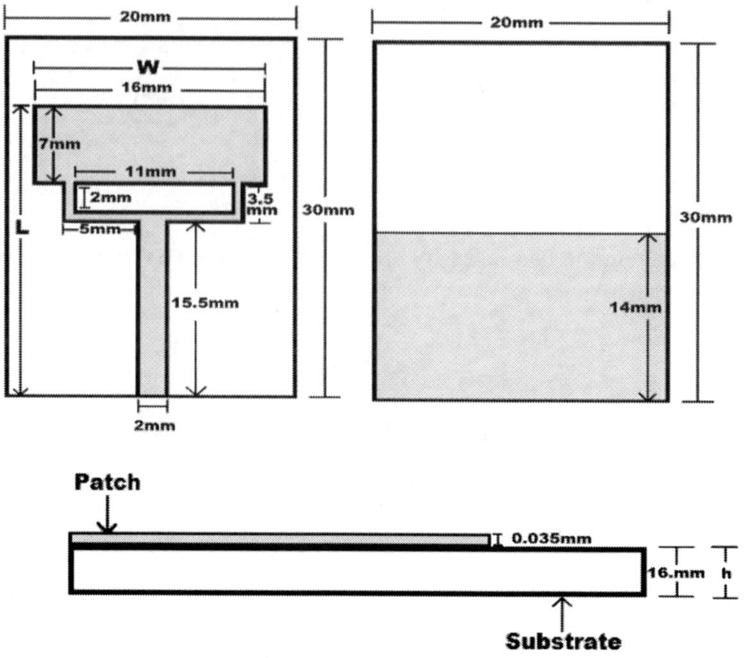

Fig. 2. Parameter Design of Proposed Antenna.

2.2 Antenna bending

To demonstrate the bending of the antenna, Figure 3 shows a cylinder with radii of 5-10 mm. The bending function is visualized to bend on a cylindrical surface in CST software. Figure 4 shows the bending direction towards the cylindrical surface. The antenna is bend on XZ plane which bends from both sides of the antenna. The antenna is tested on different type of radii bending of the cylinder. While the bending of antenna is increased, the radii of the cylinder is decreased. The results of S parameters, VSWR and Radiation Pattern is simulated.

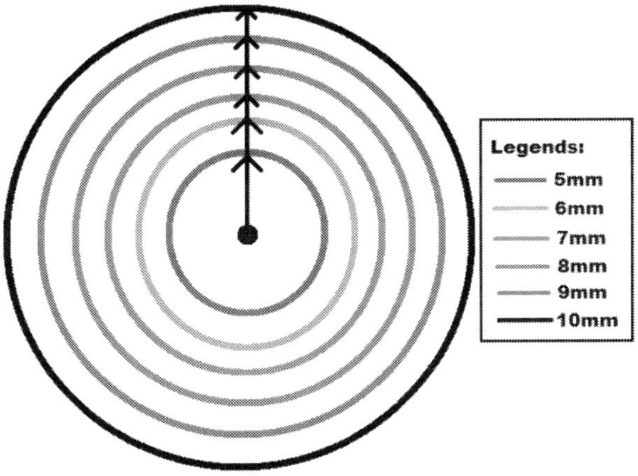

Fig. 3. Different radii of the cylinder.

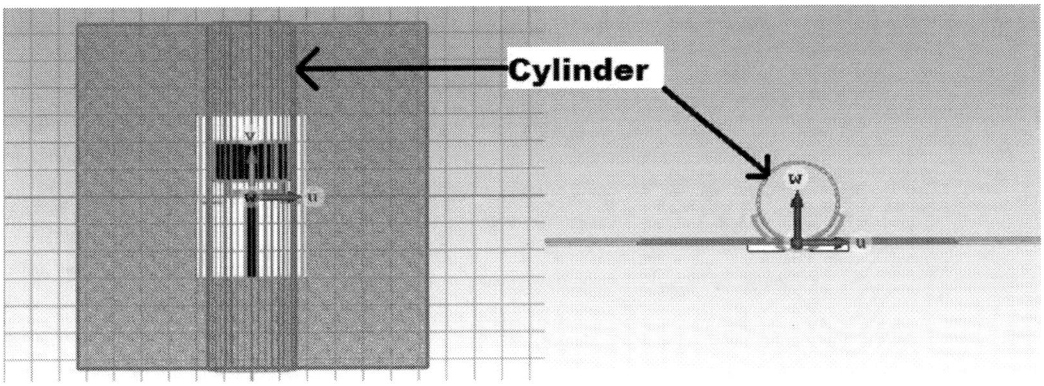

Fig. 4. Cylinder object for the antenna to bent towards in CST.

3 Result and discussion

3.1 Planar antenna performance

Simulations are carried out using Computer Simulation Software (CST) to simulate the behaviour of the antenna. Figures 5 and 6 show the simulated results in terms of return loss and VSWR of the antenna. The simulated return loss shows an impedance bandwidth of 2.67 GHz (3.08 GHz – 5.75 GHz) which is in a good operate operation. The VSWR shows a huge drop for 1 GHz to 3 GHz. One of the reason is that the frequency range is not suitable to simulate a better VSWR result since the VSWR is more than 2. The results starting from 3GHz to 8GHz has a better measurement in the VSWR which satisfies the VSWR ratio requirement of less than 2. In order to test and determine the performance of the planar antenna, the planar antenna is then bent in the XZ plane with different cylinder radii (r = 5, 6, 7, 8, 9 and 10 mm). The results of the bent design are compare with the results of planar design.

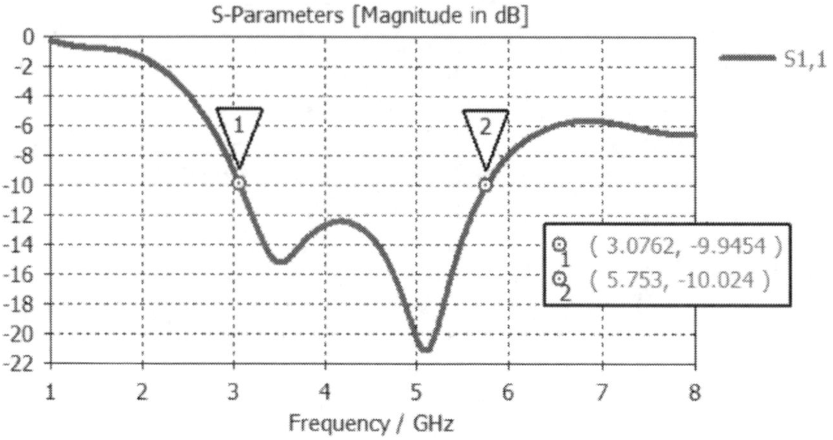

Fig. 5. Simulated S Parameter for Planar Antenna.

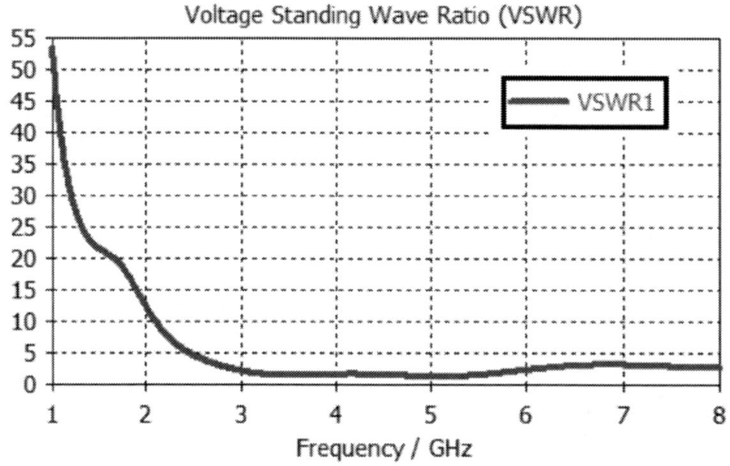

Fig. 6. Simulated VSWR for Planar Antenna.

3.2 Effects on return loss and bandwidth

Figure 7 compares planar antenna and bent antenna that is cylindrical bent with different radii of 5-10 mm. Theoretically, the antenna changes its effective length when the antenna is bent. The bending on the antenna increases, the effective length is decreases along and the resonant frequency will shift to a higher band [7]. Table 1 show that the bandwidth is large (in GHz) due to the low permittivity of the dielectric substrate [2]. This explains that lower permittivity dielectric substrate with thicker in thickness is able to produce an antenna design with high efficiency and large bandwidth. Meanwhile, dielectric substrate with high permittivity will result in a lower efficiency in the antenna design results. The return loss shows increment along as the bending increases which means the antenna is operating normally. Generally, a larger return loss results in more energy is delivered into the antenna. The antenna is to be design in wide frequency range and the wider of the bandwidth, the better in the antenna performance. The antenna operates normally within the desire range even though there is a shift in frequency. The performance of the proposed antenna is slightly decreased while the antenna bending is increased but the antenna still operates efficiently at resonant frequency of 4.591 GHz.

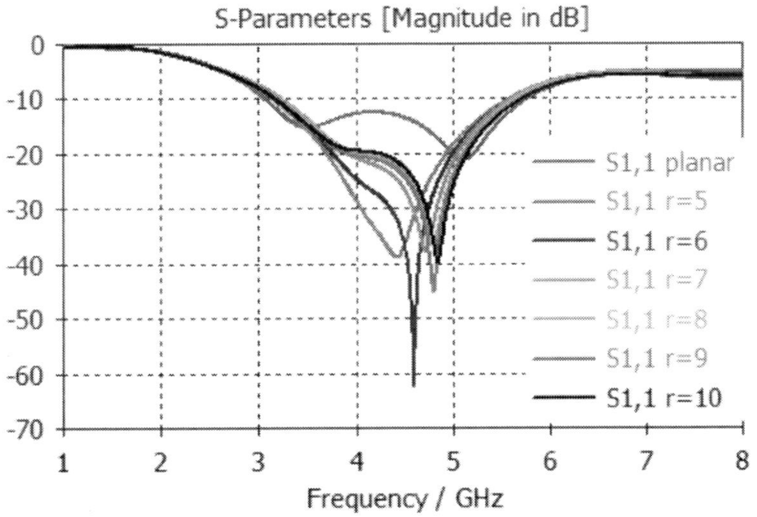

Fig. 3. Variation of return loss with bending.

Table 1. Change in radiation characteristics of antenna with bending.

Cylindrical Bend [mm]	Return Loss at 4.591 GHz [dB]	Bandwidth [GHz]
r=0 (planar)	-14.17	2.68 (5.76-3.08)
r=5	-31.56	2.40 (5.56-3.16)
r=6	-62.32	2.40 (5.59-3.19)
r=7	-30.14	2.35 (5.59-3.24)
r=8	-26.12	2.40 (5.63-3.23)
r=9	-26.78	2.50 (5.68-2.50)
r=10	-24.63	2.52 (5.70-3.18)

3.3 Effect on voltage standing wave ratio (VSWR)

Figure 8 shows the graph of VSWR comparison between both planar antenna and bent antennas. The graph shows that the bent antennas has a similar trend as the planar antenna in terms of VSWR. From the generated Voltage Standing Wave Ratio (VSWR), the results are taken from the resonant frequency at 4.591 GHz. Table 2 shows the percentage of power delivered by the receiver reflected from the antenna. The maximum power reflected is 3.8% which 96.2% of the power is delivered to the antenna. The minimum power reflected is 5.86E-05% which 99.99% of the power is delivered to the antenna. The antennas performance is considered good and little would be gained by the impedance matching since the simulated VSWR results are less than 2. The VSWR from 1 GHz to 3 GHz shows a high VSWR which means that the range of frequency is reflecting a higher power back to the antenna which may affect the antenna performance. After that, the VSWR is within the acceptable range which most of the antenna VSWR are below 2.

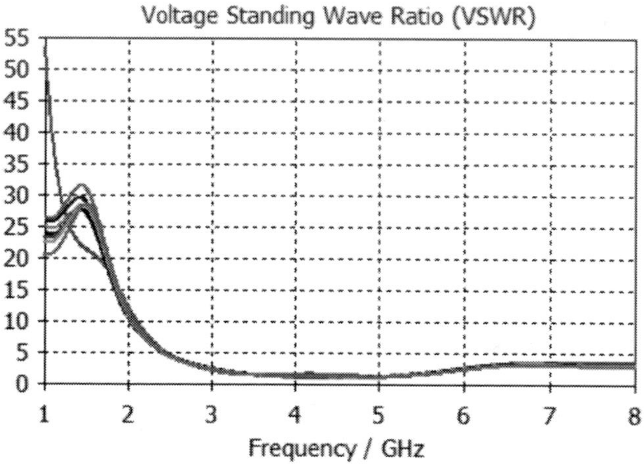

Fig. 8. Comparison of planar antenna and bent antenna in terms of Voltage Standing Wave Ratio.

Table 2. Change in VSWR of antenna with bending.

Cylindrical Bend, *r* [mm]	VSWR	Return Loss at 4.591 GHz [dB]	Reflection Coefficient, *Γ*	Reflected Power (%)
0	1.49	-14.17	0.19	3.82
5	1.05	-31.56	0.03	0.069
6	1.00	-62.32	0.0007	5.86071E-05
7	1.06	-30.14	0.031	0.10
8	1.10	-26.12	0.049	0.24
9	1.10	-26.78	0.046	0.21
10	1.12	-24.63	0.059	0.34

3.4 Effect on radiation pattern

Figure 9 shows the 2D radiation patterns for planar and bent antenna on XZ plane. Table 3 shows the angular width (3 dB) tends to decrease along with the increase in bending of the antenna. This leads to shrink in the radiation pattern in bending plane which the gain and efficiency of the antenna increases with the bending. The antenna gain, G is related to directivity, D and antenna efficiency of the antenna which the efficiency of the antenna increases along with the bending of antenna.

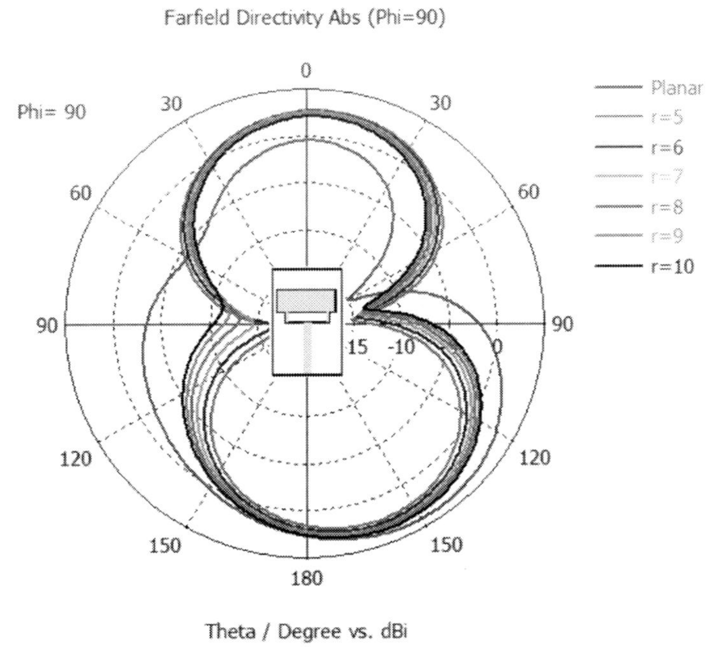

Fig. 4. Radiation pattern of planar antenna and bent antenna.

Table 3. Change in radiation pattern characteristics of antenna with bending.

Antenna Radii, *r* [mm]	Main Lobe Magnitude [dBi]	Main Lobe Direction [deg.]	Angular width (3 dB) [deg.]	Gain (dB)	Directivity (dBi)	Efficiency (%)
0 (Planar)	3.93	147.0	96.0	1.00	3.93	25.5
5	2.70	6.0	78.0	1.33	2.70	49.5
6	2.61	5.0	75.7	1.06	2.61	40.6
7	2.86	159.0	80.9	0.89	2.86	31.2
8	3.07	158.0	82.7	0.75	3.06	24.5
9	3.22	158.0	84.1	0.79	3.21	24.7
10	3.33	157.0	85.3	0.83	3.33	24.9

5 Conclusion and future work

The paper is focused on the limitation of planar antenna performances when the antenna design is bent at certain degrees at given frequency range (1-8 GHz). The design of planar antenna is executed in the Computer Simulation Technology (CST) Software. In this research, the result shows that the performance of the proposed antenna is slightly decreased while the antenna bending is increased but the antenna still operates efficiently on the resonant frequency of 4.591 GHz. The results of VSWR between both designs are within the acceptable range after 3 GHz onwards. The angular width decreases in the plane of bending which results in increase in the antenna gain. The antenna still operates efficiently from 25.5% to 49.5% with bending on the resonant frequency at 4.591 GHz even after bending. Based on the obtained conclusion, the antenna design can be implemented as a Wearable Antenna application. In future work, the performance analysis of the antenna in different angle will be done as well as trying on the different type of material used for the antenna substrate.

Reference

[1] N. A. ELIAS, N. A. Samsuri, M. K. A. Rahim, C. J. Panagamuwa, and W. G. Whittow, *J. Teknol.,* **77** (2015)

[2] S. Dey, N. Saha, and A. Alomainy, *Antennas and Propagation Conference (LAPC)* (2011)

[3] "S-parameter measurement Basics for High-Speed Digital Engineers." [Online]. Available: http://literature.cdn.keysight.com/litweb/pdf/5991-3736EN.pdf. [Accessed: 20-Oct-2017].

[4] Y. Ahmed, Y. Hao, and C. Parini, *Int. J. Antennas Propag.*, **2008** (2008).

[5] V. Bernard, J. Paul, and I. Iloh, *Microstrip Antenna Des. Using Transm. Line Mode*l, **3** (2013).

[6] M. A. Afridi, *Biol. Chem. Res.*, **2015** (2015).

[7] M. Rizwan, L. Sydänheimo, and L. Ukkonen, *Prog. Electromagn. Res. Symp.*, **2015-January** (2015).

Prediction of Cyclists Movement in Different Terrain Conditions

Mahesh A/L V.Nagarrettinam[1], *Aravind CV*[1,*], *Mohsen Nabi Poor*[1]

[1]School of Engineering, Taylor's University, Malaysia

Abstract. In Malaysia, most of the accidents involving a bicycle and another vehicle are due to either the driver or rider 'failing to look properly'. This is more significant with the government initiatives to support the use of bicycle making the carbon-free environment, a vision of TN50. This research addresses the safety aspect of the cyclists in terms of the driver's point of view which improves cyclist visibility during driving. The proposed helmet system implements a rule-based algorithm which predicts the turning and braking movement of the cyclists. With this system, additional illumination and signaling are provided for the cyclists. The major challenge faced is the implementation of an algorithm for various situations of cycling. To ensure the system could be used on the road, the accuracy and speed of the automatic signaling system need to adhere. Situations that affects the output of the indicators include bicycle speed, the angle of turning, body tilt, duration of turn and random body movements. This paper implements a 3-axis accelerometer and a microcontroller in a data logger to acquire the required data which are analyzed in MATLAB. Using filtering technique, the acquired data are then be cleaned to remove noise due to vibration during cycling. The characteristics of braking and turning are then analyzed in the time domain as well as frequency domain to ensure the optimum algorithm used for gesture recognition and movement prediction. The algorithm is based on sliding window, FFT and threshold-based rule algorithm. The output based on the rule-based algorithm then illuminate the corresponding signals which provide the safety feature of the system.

1 Introduction

70-80% of accidents between a bicycle and another vehicle is found to be mainly due to human error. Through initiatives raised from various governing bodies, the improved usage of the bicycle is supported to reduce carbon emissions from the motorized vehicle [1]. Therefore the safety aspect of a cyclist must be improved to encourage usage of a bicycle while reducing the statistic of accidents. Automated helmet signals are proposed to improve the safety aspect of. This is due to an observation that road users tend to neglect the usage of safety signals which could lead to serious repercussions.

[1] Corresponding author: aravindcv@ieee.org

At the initial stages, as the technology for automation is still improving, the addition of automated safety signals on the helmet is affected in terms of speed and accuracy due to low speeds of cycling as compared to a motorized vehicle. Therefore, this research intends to provide automatic signaling and illumination such as brake signal and turning signal acquired from inertial sensors mounted on the helmet. The main challenge of the research is to implement movement prediction algorithm. This is implemented by using an accelerometer. The accelerometer has an advantage over other inertial measurement unit are cheaper and accurate as compared to gyroscope and magnetometer with a wide frequency range. This effect is demonstrated in Siti Sarah et. al. (2017) [2]. The accelerometer directly measures the movement of the cyclists, but this recorded raw data has limitations in which if there is excessive noise in the signal, the analysis of the raw data in the time domain is used. Therefore, a reliable prediction technique is required if the system is to be implemented on the road. An accelerometer-only system is designed due to its low power consumption, low cost, and ease of usage. This is further backed by its capability to measure acceleration, orientation, rotation, and motion. By combining the stated characteristics, an accurate and robust prediction system could be achieved with a specifically designed algorithm for prediction. Figure 1 shows a typical 3-axis accelerometer. The efficiency of an accelerometer-only solution is explored from its ability to measure movement, orientation and rotation changes by O. Incel. The efficiency is then compared with a combination of accelerometer, gyroscope and magnetic field sensor. It is found that the performance of an accelerometer-only solution can achieve up to an 80% recognition which doesn't differ much from a combination of accelerometer, gyroscope and magnetic field sensor. When the gyroscope and magnetic field sensor are used alone, they do not reach the capabilities of the accelerometer. It is also found that when the motion, orientation and rotation information from the accelerometer are combined in a specifically designed algorithm, it provides a much higher accuracy as compared to using the motion information alone. Therefore, the accelerometer only system is preferred due to its low power consumption and accuracy [3]. To ensure minimal noise affecting the output of the accelerometer due to vibration during cycling, low pass filter is introduced to be incorporated to the measurement system. The low-pass filter functions by smoothing the measured signals to clean and remove noise while performing data averaging. A method for cleaning of noisy signal through low pass filter is explained in Suresh Manic et. al (2014) [4]. Furthermore, important patterns and trend within a signal could be made more vivid as the frequencies due to noise are removed. The filter functions by attenuating signals above a predetermined cut-off frequency which is manually set after obtaining necessary data from tests done. Figure 2(a) shows an example of the low pass filter applied to a noisy signal. The noisy signal is denoted by the green line whereas the low pass filtered signal is shown by the blue line. [5] Implementation of sliding window function applied for easier extraction of the acceleration data in time series. The sliding window is a continuous function whereby the window length is determined after analyzing the data sets recorded by the data logger. An example of a sliding window is shown in figure 2 (b) [6].

Fig. 1. 3-axis accelerometer

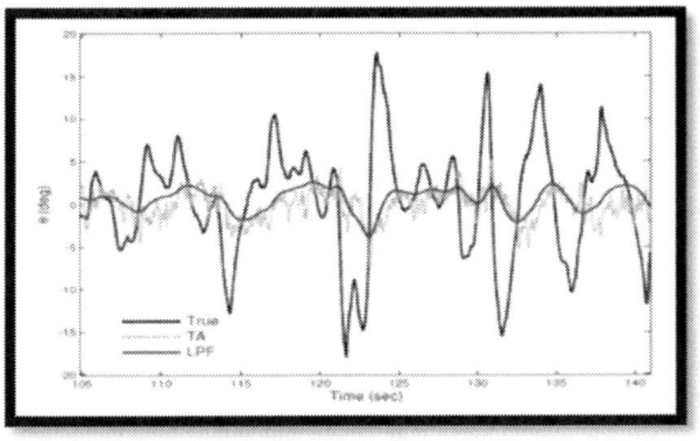

(a) Comparison of traditional attitude and low-pass filtered signal

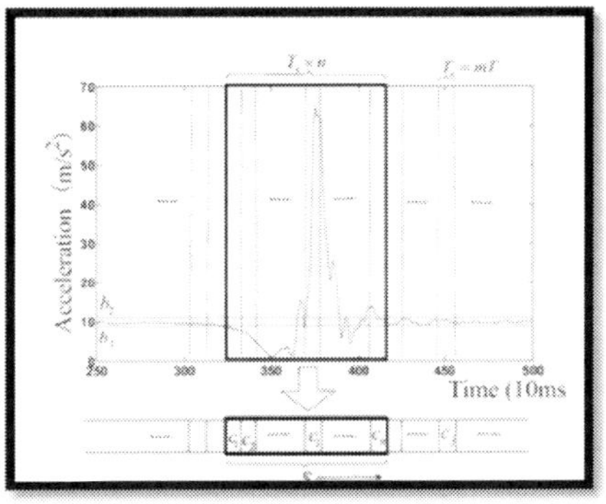

(b) Sliding window example with window length of 10 secs

Fig. 2. Comparison of low-pass filtered signal and the sliding window of 10 secs.

An implementation of FFT analysis is used next to determine the frequency domain of the data. From the frequency domain features, the frequency and frequencies amplitude of data of the accelerometer could be then calculated. The analysis of repetitive signals obtained from vibrations along with any harmonics within the signal could then be obtained. An example of FFT implementation with a time-series signal is seen in Fig.1.3 (a) [7]. A rule-based system consists of if-then statements whereby the rules are implemented to act upon specific proclamations. The rule-based system often provides answers determined by validation and verification of each manually constructed rules. Such systems provide the output for any given problems automatically in place of human experts. The data is analyzed in MATLAB to determine peak accelerations. From the peak accelerations, rules are introduced using threshold techniques. As seen in Fig. 1.3(b), the amplitude of acceleration for the movement analyzed is large for a typical fall. The threshold for the fall is determined by the green dotted line whereby the rule is set to be above a predetermined threshold [8-9]. There are variations of helmet implemented with inertial sensors, but it is proven to be bulky due to multiple inertial sensors and suitable for usage in the field. Therefore, this paper addresses the gap by implementing an accelerometer on the helmet. From the data measure real-time, sliding window function is applied to the data whereby FFT is implemented in the window length. From the frequency domain, a threshold based rule algorithm implements the output related to the rule. The development of the movement prediction system incorporates a single 3-axis accelerometer in the helmet to estimate the orientation or position of a body. Because of the advantages of using an accelerometer has over other Inertial Measurement Unit (IMU) which are its low power consumption, low cost, and small size, the accelerometer is used. For problems related to measurement biasing and noise, it is determined from above reviews that low-pass filter is easy to implement and accurate. Various limitations such as memory constraint, sensor constraint, cost constraint and microcontroller constraint are determined.

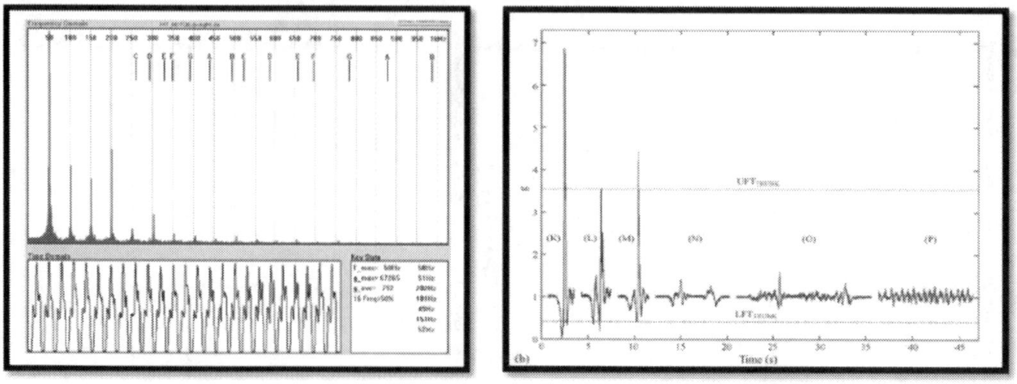

(a) Comparison of accelerometer data in time domain and frequency domain

(b) Threshold example applied to accelerometer data

Fig. 3. Time domain and frequency domain and the application of a threshold to the data.

2 Methodology

Figure 4 shows the methodology chart which determines the overall steps throughout the research. Based on the outcome of each stage, the performance of the next stage is adjusted accordingly.

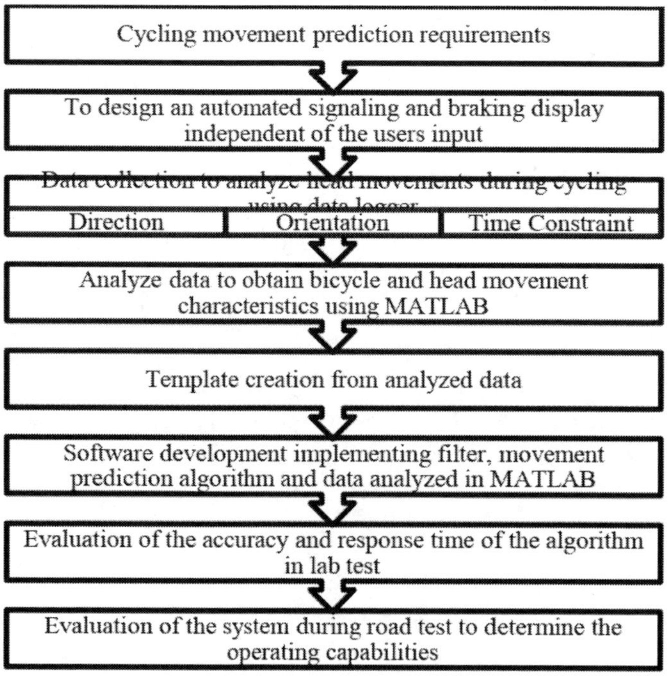

Fig. 4. Methodology

2.1 Data Logging

The data logger is built to record the head movement in conjunction with the bicycle movement of the rider. The system includes an accelerometer, SD module, microcontroller and power supply mounted to a helmet. For bicycle movement logging, a switch and potentiometer are connected to the bicycle handle and is wired to the microcontroller on the helmet. The system connection is represented and connected based on Figure 5 which shows the outline of the system. The data recorded includes three outputs of the accelerometer an output of the switch and the potentiometer output which is then be stored in the SD card and further analyzed in MATLAB. The schematic design of the data logger is as seen in Figure 6. The microcontroller is powered by a 9V battery whereby it provides a 5V to the SD card module and 3.3V to the accelerometer. For turning prediction, two types data are recorded which are the head movement acquired from the accelerometer and handle angle movement which is obtained from the potentiometer connected to the pivot of the handle bar. The handle movement data obtained is from the variation of voltages based on the turning of a potentiometer which is due to the turn of the bicycle handles. The working operation is by setting the middle of the potentiometer in position with the center of the handle angle. During straight cycling movement, the voltage is half of the supply voltage. As the bicycle turns left or right, the voltage decreases or increase corresponding to the turn. This value is then recorded in conjunction with the recorded head movement of the accelerometer.

For braking prediction, two types of data are recorded as well, which are the head movement obtained throughout the cycle and position of brake handles which is obtained from connected switches. During braking, the output of the switch is high and vice versa. The accelerometer data, handle movement and braking movement is recorded simultaneously to ensure the analysis in MATLAB is much easier and smoother. Figure 7 (a) shows the connected data logger is installed on a helmet. The reason of applying the data logger on the helmet is due to the need of analysing head movement during cycling. As there is a right turn, left turn or brake, the head move accordingly, therefore the data recorded then be analysed to determine the characteristics of the data. For the logging of movement of bicycle, the system designed and is shown in Figure 7 (b).

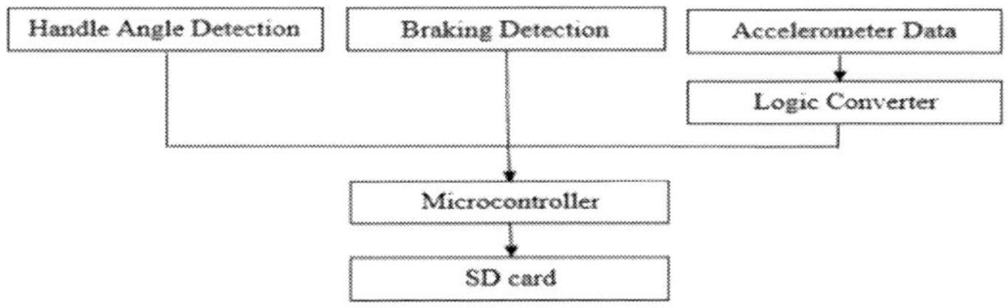

Fig. 5. System connection of the data logger

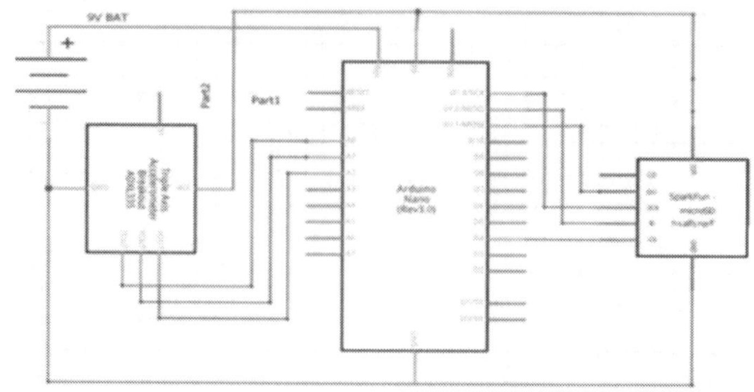

Fig. 6. Schematic design of the Data Logger

(a) Helmet Data Logger (b) Bicycle Connection of Data Logger

Fig. 7. Helmet and Bicycle Data Logger Connection

2.2 Analysis of accelerometer data

The analysis of the raw data obtained from the data logger is done in MATLAB which provides all the required functions and algorithm along with an extensive online support. The initial stage is to plot the raw data and analyse the characteristics of the braking and turning raw data. For the braking data, it can be seen in figure 8(a). A vivid indication of the braking is seen along the X-axis whereby there is sharp drop in the data. Not much changes occur in the Y and Z axis during braking. The braking signal is provided by the connected switch and indicates a high when braking and low in non-braking conditions. visible change is caused by the vibration during stopping. The turning raw data as in figure 8(b) and figure 9(a) whereby they represent left turn and right turn respectively. During turning, the most affected axis is the Y-axis. When turning to the left, the voltage of potentiometer increases while the Y-axis signal increase towards the negative Y-axis. When turning to the right, the voltage of the potentiometer decrease while the Y-axis signal increases towards the positive Y-axis. Figure 9 (b) shows the sliding window function by 2 parallel red lines. The window length of the sliding window is 150 which equates to 5 seconds. With this function, the data points is analysed easily with the implementation of FFT within the window. As seen, the frequency domain is extracted from the two-different window time. By analysing the time domain graph, it is seen that from 261 - 265 secs, a change in amplitude occurs which is generated due to turning or braking. For this set of data, from 261 – 265 secs braking occurs and from 252 -257.03 secs the cyclist is in a constant straight motion. Figure 2.7, indicates the application of FFT for two different sets of data which are from 261 – 265 secs and 252 – 257.03 sec. Figure 10 shows the frequency spectrum within the sliding window. Furthermore, for braking the FFT mainly applied at the X-axis as the change in braking is more prominent. Therefore, by analysing the frequency domain, there is a peak in the braking signal at around 2 Hz. During constant straight cycling which is from 252- 257.03 sec, it is seen that there is no peak in the frequency domain.

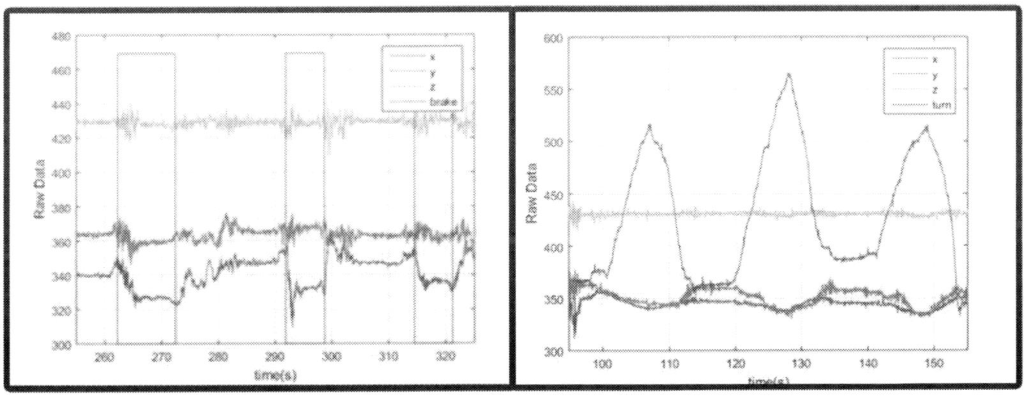

(a) X, Y, Z axis and brake raw data (b) X, Y, Z axis and left turn raw data

Fig. 8. X, Y, Z axis brake raw data and left turn raw data

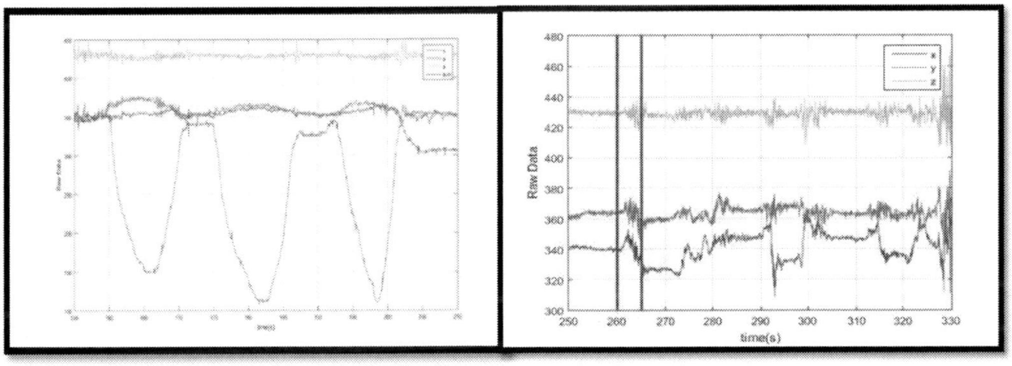

(a) X, Y, Z axis and right turn raw data (b) Sliding window applied to raw data

Fig. 9. X, Y, Z axis right turn raw data and the sliding window application.

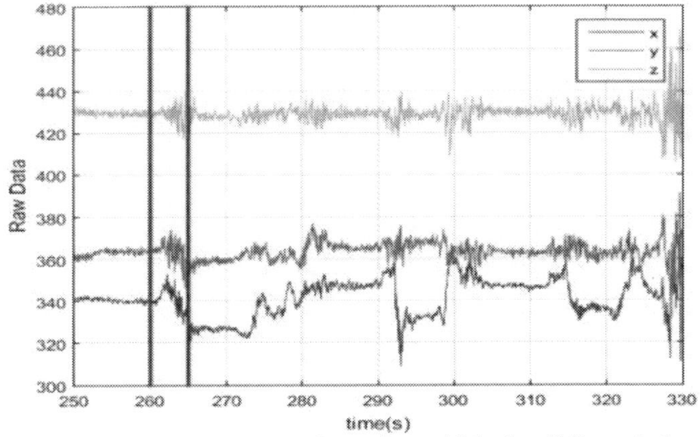

Fig. 10. Frequency Spectrum of raw data within the sliding window

3 Results and Discussion

From the methods stated, the results are processed up to the implementation of threshold-based rule algorithm for braking analysis and turning analysis.

3.1 Braking analysis

In table 1, a braking signal recorded which is from the X-axis. The data are then converted to the time domain and frequency domain. In the time domain, the analysis of impulse and transient accelerometer signals is possible. With the addition of external noise such as vibrations, it is more difficult. Therefore, FFT algorithm is applied to the time domain signal to convert it into the frequency domain for spectrum analysis. In the frequency domain, each individual frequency and its amplitude is displayed which enable the analysis of repetitive signals. As seen, the peak n frequency is at 2hz. This shows that the changes in frequency during braking is very small but at a high rate. In Table 2, a data for cycling on a road bump is recorded. It is found that by plotting the data, the main axis which displays a vivid road bump cycling is mainly at the Z-axis. The data are then converted to the time domain and frequency domain. In the time domain, it is seen that the signal peak followed by an instantaneous drop and then return to its previous position. This shows that the accelerometer moves up and down along the z-axis while in this cycling movement. When FFT algorithm is applied to the time domain and the signal is converted to the frequency domain, the frequency cluster at an average of 10 Hz. In table 3, a data for constant cycling are recorded. During constant cycling, not many changes occur on all 3 axes. The data sets are then converted to the time domain and frequency domain. In the time domain, it is seen that the signal maintains a constant level throughout the duration When FFT algorithm is applied to the time domain and the signal is converted to the frequency domain, the frequency is found to be at a very low amplitude. As seen in the y-axis of the frequency domain, the magnitude doesn't exceed $1x10^{-4}$.

Table 1. Comparison of brake data in time domain and frequency domain

Brake data in time domain	Brake data in frequency domain
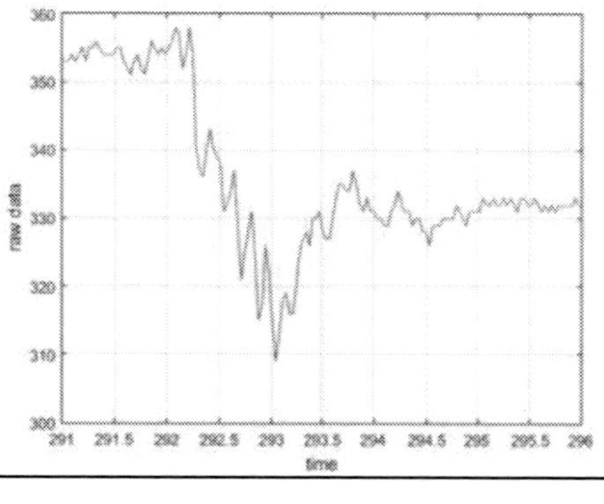	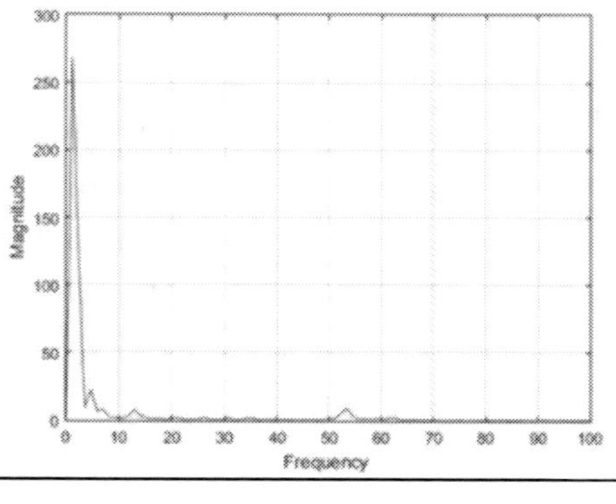

Table 2. Comparison of road bump data in time domain and frequency domain

Road bump data in time domain	Road bump data in frequency domain
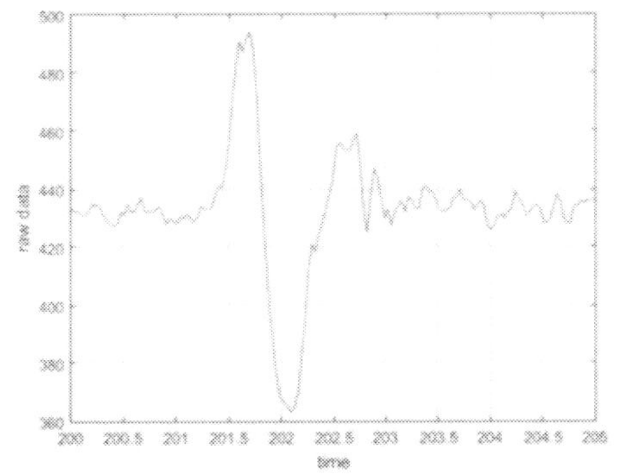	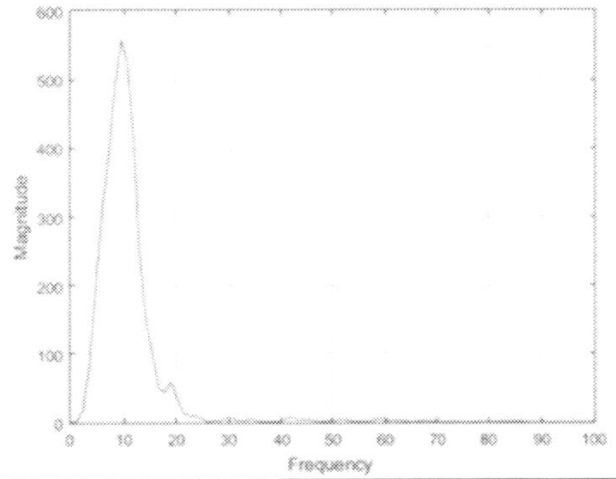

Table 3. Comparison of constant cycling data in time domain and frequency domain

Constant cycling data in time domain	Constant cycling data in frequency domain
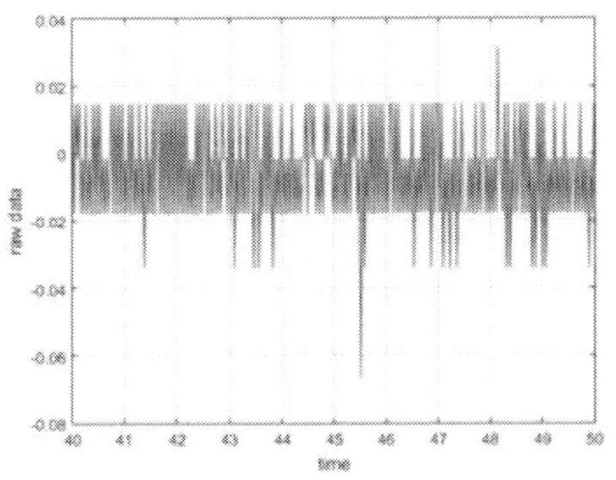	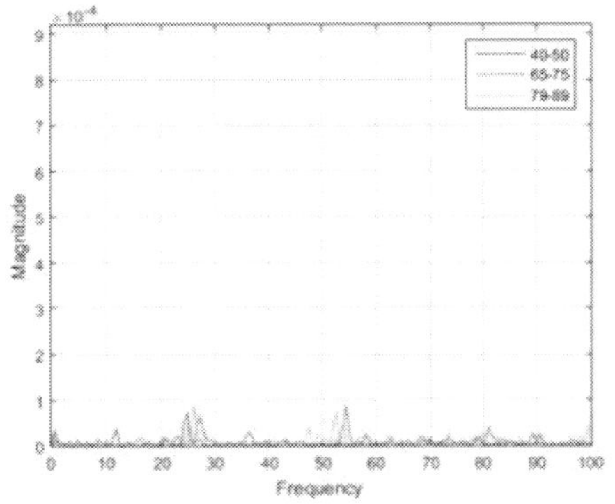

3.4 Turning analysis

Compared to the braking implementation of prediction and gesture recognition, the turning implementation is significantly harder. This is due to various situations such as duration of the turn, speed during a turn and head movement during the turn. In Table 4, a data set for left turn cycling are recorded. It is found that when the time domain of the left cycling data is converted to the frequency domain, there is not much difference with the braking frequency domain. This is due to the change in frequency during turning is too small, therefore the data sets are clustered around the 1Hz frequency. The only difference this frequency domain has over braking frequency domain is that braking frequency is derived from the X-axis while turning frequency is derived from Y-axis. By analyzing the time domain data which is in Y-axis, an initial rule for turning is derived. The rule is derived when sliding window is applied on all 3-axis separately. When there is a large change in the frequency domain for Y-axis as compared to the X and Z-axis, the algorithm can output an initial turning condition. In table 5, a data for right turn is shown. Similarly, for a left turn, right turn output the same frequency in the frequency domain. By following the rule stated earlier which is to apply sliding window on all 3-axis and then analyze the changes in the axis. If the change in the frequency domain for Y-axis is much larger than the other axis, a turn is occurring.

Table 4. Comparison of left turn cycling data in time domain and frequency domain

<table>
<tr><th>Left turn cycling data in time domain</th><th>Left turn cycling data in frequency domain</th></tr>
<tr><td>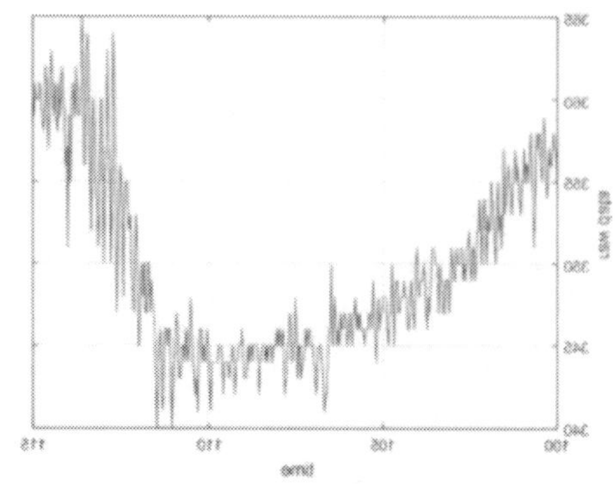</td><td>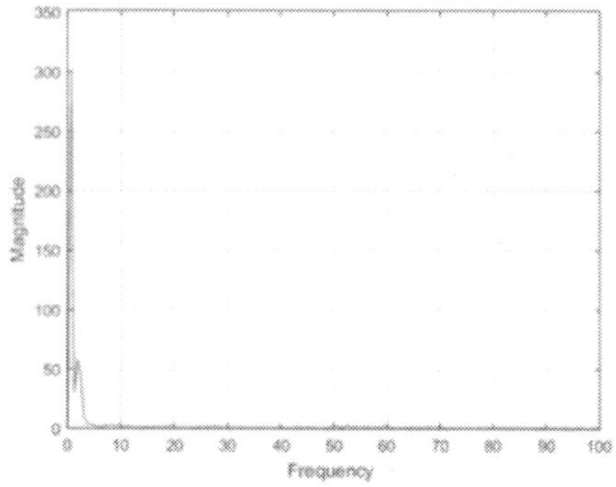
</td></tr>
</table>

Table 5. Comparison of constant cycling data in time domain and frequency domain

<table>
<tr><th>Right turn cycling data in time domain</th><th>Right cycling data in frequency domain</th></tr>
<tr><td>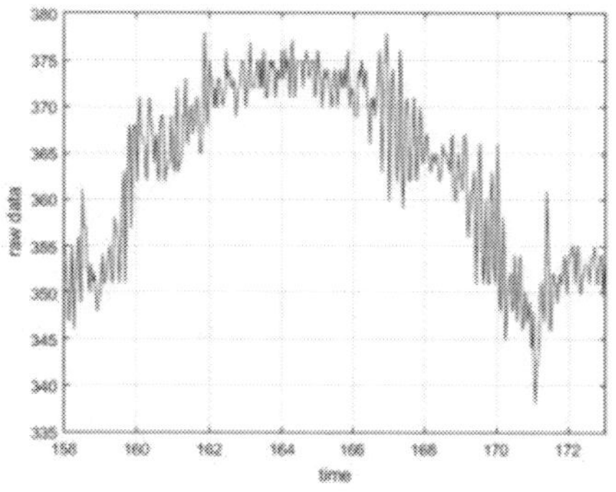
</td><td>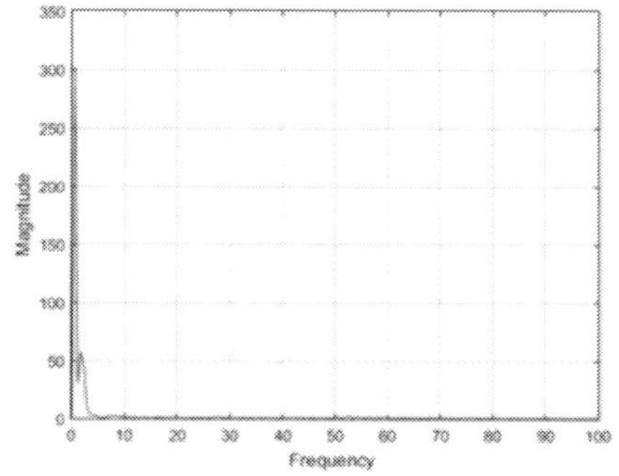
</td></tr>
</table>

4 Conclusion

A concept for movement prediction is presented using an accelerometer. The algorithm is based on the sliding window function, FFT and threshold-based rule algorithm. In terms of the accuracy for the turning indication, it is not up to the safety standards yet. The processing speed is not to be tested as the suitable microcontroller for the final product is not implemented. With the completion of this research, a significant improvement in the safety standards related to cycling is achieved. For future work, the improvement in algorithm is a must to keep up with the safety and automated standards in the society. To further increase the efficiency of the system, a GPS and mapping software is implemented alongside this system. Another method to improve this system is to implement the algorithm within a smart phone due to it containing all the necessary IMU such as an accelerometer, gyroscope, GPS, magnetometer etc. With the constant improvement and advancement in technology, the efficiency of this IMU within the smart phone keeps improving as well. This reduces the cost of the overall system and implementation will be easier due to decrease in components.

References

1. Cycling Accidents, *Road* Safety Factsheet, June (2017) [Available online: https://www.rospa.com/rospaweb/docs/advice-services/road-safety/cyclists/cycling-accidents-factsheet.pdf]
2. S. Mohd. Ariff, M. Hosseini Fouladi, N. Krishnasamy, S. Namasivayam, M. Nabi Poor, P. Madhavan, W. Eng Hwa, Diagnosis of Lung Related Diseases by Assessing Vibration Patterns of Structures at the Thoracic Cavity, Journal of Engineering Science and Technology, **6**, no. 1, pp. 15-28, (2017).
3. O. D. Incel, Analysis of movement, orientation and rotation-based sensing for phone placement recognition, Sensors (Switzerland), **15**, no. 10, pp. 25474–25506, (2015).
4. K. Suresh Manic, C. Aravind, A. Saadha, K. Pirapaharan, Implementation of Real-Time Brainwave Characterization and Visualization, Journal of Engineering Science and Technology, **2**, no. 1, pp. 50-59, (2015).
5. R. C. Leishman, J. C. MacDonald, R. W. Beard, T. W. McLain, Quadrotors and accelerometers: State estimation with an improved dynamic model, IEEE Control Syst., **34**, no. 1, pp. 28–41, (2014)
6. L. Tong, Q. Song, Y. Ge, M. Liu, HMM-based human fall detection and prediction method using the tri-axial accelerometer, IEEE Sens. J., **13**, no. 5, pp. 1849–1856, (2013).
7. K. Tuck, Frequency Analysis in the Industrial Market Using Accelerometer Sensors, Sensors (Peterborough, NH), pp. 1–13, (2008).
8. L. Lengyel, Validating rule-based algorithms, Acta Polytech. Hungarica, **12**, no. 4, pp. 59–75, (2015).
9. A. K. Bourke, J. V. O'Brien, G. M. Lyons, Evaluation of a threshold-based tri-axial accelerometer fall detection algorithm, Gait Posture, **26**, no. 2, pp. 194–199, (2007).

Design Analysis of a Dual Rotor Permanent Magnet Machine driven Electric Vehicle

Mohd Izzat Bin Zainuddin[1], *Aravind* CV[1,*]

[1]School of Engineering, Taylor's University, Malaysia

Abstract. Electric bike in urban countries such as Europe and China commonly used the brushless direct current machine (BLDC) as it able to produce high torque to transport the user from one place to another. However, BLDC torque density can't be improving due to limitation magnetic flux generated by the permanent magnet. Therefore, the performance of electric bike can't be improved. Outer rotor BLDC machine design able to improve the torque density of the motor due to increase radius of the motor which can be explained by simple physics equation (Torque = Force x radius). However, an outer rotor machine only generates constant speed, which is not suitable for operating under tractive load condition, especially electric bike. The proposed model is a new novel of double layer outer rotor BLDCPM machine which able to amplify the magnetic flux density and improve the torque density of the machine. The mutual magnetic coupling between the inner and outer rotor of the proposed model increase the magnetic flux intensity as both of them acts as individual parts. Thus, the magnetic flux generated by both rotors are double which resulted in improving the performance of the E-bike. Designing parameters and analysing the performance of the proposed 2D model is done using FEA tools. Evaluation of the conventional and proposed model by comparing torque performance, magnetic flux density and motor constant square density. Other than that, speed torque graph also is evaluated to justify either it can operate similarly to ICE engine with gears. Two model is designed which is Single Outer Rotor Brushless Direct Current (SORBLDC) and Double Outer Rotor Brushless Direct Current (DORBLDC) operated with the same cases of 27 Amp current supplied to it and operate under various speed from 500 rpm to 2000 rpm. The average torque produce by the conventional and proposed model are 2.045439 Nm and 3.102648 Nm. Furthermore, improvement of the proposed model to conventional model in terms of motor constant square density by 24.92%. Therefore, the proposed model able to improve the magnetic flux by amplifying which resulted to increase the torque density of the machine. Furthermore, the speed-torque graph of the proposed machine shows similarity with speed torque graph of ICE engine. Thus, the proposed machine is suitable to operate for bike application

[1] Corresponding author: aravindcv@ieee.org

1 Introduction

BLDC motor consist rotor embedded with permanent magnet and stator coils to produce high torque density. Furthermore, the rotor of the BLDC motor is supplied with direct current (DC) supply and it generates AC voltage waveform with a trapezoidal shape which reduces the total harmonic distortion (THD) in the machine[1]. The reason BLDC machine used for the electric vehicle due to the power supplied to the machine have lesser noise and provide better performance of the motor. However, room for improvement in terms of torque density of the machine need to be done due to the magnetic flux generated by the machine is fixed. In order to increase the magnetic flux the ratio of the permanent magnet needed to be doubled [2]. To achieve more powerful motor, the design of the motor and position along with the angle of the permanent magnet is research in order to obtain better torque density. Having a good design which included the position and angle of the interior permanent magnet in rotor section contribute to increasing the power density of the motor[3]. However, big spaces needed to insert the slanted PM even though an outer rotor machine. Other than that, the torque density of a motor is improved by increasing the radius of the rotor section. The performance of the motor can also be increased by designing new motor structure with double layer outer rotor PM. Mutual magnetic coupling between inner rotor with outer rotor improves magnetic flux intensity by making both of them as an individual part of the machine[4]. The magnetic loading of the machine increased as the flux generated by the permanent magnet is amplified. Thus, the torque density of the motor is increased. Therefore, the objective is to experiment double outer rotor layer machine that amplifies the magnetic flux which configuration on the magnetic flux density to determine the torque density of the motor.

1.1 Design Concept

A load of the machine for this research is a tractive load which to be specified is the electric bike. Tractive study is applied in order to analyze and predict the actual speed of the machine for many types of road conditions. In order to reach full capacity in working principle of an electric or hybrid vehicle, electric motor and generator machine are optimized fully in several road conditions. There are 6 conditions faced by the electric vehicle especially electric bike which are an early start, normal speed, acceleration, high speed, deceleration and stop position. Figure 1 shows the diagram of operating conditions faced by the machine[5].

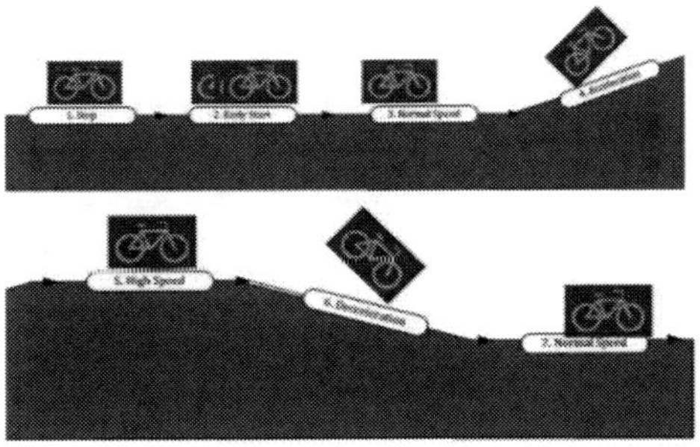

Fig. 1. Operating characteristic of E-bike for all conditions

E-bike is small transportation vehicle where the load on the propulsion unit is traction load. The loading characteristics of the proposed machine must show similarities with ICE engine. The speed-torque graph of ICE engine contains 4 parts which each of the section has its own torque and speed characteristics. Figure 2 shows the speed torque graph and speed power graph of ICE engine.

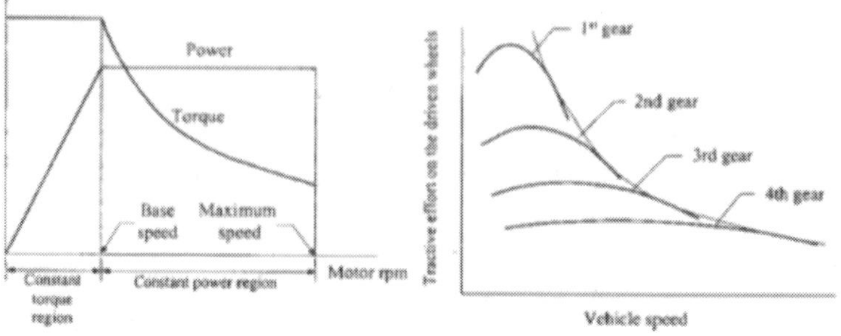

Fig. 2. Speed torque graph and speed power graph of ICE engine

In order to identify the research gap, few motor design had been researching and each of them has its own benefit and drawback. Induction motor widely uses for electric vehicle application high efficiency and low cost. The desired torque of certain speed can be controlled with an inverter can be achieved. However, control schemes of the torque are relatively difficult to control and the constant power range is also limited[6]. Permanent Magnet Synchronous Motor (PMSM) machine has a similar structure to BLDC and induction machine. Advantages PMSM machine are eddy current loss is reduced due to the segmented magnet at the rotor section thus torque density is improved. However, in order for improvement torque density of the machine, the volume of the machine needed to increased especially ratio of the segmented magnet in the rotor section[7]. Another design machine, using the in-wheel motor to SORBLDC machine to improve the magnetic flux by applying the concept of non-contact magnetic planetary gear with the permanent magnet outer rotor BLDC machine. With this method, the transmission torque is maximized resulted to increase the performance of the machine[8]. Coil Testing Study is also done on the SORBLDC in order to study electromagnetic flux generated by the coil windings which effect torque density of the machine. There are two types of armature coil winding implemented on the outer rotor synchronous machine which are concentrated winding and distributed winding. Finite element Analysis using JMAG application is done to simulate both coil winding technique on the conventional model for EV application. The output of the machine for each type of windings are analyzed using FEA tool in order to observer the magnetic field flow and flux density. The winding method that increases the performance of the machine in terms of magnetic field density is chosen. Concentrated windings generate higher peak and saturated flux across the model [9]. Thus, the torque density can be increase due to the magnetic flux generated by the permanent magnet and coil winding in the stator slot. Therefore, magnetic loading is focused for the proposed model on increasing the performance of the machine without changing the output rating of the motor for electric bike application. The proposed design is conducted by applying extra rotor layer to SORBLDC in order to create mutual magnetic coupling between outer rotor with the inner rotor to improve magnetic flux intensity by making both of them as an individual part of the machine. Therefore, the magnetic flux generated by both of individual part is amplified and the performance of the machine improved. Other than that, the proposed model of this research study able to achieve various speed with constant power and torque region[10].

The design proposed model is obtained by calculating parameters design formulas in order to know the size of each part. The magnetic circuit is redesigned in order to increase the magnetic flux in order to improve the torque density of the machine. By using FEA tools, design structure of conventional and proposed model is done. Other than that, evaluation of conventional and proposed model is done in terms of magnetic flux density, torque performance, motor constant square density, and power efficiency are done in FEA tools. Speed torque graph and speed power graph of both model is compared to ICE engine application in order to check either they are suitable for bike application. CAD tools are used to design the proposed model so that fabrication of the model can be done.

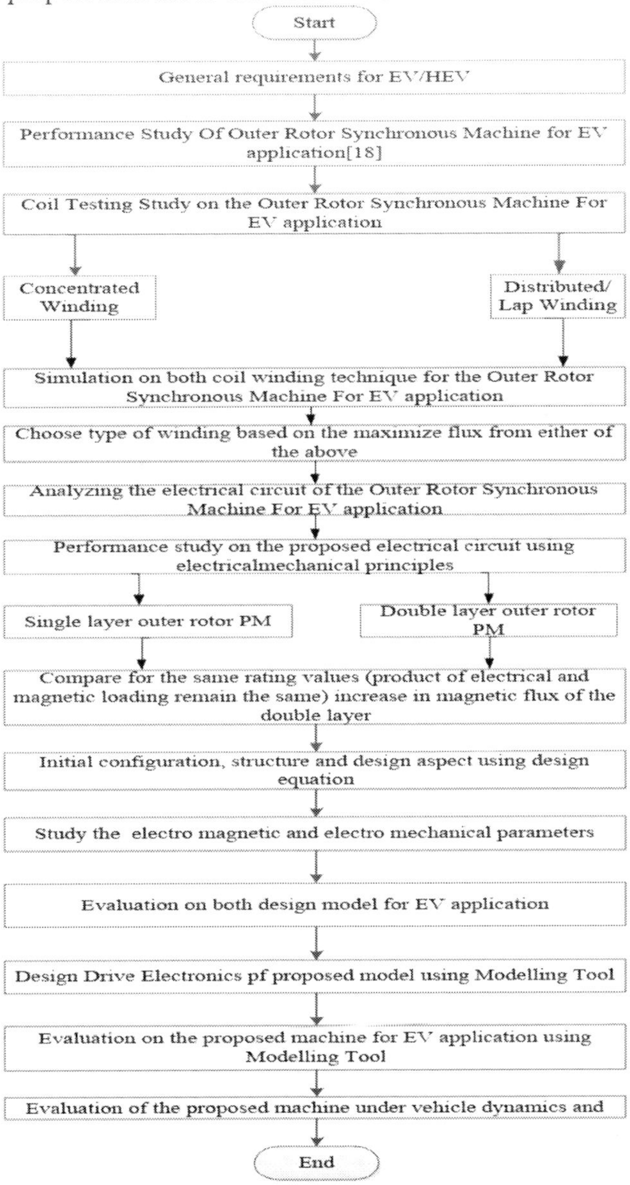

Fig. 4. Methodology used in this research

1.2 Proposed Design

Figure 5(a) shows the 2D proposed DORBLDC model design using JMAG software and Figure 6 shows the exploded view of the proposed DORBLDC model for electric bike application using Solidworks software. The proposed machine contains double layer rotor with a surface permanent magnet that supplied with 3 phase armature coil concentrated windings, 20 permanent magnets mounted on the inner surface rotor and 40 permanent magnets mounted on the outer surface rotor. The proposed model has more benefits than the conventional model as the flux is amplified which resulted to improve flux density across the machine. Table 1 shows the parameter dimensions of the proposed DORBLDC model where the outer dimension of both models are kept the same in order to compare the motor constant square density, magnetic flux density, and torque density

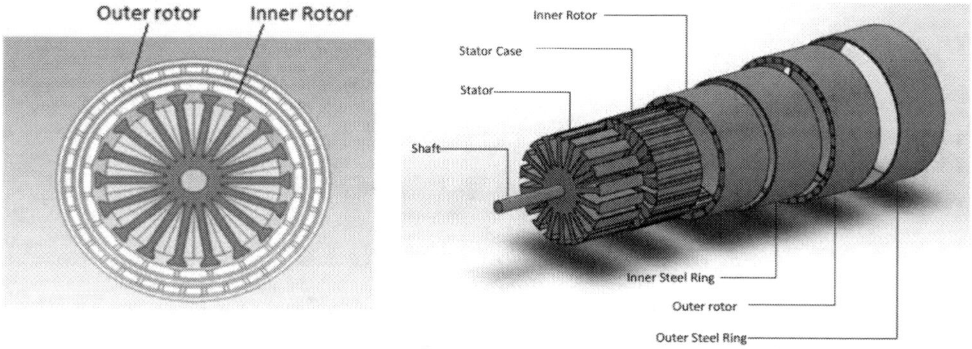

(a) Proposed 2D model (b) Exploded view

Fig. 5. The Proposed 2D model

Table 1. Specification of the proposed model

Parameter	Value (mm)
Diameter of the machine	50
Stack length	50
Number of stator poles	18
Stator Diameter	36
Physical Radial Air gap	1
Outer Ring Diameter	1.7
Inner ring diameter	1.7
Inner rotor diameter	3.523
Outer rotor diameter	3
Inner PM diameter	3
Outer PM diameter	3
Number of poles Inner magnet	18
Number of poles Outer magnet	36

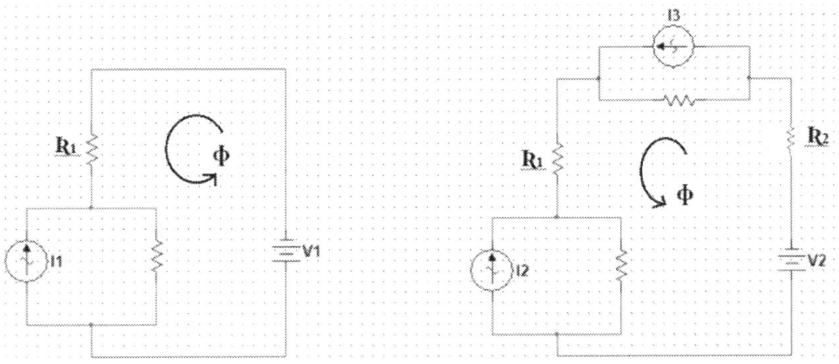

a)Conventional Model b)Proposed Model

Fig. 6. Magnetic Circuit of Conventional and Proposed Model

$$\phi = (mmf+Fm) / R \tag{1}$$

$$\phi = (mmf+Fm1+Fm2) / (R1+R2) \tag{2}$$

Fm = Flux generated in inner magnet SORBLDC
Fm1 = Flux generated in inner magnet of DORBLDC
Fm2 = Flux generated in outer magnet of DORBLDC
R = Rotor Reluctance of SORBLDC
R_1 = Inner Rotor Reluctance of DORBLDC
R_2 = Outer Rotor Reluctance of DORBLDC

The magnetic equivalent circuit shows components inside conventional and proposed model by changing them into voltage source, current source, and resistor. Figure 6 shows the equivalent magnetic circuit of conventional model and proposed a model. Based on figure 6, we can see that proposed model has higher generated flux due to mutual magnetic coupling created between outer rotor with the inner rotor to improve magnetic flux intensity by making both of them as an individual part of the machine. Therefore, the magnetic flux generated by both of individual part is amplified.

1.3 Finite Element Analysis (FEA) Method

Designing of the proposed model in JMAG Geometry Editor is the first step applied in the simulation of FEA analysis. Each part of the model is designed based on its parameters set for the machine. Model of the proposed machine updated in the JMAG software and configuration material for each part is done such as the type of material and its properties. Several conditions applied to each component such as rotational motion, torque nodal force, and FEM coil. Then, FEM coil circuit is constructed by setting the parameters such as resistance and voltage or current source. The setting of the windings and resistance of each coil followed with size and type of mesh to the model. Finally, simulation of the model is initialized as the steps are completed. Figure 7 shows the steps taken in FEA simulation.

 https://doi.org/10.1051/matecconf/201815203004

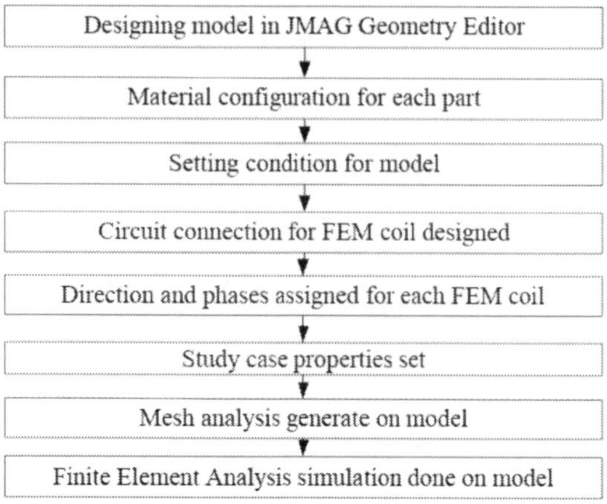

Fig. 7. Finite Element Analysis Flowchart

2 Result and Discussion

2.1 Magnetic Flux Analysis between SORBLDC and DORBLDC

Based on figure 8, the magnetic flux of amplitude for proposed model is higher than a conventional model. This shows that the proposed model has the highest magnetic flux density in the machine structure even the volume is kept the same with the conventional model. Furthermore, both of the models generates its magnetic flux pattern closing to the sinusoidal waveform which shows similarity to three-phase power supply wave. Other than that, generating magnetic flux waveform closer to sinusoidal pattern shows that harmonic occur in the machine is minimum. Torque performance analysis is done check whether the new proposed model improved the machine performance in terms of peak amplitude value. Having high generated torque of a machine enables the bike to operate under harsh road conditions such as high slope road that requires a big amount of energy to support the weight of the user and bike to accelerate uphill. In this research the torque evaluation is conducted to both models. Figure 9 shows the generated torque by both models. As shown in the graph, the proposed model generate the highest peak torque and conventional show the lower. This shows that by increasing the magnetic flux in the machine, the torque performance of the motor which increases the performance of the bike. E-bike application requires a motor that able to operate constant torque and power which show the same operation as an ICE engine with gears. The speed-torque graph of ICE engine has high starting torque and decrease as speed increases. Therefore, speed torque evaluation is done to both models in order to check whether the machine can operate same as ICE engine. Figure 10 shows the speed torque graph of the conventional and proposed model. The speed-torque graph of both models shows similarities to ICE engine speed torque. Therefore, the machine can operate for e-bike application. Based on Figure 10, the torque produced by proposed model is stable torque-speed characteristics than a conventional model. Even though SORBLDC produces high starting torque than DORBLDC, the proposed model generate higher and more stable torque-speed characteristics as the motor operate at higher speed. Therefore, the high torque produced for tougher road conditions such as uphill can be done by the proposed motor and improve the performance of the bike.

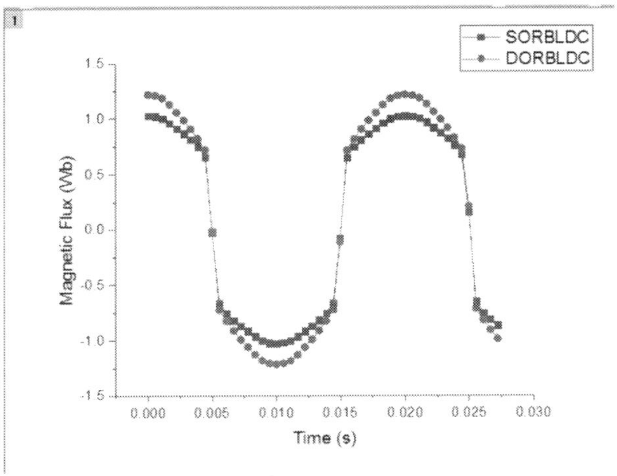

Fig. 8. Magnetic flux output of both model

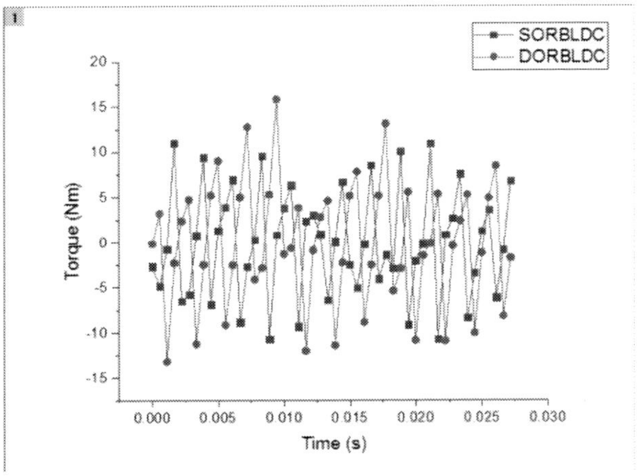

Fig. 9. Torque output of SORBLDC and DORBLDC

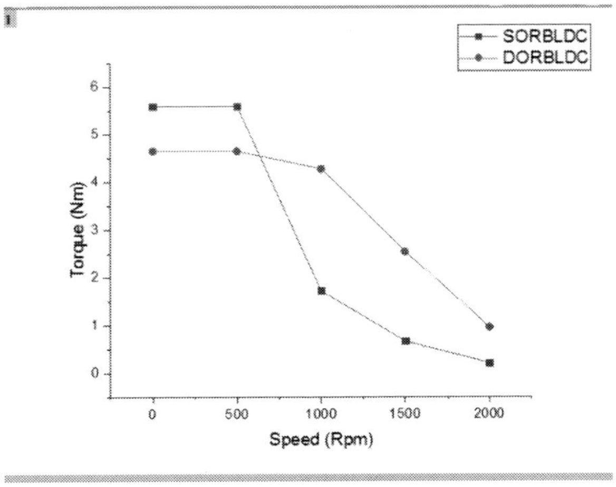

Fig. 10. Speed Torque Analysis of SORBLDC and DORBLDC

In order to evaluate constant torque and power region of a machine, speed power analysis is conducted. The maximum speed of a machine can be identified through constant power region. Figure 11 shows speed power efficiency of SORBLDC and DORBLDC. Based on Figure 11, the speed power graph of both models shows similarities to ICE engine speed power. Therefore, the machine can operate for e-bike application. SORBLDC model shows higher power efficiency during starting speed of the machine that DORBLDC model. However, DORBLDC model shows longest constant power region which means that the machine able to operate effectively at higher speed range before losing its power at the machine speed limit which is 2000 rpm.

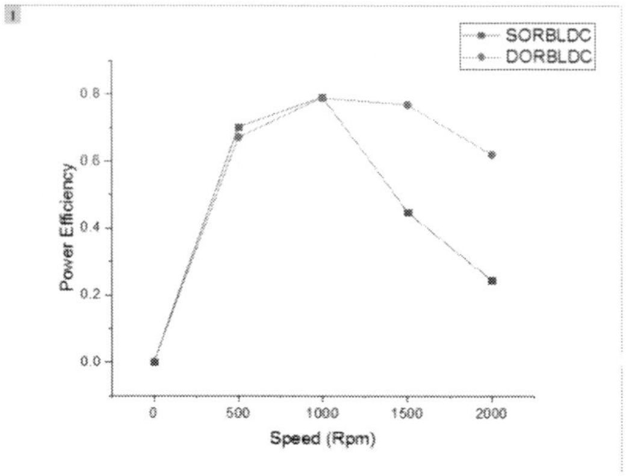

Fig. 11. Speed Power Efficiency of SORBLDC and DORBLDC

2.2 Comparative Motor Constant Square Density Evaluation

In this section, both models are being compared by evaluating each G value. The volume of both models are calculated by multiplying the cross-sectional area of the machine with the stack length set. In the Motor Constant Square, Density equation contains parameters of a machine such as the power of the motor, torque constant, motor constant and volume of the machine. All of the parameters are calculated and tabulated in Table 2. All of the variables that needed to use to calculate the G value is formed in Table 3. From the calculated and analyzed result in Table 3, it shows that double outer rotor BLDC motor (DORBLDC) produces higher motor constant square density than single rotor BLDC motor (SORBLDC). To avoid any error in evaluating G value of both model, the volume and outer radius of the machine are kept the same value. This is to ensure that the performance of both machines can be compared to the same size. From the table 3, we can see that motor constant square density of the proposed model is higher than a conventional model. The proposed model has higher G value than conventional because of having more magnet reluctance in parallel in the magnetic circuit that causes improvement flux density across the machine. It shows that the proposed model design improves the torque density of the machine whereby the performance of the motor increase as the speed of the motor increased. The difference G between DORBLDC and SORBLDC at 2000 rpm is:

Table 2. Parameters of both model

Parameter	Conventional Model	Proposed model
Stack length (mm)	50	50
Volume(mm^3)	356,415.35	311,794.70
Max. Current (A)	27	27

Table 3. G value of both model

Speed (rpm)	Conventional Model				Proposed Model			
	T_{avg}	K_t	K_m	G	T_{avg}	K_t	K_m	G
0	5.595685	0.155436	4.90842E-05	1.37716E-10	4.64406	0.129002	3.38089E-05	1.08433E-10
500	5.595685	0.155436	4.90842E-05	1.37716E-10	4.64406	0.129002	3.38089E-05	1.08433E-10
1000	1.718549	0.047737	4.62976E-06	1.29898E-11	4.276233	0.118784	2.86654E-05	9.19367E-11
1500	0.66153	0.018376	6.86016E-07	1.92477E-12	2.53823	0.070506	1.00994E-05	3.23913E-11
2000	0.20599	0.005722	6.65162E-08	1.86625E-13	0.95207	0.026446	1.42093E-06	4.55726E-12

G value improvement between SORBLDC and DORBLDC = 4.55726E-12/ 1.86625E-13

= 24.41929702 %

3 Conclusions

In the conventional model, the generated magnetic flux across the machine can't improve as the structure of the model are kept the same. Although, V shape PM improved reluctance torque of the machine, the required space needed to mount the PM is high. Therefore, new proposed design by applying extra outer rotor layer to the conventional model able to reduce space for the magnet and produced high magnetic flux density due to amplifying generate flux produced by the inner rotor. In this research project, the proposed DORBLDC with mounted PM operate more effectively than conventional model such as improvement the motor constant square density by 24.42%. Although the proposed design has few weaknesses in terms of peak torque in speed torque graph, starting power amplitude in speed power graph. However, the benefit of the proposed model such as higher magnetic flux density and longest constant torque speed characteristics than the conventional model. This shows that the proposed model is suitable for E-bike application.

References

1. J. Zhao, Y. Yangwei, Brushless DC Motor Fundamentals Application Note, MPS, Futur.

Analog IC Technol., pp. 7–8, (2011)

2. C. Aravind, L. Pin Wen, Design Evaluations of the New Double Rotor Interior Permanent Magnet Machine, Journal of Engineering Science and Technology, **2**, no. 2, pp. 73-84, (2015)

3. C. C. Hwang, S. M. Chang, C. T. Pan, T. Y. Chang, Estimation of parameters of interior permanent magnet synchronous motors, J. Magn. Magn. Mater., **239**, no. 1–3, pp. 600–603, 2002

4. R. Zhang, X. Zhu, L. Chen, Design of a new magnetic-planetary-geared outer-rotor permanent-magnet brushless motor for electric vehicles, 2014 17th Int. Conf. Electr. Mach. Syst. ICEMS 2014, pp. 658–663, (2015)

5. C. M. Chang, K. S. Yang, System integration and power flow management for the engine-generator operation of a range-extended electric vehicle, 2013 World Electric. Veh. Symp. Exhib. EVS 2014, pp. 1–10, (2014)

6. M. Cheng, C. C. Chan, General Requirement of Traction Motor Drives, Encycl. Automotive. Eng., pp. 1–18, (2014).

7. M.Norrhisam, Aravind CV, S.Khodijah, N.F.Mailah, Unified control structure of multi-type interior permanent magnet motor, Journal of Engineering Sciecnes and Technology, 10, no.3 pp.322-339

8. B. Marthandam, K. Latha, Analysis of V-Shaped Interior Permanent Magnet Synchronous Motor for Low Torque Ripple, **3**, no. 3, pp. 498–501, (2014)

9. C. Aravind, A. Saja, Vijayashankar, Design Evaluation of Double Rotor Switched Reluctance Machine, Journal of Engineering Science and Technology, **3**, no. 1, pp. 16-26, (2016)

10. C. V. Aravind, I. Grace, T. Rozita, R. Rajparthiban, R. Rajprasad Y. V. Wong, Universal computer aided design for electrical machines, IEEE 8th International Colloquium on Signal Processing and its Applications, Melaka, 2012, pp. 99-104, (2012) doi: 10.1109/CSPA.2012.6194699

Force Characteristics Analysis for Linear Machine with DC Field Excitations

Preshant A/L Krishna[1], *Aravind* CV[1,*], *Aminath* Saadha[1], and *F.* Azhar[2]

[1]School of Engineering, Taylor's University, Malaysia
[2]Faculty of Engineering, Universiti Teknikal Malaysia Melaka, Malaysia

Abstract. In urban regions and particularly in developing countries such as Malaysia with its ever-growing transport sector, there is the need for energy efficient systems. In urban railway systems there is a requirement of frequent braking and start/stop motion, and energy is lost during these processes. To improve the issues of the conventional braking systems, particularly in Japan, they have introduced linear induction motor techniques. The drawbacks of this method, however, is the use of permanent magnets, which not only increase the weight of the entire system but also increases magnetic cogging. Hence an alternative is required which uses the same principles as Magnetic-Levitation but using a magnet-less system. Therefore, the objective of this research is to propose an electromagnetic rail brake system and to analyze the effect of replacing permanent magnets with a magnet-less braking systems to produce a significant amount of brake thrust as compared with the permanent magnet system. The modeling and performance analysis of the model is done using Finite Element Analysis (FEA). The mechanical aspects of the model are designed on Solidworks and then imported to JMAG Software to proceed with the electro-magnetic analysis of the model. There are 3 models developed: Base Model (steel), Permanent Magnet (PM) Model and DC Coil Model. The performance of the proposed 2D models developed is evaluated in terms of average force produced and motor constant square density. By comparing the values for the 3 models for the same case of 9A current supplied for a 0.1mm/s moving velocity, the base model, permanent magnet model and DC coil model produced an average force of 7.78 N, 7.55 N, and 8.34 N respectively, however, with increase in DC current supplied to the DC coil model, the average force produced is increased to 13.32 N. Thus, the advantage of the DC coil (magnet-less) model, is, that the force produced can be controlled by varying the number of turns in the coil (N) or the current supply to the coil (I) given by the which is the simple principles of a solenoid: Force(mmf)=NI.

[1] Corresponding author: aravindcv@ieee.org

1 Introduction

Conventionally brake systems use friction or mechanical traction to slow down or stop a moving object, but friction produces heat and thus heat losses. Friction generates heat, thus losing energy, and the frequent use of these type of friction based braking leads to wear and tear of brake pads, wheels, and tracks leading to the requirement of frequent maintenance of the system[1]. Disc brakes require extra materials such as oil or compressed air to force the brake lining on the wheels, reduce the kinetic force of the system. Apart from the mechanical stress on the system, installing disc brakes increases the weight of the entire system, but the application is useful only during braking conditions [2]. Figure 1 shows the example of Wheel-Thread Brake System and Axle-Mounted Disc Brake System mechanisms which are of adhesion types:

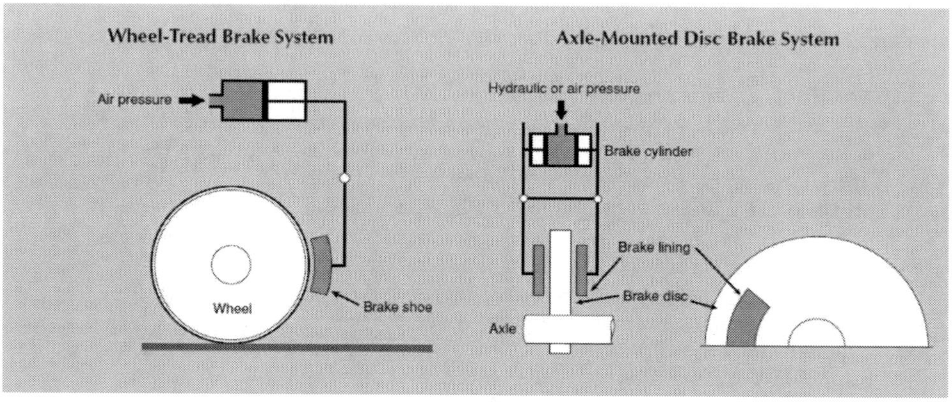

Fig. 1. Wheel-Thread Brake System and Axle-Mounted Disc Brake System [2]

To overcome this, electromagnetic brake systems that do not rely on wheel adhesion such as eddy current braking systems are introduced using electromagnetic induction principles, with the help of permanent magnets [3]. However, the drawback of this method is that, at lower speeds, the brake force is reduced without the ability to magnetically lock the load at a standstill, which poses a safety issue, and thus the conventional friction braking system has to be incorporated as well. Furthermore, the heat generated from the braking of these types of systems result in the overheating of the rail tracks causing the tracks to bend or deform. Another drawback with the use of permanent magnets is the increase in weight of the system as well as the cogging force induced. [1, 4]. Even hybrid linear motors have been researched and the effect of the core losses have been analyzed to be much more when there are permanent magnets introduced to the system[5]. Although research has been on the losses of permanent magnet systems, there is not a sufficient amount of literature and research data available for the application of magnet-less systems for railway braking aspects. [6]. Therefore, the objectives of this research are to propose an electromagnetic rail brake system through fixed DC excitation for rail brake systems which produce a significant amount of brake thrust and to analyze the effect of replacing permanent magnets with a magnet-less braking. Moreover, to compare the machine characteristics such as motor constant square density of the proposed model to the permanent magnet model, with respect to the conventional base model.

2 Research Methodology

The research encompasses the design, optimization and prototype fabrication of linear electromagnetic braking using levitating principles for railway braking systems. Then the design is simulated and analyzed in software such as JMAG for electromechanical analysis and Finite Element Analysis and Solidworks for the mechanical modeling and analysis. The limitations and scope of the research are due to the restriction to the lab environment and the specifically fixed loading capacities of the prototype model. The flowchart of the flow of research methodology is given below in Figure 2.

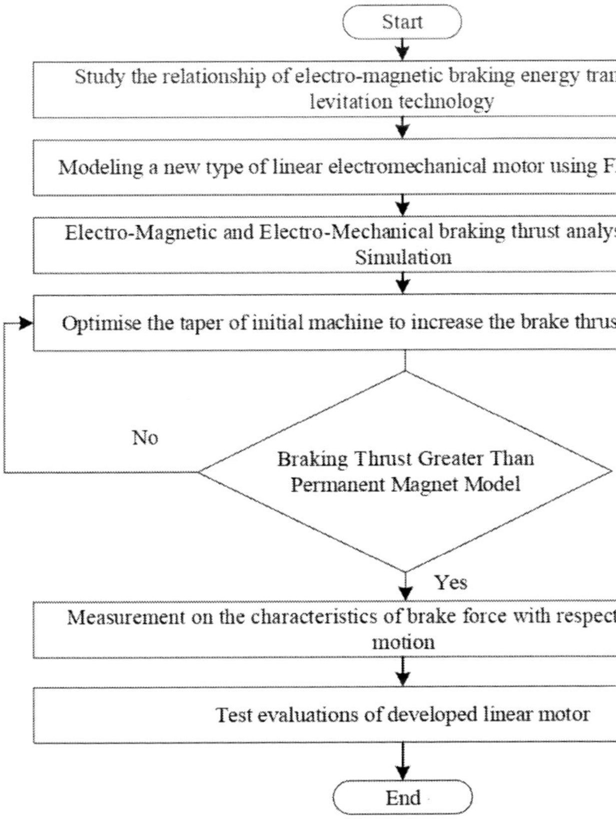

Fig. 2. Flowchart of Methodology

The relationship between the electrical and mechanical energy to the braking performance of the motor technology is mathematically developed. The purpose the study is to investigate the performance of the overall linear electromechanical technology for braking applications in railways. A set of coils energized attract steel and the flux flow in the air gap holds the rail and the track to be magnetically locked, which is the principle of the motor as seen in Figure 3. The flowchart for the overview of the design and development stages is shown below in Figure 4.

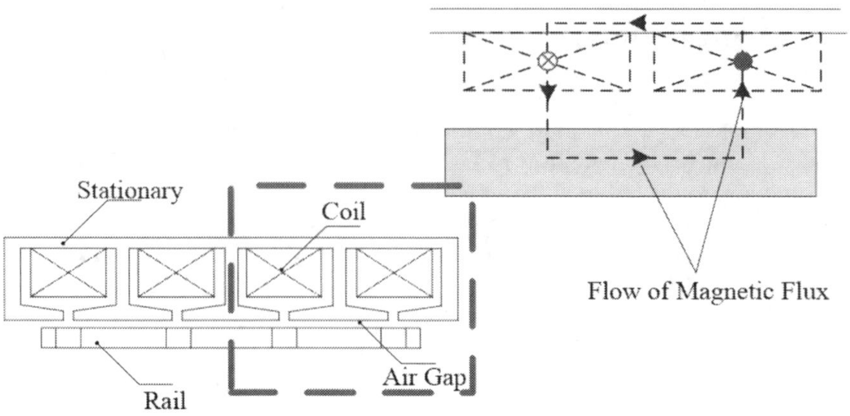

Fig. 3 Concept of Linear Switched Motor

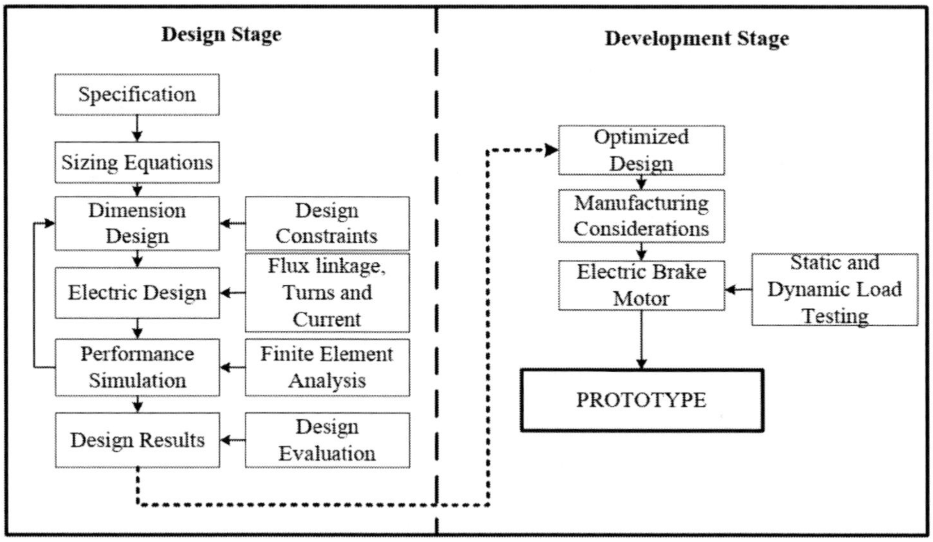

Fig. 4. Overview of Design and Development Stages

After the relationship is established, the next step is to model the new type of linear machine using numerical analysis tool. Design details are done analytically using calculations, before the modeling process. For the modeled prototype, the motor should produce maximum braking torque. The limits are set based on the sizing (height and width) calculations of the motor. Based on the analysis of the study and optimization done, the linear electromechanical motor prototype is fabricated to validate the proof of concept. Experimental analysis for the final prototype of the linear motor and the characteristic of the motor torque with a change in velocity is tested. Experimental analysis for the final prototype for the characteristics of the motor torque and speed is tested and compared to that of a linear permanent magnet motor.

2.1 Magnetic Circuit Analysis

The magnetic flux density is given by the Eq. (1) [8,9].

$$B = \mu_o H = \mu_o (\frac{NI}{x}) \quad (1)$$

whereby, B is flux density, μo is Permeability of free space (air), H is Magnetic flux intensity, N is Number of turns of the coil and I is the current carried by the coil.

The linkage flux is given by Eq. (2) [8,9].

$$\psi = \phi N = (\frac{\mu_o A N^2 I}{x}) \quad (2)$$

The self-inductance L, is given by Eq. (3) [9].

$$L = (\frac{\mu_o A N^2}{x}) \quad (3)$$

Energy stored in a current carrying coil with the self-inductance of the coil L, and if there are no other current carrying coils in the system then magnetic flux ϕ, is given by the Eq.(4) [8].

$$\phi = IL \quad (4)$$

Therefore the energy stored W, can be written as Eq. (5) below:[8]

$$W = \frac{1}{2} L I^2 \quad (5)$$

If a system consists or more than one current-carrying coils that are magnetically linked to each other, the total force of the system is the sum of all the forces acting on the coils. In a multiple loop system, the force F, associated with linear motion is defined by a variable x, and is given as Eq. (6) [8,9].

$$F = \frac{\partial W}{\partial x} \quad (6)$$

This can be substituted with Eqs.(1), (3), and(5) and rewritten to form the Eq.(7) [8,9].

$$F = \frac{d}{dx}\left(\frac{1}{2} I^2 L\right)$$

$$F = \frac{dL}{dx}\left(\frac{1}{2} I^2\right)$$

$$F = -\frac{1}{2} I^2 \left(\frac{\mu_o A N^2}{x^2}\right)$$

$$F = -\frac{1}{2\mu_o}\left(\frac{\mu_o N I}{x}\right)^2 A$$

$$F = -\frac{B^2 A}{2\mu_o} \quad (7)$$

whereby, A is the cross section in the air gap, B is the magnetic flux density and μ_o is the permeability. The negative sign indicates the fore deceases to x, thus it is an attractive force. This force is also knows as normal force which is the force of attraction between the armature core and the reaction rail of a linear motor through air gap. [8-10] Eq. (7) can also be written as Eq. (8) [8-10].

$$F = \frac{B.B.A}{2\mu_o} \tag{8}$$

Since $\phi = BA$, for a region where the magnetic field is uniform and assuming the flux density has only one directional component B with a uniform cross section A, by taking the perpendicular direction of B we get the Eq.(9) below: [8,9]

$$F = \frac{B\,\phi}{2\mu_o} \tag{9}$$

Thus the relationship can be established as Eq.(10) below:[8,9]

$$\varphi = \frac{F_{mmf}}{\mathcal{R}} = \frac{NI}{\mathcal{R}} \tag{10}$$

Thus is can be seen that keeping the area of cross section A constant, the force can be increased by increasing magnetic flux ϕ. This can be achieved by increasing the number of turns of the coil N, increasing the current I, or by decreasing the reluctance $\mathcal{R}$. Reluctance is essentially, the amount resistance to the flow of the magnetic flux through the air gap. The effects of decreasing the reluctance can be more vividly visualized by rearranging the Eq. (9) to Eq. (10), to analyze the methods of manipulating the reluctance of the system [9].

$$\mathcal{R} = \frac{mmf}{\varphi} = \frac{NI}{B\,A} \tag{11}$$

3 Design Aspects

The components of the linear motor that is designed are the mover, stator, and armature coil. The armature coils are supplied by AC current supply and produce magnetic flux which produces a force when acting on the mover. The armature coil is made up of copper windings, while the mover and stator are steel. The mover and the stator are separated by an air gap. For the Permanent Magnet model, part of the steel is modeled to be cut out to fit permanent magnets on the stator. The same dimensions are used for the DC coil model, but instead of permanent magnets, copper coils are selected as the material. Figure 5(a) shows the front view of the base model and Figure 5(b) shows the PM/DC Coil model designed on Solidworks. The dimensions of the designed model for the permanent magnet/DC coil are given in Figure 6(a) and Figure 6(b) respectively for the front view and the auxiliary view. To compare the analysis between the comparison of the system with the permanent magnet and a magnet-less system using only coils, the dimensions for the permanent magnet slots are kept the same, and replaced by coils, for the DC coil model. The 3D models of the linear motor are designed on Solidworks and then imported to JMAG for the FEM analysis. Finite Element software is capable of accurately calculating magnetic fields and the related parameters for motor design. The analysis done is shown in Figure 7(a) and Figure 7(b) for the magnetic flux density and the magnetic field lines respectively:

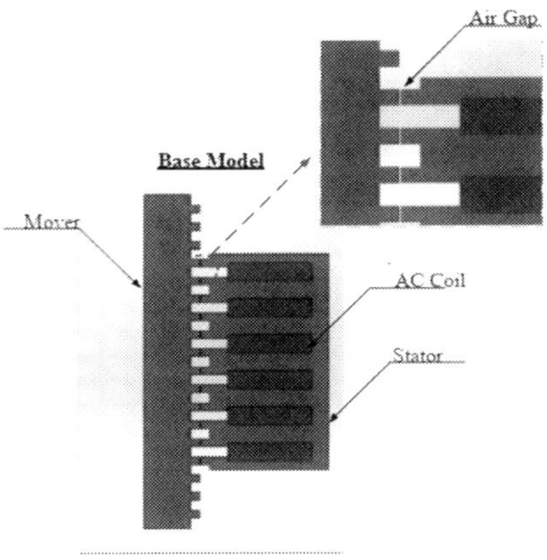

Fig. 5(a). Base Model Design

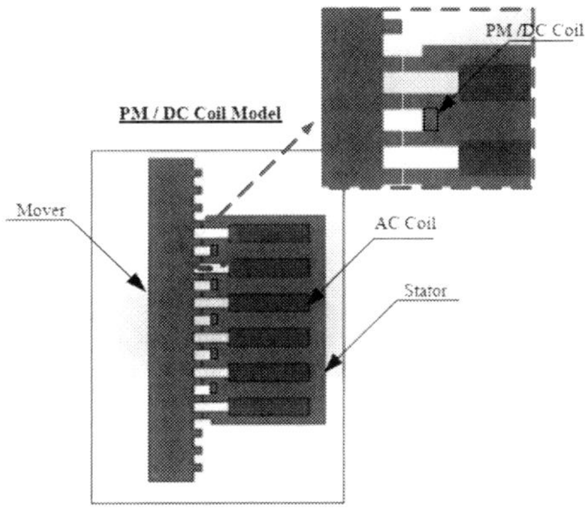

Fig. 5(b). PM / DC Coil Design

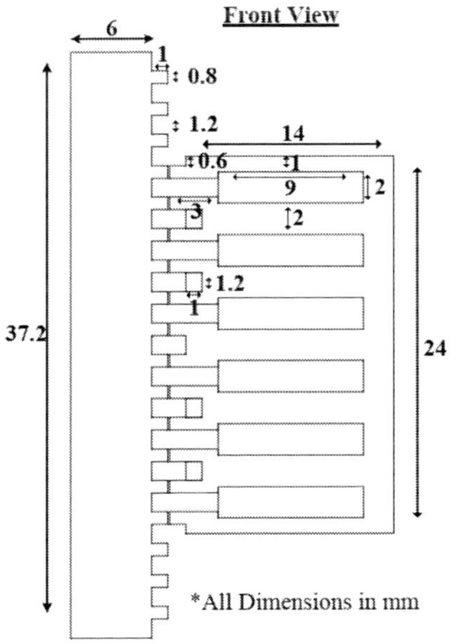

Fig. 6(a). Front View of Dimensions

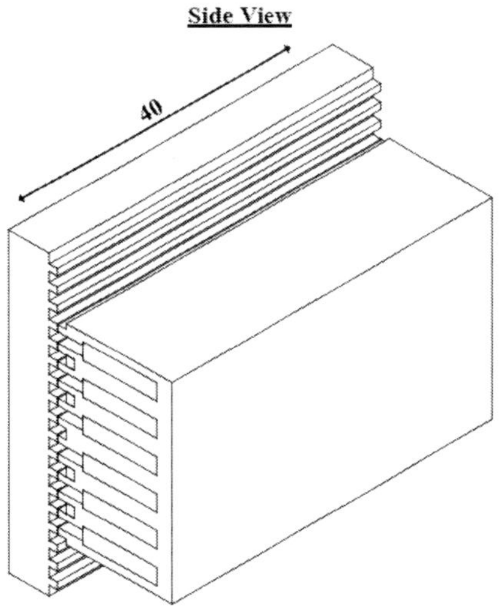

Fig. 6(b). Auxiliary View of Dimensions

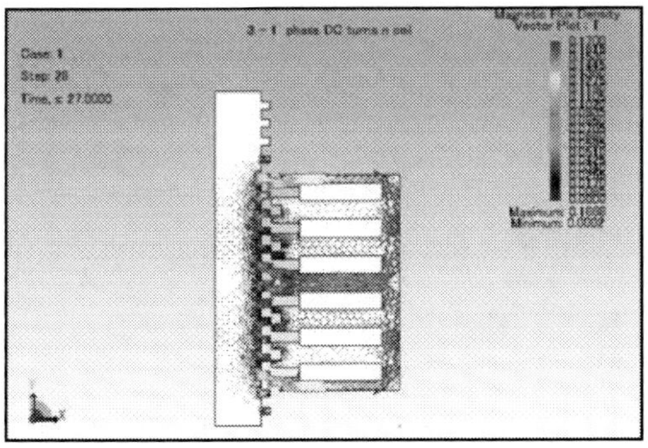

Fig. 7(a) Magnetic Flux Density

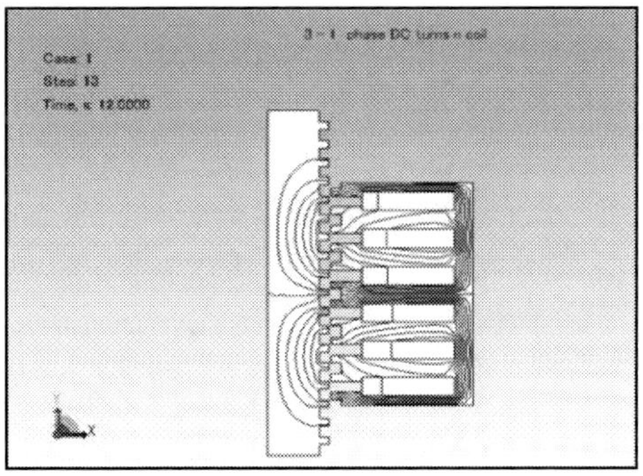

Fig. 7(b). Magnetic Flux Lines

Thus, with these tools, the performance of the model is analyzed by analyzing the flow of the magnetic field lines and flux density. It is noted that the flux originates around the armature coil and flow through the mover, back into the stator, thus producing a force on the mover. With the step feature, the movement of the mover with each step with respect to time is be visualized and the change in magnetic flux is seen. The comparison FEM analysis is done for three models of linear motors: Base Model (steel), Permanent Magnet, Model and DC Coil Model (electromagnet). For the model with the DC coil, different values of current supply are also used at 1A, 3A and 5A current for the DC coil, while the main AC supply to the stator is kept constant for all 3 models with 5A, 7A, and 9A current. For the 3 models, the average force values are tabulated for the different current values at various speeds with respect to time. The displacement values selected are 0.1mm/s, 0.3mm/s, 0.5mm/s. The force graph is obtained using JMAG and the values are tabulated and analyzed on the Origin software.

3.1 Performance Index Calculation

Thrust force produced by a machine is proportional to input power 'P', current 'I' and volume 'V'. If the volume of a motor is larger, it has a larger area to accommodate a number of coil turns and/or coils of larger size, therefore more current and power can be injected into the machine to produce a higher thrust force. [11] Thus motor constant square density G, is used as one of the performance index in order to make valid comparisons between different models. This normalizes the factors such as volume, current and input power, to make a proper unbiased comparison between models. Thrust constant k_f, and motor constant k_m, are used as performance indices to normalize the current and input power of a motor respectively. The performance indexes are calculated using the equations Eq. (12) to Eq.(15) given below:[11]

$$k_f = \frac{F}{I} \tag{12}$$

Where k_f is the thrust constant in (N/A), F is the average thrust in (N), and I is the current injected to each coil (A). [11]

$$P = I^2 R \tag{13}$$

Where P is the input power in (W) and R is the resistance of the coil.[11]

$$k_m = \frac{F}{\sqrt{P}} \tag{14}$$

Where k_m is the motor constant in (N/$\sqrt{W}$).[11]

$$G = \frac{F^2}{P\,V} \tag{15}$$

Where G is the motor constant square density in ($N^2/(Wm^3)$), and V is the volume of the motor in (m^3).[11]

4 Results and Discussion

To find the average force value for each case, from the data obtained in JMAG, the force graph for each case is integrated to find the area under the graph. The integrated values for each case in the 3 models are then tabulated and plotted on the same graph, thus the average force values for each model and each case is comparatively analyzed. Table 1 shows the data collected from the simulation done on the average force produced for each model for various cases:

Table 1. Average force produced for various cases of different models

AC Supply		**5A**			**7A**			**9A**		
Velocity (mm/s)		**0.1**	**0.3**	**0.5**	**0.1**	**0.3**	**0.5**	**0.1**	**0.3**	**0.5**
Average Force (N)	**Base**	5.87	5.87	5.36	7.10	7.10	6.46	7.78	7.75	7.08
	PM	5.65	5.65	5.15	6.87	6.87	6.24	7.54	7.54	6.85
	Coil 1A	6.57	6.57	6.00	7.71	7.72	7.02	8.34	8.34	7.57
	Coil 3A	8.85	8.85	8.08	9.82	9.82	8.96	10.3	10.38	9.48
	Coil 5A	12.03	12.04	11.03	12.82	12.83	11.78	13.32	13.32	12.24

Table 2 shows the graphical comparison of Velocity Vs Average Force produced by the 3 models for the supply AC current of 5A, 7A, and 9A:

Table 2. Comparison of Velocity Vs Force for Different Models

Supply Current (AC)	**Comparison of Velocity Vs Force graph for 3 models: Base, PM, and DC Coil (1A, 3A, 5A)**
5A	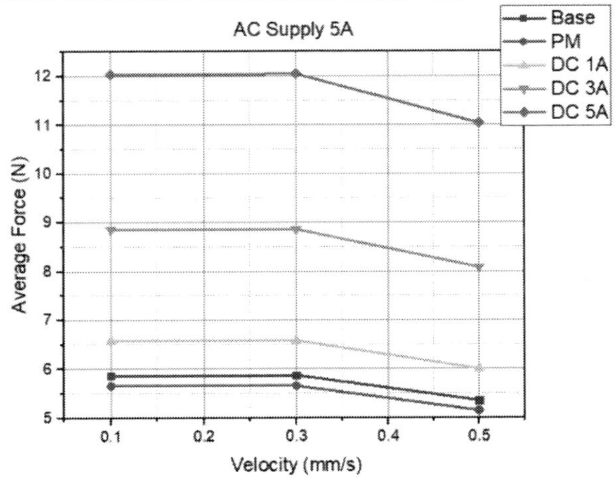

7A

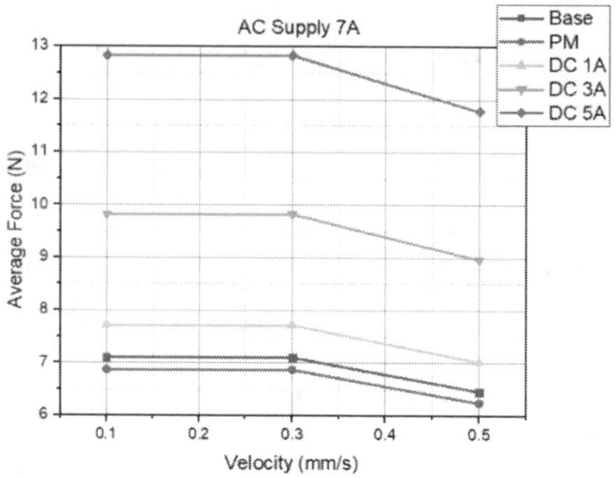

9A

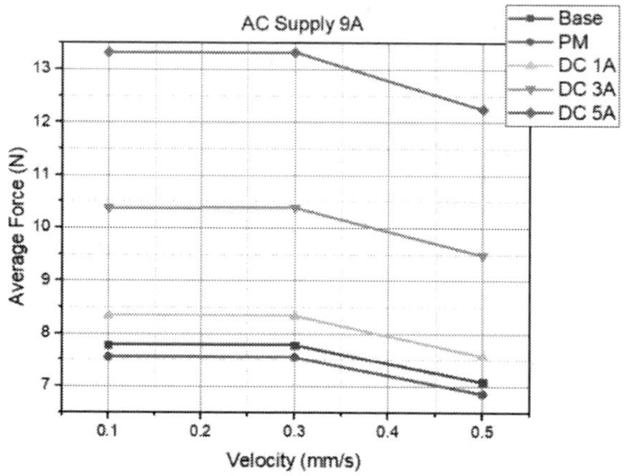

It is realized that, with varying the current in the DC coil, the brake force produced can be altered, without disrupting the main AC supply. Thus this gives a better range of force application that can be used for various brake force requirements. DC coils are used instead of permanent magnets for the proposed model, to curb the issues with magnetic cogging and to reduce the weight of the system. It also provides a wider range of operability compared to the permanent magnet as the magnetic field intensity of the coil can be increased or decreased depending on brake force requirement, but for permanent magnets, the force produced is constant and deteriorating over time. However, AC coil main supply to the system is still needed, to operate the railway system, as any unforeseen disruptions in the supply can be recovered by the AC supply, instead of affecting the entire system. The forces produced, however, are for the motoring condition, but if the polarity is reversed, the values obtained are for the braking condition. Since, the positive values indicate accelerating velocity and the negative values indicate decelerating velocity, hence braking condition. By combining the two conditions for the motoring and braking applications, the speed–force characteristics of the models is obtained. The speed-force characteristics of the different models are shown in Figure 8. The DC coil model produced an over higher average force when compared with the other models. As seen in the graph, by increasing the current value of the coil, there is an increase in the average force. Table 3 below shows the performance indexes for the research

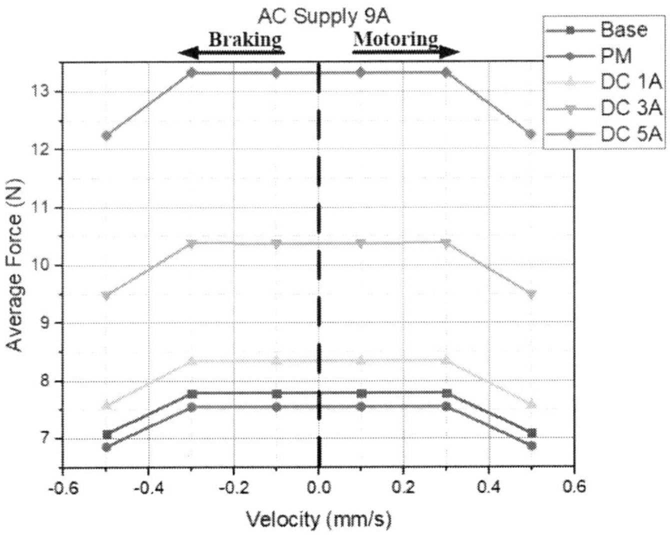

Fig. 8 Speed-Force Characteristic of the Different Models

Table 3. Parameters of Models

Par.	Unit	Base Model	PM Model	DC Coil Model		
				1A	3A	5A
k_f	N/A	0.86	0.83	0.92	1.15	1.48
k_m	$N/A/\sqrt{W}$	0.96	0.93	1.03	1.28	1.65
G	$(N/A/\sqrt{W})^2$/ mm3	4.63E-05	4.36E-05	5.32E-05	8.23E-05	1.36E-04

5 Conclusion

In conclusion,3 models are developed: Base model, Permanent Magnet model, and the DC Coil model, whereby the average force produced and motor constant square density for each model for various moving velocities are compared and analyzed. The models produced an average force of 7.78 N, 7.55 N, and 8.34 N for the base model, permanent magnet model, and DC coil model respectively. Comparing the characteristics of the models with respect to the base model, the permanent magnet model produced 5.98% decrease, while there is an increase of the DC coil model ranging from 14% to 192% with the increase of DC current supplied, for the motor constant square density.

Acknowledgments

The authors wish to express financial support of the project through the Taylor's Research Grant Scheme TRGS/MFS/1/2016/SOE/005.

References

[1] R. C. Sharma, M. Dhingra, Braking Systems in Railway Vehicles, Int. J. Eng. Res. Technol., **4**, no. 1, pp. 206–211, 2015

[2] I. Hasegawa and S. Uchida, Braking Systems, Japan Ralw. Transp. Rev. 20, **vol. 7**, no. June, pp. 52–59, (1999)

[3] C. Aravind, I. Khumira, R.N. Firdaus, Fairul.A, Single phase flux switching DC linear Actuator, Special issue pp.28-39 Journal of Science and Technology, **1**, pp. 28-39, (2013)

[4] Y. Sakamoto, T. Kashiwagi, T. Sasakawa, N. Fujii, Linear eddy current brake for railway vehicles using dynamic braking, Proc. 2008 Int. Conf. Electr. Mach. ICEM'08, pp. 1–6, (2008).

[5] A. Aravind, V. R. Akilesh, S. Gunaseelan, S. Ganesh, Eddy Current Embedded Conventional, Natl. Conf. Trends Automotive. Parts Syst. Appl., **5**, no. 7, pp. 7–12, (2016)

[6] M. Manna, S. Marwaha, M. Marwaha, Performance Optimization of Linear Induction Motor by Eddy Current and Flux Density Distribution Analysis, Journal of Engineering Science and Technology, **6**, no. 6, pp. 769-776, 2011

[7] Shibly Ahmed Al-samarraie, Yasir Khuhair Abbas, Manifold based controller design for linear systems, Journal of Engineering Science and Technology, **8**, no.6, 723-740 (2013)

[8] Q. A. Hasan, W. H. Wan Badaruzzaman, Ahmed W. Al-Zand, Azrul A. Mutalib, Finite element analysis of tapered composite plate girder with a non-linear varying web depth, Journal of Engineering Science and Technology, **12**, no.11, 2839-2854 (2017)

[9] Y. Wang, J. Miao, Y. Wei, The research of traction motor energy-saving regenerative braking control technology, 2010 Int. Conf. Intell. Comput. Technol. Autom. ICICTA 2010, **3**, pp. 930–933, (2010)

[10] I. Boldea, S. Nasar, Linear Electric Actuators and Generators. Cambridge, GBR: Cambridge University Press, (2009)

[11] H. Toliyat, G. Kliman, Handbook of Electrical Motors. Electrical and Computer Engineering. CRC Press, pp. 1-24, (2004)

[12] Rajesh V.R., Biju T.Kuzhiveli, Modelling and failure analysis of flexure springs for a stirling cryocooler, Journal of Engineering Science and Technology, **12**, no.4, 888-897 (2017)

[13] F. Azhar bin Abdul Shukor, H. Wakiwaka, K. Tashiro, M. Nirei, Design and Performance Index comparison of the Permanent Magnet Linear Motor, Progress In Electromagnetics Research M, **43**, pp. 101-108, (2015)

Thermal Energy Recovery through Optimal Salt concentration in a Parabolic Trough Systems

Rikesh Ramsurn[1], and *Aravind* CV[1,*]

School of Engineering, Taylor's University, Malaysia.

Abstract. Making a PVT system hybrid is to support the use of thermal and electrical energy simultaneously or independently, to control the thermal effect to improve electrical efficiency or vice-versa. This project makes use of the Parabolic Trough design with emphasis on making the system to be sustainable and also increasing the thermal efficiency of the system. Polystyrene and acrylic foam is utilized to maximize the heat retention capability of the system. To power, the pump that moves the heat transfer fluid (tested with salt water proportions) within the copper tube, a set of solar PV panel is to support its power demand making it sustainable. The closed loop setup designed achieved an improved thermal efficiency level of 66.2%, which contributes to having a reliable heat energy source for applications such as hot showers. The novel setup design also makes use of PV cells to support other energy demands through power electronic control designs. Using a similar heat dissipation technique, a novel setup has been designed to improve the voltage supply by making use of liquid cooling and translucent glass PV panels. Cooling the PV panel restored up to 11.7% of its rated voltage supply. This is achieved by keeping the PV panels within its best thermal operating conditions using an energy efficient electronically controlled cooling system.

1 Introduction

Solar energy is considered as the cleanest and most abundant source of renewable energy harvested from sunlight [1]–[3]. Its net input to Earth is considered as 8000 times the rate of fossil and nuclear fuels currently being used by humanity on a worldwide scale. Methods of harvesting solar energy vary a lot in design but are essentially made up of photovoltaic cells when it comes to generating electricity. Due to its solid-state nature, silicon-based PV cells have an expectancy of operating for an indefinite amount of time without wearing out [4]. However, the efficiency of PV cells is still subjected to research and development with the aim of achieving sustainability in terms of electricity generation and consumption. The efficiency of PV cells is known to be affected by heat during its exposure to solar radiation while in operation. Hence methods are being devised to improve the efficiency of PV cells through heat dissipation [5]. With energy sustainability in mind, the heat dissipated is reused for applications with heat demand such as a hot water supply.

[1] Corresponding author: aravindcv@ieee.org

A popular design currently being researched and developed for solar energy harvesting is the Parabolic Trough PVT design. Using light concentration concept, such as concentrating sunlight on an object using a magnifying glass causing it to melt or burn, the Parabolic Trough PVT design consists of a reflective sheet of metal (e.g. stainless steel) bent into a circular shape in order to reflect light rays from the sun onto a line of incidence [6]. The main application of this line of incidence is to convert the concentrated light rays into heat energy, thus contributing to the thermal aspect of the PVT design. Heat energy is collected by a copper pipe placed at the line of incidence of the reflected light rays. The heat energy collected is then be used for applications such as a hot water shower by transferring the heat to a water tank using a closed loop fluid flow circuit and a heat exchanger. The closed loop may contain water or other fluids which is facilitated faster and more efficient heat removal from the copper pipe and heat transfer to the water tank through the heat exchanger.

Following the heat dissipation concept, there is a possibility of improving the electrical efficiency of PVT systems by removing the heat build-up on PV panels/cells while reusing the heat for thermal applications. This improvement is brought through the restoration of the voltage supply which normally drops as the temperature of the panels rises [7]. A cooling technique which may be used to cool PV cells is the use of heatsinks. The heat build-up on the PV cells is transferred to the heatsinks through conduction while the latter may be air-cooled or liquid-cooled. The cooling method used may vary depending on the type of PV panel used and how the heat shall be reused [8]. Hence, it is important to further investigate and develop PVT systems such as the Parabolic Trough to achieve reliability and sustainability. In this work, a parabolic concentrating trough is experimentally studied and performance evaluated with phase layered heat removal strategy.

2 Methodology

2.1 Experimental System Setup

Heat Transfer Fluids (HTF) used in this research includes water, water with 10g/L of salt and water with 20g/L of salt. Figure 1 shows the flowchart of the system's thermal optimization implemented and experimented. The most important feature of the thermal optimization implemented is the use of a closed loop HTF circuit which involves the use of a copper coil heat exchanger for effective heat dissipation and a pump for liquid movement around the circuit. The size of the copper heat exchanger is selected according to the container size it will be in. The maximum volume the heat exchanger occupy in the container favour more effective heat dissipation. As for the pump, its specification depend on the amount of liquid to be moved around the circuit as well as the flow rate at which it shall be moved. Effective heat dissipation in a closed loop circuit requires in-depth flow rate calibration. For the experiments conducted during this research, flow rate calibration is performed experimentally by controlling the DC pump's voltage. The lower the voltage, the slower the flow rate and vice versa. Thermocouples are used to measure the various temperatures involved in this research. Experiments are conducted for each HTF for around 30 minutes under a specific weather or until the system reached a thermal settling point for the meteorological conditions, it is being experimented under. The closed loop is flushed with clean water before testing with each HTF. The recorded data are then analysed graphically. The principle of operation of the Phase Layered Heat Removal for thermal efficiency improvement is as demonstrated in Fig. 1.

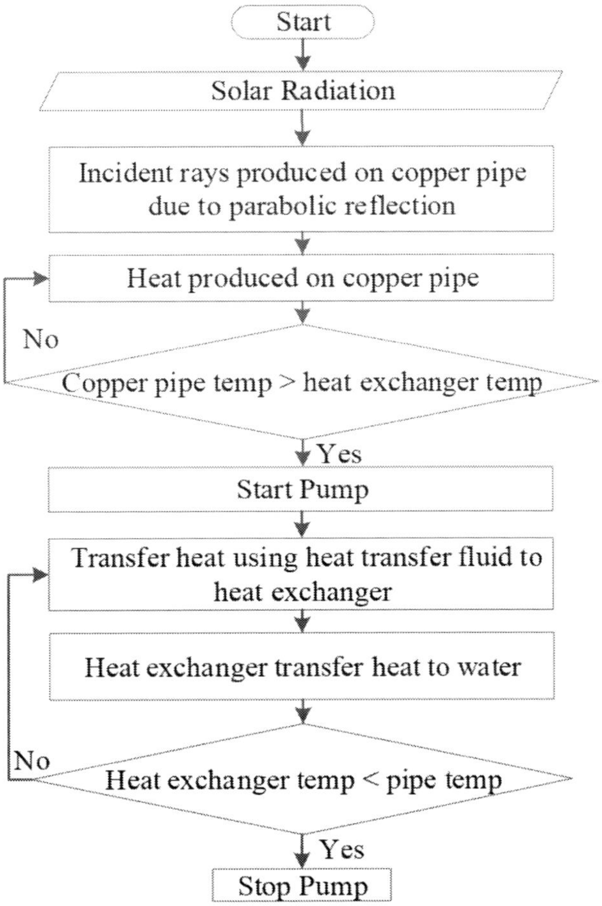

Fig. 1. Closed loop heat transfer topology.

2.2 Electrical Characteristics Study

When it comes to the electrical aspect, the cooling methods can either be active or passive. The advantage of active cooling is that it is rather faster than passive cooling. However, active cooling also affects the electrical efficiency with its power requirement [9]. Hence, the setup designed for this research is a combination of the two, whereby electricity is used by the cooling system only under specific conditions. The type of solar panel used for this research is a Silicon Thin Film of 20% Transparency. The cooling design concept involves placing the panel in between two layers of liquid for the top and bottom surface heat removal. The cooling setup designed is connected as a closed loop, making use of a pump and a heat exchanger, or as an open loop connected to the main water supply pipe of the house or building where the system is placed. As an open loop setup, electronically controlled valves are used to control the liquid flow in case the flow from the mains is not enough to cool the panel, which in case a backup water tank is used to make up for the flow through gravitational difference.

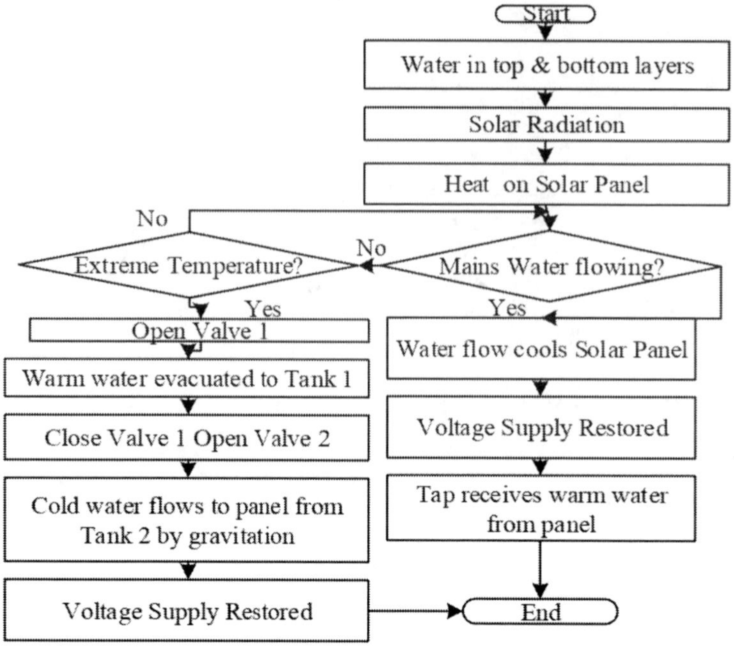

Fig. 2. Open loop heat removal topology.

The heat transfer topology for closed loop is similar to the one shown in Fig. 1 where the heat is produced on the panel instead of a copper pipe. For the open loop setup, the heat removal topology is as shown in Fig. 2. The PV panel cooling technique demonstrated by Fig. 2 is implemented at Taylor's University by connecting the water tap from the Engineering Workshop to the panel's inlet. The panel's outlet replaced the role of the water tap. The PV panel is placed under the sun and started heating up. Whenever a student is to use the water tap, the water from the tap flow into the panel's cooling layers to cool the panel. The heat from the panel is transferred to the water and leave the panel's cooling layers as warm water which the student using the tap could use. A warm water supply is appreciated at the Engineering Workshop for washing hands better than with cold water. If ever the panel's temperature get too high, an electronically controlled valve is opened to evacuate the hot water into a backup tank for later usage. The same principle is applied to a home or any other building with a water supply. While the panels are cooled, warm water is also being produced for instant use or to be stored for later usage. Fig. 3 shows the cooling setup design concept.

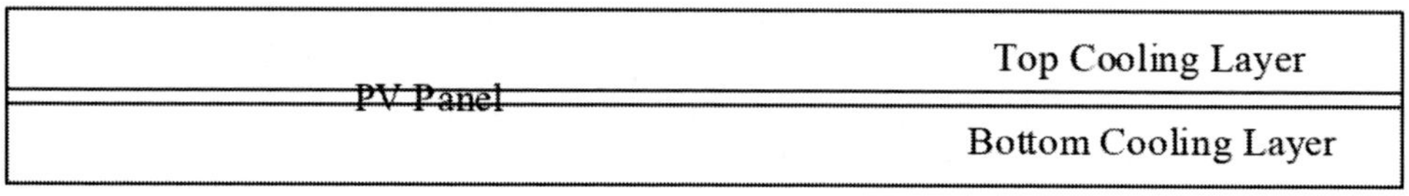

Fig. 3. Top and bottom liquid containers made of Perspex for layered cooling.

3 Results and Discussion

3.1 Thermal Efficiency

The general equation for instantaneous thermal efficiency η_T is as shown in Eq. 1 [3].

$$\eta_T = F_R(n_o - \frac{U_L}{C}(\frac{T_{out} - T_{air}}{G_b})) \quad (1)$$

Using Eq. 1 and experimental or simulation results obtained for optical efficiency [8] and heat loss coefficient [10] of the parabolic trough design, the instantaneous thermal efficiency for this research's PVT system has been derived and simplified into Eq. 2.

$$\eta_T = 0.667074 - 4.124820(\frac{T_{out} - T_{air}}{G_b}) \quad (2)$$

Where η_T = instantaneous thermal efficiency; F_R = heat removal factor; n_o = optical efficiency; U_L = overall heat loss coefficient, $W/m^2.K$; C = concentration ratio; T_{out} = temperature of HTF leaving heat exchanger, K; T_{air} = ambient temperature, K; G_b = direct solar radiation, W/m^2. Figure 4 to Figure 8 show the graphical results of the experiments conducted using the different HTFs and insulation.

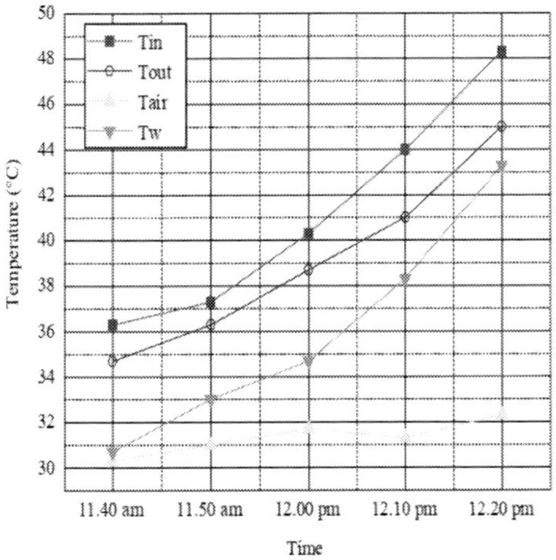

Fig. 4. Water as HTF.

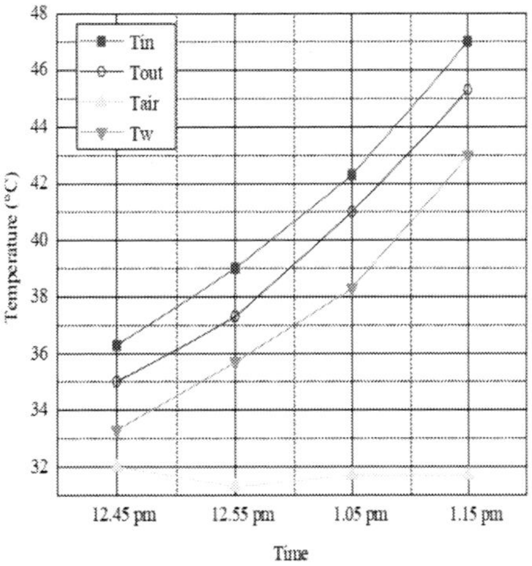

Fig. 5. Salt water at 10g/L as HTF.

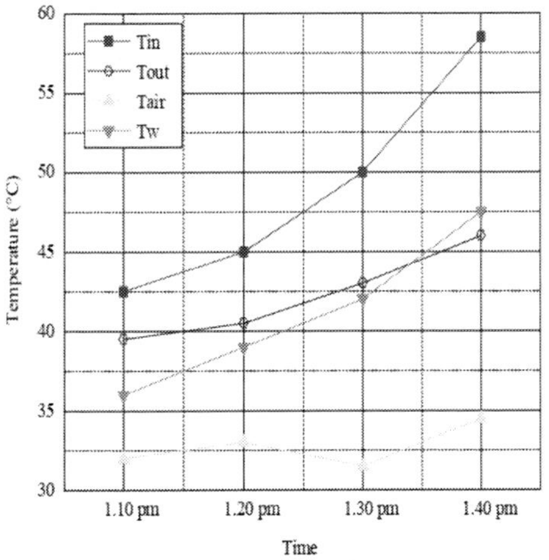

Fig. 6. Salt water at 20g/L as HTF.

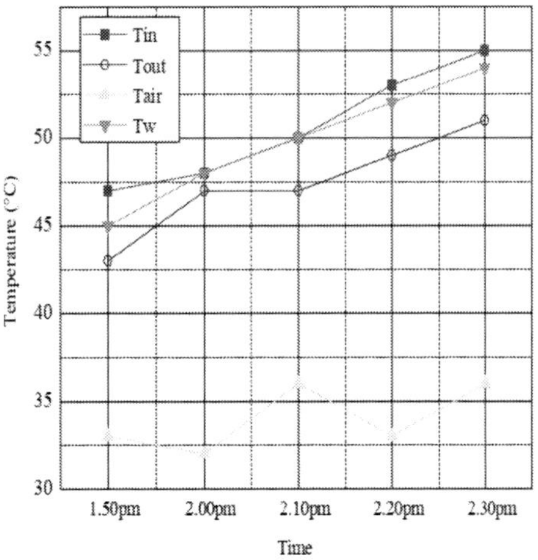

Fig.7. Salt water at 20g/L as HTF with polystyrene as pipe insulation.

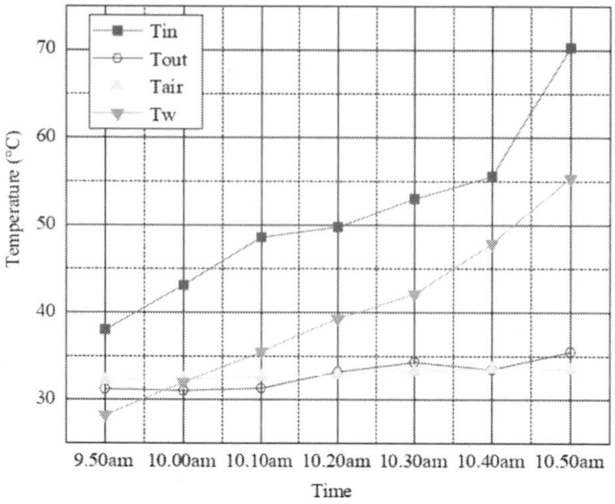

Fig. 8. Saltwater at 20g/L as HTF with acrylic foam as pipe insulation.

As seen from Fig. 4, water produces some appreciable results as an HTF but still is not good enough since it is observed from the graph that T_{in} and T_{out} are quite close. This indicates that the rate of heat transfer to the water in the tank is not that favorable. The results as shown in Fig. 5 demonstrate that Salt Water at 10g/L performs almost similarly as water thermally when compared to the results of water as HTF as seen in Fig. 4. It is observed that Salt Water at 20g/L is able to achieve much higher input temperatures as compared to water and the 10g/L salt solution. The temperature of the HTF leaving the heat exchanger is much lower, indicating good heat transfer properties of the liquid. From Fig. 7 and Fig. 8, the temperature of the water inside the tank quickly rises to the temperature of the HTF entering the heat exchanger. This indicates that the polystyrene and acrylic insulations helped reduce some heat loss from the copper pipe and hence the HTF flows with more heat capacity to be released into the water having a lower temperature. It is also observed that the temperature of the HTF leaving the heat exchanger is lower than the temperature of water, indicating that the salt water is able to transfer the heat to the water effectively. Almost all the graphs for the different configurations show a similar trend where T_{air} remains relatively constant throughout the experiment while T_{in}, T_{out}, and T_w continue to increase as the time goes by in the presence of the sun. T_{out} is always lower than T_{in} and T_w. This is because as the water enters the heat exchanger, the heat is absorbed by the water inside the tank causing T_w to increase while temperature leaving the heat exchanger decreases. Therefore, T_{out} is always lower than T_{in}. Also, T_w follows the same trend as T_{in} since the temperature of the water in the water tanks depends on the temperature of the HTF entering the heat exchanger. However, from the results obtained, Tin is always higher than T_w due to the closed loop not operating in an ideal scenario as there is heat loss to the environment. Therefore, the heat transfer between the HTF and the water inside the tank is not 100% efficient. Table 1 shows the efficiency based on Eq. 2 mentioned above and solar radiation, G_b averaging at 982W/m^2 for the first three results and 710.3W/m^2 for the last (for solar radiation data measured using a pyranometer).

Table 1. Efficiency of each experiment

HTF	T_{out} (°C)	T_{air} (°C)	η_T (%)
Water	43	31.3	61.8%
Saltwater at 10g/L	43.2	31.7	61.9%
Saltwater at 20g/L	44.5	33.8	62.2%
Saltwater at 20g/L with polystyrene	46.3	36	62.4%
Saltwater at 20g/L with acrylic foam	34.1	33.3	66.2%

The efficiency calculation has been optimized to reflect more accurate results compared to the results previously presented in paper [1]. The main difference between the results of the previous paper and this one is that only stabilized experimental data have been considered in the efficiency calculation. By considering only the experimental results of a system which has been thermally stabilized eliminates cold start results of the system which affects the accuracy of efficiency calculation results. The thermal efficiency equation used previously optimized for a closed loop system with heat exchanger, while the efficiency calculation of this paper uses Eq. 2 which has been optimized for PVT systems with cylindrical heat exchangers. Comparing the results obtained, salt water at 20g/L with acrylic foam as pipe insulation is found to have the best thermal efficiency with a 4.4% improvement from using only water as HTF, a 4% improvement from using salt water at 20g/L without insulation and a 3.8% improvement from using polystyrene as insulation. Adding 10g of salt per liter of water slightly improves the thermal efficiency by 0.1% and is observed to absorb heat faster when compared to water, taking only 30 minutes to reach the maximum temperature for T_{in} while water takes 40 minutes, as observed in Fig. 5. Hence, adding salt helps dissipate heat faster from the system but 10g/L is not enough for an effective heat transfer to the water tank. This is improved by increasing the salt concentration to 20g/L which yields an efficiency improvement of 0.3% and is able to achieve much higher T_{in} temperatures while efficiently transferring the heat to the water tank, thus keeping low T_{out} temperatures. Compared to the 10g/L concentration, the 20g/L concentration improves T_{in} by 25.5%.

3.2 Voltage Characteristics

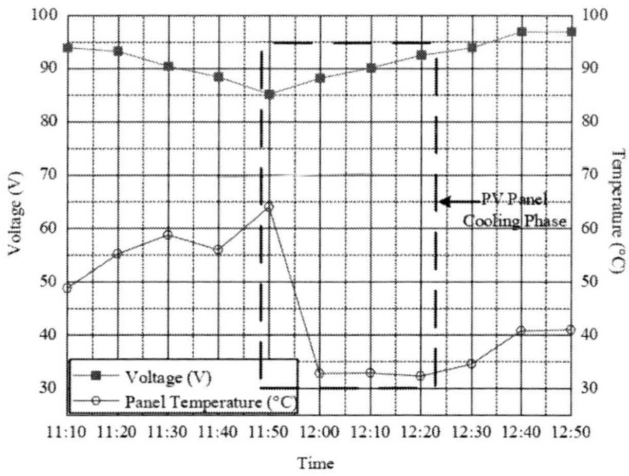

Fig. 10. PV voltage compared with PV temperature with cooling enabled.

Fig.10 shows an insight of how temperature affected the panel and the cooling period during which voltage is restored. From Fig. 10 it is observed that as the temperature of the PV panel increases, the voltage decreases. The voltage is restored by cooling the panel as highlighted by the dashed lines showing the PV panel's cooling phase. Table 2 below also shows that the voltage preserves its restored state even under lower sunlight radiation when under a low operating temperature after being cooled. The cooling phase is from around 11.50am until 12.20pm.

Table 2. Voltage Supply compared with Solar Radiation and Temperature.

Time	Voltage (V)	Solar Radiation (W/m^2)	Panel Temperature ($^{\circ}C$)
11:09:57	93.9	936	48.8
11:19:57	93.2	1005	55.2
11:29:57	90.4	1040	58.8
11:39:57	88.4	1037.1	56
11:49:57	85.2	1118.9	64
11:59:57	88.2	132.2	32.8
12:09:57	90.1	107.5	32.9
12:19:57	92.5	205.9	32.3
12:29:57	93.9	889.3	34.6
12:39:57	96.9	648.8	40.8
12:49:57	96.9	983.8	41

Considering the maximum and minimum voltage difference before and after cooling, a maximum improvement of 11.7% is noted when the panel's temperature is reduced by 24.1°C. This enabled the PV panel to achieve 96.9% of its rated maximum peak voltage even under lower solar radiation conditions (measured using a pyranometer) which is not possible without cooling the panel. However, when the PV panel is reintegrated onto the Parabolic Trough as per the experimental setup of [11] for a general comparison, the thermal efficiency of the Parabolic Trough drops by almost 40%.

4 Conclusion

Applying insulation to the concentrator pipe yielded higher thermal efficiencies. The acrylic foam performed better than the polystyrene insulation due to its advantageous thermal properties. Saltwater at 20g/L with acrylic foam as pipe insulation is found to have the best thermal efficiency with a 4.4% improvement from using only water as HTF, a 4% improvement from using salt water at 20g/L without insulation and a 3.8% improvement from using polystyrene as insulation. The cooling technique used for restoring the voltage drop of the PV panel due to heat in this research showed its potential as a hybrid PVT system where the heat build-up is reused or stored. The experiments are performed under no load as a study of the maximum voltage the panel reach compared to its rated voltage.

Acknowledgements

This research is partially supported from the Taylor's University Flagship research grant TUFR/2017/001/01.

References

[1] Paul Kramer, Why are solar cookers still unpopular among development experts, Journal of Engineering Science and Technology, **5**, no.1 75-84, (2010)

[2] M.F.HJ.Mohd. Amin, C.K.H. Chin, V.Garaniya, Application of alternatice energies in the australian offshore sector, Journal of Engineering Science and Technology, **11**, no.9,1296-1310, (2016)

[3] D.F.A.L.Riza, S.I.U.H Gilani, M.S. Aris, Standalone photovoltaic system sizing optimization using design space approach: case study for residential lighting load, Journal of Engineering Science and Technology, **10**, no.7,943-957 (2015)

[4] Vijayakumar G, Anita R, Photovoltaic based shunt active filter for power quality improvement using Icosϕ theory, Journal of Engineering Science and Technology, **10**, no.11,1422-1440, (2015)

[5] H. Machrafi, Enhancement of a photovoltaic cell performance by a coupled cooled nanocomposite thermoelectric hybrid system, using extended thermodynamics, Curr. Appl. Phys., **17**, no. 6, pp. 890–911, 2017.

[6] K.Syed Jafar, B.sivaraman, Performance characteristics of parabolic solar collector water heater system fitted with nail twisted tapes absorber, Journal of Engineering Science and Technology, **12**, no. 3, pp. 608-621, (2017)

[7] Aravind C.V., M. Al-Atabi, J. Ravishankar, A. Malik, Arkar, E. Ambikairajah, Eco-Tourism Sustainability through PV Technology: A Comprehensive View, Journal of Engineering Science and Technology, **8**, no. 6, pp. 654-669, (2013)

[8] K. Fuji, Y. Shirai, Y. Mitani, M. Watanabe, T. Kawagoe, A. Shiota, A Research of Basic Energy Reduction Approach Using PV Power Smoothing EV Storage System and LED Simple Circuit for Create Green Innovation, Journal of Engineering Science and Technology, **2**, no. 1, pp. 76-89, (2014)

[9] A. Shukla, K. Kant, A. Sharma, P. H. Biwole, Cooling methodologies of photovoltaic module for enhancing electrical efficiency: A review, Sol. Energy Mater. Sol. Cells, **160**, pp. 275–286, (2017).

[10] R. Wang, W. Qu, J. Sun, H. Hong, An On-site Test Method for Optical Efficiency of Large-size Parabolic trough Collectors, Energy Procedia, **105**, pp. 486–491, (2017)

[11] D. Kumar, S. Kumar, Simulation Analysis of Overall Heat Loss Coefficient of Parabolic Trough Solar Collector at Computed Optimal Air Gap, Energy Procedia, **109**, no. pp. 86–93, (2017)

[12] D.Prakash, A review on heat dissipiating passive cooling techniques for residential buildings at tropical region, Journal of Engineering Sciences and Technology, **12**, no.8, 2120-2140, (2017)

Magnetic Geared Radial Axis Vertical Wind Turbine for Low Velocity Regimes

Teow Wei Wei[1] and *Aravind* CV[1,*]

School of Engineering, Taylor's University, Malaysia.

Abstract. In the 21st century, every country is seeking an alternative source of energy especially the renewable sources. There are considerable developments in the wind energy technology in recent years and in more particular on the vertical axis wind turbine (VAWT) as they are modular, less installation cost and portable in comparison with that of the horizontal axis wind turbine (HAWT) systems. The cut-in speed of a conventional wind turbine is 3.5 m/s to 5 m/s. Mechanical geared generators are commonly found in wind technology to step up power conversion to accommodate the needs of the generator. Wind turbine gearboxes suffer from overload problem and frequent maintenance in spite of the high torque density produced. However, an emerging alternative to gearing system is Magnetic Gear (MG) as it offers significant advantages such as free from maintenance and inherent overload protection. In this project, numerical analysis is done on designed magnetic gear greatly affects the performance of the generator in terms of voltage generation. Magnetic flux density is distributed evenly across the generator as seen from the uniform sinusoidal output waveform. Consequently, the interaction of the magnetic flux of the permanent magnets has shown no disturbance to the output of the generator as the voltage generated shows uniform waveform despite the rotational speed of the gears. The simulation is run at low wind speed and the results show that the generator starts generating a voltage of 240 V at a wind speed of 1.04 m/s. This shows great improvement in the operating capability of the wind turbine.

1 Introduction

As oil reservoirs are running low and the increase in human population with demand, every country is seeking for an alternative source of energy especially the renewable energy sources. Due to the concerns on environmental issues, the development and application of renewable and clean energy are gaining more attention among the society. The wind technology and photovoltaic (PV) systems are found to be part of the two fast-growing technologies in the effort of generation of cleaner energy.

[1] Corresponding author: aravindcv@ieee.org

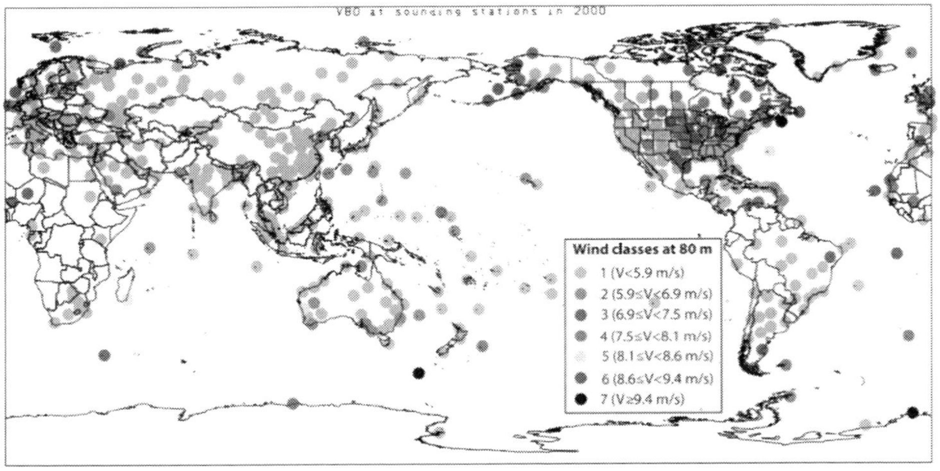

Fig. 1. World map of wind speed extrapolated to 80 m and the average for the year 2000 [2].

The only few countries with substantial wind speed are Japan, a few islands in China Sea, Guam and Mariana Islands. The majority of these areas are not suitable for wind power generation. Generally, Southeast Asia has poor wind potential on over 80% of the territory [2]. As seen in Figure 1.1, countries situated near or on the equator experience low wind speed due to the different direction of the wind that cancels on each other at the equator. Based on the wind speed characteristic of Malaysia recorded by the Malaysian Meteorological Department, typical wind speed ranges from as low as 0 m/s to 13 m/s with the highest mean daily wind speed of 3.8 m/s recorded in Mersing, Johor [3]. Wind turbines in Mersing with a cut-in speed of 2.5 m/s operate for 5084 hours a year [4]. Technically, the wind energy potential in Malaysia is very limited due to the low wind speed condition.

Industrial wind turbines fall into two general categories based on the axis of rotation of the wind turbine with respect to the ground surface and they are (a) horizontal axis wind turbine (HAWT), and (b) vertical axis wind turbine (VAWT). A conventional wind turbine has a cut-in speed of 3.5 m/s to 5 m/s and this is the minimum speed required to move the wind turbine blades into motion due to its inertia. At the rated speed of 12 m/s, the generator is operating in its ideal condition with the highest efficiency of power generation. When the wind speed is very high (above cut-out speed of 25 m/s), the wind turbine must be shut down due to mechanical constraints, while when the speed is very low (below cut-in speed), the operation is also in need of cutting off due to the variable low wind speed which is unable to move the generator at a constant speed. Many wind turbine system works on the principle of the wind energy conversion system (WECS) which can be divided into the aspects of aerodynamic, mechanical and electrical [5]. Critically optimizing each aspect helps in producing the most efficient output possible. WECS is projected and simplified in the form of block diagram as shown in Figure 2.

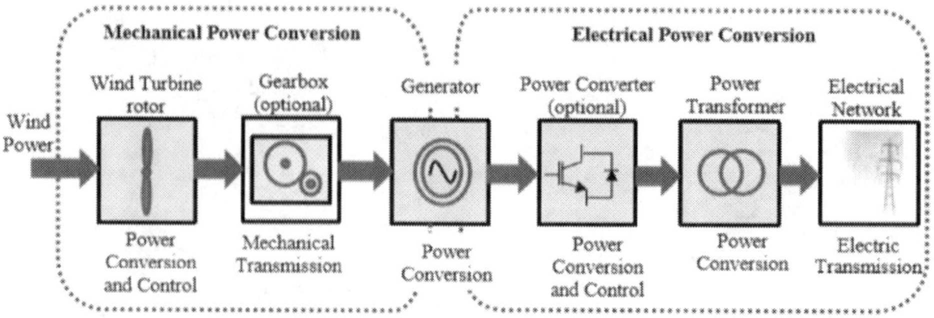

Fig. 2. Block diagram of WECS components viewed in aspects of mechanical and electrical [5].

High-speed electric generators are commonly found in many conventional wind turbines typically to step up the relatively low rotational speed of the rotor blades to accommodate the needs of the generator [8]. As a result, a gearbox is often used in the system to perform this function and since the speed increase needed is important, planetary mechanical gear type of configuration is oftentimes used to design the gearboxes of wind turbine [8]. Mechanical gears and gearboxes are commonly employed for speed change and torque transmission in various industrial applications [9], [10]. Many wind turbine gearboxes suffer from overload problem and frequent maintenance in spite of the high torque density produced by mechanical gears [9]. In contrast, an emerging alternative to gearing system is Magnetic Gear (MG) as it offers significant pros such as reduced noise and vibration, free from maintenance, and inherent overload protection [11]. In recent years, MGs have been widely used in the field of aerospace, marine-propulsion, and renewable energy [12]. Inspired by MGs, many Permanent Magnet (PM) machines with the integration of MGs are proposed to produce a higher power density than the conventional PM machines integrated with mechanical gears [12]–[14]. Nevertheless, MGs have yet to be seen to replace mechanical gears in the wind energy technology despite it being a promising technology for various torque-demanding applications due to the cost of permanent magnets[15-16].

2 Theoretical Framework

2.1 Design Concept

The conventional cut-in speed of a wind turbine is about 3.5m/s - 5m/s. A typical wind generator operates on the principle of motoring and generating. At speed lower than the cut-in, the generator operates as a motor with the help of pitch control in order to put the stationary wind turbine into motion due to the naturally occurring inertia of turbine blades. Once it reaches the cut-in speed, the generator switches into two modes, generating and motoring. In this region, the motor continues to operate until the wind turbine acquires enough momentum to match the force provided by the wind speed. The machine then switches to full generation mode as the wind turbine is capable of rotating under the wind speed (also known as rated wind speed). The operational modes discussed above is portrayed below as shown in Figure 3.

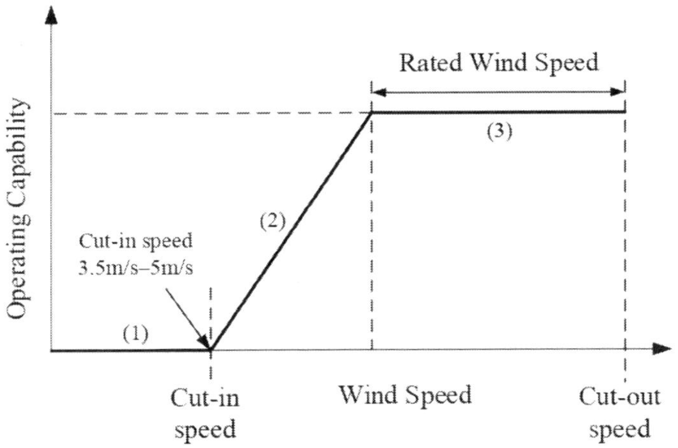

Fig. 3. Operational modes of a typical wind turbine.

Optimisation of the wind turbine system in the sense to increase the power generation capability on low wind speed condition is explained in Table 1 below.

Table 1. Optimised modes of wind turbine system.

Mode	Description
Motoring & Generating	Power generation capacity is increased with the proposed magnetic gear even under a low wind operating conditions with constant output.
Generating	In this mode, the operating mode is ideal and with the permanent magnet used to integrate into magnetic gear configuration, constant speed of the generator is able to be maintained.

The conventional mechanical structure of wind turbine system involves the use of mechanical gears and this is causing one of the major issues in the free revolution of turbine structure as the coupling between the gears induces some inertia to the system. Also, this results in the mechanical losses, friction, wear and tear due to which the gears are in contact. A magnetic gear, separated between the static structure and the revolving turbine, is proposed to address this concern. Figure 4 below shows the magnetic circuit of the ORIPMS generator and geared ORIPMS generator.

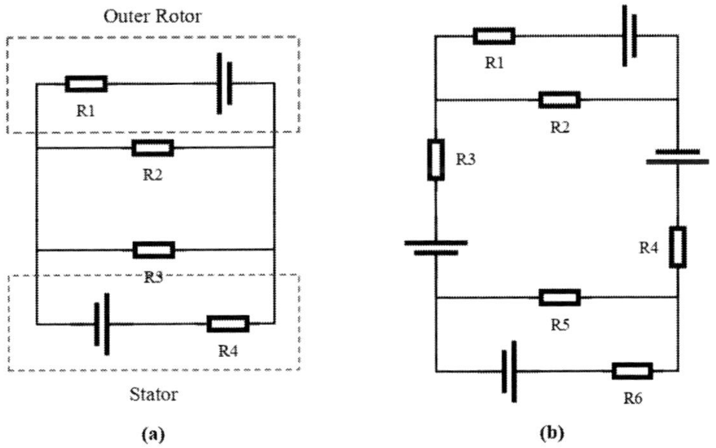

Fig. 4 (a) Equivalent magnetic circuit of ORIPMS. (b) Geared ORIPMS.

2.2 Magnetic Gear Ratio Design Concept

In a geared wind turbine mechanism, mechanical gear ratio or transmission ratio is the ratio of a number of driver gear (attached to wind turbine) over the number of driven gear (attached to the generator). The magnetic gear transmission ratio is analogous to mechanical gear principle and is discussed in the following. The magnetic gear ratio is expressed in Equation (1).

$$Gear\ ratio, \qquad G_r = \frac{V_{high}}{V_{low}} = \frac{T_{driving\ gear}}{T_{driven\ gear}} = -\frac{N_{low}}{N_{high}} \tag{1}$$

where V_{high} and V_{low} are rotational speed of high-speed gear and low-speed gear, respectively. $T_{driver\ gear}$ and $T_{the\ driven\ gear}$ represent a number of mechanical teeth of driver gear and driven gear, respectively. N_{high} and N_{low} are a number of pole-pairs of high-speed gear and low-speed gear, respectively. In order to attain the highest torque density, the number of pole pairs on each of the rotors should satisfy the following relation as in Equation (2).

$$N_{low} = N_{ferro} - N_{high} \tag{2}$$

where N_{ferro} is the number of ferromagnetic pole-pieces.

The selection of an appropriate combination of pole-pairs for the two rotating gears to produce a required gear ratio should take into consideration the torque ripples that may affect the performance of the magnetic gear. Torque ripple or cogging torque is caused by the interaction of the permanent magnets with the ferromagnetic pole-pieces. Therefore, cogging torque factor is defined to estimate the level of torque ripple on the performance (as shown in Equation (3)). The higher the factor, the higher the cogging torque is. The minimum value of this factor is unity ($C_T = 1$), which initiates a possible for good design selection as in Equation (3).

$$C_T = \frac{N_{high} N_{ferro}}{LCM(N_{high}\ ,\ N_{ferro})} \tag{3}$$

3.5 Relationship between the mechanical and electrical power in machine

The mechanical power generated, $P_{gen,low}$ by the low-speed gear is proportional to torque and speed of the gear, as shown in Equation (4).

$$P_{gen,low} = T_{low} \cdot \omega_{low} \tag{4}$$

where T_{low} is a torque generated by low-speed gear and ω_{low} is the angular velocity of the low-speed gear. As the speed the lower speed gear is proportional to the speed of higher speed gear with respect to the gear ratio, the relationship of the speeds affected by the gear ratio, N is as shown in Equations (5) and Equation (6).

$$v_{low} \propto v_{high} \tag{5}$$

$$v_{low} = G_r \cdot v_{high} \tag{6}$$

As the speed is proportional to angular velocity, the mechanical power generated by the ring magnetic gear, $P_{gen,high}$ is correlated to the gear ratio as shown in Equations (7) and Equation (8).

$$P_{gen,high} = T_{high} \cdot \omega_{high} \quad (7)$$

$$P_{gen,high} = T_{high} \cdot G_r \cdot \omega_{low} \quad (8)$$

where T_{high} is a torque generated by high-speed gear and ω_{high} is the angular velocity of the high-speed gear. Electrical power generated by the generator is expressed as in Equation (9).

$$P = V \cdot I \cos\theta \quad (9)$$

Mechanical power from the high-speed gear is converted directly to electrical power generated by the generator. With this, we can say that mechanical power is equal to the electrical power output and the relationship is expressed as in Equation (10).

$$P_{gen,high} = P_{elec} \quad (10)$$

By equating Equation (7) and Equation (8), we get power as in Equation (11).

$$T_{high} \cdot G_r \cdot \omega_{low} = V \cdot I \cos\theta \quad (11)$$

Therefore, by increasing the gear ratio of the magnetic gears, the electrical power output increase relatively. Outer Rotor Internal Permanent Magnet Synchronous (ORIPMS) generator (as shown in Figure 5) is used employed in this project due to its minimal ripple harmonics and a satisfactory amount of power produced. To further increase the efficiency of power produced by this generator, magnetic gear is introduced by increasing the flux density of the machine. FEM analysis is shown and analyzed in this section below. Comparison between the two models is done to observe the deviation of the results of the machines.

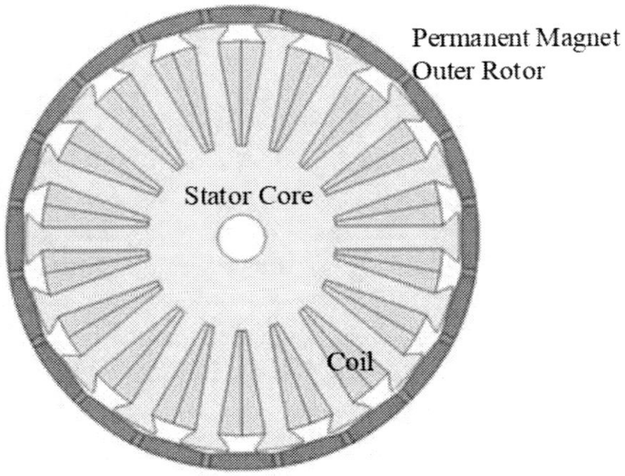

Fig. 5. Outer Rotor Internal Permanent Magnet Synchronous (ORIPMS) Generator.

Table 2. Parameters of ORIPMS Generator.

Parameter	Value
Number of poles on high-speed gear	100
Number of ferromagnetic pole-pieces	24
Number of poles on low-speed outer rotor	20
Outer diameter of rotor	50.0 mm
Inner diameter of rotor	46.5 mm
Outer diameter of PM	50.0 mm
Inner diameter of PM	47.0 mm
Width of stator pole	5.0 mm
Width of stator teeth	6.968 mm
Length of stator pole	22.5 mm
Diameter of shaft	5.0 mm
Length of air gap	0.5 mm
Stack length	50.0 mm
Outer diameter of stator	46.0 mm
Inner diameter of stator	5.0 mm
Outer diameter of coils	42.5 mm
Inner diameter of coils	20.0 mm
PM pole-arc	2 degrees
Rotor pole-arc	16 degrees
Stator Teeth pole-arc	13 degrees

3 Results and Discussion

Magnetic flux density behavior of any generator is the primary evaluation that should be considered in designing a machine as this analysis fundamentally determines the output of the machine. In this project, the proposed magnetic gear has a direct interaction with the outer rotor permanent magnets and this might affect the output quality of the machine negatively. Through JMAG simulation, the magnetic flux density of the generator rotating at 500rpm and magnetic gear rotating at 100rpm with a gear ratio of 1:5 is shown in Figure 6. Fundamentally, the operation principle of a generator is defined by Faraday's Law, which states that electromotive force induced in a coil is proportional to the rate of change of the number of magnetic flux lines passing through the surface enclosed by the coil. Therefore, with this attribute, a sinusoidal induced output voltage is expected consequently and it is proven in the following section. The output voltage generated is presented with respect to different speed from 100 rpm to 700 rpm. The voltage produced increases with the wind speed as graphed in Figure 7. Also, the generator starts generating voltage even at speed lower than the conventional cut-in speed of 3.5 m/s to 5 m/s. The generator achieves a voltage of 240 V at speed of 200 rpm or equivalent wind speed of 1.04 m/s in which this is the standard voltage supply of household in Malaysia. This means that generating a voltage at low wind speed condition is achievable. As expected based on theoretical knowledge, the voltage generated increases with wind speed. The graph is increasing linearly from 0 m/s to 2 m/s but the graph deviates a little as the wind speed continues to increase. This is possible because the generator is starting to fall into the saturation region as the rated power of this machine is only 500W.

Therefore, for this chosen generator, the size of it is only capable of low rotational speed. However, to increase its capacity, a small increment in diameter of the machine results in squared of the rated power. Based on the evaluation made above, it can be viewed that the operating range of wind turbine is improved on the low wind speed range. With this, the magnetic-geared generator enables the wind turbine to operate before the typical cut-in speed of 3.5 m/s to 5 m/s. The conventional and optimized operating capability are illustrated in the following.

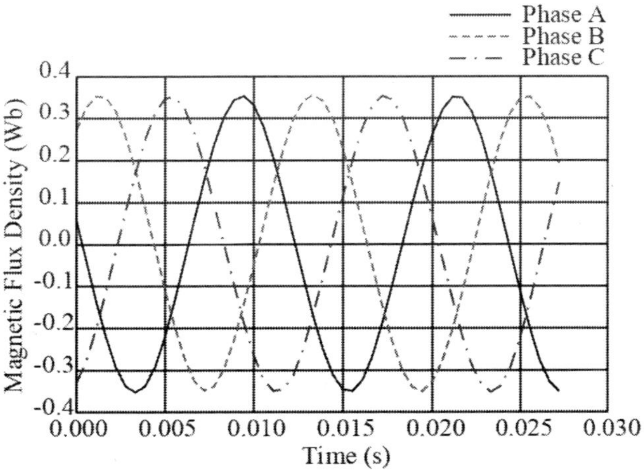

Fig. 6. The magnetic flux density of magnetic gear rotating at 100 rpm.

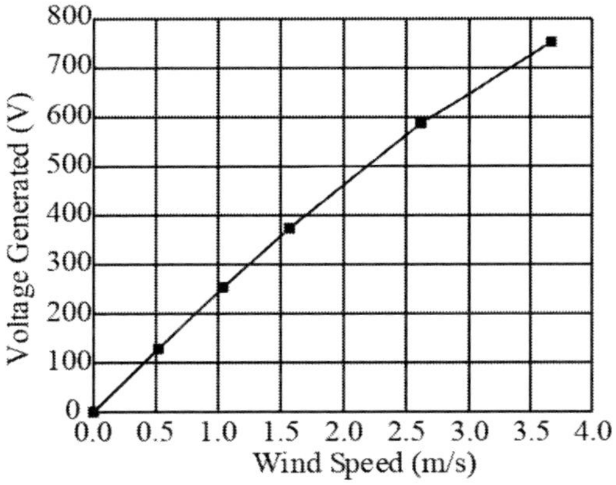

Fig. 7 Voltage generated at various wind speeds.

Table 3. Various speed versus voltage generated (rms).

Speed (rpm)		Equivalent wind speed (m/s)	Voltage Generated (V_{rms})			Average RMS Voltage (V_{rms})
Magnetic Gear	Generator Outer Rotor		Phase A	Phase B	Phase C	
100	500	0.52	129.801	123.860	130.585	128.082
200	1000	1.04	253.614	253.385	253.883	253.627
300	1500	1.57	375.507	370.097	375.525	373.710
500	2500	2.62	588.734	585.307	589.443	587.828
700	3500	3.67	752.875	751.108	753.968	752.650

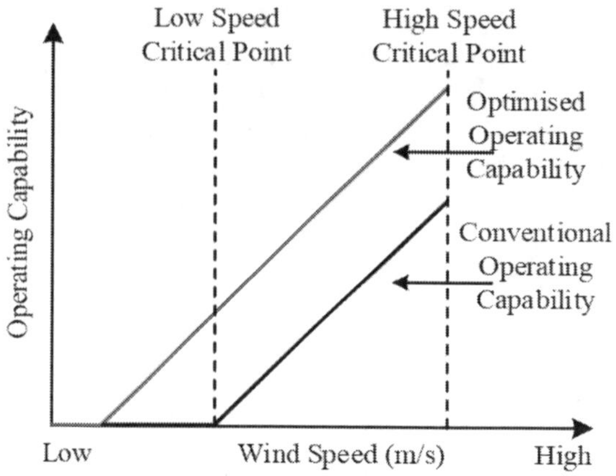

Fig. 4.3: Illustration of optimized operating capability

4 Conclusion

Magnetic flux density is distributed evenly across the generator as seen from the uniform sinusoidal output waveform. Consequently, the interaction of the magnetic flux of the permanent magnets has shown no disturbance to the output of the generator as the voltage generated shows uniform waveform despite the rotational speed of the gears. The simulation is run at low wind speed and the results show that the generator starts generating a voltage of 240 V at a wind speed of 1.04 m/s. This shows great improvement in the operating capability of the wind turbine and the objective of this project has been achieved.

References

[1] J. Damota, I. Lamas, A. Couce, J. Rodríguez. Vertical Axis Wind Turbines : Current Technologies and Future Trends, Int. Conf. Renew. Energies Power Qual. - La Coruña (Spain), 25th to 27th March, 2015, **1**, no. 13, p. 6, (2015)

[2] C. L. Archer. Evaluation of global wind power, J. Geophys. Res., **110**, D12, D12110 (2005)

[3] General Climate Information - Malaysian Meteorological Department, 2016 [Available online http://www.met.gov.my/in/web/metmalaysia/climate/climatechange/climateinformation. [Accessed: 14-Nov-2017].

[4] A. Belhamadia, M. Mansor, M. A. Younis. A study on wind and solar energy potentials in Malaysia, Int. J. Renew. energy Res., **4**,no. 4, pp. 1042–1048 (2014)

[5] M. R. M. Altimania, Modeling of Doubly-Fed Induction Generators Connected to Distribution System based on eMEGASim® Real-Time Digital Simulator, J. Chem. Inf. Model., **53**, no. 9, pp. 1689–1699 (2013)

[6] G. Semrau, S. Rimkus, T. Das, Nonlinear Systems Analysis and Control of Variable Speed Wind Turbines for Multiregime Operation, J. Dyn. Syst. Meas. Control, **137**, no. 4 (2015)

[7] M. H. Chiang, A novel pitch control system for a wind turbine driven by a variable-speed pump-controlled hydraulic servo system, Mechatronics, **21**, no. 4, pp. 753–761, (2011)

[8] C. Nutakor, A. Klodowski, J. Sopanen, A. Mikkola, J. I. Pedrero, Planetary gear sets power loss modeling: Application to wind turbines, Tribol. Int., **105**, pp.42-54 (2016)

[9] K. Atallah, D. Howe, A novel high-performance magnetic gear, IEEE Trans. Magn., **37**, no. 4, pp. 2844–2846 (2001)
[10] K. Atallah, J. Wang, D. Howe, A high-performance linear magnetic gear, J. Appl. Phys., **97**, no. 10, pp. 26–29 (2005)
[11] W. H. X. Li, K.-T. Chau, M. Cheng, Comparison of magnetic-geared permanent magnet machines, Prog. Electromagn. Res., **133**, pp. 177–198 (2013).
[12] S. Niu, Y. Mao, A Comparative Study of Novel Topologies of Magnetic Gears," Energies, **9**, no. 10, pp.no. 773, (2016)
[13] W. Li, K. T. Chau, J. Z. Jiang, Application of linear magnetic gears for pseudo-direct-drive oceanic wave energy harvesting, IEEE Trans. Magn., **47**, no. 10, pp. 2624–2627 (2011)
[14] S. Niu, S. L. Ho, W. N. Fu, Performance analysis of a novel magnetic-geared tubular linear permanent magnet machine, IEEE Trans. Magn., **47**, no. 10, pp. 3598–3601 (2011)
[15] M. M. Moh Saad, N. Asmuin, Comparison of Horizontal Axis Wind Turbines and Vertical Axis Wind Turbines, IOSR J. Eng., **4**, no. 8, pp. 2250–3021, (2014)
[16] M. Mohd Amin, C. Chin, V. Garaniya, Application of Alternative Energies in the Australian Offshore Sector, Journal of Engineering Science and Technology, **11**, no. 9, pp. 1296 - 1310 (2016)
[17] P. Sharma, T. Bhatti, K. Ramakrishnan, Permanent-Magnet Induction Generators: an Overview, Journal of Engineering Science and Technology, **6**, no. 3, pp. 332 - 338 (2011)

A Study on the Used of Bluetooth Detectors at a Fix Location for Vehicle Speed Estimation

Way Shyun Neow[1] and *Edwin C.Y.* Chung[1,*]

[1]School of Engineering, Taylor's University, 1 Jln Taylor's, 47500 Subang Jaya, Selangor, Malaysia

Abstract. Usage of Bluetooth for traffic monitoring has been widely researched and well understood. By placing a Bluetooth detector at each end of a segment of a road; the time difference when a Bluetooth device, traveling in a vehicle through that segment, is observed by these detectors is taken to be the time taken by the vehicle to travel along that known segment. The known distance between the detectors and the time it takes to traverse this distance makes it possible to estimate the average speed of the vehicle along that segment. The possibility of estimating the average speed of vehicles within the radio range of a Bluetooth detector, however, has not be carried out until now. The premise of this research is to determine the time a Bluetooth device takes to traverse through the radio range of a detector and from this estimation of the speed of the vehicle traveling through the radio range of the detector. The usage of multiple detectors at the same point to improve the accuracy of the estimates is also investigated. The results of our experiments and what is need to realise such a system are discussed.

1 Introduction

The ability to determine and disseminating accurate traffic conditions to road users in a timely manner has many advantages. Technology for traffic monitoring is a widely-researched topic and are based on technologies ranging from image processing, radio technology such as RFID & Bluetooth to GPS.

A method in image processing is to estimate traffic density through images captured [1]. The disadvantages of this method are that image processing involves complex algorithm and extensive computing infrastructure. RFID, on the other hand, involve the use of RFID sensors placed along a road which is used to detect the RFID attached to the vehicle number plate. The vehicle average speed between any 2 detectors is then estimated based on the time it takes to traverse that section of the road [2]. The main disadvantage of this method is the need to introduce RFID tagging of vehicles. Ahmad et al. [3] indicated that RFID tags can be linked to personal information and may be a barrier to implementation. Bluetooth, on the other hand, is a technology that has lots of potential in traffic monitoring. Nowadays, devices such as car radio, smart phone, laptop and various other gadgets are Bluetooth enabled. Bluetooth is in many ways no different from RFID with the exemption

* Corresponding author: Edwin.Chung@taylors.edu.my

that they are already available in vehicles traveling along our roads. It is relatively low costs and has low energy consumption.

Current Bluetooth traffic monitoring systems utilise Bluetooth detector placed at each end of a segment of a road. These detectors will intercept the Media Access Control Identity (MAC ID) of the Bluetooth devices in the vehicle and the average speed of the vehicle travelling through this segment of the road is then estimated based on the time difference a MAC ID was detected at these detectors shown in **Fig. 1.**

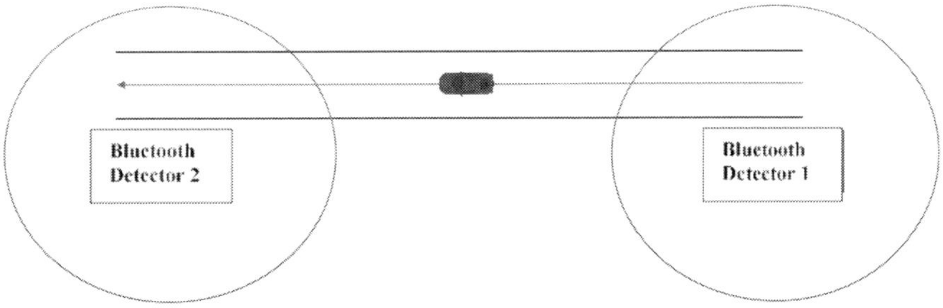

Fig. 1. Two Bluetooth detectors demarking a segment of a road being monitored.

Current research on traffic monitoring using Bluetooth technology are based on this configuration of having two detectors at each end of a segment of a road. There is, thus far, no study on the potential use of one or more Bluetooth detectors at a fix location for the estimation of vehicle speed within the radio range of the detector(s). An advantage of such a system is the reduction in the time difference between the indicated and the actual traffic. It has the potential to be able to determine the traffic condition at a round-about.

2 Bluetooth inquiry process

Bluetooth utilises an inquiry process to acquire the MAC ID of other Bluetooth devices within its radio range. During the process, the Bluetooth detector will send an inquiry packet (ID packet) via one of the 32 predefined inquiry channel. Once the vehicle Bluetooth device receive the ID packet, it will send a Frequency Hop Synchronization (FHS) packet back to the detector. **Fig. 2.** shows the format of a MAC ID[4], [5] consisting of 2, 1 and 3 bytes of non-significant address part (NAP), upper address part (UAP) and lower address part (LAP) respectively.

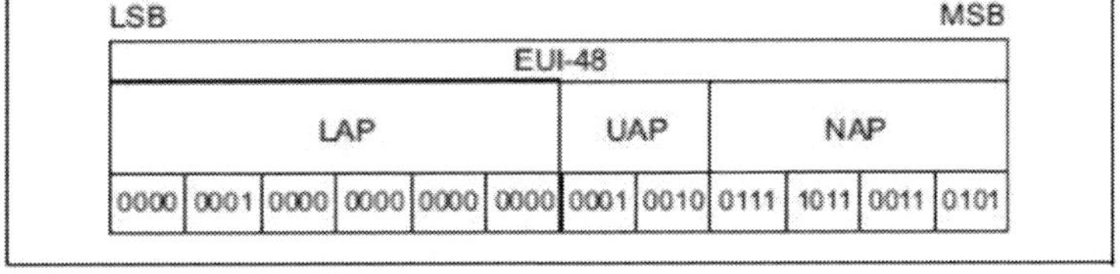

Fig. 2. MAC ID of Bluetooth devices [4].

Bluetooth utilises a frequency hopping spread spectrum technology consisting of 79 frequency channels. Of these, 32 are used as inquiry channels, 16 of these in the A train and the remaining 16 in the B train. At two consecutive 312.5μs, the Bluetooth detector will send an ID packet to two inquiry channels and listen for incoming reply at the next two consecutive 321.5μs. This is repeated 256 times for each train. In addition, one frequency will be changed between the two trains every 1.28s. **Fig. 3** shows the Inquiry process [4], [5].

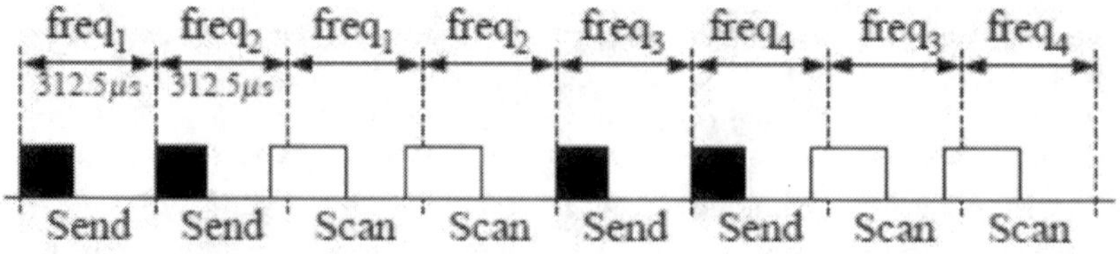

Fig. 3. Bluetooth inquiry process [5].

Franssens et al. [6] conduct research on having multiple Bluetooth inquirers in Bluetooth Discovery Process. The first study is about having multiple Bluetooth detectors present which is able to enhance the discovery of nearby Bluetooth devices. From **Fig. 4** shows having 5 inquirers (Bluetooth detectors) perform better than 1 inquirer and 3 inquirers.

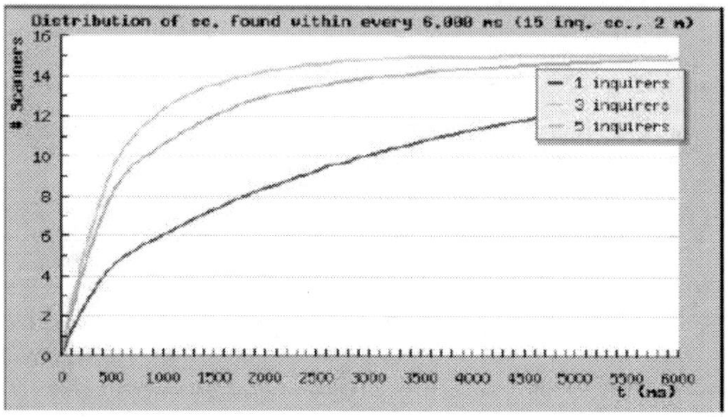

Fig. 4. Comparison of time taken for 1, 3 and 5 inquirers scan to 15 inquiry scanning device.

3 Design of the Bluetooth detector

3.1 Hardware design

3.1.1 HC-05 [7]

The HC-05, a class 2 Bluetooth module with a communication range of approximate 15 meters, is used in this research. The specification of the HC-05 is shown in **Tab. 1**. The function of pin is shown in **Tab. 2**.

Tab. 1. HC-05 specification.

Parameter	Value	Unit
Sensitivity	-80	dBm
Transmit power	4	dBm
Operating voltage	1.8 – 3.6	V

Tab. 2. HC-05 pin description and function.

Pin	Description	Function
EN	Disable module	Disable module with LOW input
+5V	Power supply	Supply voltage to the module
GND	Ground	Ground of power source
TX	Transmit	Connect to receive of micro-controller
RX	Receive	Connect to transmitter of micro-controller
STATE	Indication	To determine if the Bluetooth module is connected. HIGH when it is connected. LOW when it is not connected
Pin 34	Command mode	Provide HIGH (3.3V) input to enter command mode/AT mode

3.1.2 Data logger shield [8]

An Adafruit data logger shield is used in this research. It has a built-in Real Time Clock (RTC) module and a Secure Digital (SD) card module. The RTC module is used as the system clock and the SD card module is used for storing data. As there is no intention to build a complete system in this research, data collected are only analysed offline, hence the need for an SD card module for data storage.

3.1.3 Microcontroller development platform [9], [10]

The Arduino UNO, powered by an ATmega328 microcontroller, is used in this research. The specification of the Arduino UNO is shown in **Tab. 3**.

Tab. 3. Arduino ATmega328 MCU specification

Description		Arduino Uno
Input/output pin	Digital pin	14
	Analog pin	6
Frequency clock		16MHz
Input power		7V – 12V
Flash memory		32KB
SRAM		2KB
EEPROM		1KB

3.1.4 Power supply

A breadboard power supply is used to provide a steady voltage source (5V) to the components. It also acts as an overcurrent protection for the components. It must be noted that during this research, the Arduino UNO is power through the USB cable connecting it to the laptop and not the breadboard power supply unit. It is observed that the RTC module on data logger shield will malfunction when it is powered using the breadboard power supply.

3.1.5 Schematic design

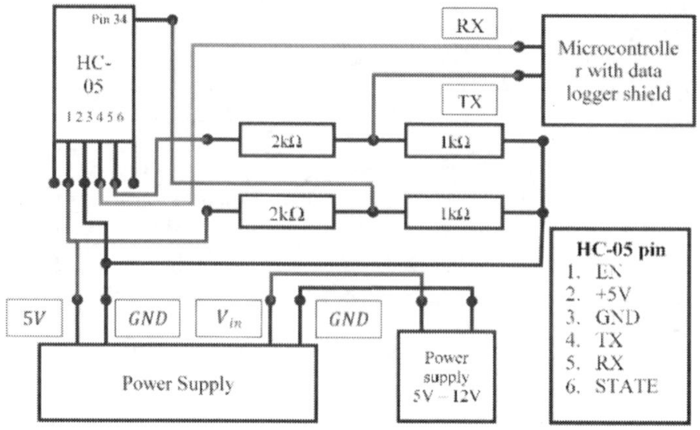

Fig. 5. Schematic diagram of a single Bluetooth detector.

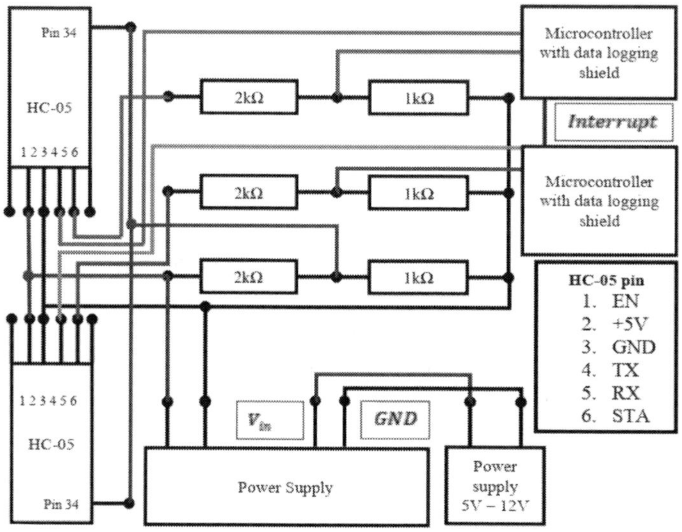

Fig. 6. Schematic diagram of a dual Bluetooth detector.

Tab. 4. Bluetooth module and microcontroller communication.

Bluetooth module		Microcontroller	
Pin 4	Transmit	Pin 8	Receive (Assigned)
Pin 5	Receive	Pin 9	Transmit (Assigned)

From the schematic diagrams above, the serial communication between the HC-05 and the UNO is through the transmit and receive pins. The connection between the communication pins are shown in **Tab. 4**. The UNO will have to assign its transmit and receive pin as activate. The transmit pin of the HC-05 is directly connected to the microcontroller which the microcontroller can treat a 3.3V digital signal it as HIGH. In contrast, microcontroller transmitter need to be step down from 5V to 3.3V as the HC-05 cannot receive signal voltage higher than 3.3V. The 5V signal is step down with using voltage division rule with resistor (1kΩ and 2kΩ). **Fig. 5** shows the design of a single Bluetooth detector and **Fig. 6** shows the design for a dual Bluetooth detectors. In the dual detector design, one of the UNO plays the role of an initiator and will interrupt the other

UNO to initiate the start of another inquiry process in the second UNO after a pre-set delay interval. It is via configuration that two independent inquiry sequence can be initiated with a pre-set delay apart. The prototype as shown in **Fig. 7** is built according to design in **Fig. 6**. One of the Bluetooth detector is switched off for a single detector configuration during testing.

Fig. 7. Prototype.

3.2 Software design

3.2.1 Flow chart

The functionality of the software embedded in the UNO is as detailed in the flowchart in **Fig. 8**. For ease of readability, the AT command detailed in this flowchart is summarised in **Tab. 5**. The reference to 'a' in the flowchart refers to a variable which record the return value from the Bluetooth module.

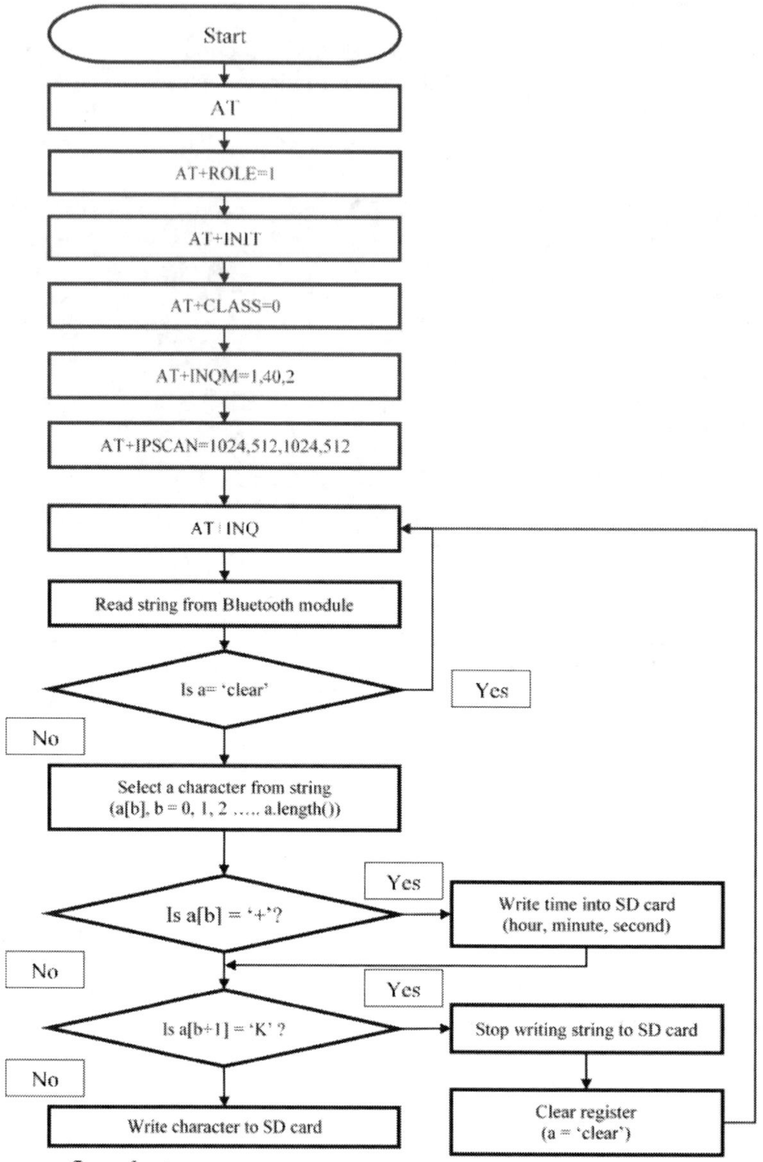

Fig. 8. Program flow chart

3.2.2 Bluetooth detector AT command explain [7]

Tab. 5. AT commands.

AT	A test command. A response "OK\r\n" is received if the Bluetooth module is operating properly.
AT+ROLE=<Param>	Set the Blutooth module mode. <Param> has the value 0, 1 or 2 for Slave, Master and Slave-Loop mode respectively.
AT+INIT	Initialise the SPP profile library. It returns an "OK\r\n" if the command is executed successfully.
AT+CLASS=<Param> AT+CLASS?	Set/Inquire device class. When setting the device class, <Param> is set to the value 0 for default class.

AT+INQM=<Param1>, <Param2>,<Param3> AT+INQM?	Set/Inquire Query access patterns. It returns an "OK\r\n" if the command is executed successfully. <Param1> = 0 for inquiry_mode_standard or 1 for inquiry_mode_rssi <Param2> = maximum number of Bluetooth devices to respond to <Param3> = maximum of limited inquiring time with value ranging from 1 to 48 corresponding to 1.28s to 61.44s.
AT+IPSCAN=<Param1>, <Param2>,<Param3>, <Param4> AT+IPSCAN?	Set/Inquire scan parameter. It returns an "OK\r\n" if the command is executed successfully. <Param1> = Query time interval <Param2> = Query duration <Param3> = Paging interval <Param4> = Call duration Default values: 1024,512,1024,512
AT+INQ	Inquire Bluetooth device with the radio range. It returns +INQ:<Param1>,<Param2>,<Param3> ... OK Where <Param1> is the Bluetooth address, <Param2> the device type and <Param3> the RSSI signal intensity.

4 Prototype testing

4.1 Static test

The objective is to identify the maximum detecting range of the Bluetooth detector. Furthermore, to justify that class 2 Bluetooth module have a detection range of 15 meters radius. This static test is conducted by having a smart phone with Bluetooth enable being placed at a known distance away from the Bluetooth detector. In the initial test, the smart phone was placed 10 meters away from the Bluetooth detector. Subsequent data points were taken with the distance between the smart phone and the detector increased in increment of 10 meters. This test is repeated until the Bluetooth detector is unable to detect the smart phone. The setting of the experiment is as shown in **Fig. 9**. Each test will last for 1 minute as the Bluetooth detector is activated. A total of 5 test is conducted for each distance. The setting of the Bluetooth detector inquiry parameters is AT+INQM = 1,1,1 throughout the test. Other than a smart phone, this static test was also conducted with a car Bluetooth enabled radio.

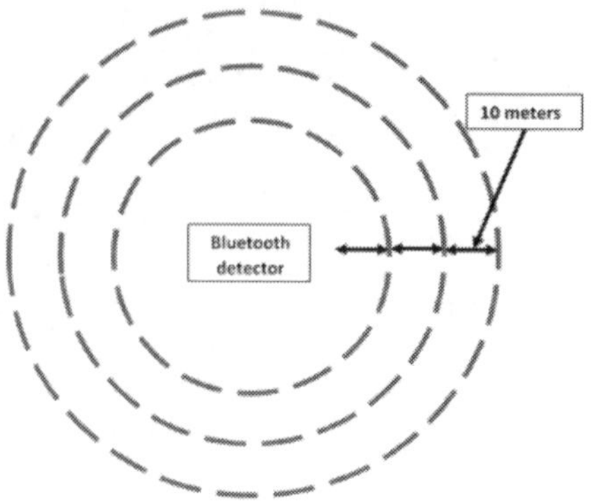

Fig. 9. Static test

4.2 Dynamic test

The purpose of the dynamic test is to collect data from the Bluetooth detector so as to determine its ability to estimate the speed of the moving Bluetooth device as well as to evaluate its performance. The test cases are as shown in **Tab. 6**. Starting from the maximum radio range of the detector (determined from the static test), a Bluetooth device used in the test is moved in a straight line to the opposite side of the radio range as shown in **Fig. 10**. Data collected from the different scenarios are then compared to determine what is the best configuration for this system.

In order to provide some insight into the test cases detailed in **Tab. 7**, the objectives of these test cases are described in the following sub-sections.

Tab. 6. Dynamic test cases.

Bluetooth module in detector		1		2		2		2	
Bluetooth device traveling		1	2	1	2	1	2	1	2
Delay				No delay		Delay by half of total inquiring time		Delay by the total inquiring time	
Inquiry parameter	1, 1, 1	√	√	√		√			
	1, 1, 2	√						√	
	1, 40, 4					√	√		

4.2.1 Effect of inquiry time

The effect of inquiry time on the performance of the Bluetooth detector is conducted by varying the maximum inquiry time limit, <Param3> of the AT+INQM command, between 1.28s and 2.56s. Accordingly, the 2 experimental runs were conducted using the following query settings

- AT+INQM=1,1,1
- AT+INQM=1,1,2

4.2.2 Effect of multiple Bluetooth detectors with different query settings and delay between their inquiry cycles

The effect of the various query settings (the same settings will be used on both detectors) and the time difference between the inquiry cycle of the 2 detectors as well as the impact on the performance of the detector was conducted. The query settings for the 2 detectors and the delay used were as follows. The delay between the 2 inquiry cycles is necessary to avoid collision.

- AT+INQM=1,1,1 with no delay
- AT+INQM=1,10,1 with 5000 milliseconds delay
- AT+INQM=1,1,1 with 0.64 second delay
- AT+INQM=1,1,2 with 2.56 second delay
- AT+INQM=1,40,4 with 2.56 second delay

4.2.3 Effect of multiple moving Bluetooth devices

The purpose of these tests is to study the performance of the system where multiple Bluetooth devices are traversing across the radio range of the detector. Two smart phones are used in these experiments and the following settings were used.

- AT+INQM=1,1,1 with a single detector
- AT+INQM=1,40,4 with 2 detectors & 2.56s gap between their inquiry cycles

The purpose of the experiment with the AT+INQM=1, 40, 4 settings and a 2.56s delay between the 2 inquiry cycles is for confirmation that it this would all the scanning of the 2 trains of 16 frequency channels concurrently.

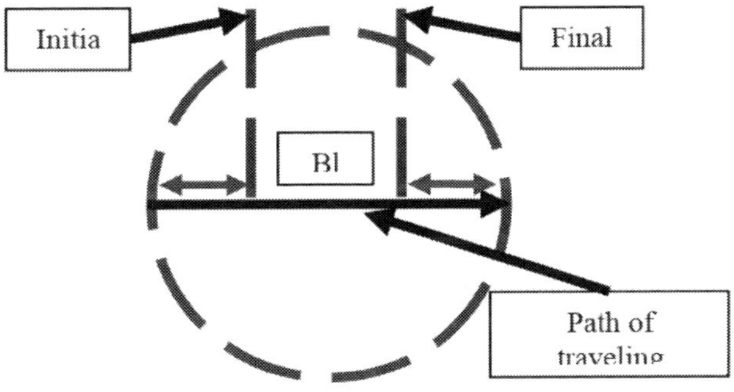

Fig. 10. Dynamic test.

In the dynamic test experiments, the average speed of a Bluetooth device traversing though the radio range of the detector is measure by taking the time it takes (t) to move across a known distance (s) using a stopwatch. The actual average speed is then computed using Equation (1).

$$v_{av,actual} = \frac{s}{t} \tag{1}$$

The duration a Bluetooth device is within the radio range of a detector can be determine from final time it was detected by the detector, t_{final}, and the time it was initially detected by the detector, $t_{initial}$. The predict average speed that is then calculated with Equation (2).

In this research, the known distance, s in Equation (2), is taken to be 2 times the radio range of the detector.

$$v_{av,predict} = \frac{s}{t_{final} - t_{initial}} \quad (2)$$

The performance of the system is then analysed by plotting the predicted average speed versus its actual average speed. As one would expect, a perfectly accurate system will produce a straight-line plot where the predicted speed equals the actual speed. As it is not plausible that that a Bluetooth device is detected the very moment it enters the radio range of the detector nor would its last detection be right at the edge as it was about to leave the radio range of the detector, the actual distance traversed by the device is a value smaller than s. As such, even if the detector is working perfectly, the predicted average speed is expected to be higher than the actual average speed. In other words, a plot of the predicted average speed versus the actual speed of a functioning system will produce a line above and parallel to the results for a perfectly accurate system. The more accurate the functional system is, the closer its plot will be to that for the perfect system.

6 Result

6.1 Static test

6.1.1 First test

The target device for this test are car Bluetooth radio and smart phone. Initial static test is conducted with car Bluetooth radio. The result is shown in **Tab. 7**.

Tab. 7. Static test with car Bluetooth radio detection.

Distance (m)	Trial 1			Trial 2		
	Time		Sample	Time		Sample
	Start time	End time		Start time	End time	
2	15:01:40	15:02:40	7	15:09:40	15:10:40	7
	15:01:58	15:02:05		15:09:49	15:10:00	
1	15:24:50	15:25:50	11	-		
	15:25:04	15:25:12				

From the initial static test results, it noted that car radio Bluetooth can only be detected over a very short distance and the samples taken are low even though the HC-05 is able to detect a smartphone Bluetooth signal over a much longer range.

6.1.2 Second test

The second static test was conducted with a smartphone over a range from 10m up to 40m in increment of 10m. At each distance, a total of five tests were conducted with each test running for one minute. However, from the results of these tests, it was noticed that the first detection occurs approximate 14 seconds after powering up. The 14s delay is most likely due to coding and hardware design that can be compensated for in a proper implementation. **Tab. 8** below shows the results of these experiments. Noticed that the number of detection within the one minute decreases as the distance between the device and the detector increases.

Tab. 8. Static test with smart phone Bluetooth.

Distance (m)	Sample	Duration (s)	Average time per sample (s)
10	25	46	1.840
	17	46	2.706
	37	46	1.243
	35	46	1.314
	34	46	1.353
20	19	46	2.421
	17	46	2.706
	26	46	1.769
	20	46	2.300
	13	46	3.538
30	21	46	2.190
	17	46	2.706
	16	46	2.875
	18	46	2.556
	19	46	2.421
40	2	46	23.000
	0	46	–
	1	46	46.000
	1	46	46.000
	1	46	46.000

6.2 Dynamic test

6.2.1 The effect of increasing inquiry time

The effect of inquiry time is examined by running the dynamic test, first with the detector settings using "AT+INQM=1, 1, 1" and subsequently using "AT+INQM=1, 1, 2." The resultant predicted average speed for both these settings are as shown in **Fig. 11**. The first observation from this result is that the accuracy of the predicted average speed decreases as the traveling speed increases. A longer inquiry time is generally more accurate as the detector has a higher chance of detecting the device within its radio range.

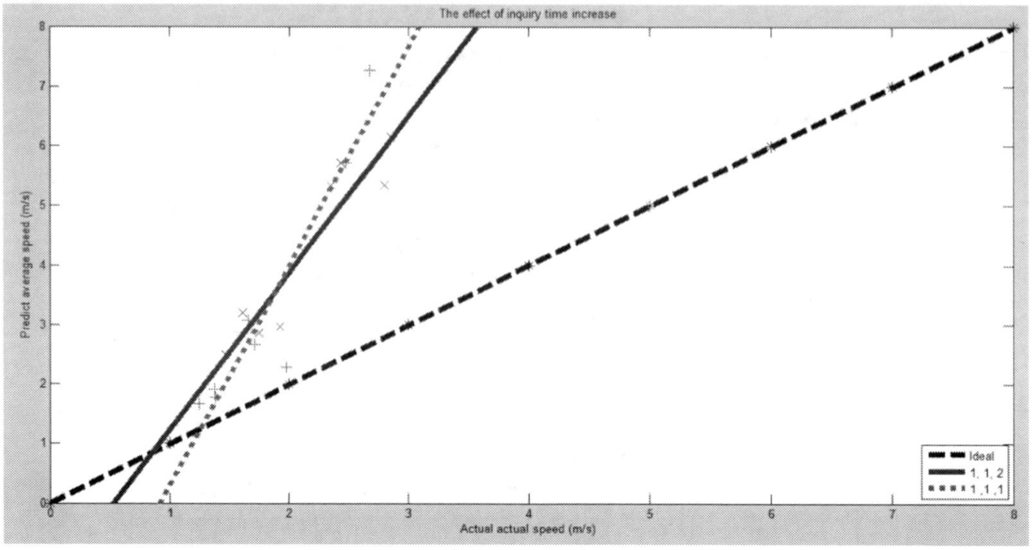

Fig. 11. The effect of increasing inquiry time.

7.2.2 The impact of having two Bluetooth detectors

The results for the experiment using two detectors is summarised in **Fig. 13**. Of the 5 settings, AT+INQM=1, 10, 1 with a 5-millisecond delay between the inquiry cycles of the 2 detectors has the worst performance. Raw data for this experiment seems to shows an anomaly where the Bluetooth module with a delay inquiry sequence received more FHS packets than the ID packets it sends out. Note that a 5-millisecond delay between the 2 inquiry sequences would have the 2 Bluetooth modules only shift the inquiry sequence by half a train as shown in **Fig. 12**.

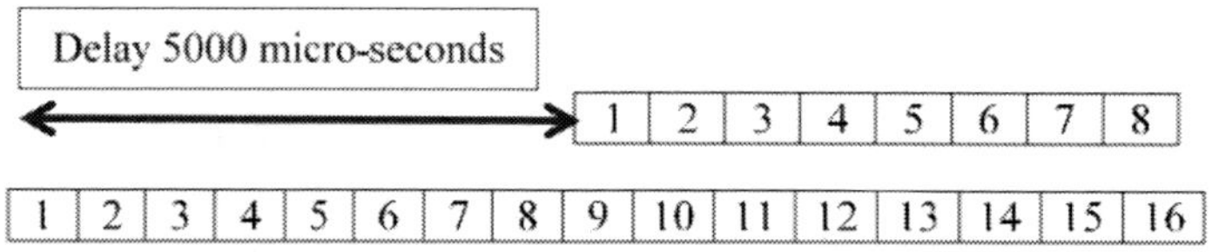

Fig. 12. AT+INQM=1, 10, 1 with a delay by 5-millisecond delay between the 2 inquiry-sequence.

The results in **Fig. 13** shows that the AT+INQM=1,1,1 query setting with no delay[†] seems to have the best performance. The results as shown in **Tab. 10** shows the Bluetooth module with the leading inquiry sequence is able to capture more data. The additional data provided by the other Bluetooth module was found to not be able to improve the accuracy of the speed estimation.

In these experiments, AT+INQM=1,40,4 with a delay of 2.56s was found to have three extreme data as detailed in **Tab. 9**. The removal of these extreme data points would produce a result as shown in **Fig. 14**. This lead us to belief that there exists a possibility of a

[†] It is important to note that though no delay is set, the execution of the command by the microcontroller will inherently introduce a delay between the 2 inquiry sequences.

way to improve the predicted average speed if we could find a way to exclude such extreme data points.

Tab. 9. Extreme data from AT+INQM=1,40,4 & a delay of 2.56 s between sequence

Actual average traveling speed	**Predict average traveling speed**
1.83	6.15
2.88	10.00
2.96	16.00

Tab. 10. Setting of AT+INQM=1, 1, 1 with no delay between the 2 inquiry-sequence.

Average speed with stopwatch			Average speed with Bluetooth Detector			Number of samples collected by Bluetooth module with leading and lagging inquiry sequence		
Distance (m)	Time (s)	Speed (m/s)	Distance (m)	Duration (s)	Speed (m/s)	Leading	lagging	total
20	24.88	0.80	20	10	2.00	6	0	6
20	23.72	0.84	20	19	1.05	14	1	15
20	23.00	0.87	20	13	1.54	15	4	19
20	20.78	0.96	20	16	1.25	6	4	10
20	18.12	1.10	20	13	1.54	9	3	12
20	13.05	1.53	20	4	5.00	6	1	7
20	11.16	1.79	20	11	1.82	9	1	10

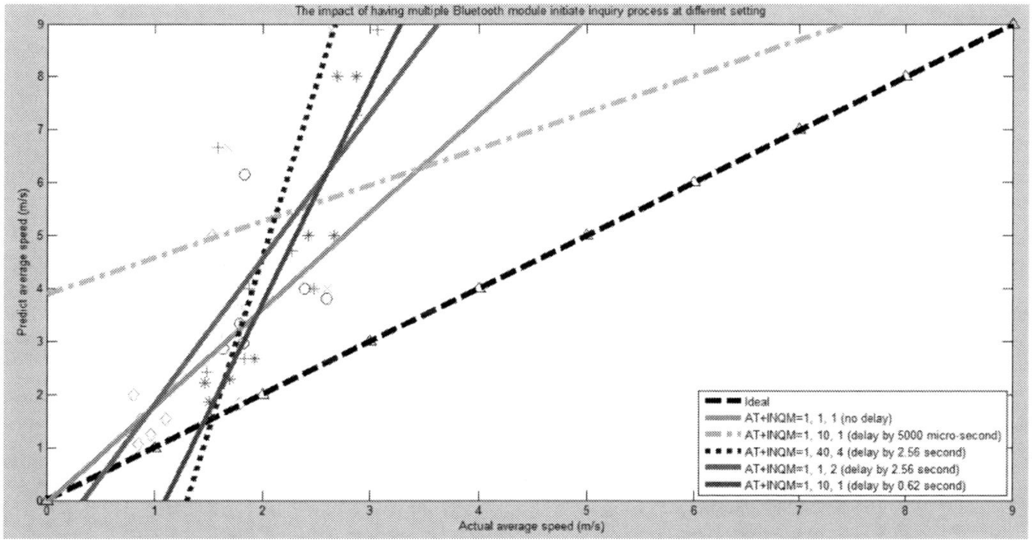

Fig. 13. The impact of having multiple Bluetooth module initiate inquiry process at different setting

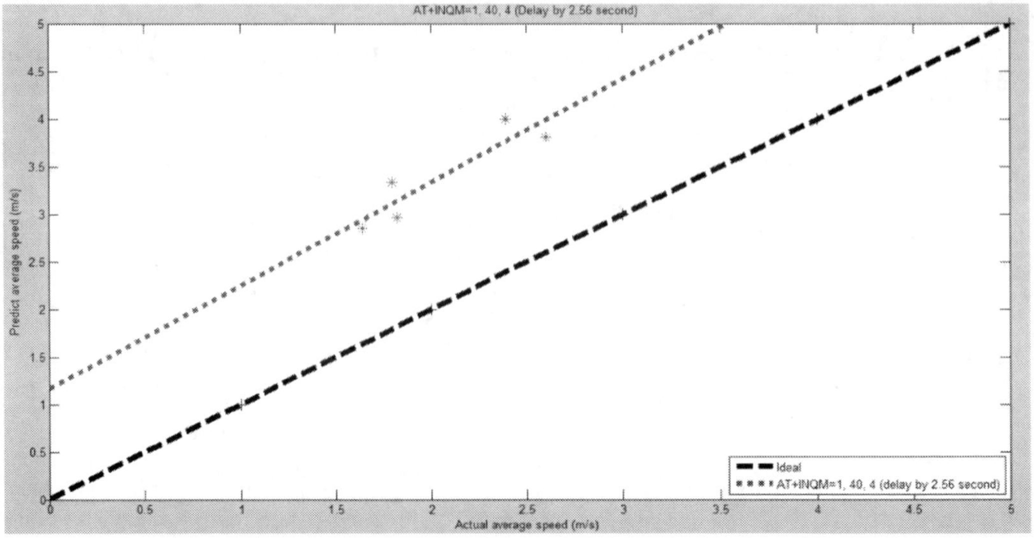

Fig. 14. AT+INQM=1,40,4 with extreme data points removed.

7.2.3 The impact of having multiple traveling Bluetooth device

Comparing **Fig. 16** and **Fig. 17** it is observed that a single Bluetooth detector with a AT+INQM=1,1,1 setting performs poorly as compared to a configuration with 2 Bluetooth detectors with a AT+INQM=1,40,4 setting and a 2.56s delay. The single detector configuration even fail to detect one of the device in a number of trials. The single Bluetooth detector is also unable to pick up sufficient data to determine the traveling speed of the device. On the other hand, a 2 Bluetooth module configuration with a AT+INQM=1,40,4 settings and a 2.56s delay between their inquiry sequence is able to pick up sufficient data to determine the average speed of both devices. This shows that with proper delay, the Bluetooth detector is able perform well with two Bluetooth modules. By allowing the two Bluetooth modules to scan at different frequency train at the same time increases the efficiency of the Bluetooth detector. The scanning sequence of the Bluetooth detector is with a 2.56s delay is as shown in **Fig. 15**.

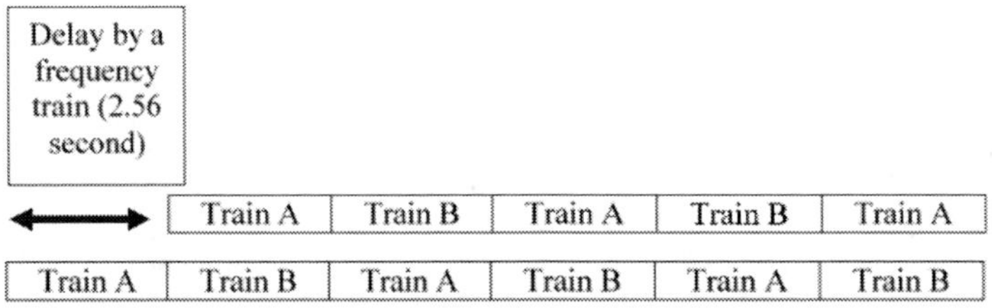

Fig. 15. Inquiry sequence with a 2.56 second delay between them.

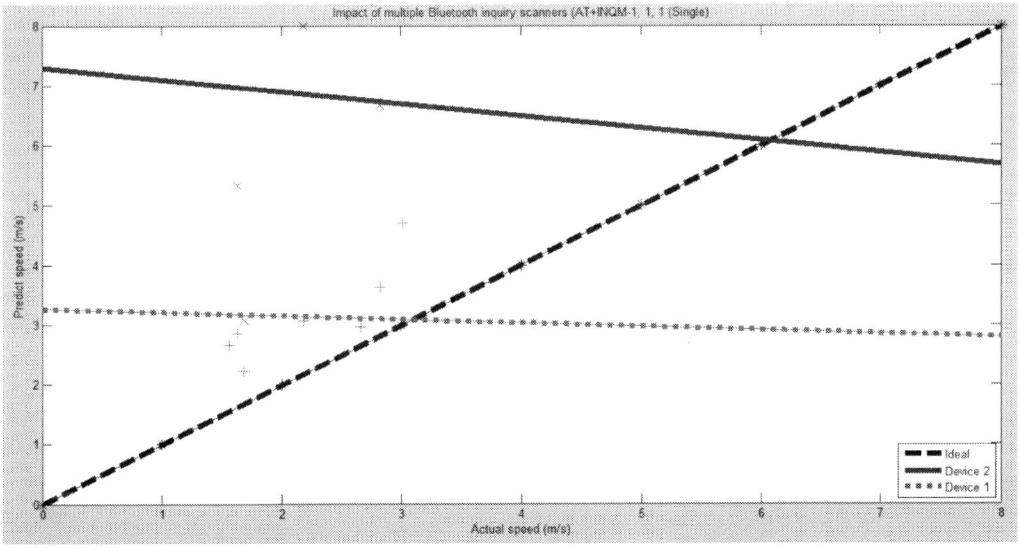

Fig. 16. Impact of two Bluetooth devices on a detector using a single Bluetooth module and a query setting of AT+INQM=1,1,1

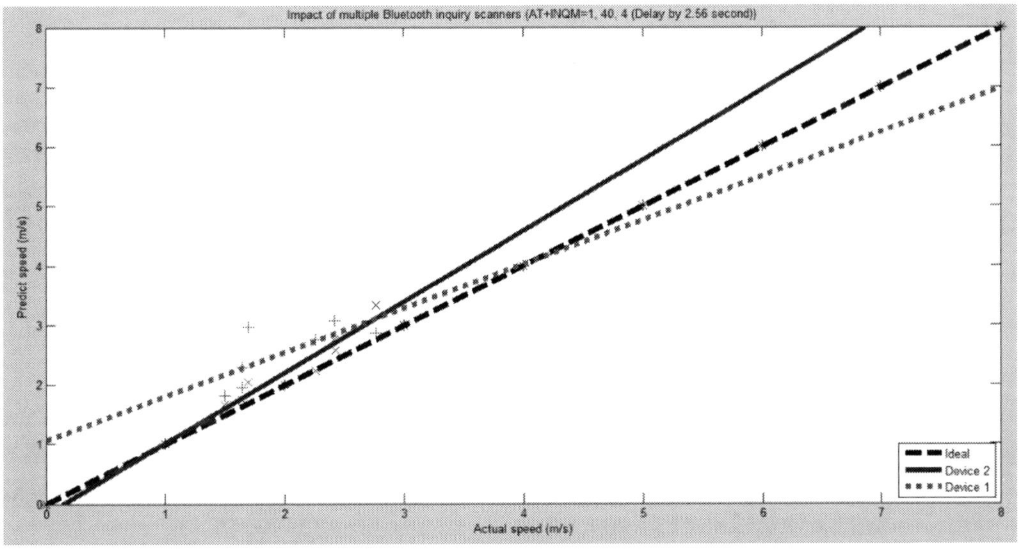

Fig. 17. Impact of two Bluetooth devices on a detector with 2 Bluetooth modules and a query setting of AT+INQM=1,40,4 and a 2.56s delay between the inquiry sequence.

8 Conclusion

Bluetooth traffic monitoring systems has advantages but there are challenges to overcome. The first challenge would be the detection range. Bluetooth detectors will have different detection range depending on the Bluetooth devices it is trying to detect. Furthermore, the power class of traveling Bluetooth device will have effect on the detection range of Bluetooth detectors. From the first dynamic test, it shows that having longer scanning time, the Bluetooth detector has higher chance of detecting device at further distance. The second dynamic test result shows that inquiry parameter AT+INQM=1,40,4 have better speed estimation compare to other setting when extreme data is removed. At the last dynamic test, it shows that with proper delay and setting, having two Bluetooth modules can outperform a single Bluetooth detector. From the dynamic test, it proves that having more Bluetooth detectors at a fix point has a better speed estimation compare to a single Bluetooth detector. However, the challenge is to avoid frequency channel interference between the two devices.

9 References

[1] P. Khanke and P. P. S. Kulkarni, "A Technique on Road Traffic Analysis using Image Processing," *Int. J. Eng. Res. Technol.*, vol. **3**, no. 2, pp. 2769–2772, (2014).

[2] F. Aloul, A. Sagahyroon, A. Nahle, M. A. Dehn, and R. Al Anani, "GuideME : An Effective RFID-Based Traffic Monitoring System," in *IASTED International Conference*, (2012), pp. 353–359.

[3] S. Ahmad, R. Lu, and M. Ziaullah, "Bluetooth an Optimal Solution for Personal Asset Tracking : A Comparison of Bluetooth, RFID and Miscellaneous Anti-lost Traking Technologies," *Int. J. u- e- Serv. Sci. Technol.*, vol. **8**, no. 3, pp. 179–188, (2015).

[4] Bluetooth SIG Proprietary, "Bluetooth Core Specification Master Table of Contents & Compliance Requirements," *Bluetooth SIG*, (2016). [Online]. Available: https://www.bluetooth.com/specifications/bluetooth-core-specification.

[5] M. Duflot, M. Kwiatkowska, G. Norman, and D. Parker, "A Formal Analysis of Bluetooth Device Discovery," *Int. J. Softw. Tools Technol. Transf.*, vol. **8**, no. 6, pp. 621–632, (2006).

[6] A. Franssens, "Impact of multiple inquires on the bluetooth discovery process : and its application to localization," UNIVERSITY OF TWENTE, (2010).

[7] Itead Studio, "HC-05 User's Manual," (2010). [Online]. Available: http://www.electronicaestudio.com/docs/istd016A.pdf.

[8] B. E. Last and P. M. Utc, "Adafruit Data Logger Shield," *Adafruit Industries*, (2017). [Online]. Available: https://cdn-learn.adafruit.com/downloads/pdf/adafruit-data-logger-shield.pdf.

[9] Arduino, "ARDUINO PRO," (2017). [Online]. Available: https://store.arduino.cc/usa/arduino-pro.

[10] J. C. Quer, "Arduino UNO datasheet," (2014). [Online]. Available: datasheet.octopart.com/A000066-Arduino-datasheet-38879526.pdf.

Design Analysis of Taper Width Variations in Magnetless Linear Machine for Traction Applications

Aminath Saadha[1], *Aravind* CV[1,*], *Preshant* Krishna[1], and *F.* Azhar[2]

[1]School of Engineering, Taylor's University, Selangor, Malaysia
[2]Faculty of Engineering, Universiti Teknikal Malaysia, Melaka Malaysia

Abstract. Linear motors are being used in a different application with a huge popularity in the use of transport industry. With the invention of maglev trains and other high-speed trains, linear motors are being used for the translation and braking applications for these systems. However, a huge drawback of the linear motor design is the cogging force, low thrust values, and voltage ripples. This paper aims to study the force analysis with change in taper/teeth width of the motor stator and mover to understand the best teeth ratio to obtain a high flux density and a high thrust. The analysis is conducted through JMAG software and it is found that the optimum teeth ratio for both the stator and mover gives an increase of 94.4% increases compared to the 0.5mm stator and mover width.

1 Introduction

Electric motors are devices that convert the electrical energy to magnetic energy and ultimately into mechanical energy. Electromagnetism is the basis of motor operation in order to produce the magnetic forces required for the production of rotational or linear motion[1]. As one of the fastest growing industries, electrical motor manufacturing represents a major industry worldwide, where electric motor driven systems account for approximately 45% of total global electricity consumption and are expected to rise to 13,360 terawatt hour (TWh) by 2030[2]. Recent developments in the industry have brought forth linear machines that provide mechanical translation without intermediate gears, screws or crankshafts. In a linear motor, either the moving or stationary member must extend over the entire range of motion of the moving member[3]. The motion occurs because of the electromagnetic force developed in the actuator. All types of motor configurations and topologies can be produced in the linear fashion, i.e. dc, induction, synchronous and reluctance[4]. Linear electric machines are associated with long linear progressive motion, such as transportation and other similar applications. The most prominent application of linear motors is the utilization of these motors in transport systems, specifically railway system in the use of train thrust and in braking [5]. Linear reluctance motors are popularly used in electrodynamic braking of high-speed trains. These brakes

[1] Corresponding author: aravindcv@ieee.org

operate on the principle of converting the traction motors of the train to a generator that convert the kinetic energy of the train into electrical power [6]. To overcome the problem of adhesion between the wheel tread and the rail, the braking force in electromagnetic brakes are achieved through strong magnetic forces that are induced with large electromagnets attached to the vehicle over the top of the rails [7]. In Electromagnetic rail brakes, the frictional forces are produced between the electromagnets and the rail [8]. If the electromagnetic fields generate eddy currents in the rails, creating forces acting in the opposite direction of the movement of the train, it is a linear eddy current brake system. However, these brake systems are still susceptible to the cogging effects of the permanent magnets due to the magnetic retention, which greatly reduces the efficiency of the motor [9]. A linear reluctance motor that does not utilize permanent magnets have been introduced, thus minimizing the cogging force and decreasing the weight of the linear motor. The objective of this paper is to study the different teeth ratios of the linear reluctance motor and to propose a ratio for the best thrust/force characteristic.

2 Literature Review

Linear Electric Machines are in general three-phase machines fed through power electronic controllers and taking advantage of regenerative electric braking for fast, robust, and precise thrust, speed, or position control. The topology of a linear switched reluctance motor is similar to that of a stepping motor with variable reluctance platen. In addition, it is equipped with position sensors [10]. The turn-on and turn-off instant of the input current is synchronized with the position of the moving part. The thrust is very sensitive to the turn-on and turn-off instant. In these type of motors, the rotor tends to move to a position where the inductance of the excited winding is maximized or the reluctance is minimized.

In the case of a linear stepping or linear switched reluctance motor, the speed υ of the moving part is as in Equation (1).

$$v = v_s = f_{sw}\tau \tag{1}$$

Where f_{sw} is the fundamental switching frequency in one armature phase winding and τ is the pole pitch of the reaction rail. For a rotary stepping or switched reluctance motor $f_{sw} = 2p_r n$ where $2p_r$ is the number of rotor poles and n is rotational speed in rev/s. The fundamental energy conversion equation for linear electric machines is as in Equation (2).

$$F\dot{x} = vi \tag{2}$$

Where F is mechanical force (N), $\dot{x}$ is mechanical velocity (m/s), v is the Voltage (V), and i is current (A). It is assumed that F and $\dot{x}$ are in the same direction. Numerous studies have been conducted into increasing the thrust of these linear motors such as, [11] where slots are used to decrease harmonic components of the thrust force. These slots can be in the form of skewed or fractional slot [12]. This allows the analysis of rated performance calculation and effects such as cogging torque, ripple torque, back-emf form prediction. Analysis which combines the orthogonal optimization algorithm for tooth shifting and pole shifting of a double-sided slotted permanent magnet synchronous linear motor are carried out in [13]. This study achieves a model with a suppressed thrust ripple which in turn increased the efficiency. A 9-pole 10-slot structure of a short primary permanent magnet

LSM is proposed in N. Bataar et.al. which is focused on reducing the cogging force through detent force minimization by optimizing the length of armature core and shape of the exterior teeth simultaneously by using (1+lambda) evolution strategy coupled with response surface method using multi-quadric radial basis function [14]. Other studies are focused on establishing a platform for the analysis and test of characteristics such as thrust ripples and detent force of the permanent magnet linear motor [15]. It is predicted that thrust ripples are generated by the distortion of the stator flux linkage distribution, reluctance force due to the relative position between the mover and stator, Y. Zhu et.al. aims to rectify this through the utilization of predictive control algorithm [16], which result in the minimization of the voltage ripple by high precision control. The same result is obtained through moving node techniques in research conducted by Ki-Chae et.al [17]. Linear Motor Systems which has electromechanical multi-parameter and strong coupling is difficult to control. W. Ai et.al proposes an analysis of relative weights of various factors for thrust fluctuations in permanent magnet LM [18], using the fuzzy analytic hierarchy process to reduce these fluctuations in the thrust. It can be seen through these research that the cogging force deteriorates the performance and even excite the mechanical resonance thus decreasing the efficiency and also the lifespan of the motor [19]. This paper look into the teeth size variation for obtaining the highest force in the mover and the stator of the linear motor [20]. The paper outline the method of analysis and then propose the results of the study, finally discussing the results and propose a ratio for the best stator and mover teeth gap for a linear reluctance motor based on the results of the study.

3 Methodology

Three basic structures; the stator, which is the stationary nonmoving part, the mover, which consists the coil and is the linear translation component are studied. Figure 1 shows the arrangement of the structure. The parameter that is to be studied is the teeth gap of both the stator and mover to show which is the best ratio of teeth width gap for obtaining the highest thrust from the linear motor. Figure 2 below shows the parameter to be studied while figure 3 gives an in-depth analysis for obtaining the ratio.

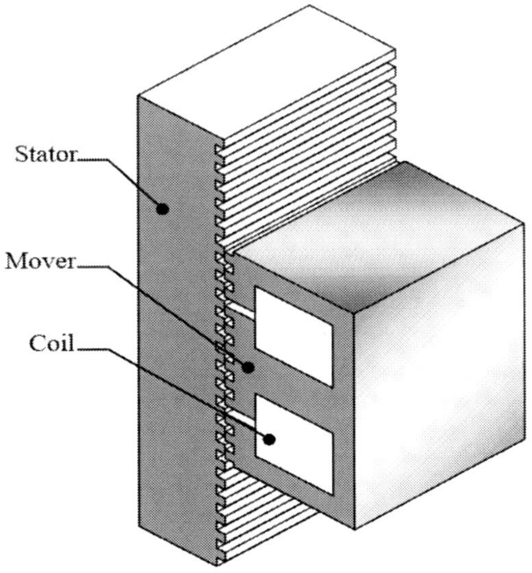

Fig. 1. The Linear Reluctance Motor.

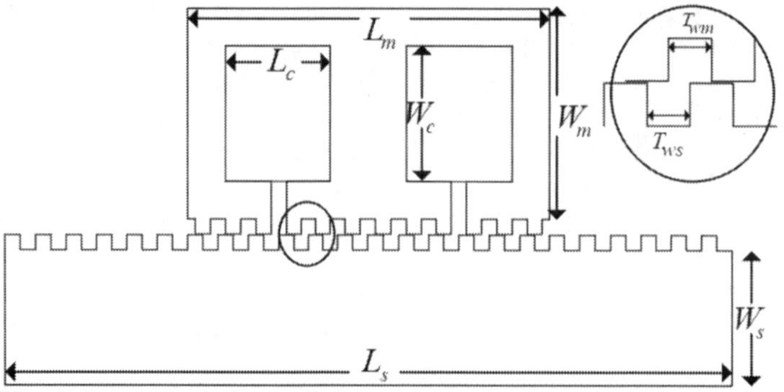

Fig. 2. Parameter to be studied

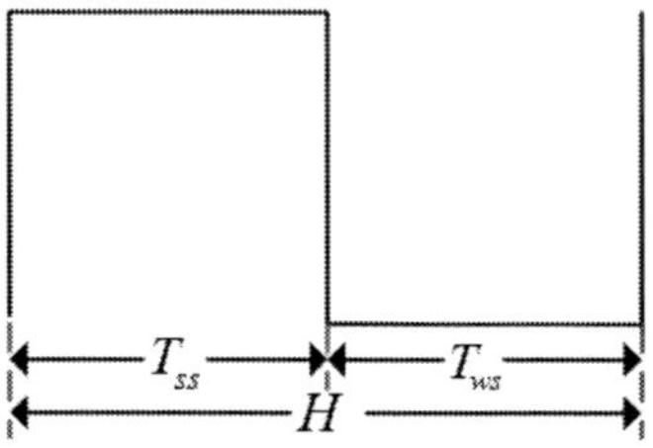

Fig. 3. Teeth ratio

As seen from figure 3, the total pole width of the motor is to be kept constant at 4mm therefore, the teeth gap for the stator is denoted in equation 3.

$$H_s = T_{ss} + T_{ws} \tag{3}$$

where, H_s is the total width of the pole and T_{ss} is the gap where the tooth protrudes and T_{ws} is the dip in the teeth. The parameter T_{ws} is then changed while keeping H_s constant. The same parameter for the mover's teeth are also varied to obtain the best ratio for the teeth and the combination of the ratio of the teeth in both stator and mover to obtain the highest thrust. The dimensions of the stator and mover are summarized in Table 1. The simulations are run on FEM software JMAG. The parameters used for the coil is 330 turns with a total resistance of 4.6Ω. The material and circuit parameters are kept constant for the whole analysis. A total of 49 simulations are carried out with different combinations of T_{ws} and T_{wm}

Table 1. Parameter List

Parameter	Sign	units	Value
Length of stator	$\boldsymbol{L_s}$	(mm)	96
Width of stator	$\boldsymbol{W_s}$	(mm)	18
Teeth height of stator		(mm)	2
Length of mover	$\boldsymbol{L_m}$	(mm)	48
Width of mover	$\boldsymbol{W_m}$	(mm)	28
Teeth height of mover		(mm)	2

Length of coil	L_c	(mm)	14
Width of coil	W_c	(mm)	18
Number of turns in coil			330
Resistance of coil		Ω	4.6

4 Results and Discussion

The analysis is run using JMAG software, in order to obtain the flux lines, magnetic flux density, an absolute force of translations and other parameters. For this study, the force in the Y direction of the motor is considered, since the increase in the force shows the decrease in cogging force which shows an increase in the efficiency of the motor. As seen in figure 4, an increase in the teeth width of both the stator and the mover, increases the absolute force to a certain point, however, further increase shows a decrease in the force that is generated. This can be seen in all of the stator teeth width measurements from $0.5T_{ws}$ to $3.5T_{ws}$. The force in Y direction of $0.5T_{ws}$ increases from 24N to 63N at $2.5T_{ws}$ and then decreases to 52N at $3.5T_{ws}$. This effect is better illustrated in Figure 5. As seen the force increases with an increase in the teeth gap showing that the magnetic flux density increases. This is because as the teeth gap T_{ws} increases the, extruded part of the teeth which is the T_{ss} decreases, since total teeth width H_s is kept constant. This decrease in T_{ss} decreases the surface area of extrusion, thus forcing the magnetic flux to be concentrated, increasing the density and hence increasing the force. However, a further increase of the gap from $3.0T_{ws}$ decreases the force since the extruded part of the teeth T_{ss}, is too small and the magnetic flux leaks from the teeth gap as shown in figure 6. This leakage decreases the density of the magnetic flux thus decreasing the force of the motor. The same effect can be seen for the mover teeth width T_{wm}, for which the force increases from 47N at 3.5 T_{wm} to 52N at 3.0 T_{wm} and then decreases to 19N at $0.5T_{wm}$ at a constant stator teeth width of 3.5 T_{ws}. Based on these results it is best to identify which ratio of teeth width for both the stator and the mover provides the highest force, thus increasing the efficiency of the motor. The highest force obtained for each stator width is illustrated in figure 7. As shown by the curve in figure 7, the highest absolute force for the motor design is at 463N which is achieved at $2.5T_{ws}$. The graph also shows that the best teeth width for the mover is at 3.0 T_{wm}, since all of the maximum force for the stator is achieved at this teeth width. However, motor the design should also take into consideration the manufacturing process and the materials required. The 2.5mm teeth width is better since this decrease the material costs for the motor production.

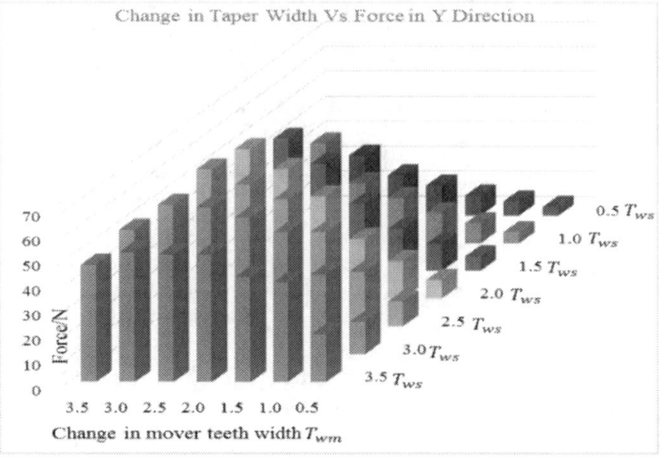

Fig. 4. Analysis of the taper width of stator and mover based on absolute force

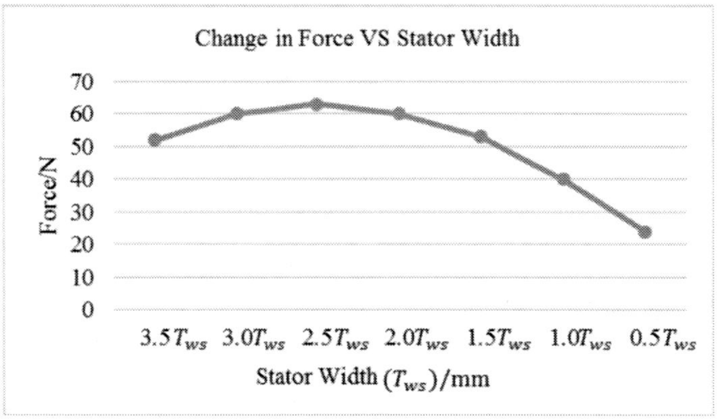

Fig. 5. Change in Force with Change in Teeth Width

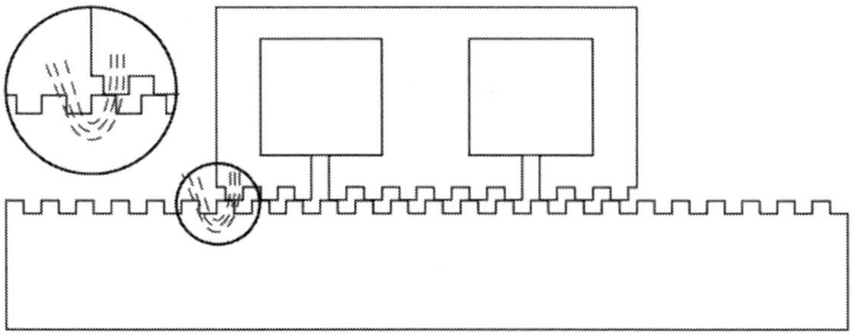

Fig. 6. Leakage of magnetic flux

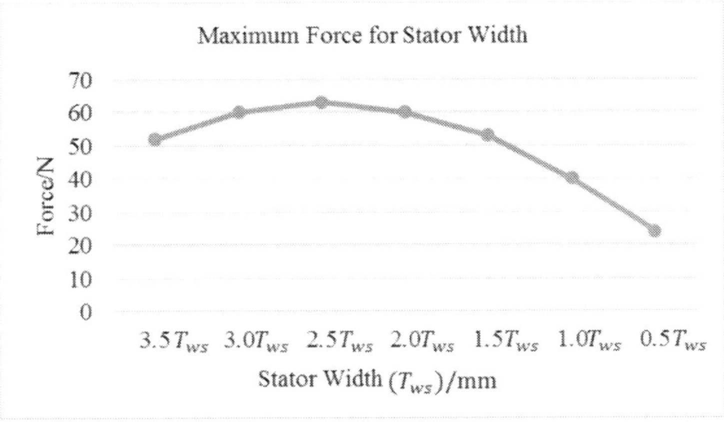

Fig. 7. Maximum force for each stator width.

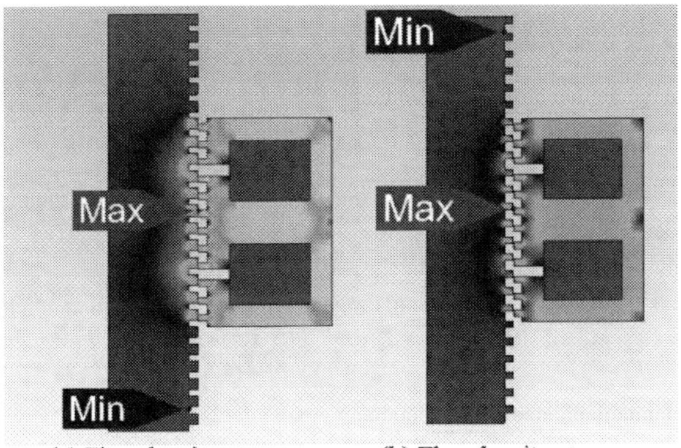

(a) Flux density at $1.5T_{ws}$ at $3.0T_{wm}$ (b) Flux density at $2.5T_{ws}$ at $3.0T_{wm}$

Fig. 8. Flux Distribution for 1.5 T_{ws} and 2.5 T_{ws}

The flux density for this size is also the best since there the distribution is more uniform and the motor shows areas of lower flux concentration which shows that continuous operation of the motor is stalled due to excessive heat buildup in the body which result in a lower life span for the motor. The comparison of the flux density for the motor is shown in figure 8. Based on these results it can be seen that the best stator tooth width is 2.5mm for T_{ws} and the mover teeth width is 3.0mm for T_{wm}. Both of these ratios provide an increase in 94.4% force compared to 0.5mm for both the stator and mover, and a 25.4% increase compared to 3.5 mm teeth width for both the stator and mover.

5 Conclusion

Linear motors are used for different applications and are gaining popularity in the transport industry mainly in railway systems. The main disadvantage of the current motor types is the cogging force by the use of permanent magnets, and voltage ripple which decreases the thrust force of the linear motor. This study aims to analyze the effects of the linear reluctance motor's taper width of both the stator and mover to identify the best ratio of teeth gap to obtain the highest force values. The analysis is done through the software JMAG, by changing the parameters of the teeth gap while keeping the other parameters such as motor dimensions, materials, circuit, coil resistance and a number of turns constant. The teeth gap of stator T_{ws} and mover T_{wm} is changed to obtain the results. As seen from the analysis the best results for the teeth gap can be found at $2.5T_{ws}$ and 3.0 T_{wm} which gives an increase of in 94.4% force compared to 0.5mm for both which is the lowest teeth width that both the stator and mover can achieve. Further analysis and studies can be conducted by changing the teeth shape and obtaining the optimum teeth shape for increased thrust, a decrease in cogging force, and higher efficiency.

Acknowledgment

This work is supported by Taylor's University through its TAYLOR'S RESEARCH SCHOLARSHIP programme. The authors wish to express financial support of the project through the Taylor's Research Grant Scheme TRGS/MFS/1/2016/SOE/005

References

[1] L. Jing Jin Mei-shan, Chen Da-Chuan, Li Jing, Qiu Chang-li, Emulated Analysis in the Performance of Linear Introduction Motor, Energy Procedia, **17**, pp. 79–83, Jan. (2012).

[2] Q. A. Hasan, W. H. Wan Badaruzzaman, Ahmed W. Al-Zand, Azrul A. Mutalib, Finite element analysis of tapered composite plate girder with a non-linear varying web depth, Journal of Engineering Science and Technology, **12**, no.11, 2839-2854 (2017)

[3] C. Aravind, I. Khumira, R.N. Firdaus, Fairul.A, Single phase flux switching DC linear Actuator, Special issue pp.28-39 Journal of Science and Technology, **1**, pp. 28-39, (2013)

[4] X. L. Qibao Lv, Zhiyuan Yao, Contact analysis and experimental investigation of a linear ultrasonic motor, Ultrasonics, **81**, pp. 32–38, Nov. (2017).

[5] Shibly Ahmed Al-samarraie, Yasir Khuhair Abbas, Manifold based controller design for linear systems, Journal of Engineering Science and Technology, **8**, no.6, 723-740 (2013)

[6] Q. Z. Xiangyu Zhou, Yi Zhang, A novel linear ultrasonic motor with characteristic of variable mode excitation, Ceram. Int., **vol. 43**, pp. S64–S69, Aug. (2017).

[7] J.-H. Jeong, J. Lim, C.-W. Ha, C.-H. Kim, and J.-Y. Choi, Thrust and efficiency analysis of linear induction motors for semi-high-speed Maglev trains using 2D finite element models, in 2016 IEEE Conference on Electromagnetic Field Computation (CEFC), (2016), pp. 1–1.

[8] P. Z. A.R. Albrecht, P.G. Howlett, P.J. Pudney, X. Vu, Energy-efficient train control: The two-train separation problem on level track, J. Rail Transp. Plan. Manag., **5**, no. 3, pp. 163–182, (2015).

[9] S. K. Satoshi Oto, Tadashi Hirayama, "Study on Thrust Improvement and Ripple Suppression of HTS Linear Switched Reluctance Motor with Coreless HTS Excitation Windings," Phys. Procedia, **81,** pp. 178–181, (2016).

[10] M. Manna, S. Marwaha, M. Marwaha, Performance Optimization of Linear Induction Motor by Eddy Current and Flux Density Distribution Analysis, Journal of Engineering Science and Technology, **6**, no. 6, pp. 769-776, (2011)

[11] Rajesh V.R., Biju T.Kuzhiveli, Modelling and failure analysis of flexure springs for

a stirling cryocooler, Journal of Engineering Science and Technology, **12**, no.4, 888-897 (2017)

[12] R. Y. Mingyi Wang, Liyi Li, Overview of thrust ripple suppression technique for linear motors - CMP Journals & Magazine, Chinese J. Electr. Eng., **2**, no. 1, pp. 77–84, (2016).

[13] T. Inagaki, K. Suzuki, and H. Dohmeki, The study of parallel synchronous drive in Permanent magnet linear synchronous motor, in Linear Drives for Industry Applications (LDIA), 2017 11th International Symposium on, Osaka, Japan, (2017).

[14] N. Baatar, Hee-Sung Yoon, Minh Trien Pham, Pan Seok Shin, and Chang Seop Koh, Shape Optimal Design of a 9-pole 10-slot PMLSM for Detent Force Reduction Using Adaptive Response Surface Method, IEEE Trans. Magn., **45**, no. 10, pp. 4562–4565, (2009).

[15] L. Zhang, B. Kou, Y. Zhang, Y. Jin, Thrust characteristic analysis and test of the synchronous permanent magnet linear motor, in 2014 17th International Conference on Electrical Machines and Systems (ICEMS), pp. 1733–1737, (2014)

[16] Y. Zhu and Y.-H. Cho, Thrust Ripples Suppression of Permanent Magnet Linear Synchronous Motor, IEEE Trans. Magn., **43**, no. 6, pp. 2537–2539, (2007)

[17] Ki-Chae Lim, Joon-Keun Woo, Gyu-Hong Kang, Jung-Pyo Hong, Gyu-Tak Kim, Detent force minimization techniques in permanent magnet linear synchronous motors, IEEE Trans. Magn., **38**, no. 2, pp. 1157–1160, (2002)

[18] P. Sharma, T. Bhatti, K. Ramakrishnan, Permanent-Magnet Induction Generators: An Overview, Journal of Engineering Science and Technology, **6**, no. 3, pp. 332 - 338, (2011)

[19] S. K. Tadashi Hirayama, Satoshi Oto, Atsushi Higashijima, Experimental Manufacture and Performance Evaluation of Linear Switched Reluctance Motor with HTS Excitation Windings, Phys. Procedia, **81**, pp. 174–177, Jan. (2016)

[20] C. Aravind, I. Khumira, R. Firdaus, A. Fairul, Single Phase Flux Switching DC Linear Actuator, Journal of Engineering Science and Technology, **1**, no. 1, pp. 28-39, (2013)

Proposed Framework of Hybrid Photovoltaic Thermal Systems to Ecotourism in Malaysia

Mohammad Taghi Hajibeigy[1,*], *Aravind* CV[1], and *Rashmi* Gangasa Walveka[1]

[1]School of Engineering, Taylor's University, 47500 Subang Jaya, Selangor, Malaysia

Abstract. Ecotourism is widely recognized for its positive impacts to the environment. Ecotourism is defined as nature-based tourist experiences, which visitors travel to regions for the purpose of appreciating their natural beauty, relaxation and being harmonious with Mother Nature. Ecotourism probably had its foundations in the ethics of conservation, which can help maintain what's left of nature and fosters environmental education. Employing the sustainable energy, paying more attention to the natural resources and developing existing natural resources will have a direct impact to healthier sustainable environment for the inhabitant. If not all, most of the ecotourism sites are located in the remote area in which to utilize electrical energy from the normal grid is difficult or very costly. On the irony, utilizing the conventional non-renewable energy have negative effect and ruins the ecotourism sites. Therefore, utilizing Hybrid Photovoltaic Thermal (PVT) systems is a fit alternative energy source to serve the limited energy demands of ecotourists sites. The amount of the energy production depends on the visitors' capacity of the site. Since the rural area enjoys the cheap and large field, it is easier to plan and assemble the PVT systems to suffice the needed energy. In this research the sustainability of local environment along with the sustainable usage of energy were considered. This paper presents the sustainable energy usage of tourism in Malaysia and aims to address the impact of sustainable and non-sustainable on the rural area of ecotourism. This paper also addresses some of the emission, carbon reduction and climate change issues that the world is expected to meet soon.

Keywords: Ecotourism, Sustainable, Environmental Impacts, Hybrid, Sustainable energy

Abbreviations

CO_2	Carbone dioxide
IUCN	International Union for Conservation of Nature
PVT	Photovoltaic Thermal

1 Introduction

Tourism is a potential matter in environmental, social and economic levels of many government agendas. Government sees tourism as a tool of development giving priority to protect the environments and traditions with minimum negative impacts [1]. Malaysia being the major attraction for the tourist destination in this part of the world enjoys a bigger share of tourist in the entire year [2]. Among the tourism, ecotourism is one of the rapid growing movements in the worldwide tourism industry. Ecotourism is a method of unconventional tourism that aims to achieve economic gain locally by development through natural resource protection. The environment has an inherent value which prevails its value as a tourism asset. Ecotourism relays to the quality of the environment which must be carefully safeguarded from the effect of the activity and type of energy utilizations in its vicinity. The International Union for Conservation of Nature (IUCN) defined ecotourism as "environmentally responsible travel and visitation to relatively undisturbed natural areas in order to enjoy and appreciate nature (and any accompany in cultural features-both past and present) that promotes conservation, has low negative visitor impact and provides for beneficially active socioeconomic involvement of local populations"[3]. World Commission on Environment and Development define sustainable as "Development that meets the needs of the present without compromising the ability of future generations to meet their own needs" [4]. The above definition implies that the important aspect of the definition is to appreciate and protect the nature at the same time an instrument for conservation and sustainable development. In order to comply with the above definition and guard the nature, the source of energy usage must be carefully considered and prohibit the usage of any source of energy which produces environmentally hazard gasses such as carbon dioxide (CO_2), carbon monoxide (CO), nitrogen dioxide (NO_2) and phosgene (PO_4).

Renewable energy resource such as sun is an extremely important part of generating energy needed in the natural rural area since Sun is the most abundance, sustainable and exist everywhere without any cost. It aims to assist in the reduction of the emission of greenhouse gases which are produced by non-renewable energy sources such as fossil fuel. However, the newly designed hybrid PVT systems can produce both electrical energy and the same time high thermal energy. Since the heat accumulated on the PV reduces the PV efficiency, the removal of the large amount of thermal energy from the PV not only help to maintain the PV electrical efficiency but also to serve as the useful thermal energy needed in the rural area. The aims of this paper is to investigate environmental impacts of energy source usage in relation to ecotourism and focuses on environmental impact of the energy source usage on ecotourism in Malaysia.

1.1 Ecotourism in Malaysia

Malaysia is an exceptional country in terms of its tourist attractions and rich natural resources. The utmost important characteristics of the country are its ethical and cultural diversity, ecotourism attractions, and diverse natural resources. The Belum Temengor Forest Complex is reputed to be more than 130 million years old. This is the oldest in the world, and even older than the Amazon forest and Congo basin. Ten top attractions for Malaysian ecotourism which have outstanding natural assets and present opportunities for local people to involve in ecotourism development are shown in Table 1 in rough order of priority [5].

Table 1. Ten Top Attractions for Malaysian Ecotourism and their Natural Assets [5].

No.	Place	Natural assets
1	Lower Kinabatangan River, Sabah	Proboscis monkeys, river, wildlife
2	Wang Kelian, Perlis	Limestone, caves and forest
3	Kenong Forest Park, Pahang	Elephants, caves, streams
4	Belum / Temenggor, Perak	Wildlife, forest and lake
5	GunungStong, Kelantan	Mountain,waterfall, rockfaces
6	Ulu Muda, Kedah	Forest, lake and sandstone
7	TasikBera, Pahang	Lake, culture, fish
8	Hose Mountains, Sarawak	Forest, trekking, birdwatching
9	TasikKenyir\, Terengganu	Lake, boating, trekking, fishing
10	PulauKuku, Johor	Mangroves, wildlife, seafood

1.2 Environmental impact of ecotourism

Ecotourism differs from conventional tourism because it mainly concerns the perceptions and values of individuals which influenced a great portion appreciation of environment. Therefore, it reduced the scale of exploitation of these natural resources [6]. In comparison to other type of tourism, manners, such as the proper management of waste at the sites and recycling are stressed in ecotourism. Ecotourism is specific in nature and it impacts the environment as well as the changes in people's attitudes in a positive sense. Hence, it is distinguished from other tourism forms owing to its natural area surroundings [7]. On the other hand, sustainable ecotourism objective is to address the needs of to sustain the visited environments, help to educate the local communities in regards to their natural surrounding and to preserve the virginity of the nature [8]. However, the impact of the tourism on the destination could be negative. Among the negative impact is the generating Carbon dioxide (CO_2) which is produced by using non-renewable energy source such as fossil fuel.

1.3 Carbon dioxide (CO_2) damage

Carbon emissions, most particularly CO_2, are part of a group of gases that harmfully impact the quality of our air and raise the greenhouse effect. Real world activities such as energy consumption, waste disposal and transportation contribute towards carbon footprint as these activities contribute towards CO_2 Ecological Footprint is a common baseline for Carbon Footprint - emission of gases contributing to climate change, associated with human production or consumption activities. In a rural tourism setting, all service providers to tourists may contribute to the deterioration of the environment in the course of their tourism activities. When consumption (operational practices and activities of tourists) supersedes carbon offset initiatives, mitigation approaches are needed [9].

1.4 Carbon Offset Plan

In order to have an effective offset plan, the true measures are vital to evaluate proper offset methods.

A carbon tax on aviation fuel would particularly affect long-haul flights, because of high emissions, and short-haul flights and the emission during take-off and landing [10].Taxation should be in accordance with the amount of pollution they generate and use the fund to reverse the pollution by implanting trees, promote a healthier environment, promote sustainable energy, and other forms of contribution to the Mother Nature on the location sites. The plan should be able to calculate the carbon emission for all the activities and offer offset price that is tied to a chosen project before the payment method is determined. The mentioned process is shown in the Figure 1.

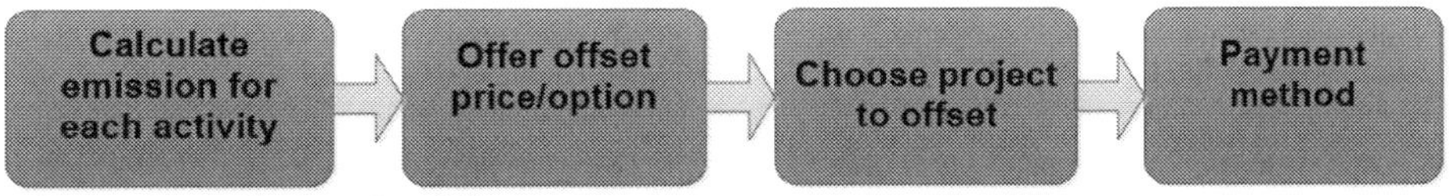

Fig.1. Offset carbon mission plan [9].

It is recommended that the vehicle operates on the site will be battery operated and the energy for using to recharge battery comes from the PVT source. The usage of renewable energy source without proper plan hinder and cause limitation on the growth of the area. The limitation in any aspect prevents growth in any form and manner such as financial impact on the local residence as well as on a national level and impact on the development of natural resources. The fund generated from the tourists can help the local community's living standard, physical, economic and social environments of destinations which solely depend on the quantity and the quality of the ecotourists. There are several ways to minimize the limitations for ecotourism. One approach is to plan for the usage of sustainable energy in place of the pollution sources energy such as fossil fuels that is non-renewable, hazardous and unnatural phenomenon in the earth's atmosphere like Global warming. Most clean, natural and renewable energy comes either directly or indirectly from the Sun. Therefore, the best answer that could be uttered out lies in the Sun. The Sun stands as a reliable renewable source of energy.

1.5 Energy Demands at the ecotourists Site

As the number of ecotourists increases the energy demand proportionally will increase as well. The PVT systems can be used to supply the demands and replaces the existing fossil fuel energy. The eco sites normally located in the rural areas and are not comfortable to transfer the energy from the normal conventional source via grid to these areas. Besides, the conventional method is not in line with the concept of ecotourism and in fact is hazardous to the Mother Nature which will ruins the eco-system. The fit alternative method is the solar energy. The solar energy is present everywhere, renewable, the most abundant, free and environmentally friendly [11]. Solar energy can cover a significant percentage of tourism infrastructures. The PVT systems can cover both heat demand and electrical demand. The proposed systems include a PV array, enhanced solar collectors, solar charger for optimized energy harvesting, and a converter-inverter module to supply ac and dc electrical energy. The PVT systems deliver both electrical energy to suffice the electrical energy need while the thermal system produces thermal energy which can heat water to be used for shower, pool or to be used in radiator to warm the rooms and offices. The application of PVT systems in the industry is suggested as a reliable solution for a wider use of solar energy systems [12]. Globally the usage of sustainable and renewable energy source has been increased. Figure 2 shows the renewable energy consumption by region in the word.

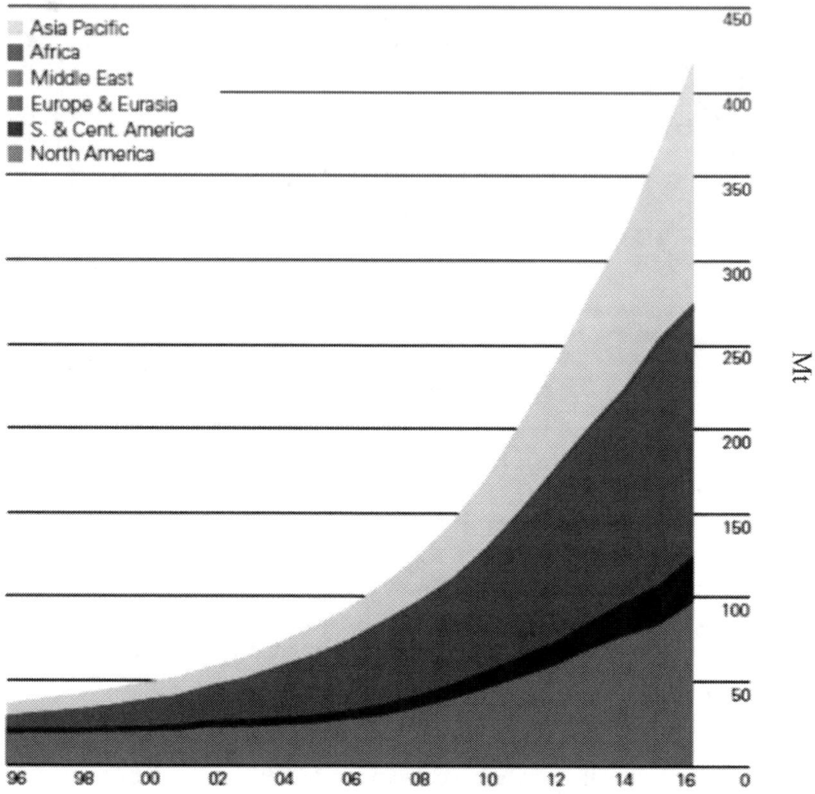

Fig. 2. Renewable energy consumption by region, year 1996 – 2016 vs. Million tonnes oil equivalent (Mtoe) [13].

The graph shows the exponential increase of renewable energy consumption in different continent. This is in favour of the natural area and better sustain the eco system from ecological effect caused by CO_2 which is generated by producing energy from non-renewable energy fossil fuel. Implementing PVT systems in rural area will boost the total renewable energy consumption of Malaysia which in turn help the ecological and sustain the ecotourists area. Generating 1kW electrical energy by PV system producing 150 kWh each month prevents 75 kg of fossil fuel from being extracted. It avoids 150 kg of CO_2 from entering the atmosphere and keeps 473 L of water from being consumed [14]. Figure 3 shows Malaysia energy consumption for the past 26 years. The graph shows a steady increase on the yearly energy consumption of Malaysia. This trend is alarming to the ecological system which put more pressure to use alternative method to non-renewable energy source to protect ecological infrastructure.

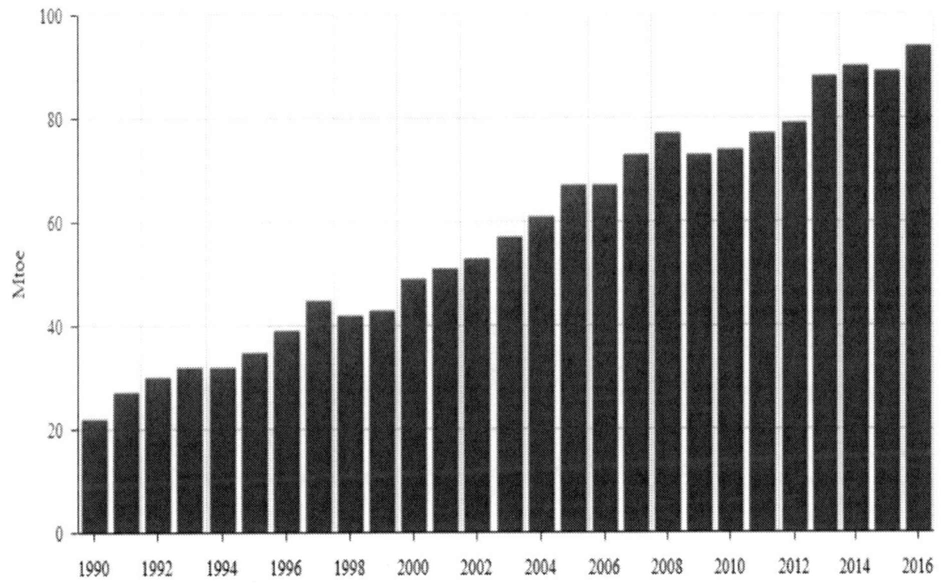

Fig. 3. Malaysian energy consumption, year 1996 – 2016 vs. million tonnes oil equivalent (Mtoe) [13].

2 Energy Requirement at the Ecotourists Site

It is highly important and critical to calculate and analyse the energy required for the ecotourism sites. Malaysia being a tropical country average sun insolation ranges about 4.8 $kWh/m^2/day$ – 5.8 $kWh/m^2/day$. Photovoltaic (PV) system is well recognized and widely utilized to convert the solar radiation to electric energy. The normal factor influence the energy management is the active energy utilization (KWh), the reactive energy consumption (KVARh) and the peak demand (KW). Conventionally the utility system put their effort on the reduction of KWh consumption and on addressing the required reactive energy to improvise the power factor [15]. Nadarajan and Aravind [2] use the below procedure for the demand analysis and the net KW demand.

Let the contracted power be (PC), the maximum demand be (PD) the power used in excess (PE) computed as:

$$PE = PC - PD \tag{1}$$

$$PD = PDm \; x \; KD \tag{2}$$

where *PDm* is the actual peak demand value from the maximum demand meter and *KD* is the demand factor.

The penalty by the supplier to the utility is computed as:

$$PP = (PC - PD) \; x \; KP \tag{3}$$

where *KP* is the penalty factor. Therefore the actual KW value (*PA*) computed is given by

$$PA = [(PC - PD) \; x \; KP] + [PDm \; x \; KD] \tag{4}$$

2.1 Energy production by utilizing PV

The generation of electrical energy solely relies on the solar radiation, the total surface area of the PV being utilized and the efficiency of the PV itself. Figure 4 shows annual average solar radiation energy for different city in Malaysia.

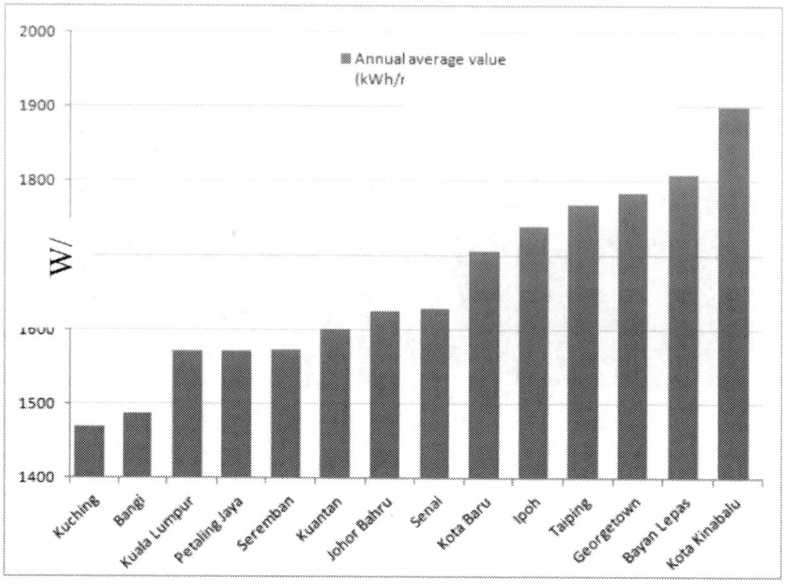

Fig. 4. Aannual average sun radiation energy watt (W/m^2) for different city in Malaysia [16].

Figure 4 above shows that for every square meter (1m^2) of PV produces 1900 W of energy per hour in Kota Kinabalu, compared to only 1571 W/h in Kuala Lumpur; assuming 100% solar cell efficiency. The electrical energy (Watt) obtained can be calculate from equation (5).

$$KWh = H\ x\ G\ x\ \eta\ x\ S \quad (5)$$

Where H is the total hours of sun light, G is the sun radiation (W/m^2), η is the efficiency of PV module and S is the total surface area of the PV. Using equation (5) a solar panel with efficiency of 15% only produces 285 Wh in Kota Kinabalu and 235 Wh in Kuala Lumpur. With 5 hours of total solar radiation per day, and with an installation of 10 m^2 of solar PV can produce 1.425 kWh in Kota Kinabalu, and 1.178 kWh in Kuala Lumpur daily [16]. Table 2 shows the estimated total electrical power generation by different PV surface area coverage with the efficiency of 15% for a full day assuming 5 hours of full solar radiation without cloud. The table was generated based on the equation (5) and the yearly average solar radiation on each location.

Table 2. Electrical Energy generated by the PV for one day for different PV surface area in different locations in Malaysia.

Locations in Malaysia	(10 m^2) KWh/ Day	(50 m^2) KWh/ Day	(200 m^2) KWh/ Day	(500 m^2) KWh/ Day	(1000 m^2) KWh/ Day	(5000 m^2) KWh/ Day
Kota Kinabalu	14.25	71.25	285.00	712.50	1425.00	7125.00
Bayan Lepass	13.58	67.88	271.50	678.75	1357.50	6787.50
Georgetown	13.35	66.75	267.00	667.50	1335.00	6675.00
Ipoh	13.05	65.25	261.00	652.50	1305.00	6525.00
Kota Baru	12.83	64.13	256.50	641.25	1282.50	6412.50
Johor Baru	12.23	61.13	244.50	611.25	1222.50	6112.50
Kuantan	12.00	60.00	240.00	600.00	1200.00	6000.00
Seremban	11.78	58.88	235.50	588.75	1177.50	5887.50
Bangi	11.10	55.50	222.00	555.00	1110.00	5550.00
Kuching	10.95	54.75	219.00	547.50	1095.00	5475.00

The electrical and thermal efficiencies of the PVT systems are independent of location. However, the solar radiation and also, to a lesser extent, ambient conditions such as temperature and wind speed will strongly influence the total energy generation. The electrical and thus thermal efficiencies are considered to vary in the ranges 18.0–15.3% and 60–50%, respectively, for operating temperatures between 25 C° and 85 C° [17]. With the thermal efficiency given above the total energy of the PVT systems can be higher than given calculated value. For a PV module in air, the total convective energy (q_{conv}) exchange from a module surface is shown in Equation (6).

$$q_{conv} = h_c \, . A. (T_{module} - T_{ambient}) \quad (6)$$

where hc is the convective heat transfer coefficient, A is the surface area of the PV module, T_{module} is the temperature of the module and $T_{ambient}$ is the ambient temperature surrounding the systems [18].

3 Modelling Framework and Methodology

3.1 PVT Hybrid Systems Design

PVT hybrid systems consisted of two main parts, electrical system and thermal system. The PV module conventionally produces electrical energy while the thermal unit produces thermal energy. The electrical system consisted of PV module, charge controller, battery, convertor and invertor for ac and dc voltage usage. The thermal system consisted of aluminium plate, heatsinks, thermal container, thermal transfer fluid, heat exchanger, thermal storage tank and pump. The control unit regulate the speed and operation of pump based on the temperature difference of thermal fluid in the thermal container and the water in the thermal storage tank. The proposed entire structure of the hybrid PVT systems is shown in Figure 5.

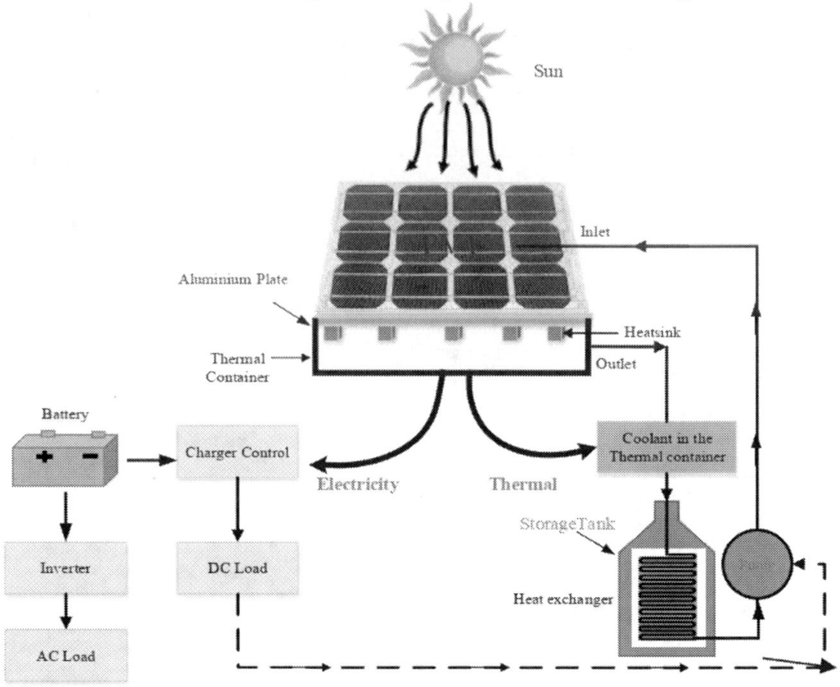

Fig. 5. Photovoltaic Thermal Hybrid PVT Systems.

3.2 Systems Configurations

The designed hybrid PVT systems consisted PV module, aluminum plate, heatsinks and thermal container. The aluminum plate is directly attached below the PV module via thermal glue to absorb the thermal energy from the PV module. The heatsinks are welded on the bottom layer of the aluminum plate to further remove the heat from the aluminum plate. The thermal container is made of acrylic which holds coolant. The coolant used as the transfer medium. The assembled unit is placed inside a thermal container, which the unit sat on the top of the thermal container. The heatsinks also fit inside the container which is immersed in the coolant. Each layer of the Hybrid PVT systems is shown in Figure 6.

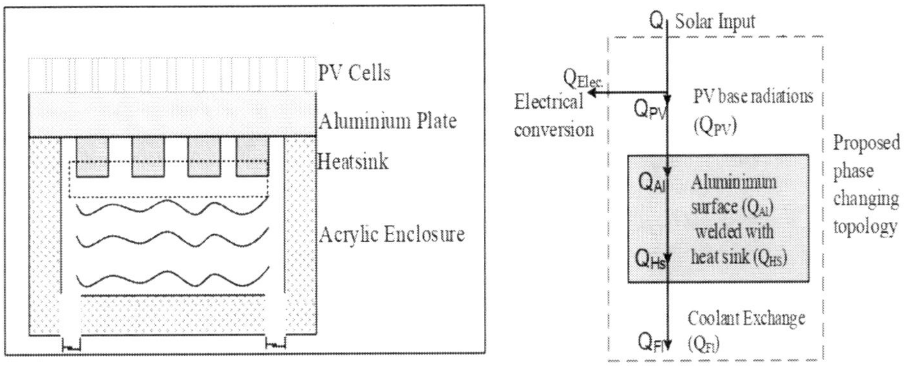

(a) Phase layered system setup (b) Stage changing topology

Fig. 6. Hybrid PVT Systems with thermal container [19].

4 Conclusions

It is evident that ecotourism has positive impacts on the environment and economy of the local and the country as whole. The development of ecotourism can support both in preservation and conservation and of the environment at the destination while the negative effect cannot be ignored. The negative effect of ecotourism mostly comes from creating CO_2 by using fossil fuel energy source. However, there are advanced technologies and designs available for hybrid PVT systems to meet the specific demands and characteristics of the tourism facilities. In order to develop the harness and the abundant amount of solar energy radiation by PVT, the systems with the hybrid controller is suggested. In order to effectively examine the sustainable development of the tourism destination in Malaysia a system to assess sustainability from the integrated perspective need to be in placed in order to evaluate the existing effect of the energy source usage.

Acknowledgement

This research was supported by Taylor's SoE Research Budget scheme towards the measurement and testing equipment.

References

1. A. Liu and G. Wall, "Planning tourism employment: a developing country perspective," *Tourism Management,* vol. **27**, pp. 159-170, (2006)
2. C.V. Aravind, M. AL-Atabi, J. Ravishankar, A. Malik, and E. T. Ambikarajah, "Eco-Tourism Sustainability Through PV Technology: A Comprehensive Review," *Journal of Engineering Science and Technology,* vol. **8**, pp. 654-669, (2013)
3. C. Lascurain. (1996) *World Conservation Union*. Available: http://www.iucnworldconservationcongress.org
4. NGO and C. o. Education, "Our Common Future," U. Nations, Ed., ed. U.S.A.: WCED, (1987)
5. I. Louis, G. Davison, J. Payne, W. Yap, P. Fletcher, and E. Lina, "Malaysian National Ecotourism Plan," vol. Excecutive Summary, M. o. Culture, Ed., ed. Malaysia, pp. 41-48, (1996)

6. I. Thomas, "Defining Tourism," *USA:Rosen Publishing Group Inc,* (2004)
7. D. N. Ravens, *The Natural and Human Environment Combined*. New York: Children's Press, (1996)
8. M. Briffa, & Le,G. *Conventional Links to Sustainable Development* Nerd Press,(2004)
9. V. Nair, "Conceptualising Carbon Footprint and Offset Measures for Rural Tourism Destinations: A Possible Plan for Himalaya? ," Leh, Ladakh June (2014)
10. R. S. J. Tol, "The impact of a carbon tax on international tourism," *Transport and Environment,* vol. **12**, pp. 129-142, (2007)
11. B. K. Maskey, "Development Governance. Agenda for Action. Centre for Development and Governance,," ed. Kathmandu, (2000)
12. G. Peterson and L. D., "People and their behavior: A challenge for recreation management," *Journal of Forestry,* vol. **77**, pp. 343-346, (1979)
13. (4 November 2017). *Global Energy Statisctical Yearbook*. Available: https://yearbook.enerdata.net/total-energy/world-consumption-statistics.html
14. Leung, D. Y. C, Yang, and Y., "Wind energy development and its environmental impact: A review," *Renewable and sustainable energy review,* vol. **16**, pp. 1031-1039, (2012)
15. Nadarajan and C.V. Aravind, "Peak Demand Analysis for the Look-ahead Energy Management System," presented at the eureca, Taylor's University, (2013)
16. (2017, 5 November) *Sustainable Energy and Environment Forum*. Available: http://www.thegreenmechanics.com/2013/08/annual-solar-radiations-in-different.html
17. A. Ramos, M. A. Chatzopoulou, I. Guarracino, J. Freeman, and C. N. Markides, "Hybrid photovoltaic-thermal solar systems for combined heating, cooling and power provision in the urban environment," *Energy Conversion and Management,* vol. **150**, pp. 838-850, 15 October (2017)
18. A.D. Jones and C.P. Underwood, "Thermal Model For Photovoltaic Systems," *Solar Energy,* vol. **70**, pp. 349-359, (2001)
19. M. T. Hajibeigy, C.V. Aravind, M. Al-Atabi, and P. Hoole, "Heat Response Model for Phase Layered Topology in a Photovoltaic Thermal System,"*Indonesian Journalof Electrical Engineering and Computer Science,* vol. **7,** pp. 52-60, July (2017)

Linking Liberatory Pedagogy to Engineering and Sustainable Development

Jayasubamani Arvi S. Moganakrishnann[1,*], *Satesh Narayana* Namasivayam[2] , and *NurHazwani* Ismail[3]

[1,2,3]Taylor's School of Engineering, Taylor's University Lakeside Campus, No. 1, Jalan Taylor's, 47500 Subang Jaya , Selangor Darul Ehsan, Malaysia

Abstract. This paper is written to link liberatory pedagogy with engineering education and sustainable development Its application to specifically, design education for community is explored through deconstructing of engineering student narratives to place understanding of power and privilege in a specific learning setting. The use of liberatory pedagogy for design education and SD is attempted where there is need of changes in understanding of learning outcomes for SD aimed at producing of sustainable physical products to address SD problems in community-based projects. We argue against the simple-minded approach of producing physical products and undermine the intersections of other knowledge forms within the community context. Without pedagogic intervention, we undermine other knowledge forms and cultures in its importance amidst technical knowledge seen as which is valid for SD. Extending liberatory pedagogy to constructivists' perspective we discuss the importance of intervention in the broader image of the engineer as the technical expert who share knowledge in partnership for SD.

1 Introduction

Engineering education and sustainable development (SD) are linked by many pedagogies which has brought possibilities for development of young engineers to acknowledge the agenda as imperative for engineering and society in the future[1]–[3]. These pedagogies have been highlighted in its benefit in bringing understanding of concepts, practice and values in relation to SD which stems from the educational efforts within Educational research undertaken particularly, from Environmental Education and Education for Sustainable Development (ESD)[4], [5]. University curricula for engineering has contributed to the development as well, framed under the Engineering Education for Sustainable Development initiative following the Barcelona Declaration and the Decade of Education for Sustainable Development (2005-2014)[6] for implementation of educational programs in link with SD .In terms of engineering education, incorporation of sustainability within the curriculum has led to exploration of relevant module to discuss sustainability as

* Corresponding author: jayasubamani@gmail.com

well as unsustainability as that which needs to be addressed particularly for professional development[2], [6]–[8]. Within engineering education, design education has received specific attention in its potential to be a constructive activity for the application of highly cognitive skills for learning of producing of solutions in the form of physical products[9]–[11] . A trend of changes to design education is noted within the literature which emphasizes on embracing of multidisciplinary towards transdisciplinary approaches which involves a more comprehensive approach towards understanding of problems solving particularly the problems raised as affecting human well-being and environment [6], [10], [12].

On the other hand, SD becomes understood as development practices which involves problem solving and decision making which are multidimensional, diverse and complex[2], [13]–[15]. This is due to involvement of people, culture and environment, pedagogies which are in support of transformative approaches are relevant with critical implementation of collaborative, intersubjective approaches seeking partnership of others involved[16]–[18]. Transformative approaches are approaches which deals with changing of perspectives, particularly relevant with regards to understanding how people, culture and environment interact in resulting sustainability and unsustainability and our changing roles in addressing them. One area of such transformation is working across discipline, being transgressive towards addressing of complexity in problem [19], [20]. Such transformation is detailed by many literatures seeking transdisciplinary imagination to build our capacity towards addressing complex problems in a . Transdisciplinary initiatives would require us to open our design space in a synergistic approach in which we could understand how we could understand contributing as part to the fresh new 'whole'. With relevance to engineering design education, bringing design into the societal domain with the current mainstream construction for such transdisciplinary endeavors may be problematic in terms of the ideals and mechanistic pathways of simplified education we prescribe to.

Reflection of such critical experiences within engineering education brings test to assumption that educational process which are constructed can be neutral, performed in positivic tradition of achieving outcomes as conceptualized without understanding society. A pedagogy which can support critical study of educational implementations for the students is critical pedagogy, aimed at deconstructing power and privileges within the narratives which inform the learning process and become dominating pathways which eliminates others and their ways of knowledge .

This paper aims to construct the liberatory pedagogy as concept, its origin as well as its worth to problematize current prevailing ideology for design education for EESD. It hopes to address the needed changes to bring possibilities for engineering students to take active roles in society with understanding of power and privileges which leads to oppression and unsustainability.

2 What is liberatory pedagogy?

Critical pedagogy was introduced by Paulo Friere through his seminal work, Pedagogy of the Oppressed published in 1968[16], [17]. It is a teaching approach informed by his work as well as other radical philosophies focusing on helping of students deal with the problematic world requiring social justice through questions. These questions are constructed to challenge dominant views which also undermine other understanding, practices and privileges certain views and knowledge. Given these premises, it becomes an approach to help students achieve critical consciousness which relates to critical thinking in reflexivity of their social and cultural positions. In terms of engineering, it is particularly relevant amidst their professional roles within the societal context such as designing for community.

In this tradition, the focus is on the educator and how efforts could be channeled towards directing the learners to question. The learners question the ideologies they prescribe to which dominate over other ideologies in addressing a whole. It refers to not considering whose interest are they representing for, in which they as diverse participants are participating in despite a homogenous intervention to gather them in a learning setting. Some of their views traditionally omitted or dismissed as they are taken for granted, even within their own institutional setting. Instead of imposing such views and expect conforming attitudes from the oppressed, the narratives spoken about each other or when they 'other' them, are deconstructed in terms of what they mean towards each other despite their participation for the whole. The deconstruction is led in a way that a legitimate, liberatory collective response through dialogue could be gathered as a fresh new perspective on the way they perceive each other as valued contributors to the whole.

Within the classroom setting, students learning begins to critically study the process in which they can begin to view that a practice in society deeply problematic. For example, in the way aboriginal groups are perceived as incompetent, unruly and uncivilized or women as incapable of technicalities or provide significant idea in equal to men. Upon presenting these views, they attempt to change the perceived oppression through constructive methods which allows them to act in direction of a collective response, in respect of those who have been oppressed by the system.

Within the classroom setting, students' learning begins to critically study the process in which they can begin to view that a practice in society deeply problematic. For example, narratives presented in the way we perceive aboriginal groups as incompetent, unruly and uncivilized or women as incapable of technicalities or provide significant ideas in equal to men. Upon presenting these views, they attempt to change the perceived oppression through constructive methods (i.e dialogue) which allows them to act in direction of a collective response, in respect of those who have been oppressed by the system.

Within sustainable development, such dialogue-oriented platforms have been identified as pertinent in bringing a more democratic and inclusive approach to bringing of learning among those involved for development of partnership and collaboration[21] . Dialogues are important communicative platforms in which we could address our differences constructively and reflexively in effort to build a collective through learning[17]. Critical pedagogy aims at providing such dialogue-oriented platforms within the curriculum, which humanizes the unrepresented and brings dignity to their views to be included within context diversified. As a critical approach, it allows educators to bring changes in building collective response in addressing issues found problematic in society particularly sustainability which is informed both socially and culturally in an overlap with .

Critical pedagogy has been explored mainly through the post formal educational thinking which champions education which emancipates the thinking of the unrepresented[3], [22]. This includes gender, religion, Eastern culture as well different abilities in challenged individuals which remain unaddressed in an education system which is mainly colonial, Western, abled and male. Although it is also argued that the approach has been championed towards transformation in context which are predominantly Western, progressive in nature and thus, may in fact destruct any traditionalist view which may be conserving the communities which it intends to represent. However, within the context of engineering education, its use may find purpose in challenging its conservative approach which limits participation and opens the cultural context to not null others and state the engineer as fixed in his image on being technoeconomic but build the reflexive engineer in support of sustainable community development.

Within engineering education, it remains rare although recent call from the engineering education research has highlighted the need to look in specific towards critical experiences in . Given the attention, certain perspectives through critical pedagogy has been explored. Using feminism in an objectivized, Western, male oriented subject such as thermodynamics is an eye opener towards use of the pedagogy in representing meaningful engineering education for women as well as other participants like international students from the East[23]–[25]. Riley used feminism in way to identify use of thermodynamics principles in applications across a diverse selection which includes inventions from untypical sources such as invented for cooking or invented in the East[24]. The use of introducing sustainable technologies not only from the Western male society but also the Eastern as well as from women inventors such as the bain-marie (double boiler) which has failed to be mentioned in most thermodynamic courses[23]. Riley also denotes the critical situation of us as educators linking critical experiences in engineering education limitedly framed to represent highly cognitive efforts and not which affects the development of the self, in our reflexive position in society[23]. However, other areas of engineering education particularly design context is yet to be discussed although the context in link with SD warrants the need for participatory context.

Understanding on the reflexivity and engineer has been discussed in terms of involvement in community development which seeks greater representation of the community within any intervention for their development as discussion of power and privileges of the engineer in their position in society. Robbins [26] discuss the perception engineers working in the context of community of their own position which are controlling in nature and inappreciative of the local knowledge which affects their opinion of the image of the engineer. Engineers discussed seem to find it difficult to accept representation of others as content of knowledge providers, potentially fits them under the label of providers of valuable information and dismiss their engagement as imperative as to technical knowledge. With regards to engineering education, the lack of community representation within the context of pedagogical approach for learning of SD become the discussion point of [27], [28]. The articles highlight the need for critical approaches within design in link with sustainable development for engineering education. In terms of critical pedagogy, it is a useful pathway in the interest of developing communicative platform from the classroom to problematize society with facilitation of engineering educators, who in their position understand critically of their changing roles both socially and culturally in facilitating development of active roles of engineers for community.

The following section will depict how critical pedagogy could be explored within the context of design education for engineers and sustainable development through educators perspective. It details examples of participatory context and descriptions of participating diverse partners involving engineers as well as narratives identified within the teaching setting of the authors which require further discussion on the link of critical pedagogy with design and sustainable development for engineering education.

3 Liberatory Pedagogy for Design

Design education is an important element of the education of engineers which has link to development of expertise as well as identity as designers. Through design education, they learn the trajectory of responding to client requirement and explore the meaning of client-designer relationship through team work of many students in a team[29]. However, with regards to SD and education which works on building of capacity to address complex problems, incorporation of more comprehensive approaches towards inquiry of problems are used[6]. This includes multidisciplinary, interdisciplinary and transdisciplinary approaches which are implemented towards transforming the disciplinary mindset, working away from silos towards a transgressive mindset of crossing boundaries in view of the common problem which affects the interest of various parties[6], [30]–[32] .

However, mainstream design education for sustainable development in its mechanistic approach addressing objectivity in design may promote a neutral perspective to a product-oriented design learning process even for projects intended for a social or environmental need of communities. This process in turn creates unhealthy power relationship □□between student and potential community of address as well as cognitive dissonance [33] with regards to the outcome which are reduced to a product which is not aligned to complex intersubjective experience [16], [34] of learning about people, space and technology through design and the issues of unsustainability.

Engineering design is a process of proceeding from problem defining based on the problems identified or explained by the client towards problem solving in the form of producing technical solution . As education, it does not allow us to understand intersubjectivity of the experience which involves people, space and technology. The values considered within the objectives of the process is depicted as static references to a social or environmental problem which may seems to correspond to a social need or environmental need. These values are not considered in a way which allows us to understand our views as intervening designers with the intention that our act of designing a physical product may solve a social or environmental problem in the living. As students who enters the intersubjective realm of complex relationships to a social and environmental problem, it does not allow us to see ourselves as emerging learning selves. The outcome which become simplified, reduced to a product that either satisfies the requirements of client or create a niche technology suggesting a technical perspective to a much more complex, multidimensional problem.

Student experience within involves projection of language which shows no reflexivity of privilege and power within the conversation which will obstruct the potential for these projects to reflect a partnership with community that they are addressing.

Within a project for the aboriginal community that was intended for providing access to clean water, discussion with educator reveals his views with dialogue entry from the educator. Students within this project have been able to conceptualize a rain water harvesting project proposal to provide access of water. However, when discussing questions to be asked to villagers on the details on the living of the aboriginal and how the project would intervene with their lifestyle, this conversation was noted

Educator : With regards to the project, we are going to visit the kampung (village) for you to talk to the villagers. Before we proceed, I would like to know what do you know about our Orang Asli (aboriginal)?

Student	:	They are uneducated, and live in uncivilized manner.
Educator	:	Why do you think so?
Student	:	(Showing a disapproving body language)..They don't have much facilities around them. Therefore, they are fine with living in the poor way
Educator	:	What about the rest, what do you think?
Student	:	(Silent a while)... They don't have much knowledge
Educator	:	Do you need more knowledge to live here or to live within the forest?
Student	:	Silent....wondering...showing facial expressions reflecting the irrelevance of the question

The depiction of conversation above is an insight to into the conversations between an educator involved in constructing education linking designing to sustainable development and engineering design students working on a project for the aboriginals. The conversation depicts the thinking of engineering student lack of awareness coming from a privileged background to understand the difficult sociohistorical conditions of the aboriginal which results in a discriminatory stance against the aboriginals they are intending to design for. The educator questioning is intended to reflect on the educational practice which incorporates the aboriginals as clients requiring a solution for clean water provision. As the current design pedagogy only requires input of problem definition without understanding sociohistorical conditions and its impact towards community deriving solutions put the engineering students in an unhealthy relationship towards building of partnership between the community they work for and themselves.

As educators, our understanding of aboriginal condition in our countries helps to lead discussion in a communicative way to help students to deconstruct their views to respond collectively as reflexive engineers who address the situation with empathy. The aboriginal living conditions around the world have become a focus [35], [36]particularly within sustainable development agenda where these communities are fighting back for their privilege towards an education that does not marginalize and access to their deprived land[37]. Within Malaysia, the aboriginal community condition has become a political focus in the recent within the National Transformational Plan[37]. The aboriginals living conditions have become focus particularly with regards to their education as dignified native citizens of Malaysia who have been marginalized by the current mainstream education . The lack of access to resources like water resulting in a socio environmental problem is also reflected within their community. The aboriginals' identification with these resources through their lifestyle such as catching fish and diving reflects a closer relationship but have been limited due to development of the mountains and hills where these resources are found. Despite having profound knowledge of the forest, their conditions which in turn has led them to reservation lands has made it difficult to access of water and fulfilled, dignified living.

Our position as designers are to be considered in terms of understanding of our power relationships when working with these communities. In partnership, a designer is sharing his or her knowledge and in turn is informed of other kind of knowledge within a process of collective inquiry . Our lending of expertise is among others who also contribute content through personal experiences, mutual place-based experience as well as holistic knowledge such as art and indigenous technology . We also recognize our knowledge among others in a synoptic process[19] which in turns builds the bigger picture, the picture of the whole in relevance to a collective response to a condition of unsustainability.

The use of liberatory pedagogy within this context would challenge the mainstream design education which simplifies the design process as that which involves technicalities as outcome. Liberatory pedagogues help students to pose questions in view of the domination of practices that does not permit learning in a critical and reflexive way. The use of liberatory pedagogy in design would involve reflexive process of the intersubjective experience in understanding knowledge of the people in informing the unsustainability in the living. For example, understanding of the wells and streams used by the aboriginals to gain water and their introspective experience of how it might be changing and affecting their lives. With regards to access to clean water, with the knowledge shared, we might be able to decide if the critical condition is to be solved with what kind of social environmental intervention. For example, do we have a problem of water availability or water treatment which provide a more socially robust knowledge from our understanding informed of socio-environmental knowledge as well as technical knowledge.

Liberatory pedagogy could also help to build our understanding on the emerging our new, learning selves from the intersubjective process of being affected by the situation resulting in an empathic stance instead of technicality[16]. Other than our technical selves, our social and cultural roles which emerges because of the interaction between people, technology and space results in a more compassionate engineer who values learning as part of the development process. Vare and Scott [38] emphasizes such outcomes in their reference within ESD as that which helps towards building of our capacity to deal with unsustainability, in reference towards learning as the main component of sustainable development. McKenzie relates to the intersubjectivity as a critical pathway towards participation for social action, a bottom up approach rather than the top down technological implementation approach. With relevance to these perspectives is Ekins[39] view of sustainable development efforts within developing countries which would involve technology transfer as its main phase of change which mainly involves changes in the field and not in the top level of planning and management.

Potential criticism of this perspective could stem from the deviation of building expertise, resulting in dilution of competency as well as subjectivity which introduces uncertainty within the philosophical tradition of objectivity and functionality. Molinga [40] denotes with regards water professionals fear of dilution of expertise in the progressive change being highlighted for engineers to become more transdiscipliuary, crossing boundaries in order to solve problems in taking various other roles. However, Brown [19], [20] asserts through her work on the development of the collective mind, the need to understand the self among the collective and how individuals contribute. Thus, with these arguments, liberative pedagogy is seen relevant in broadening the image of the engineer designer as one who is not limited to technicalities but is in the present as emerging in their learning as part of the collective response towards sustainability in society.

4 Conclusion

Thus, with linking liberatory pedagogy to design and sustainable development has been a constructionist's approach in the critical way for engineering education. This paper was written to be an eye opener of educational approaches within the critical pathways focusing on student experience of learning. As education have become a foremost proponent targeted with regards to sustainable development and building of our future, the engineer as the technical expert is limiting of an image in a rather complex platform requiring the play of their role.

Within this paper, liberatory pedagogy has been discussed in its use and its particular relevance to the engineering field based on efforts by educational pedagogues exploring opportunities to include them within engineering education. In light of this, the potential use of liberatory pedagogy was discussed within design education with arguments provided for challenging of the mainstream design education. Student narrative was used to discuss the context of delivering these approaches and its useful relevance in building the capacity of the engineer more than technical expert but in reflexive of their emerging social and cultural role when learning with others.

Potential criticism of taking such perspectives for design education was also discussed with iteration of the pedagogy as that which emancipates and not that limits the identity of the engineer.

References

1. Y. Sipos, B. Battisti, and K. Grimm, "Achieving transformative sustainability learning: engaging head, hands and heart," *Int. J. Sustain. High. Educ.*, **vol. 9,** no. 1, pp. 68–86, (Jan. 2008).

2. D. Ferrer-Balas, J. Bruno, M. De Mingo, and R. Sans, "Advances in education transformation towards sustainable development at the Technical University of Catalonia, Barcelona," *Int. J. Sustain. High. Educ.*, **vol. 5**, no. 3, pp. 251–266, (Sep. 2004).

3. I. Thomas, "Critical Thinking, Transformative Learning, Sustainable Education, and Problem-Based Learning in Universities," *J. Transform. Educ.*, **vol. 7**, no. 3, pp. 245–264, (Jul. 2009).

4. M. Pavlova, "Towards using transformative education as a benchmark for clarifying differences and similarities between Environmental Education and Education for Sustainable Development," *Environ. Educ. Res.*, **vol. 19**, no. 5, pp. 656–672, (Nov. 2012).

5. L. C. Palma *et al.*, "Transformative learning to promote sustainability: inserting the third level of learning in management programs," *Brazilian J. Sci. Technol.*, **vol. 3**, no. 1, p. 9, (Dec. 2016).

6. K. Kastenhofer, A. Lansu, R. van Dam-Mieras, and M. Sotoudeh, "The Contribution of University Curricula to Engineer ing Education for Sustainable Development," *GAIA-Ecological Perspect. Sci. Soc.*, **vol. 19(1)**, pp. 44–51, (2010).

7. K. W. Chau, "Incorporation of Sustainability Concepts into a Civil Engineering Curriculum," *J. Prof. Issues Eng. Educ. Pract.*, **vol. 133**, no. 3, pp. 188–191, (2007).

8. B. Gagnon and R. S. Leduc, "Sustainable Development in Engineering: A Review of Principles and Definition of a Conceptual Framework.," *Environ. Eng. Sci.*, **vol. 26**, no. 10, pp. 1459–1472, (2009).

9. A. I. Schäfer and B. S. Richards, "From concept to commercialisation: student learning in a sustainable engineering innovation project," *Eur. J. Eng. Educ.*, **vol. 32**, no. 2, pp. 143–165, (2007).

10. D. Tate and T. E. Maxwell, "Transdisciplinary Approaches for Teaching and Assessing Sustainable Design.," *Int. J. Eng. Educ.*, **vol. 26**, no. 2, pp. 418–429, (2010).

11. J. H. Spangenberg, A. Fuad-Luke, and K. Blincoe, "Design for Sustainability (DfS): The interface of sustainable production and consumption," *J. Clean. Prod.*, **vol. 18**, no. 15, pp. 1483–1491, (2010).

12. D. S. Strong and O. C. Kl, "Towards Effective Multidisciplinary Engineering Education : The Multidisciplinary Design Stream at Queen ' s University Professor and NSERC Chair in Design Engineering," *Proceedings of the Canadian Engineering Education Association.* (2002).

13. P. Vare and W. Scott, "Learning for a Change: Exploring the Relationship Between Education and Sustainable Development," *J. Educ. Sustain. Dev.*, **vol. 1**, no. 2, pp. 191–198, Sep. (2007).

14. S. R. Sterling, *Sustainable education: Revisioning learning and change*. United Kingdom: Green Books for the Schumacher Society, (2001).

15. C. Hopkins and R. McKeown, "Education for Sustainable Development: past experience., present action and future prospects," *Educ. Philos. Theory*, **vol. 33**, no. 2, pp. 231–244, (2001).

16. M. McKenzie, "The places of pedagogy: or, what we can do with culture through intersubjective experiences," *Environ. Educ. Res.*, (2008).

17. E. Ellsworth, "Why doesn't this feel empowering? Working through the repressive

myths of critical pedagogy," *Harv. Educ. Rev.*, (1989).

18. E. Ellsworth, *Places of learning: Media, architecture, pedagogy*. Routledge, (2005).

19. V. A. Brown, "Collective inquiry and its wicked problems," *Tackling wicked Probl. through Transdiscipl. Imagin.*, pp. 61–83, (2010).

20. V. A. Brown, "Utopian thinking and the collective mind: Beyond transdisciplinarity," *Futures*, **vol. 65**, pp. 209–216, (2015).

21. T. Meppem and R. Gill, "Planning for sustainability as a learning concept," *Ecol. Econ.*, **vol. 26**, no. 2, pp. 121–137, (Aug. 1998).

22. J. L. Kincheloe, *Critical constructivism primer*, vol. 2. Peter Lang, (2005).

23. D. Riley, "Employing liberative pedagogies in engineering education," *J. Women Minor. Sci. Eng.*, **vol. 9**, no. 2, (2003).

24. D. Riley, A. L. Pawley, J. Tucker, and G. D. Catalano, "Feminisms in engineering education: Transformative possibilities," *NWSA J.*, **vol. 21**, no. 2, pp. 21–40, (2009).

25. L. Claris and D. Riley, "Situation critical: critical theory and critical thinking in engineering education," *Eng. Stud.*, **vol. 4**, no. 2, pp. 101–120, (2012).

26. P. T. Robbins, "The reflexive engineer: perceptions of integrated development," *J. Int. Dev.*, **vol. 19**, no. 1, pp. 99–110, (2007).

27. J. Schneider and J. A. . L. Leydens, "Where is 'Community'?: Engineering education and sustainable community development.," *Eur. J. Eng. Educ.*, **vol. 33**, no. 3, pp. 307–319, (2008).

28. J. Lucena and J. Schneider, "Engineers, development, and engineering education: From national to sustainable community development.," *Eur. J. Eng. Educ.*, **vol. 33**, no. 3, pp. 247–257, (2008).

29. C. L. Dym, A. M. Agogino, O. Eris, D. D. Frey, and L. J. Leifer, "Engineering design thinking, teaching, and learning," *J. Eng. Educ.*, **vol. 94**, no. 1, pp. 103–120, (2005).

30. A. M. Lopes, D. Fam, and J. Williams, "Designing sustainable sanitation: Involving design in innovative, transdisciplinary research," *Des. Stud.*, **vol. 33**, no. 3, pp. 298–317, (2012).

31. A. Jakobsen and L. L. Bucciarelli, "Transdisciplinary variation in engineering curricula. Problems and means for solutions.," *Eur. J. Eng. Educ.*, **vol. 32,** no. 3, pp. 295–301, (2007).

32. L. Vanasupa, K. E. McCormick, C. J. Stefanco, R. J. Herter, and M. McDonald, "Challenges in Transdisciplinary, Integrated Projects: Reflections on the Case of Faculty Members' Failure to Collaborate," *Innov. High. Educ.*, **vol. 37**, no. 3, pp. 171–184, (2012).

33. N. Chabrak and R. Craig, "Student imaginings, cognitive dissonance and critical thinking," *Crit. Perspect. Account.*, **vol. 24**, no. 2, pp. 91–104, (2013).

34. S. S. Chave, "Education, sustainability and intersubjectivity: exploring the possibility of the emergence of new ways of knowing, being and acting in the world." University of Exeter, (2017).

35. J. Davies, J. White, A. Wright, Y. Maru, and M. LaFlamme, "Applying the sustainable livelihoods approach in Australian desert Aboriginal development," *Rangel. J.*, **vol. 30**, no. 1, pp. 55–65, (2008).

36. J. Altman and J. Finlayson, "Aborigines, tourism and sustainable development," *J. Tour. Stud.*, **vol. 4**, no. 1, pp. 38–50, (1993).

37. G. P. Means, "The Orang Asli: Aboriginal Policies in Malaysia," *Pac. Aff.*, **vol. 58**, no. 4, pp. 637–652, (1985).

38. P. Vare and W. Scott, "Learning for a Change: Exploring the Relationship Between Education and Sustainable Development opinion essay," **vol. 1**, no. 2, pp. 191–198, (2015).

39. P. Ekins, "Making development sustainable," *Glob. Ecol. A new arena Polit. Confl.*, pp. 91–103, (1993).

40. P. P. Mollinga, "Towards the transdisciplinary engineer: Incorporating ecology, equity and democracy concerns into water professionals' attitudes, skills and knowledge," *Irrig. Drain.*, **vol. 58**, no. S2, pp. 195–204, (2009).

AUTHOR INDEX